Universitext

Claudio Procesi

Lie Groups

An Approach through Invariants and Representations

Springer

Claudio Procesi
Dipartimento di Matematica
"G. Castelnouvo"
Universita di Roma "La Sapienza"
00185 Rome
Italy
processi@mat.uniroma l.it

Mathematics Subject Classification 2000: Primary: 22EXX, 14LXX

Library of Congress Cataloging-in-Publication Data

Processi, Claudio.
An Approach to Lie Theory through Invariants and Representations / Claudio Procesi.
p. cm. – (Universotext)
Includes bibliographical references and index.
ISBN-13: 978-0-387-26040-2
ISBN-10: 0-387-26040-4
1.Lie groups. 2. Invariants. 3. Representations of algebras. I. Title.
QA387.P76 2005
512'482—dc22 2005051743
ISBN-10: 0-387-26040-4 e-ISBN: 0-387-28929-1
ISBN-13: 978-0-387-26040-2

Procesi: Lie Groups

First Indian Reprint 2010

ISBN 978-81-8489-650-3

This edition is published by Springer (India) Private Limited, (a part of Springer Science+Business Media), Registered Office: 3rd Floor, Gandharva Mahavidyalaya, 212, Deen Dayal Upadhyaya Marg, New Delhi – 110 002, India

Printed in India by Rashtriya Printers, Delhi.

To the ladies of my life,
Liliana, Silvana, Michela, Martina

Contents

Introduction

The subject of Lie groups, introduced by Sophus Lie in the second half of the nineteenth century, has been one of the important mathematical themes of the last century. Lie groups formalize the concept of continuous symmetry, and thus are a part of the foundations of mathematics. They also have several applications in physics, notably quantum mechanics and relativity theory. Finally, they link with many branches of mathematics, from analysis to number theory, passing through topology, algebraic geometry, and so on.

This book gives an introduction to at least the main ideas of the theory. Usually, there are two principal aspects to be discussed. The first is the description of the groups, their properties and classifications; the second is the study of their representations.

The problem that one faces when introducing representation theory is that the material tends to grow out of control quickly. My greatest difficulty has been to try to understand when to stop. The reason lies in the fact that one may represent almost any class of algebraic if not even mathematical objects. In fact it is clear that even the specialists only master part of the material.

There are of course many good introductory books on this topic. Most of them however favor only one aspect of the theory. I have tried instead to present the basic methods of Lie groups, Lie algebras, algebraic groups, representation theory, some combinatorics and basic functional analysis, which then can be used by the various specialists. I have tried to balance general theory with many precise concrete examples.

This book started as a set of lecture notes taken by G. Boffi for a course given at Brandeis University. These notes were published as a "Primer in Invariant Theory" [Pr]. Later, H. Kraft and I revised these notes, which have been in use and are available on Kraft's home page [KrP]. In these notes, we present classical invariant theory in modern language. Later, E. Rogora and I presented the combinatorial approach to representations of the symmetric and general linear groups [PrR]. In past years, while teaching introductory courses on representation theory, I became convinced that it would be useful to expand the material in these various expositions to give an idea of the connection with more standard classical topics, such as the theory of Young symmetrizers and Clifford algebras, and also not to restrict to classical groups but to include general semisimple groups as well.

The reader will see that I have constantly drawn inspiration from the book of H. Weyl, *Classical Groups* [W]. On the other hand it would be absurd and quite impossible to *update* this classic.

In his book Weyl stressed the relationship between representations and invariants. In the last 30 years there has been a renewed interest in classical methods of invariant theory, motivated by problems of geometry, in particular due to the ideas of Grothendieck and Mumford on moduli spaces. The reader will see that I do not treat geometric invariant theory at all. In fact I decided that this would have deeply changed the nature of the book, which tries to always remain at a relatively elementary level, at least in the use of techniques outside of algebra. Geometric invariant theory is deeply embedded in algebraic geometry and algebraic groups, and several good introductions to this topic are available.

I have tried to explain in detail all the constructions which belong to invariant theory and algebra, introducing and using only the essential notions of differential geometry, algebraic geometry, measure theory, and functional analysis which are necessary for the treatment here. In particular, I have tried to restrict the use of algebraic geometry and keep it to a minimum, nevertheless referring to standard books for some basic material on this subject which would have taken me too long to discuss in this text. While it is possible to avoid algebraic geometry completely, I feel it would be a mistake to do so since the methods that algebraic geometry introduces in the theory are very powerful. In general, my point of view is that some of the interesting special objects under consideration may be treated by more direct and elementary methods, which I have tried to do whenever possible since a direct approach often reveals some special features which may be lost in a general theory. A similar, although less serious, problem occurs in the few discussions of homotopy theory which are needed to understand simply connected groups.

I have tried to give an idea of how 19th-century algebraists thought of the subject. The main difficulty we have in understanding their methods is in the fact that the notion of representation appears only at a later stage, while we usually start with it.

The book is organized into topics, some of which can be the subject of an entire graduate course. The organization is as follows.

The first chapter establishes the language of group actions and representations with some simple examples from abstract group theory. The second chapter is a quick look into the theory of symmetric functions, which was one of the starting points of the entire theory. First, I discuss some very classical topics, such as the resultant and the Bézoutiant. Next I introduce Schur functions and the Cauchy identity. These ideas will play a role much later in the character theory of the symmetric and the linear group.

Chapter 3 presents again a very classical topic, that of the theory of algebraic forms, à la Capelli [Ca].

In Chapter 4, I change gears completely. Taking as pretext the theory of polarizations of Capelli, I systematically introduce Lie groups and Lie algebras and start to prove some of the basic structure theorems. The general theory is completed in

Chapter 5 in which universal enveloping algebras and free Lie algebras are discussed. Later, in Chapter 10 I treat semisimple algebras and groups. I complete the proof of the correspondence between Lie groups and Lie algebras via Ado's theorem. The rest of the chapter is devoted to Cartan–Weyl theory, leading to the classification of complex semisimple groups and the associated classification of connected compact groups.

Chapter 5 is quite elementary. I decided to include it since the use of tensor algebra and tensor notation plays such an important role in the treatment as to deserve some lengthy discussion. In this chapter I also discuss Clifford algebras and the spin group. This topic is completed in Chapter 11.

Chapter 6 is a short introduction to general methods of noncommutative algebra, such as Wedderburn's theorem and the double centralizer theorem. This theory is basic to the representation theory to be developed in the next chapters.

Chapter 7 is a quick introduction to algebraic groups. In this chapter I make fair use of notions from algebraic geometry, and I try to at least clarify the statements used, referring to standard books for the proofs. In fact it is impossible, without a rather long detour, to actually develop in detail the facts used. I hope that the interested reader who does not have a background in algebraic geometry can still follow the reasoning developed here.

I have tried to stress throughout the book the parallel theory of reductive algebraic and compact Lie groups. A full understanding of this connection is gained slowly, first through some classical examples, then by the Cartan decomposition and Tannaka duality in Chapter 8. This theory is completed in Chapter 10, where I associate, to a semisimple Lie algebra, its compact form. After this the final classification theorems are proved.

Chapter 8 is essentially dedicated to matrix coefficients and the Peter–Weyl theorem. Some elementary functional analysis is used here. I end the chapter with basic properties of Hopf algebras, which are used to make the link between compact and reductive groups.

Chapter 9 is dedicated to tensor symmetry, Young symmetrizers, Schur–Weyl duality and their applications to representation theory.

Chapter 10 is a short course giving the structure and classification of semisimple Lie algebras and their representations via the usual method of root systems. It also contains the corresponding theory of adjoint and simply connected algebraic groups and their compact forms.

Chapter 11 is the study of the relationship between invariants and the representation theory of classical groups. It also contains a fairly detailed discussion of spinors and terminates with the analytic proof of Weyl's character formula.

The last four chapters are complements to the theory. In Chapter 12 we discuss the combinatorial theory of tableaux to lead to Schützenberger's proof of the Littlewood–Richardson rule.

Chapter 13 treats the combinatorial approach to invariants and representations for classical groups. This is done via the theory of standard monomials, which is developed in a characteristic-free way, for some classical representations.

Chapter 14 is a very short glimpse into the geometric theory, and finally Chapter 15 is a return to the past, to where it all started: the theory of binary forms.

Many topics could not find a place in this treatment. First, I had to restrict the discussion of algebraic groups to a minimum. In particular I chose giving proofs only in characteristic 0 when the general proof is more complicated. I could not elaborate on the center of the universal enveloping algebra, Verma modules and all the ideas relating to finite and infinite-dimensional representations. Nor could I treat the conjugation action on the group and the Lie algebra which contains so many deep ideas and results. Of course I did not even begin to consider the theory of real semisimple groups. In fact, the topics which relate to this subject are so numerous that this presentation here is just an invitation to the theory. The theory is quite active and there is even a journal entirely dedicated to its developments.

Finally, I will add that this book has some overlaps with several books, as is unavoidable when treating foundational material.

I certainly followed the path already taken by others in many of the basic proofs which seem to have reached a degree of perfection and upon which it is not possible to improve.

The names of the mathematicians who have given important contributions to Lie theory are many, and I have limited to a minimum the discussion of its history. The interested reader can now consult several sources like [Bor2], [GW].

I wish finally to thank Laura Stevens for carefully reading through a preliminary version and helping me to correct several mistakes, and Alessandro D'Andrea, Figà Talamanca and Paolo Papi for useful suggestions, and Ann Kostant and Martin Stock for the very careful and complex editing of the final text.

Claudio Procesi
Università di Roma La Sapienza
July 2006

Conventional Notations

When we introduce a new symbol or definition we will use the convenient symbol $:=$ which means that the term introduced on its left is defined by the expression on its right.

A typical example could be $P := \{x \in \mathbb{N} \mid 2 \text{ divides } x\}$, which stands for *$P$ is by definition the set of all natural numbers x such that 2 divides x.*

The symbol $\pi : A \to B$ denotes a mapping called π from the set A to the set B.

Most of our work will be for algebras over the field of real or complex numbers. Sometimes we will take a more combinatorial point of view and analyze some properties over the integers. Associative algebras will implicitly be assumed to have a unit element. When we discuss matrices over a ring A we always identify A with the scalar matrices (constant multiples of the identity matrix).

We use the standard notations:

$$\mathbb{N},\ \mathbb{Z},\ \mathbb{Q},\ \mathbb{R},\ \mathbb{C}$$

for the natural numbers (including 0), the integers, rational, real and complex numbers.

1

General Methods and Ideas

Summary. In this chapter we will develop the formal language and some general methods and theorems. To some extent the reader is advised not to read it too systematically since most of the interesting examples will appear only in the next chapters. The exposition here is quite far from the classical point of view since we are forced to establish the language in a rather thin general setting. Hopefully this will be repaid in the chapters in which we will treat the interesting results of Invariant Theory.

1 Groups and Their Actions

1.1 Symmetric Group

In our treatment groups will always appear as transformation groups, the main point being that, given a set X, the set of all bijective mappings of X into X is a group under composition. We will denote this group $S(X)$ and call it *the symmetric group* of X.

In practice, the full symmetric group is used only for X a finite set. In this case it is usually more convenient to identify X with the discrete interval $\{1, \dots, n\}$ formed by the first n integers (for a given value of n). The corresponding symmetric group has $n!$ elements and it is denoted by S_n. Its elements are called *permutations*.

In general, the groups which appear are subgroups of the full symmetric group, defined by special properties of the set X arising from some extra structure (such as from a topology or the structure of a linear space, etc.). The groups of interest to us will usually be symmetry groups of the structure under consideration. To illustrate this concept we start with a definition:

Definition. A partition of a set X is a family of nonempty disjoint subsets A_i such that $X = \cup_i A_i$.

A partition of a number n is a (non-increasing) sequence of positive numbers:

$$m_1 \geq m_2 \geq \cdots \geq m_k > 0 \text{ with } \sum_{j=1}^{k} m_j = n.$$

Remark. To a partition of the set $[1, 2, \dots, n]$ we can associate the partition of n given by the cardinalities of the sets A_i.

We will usually denote a partition by a greek letter $\lambda := m_1 \geq m_2 \geq \cdots \geq m_k$ and write $\lambda \vdash n$ to mean that it is a partition of n.

We represent graphically such a partition by a *Young diagram*. The numbers m_i appear then as the lengths of the rows (cf. Chapter 9, 2.1), e.g., $\lambda = (8, 5, 5, 2)$:

$\lambda =$  $\tilde{\lambda} =$ 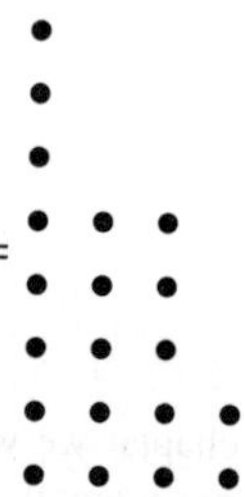

Sometimes it is useful to relax the condition and call a *partition of n* any sequence $m_1 \geq m_2 \geq \cdots \geq m_k \geq 0$ with $\sum_{j=1}^{k} m_j = n$. We then call the *height* of λ, denoted by $ht(\lambda)$, the number of nonzero elements in the sequence m_i, i.e., the number or rows of the diagram.

We can also consider the columns of the diagram which will be thought of as rows of the *dual partition*. The dual partition of λ will be denoted by $\tilde{\lambda}$. For instance, for $\lambda = (8, 5, 5, 2)$ we have $\tilde{\lambda} = (4, 4, 3, 3, 3, 1, 1, 1)$.

If $X = \cup A_i$ is a partition, the set

$$G := \{\sigma \in S_n | \sigma(A_i) = A_i, \ \forall i\},$$

is a subgroup of $S(X)$, isomorphic to the product $\prod S(A_i)$ of the symmetric groups on the sets A_i. There is also another group associated to the partition, the group of permutations which preserves the partition without necessarily preserving the individual subsets (but possibly permuting them).

1.2 Group Actions

It is useful at this stage to proceed in a formal way.

Definition 1. An action of a group G on a set X is a mapping $\pi : G \times X \to X$, denoted by $gx := \pi(g, x)$, satisfying the following conditions:

(1.2.1) $$1x = x, \quad h(kx) = (hk)x$$

for all $h, k \in G$ and $x \in X$.

The reader will note that the above definition can be reformulated as follows:

(i) The map $\varrho(h) := x \mapsto hx$ from X to X is bijective for all $h \in G$.
(ii) The map $\varrho : G \to S(X)$ is a *group homomorphism*.

In our theory we will usually fix our attention on a given group G and consider different actions of the group. It is then convenient to refer to a given action on a set X as a *G-set.*

Examples.

(a) The action of G by left multiplication on itself.
(b) For a given subgroup H of G, the action of G on the set $G/H := \{gH | g \in G\}$ of left cosets is given by

$$a(bH) := abH. \tag{1.2.2}$$

(c) The action of $G \times G$ on G given by *left and right translations:* $(a, b)c := acb^{-1}$.
(d) The action of G by conjugation on itself.
(e) The action of a subgroup of G induced by restricting an action of G.

It is immediately useful to use *categorical* language:

Definition 2. Given two G-sets X, Y, a G-equivariant mapping, or more simply a *morphism*, is a map $f : X \to Y$ such that for all $g \in G$ and $x \in X$ we have

$$f(gx) = gf(x).$$

In this case we also say that f *intertwines* the two actions. Of course if f is bijective we speak of an *isomorphism* of the two actions. If $X = Y$ the isomorphisms of the G-action X also form a group called the *symmetries of the action.*

The class of G-sets and equivariant maps is clearly a *category.*[1]

Remark. This is particularly important when G is the homotopy group of a space X and the G-sets correspond to covering spaces of X.

Example. The equivariant maps of the action of G on itself by left multiplication are the right multiplications. They form a group isomorphic to the *opposite* of G (but also to G). Note: We recall that the *opposite* of a multiplication $(a, b) \mapsto ab$ is the operation $(a, b) \mapsto ba$. One can define opposite group or opposite ring by taking the opposite of multiplication.

More generally:

Proposition. *The invertible equivariant maps of the action of G on G/H by left multiplication are induced by the right multiplications with elements of the normalizer $N_G(H) := \{g \in G \,|\, gHg^{-1} = H\}$ of H (cf. 2.2). They form a group Γ isomorphic to $N_G(H)/H$.*

---

[1] Our use of categories will be extremely limited to just a few basic functors and universal properties.

Proof. Let $\sigma : G/H \to G/H$ be such a map. Hence, for all $a, b \in G$, we have $\sigma(a\,bH) = a\sigma(bH)$. In particular if $\sigma(H) = uH$ we must have that

$$\sigma(H) = uH = \sigma(hH) = huH, \ \forall h \in H, \implies Hu \subset uH.$$

If we assume σ to be invertible, we see that also $Hu^{-1} \subset u^{-1}H$, hence $uH = Hu$ and $u \in N_G(H)$. Conversely, if $u \in N_G(H)$, the map $\sigma(u) : aH \to auH = aHu$ is well defined and an element of Γ. The map $u \to \sigma(u^{-1})$ is clearly a surjective homomorphism from $N_G(H)$ to Γ with kernel H. □

Exercise. Describe the set of equivariant maps $G/H \to G/K$ for 2 subgroups.

2 Orbits, Invariants and Equivariant Maps

2.1 Orbits

The first important notion in this setting is given by the following:

Consider the binary relation R in X given by xRy if and only if there exists $g \in G$ with $gx = y$.

Proposition. *R is an equivalence relation.*

Definition. The equivalence classes under the previous equivalence are called G-orbits (or simply orbits). The orbit of a given element x is formed by the elements gx with $g \in G$ and is denoted by Gx. The mapping $G \to Gx$ given by $g \mapsto gx$ is called the *orbit map*.

The orbit map is equivariant (with respect to the left action of G). The set X is partitioned into its orbits, and the set of all orbits (quotient set) is denoted by X/G.

In particular, we say that the action of G is *transitive* or that X is a *homogeneous space* if there is a unique orbit.

More generally, we say that a subset Y of X is *G stable* if it is a union of orbits. In this case, the action of G on X induces naturally an action of G on Y. Of course, the complement $C(Y)$ of Y in X is also G stable, and X is decomposed as $Y \cup C(Y)$ into two stable subsets.

The finest decomposition into stable subsets is the decomposition into orbits.

Basic examples.

i. Let $\sigma \in S_n$ be a permutation and A the cyclic group which it generates. Then the orbits of A on the set $\{1, \dots, n\}$ are the *cycles* of the permutation.
ii. Let G be a group and let H, K be subgroups. We have the action of $H \times K$ on G induced by the left and right action. The orbits are the *double cosets*. In particular, if either H or K is 1, we have left or right cosets.
iii. Consider G/H, the set of left cosets gH, with the action of G given by 1.2.2. Given a subgroup K of G, it still acts on G/H. The K orbits in G/H are in bijective correspondence with the double cosets KgH.

iv. Consider the action of G on itself by conjugation $(g, h) \to ghg^{-1}$. Its orbits are the *conjugacy classes.*
v. An action of the additive group $\mathbb{R}_+$ of real numbers on a set X is called a 1-*parameter group of transformations*, or in more physical language, a *reversible dynamical system.*

In example (v) the parameter t is thought of as *time,* and an orbit is seen as the time evolution of a physical state. The hypotheses of the group action mean that the evolution is reversible (i.e., all the group transformations are invertible), and the *forces* do not vary with time so that the evolution of a state depends only on the time lapse (group homomorphism property).

The previous examples also suggest the following general fact:

Remark. Let G be a group and K a normal subgroup in G. If we have an action of G on a set X, we see that G acts also on the set of K orbits X/K, since $gKx = Kgx$. Moreover, we have $(X/K)/G = X/G$.

2.2 Stabilizer

The study of group actions should start with the elementary analysis of a single orbit. The next main concept is that of *stabilizer*:

Definition. Given a point $x \in X$ we set $G_x := \{g \in G | gx = x\}$. G_x is called the *stabilizer* (or *little group*) of x.

Remark. The term *little group* is used mostly in the physics literature.

Proposition. *G_x is a subgroup, and the action of G on the orbit Gx is isomorphic to the action on the coset space G/G_x.*

Proof. The fact that G_x is a subgroup is clear. Given two elements $h, k \in G$ we have that $hx = kx$ if and only if $k^{-1}hx = x$ or $k^{-1}h \in G_x$.

The mapping between G/G_x and Gx which assigns to a coset hG_x the element hx is thus well defined and bijective. It is also clearly G-equivariant, and so the claim follows. □

Example. For the action of $G \times G$ on G by left and right translations (Example (c) of 1.2), G is a single orbit and the stabilizer of 1 is the subgroup $\Delta := \{(g, g) | g \in G\}$ isomorphic to G embedded in $G \times G$ diagonally.

Example. In the case of a 1-parameter subgroup acting continuously on a topological space, the stabilizer is a closed subgroup of $\mathbb{R}$. If it is not the full group, it is the set of integral multiples $ma, m \in \mathbb{Z}$ of a positive number a. The number a is to be considered as the first time in which the orbit returns to the starting point. This is the case of a *periodic orbit.*

Remark. Given two different elements in the same orbit, their stabilizers are conjugate. In fact, $G_{hx} = hG_xh^{-1}$. In particular when we identify an orbit with a coset space G/H this implicitly means that we have made the choice of a point for which the stabilizer is H.

Remark. The orbit cycle decomposition of a permutation π can be interpreted in the previous language. Giving a permutation π on a set S is equivalent to giving an action of the group of integers $\mathbb{Z}$ on S.

Thus S is canonically decomposed into orbits. On each orbit the permutation π induces, by definition, a *cycle*.

To study a single orbit, we need only remark that a finite orbit under $\mathbb{Z}$ is equivalent to the action of $\mathbb{Z}$ on some $\mathbb{Z}/(n)$, $n > 0$. On $Z/(n)$ the generator 1 acts as the cycle $\overline{x} \to \overline{x} + \overline{1}$.

Fixed point principle. *Given two subgroups H, K of G we have that H is conjugate to a subgroup of K if and only if the action of H on G/K has a fixed point.*

Proof. This is essentially tautological, $H \subset gKg^{-1}$ is equivalent to saying that gK is a fixed point under H. □

Consider the set of all subgroups of a group G, with G acting on this set by conjugation. The orbits of this action are the *conjugacy classes of subgroups*. Let us denote by $[H]$ the conjugacy class of a subgroup H.

The stabilizer of a subgroup H under this action is called its *normalizer.* It should not be confused with the *centralizer* which, for a given subset A of G, is the stabilizer under conjugation of all of the elements of A.

Given a group G and an action on X, it is useful to introduce the notion of *orbit type*.

Observe that for an orbit in X, the conjugacy class of the stabilizers of its elements is well defined. We say that two orbits are of the same *orbit type* if the associated stabilizer class is the same. This is equivalent to saying that the two orbits are isomorphic as G-spaces. It is often useful to partition the orbits according to their orbit types.

Exercise. Determine the points in G/H with stabilizer H.

Exercise. Show that the group of symmetries of a G action permutes transitively orbits of the same type.

Suppose that G and X are finite and assume that we have in X, n_i orbits of type $[H_i]$. Then we have, from the partition into orbits, the formula

$$\frac{|X|}{|G|} = \sum_i \frac{n_i}{|H_i|},$$

where we denote by $|A|$ the cardinality of a finite set A.

The next exercise is a challenge to the reader who is already familiar with these notions.

Exercise. Let G be a group with $p^m n$ elements, p a prime number not dividing n. Deduce the theorems of Sylow by considering the action of G by left multiplication on the set of all subsets of G with p^m elements ([Wie]).

Exercise. Given two subgroups H, K of G, describe the orbits of H acting on G/K. In particular, give a criterion for G/K to be a single H-orbit. Discuss the special case $[G : H] = 2$.

2.3 Invariants

From the elements of X, we may single out those for which the stabilizer is the full group G. These are the *fixed points* of the action or *invariant points*, i.e., the points whose orbit consists of the point alone.

These points will usually be denoted by X^G.

$$X^G := \{\textit{fixed points} \text{ or } \textit{invariant points}\}.$$

We have thus introduced in a very general sense the notion of *invariant.* Its full meaning for the moment is completely obscure; we must first proceed with the formal theory.

2.4 Basic Constructions

One of the main features of set theory is the fact that it allows us to perform constructions. Out of given sets we construct new ones. This is also the case of G-sets. Let us point out at least two constructions:

(1) Given two G-sets X, Y, we give the structure of a G-set to their disjoint sum $X \sqcup Y$ by acting separately on the two sets and to their product $X \times Y$ by

$$g(x, y) := (gx, gy), \tag{2.4.1}$$

(i.e., once the group acts on the elements it acts also on the pairs.)

(2) Consider now the set Y^X of all maps from X to Y. We can act with G (verify this) by setting

$$(gf)(x) := gf(g^{-1}x). \tag{2.4.2}$$

Notice that in the second definition we have used the action of G twice. The particular formula given is justified by the fact that it is really the only way to get a group action using the two actions.

Formula 2.4.2 reflects a general fact well known in category theory: maps between two objects X, Y are a covariant functor in Y and a contravariant functor in X.

We want to immediately make explicit a rather important consequence of our formalism:

Proposition. *A map $f : X \to Y$ between two G-sets is equivariant (cf. 1.2) if and only if it is a fixed point under the G-action on the maps.*

Proof. This statement is really a tautology, but nevertheless it deserves to be clearly understood. The proof is trivial following the definitions. Equivariance means that $f(gx) = gf(x)$. If we substitute x with $g^{-1}x$, this reads $f(x) = gf(g^{-1}x)$, which, in functional language means that the function f equals the function gf, i.e., it is invariant. □

Exercises.

(i) Show that the orbits of G acting on $G/H \times G/K$ are in canonical 1-1 correspondence with the double cosets HgK of G.
(ii) Given a G equivariant map $\pi : X \to G/H$ show that:
 (a) $\pi^{-1}(H)$ is stable under the action of H.
 (b) The set of G orbits of X is in 1-1 correspondence with the H-orbits of $\pi^{-1}(H)$.
 (c) Study the case in which $X = G/K$ is also homogeneous.

2.5 Permutation Representations

We will often consider a special case of the previous section, the case of the trivial action of G on Y. In this case of course the action of G on the functions is simply

$$^{g}f(x) = (gf)(x) = f(g^{-1}x). \tag{2.5.1}$$

Often we write ${}^{g}f$ instead of gf for the function $f(g^{-1}x)$. A mapping is equivariant if and only if it is constant on the orbits. In this case we will always speak of an *invariant function*. In view of the particular role of this idea in our treatment, we repeat the formal definition.

Definition 1. A function f on a G-set X is called *invariant* if $f(g^{-1}x) = f(x)$ for all $x \in X$ and $g \in G$.

As we have just remarked, a function is invariant if and only if it is constant on the orbits. Formally we may thus say that the quotient mapping $\pi := X \to X/G$ is an invariant map and any other invariant function factors as

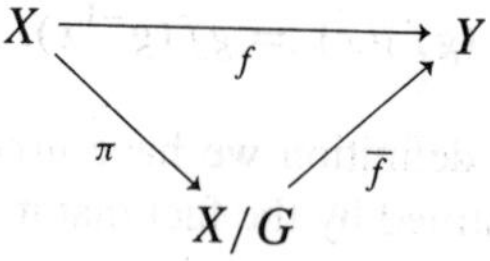

We want to make explicit the previous remark in a case of importance.

Let X be a finite G-set. Consider a field F (a ring would suffice) and the set F^X of functions on X with values in F.

An element $x \in X$ can be identified with the characteristic function of $\{x\}$. In this way X becomes a basis of F^X as a vector space.

The induced group action of G on F^X is by linear transformations which permute the basis elements.

F^X is called a *permutation representation,* and we will see its role in the next sections. Since a function is invariant if and only if it is constant on orbits we deduce:

Proposition. *The invariants of G on F^X form the subspace of F^X having as a basis the characteristic functions of the orbits.*

In other words given an orbit $\mathcal{O}$ consider $u_{\mathcal{O}} := \sum_{x\in\mathcal{O}} x$. The elements $u_{\mathcal{O}}$ form a basis of $(F^X)^G$.

We finish this section with two examples which will be useful in the theory of symmetric functions.

Consider the set $\{1, 2, \dots, n\}$ with its canonical action of the symmetric group S_n. The maps from $\{1, 2, \dots, n\}$ to the field $\mathbb{R}$ of real numbers form the standard vector space $\mathbb{R}^n$. The symmetric group acts by permuting the coordinates, and in every orbit there is a unique vector $(a_1, a_2, \dots, a_n)$ with $a_1 \geq a_2 \geq \cdots \geq a_n$.

The set of these vectors can thus be identified with the orbit space. It is a convex cone with boundary, comprising the elements in which at least two coordinates are equal.

Exercise. Discuss the orbit types of the previous example.

Definition 2. A function $M : \{1, 2, \dots, n\} \to \mathbb{N}$ (to the natural numbers) is called a *monomial.* The set of monomials is a semigroup by the addition of values, and we indicate by x_i the monomial which is the characteristic function of $\{i\}$.

As we have already seen, the symmetric group acts on these functions by $(\sigma f)(k) := f(\sigma^{-1}(k))$, and the action is compatible with the addition. Moreover $\sigma(x_i) = x_{\sigma(i)}$.

Remark. It is customary to write the semigroup law of monomials multiplicatively. Given a monomial M such that $M(i) = h_i$, we have $M = x_1^{h_1} x_2^{h_2} \cdots x_n^{h_n}$. The number $\sum_i h_i$ is the *degree* of the monomial.

Representing a monomial M as a vector $(h_1, h_2, \dots, h_n)$, we see that every monomial of degree d is equivalent, under the symmetric group, to a unique vector in which the coordinates are non-increasing. The nonzero coordinates of such a vector thus form a partition $\lambda(M) \vdash d$, with at most n parts, of the degree of the monomial.

The permutation representation associated to the monomials with coefficients in a commutative ring F is the *polynomial ring* $F[x_1, \dots, x_n]$ in the given variables.

The invariant elements are called *symmetric polynomials.*

From what we have proved, a basis of these symmetric polynomials is given by the sums

$$m_\lambda := \sum_{\lambda(M)=\lambda} M$$

of monomials with exponents the integers h_i of the partition λ. m_λ is called a *monomial symmetric function.*

Exercise. To a monomial M we can also associate a partition of the set $\{1, 2, \dots, n\}$ by the equivalence $i \cong j$ iff $M(i) = M(j)$. Show that the stabilizer of M is the group of permutations which preserve the sets of the partition (cf. 1.1). If F is a field of characteristic 0, given M with $\lambda(M) = \lambda = \{h_1, h_2, \dots, h_n\}$, we have

$$m_\lambda = \frac{1}{\prod_i h_i!} \sum_{\sigma \in S_n} \sigma(M).$$

2.6 Invariant Functions

It is time to develop some other examples. First, consider the set $\{1, \dots, n\}$ and a ring A (in most applications A will be the integers or the real or complex numbers).

A function f from $\{1, \dots, n\}$ to A may be thought of as a vector, and displayed, for instance, as a row with the notation $(a_1, a_2, \dots, a_n)$ where $a_i := f(i)$. The set of all functions is thus denoted by A^n. The symmetric group acts on such functions according to the general formula 2.5.1:

$$\sigma(a_1, a_2, \dots, a_n) = (a_{\sigma^{-1}1}, a_{\sigma^{-1}2}, \dots, a_{\sigma^{-1}n}).$$

In this simple example, we already see that the group action is linear. We will refer to this action as the *standard permutation action.*

We remark that if $\underline{e}_i$ denotes the canonical basis vector with coordinates 0 except 1 in the ith position, we have $\sigma(\underline{e}_i) = \underline{e}_{\sigma(i)}$. This formula allows us to describe the matrix of σ in the given basis: it is the matrix $\delta_{\sigma^{-1}(j),i}$. These matrices are called *permutation matrices.*

If we consider a G-set X and a ring A, the set of functions on X with values in A also forms a ring under pointwise sum and multiplication, and we have:

Remark. The group G acts on the functions with values in A as a group of ring automorphisms.

In this particular example it is important to proceed further. Once we have the action of S_n on A^n we may continue and act on the functions on A^n! In fact let us consider the *coordinate functions*: $x_i : (a_1, a_2, \dots, a_n) \to a_i$. It is clear from the general formulas that the symmetric group permutes the coordinate functions and $\sigma(x_i) = x_{\sigma(i)}$. The reader may note the fact that the inverse has now disappeared.

If we have a ring R and an action of a group G on R as ring automorphisms it is clear that:

Proposition. *The invariant elements form a subring of R.*

Thus we can speak of *the ring of invariants* R^G.

2.7 Commuting Actions

We need another generality. Suppose that we have two group actions on the same set X, i.e., assume that we have two groups G and H acting on the same set X.

We say that the two actions commute if $gh(x) = hg(x)$ for all $x \in X$, $g \in G$ and $h \in H$.

This means that every element of G gives rise to an H equivariant map (or we can reverse the roles of G and H). It also means that we really have an action of the product group $G \times H$ on X given by $(g, h)x = ghx$.

In this case, we easily see that if a function f is G-invariant and $h \in H$, then hf is also G-invariant. Hence H acts on the set of G-invariant functions.

More generally, suppose that we are given a G action on X and a normal subgroup K of G. It is easily seen that the quotient group G/K acts on the set of K-invariant functions and a function is G-invariant if and only if it is K and G/K-invariant.

Example. The right and left actions of G on itself commute (Example 1.2c).

3 Linear Actions, Groups of Automorphisms, Commuting Groups

3.1 Linear Actions

In §2.5, given an action of a group G on a set X and a field F, we deduced an action over the set F^X of functions from X to F, which is linear, i.e., given by linear operators.

In general, the groups G and the sets X on which they act may have further structures, as in the case of a topological, or differentiable, or algebraic action. In these cases it will be important to restrict the set of functions to the ones compatible with the structure under consideration. We will do it systematically.

If X is finite, the vector space of functions on X with values in F has, as a possible basis, the characteristic functions of the elements. It is convenient to identify an element x with its characteristic function and thus say that our vector space has X as a basis (cf. §2.5).

A function f is thus written as $\sum_{x \in X} f(x)x$. The linear action of G on F^X induces on this basis the action from which we started. We call such an action a *permutation representation*.

In the algebraic theory we may in any case consider the set of all functions which are finite sums of the characteristic functions of points, i.e., the functions which are 0 outside a finite set.

These are usually called *functions with finite support*. We will often denote these functions by the symbol $F[X]$, which is supposed to remind us that its elements are linear combinations of elements of X.

In particular, for the left action of G on itself we have the *algebraic regular representation* of G on $F[G]$. We shall see that this representation is particularly important.

Let us stress a feature of this representation.

We have two actions of G on G, the left and the right action, which commute with each other. In other words we have an action of $G \times G$ on G, given by $(h, k)g = hgk^{-1}$ (for which $G = G \times G/\Delta$ where $\Delta = G$ embedded diagonally; cf. 1.2c and 2.2).

Thus we have the corresponding two actions on $F[G]$ by $(h, k)f(g) = f(h^{-1}gk)$ and we may view the right action as symmetries of the left action and conversely. Sometimes it is convenient to write ${}^h f^k = (h, k)f$ to stress the left and right actions.

After these basic examples we give a general definition:

Definition 1. Given a vector space V over a field F (or more generally a module), we say that an action of a group G on V is linear if every element of G induces a linear transformation on V. A linear action of a group is also called a *linear representation*;[2] a vector space V that has a G-action is called a *G-module*.

In different language, let us consider the set of all linear invertible transformations of V. This is a group under composition (i.e., it is a subgroup of the group of all invertible transformations) and will be called the *general linear group* of V, denoted by the symbol $GL(V)$.

If we take $V = F^n$ (or equivalently, if V is finite dimensional and we identify V with F^n by choosing a basis), we can identify $GL(V)$ with the group of $n \times n$ invertible matrices with coefficients in F, denoted by $GL(n, F)$.

According to our general principles, a linear action is thus a homomorphism ϱ from G to $GL(V)$ (or to $GL(n, F)$).

When we are dealing with linear representations we usually also consider equivariant linear maps between them, thus obtaining a category and a notion of isomorphism.

Exercise. Two linear representations $\rho_1, \rho_2 : G \to GL(n, F)$ are isomorphic if and only if there is an invertible matrix $X \in GL(n, F)$ such that $X\rho_1(g)X^{-1} = \rho_2(g)$ for all $g \in G$.

Before we proceed any further, we should remark on an important feature of the theory.

Given two linear representations U, V, we can form their direct sum $U \oplus V$, which is a representation by setting $g(u, v) = (gu, gv)$. If $X = A \sqcup B$ is a G-set, where A and B are two disjoint G stable subsets, we clearly have $F^{A \sqcup B} = F^A \oplus F^B$. Thus the decomposition into a direct sum is a generalization of the decomposition of a space into G-stable sets.

If X is an orbit it cannot be further decomposed as a set, while F^X might be decomposable. The simplest example is $G = \{1, \tau = (12)\}$ the group with two elements of permutations of $[1, 2]$. The space F^X decomposes ($\mathrm{char} F \neq 2$), setting

[2] Sometimes we drop the term *linear* and just speak of *a representation*.

$$u_1 := \frac{e_1 + e_2}{2}, \quad u_2 := \frac{e_1 - e_2}{2};$$

we have $\tau u_1 = u_1,\ \tau(u_2) = -u_2$.

We have implicitly used the following ideas:

Definition 2.

(i) Given a linear representation V, a subspace U of V is a *subrepresentation* if it is stable under G.
(ii) V is a *decomposable representation* if we can find a decomposition $V = U_1 \oplus U_2$ with the U_i proper subrepresentations. Otherwise it is called *indecomposable*.
(iii) V is an *irreducible* representation if the only subrepresentations of V are V and 0.

We will study in detail some of the deep connections between these notions.

We will stress in a moment the analogy with the abstract theory of modules over a ring A. First we consider two basic examples:

Example. Let A be the group of all invertible $n \times n$ matrices over a field F.

Consider B^+ and B^-, the subgroups of all upper (resp., lower) triangular invertible matrices. Here "upper triangular" means "with 0 below the diagonal."

Exercise. The vector space F^n is irreducible as an A module, indecomposable but not irreducible as a B^+ or B^- module.

Definition 3. Given two linear representations U, V of a group G, the space of G-equivariant linear maps is denoted by $\hom_G(U, V)$ and called the space of *intertwining* operators.

In this book we will almost always treat finite-dimensional representations. Thus, unless specified otherwise, our vector spaces will always be assumed to be finite dimensional.

3.2 The Group Algebra

It is quite useful to rephrase the theory of linear representations in a different way. Consider the space $F[G]$:

Theorem.

(i) The group multiplication extends to a bilinear product on $F[G]$ under which $F[G]$ is an associative algebra with 1 called ***the group algebra****.*
(ii) Linear representations of G are the same as $F[G]$-modules.

Proof. The first part is immediate. As for the second, given a linear representation of G we have the module action $\left(\sum_{g\in G} a_g g\right)v := \sum_{g\in G} a_g(gv)$. The converse is clear. □

It is useful to view the product of elements $a, b \in F[G]$ as a *convolution of functions*:

$$(ab)(g) = \sum_{h,k\in G \mid hk=g} a(h)b(k) = \sum_{h\in G} a(h)b(h^{-1}g) = \sum_{h\in G} a(gh)b(h^{-1}).$$

Remark. Convolution can also be defined for special classes of functions on infinite groups which do not have finite support. One such extension comes from functional analysis and it applies to L^1-functions on locally compact groups endowed with a Haar measure (Chapter 8). Another extension comes in the theory of reductive algebraic groups (cf. Chapter 7).

Remark. (1) Consider the left and right action on the functions $F[G]$.

Let $h, k, g \in G$ and identify g with the characteristic function of the set $\{g\}$. Then ${}^h g^k = hgk^{-1}$ (as functions).

The space $F[G]$ as a $G \times G$ module is the permutation representation associated to $G = G \times G/\Delta$ with its $G \times G$ action (cf. 2.2 and 3.1). Thus a space of functions on G is stable under left (resp. right) action if and only if it is a left (resp. right) ideal of the group algebra $F[G]$.

(2) Notice that the direct sum of representations is the same as the direct sum as modules. Also a G-linear map between two representations is the same as a module homomorphism.

Example. Let us consider a finite group G, a subgroup K and the linear space $F[G/K]$, which as we have seen is a permutation representation.

We can identify the functions on G/K with the functions

(3.2.1) $$F[G]^K = \{a \in F[G] | ah = a, \ \forall h \in K\}$$

on G which are invariant under the right action of K. In this way the element $gK \in G/K$ is identified with the characteristic function of the coset gK, and $F[G/K]$ is identified with the left ideal of the group algebra $F[G]$ having as basis the characteristic functions χ_{gK} of the left cosets of K.

If we denote by $u := \chi_K$ the characteristic function of the subgroup K, we see that $\chi_{gK} = gu$ and that u generates this module over $F[G]$.

Given two subgroups H, K and the linear spaces $F[G/H], F[G/K] \subset F[G]$, we want to determine their intertwiners. We assume char $F = 0$.

For an intertwiner f, and $u := \chi_H$ as before, let $f(u) = a \in F[G/K]$. We have $hu = u, \ \forall h \in H$ and so, since f is an intertwiner, $a = f(u) = f(hu) = ha$. Thus we must have that a is also left invariant under H. Conversely, given such an a, the map $b \mapsto \frac{ba}{|H|}$ is an intertwiner mapping u to a. Since u generates $F[G/H]$ as a module we see that:

Proposition. *The space* $\hom_G(F[G/H], F[G/K])$ *of intertwiners can be identified with the (left) H-invariants of $F[G/K]$, or with the H-K invariants ${}^H F[G]^K$ of $F[G]$. It has as basis the characteristic functions of the double cosets HgK.*

In particular, for $H = K$ we have that the functions which are biinvariants under H form under convolution the endomorphism algebra of $F[G/H]$. Since we identify $F[G/H]$ with a subspace of $F[G]$, the claim is that multiplication on the left by a function that is biinvariant under H maps $F[G/H]$ into itself and identifies this space of functions with the full algebra of endomorphisms of $F[G/H]$.

These functions have as basis the characteristic functions of the double cosets HgH; one usually indicates by $T_g = T_{HgH}$ the corresponding operator. $\mathrm{End}(F[G/H])$ and T_g are called the *Hecke algebra* and *Hecke operators*, respectively. The multiplication rule between such operators depends on the multiplication on cosets $HgHHkH = \cup Hh_iH$, and each double coset appearing in this product appears with a positive integer multiplicity so that $T_gT_h = \sum n_i T_{h_i}$.[3]

There are similar results when we have three subgroups H, K, L and compose

$$\hom_G(F[G/H],\ F[G/K]) \times \hom_G(F[G/K], F[G/L]) \xrightarrow{\circ} \hom_G(F[G/H],\ F[G/L]).$$

The notion of permutation representation is a special case of that of *induced representation*. If M is a representation of a subgroup H of a group G we consider the space of functions $f : G \to M$ with the constraint:

$$\text{(3.2.2)} \qquad \mathrm{Ind}_H^G M := \{f : G \to M \mid f(gh^{-1}) = hf(g),\ \forall h \in H,\ g \in G\}.$$

On this space of functions we define a G-action by $(gf)(x) := f(g^{-1}x)$. It is easy to see that this is a well-defined action. Moreover, we can identify $m \in M$ with the function f_m such that $f_m(x) = 0$ if $x \notin H$ and $f_m(h) = h^{-1}m$ if $h \in H$. Now $M \subset \mathrm{Ind}_H^G M$.

Exercise. Verify that by choosing a set of representatives of the cosets G/H, we have the vector space decomposition

$$\text{(3.2.3)} \qquad \mathrm{Ind}_H^G M := \oplus_{g \in G/H}\, gM.$$

3.3 Actions on Polynomials

Let V be a G-module. Given a linear function $f \in V^*$ on V, by definition the function gf is given by $(gf)(v) = f(g^{-1}v)$ and hence it is again a linear function.

Thus G acts dually on the space V^* of linear functions. It is clear that this is a linear action, which is called the *contragredient* action.

In matrix notation, using dual bases, the contragredient action of an operator T is given by the inverse transpose of the matrix of T.

---

[3] It is important in fact to use these concepts in a much more general way as done by Hecke in the theory of modular forms. Hecke studies the action of $Sl(2, Z)$ on $M_2(Q)$ the 2×2 rational matrices. In this case one has also double cosets, a product structure on $M_2(Q)$ and the fact that a double coset is a finite union of right or left cosets. These properties suffice to develop the Hecke algebra. In this case this algebra acts on a different space of functions, the modular forms (cf. [Ogg]).

We will use the notation $\langle\varphi|v\rangle$ for the value of a linear form on a vector, and thus we have (by definition) the identity

(3.3.1) $$\langle g\varphi|v\rangle = \langle\varphi|g^{-1}v\rangle.$$

Alternatively, it may be convenient to define on V^* a *right action* by the more symmetric formula

(3.3.2) $$\langle\varphi g|v\rangle = \langle\varphi|gv\rangle.$$

Exercise. Prove that the dual of a permutation representation is isomorphic to the same permutation representation. In particular, one can apply this to the dual of the group algebra.

In the set of all functions on a finite-dimensional vector space V, the polynomial functions play a special role. By definition a polynomial function is an element of the subalgebra (of the algebra of all functions with values in F) generated by the linear functions.

If we choose a basis and consider the coordinate functions $x_1, x_2, \dots, x_n$ with respect to the chosen basis, a polynomial function is a usual polynomial in the x_i. If F is infinite, the expression as a polynomial is unique and we can consider the x_i as given variables.

The ring of polynomial functions on V will be denoted by $P[V]$ and the ring of formal polynomials by $F[x_1, x_2, \dots, x_n]$.

Choosing a basis, we always have a surjective homomorphism $F[x_1, x_2, \dots, x_n] \to P[V]$ which is an isomorphism if F is infinite.

Exercise. If F is a finite field with q elements, prove that $P[V]$ has dimension q^n over F, and that the kernel of the map $F[x_1, x_2, \dots, x_n] \to P[V]$ is the ideal generated by the elements $x_i^q - x_i$.

Since the linear functions are preserved under a given group action we have:

Proposition. *Given a linear action of a group G on a vector space V, G acts on the polynomial functions $P[V]$ as a group of ring automorphisms by the rule $(gf)(v) = f(g^{-1}v)$.*

Of course, the full linear group acts on the polynomial functions. In the language of coordinates we may view the action as a linear change of coordinates.

Exercise. Show that we always have a linear action of $GL(n, F)$ on the formal polynomial ring $F[x_1, x_2, \dots, x_n]$.

3.4 Invariant Polynomials

We assume the base field to be infinite for simplicity although the reader can see easily what happens for finite fields. One trivial but important remark is that the group action on $P[V]$ preserves the degree.

Recall that a function f on V is *homogeneous of degree* k if $f(\alpha v) = \alpha^k f(v)$ for all α and v.

The set $P_q[V]$ of homogeneous polynomials of degree q is a subspace, called in classical language the space of *quantics*. If $\dim(V) = n$, one speaks of *n-ary* quantics.[4]

In general, a direct sum of vector spaces $U = \oplus_{k=0}^{\infty} U_k$ is called a *graded vector space*. A subspace W of U is called *homogeneous*, if, setting $W_i := W \cap U_i$, we have $W = \oplus_{k=0}^{\infty} W_k$.

The space of polynomials is thus a graded vector space $P[V] = \oplus_{k=0}^{\infty} P_k[V]$. One has immediately $(gf)(\alpha v) = f(\alpha g^{-1} v) = \alpha^k (gf)(v)$, which has an important consequence:

Theorem. *If a polynomial f is an invariant (under some linear group action), then its homogeneous components are also invariant.*

Proof. Let $f = \sum f_i$ be the decomposition of f into homogeneous components, $gf = \sum gf_i$ is the decomposition into homogeneous components of gf. If f is invariant $f = gf$, then $f_i = gf_i$ for each i since the decomposition into homogeneous components is unique. □

In order to summarize the analysis done up to now, let us also recall that an algebra A is called a *graded algebra* if it is a graded vector space, $A = \oplus_{k=0}^{\infty} A_k$ and if for all h, k we have $A_h A_k \subset A_{h+k}$.[5]

Proposition. *The spaces $P_k[V]$ are subrepresentations. The set $P[V]^G$ of invariant polynomials is a graded subalgebra.*

3.5 Commuting Linear Actions

To some extent the previous theorem may be viewed as a special case of the more general setting of commuting actions.

Given two representations $\varrho_i : G \to GL(V_i)$, $i = 1, 2$, consider the linear transformations between V_1 and V_2 which are G equivariant. It is clear that they form a linear subspace of the space of all linear maps between V_1 and V_2.

The space of all linear maps will be denoted by $\hom(V_1, V_2)$, while the space of equivariant maps will be denoted by $\hom_G(V_1, V_2)$. In particular, when the two spaces coincide we write $\mathrm{End}(V)$ or $\mathrm{End}_G(V)$ instead of $\hom(V, V)$ or $\hom_G(V, V)$.

These spaces $\mathrm{End}_G(V) \subset \mathrm{End}(V)$ are in fact now algebras, under composition of operators. Choosing bases we have that $\mathrm{End}_G(V)$ is the set of all matrices which commute with all the matrices coming from the group G.

Consider now the set of invertible elements of $\mathrm{End}_G(V)$, i.e., the group H of all linear operators which commute with G.

By the remarks of §3.3, H preserves the degrees of the polynomials and maps the algebra of G-invariant functions into itself. Thus we have:

---

[4] E.g., quadratics, cubics, quartics, quintics, etc., $q = 2, 3, 4, 5, \dots$.

[5] We are restricting to $\mathbb{N}$ gradings. The notion is more general: any monoid could be used as a set of indices for the grading. In this book we will use only $\mathbb{N}$ and $\mathbb{Z}/(2)$ in Chapter 5.

Remark. H induces a group of automorphisms of the graded algebra $P[V]^G$.

We view this remark as a generalization of Theorem 3.4 since the group of scalar multiplications commutes (by definition of linear transformation) with all linear operators. Moreover it is easy to prove:

Exercise. Given a graded vector space $U = \oplus_{k=0}^{\infty} U_k$, define an action ϱ of the multiplicative group F^* of F setting $\varrho(\alpha)(v) := \alpha^k v$ if $v \in U_k$. Prove that a subspace is stable under this action if and only if it is a graded subspace (F is assumed to be infinite).

2

Symmetric Functions

Summary. Our aim is to alternate elements of the general theory with significant examples. We deal now with symmetric functions.

In this chapter we will develop some of the very basic theorems on symmetric functions, in part as a way to give a look into 19th century invariant theory, but as well to establish some useful formulas which will show their full meaning only after developing the representation theory of the linear and symmetric groups.

1 Symmetric Functions

1.1 Elementary Symmetric Functions

The theory of symmetric functions is a classical theory developed (by Lagrange, Ruffini, Galois, and others) in connection with the theory of algebraic equations in one variable and the classical question of resolution by radicals.

The main link is given by the formulas expressing the coefficients of a polynomial through its roots. A formal approach is the following.

Consider polynomials in variables $x_1, x_2, \ldots, x_n$ and an extra variable t over the ring of integers. The *elementary symmetric functions* $e_i := e_i(x_1, x_2, \ldots, x_n)$ are implicitly defined by the formula

$$(1.1.1) \qquad p(t) := \prod_{i=1}^{n}(1 + tx_i) := 1 + \sum_{i=1}^{n} e_i t^i.$$

More explicitly, $e_i(x_1, x_2, \ldots, x_n)$ is the sum of $\binom{n}{i}$ terms: the products, over all subsets of $\{1, 2, \ldots, n\}$ with i elements of the variables with indices in that subset. That is,

$$(1.1.2) \qquad e_i = \sum_{1 \le a_1 < a_2 < \cdots < a_i \le n} x_{a_1} x_{a_2} \cdots x_{a_i}.$$

If σ is a permutation of the indices, we obviously have

$$\prod_{i=1}^{n}(1+tx_i)=\prod_{i=1}^{n}(1+tx_{\sigma i}).$$

Thus the elements e_i are invariant under permutation of the variables.

Of course the polynomial $t^n p\big(-\frac{1}{t}\big)$ has the elements x_i as its roots.

Definition. A polynomial in the variables $(x_1, x_2, \dots, x_n)$, invariant under permutation of these variables, is called a symmetric function.

The functions e_i are called *elementary symmetric functions*.

There are several obviously symmetric functions, e.g., the *power sums* $\psi_k := \sum_{i=1}^{n} x_i^k$ and the functions S_k defined as the sum of all monomials of degree k. These are particular cases of the following general construction.

Consider the basis of the ring of polynomials given by the monomials. This basis is permuted by the symmetric group. By Proposition 2.5 of Chapter 1 we have:

A basis of the space of symmetric functions is given by the sums of monomials in the same orbit, for all orbits.

Orbits correspond to non-increasing vectors $\lambda := (h_1 \geq h_2 \geq \cdots \geq h_n), h_i \in \mathbb{N}$, and we have set m_λ to be the sum of monomials in the corresponding orbit.

As we will soon see there are also some subtler symmetric functions (the Schur functions) indexed by partitions, and this will play an important role in the sequel. We can start with a first important fact, the explicit connection between the functions e_i and ψ_k. To see this connection, we will perform the next computations in the ring of formal power series, although the series that we will consider also have meaning as convergent series.

Start from the identity $\prod_{i=1}^{n}(tx_i+1)=\sum_{i=0}^{n} e_i t^i$ and take the logarithmic derivative (relative to the variable t) of both sides. We use the fact that such an operator transforms products into sums to get

$$\sum_{i=1}^{n}\frac{x_i}{(tx_i+1)}=\frac{\sum_{i=1}^{n} ie_it^{i-1}}{\sum_{i=0}^{n} e_it^i}.$$

The left-hand side of this formula can be developed as

$$\sum_{i=1}^{n} x_i\sum_{h=0}^{\infty}(-tx_i)^h=\sum_{h=0}^{\infty}(-t)^h\psi_{h+1}.$$

From this we get the identity

$$\left(\sum_{h=0}^{\infty}(-t)^h\psi_{h+1}\right)\left(\sum_{i=0}^{n} e_it^i\right)=\sum_{i=1}^{n} ie_it^{i-1}$$

which gives, equating coefficients:

$$(1.1.3)\quad (-1)^m\psi_{m+1}+\sum_{i=1}^{m}(-1)^i\psi_ie_{m+1-i}=\sum_{i+j=m}(-1)^i\psi_{i+1}e_j=(m+1)e_{m+1}$$

where we take $e_i = 0$ if $i > n$.

It is clear that these formulas give recursive ways of expressing the ψ_i in terms of the e_j with integral coefficients. On the other hand, they can also be used to express the e_i in terms of the ψ_j, but in this case it is necessary to perform some division operations; the coefficients are rational and usually not integers.[6]

It is useful to give a second proof. Consider the map:

$$\pi_n : \mathbb{Z}[x_1, x_2, \dots, x_n] \to \mathbb{Z}[x_1, x_2, \dots, x_{n-1}]$$

given by evaluating x_n at 0.

Lemma. *The intersection of* $\mathrm{Ker}(\pi_n)$ *with the space of symmetric functions of degree* $< n$ *is reduced to* 0.

Proof. Consider $m_{(h_1,h_2,\dots,h_n)}$, a sum of monomials in an orbit. If the degree is less than n, we have $h_n = 0$. Under π_n we get $\pi_n(m_{(h_1,h_2,\dots,h_{n-1},0)}) = m_{(h_1,h_2,\dots,h_{n-1})}$. Thus if the degree is less than n, the map π_n maps these basis elements into distinct basis elements.

Now we give the second proof of 1.1.3. In the identity $\prod_{i=1}^n (t - x_i) := \sum_{i=0}^n (-1)^i e_i t^{n-i}$, substitute t with x_i, and then summing over all i we get (remark that $\psi_0 = n$):

$$0 = \sum_{i=0}^{n} (-1)^i e_i \psi_{n-i}, \text{ or } \psi_n = \sum_{i=1}^{n} (-1)^{i-1} e_i \psi_{n-i}.$$

By the previous lemma this identity also remains valid for symmetric functions in more than n variables and gives the required recursion.

1.2 Symmetric Polynomials

It is actually a general fact that symmetric functions can be expressed as polynomials in the elementary symmetric functions. We will now discuss an algorithmic proof.

To make the proof transparent, let us also stress in our formulas the number of variables and denote by $e_i^{(k)}$ the ith elementary symmetric function in the variables $x_1, \dots, x_k$. Since

$$\left(\sum_{i=0}^{n-1} e_i^{(n-1)} t^i\right)(tx_n + 1) = \sum_{i=0}^{n} e_i^{(n)} t^i,$$

we have

$$e_i^{(n)} = e_{i-1}^{(n-1)} x_n + e_i^{(n-1)} \quad \text{or} \quad e_i^{(n-1)} = e_i^{(n)} - e_{i-1}^{(n-1)} x_n.$$

In particular, in the homomorphism $\pi : \mathbb{Z}[x_1, \dots, x_n] \to \mathbb{Z}[x_1, \dots, x_{n-1}]$ given by evaluating x_n at 0, we have that symmetric functions map to symmetric functions and

$$\pi(e_i^{(n)}) = e_i^{(n-1)}, \; i < n, \quad \pi(e_n^{(n)}) = 0.$$

---

[6] These formulas were found by Newton, hence the name *Newton functions* for the ψ_k.

Given a symmetric polynomial $f(x_1, \dots, x_n)$ we evaluate it at $x_n = 0$. If the resulting polynomial $f(x_1, \dots, x_{n-1}, 0)$ is 0, then f is divisible by x_n.

If so, by symmetry it is divisible by all of the variables and hence by the function e_n. We perform the division and move on to another symmetric function of lower degree.

Otherwise, by recursive induction one can construct a polynomial p in $n-1$ variables which, evaluated in the $n-1$ elementary symmetric functions of $x_1, \dots, x_{n-1}$, gives $f(x_1, \dots, x_{n-1}, 0)$. Thus $f - p(e_1, e_2, \dots, e_{n-1})$ is a symmetric function vanishing at $x_n = 0$.

We are back to the previous step.

The uniqueness is implicit in the algorithm which can be used to express any symmetric polynomial as a unique polynomial in the elementary symmetric functions.

Theorem 1. *A symmetric polynomial $f \in \mathbb{Z}[x_1, \dots, x_n]$ is a polynomial, in a unique way and with coefficients in $\mathbb{Z}$, in the elementary symmetric functions.*

It is quite useful, in view of the previous lemma and theorem, to apply the same ideas to symmetric functions in larger and larger sets of variables. One then constructs a *limit ring*, which one calls just the *formal ring of symmetric functions* $\mathbb{Z}[e_1, \dots, e_i, \dots]$. It can be thought of as the polynomial ring in infinitely many variables e_i, where formally we give degree (or weight) i to e_i. The ring of symmetric functions in n variables is obtained by setting $e_i = 0, \ \forall i > n$. One often develops formal identities in this ring with the idea that, in order to verify an identity which is homogeneous of some degree m, it is enough to do it for symmetric functions in m variables.

In the same way the reader may understand the following fact. Consider the $n!$ monomials

$$x_1^{h_1} \cdots x_{n-1}^{h_{n-1}}, \ 0 \leq h_i \leq n - i.$$

Theorem 2. *The above monomials are a basis of $\mathbb{Z}[x_1, \dots, x_n]$ over $\mathbb{Z}[e_1, \dots, e_n]$.*

Remark. The same theorem is clearly true if we replace the coefficient ring $\mathbb{Z}$ by any commutative ring A. In particular, we will use it when A is itself a polynomial ring.

2 Resultant, Discriminant, Bézoutiant

2.1 Polynomials and Roots

In order to understand the importance of Theorem 1 of 1.2 on elementary symmetric functions and also the classical point of view, let us develop a geometric picture.

Consider the space $\mathbb{C}^n$ and the space $P_n := \{t^n + b_1 t^{n-1} + \cdots + b_n\}$ of monic polynomials (which can be identified with $\mathbb{C}^n$ by the use of the coefficients).

Consider next the map $\pi : \mathbb{C}^n \to P_n$ given by

$$\pi(\alpha_1, \dots, \alpha_n) := \prod_{i=1}^{n} (t - \alpha_i).$$

We thus obtain a polynomial $t^n - a_1 t^{n-1} + a_2 t^{n-2} + \cdots + (-1)^n a_n = 0$ with roots $\alpha_1, \dots, \alpha_n$ (and the coefficients a_i are the elementary symmetric functions in the roots). Any monic polynomial is obtained in this way (Fundamental Theorem of Algebra).

Two points in $\mathbb{C}^n$ project to the same point in P_n if and only if they are in the same orbit under the symmetric group, i.e., P_n parameterizes the S_n-orbits.

Suppose we want to study a property of the roots which can be verified by evaluating some symmetric polynomials in the roots (this will usually be the case for any condition on the set of all roots). Then one can perform the computation without computing the roots, since one has only to study the formal symmetric polynomial expression and, using the alogrithm discussed in §1.2 (or any equivalent algorithm), express the value of a symmetric function of the roots through the coefficients.

In other words, a symmetric polynomial function f on $\mathbb{C}^n$ factors through the map π giving rise to an effectively computable[7] polynomial function $\overline{f}$ on P_n such that $f = \overline{f}\pi$.

A classical example is given by the *discriminant.*

The condition that the roots be distinct is clearly that $\prod_{i<j}(\alpha_i - \alpha_j) \neq 0$. The polynomial $V(x) := \prod_{i<j}(x_i - x_j)$ is in fact not symmetric. It is the value of the Vandermonde determinant, i.e., the determinant of the matrix:

$$A := \begin{pmatrix} x_1^{n-1} & x_2^{n-1} & \dots & x_n^{n-1} \\ \vdots & \vdots & \ddots & \vdots \\ x_1^2 & x_2^2 & \dots & x_n^2 \\ x_1 & x_2 & \dots & x_n \\ 1 & 1 & \dots & 1 \end{pmatrix}. \tag{2.1.1}$$

Proposition 1. *$V(x)$ is antisymmetric, i.e., a permutation of the variables results in the multiplication of $V(x)$ by the sign of the permutation.*

Remark. The theory of the sign of permutations can be deduced by analyzing the Vandermonde determinant. In fact, since for a transposition τ it is clear that $V(x)^\tau = -V(x)$, it follows that $V(x)^\sigma = V(x)$ or $-V(x)$ according to whether σ is a product of an even or an odd number of transpositions. The sign is then clearly a homomorphism.

We also see immediately that V^2 is a symmetric polynomial. We can compute it in terms of the functions ψ_i as follows. Consider the matrix $B := AA^t$. Clearly in the i, j entry of B we find the symmetric function $\psi_{2n-(i+j)}$, and the determinant of B is V^2.

[7] I.e., computable without solving the equation, usually by polynomial expressions in the coefficients.

The matrix B (or rather the one reordered with ψ_{i+j-2} in the i, j position) is classically known as the *Bézoutiant*, and it carries some further information about the roots. We shall see that there is a different determinant formula for the determinant of B directly involving the elementary symmetric functions.

Let $D(e_1, e_2, \dots, e_n)$ be the expression for V^2 as a polynomial in the elementary symmetric functions (e.g., $n = 2$, $D = e_1^2 - 4e_2$).

Definition. The polynomial D is called the discriminant.

Since this is an interesting example we will pursue it a bit further.

Let us assume that F is a field, and $f(t)$ is a monic polynomial (of degree n) with coefficients in F, and let $R := F[t]/(f(t))$. We have that R is an algebra over F of dimension n.

For any finite-dimensional algebra A over a field F we can perform the following construction.

Any element a of A induces a linear transformation $L_a : x \to ax$ on A by left multiplication (and also one by right multiplication). We define $\text{tr}(a) := \text{tr}(L_a)$, the trace of the operator L_a.

We consider next the bilinear form $(a, b) := \text{tr}(ab)$. This is the *trace form* of A. It is symmetric and *associative* in the sense that $(ab, c) = (a, bc)$.

We compute it first for $R := F[t]/(t^n)$. Using the fact that t is nilpotent we see that $\text{tr}(t^k) = 0$ if $k > 0$. Thus the trace form has rank 1 with kernel the ideal generated by t.

To compute for the algebra $R := F[t]/(f(t))$ we pass to the algebraic closure $\overline{F}$ and compute in $\overline{F}[t]/(f(t))$.

We split the polynomial with respect to its distinct roots, $f(t) = \prod_{i=1}^k (t - \alpha_i)^{h_i}$, and $\overline{F}[t]/(f(t)) = \oplus_{i=1}^k \overline{F}[t]/(t - \alpha_i)^{h_i}$. Thus the trace of an element mod $f(t)$ is the sum of its traces mod $(t - \alpha_i)^{h_i}$.

Let us compute the trace of t^k mod $(t - \alpha_i)^{h_i}$. We claim that it is $h_i \alpha_i^k$. In fact in the basis $1, (t - \alpha_i), (t - \alpha_i)^2, \dots, (t - \alpha_i)^{h_i - 1}$ (mod $(t - \alpha_i)^{h_i}$) the matrix of t is lower triangular with constant eigenvalue α_i on the diagonal, and so the claim follows.

Adding all of the contributions, we see that in $F[t]/(f(t))$, the trace of multiplication by t^k is $\sum_i h_i \alpha_i^k$, the kth Newton function of the roots.

As a consequence we see that the matrix of the trace form, in the basis $1, t, \dots, t^{n-1}$, is the Bézoutiant of the roots. Since for a given block $\overline{F}[t]/(t - \alpha_i)^{h_i}$ the ideal generated by $(t - \alpha_i)$ is nilpotent of codimension 1, we see that it is exactly the radical of the block, and the kernel of its trace form. It follows that:

Proposition 2. *The rank of the Bézoutiant equals the number of distinct roots.*

Given a polynomial $f(t)$ let $\overline{f}(t)$ denote the polynomial with the same roots as $f(t)$ but all distinct. It is the generator of the radical of the ideal generated by $f(t)$. In characteristic zero this polynomial is obtained dividing $f(t)$ by the GCD between $f(t)$ and its derivative $f'(t)$.

Let us consider next the algebra $R := F[t]/(f(t))$, its radical N and $\overline{R} := R/N$. By the previous analysis it is clear that $\overline{R} = F[t]/(\overline{f}(t))$.

Consider now the special case in which $F = \mathbb{R}$ is the field of real numbers. Then we can divide the distinct roots into the real roots $\alpha_1, \alpha_2, \dots, \alpha_k$ and the complex ones $\beta_1, \overline{\beta}_1, \beta_2, \overline{\beta}_2, \dots, \beta_h, \overline{\beta}_h$.

The algebra $\overline{R}$ is isomorphic to the direct sum of k copies of $\mathbb{R}$ and h copies of $\mathbb{C}$. Its trace form is the orthogonal sum of the corresponding trace forms. Over $\mathbb{R}$ the trace form is just x^2 but over $\mathbb{C}$ we have $\mathrm{tr}((x+iy)^2) = 2(x^2 - y^2)$. We deduce:

Theorem. *The number of real roots of $f(t)$ equals the signature*[8] *of its Bézoutiant.*

As a simple but important corollary we have:

Corollary. *A real polynomial has all its roots real and distinct if and only if the Bézoutiant is positive definite.*

There are simple variations on this theme. For instance, if we consider the quadratic form $Q(x) := \mathrm{tr}(tx^2)$ we see that its matrix is again easily computed in terms of the ψ_k and its signature equals the number of real positive roots minus the number of real negative roots. In this way one can also determine the number of real roots in any interval.

These results are Sylvester's variations on Sturm's theorem. They can be found in the paper in which he discusses the Law of Inertia that now bears his name (cf. [Si]).

2.2 Resultant

Let us go back to the roots. If $x_1, x_2, \dots, x_n; y_1, y_2, \dots, y_m$ are two sets of variables, consider the polynomial

$$A(x, y) := \prod_{i=1}^{n}\prod_{j=1}^{m}(x_i - y_j).$$

This is clearly symmetric, separately in the variables x and y. If we evaluate it in numbers, it vanishes if and only if one of the values of the x's coincides with a value of the y's. Conversely, any polynomial in these two sets of variables that has this property is divisible by all the factors $x_i - y_j$, and hence it is a multiple of A.

By the general theory A, a symmetric polynomial in the x_i's, can be expressed as a polynomial R in the elementary symmetric functions $e_i(x)$ with coefficients that are polynomials symmetric in the y_j. These coefficients are thus in turn polynomials in the elementary symmetric functions of the y_j's.

Let us denote by $a_1, \dots, a_n$ the elementary symmetric functions in the x_i's and by $b_1, \dots, b_m$ the ones in the y_j's. Thus $A(x, y) = R(a_1, \dots, a_n, b_1, \dots, b_m)$ for some explicit polynomial R.

---

[8] The Bézoutiant is a real symmetric matrix; for such a matrix the notion of *signature* is explained in Chapter 5, 3.3. There are effective algorithms to compute the signature.

The polynomial R is called the *resultant.*

When we evaluate the variables x and y to be the roots of two monic polynomials $f(t),\ g(t)$ of degrees n, m, respectively, we see that the value of A can be computed by evaluating R in the coefficients (with some signs) of these polynomials. Thus the resultant is a polynomial in their coefficients, vanishing when the two polynomials have a common root.

There is a more general classical expression for the resultant as a determinant, and we drop the condition that the polynomials be monic. The theory is the following.

Let $f(t) := a_0 t^n + a_1 t^{n-1} + \cdots + a_n,\ g(t) := b_0 t^m + b_1 t^{m-1} + \cdots + b_m$ and let us denote by P_h the $h+1$-dimensional space of all polynomials of degree $\leq h$.

Consider the linear transformation:

$$T_{f,g} : P_{m-1} \oplus P_{n-1} \to P_{m+n-1} \text{ given by } T_{f,g}(a,b) := fa + gb.$$

This is a transformation between two $n+m$-dimensional spaces and, in the bases $(1,0), (t,0), \ldots, (t^{m-1},0), (0,1), (0,t), \ldots, (0,t^{n-1})$ and $1, t, t^2, \ldots, t^{n+m-1}$, it is quite easy to write down its square matrix $R_{f,g}$:

$$\tag{2.2.1} \begin{pmatrix}
a_n & 0 & 0 & \ldots & 0 & b_m & 0 & \ldots & 0 & 0 & 0 \\
a_{n-1} & a_n & 0 & \ldots & 0 & b_{m-1} & b_m & \ldots & \ldots & \ddots & \\
a_{n-2} & a_{n-1} & a_n & 0 & \ldots & b_{m-2} & b_{m-1} & b_m & \ddots & \ddots & \vdots \\
\vdots & \vdots & \vdots & \ddots & \vdots & \vdots & \ddots & \vdots & \ddots & \ddots & \vdots \\
a_1 & a_2 & a_3 & & & & & & \ddots & \ddots & \vdots \\
a_0 & a_1 & a_2 & & & & & & & & \\
0 & a_0 & a_1 & & & & & & \ddots & \ddots & \vdots \\
\vdots & \vdots & \vdots & \ddots & \vdots & \vdots & \ddots & \vdots & \ddots & \ddots & \vdots \\
0 & \vdots & & & & b_0 & b_1 & b_2 & \ddots & \vdots & \\
0 & 0 & \ldots & & & 0 & b_0 & b_1 & \ddots & \ddots & \vdots \\
0 & 0 & & & & 0 & 0 & b_0 & \ddots & \ddots & \vdots \\
\vdots & \vdots & \vdots & \ddots & \vdots & \vdots & \ddots & \vdots & \ddots & \ddots & \vdots \\
\vdots & \vdots & \vdots & \ddots & \vdots & \vdots & \ddots & \vdots & \ddots & b_0 & \vdots \\
0 & 0 & 0 & \ldots & a_0 & 0 & \ldots & \ldots & 0 & 0 & b_0
\end{pmatrix}.$$

Proposition. *If $a_0 b_0 \neq 0$, the rank of $T_{f,g}$ equals $m+n-d$ where d is the degree of $h := \mathrm{GCD}(f,g)$*[9].

Proof. By Euclid's algorithm the image of $T_{f,g}$ consists of all polynomials of degree $\leq n+m-1$ and multiples of h. Its kernel consists of pairs $(sg', -sf')$ where $f = hf',\ g = hg'$. The claim follows.

[9] $\mathrm{GCD}(f,g)$ is the greatest common divisor of f, g.

As a corollary we have that the determinant $R(f, g)$ of $R_{f,g}$ vanishes exactly when the two polynomials have a common root. This gives us a second definition of resultant.

Definition. The polynomial $R(f, g)$ is called the *resultant* of the two polynomials $f(t), g(t)$.

If we consider the coefficients of f and g as variables, we can still think of $T_{f,g}$ as a map of vector spaces, except that the base field is the field of rational functions in the given variables. Then we can solve the equation $fa + gb = 1$ by Cramer's rule and we see that the coefficients of the polynomials a, b are given by the cofactors of the first row of the matrix $R_{f,g}$ divided by the resultant. In particular, we can write $R = Af(t) + Bg(t)$ where A, B are polynomials in t of degrees $m - 1, n - 1$, respectively, and with coefficients polynomials in the variables $(a_0, a_1, \dots, a_n, b_0, b_1, \dots, b_m)$.

This can also be understood as follows. In the matrix $R_{f,g}$ we add to the first row the second multiplied by t, the third multiplied by t^2, and so on. We see that the first row becomes $(f(t), f(t)t, f(t)t^2, \dots, f(t)t^{m-1}, g(t), g(t)t, g(t)t^2, \dots, g(t)t^{n-1})$. Under these operations of course the determinant does not change. Then developing it along the first row we get the desired identity.

We have given two different definitions of resultant, which we need to compare:

Exercise. Consider the two polynomials as $a_0 \prod_{i=1}^{n}(t - x_i), b_0 \prod_{j=1}^{m}(t - y_j)$ and thus, in R, substitute the element $(-1)^i a_0 e_i(x_1, \dots, x_n)$ for the variables a_i and the element $(-1)^i b_0 e_i(y_1, \dots, y_m)$ for b_i. The polynomial we obtain is $a_0^m b_0^n A(x, y)$.

2.3 Discriminant

In the special case when we take $g(t) = f'(t)$, the derivative of $f(t)$, we have that the vanishing of the resultant is equivalent to the existence of multiple roots. We have already seen that the vanishing of the discriminant implies the existence of multiple roots. It is now easy to connect the two approaches.

The resultant $R(f, f')$ is considered as a polynomial in the variables $(a_0, a_1, \dots, a_n)$. If we substitute in $R(f, f')$ the element $(-1)^i a_0 e_i(x_1, \dots, x_n)$ for the variables a_i we have a polynomial in the x with coefficients involving a_0 that vanishes whenever two x's coincide.

Thus $R(f, f')$ is divisible by thc discriminant D of these variables. A degree computation shows in fact that it is a constant (with respect to the x) multiple cD. The constant c can be evaluated easily, for instance specializing to the polynomial $x^n - 1$. This polynomial has as roots the nth roots $e^{2\pi i k/n}$, $0 \le k < n$ of 1. The Newton functions

$$\psi_h := \sum_{i=0}^{n-1} e^{\frac{2\pi i h k}{n}} = \begin{cases} 0 & \text{if } \ h \nmid n \\ n & \text{if } \ h \mid n; \end{cases}$$

hence the Bézoutiant is $-(-n)^n$ and the computation of the resultant is n^n, so the constant is $(-1)^{n-1}$.

3 Schur Functions

3.1 Alternating Functions

Along with symmetric functions, it is also important to discuss alternating (or skew-symmetric, or antisymmetric) functions. We restrict our considerations to integral polynomials.

Definition. A polynomial f in the variables $(x_1, x_2, \dots, x_n)$ is called an alternating function if, for every permutation σ of these variables,

$$f^\sigma = f(x_{\sigma(1)}, x_{\sigma(2)}, \dots, x_{\sigma(n)}) = \epsilon_\sigma f(x_1, x_2, \dots, x_n),$$

ϵ_σ being the sign of the permutation.

We have seen the Vandermonde determinant $V(x) := \prod_{i<j}(x_i - x_j)$ as a basic alternating polynomial. The main remark on alternating functions is the following.

Proposition 1. *A polynomial $f(x)$, in the variables x, is alternating if and only if it is of the form $f(x) = V(x)g(x)$, with $g(x)$ a symmetric polynomial.*

Proof. Substitute, in an alternating polynomial f, for a variable x_j a variable x_i for $i \neq j$. We get the same polynomial if we first exchange x_i and x_j in f. Since this changes the sign, it means that under this substitution f becomes 0.

This means in turn that f is divisible by $x_i - x_j$; since i, j are arbitrary, f is divisible by $V(x)$. Writing $f = V(x)g$, it is clear that g is symmetric.

Let us be more formal. Let A, S denote the sets of *antisymmetric* and *symmetric* polynomials. We have seen that:

Proposition 2. *The space A of antisymmetric polynomials is a free rank 1 module over the ring S of symmetric polynomials generated by $V(x)$ or $A = V(x)S$.*

In particular, any integral basis of A gives, dividing by $V(x)$, an integral basis of S. In this way we will presently obtain the *Schur functions*.

To understand the construction, let us make a fairly general discussion. In the ring of polynomials $\mathbb{Z}[x_1, x_2, \dots, x_n]$, let us consider the basis given by the monomials (which are permuted by S_n).

Recall that the orbits of monomials are indexed by non-increasing sequences of nonnegative integers. To $m_1 \geq m_2 \geq m_3 \cdots \geq m_n \geq 0$ corresponds the orbit of the monomial $x_1^{m_1} x_2^{m_2} x_3^{m_3} \cdots x_n^{m_n}$.

Let f be an antisymmetric polynomial and (ij) a transposition. Applying this transposition to f changes the sign of f, while the transposition fixes all monomials in which x_i, x_j have the same exponent.

It follows that all of the monomials which have nonzero coefficient in f must have distinct exponents. Given a sequence of exponents $m_1 > m_2 > m_3 > \cdots > m_n \geq 0$ the coefficients of the monomial $x_1^{m_1} x_2^{m_2} x_3^{m_3} \cdots x_n^{m_n}$ and of $x_{\sigma(1)}^{m_1} x_{\sigma(2)}^{m_2} x_{\sigma(3)}^{m_3} \cdots x_{\sigma(n)}^{m_n}$ differ only by the sign of σ.

It follows that:

Theorem. *The functions*

$$A_{m_1>m_2>m_3>\cdots>m_n\geq 0}(x) := \sum_{\sigma\in S_n} \epsilon_\sigma x_{\sigma(1)}^{m_1} x_{\sigma(2)}^{m_2} \cdots x_{\sigma(n)}^{m_n} \tag{3.1.1}$$

are an integral basis of the space of antisymmetric polynomials.

It is often useful when making computations with alternating functions to use a simple device. Consider the subspace SM spanned by the set of *standard monomials* $x_1^{k_1} x_2^{k_2} \cdots x_n^{k_n}$ with $k_1 > k_2 > k_3 \cdots > k_n$ and the linear map L from the space of polynomials to SM which is 0 on the nonstandard monomials and the identity on SM. Then $L\big(\sum_{\sigma\in S_n} \epsilon_\sigma x_{\sigma(1)}^{m_1} x_{\sigma(2)}^{m_2} \cdots x_{\sigma(n)}\big) = x_1^{m_1} x_2^{m_2} \cdots x_n^{m_n}$, and thus L establishes a linear isomorphism between the space of alternating polynomials and SM which maps the basis of the theorem to the standard monomials.

3.2 Schur Functions

It is convenient to use the following conventions. Consider the sequence $\varrho := (n-1, n-2, \ldots, 2, 1, 0)$. We clearly have:

Lemma. *The map*

$$\begin{aligned}\lambda = (p_1, p_2, p_3, \ldots, p_n) &\mapsto \lambda + \varrho \\ &= (p_1 + n - 1, p_2 + n - 2, p_3 + n - 3, \ldots, p_n)\end{aligned}$$

is a bijective correspondence between decreasing and strictly decreasing sequences.

We thus indicate by $A_{\lambda+\varrho}$ the corresponding antisymmetric function. We can express it also as a determinant of the matrix M_λ having the element $x_j^{p_i+n-i}$ in the i, j position.[10]

Definition. The symmetric function $S_\lambda(x) := A_{\lambda+\varrho}/V(x)$ is called the *Schur function* associated to λ.

When there is no ambiguity we will drop the symbol of the variables x and use S_λ.

We can identify λ with a *partition*, with at most n parts, of the integer $\sum p_i$ and write $\lambda \vdash \sum_i p_i$.

Thus we have (with the notations of Chapter 1, 1.1) the following:

Theorem 1. *The functions S_λ, with $\lambda \vdash m$ and $ht(\lambda) \leq n$, are an integral basis of the part of degree m of the ring of symmetric functions in n variables.*

Notice that the Vandermonde determinant is the alternating function A_ϱ and $S_0 = 1$.

---

[10] It is conventional to drop the numbers equal to 0 in a decreasing sequence.

Several interesting combinatorial facts are associated to these functions; we will see some of them in the next section. The main significance of the Schur functions is in the representation theory of the linear group, as we will see later in Chapter 9.

If a is a positive integer let us denote by $\underline{a}$ the partition $(a, a, a, \dots, a)$. If $\lambda = (p_1, p_2, p_3, \dots, p_n)$ is a partition from 3.1.1, it follows that

(3.2.1) $$A_{\lambda+\varrho+\underline{a}} = (x_1 x_2 \cdots x_n)^a A_{\lambda+\varrho}, \quad S_{\lambda+\underline{a}} = (x_1 x_2 \cdots x_n)^a S_\lambda.$$

We let n be the number of variables and want to understand given a Schur function $S_\lambda(x_1, \dots, x_n)$ the form of $S_\lambda(x_1, \dots, x_{n-1}, 0)$ as symmetric function in $n-1$ variables.

Let $\lambda := h_1 \geq h_2 \geq \cdots \geq h_n \geq 0$. We have seen that, if $h_n > 0$, then $S_\lambda(x_1, \dots, x_n) = \prod_{i=1}^n x_i S_{\overline{\lambda}}(x_1, \dots, x_n)$ where $\overline{\lambda} := h_1 - 1 \geq h_2 - 1 \geq \cdots \geq h_n - 1$. In this case, clearly $S_\lambda(x_1, \dots, x_{n-1}, 0) = 0$.

Assume now $h_n = 0$ and denote the sequence $h_1 \geq h_2 \geq \cdots \geq h_{n-1}$ by the same symbol λ. Let us start from the Vandermonde determinant $V(x_1, \dots, x_{n-1}, x_n) = \prod_{i<j\leq n}(x_i - x_j)$ and set $x_n = 0$ to obtain

$$V(x_1, \dots, x_{n-1}, 0) = \prod_{i=1}^{n-1} x_i \prod_{i<j\leq n-1} (x_i - x_j) = \prod_{i=1}^{n-1} x_i V(x_1, \dots, x_{n-1}).$$

Now consider the alternating function $A_{\lambda+\varrho}(x_1, \dots, x_{n-1}, x_n)$.

Set $\ell_i := h_i + n - i$ so that $\ell_n = 0$ and

$$A_{\lambda+\varrho}(x_1, \dots, x_{n-1}, x_n) = \sum_{\sigma \in S_n} \epsilon_\sigma x_1^{\ell_{\sigma(1)}} \cdots x_n^{\ell_{\sigma(n)}}.$$

Setting $x_n = 0$ we get the sum restricted only to the terms for which $\sigma(n) = n$ or

$$A_{\lambda+\varrho}(x_1, \dots, x_{n-1}, 0) = \sum_{\sigma \in S_{n-1}} \epsilon_\sigma x_1^{\ell_{\sigma(1)}} \cdots x_{n-1}^{\ell_{\sigma(n-1)}}.$$

Now $\ell_i = h_i + n - i = (h_i + 1) + (n-1) - i$, and so in $(n-1)$ variables,

$$A_{\lambda+\varrho}(x_1, \dots, x_{n-1}, 0) = A_{\lambda+\varrho+\underline{1}}(x_1, \dots, x_{n-1}) = \prod_{i=1}^{n-1} x_i A_\lambda(x_1, \dots, x_{n-1}).$$

It follows that $S_\lambda(x_1, \dots, x_{n-1}, 0) = S_\lambda(x_1, \dots, x_{n-1})$. Thus we see that:

Proposition. *Under the evaluation of x_n at 0, the Schur function S_λ vanishes when $ht(\lambda) = n$. Otherwise it maps to the corresponding Schur function in $(n-1)$ variables.*

One uses these remarks as follows. Consider a fixed degree n, and for any m let S_m^n be the space of symmetric functions of degree n in m variables.

From the theory of Schur functions the space S_m^n has as basis the functions $S_\lambda(x_1, \dots, x_m)$ where $\lambda \vdash n$ has height $\leq m$. Under the evaluation $x_m \mapsto 0$, we have a map $S_m^n \to S_{m-1}^n$. We have proved that this map is an isomorphism as soon as $m > n$.

We recover the lemma of Section 1.1 of this chapter and the consequence that all identities which we prove for symmetric functions in n variables of degree n are valid in any number of variables.

Theorem 2. *The formal ring of symmetric functions in infinitely many variables has as basis all Schur functions S_λ. Restriction to symmetric functions in m variables sets to 0 all S_λ with height $> m$.*

When using partitions it is often more useful to describe a partition by specifying the number of parts with 1 element, the number of parts with 2 elements, and so on. Thus one writes a partition as $1^{a_1}2^{a_2}\dots i^{a_i}\dots$.

Proposition. *For the elementary symmetric functions we have*

$$e_h = S_{1^h}. \tag{3.2.2}$$

Proof. According to our previous discussion we can set all the variables $x_i,\ i > h$ equal to 0. Then e_h reduces to $\prod_{i=1}^h x_i$ as well as S_{1^h} from 3.2.1. □

3.3 Duality

Next we want to discuss the value of $S_\lambda(1/x_1, 1/x_2, \dots, 1/x_n)$.

We see that substituting x_i with $1/x_i$ in the matrix M_λ (cf. §3.2) and multiplying the jth column by $x_j^{m_1+n-1}$, we obtain a matrix which equals, up to rearranging the rows, that of the partition $\lambda' := m_1', m_2', \dots, m_n'$ where $m_i + m_{n-i+1}' = m_1$. Thus, up to a sign,

$$(x_1x_2\cdots x_n)^{m_1+n-1}A_{\lambda+\varrho}(1/x_1,\dots,1/x_n) = A_{\lambda'+\varrho}.$$

For the Schur function we have to apply the procedure to both numerator and denominator so that the signs cancel, and we get $S_\lambda(1/x_1, 1/x_2, \dots, 1/x_n) = (x_1x_2\cdots x_n)^{-m_1} S_{\lambda'}$.

If we use the diagram notation for partitions we easily visualize λ' by inserting λ in a rectangle of base m_1 and then taking its complement.

4 Cauchy Formulas

4.1 Cauchy Formulas

The formulas we want to discuss have important applications in representation theory. For now, we wish to present them as purely combinatorial identities.

$$\prod_{i,j=1} \frac{1}{1 - x_iy_j} = \sum_\lambda S_\lambda(x)S_\lambda(y), \tag{C1}$$

where the right-hand side is the sum over all partitions.

(C2)
$$\prod_{1\le i\le j\le 2m} \frac{1}{1-x_ix_j} = \sum_{\lambda\in\Lambda_{ec}} S_\lambda(x),$$

if $n = 2m$ is even.

For all n,

(C3)
$$\prod_{1\le i<j\le n} \frac{1}{1-x_ix_j} = \sum_{\lambda\in\Lambda_{er}} S_\lambda(x).$$

Here Λ_{ec}, resp. Λ_{er}, indicates the set of diagrams with columns (resp. rows) of even length.

(C4)
$$\prod_{i=1,\ j=1}^{n,\ m} (1+x_iy_j) = \sum_{\lambda} S_\lambda(x)S_{\tilde{\lambda}}(y),$$

where $\tilde{\lambda}$ denotes the dual partition (Chapter 1, 1.1) obtained by exchanging rows and columns.

We prove the first one and leave the others to Chapter 9 and 11, where they are interpreted as character formulas. We offer two proofs:

First proof of C1. It can be deduced (in a way similar to the computation of the Vandermonde determinant) considering the determinant of the $n \times n$ matrix:

$$A := (a_{ij}),\ \text{with}\ a_{ij} = \frac{1}{1-x_iy_j}.$$

We first prove that we have

(4.1.1)
$$\frac{V(x)V(y)}{\prod_{i,j=1}^{n}(1-x_iy_j)} = \det(A).$$

Subtracting the first row from the ith, $i > 1$, one has a new matrix (b_{ij}) where

$$b_{1j} = a_{1j},\ \text{and for}\ i>1,\ b_{ij} = \frac{1}{1-x_iy_j} - \frac{1}{1-x_1y_j} = \frac{(x_i-x_1)y_j}{(1-x_iy_j)(1-x_1y_j)}.$$

Thus from the ith row, $i > 1$, one can extract from the determinant the factor $x_i - x_1$ and from the jth column the factor $\frac{1}{1-x_1y_j}$.

Thus the given determinant is the product $\frac{1}{(1-x_1y_1)}\prod_{i=2}^{n}\frac{(x_i-x_1)}{(1-x_1y_i)}$ with the determinant

(4.1.2)
$$\begin{pmatrix} 1 & 1 & 1 & \dots & 1 & 1 \\ \frac{y_1}{1-x_2y_1} & \frac{y_2}{1-x_2y_2} & \dots & \dots & & \frac{y_n}{1-x_2y_n} \\ \vdots & \vdots & \vdots & \ddots & \vdots & \vdots \\ \frac{y_1}{1-x_ny_1} & \frac{y_2}{1-x_ny_2} & \dots & \dots & & \frac{y_n}{1-x_ny_n} \end{pmatrix}.$$

Subtracting the first column from the ith we get the terms $\frac{y_i - y_1}{(1-x_j y_1)(1-x_j y_i)}$. Thus, after extracting the product $\prod_{i=2}^n \frac{(y_i - y_1)}{(1-x_i y_1)}$, we are left with the determinant of the same type of matrix but without the variables x_1, y_1. The claim follows by induction.

Now we can develop the determinant of A by developing each element $\frac{1}{1-x_i y_j} = \sum_{k=0}^{\infty} x_i^k y_j^k$, or in matrix form, each row (resp. column) as a sum of infinitely many rows (or columns).

By multilinearity in the rows, the determinant is a sum of determinants of matrices:

$$\sum_{k_1=0}^{\infty} \cdots \sum_{k_n=0}^{\infty} \det(A_{k_1,k_2,\dots,k_n}), \quad A_{k_1,k_2,\dots,k_n} := ((x_i y_j)^{k_i}).$$

Clearly $\det(A_{k_1,k_2,\dots,k_n}) = \prod_i x_i^{k_i} \det(y_j^{k_i})$. This is zero if the k_i are not distinct; otherwise we reorder the sequence k_i to be decreasing. At the same time we must introduce a sign. Collecting all of the terms in which the k_i are a permutation of a given sequence $\lambda + \rho$, we get the term $A_{\lambda+\varrho}(x) A_{\lambda+\varrho}(y)$. Finally,

$$\frac{V(x)V(y)}{\prod_{i,j=1,n}(1 - x_i y_j)} = \sum_{\lambda} A_{\lambda+\varrho}(x) A_{\lambda+\varrho}(y).$$

From this the required identity follows. □

Second proof of C1. Change the matrix to $\frac{1}{x_i - y_j}$ using the fact that

$$V(x_1^{-1}, \dots, x_n^{-1}) = \left(-\prod_i x_i\right)^{1-n} V(x_1, \dots, x_n)$$

and develop the determinant as the sum of fractions $\frac{1}{\prod (x_i - y_{\sigma(i)})}$. Writing it as a rational function $\frac{f(x,y)}{\prod_{i,j=1,n}(x_i - y_j)}$, we see immediately that $f(x, y)$ is alternating in both x, y of total degree $n^2 - n$. Hence $f(x, y) = cV(x)V(y)$ for some constant c, which will appear in the formula $C1$. Comparing in degree 0 we see that $C1$ holds. □

Let us remark that Cauchy formula C1 also holds when $m \le n$, since $\prod_{i=1}^n \prod_{j=1}^m \frac{1}{1-x_i y_j}$ is obtained from $\prod_{i=1}^n \prod_{j=1}^m \frac{1}{1-x_i y_j}$ by setting $y_j = 0, \forall m < j \le n$. From Proposition 3.2 we get

$$\prod_{i=1}^n \prod_{j=1}^m \frac{1}{1 - x_i y_j} = \sum_{\lambda \vdash n,\, ht(\lambda) \le m} S_\lambda(x_1, \dots, x_n) S_\lambda(y_1, \dots, y_m).$$

Remark. The theory of symmetric functions is in fact a rather large chapter in mathematics with many applications to algebra, combinatorics, probability theory, etc. The reader is referred to the book of I.G. Macdonald [Mac] for a more extensive treatment.

5 The Conjugation Action

5.1 Conjugation

Here we study a representation closely connected to the theory of symmetric functions.

Let us consider the space $M_n(\mathbb{C})$ of $n \times n$ matrices over the field $\mathbb{C}$ of complex numbers. We view it as a representation of the group $G := GL(n, \mathbb{C})$ of invertible matrices by conjugation: XAX^{-1}; its orbits are thus the conjugacy classes of matrices.

Remark. The scalar matrices $\mathbb{C}^*$ act trivially, hence we have a representation of the quotient group (the *projective linear group*):

$$PGL(n, \mathbb{C}) := GL(n, \mathbb{C})/\mathbb{C}^*.$$

Given a matrix A consider its characteristic polynomial:

$$\det(t - A) := \sum_{i=0}^{n} (-1)^i \sigma_i(A) t^{n-i}.$$

The coefficients $\sigma_i(A)$ are polynomial functions on $M_n(\mathbb{C})$ which are clearly conjugation invariant. Since the eigenvalues are the roots of the characteristic polynomial, $\sigma_i(A)$ is the ith elementary symmetric function computed in the eigenvalues of A.

Recall that S_n can be viewed as a subgroup of $GL(n, \mathbb{C})$ (the permutation matrices). Consider the subspace D of diagonal matrices. Setting $a_{ii} = a_i$ we identify such a matrix with the vector $(a_1, \dots, a_n)$. The following is clear.

Lemma. *D is stable under conjugation by S_n. The induced action is the standard permutation action (2.6). The function $\sigma_i(A)$, restricted to D, becomes the ith elementary symmetric function.*

We want to consider the conjugation action on $M_n(\mathbb{C})$, $GL(n, \mathbb{C})$, $SL(n, \mathbb{C})$ and compute the invariant functions. As functions we will take those which come from the algebraic structure of these sets (as affine varieties, cf. Chapter 7). Namely, on $M_n(\mathbb{C})$ we take the polynomial functions: On $SL(n, \mathbb{C})$ the restriction of the polynomial functions, and on $GL(n, \mathbb{C})$ the regular functions, i.e., the quotients f/d^k where f is a polynomial on $M_n(\mathbb{C})$ and d is the determinant function.

Theorem. *Any polynomial invariant for the conjugation action on $M_n(\mathbb{C})$ is a polynomial in the functions $\sigma_i(A)$, $i = 1, \dots, n$.*

Any invariant for the conjugation action on $SL(n, \mathbb{C})$ is a polynomial in the functions $\sigma_i(A)$, $i = 1, \dots, n-1$.

Any invariant for the conjugation action on $GL(n, \mathbb{C})$ is a polynomial in the functions $\sigma_i(A)$, $i = 1, \dots, n$ and in $\sigma_n(A)^{-1}$.

Proof. We prove the first statement of the theorem. The proofs of the other two statements are similar and we leave them to the reader. Let $f(A)$ be such a polynomial. Restrict f to D. By the previous remark, it becomes a symmetric polynomial which can then be expressed as a polynomial in the elementary symmetric functions. Thus we can find a polynomial $p(A) = p(\sigma_1(A), \dots, \sigma_n(A))$ which coincides with $f(A)$ upon restriction to D. Since both $f(A),\ p(A)$ are invariant under conjugation, they must coincide also on the set of all diagonalizable matrices. The statement follows therefore from:

Exercise. The set of diagonalizable matrices is dense.

Hint.

(i) A matrix with distinct eigenvalues is diagonalizable, and these matrices are characterized by the fact that the discriminant is nonzero on them.
(ii) For every integer k, the set of points in $\mathbb{C}^k$ where a (non-identically zero) polynomial $u(x)$ is nonzero is dense. (Take any point P and a P_0 with $g(P_0) \neq 0$, on the line connecting $P,\ P_0$ the polynomial g is not identically 0, etc.). □

Remark. The map $M_n(\mathbb{C}) \to \mathbb{C}^n$ given by the functions $\sigma_i(A)$ is constant on orbits, but a fiber is not necessarily a conjugacy class. In fact when the characteristic polynomial has a multiple root, there are several types of Jordan canonical forms corresponding to the same eigenvalues.

There is a second approach to the theorem which is also very interesting and leads to some generalizations. We omit the details.

Proposition. *For an $n \times n$ matrix A the following conditions are equivalent:*

(1) There is a vector v such that the n vectors $A^i v,\ i = 0, \dots, n-1$, are linearly independent.
(2) The minimal polynomial of A equals its characteristic polynomial.
(3) The conjugacy class of A has maximal dimension $n^2 - n$.
(4) A is conjugate to a companion matrix

$$\begin{pmatrix} 0 & 0 & 0 & \dots & 0 & 0 & a_n \\ 1 & 0 & 0 & \dots & 0 & 0 & a_{n-1} \\ 0 & 1 & 0 & \dots & 0 & 0 & a_{n-2} \\ 0 & 0 & 1 & \dots & 0 & 0 & a_{n-3} \\ \dots & & & & \dots & & \dots \\ \dots & & & & \dots & & \dots \\ 0 & 0 & 0 & \dots & 1 & 0 & a_2 \\ 0 & 0 & 0 & \dots & 0 & 1 & a_1 \\ 0 & 0 & 0 & \dots & 0 & 0 & 1 \end{pmatrix}$$

with characteristic polynomial $t^n + \sum_{i=1}^n a_i t^{n-i}$.
(5) In a Jordan canonical form distinct blocks belong to different eigenvalues.

Proof. (1) and (4) are clearly equivalent, taking as the matrix conjugate to A the one of the same linear transformation in the basis $A^i v$, $i = 0, \dots, n-1$.

(2) and (5) are easily seen to be equivalent and also (5) and (1).

We do not prove (3) since we have not yet developed enough geometry of orbits. One needs the theory of Chapter 4, 3.7 showing that the dimension of an orbit equals the dimension of the group minus the dimension of the stabilizer and then one has to compute the centralizer of a regular matrix and prove that it has dimension n.

Definition. The matrices satisfying the previous conditions are called *regular*, and their set is the *regular set* or *regular sheet.*

One can easily prove that the regular sheet is open dense, and it follows again that every invariant function is determined by the value it takes on the set of companion matrices; hence we have a new proof of the theorem on invariants for the conjugation representation.

With this example we have given a glance at a set of algebro-geometric phenomena which have been studied in depth by several authors. The representations for which the same type of ideas apply are particularly simple and interesting (cf. [DK]).

3

Theory of Algebraic Forms

Summary. This is part of classical invariant theory, which is not a very appropriate name, referring to the approach to invariant theory presented in Hermann Weyl's *The Classical Groups* in which the connection between invariant theory, tensor algebra, and representation theory is stressed. One of the main motivations of Weyl was the role played by symmetry in the developments of quantum mechanics and relativity, which had just taken shape in the 30–40 years previous to the appearance of his book.

Invariant theory was born and flourished in the second half of the 19th century due to the work of Clebsch and Gordan in Germany, Cayley and Sylvester in England, and Capelli in Italy, to mention some of the best known names. It was developed at the same time as other disciplines which have a strong connection with it: projective geometry, differential geometry and tensor calculus, Lie theory, and the theory of associative algebras. In particular, the very formalism of matrices, determinants, etc., is connected to its birth.

After the prototype theorem of I.T., the theory of symmetric functions, 19th-century I.T. dealt mostly with binary forms (except for Capelli's work attempting to lay the foundations of what he called *Teoria delle forme algebriche*). One of the main achievements of that period is Gordan's proof of the finiteness of the ring of invariants. The turning point of these developments has been Hilbert's theory, with which we enter into the methods that have led to the development of commutative algebra. We shall try to give an idea of these developments and how they are viewed today.

In this chapter we will see a few of the classical ideas which will be expanded on in the next chapters. In particular, polarization operators have a full explanation in the interpretation of the Cauchy formulas by representation theory; cf. Chapter 9.

1 Differential Operators

1.1 Weyl Algebra

On the polynomial ring $\mathbb{C}[x_1, \ldots, x_n]$ acts the algebra of *polynomial differential operators,* $W(n) := \mathbb{C}\big[x_1, \ldots, x_n, \frac{\partial}{\partial x_1}, \ldots, \frac{\partial}{\partial x_n}\big]$,[11] which is a noncommutative algebra, the multiplication being the composition of operators.

[11] This is sometimes known as the Weyl algebra, due to the work of Weyl on commutation relations in quantum mechanics.

In noncommutative algebra, given any two elements a, b, one defines their commutator $[a, b] = ab - ba$, and says that $ab = ba + [a, b]$ is a *commutation relation*.

The basic commutation relations for $W(n)$ are

$$[x_i, x_j] = \left[\frac{\partial}{\partial x_i}, \frac{\partial}{\partial x_j}\right] = 0, \quad \left[\frac{\partial}{\partial x_i}, x_j\right] = \delta_i^j. \tag{1.1.1}$$

In order to study $W(n)$ it is useful to introduce a notion called the *symbol*. One starts by writing an operator as a linear combination of terms $x_1^{h_1} \cdots x_n^{h_n} \frac{\partial^{k_1}}{\partial x_1} \cdots \frac{\partial^{k_n}}{\partial x_n}$. The symbol $\sigma(P)$ of a polynomial differential operator P is obtained in an elementary way by taking the terms of higher degree in the derivatives and substituting commutative variables ξ_j for the operators $\frac{\partial}{\partial x_j}$;

$$\sigma\left(x_1^{h_1} \cdots x_n^{h_n} \frac{\partial}{\partial x_1}^{k_1} \cdots \frac{\partial}{\partial x_n}^{k_n}\right) = x_1^{h_1} \cdots x_n^{h_n} \xi_1^{k_1} \cdots \xi_n^{k_n}.$$

In a more formal way, one can filter the algebra of operators by the degree in the derivatives and take the associated graded algebra. Let us recall the method.

Definition 1. A filtration of an algebra R consists of an increasing sequence of subspaces $0 = R_0 \subset R_1 \subset R_2 \subset \cdots \subset R_i \subset \cdots \subset$ such that

$$\bigcup_{i=0}^{\infty} R_i = R, \quad R_i R_j \subset R_{i+j}.$$

Given a filtered algebra R, one can construct the *graded algebra* $\mathrm{Gr}(R) := \oplus_{i=0}^{\infty} R_{i+1}/R_i$. The multiplication in $\mathrm{Gr}(R)$ is induced by the product $R_i R_j \subset R_{i+j}$, so that if $a \in R_i$, $b \in R_j$, the class of the product ab in R_{i+j}/R_{i+j-1} depends only on the classes of a modulo R_{i-1} and of b modulo R_{j-1}. In this way $\mathrm{Gr}(R)$ becomes a *graded algebra* with R_i/R_{i-1} being the elements of degree i. For a filtered algebra R and an element $a \in R$, one can define the symbol of a as an element $\sigma(a) \in \mathrm{Gr}(R)$ as follows. We take the minimum i such that $a \in R_i$ and set $\sigma(a)$ to be the class of a in R_i/R_{i-1}.

In $W(n)$ one may consider the filtration given by the degree in the derivatives for which

$$W(n)_i := \left\langle x_1^{h_1} \cdots x_n^{h_n} \frac{\partial}{\partial x_1}^{k_1} \cdots \frac{\partial}{\partial x_n}^{k_n} \mid \sum k_j \leq i \right\rangle. \tag{1.1.2}$$

Let us denote by $S(n)$ the resulting graded algebra and by $\xi_i := \sigma\left(\frac{\partial}{\partial x_i}\right) \in S(n)_1$ the symbol. We keep for the variables x_i the same notation for their symbols since they are in $R_0 = S_0$. From the commutation relations 1.1.1, it follows that the classes ξ_i of $\frac{\partial}{\partial x_i}$ and of x_i in $S(n)$ commute and we have:

Proposition 1. *$S(n)$ is the commutative algebra of polynomials:*

$$S(n) := \mathbb{C}[x_1, \ldots, x_n, \xi_1, \ldots, \xi_n]$$

Proof. The easy proof is left to the reader. One uses the fact that the monomials in the $\frac{\partial}{\partial x_i}$ are a basis of $W(n)$ as a module over the polynomial ring $\mathbb{C}[x_1, \dots, x_n]$. □

There is a connection between differential operators and group actions. Consider in fact the linear group $GL(n, \mathbb{C})$ which acts on the space $\mathbb{C}^n$ and on the algebra of polynomials by the formula $f^g(x) = f(g^{-1}x)$. If $g \in GL(n, \mathbb{C})$ and $P \in W(n)$, consider the operator

$$P^g(f(x)) := {}^g\big(P^{g^{-1}}(f(x))\big), \ \forall f \in \mathbb{C}[x_1, \dots, x_n].$$

In other words, if we think of $g : \mathbb{C}[x_1, \dots, x_n] \to \mathbb{C}[x_1, \dots, x_n]$ as a linear map, we have $P^g = g \circ P \circ g^{-1}$.

It is clear that the action on linear operators is an automorphism of the algebra structure, and we claim that it transforms $W(n)$ into itself. Since $W(n)$ is generated by the elements $x_i, \frac{\partial}{\partial x_i}$, it is enough to understand how $GL(n, \mathbb{C})$ acts on these elements. On an element x_i, it will give a linear combination $g(x_i) = \sum_j a_{ji} x_j$. As for $g\big(\frac{\partial}{\partial x_i}\big)$, let us first remark that this operator acts as a *derivation*.

We will investigate derivations in the following chapters but now give the definition:

Definition 2. A derivation of an algebra R is a linear operator $D : R \to R$ such that

$$D(ab) = D(a)b + aD(b). \tag{1.1.3}$$

A direct computation shows that if $\phi : R \to R$ is an algebra automorphism and $D : R \to R$ a derivation, then $\phi \circ D \circ \phi^{-1}$ is also a derivation.

In particular, we may apply these remarks to $g \circ \frac{\partial}{\partial x_i} \circ g^{-1}$. Let $g^{-1}(x_i) = \sum_j b_{ji} x_j$. We have that

$$g \circ \frac{\partial}{\partial x_i} \circ g^{-1}(x_j) = g \circ \frac{\partial}{\partial x_i}\left(\sum_h b_{hj} x_h\right) = b_{ij}.$$

Proposition 2. *We have* $g \circ \frac{\partial}{\partial x_i} \circ g^{-1} = \sum_j b_{ij} \frac{\partial}{\partial x_j}$.

Proof. A derivation of $\mathbb{C}[x_1, \dots, x_n]$ is completely determined by its values on the variables x_i. By the above formulas, $g \circ \frac{\partial}{\partial x_i} \circ g^{-1}(x_j) = b_{ij} = \sum_k b_{ik} \frac{\partial}{\partial x_k}(x_j)$; hence the statement. □

The above formulas prove that the space of derivatives behaves as the dual of the space of the variables and that the action of the group is by inverse transpose. This of course has an intrinsic meaning: if V is a vector space and $\mathcal{P}(V)$ the ring of polynomials on V, we have that $V^* \subset \mathcal{P}(V)$ are the linear polynomials. The space V can be identified intrinsically with the space spanned by the derivatives. If $v \in V$, we can define the *derivative* D_v *in the direction of* v in the usual way:

$$D_v f(x) := \frac{d}{dt} f(x + tv)_{t=0}. \tag{1.1.4}$$

Intrinsically the algebra of differential operators is generated by V, V^*, with the commutation relations:

$$(1.1.5) \qquad \phi \in V^*,\ v \in V, \quad [\phi, v] = -\langle \phi \mid v \rangle.$$

The action of $GL(V)$ preserves the filtration, and thus the group acts on the graded algebra $S(n)$. We may identify this action as that on the polynomials on $V \oplus V^*$, (cf. Chapter 9).

Remark. The algebra of differential operators is a graded algebra under the following notion of degree. We say that an operator has degree k if it maps the space of homogeneous polynomials of degree h to the polynomials of degree $h + k$ for every h. For instance, the variables x_i have degree 1 while the derivatives $\frac{\partial}{\partial x_j}$ have degree -1. In particular we may consider the differential operators of degree 0.

For all i, let $\mathbb{C}[x_1, \ldots, x_n]_i$ be the space of homogeneous polynomials of degree i.

Exercise. Prove that over a field F of characteristic 0, any linear map from the space $F[x_1, \ldots, x_n]_i$ to the space $F[x_1, \ldots, x_n]_j$ can be expressed via a differential operator of degree $j - i$.

2 The Aronhold Method, Polarization

2.1 Polarizations

Before proceeding, let us recall, in a language suitable for our purposes, the usual Taylor–Maclaurin expansion.

Consider a function $F(x)$ of a vector variable $x \in V$. Under various types of assumptions we have a development for the function $F(x+y)$ of two vector variables.

For our purposes, we may restrict our considerations to polynomials and develop $F(x + y) := \sum_{i=0}^{\infty} F_i(x, y)$, where by definition $F_i(x, y)$ is homogeneous of degree i in y (of course for polynomials the sum is really finite). Therefore, for any value of a parameter λ, we have $F(x + \lambda y) := \sum_{i=0}^{\infty} \lambda^i F_i(x, y)$.

If F is also homogeneous of degree k we have

$$\sum_{i=0}^{\infty} \lambda^k F_i(x, y) = \lambda^k F(x + y) = F(\lambda(x + y)) = F(\lambda x + \lambda y) = \sum_{i=0}^{\infty} \lambda^i F_i(\lambda x, y),$$

and we deduce that $F_i(x, y)$ is also homogeneous of degree $k - i$ in x.

Given two functions F, G, we clearly have that

$$F(x+y)G(x+y) = \sum_{i=0}^{\infty} \sum_{a+b=i} F_a(x,y)G_b(x,y)$$

is the decomposition into homogeneous components relative to y.

The operator $D = D_{y,x}$ defined by the formula $D_{y,x}F(x) := F_1(x,y)$ is clearly linear, and also by the previous formula we have $D(FG) = D(F)G + F\,D(G)$. These are the defining conditions of a *derivation*.

If we use coordinates $x = (x_1, x_2, \ldots, x_n)$ and $y = (y_1, y_2, \ldots, y_n)$, we have that $D_{y,x} = \sum_{i=1}^{n} y_i \frac{\partial}{\partial x_i}$.

Definition. The operator $D_{y,x} = \sum_{i=1}^{n} y_i \frac{\partial}{\partial x_i}$ is called a *polarization* operator.

The effect of applying $D_{y,x}$ to a bihomogeneous function of two variables x, y is to decrease by one the degree of the function in x and raise by one the degree in y.

Assume we are now in characteristic 0, we have then the standard theorem of calculus:

Theorem. $F(x+y) = \sum_{i=0}^{\infty} \frac{1}{i!} D_{y,x}^i F(x) \quad \left(= \sum_{i=0}^{\infty} F_i(x,y)\right)$.

Proof. We reduce to the one variable theorem and deduce that

$$F(x+\lambda y) := \sum_{i=0}^{\infty} \frac{\lambda^i}{i!} \frac{d^i}{d\lambda^i} F(x+\lambda y)_{\lambda=0}.$$

Then

$$F_i(x,y) = \frac{1}{i!} \frac{d^i}{d\lambda^i} F(x+\lambda y)_{\lambda=0},$$

and this is, by the chain rule, equal to $\frac{1}{i!} D_{y,x}^i F(x)$. □

2.2 Restitution

Suppose now that we consider the action of an invertible linear transformation on functions. We have $(gF)(x+y) = F(g^{-1}x + g^{-1}y)$. Hence we deduce that the polarization operator commutes with the action of the linear group. The main consequence is:

Proposition. *If $F(x)$ is an invariant of a group G, so are the polarized forms $F_i(x,y)$.*

Of course we are implicitly using the (direct sum) linear action of G on pairs of variables.

Let us further develop this idea. Consider now any number of vector variables and, for a polynomial function F, homogeneous of degree m, the expansion

$$F(x_1 + x_2 + \cdots + x_m) = \sum_{h_1, h_2, \ldots, h_m} F_{h_1, h_2, \ldots, h_m}(x_1, x_2, \ldots, x_m),$$

where $\sum h_i = m$, and the indices h_i represent the degrees of homogeneity in the variables x_i. A repeated application of the Taylor–Maclaurin expansion gives

$$F_{h_1,h_2,\dots,h_m}(x_1, x_2, \dots, x_m) = \frac{1}{h_1!h_2!\cdots h_n!} D^{h_1}_{x_1 x} D^{h_2}_{x_2 x} \cdots D^{h_m}_{x_m x} F(x). \tag{2.2.1}$$

In particular, in the expansion of $F(x_1 + x_2 \cdots + x_m)$ there will be a term which is linear in all the variables x_i.

Definition. The term $F_{1,1,\dots,1}(x_1, \dots, x_m)$ multilinear in all the variables x_i is called the *full polarization of the form F*.

Let us write $PF := F_{1,1,\dots,1}(x_1, \dots, x_m)$ to stress the fact that this is a linear operator. It is clear that if $\sigma \in S_m$ is a permutation, then

$$F(x_1 + x_2 \cdots + x_n) = F(x_{\sigma 1} + x_{\sigma 2} \cdots + x_{\sigma n}).$$

Hence we deduce that the polarized form satisfies the *symmetry* property:

$$PF(x_1, \dots, x_m) = PF(x_{\sigma 1}, \dots, x_{\sigma m}).$$

We have thus found that:

Lemma. *The full polarization is a linear map from the space of homogeneous forms of degree m to the space of symmetric multilinear functions in m (vector) variables.*

Now let us substitute for each variable x_i the variable $\lambda_i x$ (the λ_i's being distinct numbers). We obtain:

$$\begin{aligned}(\lambda_1 + \lambda_2 \cdots + \lambda_m)^m F(x) &= F((\lambda_1 + \lambda_2 \cdots + \lambda_m)x) \\ &= F(\lambda_1 x + \lambda_2 x \cdots + \lambda_m x) \\ &= \sum_{h_1,h_2,\dots,h_m} F_{h_1,h_2,\dots,h_m}(\lambda_1 x, \lambda_2 x, \dots, \lambda_m x) \\ &= \sum_{h_1,h_2,\dots,h_m} \lambda_1^{h_1} \lambda_2^{h2} \dots \lambda_m^{h_m} F_{h_1,h_2,\dots,h_m}(x, x, \dots, x).\end{aligned}$$

Comparing the coefficients of the same monomials on both sides, we get

$$\binom{m}{h_1 h_2 \cdots h_m} F(x) = F_{h_1,h_2,\dots,h_m}(x, x, \dots, x).$$

In particular,

$$m!F(x) = PF(x, x, \dots, x).$$

Since we are working in characteristic zero we can also rewrite this identity as

$$F(x) = \frac{1}{m!} PF(x, x, \dots, x).$$

For a polynomial $G(x_1, x_2, \dots, x_m)$ of m vector variables, the linear operator

$$R : G(x_1, x_2, \dots, x_m) \mapsto \frac{1}{m!} G(x, x, \dots, x)$$

is called the *restitution* in the classical literature. We have:

Theorem. *The maps P, R are inverse isomorphisms, equivariant for the group of all linear transformations, between the space of homogeneous forms of degree m and the space of symmetric multilinear functions in m variables.*

Proof. We have already proved that $RPF = F$. Now let $G(x_1, x_2, \dots, x_m)$ be a symmetric multilinear function. In order to compute PRG we must determine the multilinear part of $\frac{1}{m!} G\left(\sum x_i, \sum x_i, \dots, \sum x_i\right)$.

By the multilinearity of G we have that

$$G\left(\sum x_i, \sum x_i, \dots, \sum x_i\right) = \sum G(x_{i_1}, x_{i_2}, \dots, x_{i_m}),$$

where the right sum is over all possible sequences of indices $i_1 i_2 \cdots i_m$ taken out of the numbers $1, \dots, m$. But the multilinear part is exactly the sum over all of the sequences without repetitions, i.e., the permutations. Thus

$$PRG = \frac{1}{m!} \sum_{\sigma \in S_m} G(x_{\sigma 1}, x_{\sigma 2}, \dots, x_{\sigma m}).$$

Since G is symmetric, this sum is in fact G. □

Remark. We will mainly use the previous theorem to reduce the computation of invariants to the multilinear ones. At this point it is not yet clear why this should be simpler, but in fact we will see that in several interesting cases this turns out to be true, and we will be able to compute all of the invariants by this method. This sequence of ideas is sometimes referred to as *Aronhold's method.*

2.3 Multilinear Functions

In order to formalize the previous method, consider an infinite sequence of n-dimensional vector variables $x_1, x_2, \dots, x_k, \dots$. We usually have two conventions: each x_i is either a column vector $x_{1i}, x_{2i}, \dots, x_{ni}$ or a row vector $x_{i1}, x_{i2}, \dots, x_{in}$. In other words we consider the x_{ij} as the coordinates of the space of $n \times \infty$ or of $\infty \times n$ matrices or of the space of sequences of column (resp. row) vectors.

Let $A := \mathbb{C}[x_{ij}]$ be the polynomial ring in the variables x_{ij}. For the elements of A we have the notion of being homogeneous with respect to one of the vector variables x_i, and A is in this way a *multigraded ring.*[12]

We denote by $A_{h_1, h_2, \dots, h_i, \dots}$ the multihomogeneous part relative to the degrees $h_1, h_2, \dots, h_i, \dots,$ in the vector variables $x_1, x_2, \dots, x_i, \dots$.

[12] Notice that we are not considering the separate degrees in all of the variables x_{ij}.

We have of course the notions of multihomogeneous subspace or subalgebra. For each pair i, j of indices, we consider the corresponding polarization operator

$$(2.3.1) \qquad D_{ij} = \sum_{h=1}^{n} x_{hi} \frac{\partial}{\partial x_{hj}}, \quad \text{or} \quad D'_{ij} = \sum_{h=1}^{n} x_{ih} \frac{\partial}{\partial x_{jh}}, \text{ in row notation.}$$

We view these operators as acting as derivations on the ring $A = \mathbb{C}[x_{ij}]$.

Given a function F homogeneous in the vector variables $x_1, x_2, \ldots, x_m$ of degrees $h_1, h_2, \ldots, h_m$, we can perform the process of polarization on each of the variables x_i as follows: Choose from the infinite list of vector variables m disjoint sets X_i of variables, each with h_i elements. Then fully polarize the variable x_i with respect to the chosen set X_i.

The result is multilinear and symmetric in each of the sets X_i. The function F is recovered from the polarized form by a sequence of restitutions.

We should remark that a restitution is a particular form of polarization since, if a function F is linear in the variable x_i, the effect of the operator D_{ji} on F is that of replacing in F the variable x_j with x_i.

Definition. A subspace V of the ring A is said to be *stable under polarization* if it is stable under all polarization operators.

Remark. Given a polynomial F, F is homogeneous of degree m with respect to the vector variable x_i if and only if $D_{ii}F = mF$.

From this remark one can easily prove the following:

Lemma. *A subspace V of A is stable under the polarizations D_{ii} if and only if it is multihomogeneous.*

2.4 Aronhold Method

In this section we will use the term *multilinear function* in the following sense:

Definition. We say that a polynomial $F \in A$ is multilinear if it is homogeneous of degree 0 or 1 in each of the variables x_i.

In particular we can list the indices of the variables $i_1, \ldots, i_k$ in which F is linear (the variables which appear in the polynomial) and say that F is multilinear in the x_{i_j}.

Given a subspace V of A we will denote by V_m the set of multilinear elements of V.

Theorem. *Given two subspaces V, W of A stable under polarization and such that $V_m \subset W_m$, we have $V \subset W$.*

Proof. Since V is multihomogeneous it is enough to prove that given a multihomogeneous function F in V, we have $F \in W$. We know that F can be obtained by restitution from its fully polarized form $F = RPF$. The hypotheses imply that $PF \in V$ and hence $PF \in W$. Since the restitution is a composition of polarization operators and W is assumed to be stable under polarization, we deduce that $F \in W$. □

Corollary. *If two subspaces V, W of A are stable under polarization and $V_m = W_m$, then $V = W$.*

We shall often use this corollary to compute invariants. The strategy is as follows. We want to compute the space W of invariants in A under some group G of linear transformations in n-dimensional space. We produce a list of invariants (which are more or less obvious) forming a subspace V closed under polarization. We hope to have found all invariants and try to prove $V = W$. If we can do it for the multilinear invariants we are done.

3 The Clebsch–Gordan Formula

3.1 Some Basic Identities

We start with two sets of *binary variables* (as the old algebraists would say):

$$x := (x_1, x_2); \qquad y := (y_1, y_2).$$

In modern language these are linear coordinate functions on the direct sum of two copies of a standard two-dimensional vector space. We form the following matrices of functions and differential operators:

$$\begin{pmatrix} x_1 & x_2 \\ y_1 & y_2 \end{pmatrix} \begin{pmatrix} \frac{\partial}{\partial x_1} & \frac{\partial}{\partial y_1} \\ \frac{\partial}{\partial x_2} & \frac{\partial}{\partial y_2} \end{pmatrix} = \begin{pmatrix} \Delta_{xx} & \Delta_{xy} \\ \Delta_{yx} & \Delta_{yy} \end{pmatrix}.$$

We define

$$(x, y) := \det \begin{pmatrix} x_1 & x_2 \\ y_1 & y_2 \end{pmatrix} = x_1 y_2 - y_1 x_2$$

$$\Omega := \det \begin{pmatrix} \frac{\partial}{\partial x_1} & \frac{\partial}{\partial y_1} \\ \frac{\partial}{\partial x_2} & \frac{\partial}{\partial y_2} \end{pmatrix} = \frac{\partial}{\partial x_1}\frac{\partial}{\partial y_2} - \frac{\partial}{\partial x_2}\frac{\partial}{\partial y_1}.$$

Since we are in a noncommutative setting the determinant of the product of two matrices need not be the product of the determinants. Moreover, it is even necessary to specify what one means by the determinant of a matrix with entries in a noncommutative algebra.

A direct computation takes care of the noncommutative nature of differential polynomials and yields

$$(3.1.1) \qquad (x, y)\Omega = \det \begin{pmatrix} \Delta_{xx} + 1 & \Delta_{xy} \\ \Delta_{yx} & \Delta_{yy} \end{pmatrix} = (\Delta_{xx} + 1)\Delta_{yy} - \Delta_{yx}\Delta_{xy}.$$

This is a basic identity whose extension to many variables is the *Capelli identity* on which one can see the basis for the representation theory of the linear group.

Now let $f(x, y)$ be a polynomial in x_1, x_2, y_1, y_2, homogeneous of degree m in x and n in y. Notice that

$$\Delta_{xx} f(x, y) = mf(x, y), \quad \Delta_{yy} f(x, y) = nf(x, y), \tag{3.1.2}$$

$$\Delta_{yx} f(x, y) \text{ has bidegree } (m-1, n+1), \tag{3.1.3}$$
$$\Delta_{xy} f(x, y) \text{ has bidegree } (m+1, n-1).$$

Observe that $(\Delta_{xx} + 1)\Delta_{yy} f(x, y) = (m+1)nf(x, y)$; thus the identity

$$(x, y)\Omega f(x, y) = (m+1)nf(x, y) - \Delta_{yx}\Delta_{xy} f(x, y). \tag{3.1.4}$$

The identity 3.1.4 is the beginning of the Clebsch–Gordan expansion.

We have some commutation rules:

$$[\Omega, \Delta_{yx}] = [\Omega, \Delta_{xy}] = 0, \ [\Omega, \Delta_{xx}] = \Omega = [\Omega, \Delta_{yy}],$$
$$[(x, y), \Delta_{yx}] = [(x, y), \Delta_{xy}] = 0, \tag{3.1.5}$$
$$[(x, y), \Delta_{xx}] = -(x, y) = [(x, y), \Delta_{yy}].$$

Finally

$$\Omega(x, y) - (x, y)\Omega = 2 + \Delta_{xx} + \Delta_{yy}. \tag{3.1.6}$$

In summary: In the algebra generated by the elements

$$\Delta_{xx}, \Delta_{xy}, \Delta_{yx}, \Delta_{yy},$$

the elements $\Delta_{xx} + \Delta_{yy}$ and the elements $\Omega^k (x, y)^k$ are in the center.

The Clebsch–Gordan expansion applies to a function $f(x, y)$ of bidegree (m, n), say $n \leq m$:

Theorem (Clebsch–Gordan expansion).

$$f(x, y) = \sum_{h=0}^{n} c_{h,m,n}(x, y)^h \Delta_{yx}^{n-h} \Delta_{xy}^{n-h} \Omega^h f(x, y) \tag{3.1.7}$$

with the coefficients $c_{h,m,n}$ to be determined by induction.

Proof. The proof is a simple induction on n. Since $\Omega f(x, y)$ has bidegree $(m-1, n-1)$ and $\Delta_{xy} f(x, y)$ has bidegree $(m+1, n-1)$ we have by induction on n

$$(x, y)\Omega f(x, y) = (x, y) \sum_{h=0}^{n-1} c_{h,m-1,n-1}(x, y)^h \Delta_{yx}^{n-1-h} \Delta_{xy}^{n-1-h} \Omega^h \Omega f(x, y)$$

$$\Delta_{xy} f(x, y) = \sum_{h=0}^{n-1} c_{h,m+1,n-1}(x, y)^h \Delta_{yx}^{n-1-h} \Delta_{xy}^{n-1-h} \Omega^h \Delta_{xy} f(x, y).$$

Now using 3.1.4 and 3.1.5 we obtain

$$\sum_{h=0}^{n-1} c_{h,m-1,n-1}(x, y)^{h+1} \Delta_{yx}^{n-h-1} \Delta_{xy}^{n-h-1} \Omega^{h+1} f(x, y)$$

$$= (m+1)nf(x, y) - \sum_{h=0}^{n-1} c_{h,m+1,n-1}(x, y)^h \Delta_{yx}^{n-h} \Delta_{xy}^{n-h} \Omega^h f(x, y),$$

so

$$(m+1)nf(x, y) = \sum_{h=1}^{n} c_{h-1,m-1,n-1}(x, y)^h \Delta_{yx}^{n-h} \Delta_{xy}^{n-h} \Omega^h f(x, y)$$

$$+ \sum_{h=0}^{n-1} c_{h,m+1,n-1}(x, y)^h \Delta_{yx}^{n-h} \Delta_{xy}^{n-h} \Omega^h f(x, y)$$

$$= (m+1)nf(x, y).$$

Thus, $c_{h,m,n} = \frac{1}{(m+1)n}[c_{h-1,m-1,n-1} + c_{h,m+1,n-1}]$. □

We should make some comments on the meaning of this expansion, which is at the basis of many developments. As we shall see, the decomposition given by the Clebsch–Gordan formula is unique, and it can be interpreted in the language of representation theory.

In Chapter 10 we shall prove that the spaces P_m of binary forms of degree m constitute the full list of the irreducible representations of $Sl(2)$. In this language the space of forms of bidegree (m, n) can be thought of as the tensor product of the two spaces P_m, P_n of homogeneous forms of degree m, n, respectively.

The operators Δ_{xy}, Δ_{yx}, and Ω, (x, y) are all $Sl(2)$-equivariant.

Since $f(x, y)$ has bidegree (m, n) the form $\Delta_{xy}^{n-h} \Omega^h f(x, y)$ is of bidegree $(m+n-2h, 0)$, i.e., it is an element of P_{m+n-2h}.

We thus have the $Sl(2)$-equivariant operator

$$\Delta_{xy}^{n-h} \Omega^h : P_m \otimes P_n \to P_{m+n-2h}.$$

Similarly, we have the $Sl(2)$-equivariant operator

$$(x, y)^h \Delta_{yx}^{n-h} : P_{m+n-2h} \to P_m \otimes P_n.$$

Therefore the Clebsch–Gordan formula can be understood as the expansion of an element f in $P_m \otimes P_n$ in terms of its components in the irreducible representations P_{m+n-2h}, $0 \le h \le \min(m, n)$ of this tensor product

$$P_m \otimes P_n = \bigoplus_{h=0}^{\min(m,n)} P_{m+n-2h}.$$

At this point the discussion can branch. Either one develops the basic algebraic identities for dimension greater than two (Capelli's theory), or one can enter into the developments of the theory of binary forms, i.e., the application of what has been done in the invariant theory of the group $Sl(2)$. In the next section we discuss Capelli's theory, leaving the theory of binary forms for Chapter 15.

4 The Capelli Identity

4.1 Capelli Identity

Capelli developed the algebraic formula generalizing the Clebsch–Gordan expansion, although, as we shall see, some of the formulas become less explicit.

In order to keep the notations straight we consider in general m, n-ary variables:

$$\underline{x}_i := x_{i1}\, x_{i2}\, \cdots\, x_{in};\ i = 1, \dots, m,$$

which we can view as rows of an $m \times n$ matrix of variables:

(4.1.1)
$$X := \begin{pmatrix} x_{11} & x_{12} & \dots & x_{1n} \\ \dots & \dots & \dots & \dots \\ x_{i1} & x_{i2} & \dots & x_{in} \\ \dots & \dots & \dots & \dots \\ x_{m1} & x_{m2} & \dots & x_{mn} \end{pmatrix}.$$

In the same way we form an $n \times m$ matrix of derivative operators:

(4.1.2)
$$Y := \begin{pmatrix} \frac{\partial}{\partial x_{11}} & \frac{\partial}{\partial x_{21}} & \dots & \frac{\partial}{\partial x_{m1}} \\ \dots & \dots & \dots & \dots \\ \frac{\partial}{\partial x_{1i}} & \frac{\partial}{\partial x_{2i}} & \dots & \frac{\partial}{\partial x_{mi}} \\ \dots & \dots & \dots & \dots \\ \frac{\partial}{\partial x_{1n}} & \frac{\partial}{\partial x_{2n}} & \dots & \frac{\partial}{\partial x_{mn}} \end{pmatrix}.$$

The product XY is a matrix of differential operators. In the i, j position it has the polarization operator $\Delta_{i,j} := \sum_{h=1}^{n} x_{ih} \frac{\partial}{\partial x_{jh}}$.

Now we want to repeat the discussion on the computation of the determinant of the matrix $XY = (\Delta_{i,j})$. For an $m \times m$ matrix $\mathcal{U}$ in noncommutative variables $u_{i,j}$, the determinant will be computed multiplying from left to right as

(4.1.3)
$$\det(\mathcal{U}) := \sum_{\sigma \in \mathcal{S}_n} \epsilon_\sigma u_{1,\sigma(1)} u_{2,\sigma(2)} \cdots u_{m,\sigma(m)}.$$

In our computations we shall take the elements $u_{i,j} = \Delta_{i,j}$ if $i \neq j$, while $u_{ii} = \Delta_{ii} + m - i$. Taking this we find the determinant of the matrix

(4.1.4)
$$\begin{pmatrix} \Delta_{1,1} + m - 1 & \Delta_{1,2} & \dots & \Delta_{1,m-1} & \Delta_{1,m} \\ \Delta_{2,1} & \Delta_{2,2} + m - 2 & \dots & \Delta_{2,m-1} & \Delta_{2,m} \\ \dots & \dots & \dots & \dots & \dots \\ \dots & \dots & \dots & \dots & \dots \\ \Delta_{m-1,1} & \Delta_{m-1,2} & \dots & \Delta_{m-1,m-1} + 1 & \Delta_{m-1,m} \\ \Delta_{m,1} & \Delta_{m,2} & \dots & \Delta_{m,m-1} & \Delta_{m,m} \end{pmatrix}.$$

We let $C = C_m$ be the value of this determinant computed as previously explained and call it *the Capelli element*. We have assumed on purpose that m, n are not necessarily equal and we will get a computation of C similar to what would happen in the commutative case for the determinant of a product of rectangular matrices.

Rule. When we compute such determinants of noncommuting elements, we should remark that some of the usual formal laws of determinants are still valid:

(1) The determinant is *linear* in the rows and columns.
(2) The determinant changes sign when transposing two rows.

Note that property (2) is NOT TRUE when we transpose columns!

In fact it is useful to develop a commutator identity. Recall that in an associative algebra R the commutator $[r, s] := rs - sr$ satisfies the following (easy) identity:

(4.1.5) $$[r, s_1 s_2 \cdots s_k] = \sum_{i=1}^{k} s_1 \cdots s_{i-1}[r, s_i]s_{i+1} \cdots s_k.$$

A formula follows:

Proposition.

(4.1.6) $$[r, \det(\mathcal{U})] = \sum_{i=1}^{m} \det(\mathcal{U}_i),$$

where $\mathcal{U}_i$ is the matrix obtained from $\mathcal{U}$ by replacing its ith column c_i with $[r, c_i]$.

There is a rather peculiar lemma that we will need. Given an $m \times m$ matrix $\mathcal{U}$ in noncommutative variables $u_{i,j}$, let us consider an integer k, and for all $t = 1, \ldots, m$, let us set $\mathcal{U}_{t,k}$ to be the matrix obtained from $\mathcal{U}$ by setting equal to 0 the first k entries of the tth row.

Lemma. $\sum_{t=1}^{m} \det(\mathcal{U}_{t,k}) = (m - k)\det(\mathcal{U})$.

Proof. By induction on k. For $k = 0$, it is trivial. Remark that:

$$\det(\mathcal{U}_{t,k}) = \det(\mathcal{U}_{t,k-1}) - \det \begin{pmatrix} u_{11} & \ldots & u_{1,k-1} & 0 & u_{1,k+1} & \ldots & u_{1m} \\ \ldots & \ldots & \ldots & \ldots & & & \\ u_{t-1,1} & \ldots & u_{t-1,k-1} & 0 & u_{t-1,k+1} & \ldots & u_{t-1,m} \\ 0 & \ldots & 0 & u_{t,k} & 0 & \ldots & 0 \\ u_{t+1,1} & \ldots & u_{t+1,k-1} & 0 & u_{t+1,k+1} & \ldots & u_{t+1,m} \\ \ldots & \ldots & \ldots & \ldots & & & \\ u_{m1} & u_{m2} & \ldots & 0 & \ldots & \ldots & u_{mn} \end{pmatrix}.$$

Thus

$$\sum_{t=1}^{m} \det(\mathcal{U}_{t,k}) = \sum_{t=1}^{m} \det(\mathcal{U}_{t,k-1}) - \det(\mathcal{U}) = (m - k + 1)\det(\mathcal{U}) - \det(\mathcal{U})$$
$$= (m - k)\det(\mathcal{U}).$$ □

Theorem. *The value of the Capelli element is given by*

$$C = \begin{cases} 0 & \text{if } m > n, \\ \det(X)\det(Y) & \text{if } m = n, \\ \sum_{1 \le i_1 < i_2 < \cdots < i_m \le n} X_{i_1 i_2 \cdots i_m} Y_{i_1 i_2 \cdots i_m} & \text{if } m < n \quad \text{(Binet's form)}, \end{cases}$$

where by $X_{i_1 i_2 \cdots i_m}$, respectively $Y_{i_1 i_2 \cdots i_m}$, we mean the determinants of the minors formed by the columns of indices $i_1 < i_2 < \cdots < i_m$ of X, respectively by the rows of indices $i_1 < i_2 < \cdots < i_m$ of Y.

Proof. Let us give the proof of Capelli's theorem. It is based on a simple computational idea. We start by introducing a new $m \times n$ matrix Ξ of indeterminates $\xi_{i,j}$, disjoint from the previous ones, and perform our computation in the algebra of differential operators in the variables x, ξ. Consider now the product ΞY. It is a product of matrices with elements in a commutative algebra, so one can apply to it the usual rules of determinants to get

(4.1.7)

$$\det(\Xi Y) = \begin{cases} 0, & \text{if } m > n, \\ \det(\Xi)\det(Y) & \text{if } m = n, \\ \sum_{1 \le i_1 < i_2 < \cdots < i_m \le n} \Xi_{i_1 i_2 \cdots i_m} Y_{i_1 i_2 \cdots i_m} & \text{if } m < n \quad \text{(Binet's form)}. \end{cases}$$

The matrix ΞY is an $m \times m$ matrix of differential operators. Let us use the notations:

$$\eta_{i,j} = \sum_{h=1}^{n} \xi_{ih} \frac{\partial}{\partial x_{jh}}, \qquad \gamma_{i,j} = \sum_{h=1}^{n} x_{ih} \frac{\partial}{\partial \xi_{jh}}, \qquad D_{i,j} = \sum_{h=1}^{n} \xi_{ih} \frac{\partial}{\partial \xi_{jh}}.$$

We now apply to both sides of the identity 4.1.17 the product of commuting differential operators $\gamma_i := \gamma_{ii} = \sum_{h=1}^{n} x_{ih} \frac{\partial}{\partial \xi_{ih}}$. The operator γ_i, when applied to a function linear in $\underline{\xi}_i$, has the effect of replacing $\underline{\xi}_i$ with $\underline{x}_i$. In particular we can apply the operators under consideration to functions which depend only on the $\underline{x}$ and not on the $\underline{\xi}$. For such a function we get

$$\begin{cases} \prod_{i=1}^{m} \gamma_i \sum_{\sigma \in S_n} \epsilon_\sigma \eta_{1,\sigma(1)} \eta_{2,\sigma(2)} \cdots \eta_{n,\sigma(n)} f(\underline{x}) = 0 & \text{if } m > n, \\ \det(X)\det(Y) f(\underline{x}) & \text{if } m = n, \\ \sum_{1 \le i_1 < i_2 < \cdots < i_m \le n} X_{i_1 i_2 \cdots i_m} Y_{i_1 i_2 \cdots i_m} f(\underline{x}) & \text{if } m < n. \end{cases}$$

So it is only necessary to show that

$$\prod_{i=1}^{m} \gamma_i \sum_{\sigma \in S_n} \epsilon_\sigma \eta_{1,\sigma(1)} \eta_{2,\sigma(2)} \cdots \eta_{n,\sigma(n)} f(\underline{x}) = C f(\underline{x}).$$

We will prove this formula by induction, by introducing the determinants C_{k-1} of the matrices:

$$C_{k-1} := \begin{pmatrix} \Delta_{1,1}+m-1 & \Delta_{1,2} & \dots & \Delta_{1,m-1} & \Delta_{1,m} \\ \dots & \dots & \dots & \dots & \dots \\ \Delta_{k-1,1} & \dots & \Delta_{k-1,k-1}+m-k+1 & \dots & \Delta_{k-1,m} \\ \eta_{k,1} & \eta_{k,2} & \dots & \eta_{k,m-1} & \eta_{k,m} \\ \dots & \dots & \dots & \dots & \dots \\ \eta_{m,1} & \eta_{m,2} & \dots & \eta_{m,m-1} & \eta_{m,m} \end{pmatrix}.$$

We want to show that for every k:

$$\prod_{i=1}^{k} \gamma_i \sum_{\sigma\in\mathcal{S}_n} \epsilon_\sigma \eta_{1,\sigma(1)}\eta_{2,\sigma(2)}\cdots\eta_{n,\sigma(n)} f(\underline{x}) = C_k f(\underline{x}).$$

By induction, this means that we need to prove

$$\gamma_k C_{k-1} f(\underline{x}) = C_k f(\underline{x}). \tag{4.1.8}$$

Since $\gamma_k f(\underline{x}) = 0$ this is equivalent to proving that $[\gamma_k, C_{k-1}]f(\underline{x}) = C_k f(\underline{x})$.

By 4.1.6 we have $[\gamma_k, C_{k-1}] = \sum_{j=1}^m C_{k-1}^j$ where C_{k-1}^j is the determinant of the matrix obtained from C_{k-1} by substituting its jth column c_j with the commutator $[\gamma_k, c_j]$.

In order to prove 4.1.8 we need the following easy commutation relations:

$$[\gamma_i, \eta_{hk}] = \delta_h^i \Delta_{ik} - \delta_k^i D_{hi}, \quad [\gamma_i, \Delta_{hk}] = -\delta_k^i \gamma_{hi}.$$

Thus we have

$$[\gamma_k, c_j] = \begin{vmatrix} 0 \\ 0 \\ \vdots \\ \Delta_{kj} \\ \vdots \\ 0 \end{vmatrix}, \ j \neq k, \qquad [\gamma_k, c_k] = \begin{vmatrix} -\gamma_{1k} \\ \vdots \\ -\gamma_{k-1,k} \\ \Delta_{kk} - D_{kk} \\ -D_{k+1,k} \\ \vdots \\ -D_{mk} \end{vmatrix}.$$

By the linearity in the rows we see that $[\gamma_k, C_{k-1}]f(\underline{x}) = (C_k + U)f(\underline{x})$, where U is the determinant of the matrix obtained from C_{k-1} by substituting its kth column with

$$\begin{vmatrix} -\gamma_{1k} \\ \vdots \\ -\gamma_{k-1,k} \\ -(m-k) - D_{kk} \\ -D_{k+1,k} \\ \vdots \\ -D_{mk} \end{vmatrix}.$$

We need to show that $Uf(\underline{x}) = 0$. In the development of $Uf(\underline{x})$, the factor D_{kk} is applied to a function which does not depend on $\underline{\xi}_k$, and so it produces 0. We need to analyze the determinants ($s = 1, \dots, k-1, \quad j = 1, \dots, m-k$):

$$U_s := \begin{pmatrix} \Delta_{1,1} + m - 1 & \Delta_{1,2} & \dots & 0 & \dots & \Delta_{1,m} \\ \dots & \dots & \dots & -\gamma_{s,k} & \dots & \dots \\ \Delta_{k-1,1} & \dots & \dots & 0 & \dots & \Delta_{k-1,m} \\ \eta_{k,1} & \eta_{k,2} & \dots & 0 & \dots & \eta_{k,m} \\ \dots & \dots & \dots & 0 & \dots & \\ \eta_{m,1} & \eta_{m,2} & \dots & 0 & \dots & \eta_{m,m} \end{pmatrix},$$

$$U'_j := \begin{pmatrix} \Delta_{1,1} + m - 1 & \Delta_{1,2} & \dots & 0 & \dots & \Delta_{1,m} \\ \dots & \dots & \dots & 0 & \dots & \dots \\ \Delta_{k-1,1} & \dots & \dots & 0 & \dots & \Delta_{k-1,m} \\ \eta_{k,1} & \eta_{k,2} & \dots & 0 & \dots & \eta_{k,m} \\ \dots & \dots & \dots & -D_{k+j,k} & \dots & \\ \eta_{m,1} & \eta_{m,2} & \dots & 0 & \dots & \eta_{m,m} \end{pmatrix}.$$

For the first U_s, exchange the k and s rows, changing sign, and notice that in the development of $U_s f(\underline{x})$ the only terms that intervene are the ones in which the factor $\gamma_{s,k}$ is followed by one of the $\eta_{k,t}$ since the other terms do not depend on $\underline{\xi}_k$. For these terms $\gamma_{s,k}$ commutes with all factors except for $\eta_{k,t}$ and the action of $\gamma_{s,k}\eta_{k,t}$, is the same as of $\Delta_{s,t}$

For the last U'_j, exchange the k and $k+j$ rows, changing sign, and notice that in the development of $U'_j f(\underline{x})$, the only terms that intervene are the ones in which the factor $D_{k+j,k}$ is followed by one of the $\eta_{k,t}$, since the other terms do not depend on $\underline{\xi}_k$. For these terms $D_{k+j,k}$ commutes with all factors except for $\eta_{k,t}$, and the action of $D_{k+j,k}\eta_{k,t}$ is the same as that of $\eta_{k+j,t}$. We need to interpret the previous terms. Let $\mathcal{A}$ be the matrix obtained from C_{k-1} dropping the kth row and k^{th} column, and A its determinant. We have that

$$Uf(\underline{x}) = -(m-k)Af(\underline{x}) + \left(\sum_{s+1}^{k-1} U_s + \sum_{j=1}^{m-k} U'_j\right) f(\underline{x}).$$

We have then that the determinants U_s, U'_j are the $m-1$ terms $\det(\mathcal{A}_{t,k})$ which sum to $(m-k)A$, and thus by the previous lemma we finally obtain 0. □

In 5.2 we will see how to obtain a generalization of the Clebsch–Gordan formula from the Capelli identities.

5 Primary Covariants

5.1 The Algebra of Polarizations

We want to deduce some basic formulas which we will interpret in representation theory later. The reader who is familiar with the representation theory of semisimple Lie algebras will see here the germs of the theory of highest weights, central characters, the Poincaré–Birkhoff–Witt, the Peter–Weyl and Borel–Weil theorems.

The theory is due to Capelli and consists in the study the algebra of differential operators $\mathcal{U}_m(n)$, generated by the elements $\Delta_{ij} = \sum_{h=1}^{n} x_{ih}\frac{\partial}{\partial x_{jh}},\ i, j = 1, \dots, m$.

A similar approach can also be found in the book by Deruyts ([De]).

One way of understanding $\mathcal{U}_m(n)$ is to consider it as a subalgebra of the algebra $\mathbb{C}\big[x_{ih}, \frac{\partial}{\partial x_{jk}}\big],\ i, j = 1, \dots, m,\ h, k = 1, \dots, n$, of all polynomial differential operators. The first observation is the computation of their commutation relations:

$$[\Delta_{ij}, \Delta_{hk}] = \delta_h^j \Delta_{ik} - \delta_k^i \Delta_{hj}. \tag{5.1.1}$$

In the language that we will develop in the next chapter this means that the m^2 operators Δ_{ij} span a *Lie algebra* (Chapter 4, 1.1), in fact the Lie algebra gl_m of the general linear group.

As for the associative algebra $\mathcal{U}_m(n)$, it can be thought of as a representation of the universal enveloping algebra of gl_m (Chapter 5, §7).

If we list the m^2 operators Δ_{ij} as $u_1, u_2, \dots, u_{m^2}$ in some given order, we see immediately by induction that the monomials $u_1^{h_1} u_2^{h_2} \cdots u_{m^2}^{h_{m^2}}$ are linear generators of $\mathcal{U}_m$.

Theorem. *If $n \geq m$, the monomials $u_1^{h_1} u_2^{h_2} \cdots u_{m^2}^{h_{m^2}}$ are a linear basis of $\mathcal{U}_m(n)$.*[13]

Proof. We can prove this theorem using the method of computing the symbol of a differential operator. The symbol of a polynomial differential operator in the variables $x_i, \frac{\partial}{\partial x_j}$ is obtained in an elementary way, by taking the terms of higher degree in the derivatives and substituting for the operators $\frac{\partial}{\partial x_j}$ the commutative variables ξ_j. In a more formal way, one can filter the algebra of operators by the degree in the derivatives and take the associated graded algebra. The symbol of Δ_{ij} is $\sum_{h=1}^{n} x_{ih}\xi_{jh}$. If $n \geq m$ these polynomials are algebraically independent and this implies the linear independence of the given monomials. □

Remark. If $n < m$, the previous theorem does not hold. For instance, we have the Capelli element which is 0. One can analyze all of the relations in a form which is the noncommutative analogue of determinantal varieties.

The previous theorem implies that as an abstract algebra, $\mathcal{U}_m(n)$ is independent of n, for $n \geq m$. We will denote it by $\mathcal{U}_m$. $\mathcal{U}_m$ is the *universal enveloping algebra* of gl_m (cf. Chapter 5, §7).

5.2 Primary Covariants

Let us now introduce the notion of a *primary covariant*.[14] We will see much later that this is the germ of the theory of highest weight vectors.

Let X be as before a rectangular $m \times n$ matrix of variables.

---

[13] This implies the Poincaré–Birkhoff–Witt theorem for linear Lie algebras.

[14] The word covariant has a long history and we refer the reader to the discussion in Chapter 15.

Definition. A $k \times k$ minor of the matrix X is primary if it is extracted from the first k rows of X. A primary covariant is a product of determinants of primary minors of the matrix X.

Let us introduce a convenient notation for the primary covariants. Given a number $p \leq m$ and p indices $1 \leq i_1 < i_2 < \cdots < i_p \leq n$, denote by the symbol $|i_1, i_2, \ldots, i_p|$ the determinant of the $p \times p$ minor of the matrix X extracted from the first p rows and the columns of indices $i_1, i_2, \ldots, i_p$. By definition a primary covariant is a product of polynomials of type $|i_1, i_2, \ldots, i_p|$.

If we perform on the matrix X an elementary operation consisting of adding to the ith row a linear combination of the preceding rows, we see that the primary covariants do not change. In other words primary covariants are invariant under the action of the group U^- of strictly lower triangular matrices acting by left multiplication. We shall see in Chapter 13 that the converse is also true: a polynomial in the entries of X invariant under U^- is a primary covariant.

Then, the decomposition of the ring of polynomials in the entries of X as a representation of $GL(m) \times GL(n)$, together with the highest weight theory, explains the true nature of primary covariants.

Let $f(\underline{x}_1, \underline{x}_2, \ldots, \underline{x}_m)$ be a function of the vector variables $\underline{x}_i$, such that f is multihomogeneous of degrees $\underline{h} := (h_1, h_2, \ldots, h_m)$. Let us denote by $\mathcal{U}_m$ the algebra generated by the operators $\Delta_{i,j},\ i, j \leq m$.

Lemma. *If $i \neq j \leq p$, then Δ_{ij} commutes with $|i_1, i_2, \ldots, i_p|$ while*

$$\Delta_{ii}|i_1, i_2, \ldots, i_p| = |i_1, i_2, \ldots, i_p|(1 + \Delta_{ii}).$$

Proof. Immediate. □

Theorem (Capelli–Gordan expansion). *Given a multidegree $\underline{h}$, there exist elements $C_i(\underline{h}), D_i(\underline{h}) \in \mathcal{U}_m$ depending only on the multidegree $\underline{h}$, such that for any function f of multidegree $\underline{h}$ we have*

$$f = \sum_i C_i(\underline{h}) D_i(\underline{h}) f. \tag{5.2.1}$$

Moreover, for all i, the polynomial $D_i(\underline{h}) f$ is a primary covariant.

Proof. We apply first induction on the total degree. For a fixed total degree N, we apply induction on the set of multidegrees ordered opposite to the lexicographic order. Thus the basis of this second induction is multidegree $(N, 0, 0, \ldots, 0)$, i.e., for a function only in the variable $\underline{x}_1$ which is clearly a primary covariant. In general, assume that the degrees $h_i = 0$ for $i > k$. For $h_k = p$, we apply the Capelli identities expressing the value of the Capelli determinant C_p (associated to the $\Delta_{i,j},\ i, j \leq p$). If $p > n$, we get $0 = C_p f$, and developing the determinant C_p from right to left we get a sum

$$0 = (\Delta_{11} + p - 1) \cdots (\Delta_{pp}) f + \sum_{i<p} A_{ij} \Delta_{ip} f$$

$$= \prod_{i=1}^{p} (h_i + p - i) f + \sum_{i<p} A_{ij} \Delta_{ip} f,$$

where the A_{ip} are explicit elements of $\mathcal{U}_m$. The functions $\Delta_{ip} f$ have multidegree higher than that of f, and we can apply induction.

If $p \leq n$ we also apply the Capelli identity, but now we have

$$\sum_{1 \leq i_1 < i_2 < \cdots < i_p \leq n} X_{i_1 i_2 \cdots i_p} Y_{i_1 i_2 \cdots i_p} f = (\Delta_{11} + p - 1) \cdots (\Delta_{pp}) f + \sum_{i<p} A_{ij} \Delta_{ip} f$$

from which

$$f = \prod_{i=1}^{p} (h_i + p - i)^{-1} \Big(\sum_{1 \leq i_1 < i_2 < \cdots < i_p \leq n} X_{i_1 i_2 \cdots i_p} Y_{i_1 i_2 \cdots i_p} f - \sum_{i<p} A_{ip} \Delta_{ip} f \Big).$$

Now the induction can be applied to both the functions $\Delta_{ip} f$ and $Y_{i_1 i_2 \cdots i_p} f$. Next one has to use the commutation rules of the elements $X_{i_1 i_2 \cdots i_p}$ with the Δ_{ij}, which imply that the elements $X_{i_1 i_2 \cdots i_p}$ contribute determinants which are primary covariants. □

A classical application of this theorem gives another approach to computing invariants. Start from a linear action of a group G on a space and take several copies of this representation. We have already seen that the polarization operators commute with the group G and so preserve the property of being invariant. In Section 2.4 we have discussed a possible strategy of computing invariants by reducing to multilinear ones with polarization.

The Capelli–Gordan expansion offers a completely opposite strategy. In the expansion 5.2.1, if f is a G-invariant so are the primary covariants $D_i(\underline{h}) f$. Thus we deduce that each invariant function of several variables can be obtained under polarization from invariant primary covariants, in particular, from invariants which depend at most on n vector variables. We will expand on these ideas in Chapter 11, §1 and §5, where they will be set forth in the language of representation theory.

5.3 Cayley's Ω Process

There is one more classical computation which we shall reconsider in the representation theory of the linear group.

Let us take $m = n$ so that the two matrices X, Y defined in 4.1 are square matrices.

By definition, $D := \det(X) = |1, 2, \ldots, n|$, while $\det(Y)$ is a differential operator classically denoted by Ω and called *Cayley's* Ω *process*. We have $C = |1, 2, \ldots, n| \Omega = D\Omega$.

From Lemma 5.2 we have the commutation relations:

$$[\Delta_{ij}, D] = 0, \; i \neq j, \quad [\Delta_{ii}, D] = D. \tag{5.3.1}$$

We have a similar result for Ω:

Lemma. *If $i \neq j$, then Δ_{ij} commutes with Ω, while*

(5.3.2)
$$\Delta_{ii}\Omega = \Omega(\Delta_{ii} - 1).$$

Proof. Let us apply 4.1.6. The operator Δ_{ij} commutes with all of the columns of Ω except for the ith column ω_i with entries $\frac{\partial}{\partial x_{it}}$. Now $\left[\Delta_{ij}, \frac{\partial}{\partial x_{it}}\right] = -\frac{\partial}{\partial x_{jt}}$, from which $[\Delta_{ij}, \omega_i] = -\omega_j$. The result follows immediately. □

Let us introduce a more general determinant, analogous to a characteristic polynomial. We denote it by $C_m(\rho) = C(\rho)$ and define it as

(5.3.3)
$$\begin{pmatrix} \Delta_{1,1} + m - 1 + \rho & \Delta_{1,2} & \dots & \Delta_{1,m} \\ \Delta_{2,1} & \Delta_{2,2} + m - 2 + \rho & \dots & \Delta_{2,m} \\ \dots & \dots & \dots & \dots \\ \dots & \dots & \dots & \dots \\ \Delta_{m-1,1} & \Delta_{m-1,2} & \dots & \Delta_{m-1,m} \\ \Delta_{m,1} & \Delta_{m,2} & \dots & \Delta_{m,m} + \rho \end{pmatrix}.$$

We have now a generalization of the Capelli identity:

Proposition.

(5.3.4)
$$\Omega C(k) = C(k+1)\Omega, \quad |1, 2, \dots, m| C(k) = C(k-1)|1, 2, \dots, m|,$$

(5.3.5)
$$D^k\Omega^k = C(-(k-1))C(-(k-2)) \cdots C(-1)C, \ \Omega^k D^k = C(k)C(k-1) \cdots C(1).$$

Proof. We may apply directly 5.3.1 and 5.3.2 and then proceed by induction. □

We now develop $C_m(\rho)$ as a polynomial in ρ, obtaining an expression

(5.3.6)
$$C_m(\rho) = \rho^m + \sum_{i=1}^{m} K_i \rho^{m-i}.$$

Theorem.

(i) *The elements K_i generate the center of the algebra $\mathcal{U}_m$, generated by the elements Δ_{ij}, $i, j = 1, \dots, m$.*
(ii) *The operator $C_m(\rho)$ applied to the polynomial $\prod_{i=1}^{m} |1, 2, \dots, i-1, i|^{k_i}$ multiplies it by $\prod_j (\ell_j + m - j + \rho)$, where the numbers ℓ_j are the multidegrees of the given polynomial and $\ell_j := \sum_{i \geq j} k_i$.*

Proof. To prove (i) we begin by proving that the K_i belong to the center of $\mathcal{U}_m$. From 5.3.1, 5.3.2, and 5.3.4 it follows that, for every k, the element $C(k)$ is in the center of $\mathcal{U}_m$. But then this easily implies that all the coefficients of the polynomial are also in the center. To prove that they generate the center we need some more theory, so we postpone the proof to Chapter 5, §7.2, only sketching its steps here. Let $\sigma_{ij} := \sum_{h=1}^{m} x_{ih}\xi_{jh}$ be the symbol of Δ_{ij}. We think of the σ_{ij} as coordinates of

$m \times m$ matrices. One needs to prove first that the symbol of a central operator is an invariant function under conjugation.

Next one shows that the coefficients of the characteristic polynomial of a matrix generate the invariants under conjugation. Finally, the symbols of the Capelli elements are the coefficients of the characteristic polynomial up to a sign, and then one finishes by induction.

Now we prove (ii). We have, by the definition of polarizations, that $\Delta_{hk}|1, 2, \dots, i-1, i| = 0$ if $h < k$, or if $k > i$, while $\Delta_{hh}|1, 2, \dots, i-1, i|^j = j|1, 2, \dots, i-1, i|^j$, $h \leq i$. Therefore it follows that

$$\Delta_{hh} \prod |1, 2, \dots, i-1, i|^{k_i} = \Big(\sum_{i \geq h} k_i\Big) \prod |1, 2, \dots, i-1, i|^{k_i}$$
$$= \ell_h \prod |1, 2, \dots, i-1, i|^{k_i}.$$

The development of the determinant of 4.1.7 applied to $\prod_{i=1}^{m} |1, 2, \dots, i-1, i|^{k_i}$ gives 0 for all of the terms in the upper triangle and thus it is equal to the expression

$$\prod_{i=1}^{m} (\Delta_{ii} + m - i + \rho) \prod_{i=1}^{m} |1, 2, \dots, i-1, i|^{k_i}$$
$$= \prod_{i=1}^{m} (\ell_i + m - i + \rho) \prod_{i=1}^{m} |1, 2, \dots, i-1, i|^{k_i}. \qquad \square$$

The significance of this theorem is that the polynomials we have considered are highest weight vectors for irreducible representations of the group $GL(m)$ over which, by Schur's lemma, the operators K_i must act as scalars. Thus we obtain the explicit value of the central character (cf. Chapter 11, §5).

4

Lie Algebras and Lie Groups

Summary. In this chapter we will discuss topics on differential geometry. In the spirit of the book, the proofs will be restricted to the basic ideas. Our goal is twofold: to explain the meaning of polarization operators as part of Lie algebra theory and to introduce the basic facts of this theory. The theory of Lie algebras is presented extensively in various books, as well as the theory of Lie groups (cf. [J1], [J2], [Ho], [Kn], [Se1], [Se2], [B1], [B2], [B3], [Wa]).

We assume that the reader is familiar with the basic definitions of differential geometry.

1 Lie Algebras and Lie Groups

1.1 Lie Algebras

Polarization operators (Chapter 3, §2) are special types of *derivations*. Let us recall the general definitions. Given an associative algebra A, we define the *Lie product*

$$[a, b] := ab - ba,$$

and verify immediately that it satisfies the *Lie axioms*:

$[a, b] = -[b, a]$ (*antisymmetry*), and $[a, [b, c]] + [b, [c, a]] + [c, [a, b]] = 0$ (*Jacobi identity*).

Recall that a general *algebra* is a vector space with a bilinear product.

Definition 1. An algebra with a product $[a, b]$ satisfying the antisymmetry axiom and the Jacobi identity is called a *Lie algebra*. $[a, b]$ is called a *Lie bracket*.

Exercise. The algebra $L, [\ ,\ ]$ with the new product $\{a, b\} := [b, a]$ is also a Lie algebra isomorphic to L.

The first class of Lie algebras to be considered are the algebras $gl(U)$, the Lie algebra associated to the associative algebra $\mathrm{End}(U)$ of linear operators on a vector space U.

Given any algebra A (not necessarily associative), with product denoted ab, we define:

Definition 2. A derivation of A is a linear mapping $D : A \to A$, satisfying $D(ab) = D(a)b + aD(b)$, for every $a, b \in A$.

The main remarks are:

Proposition.

(i) In a Lie algebra L the Jacobi identity expresses the fact that the map

$$\mathrm{ad}(a) := b \mapsto [a, b]$$

is a derivation.[15]

(ii) The derivations of any algebra A form a Lie subalgebra of the space of linear operators.

Proof. The proof is by direct verification. □

In an associative algebra or a Lie algebra, the derivations given by $b \mapsto [a, b]$ are called *inner*.

The reason why Lie algebras and derivations are important is that they express infinitesimal analogues of groups of symmetries, as explained in the following sections. The main geometric example is:

Definition 3. A derivation X of the algebra $C^\infty(M)$ of C^∞ functions on a manifold M is called a *vector field*. We will denote by $\mathcal{L}(M)$ the Lie algebra of all vector fields on M.

Our guiding principle is (cf. 1.4):

$\mathcal{L}(M)$ is the infinitesimal form of the group of all diffeomorphisms of M.

The first formal property of vector fields is that they are *local*. This means that, given a function $f \in C^\infty(M)$ and an open set U, the value of $X(f)$ on U depends only on the value of f on U. In other words:

Lemma. *If $f = 0$ on U, then also $X(f) = 0$ on U.*

Proof. Let $p \in U$ and let V be a small neighborhood of p in U. We can find a C^∞ function u on M which is 1 outside of U and 0 on V.

Hence $f = uf$ and $X(f) = X(u)f + uX(f)$ which is manifestly 0 at p. □

Recall that given an n-dimensional manifold M, a *tangent vector* v at a point $p \in M$ is a linear map $v : C^\infty(M) \to \mathbb{R}$ satisfying $v(fg) = v(f)g(p) + f(p)v(g)$. The tangent vectors in p are an n-dimensional vector space, denoted $T_p(M)$, with basis the operators $\frac{\partial}{\partial x_i}$ if $x_1, \dots, x_n$ are local coordinates.

The union of the tangent spaces forms the *tangent bundle* to M, itself a manifold.

Finally, if $F : M \to N$ is a C^∞ map of manifolds and $p \in M$, we have the *differential* $dF_p : T_p(M) \to T_{F(p)}M$ which is implicitly defined by the formula

[15] ad stands for *adjoint* action.

$$(1.1.1) \qquad dF_p(v)(f) := v(f \circ F).$$

For a composition of maps $M \xrightarrow{F} N \xrightarrow{G} P$ we clearly have $d(G \circ F) = dG \circ dF$ (the chain rule).

The previous property can be easily interpreted (cf. [He]) by saying that we can consider a derivation of the algebra of C^∞ functions as a section of the tangent bundle. In local coordinates x_i, we have $X = \sum_{i=1}^n f_i(x_1, \dots, x_n)\frac{\partial}{\partial x_i}$, a *linear differential operator.*

Remark. Composing two linear differential operators gives a quadratic operator. The remarkable fact, which we have proved using the notion of derivation, is that the Lie bracket $[X, Y]$ of two linear differential operators is still linear. The quadratic terms cancel.

As for any type of algebraic structure we have the notion of *Lie subalgebra A* of a Lie algebra L: a subspace $A \subset L$ closed under bracket.

The homomorphism $f : L_1 \to L_2$ of Lie algebras, i.e., a linear map preserving the Lie product.

Given a Lie algebra L and $x \in L$ we have defined the linear operator $\mathrm{ad}(x) : L \to L$ by $\mathrm{ad}(x)(y) := [x, y]$. The operator $\mathrm{ad}(x)$ is called the *adjoint* of x.

As for the notion of ideal, which is the kernel of a homomorphism, the reader will easily verify this:

Definition 4. An ideal I of L is a linear subspace stable under all of the operators $\mathrm{ad}(x)$.

The quotient L/I is naturally a Lie algebra and the usual homomorphism theorems hold. Conversely, the kernel of a homomorphism is an ideal.

1.2 Exponential Map

The main step in passing from infinitesimal to global transformations is done by integrating a system of differential equations. Formally (but often concretely) this consists of taking the exponential of the linear differential operator.

Consider the finite-dimensional vector space F^n where F is either $\mathbb{C}$ or $\mathbb{R}$ (complex or real field), with its standard Hilbert norm. Given a matrix A we define its norm:

$$|A| := \max\left\{\frac{|A(v)|}{|v|},\ v \neq 0\right\}, \text{ or equivalently } \quad |A| = \max\{|A(v)|,\ |v| = 1\}.$$

Of course this extends to infinite-dimensional Hilbert spaces and bounded operators, i.e., linear operators A with $\sup_{|v|=1} |A(v)| := |A| < \infty$. $|A|$ is a *norm* in the sense that:

(1) $|A| \geq 0,\ |A| = 0$ if and only if $A = 0$.
(2) $|\alpha A| = |\alpha||A|,\ \forall \alpha \in F, \forall A$.
(3) $|A + B| \leq |A| + |B|$.

With respect to the multiplicative structure, the following facts can be easily verified:

Proposition 1.

(i) Given two operators A, B *we have* $|AB| \le |A||B|$.
(ii) The series $e^A := \sum_{k=0}^{\infty} \frac{A^k}{k!}$ *is totally convergent in any bounded set of operators.*
(iii) $\log(1 + A) := \sum_{k=1}^{\infty}(-1)^{k+1}\frac{A^k}{k}$ *is totally convergent for* $|A| \le 1 - \epsilon$, *for any* $1 \ge \varepsilon > 0$.
(iv) The functions e^A, $\log A$ *are inverse of each other in suitable neighborhoods of* 0 *and* 1.

Remark. For matrices (a_{ij}) we can also take the equivalent norm $\max(|a_{ij}|)$.

The following properties of the *exponential map*, $A \mapsto e^A$, are easily verified:

Proposition 2.

(i) If A, B *are two commuting operators (i.e.,* $AB = BA$*) we have* $e^A e^B = e^{A+B}$ *and also* $\log(AB) = \log(A) + \log(B)$ *if* A, B *are sufficiently close to 1.*
(ii) $e^{-A}e^A = 1$.
(iii) $\frac{de^{tA}}{dt} = Ae^{tA}$.
(iv) $Be^A B^{-1} = e^{BAB^{-1}}$.
If A is an $n \times n$ *matrix we also have:*
(v) If $\alpha_1, \alpha_2, \dots, \alpha_n$ *are the eigenvalues of A, the eigenvalues of* e^A *are* $e^{\alpha_1}, e^{\alpha_2}, \dots, e^{\alpha_n}$.
(vi) $\det(e^A) = e^{Tr(A)}$.
(vii) $e^{A^t} = (e^A)^t$.

In particular the mapping $t \mapsto e^{tA}$ is a homomorphism from the additive group of real (or complex) numbers to the multiplicative group of matrices (real or complex).

Definition. The map $t \mapsto e^{tA}$ is called the 1-*parameter subgroup* generated by A. A is called its infinitesimal generator.

Theorem. *Given a vector* v_0 *the function* $v(t) := e^{tA}v_0$ *is the solution to the differential equation* $v'(t) = Av(t)$ *with initial condition* $v(0) = v_0$.

Proof. The proof follows immediately from (iii). □

It is not restrictive to consider such 1-parameter subgroups. In fact we have:

Exercise. A continuous homomorphism $\varphi : (\mathbb{R}, +) \to Gl(n, F)$ is of the form e^{tA} for a unique matrix A, the *infinitesimal generator* of the group φ. We also have $A = \frac{d\varphi(t)}{dt}\big|_{t=0}$.

Hint. Take the logarithm and prove that we obtain a linear map. □

1.3 Fixed Points, Linear Differential Operators

Let us develop a few basic properties of 1-parameter groups. First:

Proposition 1. *A vector v is fixed under a group $\{e^{tA},\ t \in \mathbb{R}\}$ if and only if $Av = 0$.*

Proof. If $Av = 0$, then $t^k A^k v = 0$ for $k > 0$. Hence $e^{tA} v = v$. Conversely, if $e^{tA} v$ is constant, then its derivative is 0. But its derivative at 0 is in fact Av. □

Remark. Suppose that e^{tA} leaves a subspace U stable. Then, if $v \in U$, we have that $Av \in U$ since this is the derivative of $e^{tA} v$ at 0. Conversely, if A leaves U stable, it is clear from the definition that e^{tA} leaves U stable. Its action on U agrees with the 1-parameter subgroup generated by the restriction of A.

In dual bases e_i, e^j the vector $x(t)$ of coordinate functions $x_i(t) := (e^i, e^{tA} v)$ of the evolving vector satisfies the system of ordinary linear differential equations $\dot{x}(t) = A^t x(t)$.[16] Such a system, with the initial condition $x(0)$, has the global solution $x(t) = e^{tA^t} x(0)$.

Let us now consider a function $f(x)$ on an n-dimensional vector space. We can follow its evolution under a 1-parameter group and set $\varphi(t)(f) := F(t, x) := f(e^{-tA} x)$.

We thus have a 1-parameter group of linear transformations on the space of functions, induced from the action of $\varphi(t)$ on this space.

This is not a finite-dimensional space, and we cannot directly apply the results from the previous section 1.2. If we restrict to homogeneous polynomials, we are in the finite-dimensional case. Thus for this group we have $\varphi(t)(f) = e^{tD_A} f$, where D_A is the operator

$$D_A(f) = \left.\frac{dF(t, x)}{dt}\right|_{t=0}.$$

We have

$$\frac{dF(t, x)}{dt} = \sum_{i=1}^{n} \frac{\partial f(e^{-tA} x)}{\partial x_i} \frac{dx_i(t)}{dt}$$

and, since $\frac{dx(t)}{dt} = -Ax(t)$, at $t = 0$ we have

$$\left.\frac{dx_i(t)}{dt}\right|_{t=0} = -\sum_{j=1}^{n} a_{ij} x_j.$$

Hence

$$\left.\frac{dF(t, x)}{dt}\right|_{t=0} = -\sum_{i=1}^{n} \sum_{j=1}^{n} a_{ij} x_j \frac{\partial f}{\partial x_i}.$$

Thus we have found that D_A is the differential operator:

[16] $\dot{f}(t)$ is the time derivative.

(1.3.1) $$D_A := -\sum_{i=1}^{n}\sum_{j=1}^{n} a_{ij}x_j\frac{\partial}{\partial x_i}.$$

We deduce that the formula $\varphi(t)(f) = e^{tD_A}f$ is just the Taylor series:

$$F(t,x) = \sum_{k=0}^{\infty}\frac{(tD_A)^k}{k!}F(0,x) = \sum_{k=0}^{\infty}\frac{t^k}{k!}D_A^k f(x).$$

In order to better understand the operators D_A, let us compute

$$D_A x_i = -\sum_{j=1}^{n} a_{ij}x_j.$$

We see that on the linear space with basis the functions x_i, this is just the linear operator given by the matrix $-A^t$.

Since a derivation is determined by its action on the variables x_i we have:

Proposition 2. *The differential operators D_A are a Lie algebra and*

(1.3.2) $$[D_A, D_B] = D_{[A,B]}.$$

Proof. This is true on the space spanned by the x_i since $[-A^t, -B^t] = -[A,B]^t$. □

Example. $Sl(2,\mathbb{C})$. We want to study the case of polynomials in two variables x, y.

The Lie algebra of 2×2 matrices decomposes as the direct sum of the 1-dimensional algebra generated by $D = x\frac{\partial}{\partial x} + y\frac{\partial}{\partial y}$ and the 3-dimensional algebra $sl(2,\mathbb{C})$ with basis

$$H = -x\frac{\partial}{\partial x} + y\frac{\partial}{\partial y},\quad E = -y\frac{\partial}{\partial x},\quad F = -x\frac{\partial}{\partial y}.$$

These operators correspond to the matrices

$$\begin{pmatrix}-1 & 0\\ 0 & -1\end{pmatrix},\quad \begin{pmatrix}1 & 0\\ 0 & -1\end{pmatrix},\quad \begin{pmatrix}0 & 1\\ 0 & 0\end{pmatrix},\quad \begin{pmatrix}0 & 0\\ 1 & 0\end{pmatrix}.$$

We can see how these operators act on the space P_n of homogeneous polynomials of degree n. This is an $n+1$-dimensional space spanned by the monomials $u_i := (-1)^i y^{n-i}x^i$, on which D acts by multiplication by n. We have

(1.3.3) $$Hu_i = (n-2i)u_i\,,\qquad Fu_i = (n-i)u_{i+1}\,,\qquad Eu_i = iu_{i-1}.$$

The reader who has seen these operators before will recognize the standard irreducible representations of the Lie algebra $sl(2,\mathbb{C})$ (cf. Chapter 10, 1.1).

The action of $Sl(2,\mathbb{C})$ on the polynomials $\sum_i a_i(-1)^i y^{n-i}x^i \in P_n$ extends to an action on the functions $p(a_0, a_1, \dots, a_n)$ of the coordinates a_i. We have that

$$E = -\sum(i+1)a_{i+1}\frac{\partial}{\partial a_i},\quad F = -\sum(n-i+1)a_{i-1}\frac{\partial}{\partial a_i},$$

(1.3.4) $$H = -\sum(n-2i)a_i\frac{\partial}{\partial a_i}$$

are the induced differential operators. We will return to this point later.

1.4 One-Parameter Groups

Of course these ideas have a more general range of validity. For instance, the main facts about the exponential and the logarithm are sufficiently general to hold for any Banach algebra, i.e., an algebra with a norm under which it is complete. Thus one can also apply these results to bounded operators on a Hilbert space.

The linearity of the transformations A is not essential. If we consider a C^∞ differentiable manifold M, we can discuss dynamical systems (in this case also called *flows*) in 2 ways.

Definition 1. A C^∞ flow on a manifold M, or a 1-parameter group of diffeomorphisms, is a C^∞ map:

$$\phi(t, x) : \mathbb{R} \times M \to M$$

which defines an additive group action of $\mathbb{R}$ on M.

This can also be thought of as a C^∞ family of diffeomorphisms:

$$\phi_s : M \to M, \qquad \phi_s(m) := \phi(s, m), \qquad \phi_0 = 1_M, \quad \phi_{s+t} = \phi_s \circ \phi_t.$$

To a flow is associated a vector field X, called the *infinitesimal generator of the flow*.

The vector field X associated to a flow $\phi(t, x)$ can be defined at each point p as follows. Let us start from a fixed p. Denote by $\phi_p(t) := \phi(t, p)$. This is now a map from $\mathbb{R}$ to M which represents the evolution of p with time. In this notation the group property is $\phi_{\phi_p(s)}(t) = \phi_p(s + t)$. Denote the differential of ϕ_p at a point s, by $d\phi_p(s)$. Let

$$X_p := d\phi_p(0)\Big(\frac{d}{dt}\Big), \quad \text{i.e.,} \quad X_p(f) = \frac{d}{dt} f(\phi(t, p))_{t=0}, \ \forall f \in C^\infty(M).$$

Definition 2. The vector field X which, given a function f on M, produces

$$X_p(f) = X(f)(p) := \frac{d}{dt} f(\phi(t, p))_{t=0}, \tag{1.4.1}$$

is the *infinitesimal generator* of the flow $\phi(t)$.

Given a point p, the map $t \mapsto \phi_p(t) = \phi(t, p)$ describes a curve in M which is the *evolution of* p, i.e., the *orbit* under the flow. X_p is the *velocity* of this evolution of p, which by the property $\phi_{\phi_p(s)}(t) = \phi_p(s + t)$ depends only on the position and not on the time:

$$X_{\phi(s,p)} = d\phi_{\phi(s,p)}(0)\Big(\frac{d}{dt}\Big) = d\phi_p(s)\Big(\frac{d}{dt}\Big).$$

A linear operator T on a vector space V induces two linear flows in V and V^*. Identifying V with its tangent space at each point, the vector field X_A associated to the linear flow at a point v is Tv, while the generator of the action on functions is $-T^t$. In dual bases, if A denotes the matrix of T, the vector field is thus (cf. 1.3.2):

(1.4.2) $$-X_A = D_A = -\sum_{i=1}^{n}\sum_{j=1}^{n} a_{ij}x_j \frac{\partial}{\partial x_i} \implies [X_A, X_B] = X_{[B,A]}.$$

A vector field $\sum_i f_i(x_1, \dots, x_n)\frac{\partial}{\partial x_i}$ gives rise, at least locally, to a flow which one obtains by solving the linear system of ordinary differential equations

$$\frac{dx_i(t)}{dt} = f_i\big(x_1(t), \dots, x_n(t)\big).$$

Thus one has local solutions $\varphi(t)(x) = F(t, x)$, depending on the parameters x, with initial conditions $F(0, x) = x$. For a given point x^0 such a solution exists, for small values of t in a neighborhood of a given point x^0.

We claim that the property of a local 1-parameter group is a consequence of the uniqueness of solutions of ordinary differential equations. In fact we have that for given t, letting s be a new parameter,

$$\frac{dx_i(t+s)}{ds} = f_i\big(x_1(t+s), \dots, x_n(t+s)\big).$$

Remark. A point is fixed under the flow if and only if the vector field vanishes on it.

For the vector field $X_A = -D_A$, the flow is the 1-parameter group of linear operators e^{tA}. We can in fact, at least locally, linearize every flow by looking at its induced action on functions. By the general principle of (Chapter 1, 2.5.1) the evolution of a function f is given by the formula $\phi(t)f(x) = f(\phi(-t)x) = f(F(-t, x))$. When we fix x, we are in fact restricting to an orbit. We can now develop $\phi(t)f(x)$ in Taylor series. By definition, the derivative with respect to t at a point of the orbit is the same as the derivative with respect to the vector given by X, hence the Taylor series is $\phi(t)f(x) = \sum_{k=0}^{\infty}(-t)^k \frac{X^k}{k!} f(x)$.

In this sense the flow $\phi(t)$ becomes a linear flow on the space of functions, with infinitesimal generator $-X$. We have $-Xf = \frac{d\phi(t)f}{dt}(0)$, and $\phi(t) = e^{-tX}$.[17]

Of course in order to make the equality $\phi(t)f(x) = \sum_{k=0}^{\infty} t^k \frac{(-X)^k}{k!} f(x)$ valid for all x, t, we need some hypotheses, such as the fact that the flow exists globally, and also that the functions under consideration are analytic.

The special case of linear flows has the characteristic that one can find global coordinates on the manifold so that the evolution of these coordinates is given by a linear group of transformations of the finite-dimensional vector space, spanned by the coordinates!

In general of course the evolution of coordinates develops nonlinear terms. We will use a simple criterion for Lie groups, which ensures that a flow exists globally.

Lemma 1. *Suppose there is an $\epsilon > 0$, independent of p, so that the flow exists for all p and all $t < \epsilon$. Then the flow exists globally for all values of t.*

[17] If we want to be consistent with the definition of action induced on functions by an action of a group on a set, we must define the evolution of f by $f(t, x) := f(F(-t, x))$, so the flow on functions is e^{-tX}.

Proof. We have for small values of t the diffeomorphisms $\phi(t)$ which, for t, s sufficiently small, satisfy $\phi(s+t) = \phi(s)\phi(t)$. Given any t we consider a large integer N and set $\phi(t) := \phi(t/N)^N$. The previous property implies that this definition is independent of N and defines the flow globally. □

We have already seen that the group of diffeomorphisms of a manifold (as well as subgroups, as for instance a flow) acts linearly as algebra automorphisms on the algebra of C^∞ functions, by $(gf)(x) = f(g^{-1}x)$. We can go one step further and deduce a linear action on the space of linear operators T on functions. The formula is $g \circ T \circ g^{-1}$ or gTg^{-1}. In particular we may apply this to vector fields. Recall that:

(i) A vector field on M is just a derivation of the algebra of functions $C^\infty(M)$.
(ii) Given a derivation D and an automorphism g of an algebra A, gDg^{-1} is a derivation.

We see that given a diffeomorphism g, the induced action on operators maps vector fields to vector fields.

We can compute at each point $g \circ X \circ g^{-1}$ by computing on functions, and see that:

Lemma 2.

(i) At each point $p = gq$ we have $(gXg^{-1})_p = dg_q(X_q)$.
(ii) Take a vector field X generating a 1-parameter group $\phi_X(t)$ and a diffeomorphism g. Then, the vector field gXg^{-1} generates the 1-parameter group $g\phi_X(t)g^{-1}$.
(iii) The map $X \mapsto gXg^{-1}$ is a homomorphism of the Lie algebra structure.

Proof.

(i)
$$dg_q(X_q)(f) = X_q(f \circ g) = X(f(gx))(q)$$

(1.4.3)
$$= X(g^{-1}f)(q) = g(X(g^{-1}f))(p).$$

(ii) Let us compute $\frac{d}{dt} f(g\phi(t)g^{-1}p)$. The curve $\phi(t)g^{-1}p = \phi(t)q$ has tangent vector X_q at $t = 0$. The curve $g\phi(t)g^{-1}p$ has tangent vector $dg_q(X_q) = (gXg^{-1})_p$ at $t = 0$.
(iii) The last claim is obvious. □

The main result regarding evolution of vector fields is:

Theorem. *Let X and Y be vector fields, $\phi(t)$ the 1-parameter group generated by X. Consider the time-dependent vector field*

$$Y(t) := \phi(t)Y\phi(t)^{-1}, \quad Y(t) = d\phi(t)_{\phi(t)^{-1}p}(Y_{\phi(t)^{-1}p}).$$

Then $Y(t)$ satisfies the differential equation $\frac{dY(t)}{dt} = [Y(t), X]$.

Proof. Thanks to the group property, it is enough to check the equality at $t = 0$. Let f be a function and let us compute the Taylor series of $Y(t)f$. We have

$$(1.4.4) \quad \begin{aligned} Y(t)f &= \phi(t)Y\phi(t)^{-1}f = \phi(t)[Yf + tYXf + O(t^2)] \\ &= Yf + tYXf + O(t^2) - tX[Yf + tYXf + O(t^2)] + O(t^2) \\ &= Yf + t[Y, X]f + O(t^2). \end{aligned}$$

In general by $O(t^k)$ we mean a function $h(t)$ which is infinitesimal of order t^k, i.e., $\lim_{t\to 0} t^{k-1}h(t) = 0$. From this the statement follows. □

In different terminology, consider $d\phi(t)^{-1}_{\phi(t)p} Y_{\phi(t)p}$. The derivative of this field at $t = 0$ is $[X, Y]$ and it is called the *Lie derivative*. In other words $[X, Y]_p$ measures the infinitesimal evolution of Y_p on the orbit of p under the time evolution of $\phi(t)$.

Corollary 1. *The* 1*-parameter groups of diffeomorphisms generated by two vector fields X, Y commute if and only if $[X, Y] = 0$.*

Proof. Clearly $Y(t) = \phi_X(t)Y\phi_X(t)^{-1}$ is the generator of $\phi_X(t)\phi_Y(s)\phi_X(t)^{-1}$ (in the parameter s). If the two groups commute, $Y(t)$ is constant so its derivative $[Y, X]$ is 0. Conversely, if $Y(t)$ is constant, then the two groups commute. □

Remark. If X and Y are commuting vector fields and a, b two numbers, $aX + bY$ is the infinitesimal generator of the flow $\phi_X(at)\phi_Y(bt)$.

There is an important special case to notice, when p is a fixed point of $\phi_X(t)$. In this case, although p is fixed, the tangent vectors at p are not necessarily fixed, but *move* according to the linear 1-parameter group $d\phi_X(t)_p$. Thus the previous formulas imply the following.

Corollary 2. *Let X be a vector field vanishing at p, and Y any vector field. Then the value $[Y, X]_p$ depends only on Y_p. The linear map $Y_p \to [Y, X]_p$ is the infinitesimal generator of the* 1*-parameter group $d\phi_X(t)_p$ on the tangent space at p.*

1.5 Derivations and Automorphisms

Let us go back to *derivations* and *automorphisms*.

Consider an algebra A and a linear operator D on A. Assume that there are sufficient convergence properties to ensure the existence of e^{tD} as a convergent power series:[18]

Proposition. *D is a derivation if and only if e^{tD} is a group of automorphisms.*

[18] as for Banach algebras

Proof. This is again a variation of the fact that a vector v is fixed under e^{tD} if and only if $Dv = 0$. In fact to say that e^{tD} are automorphisms means that

$$\forall a, b \in A,\ e^{tD}(ab) - e^{tD}(a)e^{tD}(b) = 0.$$

Writing in power series and taking the coefficient of the linear term we get

$$D(ab) - D(a)b - aD(b) = 0,$$

the condition for a derivation.

Conversely, given a derivation, we see by easy induction that for any positive integer k,

$$D^k(ab) = \sum_{i=0}^{k} \binom{k}{i} D^{k-i}(a)D^i(b).$$

Hence

$$e^{tD}(ab) = \sum_{k=0}^{\infty} \frac{t^k D^k(ab)}{k!} = \sum_{k=0}^{\infty} t^k \sum_{i=0}^{k} \frac{1}{k!} \binom{k}{i} D^{k-i}(a)D^i(b)$$

$$= \sum_{k=0}^{\infty} \sum_{i=0}^{k} \frac{1}{(k-i)!i!} t^{k-i} D^{k-i}(a) t^i D^i(b) = e^{tD}(a)e^{tD}(b). \qquad \square$$

Our heuristic idea is that, for a differentiable manifold M, its group of diffeomorphisms should be *the group of automorphisms of the algebra of C^∞ functions*. Our task is then to translate this idea into rigorous finite-dimensional statements.

2 Lie Groups

2.1 Lie Groups

As we have already mentioned, it is quite interesting to analyze group actions subject to special structural requirements.

The structure of each group G is described by two basic maps: the multiplication $m : G \times G \to G,\ m(a, b) := ab$ and the inverse $i : G \to G,\ i(g) := g^{-1}$. If G has an extra geometric structure we require the compatibility of these maps with it. Thus we say that:

Definition. A group G is a: (1) topological group, (2) Lie group, (3) complex analytic group, (4) algebraic group, (5) affine group,

if G is also a (1) topological space, (2) differentiable manifold, (3) complex analytic manifold, (4) algebraic variety, (5) affine variety,

and if the two maps m, i are compatible with the given structure, i.e., are continuous, differentiable, complex analytic or regular algebraic.

When speaking of Lie groups we have not discussed the precise differentiability hypotheses. A general theorem (solution of Hilbert's 5th problem (cf. [MZ], [Kap])) ensures that a topological group which is locally homeomorphic to Euclidean space can be naturally given a real analytic structure. So Lie groups are in fact real analytic manifolds.

The group $GL(n, \mathbb{C})$ is an affine algebraic group (cf. Chapter 7), acting on $\mathbb{C}^n$ by linear and hence algebraic transformations. A group G is called a *linear group* if it can be embedded in $GL(n, \mathbb{C})$ (of course one should more generally consider as linear groups the subgroups of $GL(n, F)$ for an arbitrary field F).

For an action of G on a set X we can also have the same type of analysis: continuous action of a topological group on a topological space, differentiable actions of Lie groups on manifolds, etc. We shall meet many very interesting examples of these actions in the course of our treatment.

Before we concentrate on Lie groups, let us collect a few simple general facts about topological groups, and actions (cf. [Ho], [B1]). For our discussion, "connected" will always mean "arc-wise connected." Let G be a topological group.

Proposition.

(1) Two actions of G coinciding on a set of generators of G are equal.[19]
(2) An open subgroup of a topological group is also closed.
(3) A connected group is generated by the elements of any given nonempty open set.
(4) A normal discrete subgroup Z of a connected group is in the center.
(5) A topological group G is discrete if and only if 1 is an isolated point.
(6) Let $f : H \to G$ be a continuous homomorphism of topological groups. Assume G connected and there is a neighborhood U of $1 \in H$ for which $f(U)$ is open and $f : U \to f(U)$ is a homeomorphism. Then f is a covering space.

Proof.

(1) is obvious.
(2) Let H be an open subgroup. G is the disjoint union of all its left cosets gH which are then open sets. This means that the complement of H is also open, hence H is closed.
(3) Let U be a nonempty open set of a group G and H be the subgroup that it generates. If $h \in H$ we have that also hU is in H. Hence H is an open subgroup, and by the previous step, it is also closed. Since G is connected and H nonempty, we have $H = G$.
(4) Let $x \in Z$. Consider the continuous map $g \mapsto gzg^{-1}$ from G to Z. Since G is connected and Z discrete, this map is constant and thus equal to $z = 1z1^{-1}$.
(5) is clear.
(6) By hypothesis and (5), $A := f^{-1}1$ is a discrete group. We have $f^{-1}f(U) = \sqcup_{h\in A} Uh$. The covering property is proved for a neighborhood of 1. From 3) it follows that f is surjective. Then for any $g \in G$, $g = f(b)$ we have

[19] For topological groups and continuous actions, we can take topological generators, i.e., elements which generate a dense subgroup.

$$f^{-1}f(bU) = \sqcup_{h \in A} bUh. \qquad \square$$

There is a converse to point 6) which is quite important. We can apply the theory of covering spaces to a connected, locally connected and locally simply connected topological group G. Let $\tilde{G}$ be the universal covering space of G, with covering map $\pi : \tilde{G} \to G$. Let us first define a transformation group $\overline{G}$ on $\tilde{G}$:

$$\overline{G} := \{T \mid T : \tilde{G} \to \tilde{G}, \text{ continuous} \mid \exists g \in G \text{ with } \pi(T(x)) = g\pi(x)\}.$$

The theory of covering spaces implies immediately:

Theorem. *$\overline{G}$ is a group which acts in a simply transitive way on $\tilde{G}$.*

Proof. It is clear that $\overline{G}$ is a group. Given two points $x, y \in \tilde{G}$ there is a unique $g \in G$ with $g\pi(x) = \pi(y)$, and therefore a unique lift T of the multiplication by g, with $T(x) = y$. □

Therefore, given $x \in \tilde{G}$, we can identify $\overline{G}$ with $\tilde{G}$. One easily verifies that:

Corollary. *With this identification $\tilde{G}$ becomes a simply connected topological group. If $\pi(x) = 1$, the mapping π is a homomorphism.*

3 Correspondence between Lie Algebras and Lie Groups

We review here the basics of Lie theory referring to standard books for a more leisurely discussion.

3.1 Basic Structure Theory

In this section the general theory of vector fields and associated 1-parameter groups will be applied to Lie groups. Given a Lie group G, we will associate to it a Lie algebra $\mathfrak{g}$ and an exponential map $\exp : \mathfrak{g} \to G$. The Lie algebra $\mathfrak{g}$ can be defined as the Lie algebra of left-invariant vector fields on G, under the usual bracket of vector fields. The exponential map is obtained by integrating these vector fields, proving that, in this case, the associated 1-parameter groups are global.[20] A homomorphism $\phi : G_1 \to G_2$ of Lie groups induces a homomorphism $d\phi : \mathfrak{g}_1 \to \mathfrak{g}_2$ of the associated Lie algebras. In particular, this applies to linear representations of G which induce linear representations of the Lie algebra. Conversely, a homomorphism of Lie algebras integrates to a homomorphism of Lie groups, provided that G_1 is simply connected.

Before we enter into the details let us make a fundamental definition:

Definition 1. A 1-parameter subgroup of a topological group G is a continuous homomorphism $\phi : \mathbb{R} \to G$, i.e., $\phi(s + t) = \phi(s)\phi(t)$.

[20] This can also be interpreted in terms of geodesics of Riemannian geometry, for a left-invariant metric.

Remark. For a Lie group we will assume that ϕ is C^∞, although one can easily see that this is a consequence of continuity.

Let G be a Lie group. Consider left and right actions, $L_g(h) := gh$, $R_g(h) := hg^{-1}$.

Lemma. *If a transformation $T : G \to G$ commutes with the left action, then $T = R_{T(1)^{-1}}$.*

Proof. We have $T(g) = T(L_g(1)) = L_g(T(1)) = gT(1) = R_{T(1)^{-1}}(g)$. □

Definition 2. We say that a vector field X is left-invariant if $L_g \circ X \circ L_g^{-1} = X, \ \forall g \in G$.

Proposition 1.

(1) A left-invariant vector field $X := X_a$ is uniquely determined by the value $X(1) := a$. Then its value at $g \in G$ is given by the formula

$$X_a(g) = L_g X_a L_g^{-1}(g) = dL_g(X_a(1)) = dL_g(a). \tag{3.1.1}$$

(2) A tangent vector $a \in T_1(G)$ is the velocity vector of a uniquely determined 1-parameter group $t \mapsto \phi_a(t)$. The corresponding left-invariant vector field X_a is the infinitesimal generator of the 1-parameter group of right translations $\phi_{X_a}(t)(g) := g\phi_a(t)$.

Proof.

(1) 3.1.1 is a special case of formula 1.4.3.

(2) By Lemma 1.4 a left-invariant vector field is the infinitesimal generator of a 1-parameter group of diffeomorphisms commuting with left translations. By the previous lemma these diffeomorphisms must be right translations which are defined globally.

Given a tangent vector a at 1, the corresponding vector field X_a and the 1-parameter group of diffeomorphisms $\Phi_a(t) := \phi_{X_a}(t)$, consider the curve $\phi_a(t) := \Phi_a(t)(1)$. We thus have $\Phi_a(t)(g) = g\phi_a(t)$ and also $\phi_a(t+s) = \phi_{X_a}(t+s)(1) = \phi_{X_a}(t)\phi_{X_a}(s)(1) = \phi_a(s)\phi_a(t)$ is a 1-parameter subgroup of G. □

Remark. The 1-parameter group $\phi_a(t)$ with velocity a is also called the *exponential*, denoted $\phi_a(t) = e^{ta} = \exp(ta)$. The map $\exp : T_1(G) \to G$, $\exp(a) := e^a$ is called the *exponential map*.

The differential of exp at 0 is the identity, so $\exp(T_1(G))$ contains a neighborhood of 1. Therefore, if G is connected, we deduce that G is generated by $\exp(T_1(G))$.

Since applying a diffeomorphism to a vector field preserves the Lie bracket we have:

Theorem 1. *The left-invariant vector fields form a Lie subalgebra of the Lie algebra of vector fields on G, called the Lie algebra of G and denoted by L_G. Evaluation at 1 establishes a linear isomorphism of L_G with the tangent space $T_1(G)$.*

In view of the linear isomorphism $L_G \equiv T_1(G)$, it is usual to endow $T_1(G)$ with the Lie algebra structure given implicitly by $X_{[a,b]} := [X_a, X_b]$.

We could have started from right-invariant vector fields Z_a for which the expression is $Z_a(g) = dR_{g^{-1}}(a)$. It is easy to compare the two approaches from the following:

Remark 1. Consider the diffeomorphism $i : x \mapsto x^{-1}$. We have $iL_g i^{-1} = R_g$, so i transforms right into left-invariant vector fields.

We claim that the differential at 1 is $di(a) = -a$. In fact, for the differential of the multiplication map $m : G \times G \to G$ we must have $dm : (a, b) \mapsto a + b$, since composing with the two inclusions, $i_1 : g \mapsto (g, 1)$, $i_2 : g \mapsto (1, g)$ gives the identity on G. Then $m(g, i(g)) = m(i(g), g) = 1$ implies that $di(a) + a = 0$.

In summary, we have two Lie algebra structures on $T_1(G)$ induced by right and left-invariant vector fields. The map $a \mapsto -a$ is an isomorphism of the two structures. In other words, $[Z_a, Z_b] = Z_{[b,a]}$.

The reason to choose left-invariant rather than right-invariant vector fields in the definition of the Lie algebra is the following:

Proposition. *When $G = GL(n, \mathbb{R})$ is the linear group of invertible matrices, its tangent space at 1 can be identified with the space of all matrices $M_n(\mathbb{R})$. If we describe the Lie algebra of $GL(n, \mathbb{R})$ as left-invariant vector fields, we get the usual Lie algebra structure $[A, B] = AB - BA$ on $M_n(\mathbb{R})$.*[21]

Proof. A matrix A (in the tangent space) generates the 1-parameter group of right translations $X \mapsto e^{tA}$, whose infinitesimal generator is the linear vector field associated to the map $R_A : X \mapsto XA$. We have

$$[R_B, R_A](X) = (R_B R_A - R_A R_B)X$$
$$= XAB - XBA = R_{[A,B]}(X), \quad \text{i.e,} \quad [R_B, R_A] = R_{[A,B]}.$$

From 1.4.3, if X_A is the linear vector field associated to R_A we finally have $[X_A, X_B] = X_{[A,B]}$, as required. □

Remark 2. From 1.4 it follows that if $[a, b] = 0$, then $\exp(a)\exp(b) = \exp(a + b)$. In general the Lie bracket is a second-order correction to this equality.

Since right translations commute with left translations, they must map L_G into itself. We have thus a linear action of G on the Lie algebra, called the *adjoint action,* given by

$$\text{(3.1.2)} \qquad \mathrm{Ad}(g)X_a := R_g \circ X_a \circ R_g^{-1}.$$

Explicitly, since $\mathrm{Ad}(g)X_a \in L_G$, we have $\mathrm{Ad}(g)X_a = X_{\mathrm{Ad}(g)(a)}$ for some unique element $\mathrm{Ad}(g)(a)$. From formulas 1.4.3 and 3.1.1 we have

[21] With right-invariant vector fields, we would get $BA - AB$.

$$\mathrm{Ad}(g)(a) = dR_g(X_a(R_g^{-1}(1))) = dR_g(X_a(g))$$

(3.1.3) $$= dR_g(dL_g(a)) = d(R_g \circ L_g)(a).$$

The diffeomorphism $R_g \circ L_g$ is just $x \mapsto gxg^{-1}$, i.e., it is conjugation by g.

For an element $g \in G$ let us denote by $C_g : x \mapsto gxg^{-1}$ the conjugation action. C_g is an automorphism and $C_g \circ C_h = C_{gh}$. Of course $C_g(1) = 1$, hence C_g induces a linear map $dC_g : T_1(G) \to T_1(G)$. The map $g \mapsto dC_g$ is a linear representation of G on $L_G = T_1(G)$. We have just proved:

Proposition 2. *We have*

(3.1.4) $$\mathrm{Ad}(g) = dC_g, \quad \textit{differential of the adjoint map.}$$

(3.1.5) $$ge^a g^{-1} = e^{\mathrm{Ad}(g)(a)}, \ \forall g \in G, \ a \in L_G.$$

Proof. We have proved the first formula. As for the second we have the composition $ge^{ta}g^{-1} : \mathbb{R} \xrightarrow{e^{ta}} G \xrightarrow{C_g} G$ so $\mathrm{Ad}(g)(a) = dC_g(a) = d(ge^{ta}g^{-1})\left(\frac{d}{dt}\right)$ is the infinitesimal generator of the 1-parameter group $ge^{ta}g^{-1}$. □

At this point we can make explicit the Lie algebra structure on $T_1(G)$.

Let $a, b \in T_1(G)$ and consider the two left-invariant vector fields X_a, X_b so that $X_{[a,b]} := [X_a, X_b]$ by definition. By Theorem 1.4 we have that $[X_a, X_b]$ is the derivative at 0 of the variable vector field

$$R_{\phi_a(t)}^{-1} X_b R_{\phi_a(t)}.$$

At the point 1 this takes the value

$$dR_{\phi_a(-t)}(\phi_a(t))X_b(\phi_a(t)) = dR_{\phi_a(-t)}(\phi_a(t))dL_{\phi_a(t)}b = dC_{\phi_a(t)}(b) = \mathrm{Ad}(\phi_a(t))(b).$$

We deduce:

Theorem 2. *Given $a \in T_1(G)$, the linear map $\mathrm{ad}(a) : b \mapsto [a, b]$ is the infinitesimal generator of the 1-parameter group $\mathrm{Ad}(\phi_a(t)) : L_G \to L_G$.*

(3.1.6) $$\mathrm{Ad}(e^a) = e^{\mathrm{ad}(a)}, \quad e^a e^b e^{-a} = e^{\mathrm{Ad}(e^a)(b)} = e^{e^{\mathrm{ad}(a)}(b)}, \ \forall a, b \in L_G.$$

It may be useful to remark how one computes the differential of the multiplication m at a pair of points $g_0, h_0 \in G$. Recall that we have identified $T_{g_0}(G)$ with L via the linear map $dL_{g_0} : L = T_1(G) \to T_{g_0}(G)$. Therefore given two elements $dL_{g_0}(a), dL_{h_0}(b), \ a, b \in L$, computing $dm_{g_0,h_0}(dL_{g_0}(a), dL_{h_0}(b))$ is equivalent to computing $df_{1,1}(a, b)$ where $f : (x, y) \mapsto g_0 x h_0 y$. Moreover we want to find the $c \in L$ such that $df_{1,1}(a, b) = dL_{g_0 h_0}(c)$, and $c = dh_{1,1}(a, b)$ where $h : (x, y) \mapsto (g_0 h_0)^{-1} g_0 x h_0 y = \mathrm{Ad}(h_0^{-1})(x)y$; thus we get

(3.1.7) $$c = \mathrm{Ad}(h_0^{-1})(a) + b.$$

3.2 Logarithmic Coordinates

We have seen that Lie algebras are an infinitesimal version of Lie groups. In order to understand this correspondence, we need to be able to do some local computations in a Lie group G with Lie algebra L in a neighborhood of 1.

Lemma 1. *The differential at* 0 *of the map* $a \mapsto \exp(a)$ *is the identity. Therefore* exp *is a local diffeomorphism between L and G.*

We can thus use the coordinates given by $\exp(a)$ which we call *logarithmic coordinates*. Fixing an arbitrary Euclidean norm on L, they will be valid for $|a| < R$ for some R. In these coordinates, if $|a|, N|a| < R$ and N is an integer, we have that $\exp(Na) = \exp(a)^N$. Moreover, (introducing a small parameter t) the multiplication in these coordinates has the form $m(ta, tb) = ta+tb+t^2R(t, a, b)$ where $R(t, a, b)$ is bounded around 0. An essential remark is:

Lemma 2. *Given two* 1*-parameter groups* $\exp(ta)$, $\exp(tb)$ *we have*

$$\lim_{N\to\infty} [\exp\left(\frac{t}{N}a\right)\exp\left(\frac{t}{N}b\right)]^N = \exp(t(a+b)).$$

Proof. If t is very small we can apply the formulas in logarithmic coordinates and see that the coordinates of $[\exp\left(\frac{t}{N}a\right)\exp\left(\frac{t}{N}b\right)]^N$ are $ta+tb+t^2/N\, R(t, a/N, b/N)$. Since $R(t, a/N, b/N)$ is bounded, $\lim_{N\to\infty} ta + tb + t^2/N\, R(t, a/N, b/N) = t(a+b)$ as required. One reduces immediately to the case where t is small. □

We shall use the previous lemma to show that a closed subgroup H of a Lie group G is a Lie subgroup, i.e., it is also a differentiable submanifold. Moreover we will see that the Lie algebra of H is given by restricting to H the left-invariant vector fields in G which are tangent to H at 1. In other words, the Lie algebra of H is the subspace of tangent vectors to G at 1 which are tangent to H as a Lie subalgebra.

Theorem 1. *Let G be a Lie group and H a closed subgroup. Then H is a Lie subgroup of G. Its Lie algebra is* $L_H := \{a \in L_G | e^{ta} \in H,\ \forall t \in \mathbb{R}\}$.

Proof. First we need to see that L_H is a Lie subalgebra. We use the previous lemma.

Clearly L_H is stable under multiplication by scalars. If $a, b \in L_H$, we have that $\lim_{N\to\infty}[\exp\left(\frac{t}{N}a\right)\exp\left(\frac{t}{N}b\right)]^N = \exp(t(a+b)) \in H$ since H is closed; hence $a+b \in L_H$. To see that it is closed under Lie bracket we use Theorem 1.4 and the fact that $e^{ta}e^{sb}e^{-ta} \in H$ if $a, b \in L_H$. Thus, we have that $\mathrm{Ad}(e^{ta})(b) \in L_H$, $\forall t$. Finally $[a, b] \in L_H$ since it is the derivative at 0 of $\mathrm{Ad}(e^{ta})(b)$.

Now consider any linear complement A to L_H in L_G and the map $\psi : L_G = A \oplus L_H \to G$, $\psi(a, b) := e^a e^b$. The differential at 0 is the identity map so ψ is a local homeomorphism. We claim that in a small neighborhood of 0, we have $\psi(a, b) \in H$ if and only if $a = 0$; of course $e^a e^b = \psi(a, b) \in H$ if and only if $e^a \in H$. Otherwise, fix an arbitrary Euclidean norm in A. There is an infinite sequence of nonzero elements $a_i \in A$ tending to 0 and with $e^{a_i} \in H$. By compactness we can extract from this sequence a subsequence for which $a_i/|a_i|$ has as its limit a unit vector $a \in A$. We claim that $a \in L_H$, which is a contradiction. In fact, compute $\exp(ta) = \lim_{i\to\infty} \exp(ta_i/|a_i|)$. Let m_i be the integral part of $t/|a_i|$. Clearly since the a_i tend to 0 we have $\exp(ta) = \lim_{i\to\infty} \exp(ta_i/|a_i|) = \lim_{i\to\infty} \exp(m_i a_i) = \lim_{i\to\infty} \exp(a_i)^{m_i} \in H$.

Logarithmic coordinates thus show that the subgroup H is a submanifold in a neighborhood U of 1. By the group property this is true around any other element as well. Given any $h \in H$ we have that $H \cap hU = h(H \cap U)$ is a submanifold in hU. Thus H is a submanifold of G. The fact that L_H is its Lie algebra follows from the definition of L_H. □

In the correspondence between Lie groups and Lie algebras we have:

Theorem 2.

(i) A homomorphism $\rho : G_1 \to G_2$ of Lie groups induces a homomorphism $d\rho$ of the corresponding Lie algebras.
(ii) The kernel of $d\rho$ is the Lie algebra of the kernel of ρ. The map $d\rho$ is injective if and only if the kernel of ρ is discrete.
(iii) If G_2 is connected, $d\rho$ is surjective if and only if ρ is surjective. The map $d\rho$ is an isomorphism if and only if ρ is a covering.

Proof.

(i) Given $a \in L_{G_1}$, we know that $\mathrm{ad}(a)$ is the generator of the 1-parameter group $\mathrm{Ad}(\phi_a(t))$ acting on the tangent space at 1. Under ρ we have that $\rho(\phi_a(t)) = \phi_{d\rho(a)}(t)$ and $\rho \circ C_g = C_{\rho(g)} \circ \rho$. Thus $d\rho \circ \mathrm{Ad}(g) = d\rho \circ dC_g = dC_{\rho(g)} \circ d\rho = \mathrm{Ad}(\rho(g)) \circ d\rho$.

We deduce $d\rho \circ \mathrm{Ad}(\phi_a(t)) = \mathrm{Ad}(\rho(\phi_a(t))) \circ d\rho = \mathrm{Ad}(\phi_{d\rho(a)}(t)) \circ d\rho$. Taking derivatives we have $d\rho \circ \mathrm{ad}(a) = \mathrm{ad}(d\rho(a)) \circ d\rho$ and the formula follows.

(ii) Now, $d\rho(a) = 0$ if and only if $\rho(e^{ta}) = 1\ \forall t$. This means that a is in the Lie algebra of the kernel K, by Theorem 1. To say that this Lie algebra is 0 means that the group K is discrete.

(iii) If $d\rho$ is surjective, by the implicit function theorem, the image of ρ contains a neighborhood U of 1 in G_2. Hence also the subgroup H generated by U. Since G_2 is connected, $H = G_2$. If $d\rho(a)$ is an isomorphism, ρ is also a local isomorphism. The kernel is thus a discrete subgroup and the map ρ is a covering by (6) of Proposition 2.1. □

Exercise. If G is a connected Lie group, then its universal cover is also a Lie group.

The simplest example is $SU(2, \mathbb{C})$, which is a double covering of the 3-dimensional special orthogonal group $SO(3, \mathbb{R})$, providing the first example of *spin* (cf. Chapter 5, §6). Nevertheless, one can establish a bijective correspondence between Lie algebras and Lie groups by restricting to simply connected groups. This is the topic of the next two sections.

3.3 Frobenius Theorem

There are several steps in the correspondence. We begin by recalling the relevant theory. First, we fix some standard notation.

Definition 1. A C^∞ map $i : N \to M$ of differentiable manifolds is an *immersion* if it is injective and, for each $x \in N$, the differential di_x is also injective.

Definition 2. An n-dimensional distribution on a manifold M is a function that to each point $p \in M$ assigns an n-dimensional subspace $P_p \subset T_p(M)$.

The distribution is smooth if, for every $p \in M$, there is a neighborhood U_p and n-linearly independent smooth vector fields X_i on U_p, such that $X_\iota(p)$ is a basis of $P_p, \forall p \in U_p$.

An *integral manifold* for the distribution is an immersion $j : N \to M$ of an n-dimensional manifold N, so that for every $x \in N$ we have $dj_x(T_x N) = P_{j(x)}$. In other words, N is a submanifold for which the tangent space at each point x is the prescribed space P_x.

It is quite clear that in general there are no integral manifolds. Formally, finding integral manifolds means solving a system of partial differential equations, and as usual there is a compatibility condition. This condition is easy to understand geometrically since the following is an easy exercise.

Exercise. Given an immersion $j : N \to M$ and two vector fields X and Y on M tangent to N, then $[X, Y]$ is also tangent to N.

This remark suggests the following:

Definition 3. A smooth distribution on M is said to be *involutive* if, given any point p in M and a basis $X_1, \dots, X_n$ of vector fields in a neighborhood U of p for the distribution, there exist C^∞ functions $f_{i,j}^k$ on U, with $[X_i, X_j] = \sum_k f_{i,j}^k X_k$.

The prime and only example which will concern us is the distribution induced on a Lie group G by a Lie subalgebra H. In this case a basis of H gives a global basis of vector fields for the distribution. It is clear that the distribution is involutive (in fact the functions $f_{i,j}^k$ are the multiplication constants of the Lie bracket).

If $Y_1, \dots, Y_n$ is a basis of a distribution and $f_{i,j}(x)$ is an invertible $n \times n$ matrix of functions, then $Z_i = \sum_j f_{j,i} Y_j$ is again a basis. The property of being involutive is independent of a choice of the basis since $[Z_h, Z_k] = \sum_{s,t}(f_{s,h} f_{t,k}[Y_s, Y_t] + f_{s,h} Y_s(f_{t,k}) Y_t - f_{t,k} Y_t(f_{s,h}) Y_s)$.

Proposition. *Given an involutive distribution and a point $p \in M$, there exists a neighborhood U of p and vector fields $X_1, \dots, X_n$ in U such that.*

(1) *The vector fields $X_1, \dots, X_n$ are a basis of the distribution in U.*
(2) $[X_i, X_j] = 0$.

Proof. Start in some coordinates $x_1, \dots, x_m$ with some basis $Y_i = \sum_j a_{j,i}(x)\frac{\partial}{\partial x_j}$.

Since the Y_i are linearly independent, a maximal $n \times n$ minor of the matrix $(a_{j,i})$ is invertible (in a possibly smaller neighborhood). Changing basis using the inverse of this minor, which we may assume to be the first, we reduce to the case in which $a_{i,j}(x) = \delta_i^j, \forall i, j \leq n$. Thus the new basis is $X_i = \frac{\partial}{\partial x_i} + \sum_{h=n+1}^m a_{h,i}(x)\frac{\partial}{\partial x_h}$. The Lie bracket $[X_i, X_j]$ is a linear combination of the derivatives $\frac{\partial}{\partial x_h}$, $h > n$. On the other hand the assumption of being involutive means that this commutator is some linear combination $[X_i, X_j] = \sum_{k=1}^n f_{i,j}^k X_k = \sum_{k=1}^n f_{i,j}^k \frac{\partial}{\partial x_k}$ + other terms. Since the coefficients of $\frac{\partial}{\partial x_k}$, $k \leq n$ in $[X_i, X_j]$ equal 0, we deduce that all $f_{i,j}^k = 0$. Hence $[X_i, X_j] = 0$. □

Theorem (Frobenius). *Given an involutive distribution and $p \in M$, there exists a neighborhood of p and a system of local coordinates $(x_1, \dots, x_m)$, such that the distribution has as basis $\frac{\partial}{\partial x_i}$, $i = 1, \dots, n$. So it is formed by the tangent spaces to the **level manifolds** $x_i = a_i, i = n+1, \dots m$ (a_i constants). These are integral manifolds for the distribution.*

Proof. First we use the previous proposition to choose a basis of commuting vector fields X_i for the distribution. Integrating the vector fields X_i in a neighborhood of p gives rise to n commuting local 1-parameter groups $\phi_i(t_i)$. Choose a system of coordinates y_i around p so that the coordinates of p are 0 and $\frac{\partial}{\partial y_i}$ equals X_i at $p, i = 1, \dots, n$. Consider the map of a local neighborhood of 0 in $\mathbb{R}^n \times \mathbb{R}^{m-n}$ given by $\pi : (x_1, \dots, x_n, x_{n+1}, \dots, x_m) := \phi_1(x_1)\phi_2(x_2)\dots\phi_n(x_n)(0, 0, \dots, 0, x_{n+1}, \dots, x_m)$. It is clear that the differential $d\pi$ at 0 is the identity matrix, so π is locally a diffeomorphism. Further, since the groups $\phi_i(x_i)$ commute, acting on the source space by the translations $x_i \mapsto x_i + s_i$ corresponds under π to the action by $\phi_i(s_i)$:

$$\text{(3.3.1)} \qquad \begin{aligned} \pi &: (x_1, \dots, x_i + s_i, \dots, x_n, x_{n+1}, \dots, x_m) \\ &= \phi_i(s_i)\phi_1(x_1)\phi_2(x_2)\dots\phi_n(x_n)(x_{n+1}, \dots, x_m). \end{aligned}$$

Thus, in the coordinates $x_1, \dots, x_n, x_{n+1}, \dots, x_m$, we have $X_i = \frac{\partial}{\partial x_i}$. The rest follows. □

The special coordinate charts in which the integral manifolds are the submanifolds in which the last $m - n$ coordinates are fixed will be called *adapted* to the distribution.

By the construction of integral manifolds, it is clear that an integral manifold through a point p is (at least locally) uniquely determined and spanned by the evolution of the 1-parameter groups generated by the vector fields defining the distribution. It follows that if we have two integral manifolds A, B and p is a point in their intersection, then an entire neighborhood of p in A is also a neighborhood of p in B. This allows us to construct *maximal integral manifolds* as follows. Let us define a new topology on M. If U is an open set with an adapted chart we redefine the topology on U by *separating* all the level manifolds; in other words, we declare all level manifolds open, leaving in each level manifold its induced topology. The previous remarks show that if we take two adapted charts U_1, U_2, then the new topology on U_1 induces on $U_1 \cap U_2$ the same topology as the new topology induced by U_2.

Call a maximal integral manifold a connected component M_α of M under the new topology. It is clear that such a component is covered by coordinate charts, the coordinate changes are C^∞, and the inclusion map $M_\alpha \to M$ is an immersion. The only unclear point is the existence of a countable set dense in M_α. For a topological group we can use:

Lemma. *Let G be a connected topological group such that there is a neighborhood U of 1 with a countable dense set. Then G has a countable dense set.*

Proof. We may assume $U = U^{-1}$ and X dense in U and countable. Since a topological group is generated by a neighborhood of the identity, we have $G = \cup_{k=1}^{\infty} U^k$.

Then $Y := \cup_{k=1}^{\infty} X^k$ is dense and countable. □

In the case of a Lie group and a Lie subalgebra M the maximal integral manifold through 1 of the distribution satisfies the previous lemma.

In the general case it is still true that maximal integral manifolds satisfy the countability axioms. We leave this as an exercise. *Hint*: If A is a level manifold in a given adapted chart U and U' is a second adapted chart, prove that $A \cap U'$ is contained in a countable number of level manifolds for U'. Then cover M with countably many adapted charts.

Theorem 2. *The maximal integral manifold H through 1 is a subgroup of G. The other maximal integral manifolds are the left cosets of H in G. With the natural topology and local charts H is a Lie group of Lie algebra M. The inclusion map is an immersion.*

Proof. Given $g \in G$, consider the diffeomorphism $x \mapsto gx$. Since the vector fields of the Lie algebra M are left-invariant this diffeomorphism preserves the distribution, hence it permutes the maximal integral manifolds. Thus it is sufficient to prove that H is a subgroup. If we take $g \in H$, we have $g1 = g$, hence H is sent to itself. Thus H is closed under multiplication. Applying now the diffeomorphism $x \mapsto g^{-1}x$, we see that $1 \in g^{-1}H$, hence $g^{-1}H = H$ and $g^{-1} = g^{-1}1 \in H$. □

As we already remarked, H need not be closed. Nevertheless, $\overline{H}$ is clearly a subgroup. Thus we find the following easy criterion for H to be closed.

Criterion. *If there is a neighborhood A of 1, and a closed set $X \supset H$ such that $X \cap A = H \cap A$, then $H = \overline{H}$.*

Proof. Both H and $\overline{H}$ are connected and $\overline{H} \subset X$. Thus $A \cap \overline{H} = A \cap H$. By 2.1, Proposition (2), a connected group is generated by any open neighborhood of the identity, hence the claim. □

3.4 Simply Connected Groups

A given connected Lie group G has a unique universal covering space which is a simply connected Lie group with the same Lie algebra.

The main existence theorem is:

Theorem.

(i) For every Lie algebra $\mathfrak{g}$, there exists a unique simply connected Lie group G such that $\mathfrak{g} = \mathrm{Lie}(G)$.

(ii) Given a morphism $f : \mathfrak{g}_1 \to \mathfrak{g}_2$ of Lie algebras there exists a unique morphism of the associated simply connected Lie groups, $\phi : G_1 \to G_2$ which induces f, i.e., $f = d\phi_1$.

Proof. (i) We will base this result on Ado's theorem (Chapter 5, §7), stating that a finite-dimensional Lie algebra L can be embedded in matrices. If $L \subset gl(n, \mathbb{R})$, then by Theorem 2, 3.3 we can find a Lie group H with Lie algebra L and an immersion to $GL(n, \mathbb{R})$.

Its universal cover is the required simply connected group. Uniqueness follows from the next part applied to the identity map.

(ii) Given a homomorphism $f : \mathfrak{g}_1 \to \mathfrak{g}_2$ of Lie algebras, its graph $\Gamma_f := \{(a, f(a)) \mid a \in \mathfrak{g}_1\}$ is a Lie subalgebra of $\mathfrak{g}_1 \oplus \mathfrak{g}_2$. Let G_1, G_2 be simply connected groups with Lie algebras $\mathfrak{g}_1, \mathfrak{g}_2$. By the previous theorem we can find a Lie subgroup H of $G_1 \times G_2$ with Lie algebra Γ_f. The projection to G_1 induces a Lie homomorphism $\pi : H \to G_1$ which is the identity at the level of Lie algebras. Hence π is a covering. Since G_1 is simply connected we have that π is an isomorphism. The inverse of π composed with the second projection to G_2 induces a Lie homomorphism whose differential at 1 is the given f. □

One should make some remarks regarding the previous theorem. First, the fact that $f : \mathfrak{g}_1 \to \mathfrak{g}_2$ is an injective map does not imply that G_1 is a subgroup of G_2. Second, if G_1 is not simply connected, the map clearly may not exist.

When $M \subset L$ is a Lie subalgebra, we have found, by the method of Frobenius, a Lie group G_1 of Lie algebra M mapped isomorphically to the subgroup of G_2 generated by the elements $\exp(M)$. In general G_1 is not closed in G_2 and its closure can be a much bigger subgroup. The classical example is when we take the 1-parameter group $t \mapsto (e^{tr_1 i}, \dots, e^{tr_n i})$ inside the *torus* of n-tuples of complex numbers of absolute value 1. It is a well-known observation of Kronecker (and not difficult) that when the numbers r_i are linearly independent over the rational numbers the image of this group is dense.[22] The following is therefore of interest:

Proposition. *If $M \subset L$ is an ideal of the Lie algebra L of a group G, then the subgroup $G_1 \subset G$ generated by $\exp(M)$ is normal, and closed, with Lie algebra M.*

[22] This is a very important example, basic in *ergodic theory*.

Proof. It is enough to prove the proposition when G is simply connected (by a simple argument on coverings). Consider the homomorphism $L \to L/M$ which induces a homomorphism from G to the Lie group K of Lie algebra L/M. Its kernel is a closed subgroup with Lie algebra M, hence the connected component of 1 must coincide with G_1. □

Corollary. *In a connected Lie group G we have a 1-1 correspondence between closed connected normal subgroups of G and ideals of its Lie algebra L.*

Proof. One direction is the previous proposition. Let K be a closed connected normal subgroup of G and let M be its Lie algebra. For every element $a \in L$ we have that $\exp(ta)K\exp(-ta) \in K$. From 3.1, it follows that $\mathrm{ad}(a)M \subset M$, hence M is an ideal. □

These theorems lay the foundations of Lie theory. They have taken some time to prove. In fact, Lie's original approach was mostly infinitesimal, and only the development of topology has allowed us to understand the picture in a more global way.

After having these foundational results (which are quite nontrivial), one can set up a parallel development of the theory of groups and algebras and introduce basic structural notions such as solvable, nilpotent, and semisimple for both cases and show how they correspond.

The proofs are not always simple, and sometimes it is simpler to prove a statement for the group or sometimes for the algebra. For Lie algebras the methods are essentially algebraic, while for groups, more geometric ideas may play a role.

Exercise. Show that the set of Lie groups with a prescribed Lie algebra is in correspondence with the discrete subgroups of the center of the unique simply connected group. Analyze some examples.

3.5 Actions on Manifolds

Let us now analyze group actions on manifolds. If G acts on a manifold M by $\rho : G \times M \to M$, and $\phi_a(t)$ is a 1-parameter subgroup of G, we have the 1-parameter group of diffeomorphisms $\phi_a(t)m = \rho(\phi_a(t), m)$, given by the action. We call Y_a its infinitesimal generator. Its value in m is the velocity of the curve $\phi_a(t)m$ at $t = 0$, or $d\rho_{1,m}(a, 0)$.

Theorem 1. *The map $a \mapsto -Y_a$ from the Lie algebra L of G to the Lie algebra of vector fields on M is a Lie algebra homomorphism.*

Proof. Apply Theorem 1.4. Given $a, b \in L$, the 1-parameter group (in the parameter s) $\phi_a(t)\phi_b(s)\phi_a(-t)m$ (depending on t) is generated by a variable vector field $Y_b(t)$ which satisfies the differential equation $\dot{Y}_b(t) = [Y_b(t), Y_a]$, $Y_b(0) = Y_b$. Now $\phi_a(t)\phi_b(s)\phi_a(-t) = \phi_{\mathrm{Ad}(\phi_a(t))(b)}(s)$, so $[Y_b, Y_a]$ is the derivative at $t = 0$ of $Y_{\mathrm{Ad}(\phi_a(t))(b)}$. This, in any point m, is computed by $\frac{d}{dt} d\rho_{1,m}(\mathrm{Ad}(\phi_a(t))(b), 0)_{t=0}$, which equals $d\rho_{1,m}([a, b], 0)$. Thus $[Y_b, Y_a] = Y_{[a,b]}$, hence the claim. □

Conversely, let us give a homomorphism $a \mapsto Z_a$ of the Lie algebra L_G into $\mathcal{L}(M)$, the vector fields on M. We can then consider a copy of L_G as vector fields on $G \times M$ by adding the vector field X_a on G to the vector field Z_a on M. In this way we have a copy of the Lie algebra L_G, and at each point the vectors are linearly independent. We have thus an integrable distribution and can consider a maximal integral manifold.

Exercise. Use the Frobenius theorem to understand, at least locally, how this distribution gives rise to an action of G on M.

Let us also understand an *orbit map*. Given a point $p \in M$ let G_p be its stabilizer.

Theorem 2. *The Lie algebra $L(G_p)$ of G_p is the set of vectors $v \in L$ for which $Y_v(p) = 0$.*

G/G_p has the structure of a differentiable manifold, so that the orbit map $i : G/G_p \to M, i(gG_p) := gp$ is an immersion.

Proof. $v \in L(G_p)$ if and only if $\exp(tv) \in G_p$, which means that the 1-parameter group $\exp(tv)$ fixes p. This happens if and only if Y_v vanishes at p.

Let M be a complementary space to $L(G_p)$ in L. The map $j : M \oplus L(G_p) \to G,\ j(a, b) := \exp(a)\exp(b)$ is a local diffeomorphism from some neighborhood $A \times B$ of 0 to a neighborhood U of 1 in G. Followed by the orbit map we have $\exp(a)\exp(b)p = \exp(a)p$ and the map $a \mapsto \exp(a)p$ is an immersion. This gives the structure of a differentiable manifold to the orbit locally around p. At the other points, we translate the chart by elements of G. □

We want to apply the previous analysis to invariant theory.

Corollary. *If G is connected, acting on M, a function f is invariant under G if and only if it satisfies the differential equations $Y_a f = 0$ for all $a \in L_G$.*

Proof. Since G is connected, it is generated by its 1-parameter subgroups $\exp(ta)$, $a \in L_G$. Hence f is fixed under G if and only if it is fixed under these 1-parameter groups. Now f is constant under $\exp(ta)$ if and only if $Y_a f = 0$. □

For instance, for the invariants of binary forms, the differential equations are the ones obtained using the operators 1.3.4.

3.6 Polarizations

We go back to polarizations. Let us consider, as in Chapter 3, 2.3, m-tuples of vector variables $x_1, x_2, \dots, x_m$, each x_i being a column vector $x_{1i}, x_{2i}, \dots, x_{ni}$. In other words we consider the x_{ij} as the coordinates of the space $M_{n,m}$ of $n \times m$ matrices.

Let $A = F[x_{ij}]$ ($F = \mathbb{R}, \mathbb{C}$) be the polynomial ring in the variables x_{ij}, which we also think of as polynomials in the vector variables x_i given by the columns. We want to consider some special 1-parameter subgroups on $M_{n,m}$ (induced by left or right multiplications).

For any $m \times m$ matrix A we consider the 1-parameter group $X \to Xe^{-tA}$.

In particular for the elementary matrix $e_{ij},\ i \neq j$ (with 1 in the ij position and 0 elsewhere), we have $e_{ij}^2 = 0,\ e^{-te_{ij}} = 1 - te_{ij}$ and the matrix $Xe^{-te_{ij}}$ is obtained from X adding to its jth column its ith column multiplied by $-t$.

For e_{ii} we have that $Xe^{-te_{ii}}$ is obtained from X multiplying its ith column by e^{-t}. We act dually on the functions in A and the 1-parameter group acts substituting x_j with $x_j + tx_i,\ i \neq j$, resp. x_i with $e^t x_i$. By the previous sections and Chapter 3, Theorem 2.1 we see:

Proposition. *The infinitesimal generator of the transformation of functions induced by $X \to Xe^{-te_{ij}}$ is the polarization operator D_{ij}.*

We should summarize these ideas. The group $GL(m, F)$ (resp. $GL(n, F)$) acts on the space of $n \times m$ matrices by the rule $(A, X) \mapsto XA^{-1}$ (resp. $(B, X) \mapsto BX$).

The infinitesimal action is then $X \mapsto -XA := r_A(X)$ (resp. $X \mapsto BX$).

If we denote this operator by r_A, we have $[r_A, r_B] = r_{[A,B]}$. In other words, the map $A \mapsto r_A$ is a *Lie algebra homomorphism* associated to the given action.

The derivation operators induced on polynomials (by the right multiplication action) are the linear span of the polarization operators which correspond to elementary matrices.

We state the next theorem for complex numbers although this is not really necessary.

Recall that an $n \times m$ matrix X can be viewed either as the list of its column vectors $x_1, \dots, x_m$ or of its row vectors which we will call $x^1, x^2, \dots, x^n$.

Theorem. *A space of functions $f(x_1, \dots, x_m) = f(X)$ in m vector variables, is stable under polarization if and only if it is stable under the action of $GL(m, \mathbb{C})$ given, for $A = (a_{ji}) \in GL(m, \mathbb{C})$, by*

$$f^A(X) = f(XA), \quad f^A(x_1, \dots, x_m) := f\Big(\sum_j a_{j1}x_j, \sum_j a_{j2}x_j, \dots \sum_j a_{jn}x_j\Big)$$

or

$$f^A(x^1, \dots, x^n) := f(x^1A, x^2A, \dots, x^nA).$$

Proof. $GL(m, \mathbb{C})$ is connected, so it is generated by the elements e^{tA}. A subspace of a representation of $GL(m, \mathbb{C})$ is stable under e^{tA}, if and only if it is stable under A. In our case the infinitesimal generators are the polarizations D_{ij}. □

3.7 Homogeneous Spaces

We want to discuss a complementary idea, which is important in itself, but for us it is useful in order to understand which groups are simply connected. Let us explain with an example:

Example. The simplest noncommutative example of a simply connected Lie group is $SL(n, \mathbb{C})$ ($SL(n, \mathbb{R})$ is not simply connected).

One way to compute $\pi_1(G)$ and hence check that a Lie group G is simply connected is to work by induction, using the long exact sequence of homotopy for a fibration $H \to G \to G/H$ where H is some closed subgroup. In algebraic topology there are rather general definitions of fibrations which are special maps $f : X \to B$ of spaces with base point $x_0 \in X, b_0 \in B, \ f(x_0) = b_0$ and for which one considers the *fiber* $F := f^{-1}(b_0)$. One has the long exact sequence of homotopy groups (cf. [Sp]):

$$\ldots \pi_i(F) \to \pi_i(X) \to \pi_i(B) \to \ldots \to \pi_1(F)$$
$$\to \pi_1(X) \to \pi_1(B) \to \pi_0(F) \to \pi_0(X) \to \pi_0(B).$$

In our case the situation is particularly simple. We will deal with a *locally trivial fibration*, a very special type of fibration for which the long exact sequence of homotopy holds. This means that we can cover B with open sets U_i and we can identify $\pi^{-1}(U_i)$ with $U_i \times F$, so that under this identification, the map π becomes the first projection.

Theorem. *Given a Lie group G and a closed subgroup H, the coset space G/H naturally has the structure of a differentiable manifold (on which G acts in a C^∞ way).*

The orbit map $g \mapsto gH$ is a locally trivial fibration with fiber H.

Once we have established this fact we can define:

Definition. G/H with its C^∞ structure is called a *homogeneous space*.

In other words, a homogeneous space M for a Lie group G is a manifold M with a C^∞ action of G which is transitive.

The structure of a differentable manifold on G/H is quite canonical in the following sense. Whenever we are given an action of G on a manifold M and H stabilizes a point p, we have that the orbit map $\rho : G \to M, \rho(g) = gp$ factors through a C^∞ map $i : G/H \to M$. If H is the full stabilizer of p, then i is an immersion of manifolds. Thus in practice, rather than describing G/H, we describe an action for which H is a stabilizer.

Let us prove the previous statements. The proof is based on the existence of a *tubular neighborhood* of H in G. Let $H \subset G$ be a closed subgroup, let A and B be the Lie algebras of H and G. Consider a linear complement C to A in B. Consider the map $f : C \times H \to G$ given by $f(c, h) := \exp(c)h$. The differential of f at $(0, 1)$ is the identity. Hence, since the map is equivariant with respect to right multiplication by H, there is an open neighborhood U of 0 in C such that df is bijective at all points of $U \times H$. We want to see that:

Tubular neighborhood lemma. *If U is sufficiently small, we have that $f : U \times H \to G$ is a diffeomorphism to an open subset containing H (a union of cosets* $\exp(a)H$, *$a \in U$).*

Proof. Since df is bijective on $U \times H$, it is sufficient to prove that we can choose U so that f is injective.

Since H is a closed submanifold, there are neighborhoods U of 0 in C and V of 1 in H so that the map $i : (a, b) \mapsto \exp(a)b,\ U \times V \to G$ is a diffeomorphism to a neighborhood W of 1 and $\exp(a)b \in H$ if and only if $a = 0$.

We can consider a smaller neighborhood A of 0 in C so that, if $a_1, a_2 \in A$, we have $\exp(-a_2)\exp(a_1) \in W$. We claim that A satisfies the property that the map $f : A \times H \to G$ is injective. In fact if $\exp(a_1)b_1 = \exp(a_2)b_2$ we have $b := b_2 b_1^{-1} = \exp(-a_2)\exp(a_1) \in W \cap H = V$. Therefore $i(a_1, 1) = \exp(a_1) = \exp(a_2)b = i(a_2, b), b \in V$. Since the map i on $A \times V \to G$ is injective, this implies $a_1 = a_2, b = 1$. Therefore $f(A \times H) = \exp(A)H$ is the required tubular neighborhood, and we identify it (using f) to $A \times H$. □

Thus A naturally parameterizes a set in G/H. We can now give G/H the structure of a differentiable manifold as follows. First we give G/H the quotient topology induced by the map $\pi : G \to G/H$. By the previous construction the map π restricted to the tubular neighborhood $A \times H$ can be identified with the projection $(a, h) \mapsto a$. Its image in G/H is an open set isomorphic to A and will be identified with A. It remains to prove that the topology is Hausdorff and that one can cover G/H with charts gA translating A. Since G acts continuously on G/H, in order to verify the Hausdorff property it suffices to see that if $g \notin H$, we can separate the two cosets gH and H by open neighborhoods. Clearly we can find a neighborhood A' of 0 in A such that $\exp(-A')g\exp(A') \cap H = \emptyset$. Thus we see that $\exp(A')H \cap g\exp(A')H = \emptyset$. The image of $g\exp(A')H$ is a neighborhood of gH which does not intersect the image of $\exp(A')H$. Next one easily verifies that the coordinate changes are C^∞ (and even analytic), giving a manifold structure to G/H. The explicit description given also shows easily that G acts in a C^∞ way and that π is a locally trivial fibration. We leave the details to the reader.

Let us return to showing that $SL(n, \mathbb{C})$ is simply connected. We start from $SL(1, \mathbb{C}) = \{1\}$. In the case of $SL(n, \mathbb{C})$, $n > 1$ we have that $SL(n, \mathbb{C})$ acts transitively on the nonzero vectors in $\mathbb{C}^n$, which are homotopic to the sphere S^{2n-1}, so that $\pi_1(\mathbb{C}^n - \{0\}) = \pi_2(\mathbb{C}^n - \{0\}) = 0$ (cf. [Sp]). By the long exact sequence in homotopy for the fibration $H \to SL(n, \mathbb{C}) \to \mathbb{C}^n - \{0\}$, where H is the stabilizer of e_1, we have that $\pi_1(SL(n, \mathbb{C})) = \pi_1(H)$. H is the group of block matrices $\begin{vmatrix} 1 & a \\ 0 & B \end{vmatrix}$ where a is an arbitrary vector and $B \in SL(n-1, \mathbb{C})$. Thus H is homeomorphic to $SL(n-1, \mathbb{C}) \times \mathbb{C}^{n-1}$ which is homotopic to $SL(n-1, \mathbb{C})$ and we finish by induction. Remark that the same proof shows also that $SL(n, \mathbb{C})$ is connected.

Remark. There is an important, almost immediate generalization of the fibration $H \to G \to G/H$. If $H \subset K \subset G$ are closed subgroups we have a locally trivial fibration of homogeneous manifolds:

$$K/H \to G/H \to G/K. \tag{3.7.1}$$

4 Basic Definitions

4.1 Modules

An important notion is that of *module* or *representation.*

Definition 1. A module over a Lie algebra L consists of a vector space and a homomorphism (of Lie algebras) $\rho : L \to gl(V)$.

As usual one can also give the definition of module by the axioms of an *action.* This is a bilinear map $L \times V \to V$ denoted by $(a, v) \mapsto av$, satisfying the Lie homomorphism condition $[a, b]v = a(bv) - b(av)$.

Remark. Another interpretation of the Jacobi identity is that L is an L-module under the action $[a, b] = \mathrm{ad}(a)(b)$.

Definition 2. The Lie homomorphism $\mathrm{ad} : L \to gl(L)$, $x \mapsto \mathrm{ad}(x)$ is called the *adjoint representation.*

$\mathrm{ad}(x)$ is a derivation, and $x \mapsto \exp(t\,\mathrm{ad}(x))$ is a 1-parameter group of automorphisms of L (1.6).

Definition 3. The group of automorphisms of L generated by the elements $\exp(\mathrm{ad}(x))$ is called the *adjoint group*, $\mathrm{Ad}(L)$, of L, and its action on L the adjoint action.

If g is in the adjoint group, we indicate by $\mathrm{Ad}(g)$ the linear operator it induces on L.

Remark. From §3.1, if L is the Lie algebra of a connected Lie group G, the adjoint action is induced as the differential at 1 of the conjugation action of G on itself.

The kernel of the adjoint representation of L is

(4.1.1) $$Z(L) := \{x \in L \mid [x, L] = 0\}, \quad \text{the center of } L.$$

From 3.1.2 the kernel of the adjoint representation of G is made of the elements which commute with the 1-parameter groups $\exp(ta)$. If G is connected, these groups generate G, hence the kernel of the adjoint representation is the center of G.

As usual one can speak of homomorphisms of modules, or L-linear maps, of submodules and quotient modules, direct sums, and so on.

Exercise. Given two L-modules, M, N, we have an L-module structure on the vector space $\hom(M, N)$ of all linear maps, given by $(af)(m) := a(f(m)) - f(am)$. A linear map f is L-linear if and only if $Lf = 0$.

4.2 Abelian Groups and Algebras

The basic structural definitions for Lie algebras are similar to those given for groups.

If A, B are two subspaces of a Lie algebra, $[A, B]$ denotes the linear span of the elements $[a, b],\ a \in A,\ b \in B$, called the *commutator* of A and B.

Definition. A Lie algebra L is *abelian* if $[L, L] = 0$.

The first remark of this comparison is the:

Proposition. *A connected Lie group is abelian if and only if its Lie algebra is abelian. In this case the map* $exp : L \to G$ *is a surjective homomorphism with discrete kernel.*

Proof. From Corollary 1 of 1.4, G is abelian if and only if L is abelian. From Remark 2 of §3.1, then exp is a homomorphism. From point (3) of the Proposition in §2.1, exp is surjective. Finally, Lemma 1 and Theorem 2 of §3.2 imply that the kernel is a discrete subgroup. □

As a consequence we have the description of abelian Lie groups. We have the two basic abelian Lie groups: $\mathbb{R}$, the additive group of real numbers, and $S^1 = U(1, \mathbb{C}) = \mathbb{R}/\mathbb{Z}$, the multiplicative group of complex numbers of absolute value 1. This group is compact.

Theorem. *A connected abelian Lie group* G *is isomorphic to a product* $\mathbb{R}^k \times (S^1)^h$.

Proof. By the previous proposition $G = \mathbb{R}^n/\Lambda$ where Λ is a discrete subgroup of $\mathbb{R}^n$. Thus it suffices to show that there is a basis e_i of $\mathbb{R}^n$ and an $h \leq n$ such that Λ is the set of integral linear combinations of the first h vectors e_i. This is easily proved by induction. If $n = 1$ the argument is quite simple. If $\Lambda \neq 0$, since Λ is a discrete subgroup of $\mathbb{R}$, there is a minimum $a \in \Lambda,\ a > 0$. If $x \in \Lambda$ write $x = ma + r$ where $m \in \mathbb{Z},\ |r| < a$. We see that $\pm r \in \Lambda$ which implies $r = 0$ and $\Lambda = \mathbb{Z}a$. Taking a as basis element, $\Lambda = \mathbb{Z}$.

In general take a vector $e_1 \in \Lambda$ such that e_1 generates the subgroup $\Lambda \cap \mathbb{R}e_1$. We claim that the image of Λ in $\mathbb{R}^n/\mathbb{R}e_1$ is still discrete. If we can prove this we find the required basis by induction. Otherwise we can find a sequence of elements $a_i \in \Lambda,\ a_i \notin \mathbb{Z}e_1$ whose images in $\mathbb{R}^n/\mathbb{R}e_1$ tend to 0. Completing e_1 to a basis we write $a_i = \lambda_i e_1 + b_i$ where the b_i are linear combinations of the remaining basis elements. By hypothesis $b_i \neq 0,\ \lim_{i\to\infty} b_i = 0$. We can modify each a_i by subtracting an integral multiple of e_1 so to assume that $|\lambda_i| \leq 1$. By compactness we can extract from this sequence another one, converging to some vector λe_1. Since the group is discrete, this means that $a_i = \lambda e_1$ for large i and this contradicts the hypothesis $b_i \neq 0$. □

A compact connected abelian group is isomorphic to $(S^1)^h$ and is called a *compact h-dimensional torus.*

4.3 Nilpotent and Solvable Algebras

Definition 1. Let L be a Lie algebra. The *derived series* is defined inductively:

$$L^{(1)} = L, \dots, L^{(i+1)} := [L^{(i)}, L^{(i)}].$$

The *lower central series* is defined inductively:

$$L^1 = L, \dots, L^{i+1} := [L, L^i].$$

A Lie algebra is *solvable* (resp. *nilpotent*) if $L^{(i)} = 0$ (resp. $L^i = 0$) for some i.

Clearly $L^{(i)} \subset L^i$ so nilpotent implies solvable. The opposite is not true as we see by the following:

Basic example. Let B_n resp. U_n be the algebra of upper triangular $n \times n$ matrices over a field F (i.e., $a_{i,j} = 0$ if $i > j$), resp. of strictly upper triangular $n \times n$ matrices (i.e., $a_{i,j} = 0$ if $i \geq j$).

These are two subalgebras for the ordinary product of matrices, hence also Lie subalgebras. Prove that $B_n^{(n)} = 0,\ B_n^i = U_n, \forall i \geq 1, U_n^n = 0$.

B_n is solvable but not nilpotent; U_n is nilpotent.

Remark. To say that $L^i = 0$ means that for all $a_1, a_2, \dots, a_i \in L$ we have that $[a_1, [a_2, [\dots, a_i]]]] = 0$. With different notation this means that the operator

(4.3.2) $$\mathrm{ad}(a_1)\,\mathrm{ad}(a_2)\dots\mathrm{ad}(a_{i-1}) = 0, \quad \forall a_1, a_2, \dots, a_i \in L.$$

Proposition 1. *A subalgebra of a solvable (resp. nilpotent) Lie algebra is solvable (nilpotent). If L is a Lie algebra and I is an ideal, L is solvable if and only if L/I and I are both solvable. The sum of two solvable ideals is a solvable ideal.*

Proof. The proof is straightforward. □

Warning. It is *not true* that L/I nilpotent and I nilpotent implies L nilpotent (for instance B_n/U_n is abelian).

Remark. The following identity can be proved by a simple induction and will be used in the next proposition:

$$[\mathrm{ad}(b), \mathrm{ad}(a_1)\,\mathrm{ad}(a_2)\dots\mathrm{ad}(a_i)] = \sum_{h=1}^{i} \mathrm{ad}(a_1)\,\mathrm{ad}(a_2)\dots\mathrm{ad}([b, a_h])\dots\mathrm{ad}(a_i).$$

Proposition 2. *The sum of two nilpotent ideals A and B is a nilpotent ideal.*

Proof. Note first that using the Jacobi identity, for each i, A^i and B^i are ideals. Assume $A^k = B^h = 0$. We claim that $(A + B)^{h+k-1} = 0$. We need to show that the product of $h - 1 + k - 1$ factors $\mathrm{ad}(a_i)$ with $a_i \in A$ or $a_i \in B$ is 0. At least $k - 1$ of these factors are in A, or $h - 1$ of these factors are in B. Suppose we are in the first case. Each time we have a factor $\mathrm{ad}(a)$ with $a \in B$, which comes to the left of some factor in A, we can apply the previous commutation relation and obtain a sum of terms in which the number of factors in A is not changed, but one factor in B is either dropped or it moves to the right of the monomial. Iterating this procedure we get a sum of monomials each starting with a product of $k-1$ factors in A, which is thus equal to 0. □

In a finite-dimensional Lie algebra L the previous propositions allow us to define the *solvable radical* as the maximal solvable ideal of L, and the *nilpotent radical* as the maximal nilpotent ideal of L.

4.4 Killing Form

In the following discussion all Lie algebras will be assumed to be finite dimensional. If L is finite dimensional one can define a symmetric bilinear form on L:

$$(x, y) := \operatorname{tr}(\operatorname{ad}(x)\operatorname{ad}(y)) \qquad \text{the Killing form.} \tag{4.4.1}$$

It has the following *invariance* or *associativity* property $([x, y], z) = (x, [y, z])$.

Proof.

$$\begin{aligned}([x, y], z) &= \operatorname{tr}(\operatorname{ad}([x, y])\operatorname{ad}(z)) = \operatorname{tr}(\operatorname{ad}(x)\operatorname{ad}(y)\operatorname{ad}(z) - \operatorname{ad}(y)\operatorname{ad}(x)\operatorname{ad}(z))\\ &= \operatorname{tr}(\operatorname{ad}(x)\operatorname{ad}(y)\operatorname{ad}(z) - \operatorname{ad}(z)\operatorname{ad}(y)\operatorname{ad}(x))\\ &= \operatorname{tr}(\operatorname{ad}(x)\operatorname{ad}([y, z])) = (x, [y, z]).\end{aligned}$$ □

Remark. The associativity formula means also that $\operatorname{ad}(x)$ is skew adjoint with respect to the Killing form, or

$$(\operatorname{ad}(x)y, z) = -(y, \operatorname{ad}(x)z). \tag{4.4.2}$$

Definition 1. A Lie algebra L is *simple* if it has no nontrivial ideals and it is not abelian.

A finite-dimensional Lie algebra is *semisimple* if its solvable radical is 0.

Over $\mathbb{C}$ there are several equivalent definitions of semisimple Lie algebra. For a Lie algebra L the following are equivalent (cf. [Hu1], [Se2],[J1]) and the next sections.

(1) L is a direct sum of simple Lie algebras.
(2) The Killing form $(x, y) := \operatorname{tr}(\operatorname{ad}(x)\operatorname{ad}(y))$ is nondegenerate.
(3) L has no abelian ideals.

For the moment let us see at least:

Lemma 1. *If L has a nonzero solvable radical, then it also has an abelian ideal.*

Proof. Let N be its solvable radical, We claim that for each i, $N^{(i)}$ is an ideal. By induction, $[L, N^{(i+1)}] = [L, [N^{(i)}, N^{(i)}]] \subset [[L, N^{(i)}], N^{(i)}] + [N^{(i)}, [L, N^{(i)}]] \subset [N^{(i)}, N^{(i)}] = N^{(i+1)}$. Thus it is enough to take the last i for which $N^{(i)} \neq 0$, $N^{(i+1)} = 0$. □

Lemma 2. *The Killing form is invariant under any automorphism of the Lie algebra.*

Proof. Let ρ be an automorphism. We have

$$d(\rho(a)) = \rho \circ \operatorname{ad}(a) \circ \rho^{-1} \implies (\rho(a), \rho(b)) = \operatorname{tr}(\operatorname{ad}(\rho(a)) \operatorname{ad}(\rho(b)))$$

$$= \operatorname{tr}(\rho \circ \operatorname{ad}(a) \operatorname{ad}(b) \circ \rho^{-1}) = \operatorname{tr}(\operatorname{ad}(a) \operatorname{ad}(b)). \tag{4.4.3}$$

□

5 Basic Examples

5.1 Classical Groups

We list here a few of the interesting groups and Lie algebras, which we will study in the book. One should look specifically at Chapter 6 for a more precise discussion of orthogonal and symplectic groups and Chapter 10 for the general theory.

We have already seen the linear groups:

$$GL(n, \mathbb{C}), \quad GL(n, \mathbb{R}), \quad SL(n, \mathbb{C}), \quad SL(n, \mathbb{R})$$

which are readily seen to have real dimension $2n^2, n^2, 2(n^2 - 1), n^2 - 1$. We also have:

The *unitary group* $U(n, \mathbb{C}) := \{X \in GL(n, \mathbb{C}) \mid \overline{X}^t X = 1\}$.

Notice in particular $U(1, \mathbb{C})$ is the set of complex numbers of absolute value 1, the circle group, denoted also by S^1.

The *special unitary group* $SU(n, \mathbb{C}) := \{X \in SL(n, \mathbb{C}) \mid \overline{X}^t X = 1\}$.

The *complex and real orthogonal groups*

$$O(n, \mathbb{C}) := \{X \in GL(n, \mathbb{C}) \mid X^t X = 1\}, \quad O(n, \mathbb{R}) := \{X \in GL(n, \mathbb{R}) \mid X^t X = 1\}.$$

The *special complex and real orthogonal groups*

$$SO(n, \mathbb{C}) := \{X \in SL(n, \mathbb{C}) \mid X^t X = 1\}, \quad SO(n, \mathbb{R}) := \{X \in SL(n, \mathbb{R}) \mid X^t X = 1\}.$$

There is another rather interesting group called by Weyl the *symplectic group* $Sp(2n, \mathbb{C})$. We can define it starting from the $2n \times 2n$ skew symmetric block matrix $J_n := \begin{vmatrix} 0 & 1_n \\ -1_n & 0 \end{vmatrix}$ (where 1_n denotes the identity matrix of size n) as

$$Sp(2n, \mathbb{C}) := \{X \in GL(2n, \mathbb{C}) \mid X^t J_n X = J_n\}.$$

Since $J_n^2 = -1$ the last condition is better expressed by saying that $X^s X = 1$ where $X^s := -J_n X^t J_n = J_n X^t J_n^{-1}$ is the *symplectic transpose*.

Write X as 4 blocks of size n and have

$$\begin{vmatrix} A & B \\ C & D \end{vmatrix}^s = \begin{vmatrix} D^t & -B^t \\ -C^t & A^t \end{vmatrix}. \tag{5.1.1}$$

Finally the compact symplectic group:

$$Sp(n) := \{X \in U(2n, \mathbb{C}) \mid X^t J_n X = J_n\}.$$

Out of these groups we see that

$$GL(n, \mathbb{C}), \quad SL(n, \mathbb{C}), \quad SO(n, \mathbb{C}), \quad O(n, \mathbb{C}), \quad Sp(2n, \mathbb{C})$$

are *complex algebraic*, that is they are defined by polynomial equations in complex space (see Chapter 7 for a detailed discussion), while

$$U(n, \mathbb{C}), \quad SU(n, \mathbb{C}), \quad SO(n, \mathbb{R}), \quad O(n, \mathbb{R}), \quad Sp(n)$$

are *compact* as topological spaces. In fact they are all closed bounded sets of some complex space of matrices. Their local structure can be deduced from the exponential map but also from another interesting device, the *Cayley transform*.

The Cayley transform is a map from matrices to matrices defined by the formula

$$C(X) := \frac{1 - X}{1 + X}. \tag{5.1.2}$$

Of course this map is not everywhere defined but only on the open set U of matrices with $1+X$ invertible, that is without the eigenvalue -1. We see that $1+C(X) = \frac{2}{1+X}$, so rather generally if 2 is invertible (and not only over the complex numbers) we have $C(U) = U$, and we can iterate the Cayley transform and see that $C(C(X)) = X$.

The Cayley transform maps U to U and 0 to 1. In comparison with the exponential, it has the advantage that it is algebraic but the disadvantage that it works in less generality.

Some immediate properties of the Cayley transform are (conjugation applies to complex matrices):

$$C(-X) = C(X)^{-1}, \; C(X^t) = C(X)^t, \; C(\overline{X}) = \overline{C(X)}, \; C(AXA^{-1}) = AC(X)A^{-1},$$

and finally $C(X^s) = C(X)^s$.

Therefore we obtain:

$$\begin{aligned} C(X) &\in U(n, \mathbb{C}) \quad \text{if and only if} \quad -X = \overline{X}^t. \\ C(X) &\in O(n, \mathbb{C}) \quad \text{if and only if} \quad -X = X^t. \\ C(X) &\in Sp(n, \mathbb{C}) \quad \text{if and only if} \quad -X = X^s. \end{aligned}$$

There are similar conditions for $O(n, \mathbb{R})$ and $Sp(n)$.

It is now easy to see that the matrices which satisfy any one of these conditions form a Lie algebra. In fact the conditions are that $-x = x^*$ where $x \to x^*$ is a ($\mathbb{R}$-linear map) satisfying $(xy)^* = y^*x^*, \ (x^*)^* = x$, cf. Chapter 5 where such a map is called an involution.

Now we have

$$-x = x^*, -y = y^* \implies -[x, y] = -xy + yx = -x^*y^* + y^*x^*$$
$$= -(yx)^* + (xy)^* = [x, y]^*.$$

Proposition. *In an associative algebra A the space of elements which satisfy $-x = x^*$ for an involution $x \to x^*$ is closed under Lie bracket and so it is a Lie subalgebra.*

Remark. A priori it is not obvious that these are the Lie algebras defined via the exponential map. This is clear if one identifies the Lie algebra as the tangent space to the group at 1, which can be computed using any parameterization around 1.

5.2 Quaternions

Denote by $\mathbb{H}$ the algebra of quaternions.

$$\mathbb{H} = \{a + bi + cj + dk \mid a, b, c, d \in \mathbb{R}\}, \quad i^2 = j^2 = k^2 = -1, \ ij = k.$$

Proposition. *$Sp(n)$ can be defined as the set of quaternionic $n \times n$ matrices X with $\overline{X}^t X = 1$.*

In the previous formula, conjugation is the conjugation of quaternions. In this representation $Sp(n)$ will be denoted by $Sp(n, \mathbb{H})$. This is not the same presentation we have given, but rather it is in a different basis.

Proof. We identify the $2n \times 2n$ complex matrices with $n \times n$ matrices with entries 2×2 matrices and replace J_n with a *diagonal block matrix* J_n' made of n diagonal blocks equal to the 2×2 matrix J_1.

We use Cayley's representation of quaternions $\mathbb{H}$ as elements $q = \alpha + j\beta$ with $\alpha, \beta \in \mathbb{C}$ and $j\alpha = \overline{\alpha} j, \ j^2 = -1$. Setting $k := -ji$ we have $a + bi + cj + dk = (a + bi) + j(c - di)$.

Consider $\mathbb{H}$ as a right vector space over $\mathbb{C}$ with basis $1, j$. An element $\alpha + j\beta = q \in \mathbb{H}$ induces a matrix by left multiplication. We have $q1 = q, qj = -\overline{\beta} + j\overline{\alpha}$, thus the matrix

(5.2.1) $$q = \begin{vmatrix} \alpha & -\overline{\beta} \\ \beta & \overline{\alpha} \end{vmatrix}, \quad \det(q) = \alpha\overline{\alpha} + \beta\overline{\beta}.$$

From the formula of symplectic involution we see that the symplectic transpose of a quaternion is a quaternion $q^s = \overline{\alpha} - j\beta, \quad (a + bi + cj + dk)^s = a - bi - cj - dk$.

From 5.1.1 and 5.2.1 it follows that *a 2×2 matrix q is a quaternion if and only if $q^s = \overline{q}^t$.*

We define symplectic transposition using the matrix J'_n. Take an $n \times n$ matrix $X = (a_{i,j})$ of block 2×2 matrices $a_{i,j}$. We see that $X^s = -J'_n X^t J'_n = (a^s_{j,i})$ while $\overline{X}^t = (\overline{a}^t_{j,i})$. Thus $X^s = \overline{X}^t$ if and only if X is a quaternionic matrix. Thus if $X^s = X^{-1} = \overline{X}^t$ we must have X quaternionic and the proof is complete. □

5.3 Classical Lie Algebras

This short section anticipates ideas which will be introduced systematically in the next chapters. We use freely some tensor algebra which will be developed in Chapter 5.

The list of classical Lie algebras, where n is called the rank, is a reformulation of the last sections. It is, apart from some special cases, a list of simple algebras. The verification is left to the reader but we give an example:

(1) $sl(n+1, \mathbb{C})$ is the Lie algebra of the special linear group $SL(n+1, \mathbb{C})$, $\dim sl(n+1, \mathbb{C}) = (n+1)^2 - 1$. $sl(n+1, \mathbb{C})$ is the set of $(n+1) \times (n+1)$ complex matrices with trace 0; if $n > 0$ it is a simple algebra, said to be *of type* A_n.

Hint as to how to prove simplicity: Let I be an ideal. I is stable under the adjoint action of the diagonal matrices so it has a basis of eigenvectors. Choose one such eigenvector; it is either a diagonal matrix or a matrix unit. Commuting a nonzero diagonal matrix with a suitable off-diagonal matrix unit we can always find a matrix unit $e_{i,j}$, $i \neq j$ in the ideal. Then by choosing appropriate other matrix units we can find all matrices in the ideal.

(2) $so(2n, \mathbb{C})$ is the Lie algebra of the special orthogonal group $SO(2n, \mathbb{C})$, we have $\dim so(2n, \mathbb{C}) = 2n^2 - n$. In order to describe it in matrix form it is convenient to choose a hyperbolic basis $e_1, \dots, e_n, f_1, \dots, f_n$ where the matrix of the form is (cf. Chapter 5, §3.5):

$$I_{2n} := \begin{pmatrix} 0 & 1_n \\ 1_n & 0 \end{pmatrix}, \qquad so(2n, \mathbb{C}) := \{A \in M_{2n}(\mathbb{C}) \,|\, A^t I_{2n} = -I_{2n} A\}.$$

If $n > 3$, $so(2n, \mathbb{C})$ is a simple algebra, said to be *of type* D_n.

Proposition 1. *For $n = 3$ we have the special isomorphism $so(6, \mathbb{C}) = sl(4, \mathbb{C})$.*

Proof. If V is a 4-dimensional vector space, we have an action of $SL(V)$ on $\bigwedge^2 V$ which is 6-dimensional. The action preserves the symmetric pairing $\bigwedge^2 V \times \bigwedge^2 V \to \bigwedge^4 V = \mathbb{C}$. So we have a map $SL(V) \to SO(\bigwedge^2 V)$. We leave to the reader to verify that it is surjective and at the level of Lie algebras induces the required isomorphism. □

Proposition 2. *For $n = 2$ we have the special isomorphism $so(4, \mathbb{C}) = sl(2, \mathbb{C}) \oplus sl(2, \mathbb{C})$.*

Proof. Let V be a 2-dimensional space, V has the symplectic form $V \times V \to \bigwedge^2 V = \mathbb{C}$ and $V \otimes V$ has an induced symmetric form $(u_1 \otimes u_2, v_1 \otimes v_2) = (u_1 \wedge v_1)(u_2 \wedge v_2)$. Then $SL(V) \times SL(V)$ acting as a tensor product preserves this form, and we have a surjective homomorphism $SL(V) \times SL(V) \to SO(V \otimes V)$ which at the level of Lie algebras is an isomorphism.[23] □

(3) $so(2n+1, \mathbb{C})$ is the Lie algebra of the special orthogonal group $SO(2n+1, \mathbb{C})$. We have $\dim so(2n+1, \mathbb{C}) = 2n^2 + n$. In order to describe it in matrix form it is convenient to choose a hyperbolic basis $e_1, \dots, e_n, f_1, \dots, f_n, u$ where the matrix of the form is $I_{2n+1} := \begin{vmatrix} 0 & 1_n & 0 \\ 1_n & 0 & 0 \\ 0 & 0 & 1 \end{vmatrix}$ and $so(2n+1, \mathbb{C}) := \{A \in M_{2n+1}(\mathbb{C}) | A^t I_{2n+1} = -I_{2n+1} A\}$. If $n > 0$, it is a simple algebra, said to be *of type B_n for $n > 1$.*

Proposition 3. *For $n = 1$ we have the special isomorphism $so(3, \mathbb{C}) = sl(2, \mathbb{C})$.*

Proof. This can be realized by acting with $SL(2, \mathbb{C})$ by conjugation on its Lie algebra, the 3-dimensional space of trace 0, 2×2 matrices. This action preserves the form $\mathrm{tr}(AB)$ and so it induces a map $SL(2, \mathbb{C}) \to SO(3, \mathbb{C})$ that is surjective and, at the level of Lie algebras, induces the required isomorphism. □

(4) $sp(2n, \mathbb{C})$ is the Lie algebra of the symplectic group $Sp(2n, \mathbb{C})$, $\dim sp(2n, \mathbb{C}) = 2n^2 + n$. In order to describe it in matrix form it is convenient to choose a symplectic basis $e_1, \dots, e_n, f_1, \dots, f_n$, where the matrix of the form is $J := \begin{pmatrix} 0 & 1_n \\ -1_n & 0 \end{pmatrix}$ and $sp(2n, \mathbb{C}) := \{A \in M_{2n}(\mathbb{C}) | A^t J = -JA\}$. If $n > 1$ it is a simple algebra, said to be *of type C_n for $n > 1$.*

Proposition 4. *For $n = 1$ we have the special isomorphism $sp(2, \mathbb{C}) = sl(2, \mathbb{C})$. For $n = 2$ we have the isomorphism $sp(4, \mathbb{C}) = so(5, \mathbb{C})$, hence $B_2 = C_2$.*

Proof. This is seen as follows. As previously done, take a 4-dimensional vector space V with a symplectic form, which we identify with $I = e_1 \wedge f_1 + e_2 \wedge f_2 \in \bigwedge^2 V$. We have an action of $Sp(V)$ on $\bigwedge^2 V$ which is 6-dimensional and preserves the symmetric pairing $\bigwedge^2 V \times \bigwedge^2 V \to \bigwedge^4 V = \mathbb{C}$. So we have a map $Sp(V) \to SO(\bigwedge^2 V)$. The element I is fixed and its norm $I \wedge I \neq 0$, hence $Sp(V)$ fixes the 5-dimensional orthogonal complement $I^\perp$ and we have an induced map $Sp(4, \mathbb{C}) \to SO(5, \mathbb{C})$. We leave to the reader to verify that it is surjective and at the level of Lie algebras induces the required isomorphism. □

No further isomorphisms arise between these algebras. This follows from the theory of root systems (cf. Chapter 10, §2,3). The list of all complex simple Lie algebras is completed by adding the five exceptional types, called G_2, F_4, E_6, E_7, E_8.

The reason to choose these special bases is that in these bases it is easy to describe a Cartan subalgebra and the corresponding theory of roots (cf. Chapter 10).

---

23 Since there are spin groups one should check that these maps we found are in fact not isomorphisms of groups, but rather covering maps.

Remark. We have described parameterizations around the identity element 1. If we take any element g in the group, we can find a parameterization around g, remarking that the map $x \to gx$ maps a neighborhood of 1 into one of g, and the group into the group.

6 Basic Structure Theorems for Lie Algebras

6.1 Jordan Decomposition

The theory of Lie algebras is a generalization of the theory of a single linear operator. For such an operator on a finite-dimensional vector space the basic fact is the theorem of the Jordan canonical form. Of this theorem we will use the *Jordan decomposition.* Let us for simplicity work over an algebraically closed field.

Definition. A linear map a on a finite-dimensional vector space V is semisimple if it has a basis of eigenvectors. It is nilpotent if $a^k = 0$ for some $k > 0$.

A linear operator is nilpotent and semisimple only if it is 0.

Theorem. *Given a linear operator a on V, there exist unique operators a_s semisimple, a_n nilpotent, such that $a = a_s + a_n$, $[a_s, a_n] = 0$.*

Moreover a_s can be written as a polynomial without constant term in a. If $V \supset A \supset B$ are linear subspaces and $aA \subset B$ we have $a_s A \subset B$, $a_n A \subset B$.

Proof. Let $\alpha_1, \dots, \alpha_k$ be the distinct eigenvalues of a and $n := \dim V$. One can decompose $V = \oplus_i V_i$ where $V_i := \{v \in V \mid (a - \alpha_i)^n v = 0\}$. By the Chinese Remainder Theorem let $f(x)$ be a polynomial with $f(x) \equiv \alpha_i \mod (x - \alpha_i)^n$, $f(x) \equiv 0 \mod x$. Clearly $f(a) = a_s$ is the semisimple operator α_i on V_i. $a_n := a - a_s$ is nilpotent. The rest follows. □

6.2 Engel's Theorem

There are three basic theorems needed to found the theory. The first, Engel's theorem, is true in all characteristics. The other two, the theorem of Lie and Cartan's criterion, hold only in characteristic 0.

In a way these theorems are converses of the basic examples of Section 4.2.

We start with a simple lemma:

Lemma. *Let a be a semisimple matrix with eigenvalues α_i and eigenvectors a basis e_i.*

$\mathrm{ad}(a)$ *is also semisimple with eigenvalues $\alpha_i - \alpha_j$ and eigenvectors the matrix units in the basis e_i.*

If a is nilpotent, $a^k = 0$ we have $\mathrm{ad}(a)^{2k-1} = 0$.

Proof. The statement for semisimple matrices is clear. For nilpotent we use the identity $ad(a) = a_L - a_R$, $a_L(b) = ab$, $a_R(b) = ba$. Since left and right multiplications a_L, a_R are commuting operators,

$$\mathrm{ad}(a)^{2k-1} = (a_L - a_R)^{2k-1} = \sum_{i=0}^{2k-1} \binom{2k-1}{i} a_L^i a_R^{2k-1-i} = 0. \qquad \square$$

Corollary. *If $a \in gl(V)$ and $a = a_s + a_n$ is its Jordan decomposition, then we have that $\mathrm{ad}(a) = \mathrm{ad}(a_s) + \mathrm{ad}(a_n)$ is the Jordan decomposition of $\mathrm{ad}(a)$.*

Engel's Theorem. *Let V be a finite-dimensional vector space, and let $L \subset \mathrm{End}(V)$ be a linear Lie algebra all of whose elements are nilpotent. There is a basis of V in which L is formed by strictly upper triangular matrices. In particular L is nilpotent.*

Proof. The proof can be carried out by induction on both $\dim V$, $\dim L$. The essential point is to prove that there is a nonzero vector $v \in V$ with $Lv = 0$, since if we can prove this, then we repeat the argument with L acting on V/Fv. If L is 1-dimensional, then this is the usual fact that a nilpotent linear operator is strictly upper triangular in a suitable basis.

Let $A \subset L$ be a proper maximal Lie subalgebra of L. We are going to apply induction in the following way. By the previous lemma, $\mathrm{ad}(A)$ consists of nilpotent operators on $gl(V)$, hence the elements of $\mathrm{ad}(A)$ are also nilpotent acting on L and L/A.

By induction there is an element $u \in L$, $u \notin A$ with $\mathrm{ad}(A)u = 0$ modulo A or $[A, u] \subset A$. This implies that $A + Fu$ is a larger subalgebra; by maximality of A we must have that $L = A + Fu$ and also that A is an ideal of L. Now let $W := \{v \in V \mid Au = 0\}$. By induction, $W \neq 0$. We have, if $w \in W, a \in A$, that $a(uw) = [a, u]w + uaw = 0$ since $[a, u] \in A$. So W is stable under u, and since u is nilpotent, there is a nonzero vector $v \in W$ with $Aw = uw = 0$. Hence $Lw = 0$. $\square$

6.3 Lie's Theorem

Lie's Theorem. *Let V be a finite-dimensional vector space over the complex numbers, and let $L \subset \mathrm{End}(V)$ be a linear solvable Lie algebra. There is a basis of V in which L is formed by upper triangular matrices.*

Proof. The proof can be carried out by induction on $\dim L$. The essential point is again to prove that there is a nonzero vector $v \in V$ which is an eigenvector for L. If we can prove this, then we repeat the argument with L acting on V/Fv. If L is 1-dimensional, then this is the usual fact that a linear operator has a nonzero eigenvector.

We start as in Engel's Theorem. Since L is solvable any proper maximal subspace $A \supset [L, L]$ is an ideal of codimension 1. We have again $L = A + \mathbb{C}u$, $[u, A] \subset A$. Let, by induction, $v \in V$ be a nonzero eigenvector for A. Denote by $av := \lambda(a)v$ the eigenvalue (a linear function on A).

Consider the space $W := \{v \in V \mid av = \lambda(a)v, \forall a \in A\}$. If we can prove that W is stabilized by u, then we can finish by choosing a nonzero eigenvector for u in W.

Then let $v \in W$; for some m we have the linearly independent vectors $v, uv, u^2 v, \dots, u^m v$ and $u^{m+1} v$ dependent on the preceding ones. We need to prove that $uv \in W$. In any case we have the following identity for $a \in A$: $auv = [a, u]v + uav = \lambda(a)uv + \lambda([u, v])v$. We have thus to prove that $\lambda([u, v]) = 0$. We repeat $au^i v = \lambda(a)u^i v + \cdots + u[a, u]u^{i-2} v + [a, u]u^{i-1} v$, inductively $[a, u]u^{i-2} v = \lambda([a, u])u^{i-2} v + \sum_{k<i-2} c_k u^k v$. Thus a acts as an upper triangular matrix on the span $M := \langle v, uv, u^2 v, \dots, u^m v \rangle$ of the vectors $u^i v$ with $\lambda(a)$ on the diagonal. On the other hand, since M is stable under u and a we have that $[u, a]$ is a commutator of two operators on M. Thus the trace of the operator $[u, a]$ restricted to M is 0. On the other hand, by the explicit triangular form of $[u, a]$ we obtain for this trace $(m + 1)\lambda([u, v])$. Since we are in characteristic 0, we have $m + 1 \neq 0$, hence $\lambda([u, v]) = 0$. □

Corollary. *If $L \subset gl(V)$ is a solvable Lie algebra, then $[L, L]$ is made of nilpotent elements and is thus nilpotent.*

For a counterexample to this theorem in positive characteristic, see [Hu], p. 20.

6.4 Cartan's Criterion

Criterion. *Let V be a finite-dimensional vector space over $\mathbb{C}$, and let $L \subset \mathrm{End}(V)$ be a linear Lie algebra. If $\mathrm{tr}(ab) = 0$ for all $a \in L$, $b \in [L, L]$, then L is solvable.*

Proof. There are several proofs in the literature ([Jac], [Hu]). We follow the *slick proof*. The goal is to prove that $[L, L]$ is nilpotent, or using Engel's theorem, that all elements of $[L, L]$ are nilpotent. First we show that we can make the statement more abstract. Let $M := \{x \in gl(V) \mid [x, L] \subset [L, L]\}$. Of course $M \supset L$ and we want to prove that we still have $\mathrm{tr}(ab) = 0$ when $a \in M, b \in [L, L]$. In fact if $b = [x, y], x, y \in L$, then we have by the associativity of the trace form $\mathrm{tr}([x, y]a) = \mathrm{tr}(x[y, a]) = 0$ since $x \in L, [y, a] \in [L, L]$. Thus the theorem will follow from the next general lemma, applied to $A = [L, L], B = L$. □

Lemma. *Let V be a vector space of finite dimension n over the complex numbers, and let $A \subset B \subset \mathrm{End}(V)$ be linear spaces. Let $M := \{x \in gl(V) \mid [x, B] \subset A\}$. If an element $a \in M$ satisfies $\mathrm{tr}(ab) = 0$ for all $b \in M$, then a is nilpotent.*

Proof. The first remark to make is that if $a \in M$, also $a_s, a_n \subset M$. This follows from Lemma 6.2 and the properties of the Jordan decomposition. We need to show that $a_s = 0$. Let $\alpha_1, \dots, \alpha_n$ be the eigenvalues of a_s and $e_1, \dots, e_n$ a corresponding basis of eigenvectors. For any semisimple element b, which in the same basis is diagonal with eigenvalues $\beta_1, \dots, \beta_n$, if $b \in M$ we have $\sum_i \alpha_i \beta_i = 0$. A sufficient condition for $b \in M$ is that $\mathrm{ad}(b)$ is a polynomial in $\mathrm{ad}(a_s)$ without constant terms. In turn this is true if one can find a polynomial $f(x)$ without constant term with $f(\alpha_i - \alpha_j) = \beta_i - \beta_j$. By Lagrange interpolation, the only condition we need is that if $\alpha_i - \alpha_j = \alpha_h - \alpha_k$ we must have $\beta_i - \beta_j = \beta_h - \beta_k$. For this it is sufficient to choose the β_i as the value of any $\mathbb{Q}$-linear function on the $\mathbb{Q}$-vector space spanned by the α_i in $\mathbb{C}$. Take such a linear form g. We have $\sum_i \alpha_i g(\alpha_i) = 0$ which implies $\sum_i g(\alpha_i)^2 = 0$. If the numbers $g(\alpha_i)$ are rationals, then this implies that $g(\alpha_i) = 0, \forall i$. Since g can be any $\mathbb{Q}$ linear form on the given $\mathbb{Q}$ vector space, this is possible only if all the $\alpha_i = 0$. □

6.5 Semisimple Lie Algebras

Theorem 1. *Let L be a finite-dimensional Lie algebra over $\mathbb{C}$. Then L is semisimple if and only if the Killing form is nondegenerate.*

Proof. Assume first that the Killing form is nondegenerate. We have seen in §4.4 (Lemma 1) that, if L is not semisimple, then it has an abelian ideal A. Let us show that A must be in the kernel of the Killing form. If $a \in L, b \in A$ we have that $\mathrm{ad}(a)$ is a linear map that preserves A, while $\mathrm{ad}(b)$ is a linear map which maps L into A and is 0 on A.

Hence $\mathrm{ad}(a)\,\mathrm{ad}(b)$ maps L into A and it is 0 on A, so its trace is 0, and A is in the kernel of the Killing form.

Conversely, let A be the kernel of the Killing form. A is an ideal. In fact by associativity, if $a \in A, b, c \in L$, we have $(b, [c, a]) = ([b, c], a) = 0$. Next consider the elements $\mathrm{ad}(A)$. They form a Lie subalgebra with $\mathrm{tr}(ab) = 0$ for all $a, b \in \mathrm{ad}(A)$. By Cartan's criterion $\mathrm{ad}(A)$ is solvable. Finally the kernel of the adjoint representation is the center of L so if $\mathrm{ad}(A)$ is solvable, also A is solvable, and we have found a nonzero solvable ideal. □

Theorem 2. *A finite-dimensional semisimple Lie algebra is a direct sum of simple Lie algebras L_i which are mutually orthogonal under the Killing form.*

Proof. If L is simple there is nothing to prove, otherwise let A be a minimal ideal in L. Let $A^\perp$ be its orthogonal complement with respect to the Killing form. We have always by the associativity that also $A^\perp$ is an ideal. We claim that $A \cap A^\perp = 0$ so that $L = A \oplus A^\perp$. In fact by minimality the only other possibility is $A \subset A^\perp$. But this implies that A is solvable by Cartan's criterion, which is impossible. Since $L = A \oplus A^\perp$, by minimality A is a simple Lie algebra, $A^\perp$ is semisimple, and we can proceed by induction. □

6.6 Real Versus Complex Lie Algebras

Although we have privileged complex Lie algebras, real Lie algebras are also interesting. We want to make some elementary remarks.

First, given a real Lie algebra L, we can complexify it, getting $L_\mathbb{C} := L \otimes_\mathbb{R} \mathbb{C}$. The algebra $L_\mathbb{C}$ continues to carry some of the information of L, although different Lie algebras may give rise to the same complexification. We say that L *is a real form* of $L_\mathbb{C}$. For instance, $sl(2, \mathbb{R})$ and $su(2, \mathbb{C})$ are different and both complexify to $sl(2, \mathbb{C})$. Some properties are easily verified. For instance, $M \subset L$ is a subalgebra or ideal if and only if the same is true for $M_\mathbb{C} \subset L_\mathbb{C}$. The reader can verify also the compatibility of the derived and lower central series:

$$(L^{(i)})_{\mathbb{C}} = (L_{\mathbb{C}})^{(i)}, \qquad (L^i)_{\mathbb{C}} = (L_{\mathbb{C}})^i. \tag{6.6.1}$$

Therefore the notions of solvability, nilpotency, and semisimplicity are compatible with the complexification. Finally, the Killing form of $L_{\mathbb{C}}$ is just the complexification of the Killing form of L, thus it is nondegenerate if and only if L is semisimple.

In this case there is an interesting invariant, since the Killing form in the real case is a real symmetric form one can consider its signature (cf. Chapter 5, §3.3) which is thus an invariant of the real form. Often it suffices to detect the real form. In particular we have:

Exercise. Let L be the Lie algebra of a semisimple group K. Then the Killing form of L is negative definite if and only if K is compact.

In fact to prove in full this exercise is not at all easy and the reader should see Chapter 10, §5. It is not too hard when K is the adjoint group (see Chapter 10).

When one studies real Lie algebras, it is interesting to study also real representations, then one can use the methods of Chapter 6, 3.2.

7 Comparison between Lie Algebras and Lie Groups

7.1 Basic Comparisons

In this section we need to compare the concepts of nilpotency, solvability, and semisimplicity introduced for Lie algebras with their analogues for Lie groups.

For Lie groups and algebras *abelian* is the same as *commutative*.

We have already seen in §4.2 that a connected Lie group G is abelian if and only if its Lie algebra is abelian. We want to prove now that the derived group of a connected Lie group has, as Lie algebra, the derived algebra.

For groups there is a parallel theory of central and derived series.[24] In a group G the *commutator of two elements* is the element $\{x, y\} := xyx^{-1}y^{-1}$. Given two subgroups H, K of a group G, one defines $\{H, K\}$ to be the subgroup generated by the commutators $\{x, y\}, \ x \in H, \ y \in K$.

The *derived group* of a group G is the subgroup $\{G, G\}$ generated by the commutators. $\{G, G\}$ is clearly the minimal normal subgroup K of G such that G/K is abelian.

The *derived series* is defined inductively:

$$G^{(1)} = G, \dots, G^{(i+1)} := \{G^{(i)}, G^{(i)}\}.$$

The *lower central series* is defined inductively:

$$G^1 = G, \dots, G^{i+1} := \{G, G^i\}.$$

[24] In fact all these concepts were first developed for finite groups.

In a topological group we define the *derived group* to be the closed subgroup generated by the commutators, and similar definitions for derived and lower central series.[25]

One has thus the notions of solvable and nilpotent group. Let G be a connected Lie group with Lie algebra L.[26]

Proposition 1. *The derived group of G has Lie algebra* $[L, L]$.

Proof. The derived group is the minimal closed normal subgroup H of G such that G/H is abelian. Since a Lie group is abelian if and only if its Lie algebra is abelian, the proposition follows from Proposition 3.4 since subgroups corresponding to Lie ideals are closed. □

Proposition 2. *The Lie algebra of* $G^{(i+1)}$ *is* $L^{(i+1)}$. *The Lie algebra of* G^{i+1} *is* L^{i+1}.

Proof. The first statement follows from the previous proposition. For the second, let H^i be the connected Lie subgroup of Lie algebra L^i. Assume by induction $H^i = G^i$. Observe that G^{i+1} is the minimal normal subgroup K of G contained in G^i with the property that the conjugation action of G on G^i/K is trivial. Since G is connected, if G acts on a manifold, it acts trivially if and only if its Lie algebra acts by trivial vector fields. If K is a normal subgroup of G contained in G^i with Lie algebra M, G acts trivially by conjugation on G^i/K if and only if the Lie algebra of G acts by trivial vector fields. In particular the restriction of the adjoint action of G on L^i/M is trivial and so $K \supset H^{i+1}$. Conversely it is clear that $H^{i+1} \supset G^{i+1}$. □

Thus a connected Lie group G has a maximal closed connected normal solvable subgroup, the *solvable radical*, whose Lie algebra is the solvable radical of the Lie algebra of G. The nilpotent radical is defined similarly.

Proposition 3. *For a Lie group G the following two conditions are equivalent:*

(1) The Lie algebra of G is semisimple.
(2) G has no connected solvable normal subgroups.

Definition. A group G satisfying the previous two conditions is called a *semisimple* group.

Remark. G may have a nontrivial discrete center. Semisimple Lie groups are among the most interesting Lie groups. They have been completely classified. Of this classification, we will see the complex case (the real case is more intricate and beyond the purpose of this book). This classification is strongly related to algebraic and compact groups as we will illustrate in the next chapters.

---

[25] In fact for a Lie group the two notions coincide; cf. [OV].
[26] The connectedness hypothesis is obviously necessary.

5

Tensor Algebra

Summary. In this chapter we develop somewhat quickly the basic facts of tensor algebra, assuming the reader is familiar with linear algebra. Tensor algebra should be thought of as a natural development of the theory of functions in several vector variables. To some extent it is equivalent, at least in our setting, to this theory.

1 Tensor Algebra

1.1 Functions of Two Variables

The language of functions is most suitably generalized into the language of tensor algebra. The idea is simple but powerful: the dual V^* of a vector space V is a space of functions on V, and V itself can be viewed as functions on V^*.

A way to stress this symmetry is to use the bra-ket $\langle\ |\ \rangle$ notation of the physicists:[27] given a linear form $\phi \in V^*$ and a vector $v \in V$, we denote by $\langle \phi | v \rangle := \phi(v)$ the value of ϕ on v (or of v on ϕ !).

From linear functions one can construct polynomials in one or several variables. Tensor algebra provides a coherent model to perform these constructions in an intrinsic way.

Let us start with some elementary remarks. Given a set X (with n elements) and a field F, we can form the n-dimensional vector space F^X of functions on X with values in F.

This space comes equipped with a canonical basis: the characteristic functions of the elements of X. It is convenient to identify X with this basis and write $\sum_{x \in X} f(x)x$ for the vector corresponding to a function f.

From two sets X, Y (with n, m elements, respectively) we can construct F^X, F^Y, and also $F^{X \times Y}$. This last space is the space of functions in two variables. It has dimension nm.

[27] This was introduced in quantum mechanics by Dirac.

Of course, given a function $f(x) \in F^X$ and a function $g(y) \in F^Y$, we can form the two variable function $F(x, y) := f(x)g(y)$; the product of the given basis elements is just $xy = (x, y)$. A simple but useful remark is the following:

Proposition. *Given two bases $u_1, \dots, u_n$ of F^X and $v_1, \dots, v_m$ of X^Y the nm elements $u_i v_j$ are a basis of $F^{X \times Y}$.*

Proof. The elements xy are a basis of $F^{X \times Y}$. We express x as a linear combination of the $u_1, \dots, u_n$ and y as one of the $v_1, \dots, v_m$.

We then see, by distributing the products, that the nm elements $u_i v_j$ span the vector space $F^{X \times Y}$. Since this space has dimension nm the $u_i v_j$ must be a basis. □

1.2 Tensor Products

We perform the same type of construction with a tensor product of two spaces, without making any reference to a basis. Thus we define:

Definition 1. Given 3 vector spaces U, V, W a map $f(u, v) : U \times V \to W$ is *bilinear* if it is linear in each of the variables u, v separately.

If U, V, W are finite dimensional we easily see that:

Proposition. *The following conditions on a bilinear map $f : U \times V \to W$ are equivalent:*

(i) There exist bases $u_1, \dots, u_n$ of U and $v_1, \dots, v_m$ of V such that the nm elements $f(u_i, v_j)$ are a basis of W.
(ii) For all bases $u_1, \dots, u_n$ of U and $v_1, \dots, v_m$ of V, the nm elements $f(u_i, v_j)$ are a basis of W.
(iii) $\dim(W) = nm$, and the elements $f(u, v)$ span W.
(iv) Given any vector space Z and a bilinear map $g(u, v) : U \times V \to Z$ there exists a unique linear map $G : W \to Z$ such that $g(u, v) = G(f(u, v))$ (universal property).

Definition 2. A bilinear map is called a *tensor product* if it satisfies the equivalent conditions of the previous proposition.

Property (iv) ensures that two different tensor product maps are canonically isomorphic. In this sense we will speak of W as the *tensor product of two vector spaces* which we will denote by $U \otimes V$. We will denote by $u \otimes v$ the image of the pair (u, v) in the bilinear (tensor product) map.

Definition 3. The elements $u \otimes v$ are called *decomposable tensors.*

Example. The bilinear product $F \times U \to U$ given by $(\alpha, u) \mapsto \alpha u$ is a tensor product.

1.3 Bilinear Functions

To go back to functions, we can again concretely treat our constructions as follows. Consider the space $\mathrm{Bil}(U \times V, F)$ of bilinear functions with values in the field F. We have a bilinear map

$$F : U^* \times V^* \to \mathrm{Bil}(U \times V, F)$$

given by $F(\varphi, \psi)(u, v) := \langle \varphi | u \rangle \langle \psi | v \rangle$. In other and more concrete words, the product of two linear functions, in separate variables, is bilinear.

In given bases $u_1, \dots, u_n$ of U and $v_1, \dots, v_m$ of V we have for a bilinear function

$$f\left(\sum_{i=1}^{n} \alpha_i u_i, \sum_{j=1}^{m} \beta_j v_j \right) = \sum_{i=1}^{n} \sum_{j=1}^{m} \alpha_i \beta_j f(u_i, v_j). \tag{1.3.1}$$

Let e^{hk} be the bilinear function defined by the property

$$e^{hk}\left(\sum_{i=1}^{n} \alpha_i u_i, \sum_{j=1}^{m} \beta_j v_j \right) = \alpha_h \beta_k. \tag{1.3.2}$$

We easily see that these bilinear functions form a basis of $\mathrm{Bil}(U \times V, F)$, and a general bilinear function f is expressed in this basis as

$$f(u, v) = \sum_{i=1}^{n} \sum_{j=1}^{m} f(u_i, v_j) e^{ij}(u, v), \quad f = \sum_{i=1}^{n} \sum_{j=1}^{m} f(u_i, v_j) e^{ij}. \tag{1.3.3}$$

Moreover let u^i and v^j be the dual bases of the two given bases. We see immediately that $e^{hk}(u, v) = u^h(u) v^k(v)$. Thus we are exactly in the situation of a tensor product, and we may say that $\mathrm{Bil}(U \times V, F) = U^* \otimes V^*$.

In the more familiar language of polynomials, we can think of n variables x_i and m variables y_j. The space of bilinear functions is the span of the bilinear monomials $x_i y_j$.

Since a finite-dimensional vector space U can be identified with its double dual it is clear how to construct a tensor product. We may set[28]

$$U \otimes V := \mathrm{Bil}(U^* \times V^*, F).$$

1.4 Tensor Product of Operators

The most important point for us is that one can also perform the tensor product of operators using the universal property.

28 Nevertheless, the tensor product construction holds for much more general situations than the one we are treating now. We refer to N. Bourbaki for a more detailed discussion [B].

If $f : U_1 \to V_1$ and $g : U_2 \to V_2$ are two linear maps, the map $U_1 \times U_2 \to V_1 \otimes V_2$ given by $(u, v) \to f(u) \otimes g(v)$ is bilinear. Hence it factors through a unique linear map denoted by $f \otimes g : U_1 \otimes U_2 \to V_1 \otimes V_2$.

This is characterized by the property

$$(f \otimes g)(u \otimes v) = f(u) \otimes g(v). \tag{1.4.1}$$

In matrix notation the only difficulty is a notational one. Usually it is customary to index basis elements with integral indices. Clearly if we do this for two spaces, the tensor product basis is indexed with pairs of indices and so the corresponding matrices are indexed with pairs of pairs of indices.

Concretely, if $f(u_i) = \sum_j a_{ji} u'_j$ and $g(v_h) = \sum_k b_{kh} v'_k$ we have

$$(f \otimes g)(u_i \otimes v_h) = \sum_{jk} a_{ji} b_{kh} u'_j \otimes v'_k. \tag{1.4.2}$$

Hence the elements $a_{ji} b_{kh}$ are the entries of the tensor product of the two matrices.

An easy exercise shows that the tensor product of maps is again bilinear and thus defines a map

$$\hom(U_1, V_1) \otimes \hom(U_2, V_2) \to \hom(U_1 \otimes U_2, V_1 \otimes V_2).$$

Using bases and matrix notations (and denoting by $M_{m,n}$ the space of $m \times n$ matrices), we have thus a map

$$M_{m,n} \otimes M_{p,q} \to M_{mp,nq}.$$

We leave it to the reader to verify that the tensor product of the elementary matrices gives the elementary matrices, and hence that this mapping is an isomorphism.

Finally we have the obvious associativity conditions. Given

$$U_1 \xrightarrow{f} V_1 \xrightarrow{h} W_1$$

$$U_2 \xrightarrow{g} V_2 \xrightarrow{k} W_2$$

we have $(h \otimes k)(f \otimes g) = hf \otimes kg$. In particular, consider two spaces U, V and endomorphisms $f : U \to U$, $g : V \to V$. We see that:

Proposition. *The mapping $(f, g) \to f \otimes g$ is a representation of $GL(U) \times GL(V)$ in $GL(U \otimes V)$.*

There is an abstraction of this notion. Suppose we are given two associative algebras A, B over F. The vector space $A \otimes B$ has an associative algebra structure, by the universal property, which on decomposable tensors is

$$(a \otimes b)(c \otimes d) = ac \otimes bd.$$

Given two modules M, N on A, and B respectively, $M \otimes N$ becomes an $A \otimes B$-module by

$$(a \otimes b)(m \otimes n) = am \otimes bn.$$

Remark. (1) Given two maps $i, j : A, B \to C$ of algebras such that the images commute, we have an induced map $A \otimes B \to C$ given by $a \otimes b \to i(a)j(b)$.

This is a characterization of the tensor product by universal maps.

(2) If A is an algebra over F and $G \supset F$ is a field extension, then $A \otimes_F G$ can be thought of as a G algebra.

Remark. Given an algebra A and two modules M, N, in general $M \otimes N$ does not carry any natural A-module structure. This is the case for group representations or more generally for Hopf algebras, in which one assumes, among other things, to have a homomorphism $\Delta : A \to A \otimes A$ (for the group algebra of G it is induced by $g \mapsto g \otimes g$), cf. Chapter 8, §7.

1.5 Special Isomorphisms

We analyze some special cases.

First, we can identify any vector space U with $\hom(F, U)$ associating to a map $f \in \hom(F, U)$ the vector $f(1)$. We have also seen that $F \otimes U = U$, and $a \otimes u = au$.

We thus follow the identifications:

$$V \otimes U^* = \hom(F, V) \otimes \hom(U, F) = \hom(F \otimes U, V \otimes F).$$

This last space is identified with $\hom(U, V)$.

Proposition. *There is a canonical isomorphism* $V \otimes U^* = \hom(U, V)$.

It is useful to make this identification explicit and express the action of a decomposable element $v \otimes \varphi$ on a vector u, as well as the composition law of morphisms. Consider the tensor product map

$$\hom(V, W) \times \hom(U, V) \to \hom(U, W).$$

With the obvious notations we easily find:

(1.5.1) $$(v \otimes \varphi)(u) = v\langle\varphi|u\rangle, \quad w \otimes \psi \circ v \otimes \varphi = w \otimes \langle\psi|v\rangle\varphi.$$

In the case of $\mathrm{End}(U) := \hom(U, U)$, we have the identification $\mathrm{End}(U) = U \otimes U^*$; in this case we can consider the linear map $\mathrm{Tr} : U \otimes U^* \to F$ induced by the bilinear pairing given by duality

(1.5.2) $$\mathrm{Tr}(u \otimes \varphi) := \langle\varphi|u\rangle.$$

Definition. The mapping $\mathrm{Tr} : \mathrm{End}(U) \to F$ is called the *trace*. In matrix notations, if e_i is a basis of U and e^i the dual basis, given a matrix $A = \sum_{ij} a_{ij} e_i \otimes e^j$ we have $\mathrm{Tr}(A) = \sum_{ij} a_{ij}\langle e^j|e_i\rangle = \sum_i a_{ii}$.

For the tensor product of two endomorphisms of two vector spaces one has

(1.5.3) $$\mathrm{Tr}(A \otimes B) = \mathrm{Tr}(A)\,\mathrm{Tr}(B),$$

as verified immediately.

Finally, given linear maps $X : W \to U,\ Y : V \to Z,\ v \otimes \phi : U \to V$ we have

(1.5.4) $$Y \circ v \otimes \phi \circ X = Yv \otimes X^t\phi.$$

1.6 Decomposable Tensors

An immediate consequence of the previous analysis is:

Proposition. *The decomposable tensors in* $V \otimes U^* = \hom(U, V)$ *are the maps of rank* 1.

In particular this shows that most tensors are not decomposable. In fact, quite generally:

Exercise. In a tensor product (with the notations of Section 1) a tensor $\sum a_{ij} u_i \otimes v_j$ is decomposable if and only if the $n \times m$ matrix with entries a_{ij} has rank ≤ 1.

Another important case is the sequence of identifications:

$$U^* \otimes V^* = \hom(U, F) \otimes \hom(V, F) = \hom(U \otimes V, F \otimes F) = \hom(U \otimes V, F), \tag{1.6.1}$$

i.e., the tensor product of the duals is identified with the dual of the tensor product. In symbols

$$(U \otimes V)^* = U^* \otimes V^*.$$

It is useful to write explicitly the duality pairing at the level of decomposable tensors:

$$\langle \varphi \otimes \psi | u \otimes v \rangle = \langle \varphi | u \rangle \langle \psi | v \rangle. \tag{1.6.2}$$

In other words, if we think of $U^* \otimes V^*$ as the space of bilinear functions $f(u, v)$, $u \in U$, $v \in V$ the tensor $a \otimes b$ is identified, as a linear function on this space and as the evaluation $f \to f(a, b)$. The interpretation of $U \otimes V$ as bilinear functions on $U^* \times V^*$ is completely embedded in this basic pairing.

Summarizing, we have seen the intrinsic notion of $U \otimes V$ as the solution of a universal problem, as bilinear functions on $U^* \times V^*$, and finally as the dual of $U^* \otimes V^*$.[29]

1.7 Multiple Tensor Product

The tensor product construction can clearly be iterated. The multiple tensor product map

$$U_1 \times U_2 \times \cdots \times U_m \to U_1 \otimes U_2 \otimes \cdots \otimes U_m$$

is the universal multilinear map, and we have in general the dual pairing:

$$U_1^* \otimes U_2^* \otimes \cdots \otimes U_m^* \times U_1 \otimes U_2 \otimes \cdots \otimes U_m \to F$$

[29] Of these three definitions, the solution of the universal problem is the one which admits the widest generalizations.

given on the decomposable tensors by

(1.7.1) $$\langle \varphi_1 \otimes \varphi_2 \otimes \cdots \otimes \varphi_m | u_1 \otimes u_2 \otimes \cdots \otimes u_m \rangle = \prod_{i=1}^{m} \langle \varphi_i | u_i \rangle.$$

This defines a canonical identification of $(U_1 \otimes U_2 \otimes \cdots U_m)^*$ with $U_1^* \otimes U_2^* \otimes \cdots U_m^*$.

Similarly we have an identification

$$\hom(U_1 \otimes U_2 \otimes \cdots U_m, V_1 \otimes V_2 \otimes \cdots V_m) \\ \cong \hom(U_1, V_1) \otimes \hom(U_2, V_2) \otimes \cdots \otimes \hom(U_m, V_m).$$

Let us consider the self-dual pairing on $\mathrm{End}(U)$ given by $\mathrm{Tr}(AB)$ in terms of decomposable tensors. If $A = v \otimes \psi$ and $B = u \otimes \varphi$ we have

(1.7.2) $$\mathrm{Tr}(AB) = \mathrm{Tr}(v \otimes \psi \circ u \otimes \varphi) = \mathrm{Tr}(\langle \psi | u \rangle v \otimes \varphi) = \langle \varphi | v \rangle \langle \psi | u \rangle.$$

We recover the simple fact that $\mathrm{Tr}(AB) = \mathrm{Tr}(BA)$.

We remark also that this is a nondegenerate pairing, and $\mathrm{End}(U) = U \otimes U^*$ is identified by this pairing with its dual:

$$\mathrm{End}(U)^* = (U \otimes U^*)^* = U^* \otimes (U^*)^* = U^* \otimes U \cong U \otimes U^*.$$

We identify an operator A with the linear function $X \mapsto \mathrm{Tr}(AX) = \mathrm{Tr}(XA)$.

Since $\mathrm{Tr}([A, B]) = 0$, the operators with trace 0 form a Lie algebra (of the group $SL(U)$), called $sl(U)$ (and an ideal in the Lie algebra of all linear operators).

For the identity operator in an n-dimensional vector space we have $\mathrm{Tr}(1) = n$.

If we are in characteristic 0 (or prime with n) we can decompose each matrix as $A = \frac{\mathrm{Tr}(A)}{n} 1 + A_0$ where A_0 has zero trace. Thus the Lie algebra $gl(U)$ decomposes as the direct sum $gl(U) = F \oplus sl(U)$, F being identified with the *scalar matrices*, i.e., with the multiples of the identity matrix.

It will be of special interest to us to consider the tensor product of several copies of the same space U, i.e., the *tensor power* of U, denoted by $U^{\otimes m}$. It is convenient to form the direct sum of all of these powers since this space has a natural algebra structure defined on the decomposable tensors by the formula

$$(u_1 \otimes u_2 \otimes \cdots \otimes u_h)(v_1 \otimes v_2 \otimes \cdots \otimes v_k) \\ := u_1 \otimes u_2 \otimes \cdots \otimes u_h \otimes v_1 \otimes v_2 \otimes \cdots \otimes v_k.$$

We usually use the notation

$$T(U) := \bigoplus_{k=0}^{\infty} U^{\otimes k}.$$

This is clearly a graded algebra generated by the elements of degree 1.

Definition. $T(U)$ is called the *tensor algebra* of U.

This algebra is characterized by the following *universal property*:

Proposition. *Any linear mapping $j : U \to R$ into an associative algebra R extends uniquely to a homomorphism $\overline{j} : T(U) \to R$.*

Proof. The mapping $U \times U \times \cdots \times U \to R$ given by $j(u_1)j(u_2)\dots j(u_k)$ is multilinear and so defines a linear map $U^{\otimes k} \to R$.

The required map is the sum of all these maps and is clearly a homomorphism extending j; it is also the unique possible extension since U generates $T(U)$ as an algebra. □

1.8 Actions on Tensors

In particular, a linear automorphism g of U extends to an automorphism of the tensor algebra which acts on the tensors $U^{\otimes m}$ as $g^{\otimes m} := g \otimes g \otimes g \cdots \otimes g$.

Thus we have:

Proposition. *$GL(U)$ acts naturally on $T(U)$ as algebra automorphisms (preserving the degree and extending the standard action on U).*

It is quite suggestive to think of the tensor algebra in a more concrete way. Let us fix a basis of U which we think of as indexed by the letters of an *alphabet* A with n letters.[30]

If we write the tensor product omitting the symbol $\otimes$ we see that a basis of $U^{\otimes m}$ is given by all the n^m *words* of length m in the given alphabet.

The multiplication of two words is just the *juxtaposition* of the words (i.e., write one after the other as a unique word). In this language we see that the tensor algebra can be thought of as *the noncommutative polynomial ring in the variables A,* or *the free algebra on A,* or the *monoid algebra* of the free monoid.

When we think in these terms we adopt the notation $F\langle A\rangle$ instead of $T(U)$.

In this language the universal property is that of polynomials, i.e., we can evaluate a polynomial in any algebra once we give the *values* for the variables.

In fact, since A is a basis of U, a linear map $j : U \to R$ is determined by assigning arbitrarily the *values for the variables A*. The resulting map sends a *word*, i.e., a product of variables, to the corresponding product of the values. Thus this map is really the evaluation of a polynomial.

The action of a linear map on U is a special substitution of variables, a *linear substitution*.

Notice that we are working in the category of all associative algebras and thus we have to use *noncommutative* polynomials, i.e., elements of the free algebra. Otherwise the evaluation map is either not defined or not a homomorphism.

Remark. As already mentioned the notion of tensor product is much more general than the one we have given. We will use at least one case of the more general definition. If A is an algebra over a field k, M a right A-module, N a left A-module, we

---

30 Of course we use the usual alphabet, and so in our examples this restricts n artificially, but there is no theoretical obstruction to think of a possibly infinite alphabet.

define $M \otimes_A N$ to be the quotient of the vector space $M \otimes N$ modulo the elements $ma \otimes n - m \otimes an$. This construction typically is used when dealing with *induced representations*. We will use some simple properties of this construction which the reader should be able to verify.

2 Symmetric and Exterior Algebras

2.1 Symmetric and Exterior Algebras

We can reconstruct the commutative picture by passing to a quotient.

Given an algebra R, there is a unique minimal ideal I of R such that R/I is commutative. It is the ideal generated by all of the commutators $[a, b] := ab - ba$.

In fact, I is even generated by the commutators of a set of generators for the algebra since if an algebra is generated by pairwise commuting elements, then it is commutative.

Consider this ideal in the case of the tensor algebra. It is generated by the commutators of the elements of degree 1, hence it is a homogeneous ideal, and so the resulting quotient is a graded algebra, called the *symmetric algebra* on U.

It is usually denoted by $S(U)$ and its homogeneous component of degree m is called the mth *symmetric power* of the space U and denoted $S^m(U)$.

In the presentation as a free algebra, to make $F\langle A\rangle$ commutative means to impose the commutative law on the variables A. This gives rise to the polynomial algebra $F[A]$ in the variables A. Thus $S(U)$ is isomorphic to $F[A]$.

The canonical action of $GL(U)$ on $T(U)$ clearly leaves invariant the commutator ideal and so induces an action as algebra automorphisms on $S(U)$. In the language of polynomials we again find that the action may be realized by changes of variables.

There is another important algebra, the *Grassmann* or *exterior algebra*. It is defined as $T(U)/J$ where J is the ideal generated by all the elements $u^{\otimes 2}$ for $u \in U$. It is usually denoted by $\bigwedge U$.

The multiplication of elements in $\bigwedge U$ is indicated by $a \wedge b$. Again we have an action of $GL(U)$ on $\bigwedge U = \oplus_k \bigwedge^k U$ as automorphisms of a graded algebra, and the algebra satisfies a universal property with respect to linear maps. Given a linear map $j : U \to R$ into an algebra R, restricted by the condition $j(u)^2 = 0, \forall u \in U$, we have that j extends to a unique homomorphism $\bigwedge U \to R$.

In the language of alphabets we have the following description. The variables in A satisfy the rules:

$$a \wedge b = -b \wedge a, \ a \wedge a = 0.$$

We order the letters in A. A monomial M is 0 if it contains a repeated letter. Otherwise reorder it in alphabetical order, introducing a negative sign if the permutation used to reorder is odd; let us denote by $a(M)$ this value.

Consider the monomials in which the letters appear in strict increasing order, and we call these the *strict monomials* (if A has n elements we have $\binom{n}{k}$ strict monomials of degree k for a total of 2^n monomials). For example, from a basis $e_1, \dots, e_n$ of U we deduce a basis

$$e_{i_1} \wedge e_{i_2} \wedge \cdots \wedge e_{i_k}, \quad 1 \leq i_1 < i_2 < \cdots < i_k \leq n$$

of $\bigwedge U$.

Theorem. *The strict monomials are a basis of $\bigwedge U$.*

Proof. We hint at a combinatorial proof. We construct a vector space with basis the strict monomials. We then define a product by $M \wedge N := a(MN)$. A little combinatorics shows that we have an associative algebra R, and the map of A into R determines an isomorphism of R with $\bigwedge U$. □

For a different proof see Section 4.1 in which we generalize this theorem to Clifford algebras.

In particular, we have the following dimension computations. If $\dim U = n$,

$$\dim \bigwedge^k U = \binom{n}{k}, \ \dim \bigwedge U = 2^n, \ \dim \bigwedge^n U = 1.$$

Let $\dim U = n$. The bilinear pairing $\bigwedge^k U \times \bigwedge^{n-k} U \to \bigwedge^n U$ induces a linear map

$$j : \bigwedge^k U \to \hom\left(\bigwedge^{n-k} U, \bigwedge^n U\right) = \bigwedge^n U \otimes \left(\bigwedge^{n-k} U\right)^*,$$

$$j(u_1 \wedge \cdots \wedge u_k)(v_1 \wedge \cdots v_{n-k}) := u_1 \wedge \cdots \wedge u_k \wedge v_1 \wedge \cdots \wedge v_{n-k}.$$

In a given basis $e_1, \dots, e_n$ we have $j(e_{i_1} \wedge \cdots \wedge e_{i_k})(e_{j_1} \wedge \cdots \wedge e_{j_{n-k}}) = 0$ if the elements $i_1, \dots, i_k, j_1, \dots, j_{n-k}$ are not a permutation of $1, 2, \dots, n$. Otherwise, reordering, the value we obtain is $\epsilon_\sigma e_1 \wedge e_2 \dots \wedge e_n$, where σ is the (unique) permutation that brings the elements $i_1, \dots, i_k, j_1, \dots, j_{n-k}$ into increasing order and ϵ_σ denotes its sign.

In particular we obtain

Proposition. *The map $j : \bigwedge^k U \to \bigwedge^n U \otimes (\bigwedge^{n-k} U)^*$ is an isomorphism.*

This statement is a duality statement in the exterior algebra; it is part of a long series of ideas connected with duality. It is also related to the Laplace expansion of a determinant and the expression of the inverse of a given matrix. We leave to the reader to make these facts explicit (see the next section).

2.2 Determinants

Given a linear map $A : U \to V$, the composed map $j : U \to V \to \bigwedge V$ satisfies the universal property and thus induces a homomorphism of algebras, denoted by $\bigwedge A : \bigwedge U \to \bigwedge V$. For every k the map $\bigwedge A$ induces a linear map $\bigwedge^k A : \bigwedge^k U \to \bigwedge^k V$ with

$$\bigwedge^k A(u_1 \wedge u_2 \wedge \cdots \wedge u_k) = Au_1 \wedge Au_2 \wedge \cdots \wedge Au_k.$$

In this way $(\bigwedge U, \bigwedge A)$ is a *functor* from vector spaces to graded algebras (a similar fact holds for the tensor and symmetric algebras).

In particular, for every k the space $\bigwedge^k U$ is a linear representation of the group $GL(U)$.

This is the proper setting for the theory of determinants. One can define the determinant of a linear map $A : U \to U$ of an n-dimensional vector space U as the linear map $\bigwedge^n A$.

Since $\dim \bigwedge^n U = 1$ the linear map $\bigwedge^n A$ is a scalar. One can identify $\bigwedge^n U$ with the base field by choosing a basis of U; any other basis of U gives the same identification if and only if the matrix of the base change is of determinant 1.

Definition. Given an n-dimensional space V, the special linear group $SL(V)$ is the group of transformations A of determinant 1, or $\bigwedge^n A = 1$.

Sometimes one refers to a matrix of determinant 1 as *unimodular*.

More generally, given bases $u_1, \dots, u_m$ for U and $v_1, \dots, v_n$ for V, we have the induced bases on the Grassmann algebra, and we can compute the matrix of $\bigwedge A$ starting from the matrix a_j^i of A. We have $Au_j = \sum_i a_j^i v_i$ and

$$\begin{aligned}\bigwedge^k A(u_{j_1} \wedge \cdots \wedge u_{j_k}) &= Au_{j_1} \wedge \cdots \wedge Au_{j_k}\\ &= \Big(\sum_{i_1} a_{j_1}^{i_1} v_{i_1}\Big) \wedge \Big(\sum_{i_2} a_{j_2}^{i_2} v_{i_2}\Big) \wedge \dots \wedge \Big(\sum_{i_k} a_{j_k}^{i_k} v_{i_k}\Big)\\ &= \sum_{i_1,\dots,i_k} A(i_1, \dots, i_k | j_1, \dots, j_k) v_{i_1} \wedge \cdots \wedge v_{i_k}.\end{aligned}$$

Proposition. *The coefficient $A(i_1, \dots, i_k | j_1, \dots, j_k)$ is the determinant of the minor of the matrix extracted from the matrix of A from the rows of indices $i_1, \dots, i_k$ and the columns of indices $j_1, \dots, j_k$.*

Proof. By expanding the product and collecting terms. □

Given two matrices A, B with product BA, the multiplication formula of the two matrices associated to two exterior powers, $\bigwedge^k(BA) = \bigwedge^k B \circ \bigwedge^k A$, is called *Binet's formula.*

2.3 Symmetry on Tensors

The theory developed is tied with the concepts of symmetry. We have a canonical action of the symmetric group S_n on $U^{\otimes n}$, induced by the permutation action on $U \times U \times \cdots \times U$. Explicitly,

$$\sigma(u_1 \otimes u_2 \otimes \cdots \otimes u_n) = u_{\sigma^{-1}1} \otimes u_{\sigma^{-1}2} \otimes \cdots \otimes u_{\sigma^{-1}n}.$$

We will refer to this action as the *symmetry action* on tensors.[31]

[31] It will be studied intensively in Chapter 9.

Definition 1. The spaces $\Sigma_n(U)$, $A_n(U)$ of symmetric and antisymmetric tensors are defined by

$$\Sigma_n(U) = \{u \in U^{\otimes n} | \sigma(u) = u\}, \ A_n(U) = \{u \in U^{\otimes n} | \sigma(u) = \epsilon(\sigma)u, \ \forall \sigma \in S_n\},$$

($\epsilon(\sigma)$ indicates the sign of σ).

In other words, the space of symmetric tensors is the sum of copies of the trivial representation while the space of antisymmetric tensors is the sum of copies of the sign representation of the symmetric group.

One can explicitly describe bases for these spaces along the lines of §1.

Fix a basis of U which we think of as an ordered alphabet, and take for a basis of $U^{\otimes n}$ the words of length n in this alphabet. The symmetric group permutes these words by reordering the letters, and $U^{\otimes n}$ is thus a *permutation representation.*

Each word is equivalent to a unique word in which all the letters appear in increasing order. If the letters appear with multiplicity $h_1, h_2, \dots, h_k$, the stabilizer of this word is the product of the symmetric groups $S_{h_1} \times \cdots \times S_{h_k}$. The number of elements in its orbit is $\binom{n}{h_1\, h_2 \dots h_k}$.

The sum of the elements of such an orbit is a symmetric tensor denoted by $e_{h_1}, \dots, e_{h_k}$ and these tensors are a basis of $\Sigma_n(U)$. For skew-symmetric tensors we can only use words without multiplicity, since otherwise a transposition fixes such a word but by antisymmetry must change sign to the tensor. The sum of the elements of such an orbit taken with the sign of the permutation is an antisymmetric tensor, and these tensors are a basis of $A_n(U)$.

Theorem. *If the characteristic of the base field F is 0, the projections of $T(U)$ on the symmetric and on the Grassmann algebra are linear isomorphisms when restricted to the symmetric, respectively, the antisymmetric, tensors.*

Proof. Take a symmetric tensor sum of $\binom{n}{h_1\, h_2 \dots h_k}$ elements of an orbit. The image of all the elements in the same orbit in the symmetric algebra is always the same monomial.

Thus the image of this basis element in the symmetric algebra is the corresponding commutative monomial times the order of the orbit, e.g.,

$$aabb + abab + abba + baab + baba + bbaa \to 6a^2b^2.$$

In char $= 0$, this establishes the isomorphism since it sends a basis to a basis. In order to be more explicit, let us denote by $e_1, e_2, \dots, e_m$ a basis of U. The element

$$\frac{1}{n!} \sum_{\sigma \in S_n} e_{i_{\sigma(1)}} \otimes e_{i_{\sigma(2)}} \otimes \cdots e_{i_{\sigma(n)}}$$

is a symmetric tensor which, by abuse of notation, we identify with the monomial

$$e_{i_1} e_{i_2} \dots e_{i_n}$$

to which it corresponds in the symmetric algebra.

Now for the antisymmetric tensors, an orbit gives rise to an antisymmetric tensor if and only if the stabilizer is 1, i.e., if all the $h_i = 1$. Then the antisymmetric tensor corresponding to a word $a_1 a_2 \dots a_n$ is

$$\frac{1}{n!}\sum_{\sigma\in S_n} \epsilon_\sigma a_{\sigma(1)} a_{\sigma(2)} \dots a_{\sigma(n)}.$$

This tensor maps in the Grassmann algebra to

$$a_1 \wedge a_2 \wedge \dots \wedge a_n.$$

It is often customary to identify $e_{i_1} e_{i_2} \dots e_{i_n}$ or $a_1 \wedge a_2 \wedge \dots \wedge a_n$, with the corresponding symmetric or antisymmetric tensor, but of course one loses the algebra structure. □

Let us now notice one more fact. Given a vector $u \in U$ the tensor $u^{\otimes n} = u \otimes u \otimes u \otimes \cdots \otimes u$ is clearly symmetric and identified with $u^n \in S_n(U)$. If $u = \sum_k \alpha_k e_k$ we have

$$u^{\otimes n} = \sum_{h_1+h_2+\dots+h_m=n} \alpha_1^{h_1}\alpha_2^{h_2}\cdots\alpha_m^{h_m} e_{h_1,\dots,h_k}$$

$$u^n \equiv u^{\otimes n}, \quad e_{h_1,\dots,h_k} \equiv \binom{n}{h_1\, h_2 \cdots h_m} e_1^{h_1} e_2^{h_2} \cdots e_m^{h_m}.$$

We notice a formal fact. A homogeneous polynomial function f on U of degree n factors through $u \mapsto u^{\otimes n}$ and a uniquely determined linear map on $\Sigma_n(U)$. In other words, with $P_n(U)$ the space of homogeneous polynomials of degree n on U, we have:

Proposition. *$P_n(U)$ is canonically isomorphic to the dual of $\Sigma_n(U)$.*[32]

The identification is through the factorization

$$U \xrightarrow{u^{\otimes n}} \Sigma_n(U) \xrightarrow{f} F, \quad f(u^{\otimes n}) = \sum_{h_1+h_2+\dots+h_m=n} \alpha_1^{h_1}\alpha_2^{h_2}\cdots\alpha_m^{h_m} f(e_{h_1,\dots,h_k}).$$

One in fact can more generally define:

Definition 2. A *polynomial map* $F : U \to V$ between two vector spaces is a map which in coordinates is given by polynomials.

In particular, one can define homogeneous polynomial maps. We thus have that the map $U \to \Sigma_n(U)$ given by $u \mapsto u^{\otimes n}$ is a polynomial map, homogeneous of degree n and *universal*, in the sense that:

Corollary. *Every homogeneous polynomial map of degree n from U to a vector space V factors through the map $u \to u^{\otimes n}$, with a linear map $\Sigma_n(U) \to V$.*

In characteristic 0 we use instead $u \to u^n$ as a universal map.

[32] This will be useful in the next chapter.

3 Bilinear Forms

3.1 Bilinear Forms

At this point it is important to start introducing the language of bilinear forms in a more systematic way.

We have already discussed the notion of a bilinear mapping $U \times V \to F$. Let us denote the value of such a mapping with the *bra-ket* notation $\langle u|v\rangle$.

Choosing bases for the two vector spaces, the pairing determines a matrix A with entries $a_{ij} = \langle u_i|v_j\rangle$. Using column notation for vectors, the form is given by the formula

$$(u, v) := u^t Av. \tag{3.1.1}$$

If we change the two bases with matrices B, C, and $u = Bu'$, $v = Cv'$, the corresponding matrix of the pairing becomes $B^t AC$.

If we fix our attention on one of the two variables we can equivalently think of the pairing as a linear map $j : U \to \hom(V, F)$ given by $\langle j(u)|v\rangle = \langle u|v\rangle$ or $j(u) : v \mapsto \langle u|v\rangle$.

We have used the bracket notation for our given pairing as well as the duality pairing, thus we can think of a pairing as a linear map from U to V^*.

Definition. We say that a pairing is *nondegenerate* if it induces an isomorphism between U and V^*.

Associated to this idea of bilinear pairing is the notion of *orthogonality*. Given a subspace M of U, its orthogonal is the subspace

$$M^\perp := \{v \in V|\langle u|v\rangle = 0,\ \forall u \in M\}.$$

Remark. The pairing is an isomorphism if and only if its associated (square) matrix is nonsingular. In the case of nondegenerate pairings we have:

(a) $\dim(U) = \dim(V)$.
(b) $\dim(M) + \dim(M^\perp) = \dim(U)$; $(M^\perp)^\perp = M$ for all the subspaces.

3.2 Symmetry in Forms

In particular consider the case $U = V$. In this case we speak of a *bilinear form* on U. For such forms we have a further important notion, that of symmetry:

Definition. We say that a form is *symmetric*, respectively *antisymmetric* or *symplectic*, if $\langle u_1|u_2\rangle = \langle u_2|u_1\rangle$ or, respectively, $\langle u_1|u_2\rangle = -\langle u_2|u_1\rangle$, for all $u_1, u_2 \in U$.

One can easily see that the symmetry condition can be written in terms of the associated map $j : U \to U^* : \langle j(u)|v\rangle = \langle u|v\rangle$.

We take advantage of the identification $U = U^{**}$ and so we have the transpose map $j^* : U^{**} = U \to U^*$.

Lemma. *The form is symmetric if and only if $j = j^*$. It is antisymmetric if and only if $j = -j^*$.*

Sometimes it is convenient to give a uniform treatment of the two cases and use the following language. Let ϵ be 1 or -1. We say that the form is *ϵ-symmetric* if

$$\langle u_1 | u_2 \rangle = \epsilon \langle u_2 | u_1 \rangle. \tag{3.2.1}$$

Example 1. The space $\mathrm{End}(U)$ with the form $\mathrm{Tr}(AB)$ is an example of a nondegenerate symmetric bilinear form. The form is nondegenerate since it induces the isomorphism between $U^* \otimes U$ and its dual $U \otimes U^*$ given by exchanging the two factors of the tensor product (cf. 1.7.2).

Example 2. Given a vector space V we can equip $V \oplus V^*$ with a canonical symmetric form, and a canonical antisymmetric form, by the formula

$$\langle (v_1, \varphi_1) | (v_2, \varphi_2) \rangle := \langle \varphi_1 | v_2 \rangle \pm \langle \varphi_2 | v_1 \rangle. \tag{3.2.2}$$

On the right-hand side we have used the dual pairing to define the form. We will sometimes refer to these forms as *standard hyperbolic* (resp. *symplectic*) *form*. One should remark that the group $GL(V)$ acts naturally on $V \oplus V^*$ preserving the given forms.

The previous forms are nondegenerate. For an ϵ-symmetric form $(\ ,\ )$ on V we have

$$\{v \in V \mid (v, w) = 0,\ \forall w \in V\} = \{v \in V \mid (w, v) = 0,\ \forall w \in V\}.$$

This subspace is called the *kernel* of the form. The form is nondegenerate if and only if its kernel is 0.

3.3 Isotropic Spaces

For bilinear forms we have the important notion of an *isotropic* and a *totally isotropic subspace*.

Definition. A subspace $V \subset U$ is isotropic if the restriction of the form to V is degenerate and totally isotropic if the restricted form is identically 0.

From the formulas of the previous section it follows:

Proposition. *For a nondegenerate bilinear form on a space U, a totally isotropic subspace V has dimension* $\dim V \leq \dim U/2$.

In particular if $\dim U = 2m$, a maximal totally isotropic subspace has at most dimension m.

Exercise.

(1) Prove that a nondegenerate symmetric bilinear form on a space U of dimension $2m$ has a maximal totally isotropic subspace of dimension m if and only if it is isomorphic to the standard hyperbolic form.
(2) Prove that a nondegenerate antisymmetric bilinear form on a space U exists only if U is of even dimension $2m$. In this case, it is isomorphic to the standard symplectic form.

The previous exercise shows that for a given even dimension there is only one symplectic form up to isomorphism. This is not true for symmetric forms, at least if the field F is not algebraically closed. Let us recall the theory for real numbers. Given a symmetric bilinear form on a vector space over the real number $\mathbb{R}$ there is a basis in which its matrix is diagonal with entries $+1, -1, 0$. The number of 0 is the dimension of the kernel of the form. The fact that the number of $+1$'s (or of -1's) is independent of the basis in which the form is diagonal is Sylvester's law of inertia. The form is *positive* (resp. *negative*) *definite* if the matrix is $+1$ (resp. -1). Since the positive definite form is the usual Euclidean norm, one refers to such space as Euclidean space. In general the number of $+1$'s minus the number of -1's is an invariant of the form called its *signature*.

3.4 Adjunction

For a nondegenerate ϵ-symmetric form we have also the important notion of *adjunction* for operators on U. For $T \in \mathrm{End}(U)$ one defines T^*, the adjoint of T, by

$$(u, T^*v) := (Tu, v). \tag{3.4.1}$$

Using the matrix notation $(u, v) = u^t A v$ we have

(3.4.2)

$$(Tu, v) = (Tu)^t Av = u^t\, T^t Av = u^t AA^{-1}T^t Av = u^t AT^*v \implies T^* = A^{-1}T^t A.$$

Adjunction defines an involution on the algebra of linear operators. Let us recall the definition:

Definition. An involution of an F-algebra R is a linear map $r \mapsto r^*$ satisfying:

$$(rs)^* = s^*r^*, \quad (r^*)^* = r. \tag{3.4.3}$$

In other words $r \mapsto r^*$ is an isomorphism between R and its opposite R^o and it is of order 2. Sometimes it is also convenient to denote an involution by $r \to \bar{r}$.

Let us use the form to identify U with U^* as in 3.1, by identifying u with the linear form $\langle j(u)|v\rangle = (u, v)$.

This identifies $\mathrm{End}(U) = U \otimes U^* = U \otimes U$. With these identifications we have:

$$(u \otimes v)w = u(v, w), \ (a \otimes b)(c \otimes d) = a \otimes (b, c)d,$$
$$(a \otimes b)^* = \varepsilon b \otimes a, \mathrm{tr}(a \otimes b) = (b, a). \tag{3.4.4}$$

These formulas will be used systematically in Chapter 11.

3.5 Orthogonal and Symplectic Groups

Another important notion is that of the symmetry group of a form. We define *an orthogonal transformation* T for a form to be one for which

$$(3.5.1) \qquad (u, v) = (Tu, Tv), \text{ for all } u, v \in U.$$

Equivalently $T^*T = TT^* = 1$ (if the form is nondegenerate), in matrix notations. From 3.4.2 we have $A^{-1}T^tAT = TA^{-1}T^tA = 1$ or $T^tAT = A,\ TA^{-1}T^t = A^{-1}$.

One checks immediately that if the form is nondegenerate, the orthogonal transformations form a group. For a nondegenerate symmetric form the corresponding group of orthogonal transformations is called the *orthogonal group*. For a nondegenerate skew-symmetric form the corresponding group of orthogonal transformations is called the *symplectic group*.

We will denote by $O(V)$, $Sp(V)$ the orthogonal or symplectic group when there is no ambiguity with respect to the form.

For explicit computations it is useful to have a matrix representation of these groups. For the orthogonal group there are several possible choices, which for a non-algebraically closed field may correspond to non-equivalent symmetric forms and non-isomorphic orthogonal groups.

If A is the identity matrix we get the usual relation $T^* = T^t$. In this case the orthogonal group is

$$O(n, F) := \{X \in GL(n, F) \mid XX^t = 1\}.$$

It is immediate then that for $X \in O(n, F)$ we have $\det(X) = \pm 1$. Together with the orthogonal group it is useful to consider the *special orthogonal group*:

$$SO(n, F) = \{X \in O(n, F) \mid \det X = 1\}.$$

Often one refers to elements in $SO(n, F)$ as *proper orthogonal transformations* while the elements of determinant -1 are called *improper*.

Consider the case of the *standard hyperbolic form 3.2.2* where $U = V \oplus V^*$, $\dim(U) = 2m$ is even.

Choose a basis v_i in V and correspondingly the dual basis v^i in V^*. We see that the matrix of the standard hyperbolic form is $A = \begin{vmatrix} 0 & 1_m \\ 1_m & 0 \end{vmatrix}$ (note that $A = A^{-1}$).

It is useful to consider the orthogonal group for this form, which for non-algebraically closed fields is usually different from the standard form and is called the *split form* of the orthogonal group.

Similarly for the standard symplectic form we have $A = \begin{vmatrix} 0 & 1_m \\ -1_m & 0 \end{vmatrix}$. Notice that $A^{-1} = -A = A^t$. The standard matrix form of the symplectic group is

$$(3.5.2) \qquad Sp(2m, F) := \{X \in GL(m, F) \mid X^tAX = A \text{ or } XAX^t = A\}.$$

In the previous cases, writing a matrix T in block form $\begin{vmatrix} a & b \\ c & d \end{vmatrix}$, we see that

(3.5.3) $$T^* = \begin{vmatrix} d^t & b^t \\ c^t & a^t \end{vmatrix} \quad \text{(hyperbolic adjoint)}$$

(3.5.4) $$T^* = \begin{vmatrix} d^t & -b^t \\ -c^t & a^t \end{vmatrix} \quad \text{(symplectic adjoint)}$$

One could easily write the condition for a block matrix to belong to the corresponding orthogonal or symplectic group. Rather we work on the real or complex numbers and deduce the Lie algebras of these groups. We have that $(e^{tX})^* = e^{tX^*}$, $(e^{tX})^{-1} = e^{-tX}$, hence:

Proposition. *The Lie algebra of the orthogonal group of a form is the space of matrices with $X^* = -X$.*

From the formulas 3.5.3 and 3.5.4 we get immediately an explicit description of these spaces of matrices, which are denoted by $so(2n, F), sp(2n, F)$.

(3.5.5) $$so(2n, F) := \left\{ \begin{vmatrix} a & b \\ c & -a^t \end{vmatrix};\ b, c \text{ skew-symmetric} \right\}$$

(3.5.6) $$sp(2n, F) := \left\{ \begin{vmatrix} a & b \\ c & -a^t \end{vmatrix};\ b, c \text{ symmetric} \right\}.$$

Thus we have for their dimensions:

$$\dim(so(2n, F)) = 2n^2 - n, \ \dim(sp(2n, F)) = 2n^2 + n.$$

We leave it to the reader to describe $so(2n + 1, F)$. The study of these Lie algebras will be taken up in Chapter 10 in a more systematic way.

3.6 Pfaffian

We want to complete this treatment recalling the properties and definitions of the *Pfaffian* of a skew matrix.

Let V be a vector space of dimension $2n$ with basis e_i. A skew-symmetric form ω_A on V corresponds to a $2n \times 2n$ skew-symmetric matrix A defined by $a_{i,j} := \omega_A(e_i, e_j)$.

According to the theory of exterior algebras we can think of ω_A as the 2-covector[33] given by $\omega_A := \sum_{i<j} a_{i,j} e^i \wedge e^j = 1/2 \sum_{i,j} a_{i,j} e^i \wedge e^j$.

Definition. We define $Pf(A)$ through the formula

(3.6.1) $$\omega_A^n = n!\, Pf(A) e^1 \wedge e^2 \wedge \cdots \wedge e^{2n}.$$

[33] One refers to an element of $\bigwedge^k V^*$ as a k-covector.

Theorem.

(i) For any invertible matrix B we have

(3.6.2) $$Pf(BAB^t) = \det(B)Pf(A).$$

(ii) $\det(A) = Pf(A)^2$.

Proof. (i) One quickly verifies that $\omega_{BAB^t} = (\wedge^2 B)(\omega_A)$. The identity (i) follows since the linear group acts as algebra automorphisms of the exterior algebra.

The identity (ii) follows from (i) since every skew matrix is of the form BJ_kB^t, where J_k is the standard skew matrix of rank $2k$ given by the direct sum of k blocks of size 2×2 $\begin{pmatrix} 0 & 1 \\ -1 & 0 \end{pmatrix}$. The corresponding covector is $\Omega_{J_k} = \sum_{j=1}^{k} e^{2j-1} \wedge e^{2j}$,

$$\Omega_{J_k}^n = 0, \text{ if } k < n, \quad \Omega_{J_n}^n = n!\, e^1 \wedge e^2 \wedge \cdots \wedge e^{2n}, \quad Pf(J_n) = 1.$$

The identity is verified directly. □

From formula 3.6.2 it follows that *the determinant of a symplectic matrix is* 1.

Exercise. Let $x_{ij} = -x_{ji}$, $i, j = 1\dots, 2n$, be antisymmetric variables and X the *generic antisymmetric matrix* with entries x_{ij}. Consider the symmetric group S_{2n} acting on the matrix indices and on the monomial $x_{12}x_{34}\dots x_{2n-1\,2n}$. Up to sign this monomial is stabilized by a subgroup H isomorphic to the semidirect product $S_n \ltimes \mathbb{Z}/(2)^n$. Prove that

$$Pf(X) = \sum_{\sigma \in S_{2n}/H} \epsilon_\sigma x_{\sigma(1)\,\sigma(2)} x_{\sigma(3)\,\sigma(4)} \cdots x_{\sigma(2n-1)\,\sigma(2n)}.$$

Prove that the polynomial $Pf(X)$ (in the variables which are the coordinates of a skew matrix) is irreducible.

There is another interesting formula to point out. We will return to this in Chapter 13.

Let us introduce the following notation. Given X as before, $k \leq n$, and indices $1 \leq i_1 < i_2 < \cdots < i_{2k} \leq 2n$ we define the symbol $[i_1, i_2, \dots, i_{2k}]$ to be the Pfaffian of the principal minor of X extracted from the rows and the columns of indices $i_1 < i_2 < \cdots < i_{2k}$. If $\omega_X := \sum_{i<j} x_{i,j} e_i \wedge e_j$, we have

(3.6.3) $$\exp(\omega_X) = \sum_k \sum_{i_1 < i_2 < \cdots < i_{2k}} [i_1, i_2, \dots, i_{2k}] e_{i_1} \wedge e_{i_2} \wedge \cdots \wedge e_{i_{2k-1}} \wedge e_{i_{2k}}.$$

3.7 Quadratic Forms

Given a symmetric bilinear form on U, we define its associated *quadratic form* by $Q(u) := \langle u|u\rangle$. We see that $Q(u)$ is a homogeneous polynomial of degree 2. We have $Q(u+v) = \langle u+v|u+v\rangle = Q(u) + Q(v) + 2\langle u|v\rangle$ by the bilinearity and symmetry properties. Thus (if 2 is invertible):

$$\tfrac{1}{2}(Q(u+v) - Q(u) - Q(v)) = \langle u|v\rangle.$$

Notice that this is a very special case of the theory of polarization and restitution, thus quadratic forms or symmetric bilinear forms are equivalent notions (if 2 is invertible).[34]

Suppose we are now given two bilinear forms on two vector spaces U, V. We can then construct a bilinear form on $U \otimes V$ which, on the decomposable tensors, is

$$\langle u_1 \otimes v_1 | u_2 \otimes v_2\rangle = \langle u_1|u_2\rangle\langle v_1|v_2\rangle.$$

We see immediately that if the forms are ϵ_1, ϵ_2-symmetric, then the tensor product is $\epsilon_1\epsilon_2$-symmetric.

One easily verifies that if the two forms are associated to the maps

$$j : U \to U^*,\ k : V \to V^*,$$

then the tensor product form corresponds to the tensor product of the two maps. As a consequence we have:

Proposition. *The tensor product of two nondegenerate forms is nondegenerate.*

Iterating the construction we have a bilinear function on $U^{\otimes m}$ induced by a bilinear form on U.

If the form is symmetric on U, then it is symmetric on all the tensor powers, but if it is antisymmetric, then it will be symmetric on the even and antisymmetric on the odd tensor powers.

Example. We consider the classical example of binary forms ([Hilb]).

We start from a 2-dimensional vector space V with basis e_1, e_2. The element $e_1 \wedge e_2$ can be viewed as a skew-symmetric form on the dual space.

The symplectic group in this case is just the group $SL(2, \mathbb{C})$ of 2×2 matrices with determinant 1.

The dual space of V is identified with the space of linear forms in two variables x, y where x, y represent the dual basis of e_1, e_2.

A typical element is thus a linear form $ax + by$. The skew form on this space is

$$[ax + by, cx + dy] := ad - bc = \det\begin{vmatrix} a & b \\ c & d \end{vmatrix}.$$

This skew form determines corresponding forms on the tensor powers of V^*. We restrict such a form to the symmetric tensors which are identified with the space of binary forms of degree n. We obtain on the space of binary forms of even degree, a nondegenerate symmetric form and, on the ones of odd degree a nondegenerate skew-symmetric form.

The group $SL(2, \mathbb{C})$ acts correspondingly on these spaces by orthogonal or symplectic transformations.

---

[34] The theory in characteristic 2 can be developed but it is rather more complicated.

One can explicitly evaluate these forms on the special symmetric tensors given by taking the power of a linear form

$$[u \otimes u \otimes \dots u, v \otimes v \otimes \cdots \otimes v] = [u, v]^n.$$

If $u = ax + by,\ v = cx + dy$, we get

$$\left[\sum_{i=0}^{n} \binom{n}{i} a^{n-i} b^i x^{n-i} y^i, \sum_{j=0}^{n} \binom{n}{j} c^{n-j} d^j x^{n-j} y^j\right] = (ad - bc)^n.$$

Setting $u_{ij} := [x^{n-i} y^i, x^{n-j} y^j]$ we get

$$\sum_{i=0}^{n} \binom{n}{i} \sum_{j=0}^{n} \binom{n}{j} u_{ij} a^{n-i} b^i c^{n-j} d^j = \sum_{k=0}^{n} \binom{n}{k} (-1)^k (ad)^k (bc)^{n-k}.$$

Comparing the coefficients of the monomials we finally have

$$u_{ij} = 0,\ \text{if } i + j \neq n,\ u_{i,n-i} = (-1)^{n-i} \binom{n}{i}^{-1}.$$

3.8 Hermitian Forms

When one works over the complex numbers there are several notions associated with complex conjugation.[35]

Given a vector space U over $\mathbb{C}$ one defines the *conjugate space* $\overline{U}$ to be the group U with the new scalar multiplication $\circ$ defined by

$$\alpha \circ u := \overline{\alpha} u.$$

A linear map from $A : \overline{U} \to V$ to another vector space V, is the same as an *antilinear* map from U to V, i.e., a map A respecting the sum and for which $A(\alpha u) = \overline{\alpha} A(u)$.

The most important concept associated to antilinearity is perhaps that of a Hermitian form and Hilbert space structure on a vector space U.

From the algebraic point of view:

Definition. An Hermitian form is a bilinear map $U \times \overline{U} \to \mathbb{C}$ denoted by (u, v) with the property that (besides the linearity in u and the antilinearity in v) one has

$$(v, u) = \overline{(u, v)},\ \forall u, v \in U.$$

An Hermitian form is positive if $\|u\|^2 := (u, u) > 0$ for all $u \neq 0$.

A positive Hermitian form is also called a pre-Hilbert structure on U.

[35] One could extend several of these notions to automorphisms of a field, or automorphisms of order 2.

Remark. The Hilbert space condition is not algebraic, but is the completeness of U under the metric $\|u\|$ induced by the Hilbert norm.

In a finite-dimensional space, completeness is always ensured. A Hilbert space always has an orthonormal basis u_i with $(u_i, u_j) = \delta_{ij}$. In the infinite-dimensional case this basis will be infinite and it has to be understood in a topological way (cf. Chapter 8). The most interesting example is the *separable* case in which any orthonormal basis is countable.

A pre-Hilbert space can always be completed to a Hilbert space by the standard method of Cauchy sequences modulo sequences converging to 0.

The group of linear transformations preserving a given Hilbert structure is called the unitary group. In the finite-dimensional case and in an orthonormal basis it is formed by the matrices A such that $A\overline{A}^t = 1$.

The matrix $\overline{A}^t$ is denoted by A^* and called the *adjoint* of A. It is connected with the notion of adjoint of an operator T which is given by the formula $(Tu, v) = (u, T^*v)$. In the infinite-dimensional case and in an orthonormal basis the matrix of the adjoint of an operator is the adjoint matrix.

Given two Hilbert spaces H_1, H_2 one can form the tensor product of the Hilbert structures by the obvious formula $(u \otimes v, w \otimes x) := (u, w)(v, x)$. This gives a pre-Hilbert space $H_1 \otimes H_2$; if we complete it we have a *complete tensor product* which we denote by $H_1 \hat{\otimes} H_2$.

Exercise. If u_i is an orthonormal basis of H_1 and v_j an orthonormal basis of H_2, then $u_i \otimes v_j$ is an orthonormal basis of $H_1 \hat{\otimes} H_2$.

The real and imaginary parts of a positive Hermitian form $(u, v) := S(u, v) + iA(u, v)$ are immediately seen to be bilinear forms on U as a real vector space. $S(u, u)$ is a positive quadratic form while $A(u, v)$ is a nondegenerate alternating form.

An orthonormal basis $u_1, \dots, u_n$ for U (as Hilbert space) defines a basis for U as real vector space given by $u_1, \dots, u_n, iu_1, \dots, iu_n$ which is an orthonormal basis for S and a standard symplectic basis for A, which is thus nondegenerate.

The connection between S, A and the complex structure on U is given by the formula

$$A(u, v) = S(u, iv), \ \ S(u, v) = -A(u, iv).$$

3.9 Reflections

Consider a nondegenerate quadratic form Q on a vector space V over a field F of characteristic $\neq 2$. Write $\|v\|^2 = Q(v), \ (v, w) = \frac{1}{2}(Q(v + w) - Q(v) - Q(w))$.

If $v \in V$ and $Q(v) \neq 0$, we may construct the map $s_v : w \to w - \frac{2(v,w)}{Q(v)} v$. Clearly $s_v(w) = w$ if w is orthogonal to v, while $s_v(v) = -v$.

The map s_v is called the *orthogonal reflection* relative to the hyperplane orthogonal to v. It is an improper orthogonal transformation (of determinant -1) of order two ($s_v^2 = 1$).

Example. Consider the hyperbolic space V of dimension 2. In a hyperbolic basis the matrices $\begin{vmatrix} a & b \\ c & d \end{vmatrix}$ of the orthogonal transformations satisfy

$$\begin{vmatrix} a & c \\ b & d \end{vmatrix} \begin{vmatrix} 0 & 1 \\ 1 & 0 \end{vmatrix} \begin{vmatrix} a & b \\ c & d \end{vmatrix} = \begin{vmatrix} 0 & 1 \\ 1 & 0 \end{vmatrix} \implies ac = bd = 0,\ cb + ad = 1.$$

From the above formulas one determines the proper transformations $\begin{vmatrix} a & 0 \\ 0 & a^{-1} \end{vmatrix}$, and improper transformations $\begin{vmatrix} 0 & a \\ a^{-1} & 0 \end{vmatrix}$ which are the reflections $s_{(-a,1)}$.

Theorem (Cartan–Dieudonné). *If* $\dim V = m$, *every orthogonal transformation of* V *is the product of at most* m *reflections.*

Proof. Let $T : V \to V$ be an orthogonal transformation. If T fixes a non-isotropic vector v, then T induces an orthogonal transformation in the orthogonal subspace $v^\perp$, and we can apply induction.

The next case is when there is a non-isotropic vector v such that $u := v - T(v)$ is non-isotropic. Then $(v, u) = (v, v) - (v, T(v))$, $(u, u) = (v, v) - (v, T(v)) - (T(v), v) + (T(v), T(v)) = 2((v, v) - (v, T(v)))$ so that $\frac{2(u,v)}{(u,u)} = 1$ and $s_u(v) = T(v)$. Now $s_u T$ fixes v and we can again apply induction.

This already proves the theorem in the case of a definite form, for instance for the Euclidean space.

The remaining case is that in which every fixed point is isotropic and for every non-isotropic vector v we have $u := v - T(v)$ is isotropic. We claim that if $\dim V \geq 3$, then:

(1) $v - T(v)$ is always isotropic.
(2) V has even dimension $2m$.
(3) T is a proper orthogonal transformation.

Let $s := 1 - T$, and let $V_1 = \ker s$ be the subspace of vectors fixed by T.

Let v be isotropic and consider $v^\perp$ which is a space of dimension $n - 1 > n/2$. Thus $v^\perp$ contains a non-isotropic vector w, and also $\lambda v - w$ is non-isotropic for all λ. Thus by hypothesis

$$0 = Q(s(w)) = Q(s(v - w)) = Q(s(-v - w)).$$

From these equalities follows

$$0 = Q(s(v)) + Q(s(w)) - 2(s(v), s(w)) = Q(s(v)) - 2(s(v), s(w)),$$
$$0 = Q(s(v)) + Q(s(w)) + 2(s(v), s(w)) = Q(s(v)) + 2(s(v), s(w)).$$

Hence $Q(s(v)) = 0$. From $(v - T(v), v - T(v)) = 0$ for all v follows $(v - T(v), w - T(w)) = 0$ for all v, w. From the orthogonality of T follows that

$$(v, 2w) + (v, -T(w)) + (v, -T^{-1}w) = 0$$

for all v, w or $2 - T - T^{-1} = 0$. Hence $2T - T^2 - 1 = 0$ or $s^2 = (1 - T)^2 = 0$.

We have, by hypothesis, that $V_1 = \ker s$ is a totally isotropic subspace, so $2 \dim V_1 \leq \dim V$. Since $s^2 = 0$ we have that $s(V) \subset V_1$, thus $s(V)$ is made of isotropic vectors. Since $\dim V = \dim s(V) + \dim(\ker s)$, it follows that $V_1 = s(V)$ is a maximal totally isotropic subspace and V is of even dimension $2m$. We have that $T = 1 + s$ has only 1 as an eigenvalue and so it has determinant 1, and is thus a proper orthogonal transformation. If S_w is any reflection, we have that $S_w T$ cannot satisfy the same conditions as T, otherwise it would be of determinant 1. Thus we can apply induction and write it as a product of $\leq 2m$ reflections, but this number must be odd, since $S_w T$ has determinant -1. So $S_w T$ is a product of $< 2m$ reflections hence $T = S_w(S_w T)$ is the product of $\leq 2m$ reflections.

In the case $\dim V = 2$, we may assume that the space has isotropic vectors. Hence it is hyperbolic, and we have the formulas of the example. The elements $\begin{vmatrix} 0 & a \\ a^{-1} & 0 \end{vmatrix}$ are reflections, and clearly the proper transformations are products of two reflections. □

3.10 Topology of Classical Groups

Over the complex or real numbers, the groups we have studied are Lie groups. In particular it is useful to understand some of their topology. For the orthogonal groups we have just one group $O(n, \mathbb{C})$ over the complex numbers. Over the real numbers, the group depends on the signature of the form, and we denote by $O(p, q; \mathbb{R})$ the orthogonal group of a form with p entries $+1$ and q entries -1 in the diagonal matrix representation. Let us study the connectedness properties of these groups. Since the special orthogonal group has index 2 in the orthogonal group, we always have for any form and field $O(V) = SO(V) \cup SO(V)\eta$ where η is any given improper transformation. Topologically this is a disjoint union of closed and open subsets, so for the study of the topology we may reduce to the special orthogonal groups.

Let us remark that if T_1, T_2 can be joined by a path to the identity in a topological group, then so can $T_1 T_2$. In fact if $\phi_i(t)$ is a path with $\phi_i(0) = 1$, $\phi_i(1) = T_i$, we take the path $\phi_1(t)\phi_2(t)$.

Proposition. *The groups $SO(n, \mathbb{C})$, $SO(n, \mathbb{R})$ are connected.*

Proof. It is enough to show that any element T can be joined by a path to the identity. If we write T as a product of an even number of reflections, it is enough by the previous remark to treat a product of two reflections. In this case the fixed vectors have codimension 2, and the transformation is essentially a rotation in 2-dimensional space. Then for the complex groups these rotations can be identified (in a hyperbolic basis) with the invertible elements of $\mathbb{C}$ which is a connected set. $SO(2, \mathbb{R})$ is the *circle* group of rotations of the plane, which is clearly connected. □

Things are different for the groups $SO(p, q; \mathbb{R})$. For instance $SO(1, 1; \mathbb{R}) = \mathbb{R}^*$ has two components. In this case we claim that if $p, q > 0$, then $SO(p, q; \mathbb{R})$ has two components. Let us give the main ideas of the proof, leaving the details as exercise.

Suppose $p = 1$. The quadratic form can be written as $Q(x, y) := x^2 - \sum_{i=1}^q y_i^2$. The set of elements with $x^2 - \sum_{i=1}^q y_i^2 = 1$ has two connected components. The group $SO(1, q; \mathbb{R})$ acts transitively on these vectors, and the stabilizer of a given vector is $SO(q; \mathbb{R})$ which is connected.

If $p > 1$, the set of elements $\sum_{j-1}^p x_j^2 = 1 + \sum_{i=1}^q y_i^2$ is connected. The stabilizer of a vector is $SO(p-1, q; \mathbb{R})$ so we can apply induction.

Exercise. Prove that the groups $GL(n, \mathbb{C})$, $SL(n, \mathbb{C})$, $SL(n, \mathbb{R})$ are connected while $GL(n, \mathbb{R})$ has two connected components.

More interesting is the question of which groups are simply connected. We have seen in Chapter 4, §3.7 that $SL(n, \mathbb{C})$ is simply connected. Let us analyze $SO(n, \mathbb{C})$ and $Sp(2n, \mathbb{C})$. We use the same method of fibrations developed in Chapter 4, §3.7. Let us treat $Sp(2n, \mathbb{C})$ first. As for $SL(n, \mathbb{C})$, we have that $Sp(2n, \mathbb{C})$ acts transitively on the set W of pairs of vectors u, v with $[u, v] = 1$. The stabilizer of e_1, f_1 is $Sp(2(n-1), \mathbb{C})$. Now let us understand the topology of W. Consider the projection $(u, v) \mapsto u$ which is a surjective map to $\mathbb{C}^{2n} - \{0\}$ with fiber at e_1 the set $f_1 + ae_1 + \sum_{i>2} a_i e_i + b_i f_i$, a contractible space.

This is a special case of Chapter 4, §3.7.1 for the groups $Sp(2(n-1), \mathbb{C}) \subset H \subset Sp(2n, \mathbb{C})$ where H is the stabilizer, in $Sp(2n, \mathbb{C})$, of the vector e_1.

Thus $\pi_1(Sp(2n, \mathbb{C})) = \pi_1(Sp(2(n-1), \mathbb{C}))$. By induction we have:

Theorem. *$Sp(2n, \mathbb{C})$ is simply connected.*

As for $SO(n, \mathbb{C})$ we make two remarks. In Chapter 8, §6.2 we will see a fact, which can be easily verified directly, implying that $SO(n, \mathbb{C})$ and $SO(n, \mathbb{R})$ are homotopically equivalent. We discuss $SO(n, \mathbb{R})$ at the end of §5, proving that $\pi_1(SO(n, \mathbb{R})) = \mathbb{Z}/(2)$.

At this point, if we restrict our attention only to the infinitesimal point of view (that is, the Lie algebras), we could stop our search for classical groups.

This, on the other hand, misses a very interesting point. The fact that the special orthogonal group is not simply connected implies that, even at the infinitesimal level, not all the representations of its Lie algebra arise from representation of this group.

In fact we miss the rather interesting *spin representations*. In order to discover them we have to construct the *spin group*. This will be done, as is customary, through the analysis of Clifford algebras, to which the next section is devoted.

4 Clifford Algebras

4.1 Clifford Algebras

Given a quadratic form on a space U we can consider the ideal J of $T(U)$ generated by the elements $u^{\otimes 2} - Q(u)$.

Definition 1. The quotient algebra $T(U)/J$ is called the *Clifford algebra of the quadratic form.*

Notice that the Clifford algebra is a generalization of the Grassmann algebra, which is obtained when $Q = 0$. We will denote it by $Cl_Q(U)$ or by $Cl(U)$ when there is no ambiguity for the quadratic form.

By definition the Clifford algebra is the universal solution for the construction of an algebra R and a linear map $j : U \to R$ with the property that $j(u)^2 = Q(u)$. Let us denote by $(u, v) := 1/2(Q(u+v) - Q(u) - Q(v))$ the bilinear form associated to Q. We have in the Clifford algebra:

$$(4.1.1) \qquad v, w \in U, \implies vw + wv = (v+w)^2 - v^2 - w^2 = 2(v, w).$$

In particular if v, w are orthogonal they anticommute in the Clifford algebra $vw = -wv$.

If $G \supset F$ is a field extension, the given quadratic form Q on U extends to a quadratic form Q_G on $U_G := U \otimes_F G$. By the universal property it is easy to verify that

$$Cl_{Q_G}(U_G) = Cl_Q(U) \otimes_F G.$$

There are several efficient ways to study the Clifford algebra. We will go through the theory of superalgebras ([ABS]).

One starts by remarking that, although the relations defining the Clifford algebra are not homogeneous, they are of even degree. In other words, decompose the tensor algebra $T(U) = T_0(U) \oplus T_1(U)$, where $T_0(U) = \oplus_k \otimes^{2k} U$, $T_1(U) = \oplus_k \otimes^{2k+1} U$, as a direct sum of its even and odd parts. The ideal I defining the Clifford algebra decomposes also as $I = I_0 \oplus I_1$ and $Cl(U) = T_0(U)/I_0 \oplus T_1(U)/I_1$. This suggests the following:

Definition 2. A superalgebra is an algebra A decomposed as $A_0 \oplus A_1$, with $A_i A_j \subset A_{i+j}$, where the indices are taken modulo 2.[36]

A superalgebra is thus graded modulo 2. For a homogeneous element a, we set $d(a)$ to be its degree (modulo 2). We have the obvious notion of (graded) homomorphism of superalgebras. Often we will write $A_0 = A^+$, $A_1 = A^-$.

Given a superalgebra A, a *superideal* is an ideal $I = I_0 \oplus I_1$, and the quotient is again a superalgebra.

More important is the notion of a *super-tensor product* of associative superalgebras.

Given two superalgebras A, B we define a superalgebra:

$$A \hat{\otimes} B := (A_0 \otimes B_0 \oplus A_1 \otimes B_1) \oplus (A_0 \otimes B_1 \oplus A_1 \otimes B_0),$$
$$(4.1.2) \qquad (a \otimes b)(c \otimes d) := (-1)^{d(b)d(c)} ac \otimes bd.$$

It is left to the reader to show that this defines an associative superalgebra.

---

[36] Superalgebras have been extensively used by physicists in the context of the theory of elementary particles. In fact several basic particles like electrons are *Fermions*, i.e., they obey a special statistics which is suitably translated with the spin formalism. Some further proposed theories, such as supersymmetry, require the systematic use of superalgebras of operators.

Exercise. A superspace is just a $\mathbb{Z}/(2)$ graded vector space $U = U_0 \oplus U_1$, we can grade the endomorphism ring $\mathrm{End}(U)$ in an obvious way:

$$\mathrm{End}(U)_0 = \mathrm{End}(U_0) \oplus \mathrm{End}(U_1), \quad \mathrm{End}(U)_1 = \hom(U_0, U_1) \oplus \hom(U_1, U_0).$$

Prove that, given two superspaces U and V, we have a natural structure of superspace on $U \otimes V$ such that $\mathrm{End}(U \otimes V)$ is isomorphic to $\mathrm{End}(U) \hat{\otimes} \mathrm{End}(V)$ as superalgebra.

In this vein of definitions we have the notion of a *supercommutator*, which on homogeneous elements is

$$\{a, b\} := ab - (-1)^{d(a)d(b)} ba. \tag{4.1.3}$$

Definitions 4.1.2,3 are then extended to all elements by bilinearity.

Accordingly we say that a superalgebra is *supercommutative* if $\{a, b\} = 0$ for all the elements. For instance, the Grassmann algebra is supercommutative.[37]

The connection between supertensor product and supercommutativity is in the following:

Exercise. Given two graded maps $i, j : A, B \to C$ of superalgebras such that the images supercommute we have an induced map $A \hat{\otimes} B \to C$ given by $a \otimes b \to i(a)j(b)$.

Exercise. Discuss the notions of *superderivation* ($D(ab) = D(a)b + (-1)^{d(a)} aD(b)$), supermodule, and super-tensor product of such supermodules.

We can now formulate the main theorem:

Theorem 1. *Given a vector space U with a quadratic form and an orthogonal decomposition $U = U_1 \oplus U_2$, we have a canonical isomorphism*

$$Cl(U) = Cl(U_1) \hat{\otimes} Cl(U_2). \tag{4.1.4}$$

Proof. First consider the linear map $j : U \to Cl(U_1) \hat{\otimes} Cl(U_2)$ which on U_1 is $j(u_1) := u_1 \otimes 1$ and on U_2 is $j(u_2) = 1 \otimes u_2$.

It is easy to see, by all of the definitions given, that this map satisfies the universal property for the Clifford algebra and so it induces a map $\overline{j} : Cl(U) \to Cl(U_1) \hat{\otimes} Cl(U_2)$.

Now consider the two inclusions of U_1, U_2 in U which define two maps $Cl(U_1) \to Cl(U), Cl(U_2) \to Cl(U)$.

It is again easy to see (since the two subspaces are orthogonal) that the images supercommute. Hence we have a map $\overline{i} : Cl(U_1) \hat{\otimes} Cl(U_2) \to Cl(U)$.

On the generating subspaces $U, U_1 \otimes 1 \oplus 1 \otimes U_2$, the maps $\overline{j}, \overline{i}$ are isomorphisms inverse to each other. The claim follows. □

[37] Sometimes in the literature just the term commutative, instead of supercommutative, is used for superalgebras.

For a 1-dimensional space with basis u and $Q(u) = \alpha$, the Clifford algebra has basis $1, u$ with $u^2 = \alpha$.

Thus we see by induction that:

Lemma 1. *If we fix an orthogonal basis $e_1, \dots, e_n$ of the vector space U, then the 2^n elements $e_{i_1} e_{i_2} \dots e_{i_k}$, $i_1 < i_2 < \dots < i_k$ give a basis of $Cl(U)$.*

For an orthogonal basis e_i we have the defining commuting relations $e_i^2 = Q(e_i)$, $e_i e_j = -e_j e_i$, $i \neq j$. If the basis is orthonormal we have also $e_i^2 = 1$.

It is also useful to present the Clifford algebra in a hyperbolic basis, i.e., the Clifford algebra of the standard quadratic form on $V \oplus V^*$ which is convenient to renormalize by dividing by 2, so that $Q((v, \phi)) = \langle \phi \mid v \rangle$. If $V = \mathbb{C}^n$ we denote this Clifford algebra by C_{2n}.

The most efficient way to treat C_{2n} is to exhibit the exterior algebra $\bigwedge V$ as an irreducible module over $Cl(V \oplus V^*)$, so that $Cl(V \oplus V^*) = \mathrm{End}(\bigwedge V)$. This is usually called the *spin formalism.*

For this let us define two linear maps i, j from V, V^* to $\mathrm{End}(\bigwedge V)$:

$$i(v)(u) := v \wedge u,$$

$$j(\varphi)(v_1 \wedge v_2 \dots \wedge v_k) := \sum_{t=1}^{k} (-1)^{t-1} \langle \varphi | v_t \rangle v_1 \wedge v_2 \dots \check{v}_t \dots \wedge v_k, \tag{4.1.5}$$

where $\check{v}_t$ means that this term has been omitted.

Notice that the action of $i(v)$ is just the left action of the algebra $\bigwedge V$ while $j(\varphi)$ is the superderivation induced by the contraction by φ on V.

One immediately verifies

$$i(v)^2 = j(\varphi)^2 = 0, \; i(v)j(\varphi) + j(\varphi)i(v) = \langle \varphi | v \rangle. \tag{4.1.6}$$

Theorem 2. *The map $i + j : V \oplus V^* \to \mathrm{End}(\bigwedge V)$ induces an isomorphism between the algebras $Cl(V \oplus V^*)$ and $\mathrm{End}(\bigwedge V)$ (as superalgebras).*

Proof. From 4.1.6 we have that $i + j$ satisfies the universal condition defining the Clifford algebra for $1/2$ of the standard form.

To prove that the resulting map is an isomorphism between $Cl(V \oplus V^*)$ and $\mathrm{End}(\bigwedge V)$ one has several options. One option is to show directly that $\bigwedge V$ is an irreducible module under the Clifford algebra and then remark that, if $n = \dim(V)$ then $\dim Cl(V \oplus V^*) = 2^{2n} = \dim \mathrm{End}(\bigwedge V)$. The second option is to analyze first the very simple case of $\dim V = 1$ where we verify the statement by direct inspection. Next we decompose the exterior algebra as the tensor product of the exterior algebras on 1-dimensional spaces. Each such space is a graded irreducible 2-dimensional module over the corresponding Clifford algebra, and we get the identity by taking super-tensor products. □

The Clifford algebra in the odd-dimensional case is different. Let us discuss the case of a standard orthonormal basis, $e_1, e_2, \dots, e_{2n+1}$. Call this Clifford algebra C_{2n+1}.

Lemma 2. *The element* $c := e_1 e_2 \dots e_{2n+1}$ *is central and* $c^2 = (-1)^n$.

Proof. From the defining commutation relations we immediately see that.

$$e_i c = e_i e_1 e_2 \dots e_{i-1} e_i \dots e_{2n+1} = (-1)^{i-1} e_1 e_2 \dots e_{i-1} e_i^2 \dots e_{2n+1};$$
$$c e_i = e_1 e_2 \dots e_{i-1} e_i \dots e_{2n+1} e_i = (-1)^{2n+1-i} e_1 e_2 \dots e_{i-1} e_i^2 \dots e_{2n+1}.$$

As for c^2, we have again by commutation relations:

$$\begin{aligned} e_1 e_2 \dots e_{2n+1} e_1 e_2 \dots e_{2n+1} &= e_1^2 e_2 \dots e_{2n+1} e_2 \dots e_{2n+1} \\ &= e_2 \dots e_{2n+1} e_2 \dots e_{2n+1} = -e_2^2 e_3 \dots e_{2n+1} e_3 \dots e_{2n+1} \\ &= -e_3 \dots e_{2n+1} e_3 \dots e_{2n+1} = \ \dots \ = (-1)^n. \end{aligned}$$ □

Take the Clifford algebra C_{2n} on the first $2n$ basis elements. Since $e_{2n+1} = e_{2n} e_{2n-1} \dots e_1 c$, we see that $C_{2n+1} = C_{2n} + C_{2n} c$. We have proved that:

Theorem 3. $C_{2n+1} = C_{2n} \otimes_F F[c]$. *If* $F[c] = F \oplus F$, *which happens if* $(-1)^n$ *has a square root in* F, *then* C_{2n+1} *is isomorphic to* $C_{2n} \oplus C_{2n}$.

4.2 Center

One may apply the previous results to have a first study of Clifford algebras as follows. Let U be a quadratic space of dimension n over a field F and let $G = \overline{F}$ be an algebraic closure. Then $U \otimes_F \overline{F}$ is hyperbolic, so we have that $Cl_Q(U) \otimes_F G$ is the Clifford algebra of a hyperbolic form. Thus if $n = 2k$ is even, $Cl_Q(U) \otimes_F G = M_{2^k}(G)$ is the algebra of matrices over G, while if $n = 2k + 1$ is odd, we have $Cl_Q(U) \otimes_F G = M_{2^k}(G) \oplus M_{2^k}(G)$. We can draw some consequences from this statement using the following simple lemma:

Lemma. *Let* R *be an algebra over a field* F *with center* Z *and let* G *be a field extension of* F. *Then the center of* $R \otimes_F G$ *is* $Z \otimes_F G$.

Proof. Let u_i be a basis of G over F. Consider an element $s := \sum_i r_i \otimes u_i \in R \otimes_F G$. To say that it is in the center implies that for $r \in R$ we have $0 = rs - sr = \sum_i (rr_i - r_i r) \otimes u_i$. Hence $rr_i - r_i r = 0$ for all i and $r_i \in Z$ for all i. The converse is also obvious. □

As a consequence we have that:

Proposition. *The center of* $Cl_Q(U)$ *is* F *if* n *is even. If* $n = 2k + 1$, *then the center is* $F + cF$, *where* $c := u_1 u_2 \dots u_{2k+1}$ *for any orthogonal basis* $u_1, u_2, \dots, u_{2k+1}$ *of* U.

Proof. First, one uses the fact that the center of the algebra of matrices over a field is the field itself. Second, up to multiplying c by a nonzero scalar, we may assume that the basis is orthonormal. Finally we are reduced to the theory developed in the previous paragraph. □

Remark. In the case of odd dimension the center may either be isomorphic to $F \oplus F$ or to a quadratic extension field of F. This depends on whether the element $c^2 \in F^*$ is a square or not (in F^*).

4.3 Structure Theorems

It is also important to study the Clifford algebras $C(n), C'(n)$ over $\mathbb{R}$ for the negative and positive definite forms $-\sum_i^n x_i^2, \sum_i^n x_i^2$.

For $n = 1$ we get $C(1) := \mathbb{R}[x]/(x^2+1) = \mathbb{C}$, $C'(1) := \mathbb{R}[x]/(x^2-1) = \mathbb{R}\oplus\mathbb{R}$.

For $n = 2$ we get $C(2) := \mathbb{H}$, the quaternions, since setting $i := e_1, j := e_2$, the defining relations are $i^2 = -1, \ j^2 = -1, ij = -ji$.

$C'(2)$ is isomorphic to $M_2(\mathbb{R})$, the 2×2 matrices over $\mathbb{R}$. In fact in this case the defining relations are $i^2 = 1, j^2 = 1, ij = -ji$, which are satisfied by the matrices:

$$i := \begin{vmatrix} 1 & 0 \\ 0 & -1 \end{vmatrix}, \quad j := \begin{vmatrix} 0 & 1 \\ 1 & 0 \end{vmatrix}, \quad -ji = ij = \begin{vmatrix} 0 & 1 \\ -1 & 0 \end{vmatrix}.$$

To study Clifford algebras in general we make a remark. Let Q be a nondegenerate quadratic form on the space U. Decompose U into an orthogonal direct sum $U = U_1 \oplus U_2$, with $\dim U_1 = 2$. Denote by Q_1, Q_2 the induced quadratic forms on U_1, U_2.

Fix an orthogonal basis u, v of U_1 and an orthogonal basis $e_1, \dots, e_k$ of U_2. Let $u^2 = \alpha, \ v^2 = \beta, \ \lambda_i := e_i^2 = Q(e_i)$ and set $\delta := \alpha\beta \neq 0$, since the form is nondegenerate. We have the commutation relations:

$$ue_i = -e_iu, \quad ve_i = -ve_i, \quad uve_i = e_iuv, \quad uv = -vu, \quad (uv)^2 = -\delta.$$

Moreover, $uuv = -uvu$, $vuv = -uvv$. If we set $f_i := uve_i$ we deduce the following commutation relations:

$$f_if_j = -f_jf_i, \ i \neq j; \quad f_i^2 = -\delta\lambda_i; \quad uf_i = f_iu; \quad vf_i = f_iv.$$

From these commutation relations we deduce that the subalgebra $F(f_1, \dots, f_k)$ of the Clifford algebra generated by the elements f_i is a homomorphic image of the Clifford algebra on the space U_2 but relative to the quadratic form $-\delta Q_2$. Moreover, this subalgebra commutes with the subalgebra $F(u, v) = Cl_{Q_1}(U_1)$. We have thus a homomorphism:

$$Cl_{Q_1}(U_1) \otimes Cl_{-\delta Q_2}(U_2) \xrightarrow{i} Cl_Q(U).$$

Proposition. *The map i is an isomorphism.*

Proof. Since the dimensions of the two algebras are the same, it is enough to show that the map is surjective, i.e., that the elements u, v, f_i generate $C_Q(U)$. This is immediate since $e_i = -\delta^{-1}uvf_i$. □

We can apply this proposition to the Clifford algebras $C(n), C'(n)$ over $\mathbb{R}$ for the negative and positive definite quadratic form. In this case $\delta = 1$, so we get

$$C(n) = \mathbb{H} \otimes C'(n-2), \quad C'(n) = M_2(\mathbb{R}) \otimes C(n-2).$$

Iterating, we get the recursive formulas:

$$C(n) = C(4k) \otimes C(n-4k), \quad C'(n) = C'(4k) \otimes C'(n-4k).$$

In order to complete the computations we need the following simple facts:

Lemma.

(1) If A, B are two F-algebras and $M_h(A)$, $M_k(B)$ denote matrix algebras (over A, B respectively), we have

$$M_h(A) \otimes_F M_k(B) = M_{hk}(A \otimes_F B).$$

(2)
$$\mathbb{H} \otimes_{\mathbb{R}} \mathbb{H} = M_4(\mathbb{R}), \quad \mathbb{H} \otimes_{\mathbb{R}} \mathbb{C} = M_2(\mathbb{C}).$$

Proof. (1) is an easy exercise. (2) can be shown as follows. We have a homomorphism $\psi : \mathbb{H} \otimes_{\mathbb{R}} \mathbb{H} \to \operatorname{End}_{\mathbb{R}} \mathbb{H}$ given by $\psi(a \otimes b)(c) := ac\bar{b}$ which one easily verifies is an isomorphism.

For the second consider $\mathbb{C} \subset \mathbb{H}$ in the usual way, and consider $\mathbb{H}$ as vector space over $\mathbb{C}$ by right multiplication. We have a homomorphism $\phi : \mathbb{H} \otimes_{\mathbb{R}} \mathbb{C} \to \operatorname{End}_{\mathbb{C}} \mathbb{H}$ given by $\phi(a \otimes b)(c) := acb$ which one easily verifies is an isomorphism. □

We deduce the following list for $C(n)$, $n = 0, 1, 2 \dots, 8$:

$$\mathbb{R},\ \mathbb{C},\ \mathbb{H},\ \mathbb{H} \oplus \mathbb{H},\ M_2(\mathbb{H}),\ M_4(\mathbb{C}),\ M_8(\mathbb{R}),\ M_8(\mathbb{R}) \oplus M_8(\mathbb{R}),\ M_{16}(\mathbb{R})$$

and *periodicity 8:* $C(n) = M_{16}(C(n-8))$.[38]

The list of the same algebras but over the complex numbers

$$C(n) \otimes_{\mathbb{R}} \mathbb{C} := C_{\mathbb{C}}(n) = C'_{\mathbb{C}}(n) = C'(n) \otimes_{\mathbb{R}} \mathbb{C}$$

is deduced by tensoring with $\mathbb{C}$ as

$$\mathbb{C},\ \mathbb{C} \oplus \mathbb{C},\ M_2(\mathbb{C}),\ M_2(\mathbb{C}) \oplus M_2(\mathbb{C}),\ M_4(\mathbb{C})$$

and *periodicity 2:* $C_{\mathbb{C}}(n) = M_2(C_{\mathbb{C}}(n-2))$. Of course over the complex numbers the form is hyperbolic, so we get back the result we already knew by the spin formalism.

4.4 Even Clifford Algebra

It is also important to study the degree 0 part of the Clifford algebra, i.e., $Cl_Q^+(U)$, since it will appear in the definition of the spin groups. This is the subalgebra of $Cl_Q(U)$ spanned by products $u_1 u_2 \dots u_{2k}$ of an even number of vectors in U. Let $\dim U = s + 1$, and fix an orthogonal basis which we write $u, e_1, \dots, e_s$ to stress the decomposition $U = Fu \oplus U'$. Let Q' be the restriction of Q to U'. Define $f_i := ue_i$. The f_i are elements of $Cl_Q^+(U)$. Let $\delta := u^2 = Q(u)$.

We have from the commutation relations:

$$i \neq j, \quad f_i f_j = ue_i ue_j = -ue_i e_j u = ue_j e_i u = -ue_j ue_i = -f_j f_i,$$
$$f_i^2 = ue_i ue_i = -u^2 e_i^2.$$

It follows then that the elements f_i satisfy the commutation relations for the Clifford algebra of the space U' equipped with the form $-\delta Q'$. Thus we have a homomorphism $i : Cl_{-\delta Q'}(U') \to Cl_Q^+(U)$.

[38] This is related to the Bott periodicity theorem for homotopy groups and K-theory.

Proposition 1. *The map i is an isomorphism.*

Proof. Since the dimensions of the two algebras are the same it is enough to show that the map is surjective, i.e., that the elements f_i generate $C^+_Q(U)$. Let $\alpha = u^2 \neq 0$. We have that $e_i = \alpha^{-1}uf_i$. Moreover $f_iu = ue_iu = -uf_i$. Take an even monomial in the elements $u, e_1, \dots, e_s$, such that u appears h times and the e_i appear $2k - h$ times. Substitute $e_i = \alpha^{-1}uf_i$. Up to a nonzero scalar, we obtain a monomial in u, f_i in which u appears $2k$ times. Using the commutation relations we can bring u^{2k} in front and then use the fact that this is a scalar to conclude. □

If we apply the previous result to $C(n)$, we obtain $C^+(n) \equiv C(n-1)$.

As preparation for the theory of the spin group let us make an important remark.

Let us take a space V with a nondegenerate symmetric form (a, b) and let $C(V)$ be the Clifford algebra for $1/2(a, b)$.[39] Consider the space $L := \{ab \mid a, b \in V\} \subset C^+(V)$ and the map $a : \bigwedge^2 V \to C^+(V)$, $a(v \wedge w) := [v, w]/2$. Fixing an orthogonal basis e_i for V we have $a(e_i \wedge e_j) = e_ie_j$, $i < j$. From Lemma 1, 4.1 it then follows that a is injective, so we identify $\bigwedge^2 V \subset C^+(V)$.

Proposition 2. *$L = F \oplus \bigwedge^2 V$ is a Lie subalgebra of $C^+(V)$, $[L, L] = \bigwedge^2 V$. Under the adjoint action, V is an L-submodule for which $\bigwedge^2 V$ is isomorphic to the Lie algebra of $SO(V)$.*

Proof. $ab + ba = (a, b)$ means that $ab = \frac{[a,b]}{2} + (a, b)/2$ so $ab = a \wedge b + (a, b)/2$, $\forall a, b \in V$. It follows that $L = F \oplus \bigwedge^2 V$ is the span of the products ab, $a, b \in V$. Next, given $a, b, c, d \in V$ we have (applying the relations):

$$cdab = abcd + [(b, d)ac + (a, d)cb - (a, c)db - (b, c)ad].$$

$$\begin{aligned}\text{Hence}\quad [cd, ab] &= [(b, d)ac + (a, d)cb - (a, c)db - (b, c)ad] \\ &= 1/2\{(b, d)[a, c] + (a, d)[c, b] - (a, c)[d, b] - (b, c)[a, d]\}.\end{aligned}$$

So L is a Lie algebra and $[L, L] \subset \bigwedge^2 V$. Furthermore

$$(4.4.1)\quad [c \wedge d, a \wedge b] = (b, d)a \wedge c + (a, d)c \wedge b - (a, c)d \wedge b - (b, c)a \wedge d.$$

$$(4.4.2)\quad [ab, c] = (b, c)a - (a, c)b, \quad [a \wedge b, c] = (b, c)a - (a, c)b.$$

Then

$$([a \wedge b, c], d) = (b, c)(a, d) - (a, c)(b, d)$$

is skew-symmetric as a function of c, d. This shows that L acts as $so(V)$. An element of L is in the kernel of the action if and only if it is a scalar. Of course we get F from the elements a^2.

Since $\bigwedge^2 V$ and $so(V)$ have the same dimension we must have the isomorphism. □

---

[39] The normalization 1/2 is important to eliminate a lot of factors of 2 and also reappears in the spin formalism.

4.5 Principal Involution

The Clifford algebra has an *involution* (cf. 3.4.3).

Consider the embedding of V in $Cl_Q(V)^o$. This embedding still satisfies the property $i(v)^2 = Q(v)$, and hence it extends to a homomorphism $* : Cl_Q(V) \to Cl_Q(V)^o$. In other words there is an *antihomomorphism* $* : Cl_Q(V) \to Cl_Q(V)$ such that $v^* = v, \forall v \in V$. The homomorphism $r \to (r^*)^*$ is the identity on V and therefore, by the universal property, it must be the identity of $Cl_Q(V)$. This proves the existence of an involution, called the *principal involution* on $Cl_Q(V)$, such that $v^* = v, \forall v \in V$.

Remark. The subalgebra $Cl_Q^+(V)$ is stable under the principal involution.

In fact, for the elements defined in 4.4, we have $f_i^* = (ue_i)^* = e_i^* u^* = e_i u = -f_i$.

This formula could of course be defined in general; one could have defined an involution setting $v^* := -v$. For $C(1) = \mathbb{C}$ this last involution is just complex conjugation. For $C(2) = \mathbb{H}$ it is the standard quaternion involution $q = a + b\underline{i} + c\underline{j} + d\underline{k} \mapsto \overline{q} := a - b\underline{i} - c\underline{j} - d\underline{k}$.

5 The Spin Group

5.1 Spin Groups

The last of the groups we want to introduce is the spin group. Consider again the Clifford algebra of a quadratic form Q on a vector space V over a field F. Write $\|v\|^2 = Q(v)$, $(v, w) = \frac{1}{2}(Q(v + w) - Q(v) - Q(w))$.

Definition 1. The Clifford group $\Gamma(V, Q)$ is the subgroup of invertible elements $x \in Cl_Q(V)^*$ with $xVx^{-1} = V$. The Clifford group $\Gamma^+(V, Q)$ is the intersection $\Gamma^+(V, Q) := \Gamma(V, Q) \cap Cl_Q^+(V)$.

Let $x \in \Gamma(V, Q)$ and $u \in V$. We have $(xux^{-1})^2 = xu^2x^{-1} = Q(u)$. Therefore the map $u \mapsto xux^{-1}$ is an orthogonal transformation of V. We have thus a homomorphism $\pi : \Gamma(V, Q) \to O(V)$. If $v, w \in V$ and $Q(v) \neq 0$, we have that v is invertible and

$$v^{-1} = \frac{v}{Q(v)}, \qquad vw + wv = 2(v, w), \qquad vwv^{-1} = \frac{2(v, w)}{Q(v)} v - w.$$

The map $w \to w - \frac{2(v,w)}{Q(v)} v$ is the orthogonal reflection r_v relative to the hyperplane orthogonal to v. Thus conjugation by v induces $-r_v$. If $\dim V$ is even, it is an improper orthogonal transformation.

We have that $v \in \Gamma(V, Q)$ and that a product $v_1 v_2 \dots v_{2k}$ of an even number of such v induces an even product of reflections, hence a special orthogonal transformation in V. By the Cartan–Dieudonné theorem, any special orthogonal transformation can be so induced. Similarly, a non-special (or improper) orthogonal transformation is the product of an odd number of reflections. We obtain:

Proposition 1. *The image of the homomorphism* $\pi : \Gamma(V, Q) \to O(V)$ *contains* $SO(V)$. *If* $\dim V$ *is even,* π *is surjective. If* $\dim V$ *is odd, the image of* π *is* $SO(V)$.

Proof. Only the last statement has not been checked. If in the odd case π were surjective, there is an element $x \in \Gamma(V, Q)$ with $xvx^{-1} = -v, \forall v \in V$. It follows that $xcx^{-1} = -c$ which is absurd, since c is in the center. □

The kernel of π is composed of elements which commute with the elements of V. Since these elements generate the Clifford algebra, we deduce that $\operatorname{Ker}\pi$ is the set Z^* of invertible elements of the center of $Cl_Q(V)$. We have thus the exact sequence:

(5.1.1)
$$1 \to Z^* \to \Gamma(V, Q) \xrightarrow{\pi} O(V).$$

If $n = \dim V$ is even the center is the field F, otherwise it is the set $\alpha + \beta c$, $\alpha, \beta \in F$ and $c := u_1 u_2 \dots u_{2k+1}$ for any given orthogonal basis $u_1, u_2, \dots, u_{2k+1}$ of U (cf. 4.2). Let us consider now $\Gamma^+(V, Q)$, its intersection with the center is clearly F^*. Since every element of $O(V)$ is a product of reflections, we deduce that every element γ of $\Gamma(V, Q)$ is a product $\alpha v_1 v_2 \cdots v_k$, of an element $\alpha \in Z^*$ and elements $v_i \in V$. If $\dim V$ is odd, by Proposition 1 we can assume that this last product is even ($k = 2h$). If $\gamma \in \Gamma^+(V, Q)$ we deduce $\alpha \in F^*$. If $\dim V$ is even we have $Z^* = F^*$ and so again, if $\gamma \in \Gamma^+(V, Q)$ we deduce that $k = 2h$ is even. The image of γ in $O(V)$ is contained in $SO(V)$, and we have an exact sequence:

(5.1.2)
$$1 \to F^* \to \Gamma^+(V, Q) \xrightarrow{\pi} SO(V) \to 1.$$

Let us compute $N(r) := rr^*$ when $r = v_1 v_2 \cdots v_j \in \Gamma(V, Q)$. We have $r^* = v_j v_{j-1} \dots v_1$ and by easy induction we obtain

(5.1.3)
$$rr^* = Q(v_1) Q(v_2) \dots Q(v_j) \in F^*.$$

Lemma. *The map* $r \to N(r) = rr^*$ *restricted to* $\Gamma^+(V, Q)$ *is a homomorphism to* F^*.

Proof. For two elements $r = v_1 v_2 \dots v_j, s = u_1 u_2 \dots u_h$ we have $N(r) = \prod_i Q(v_i)$, $N(s) = \prod_k Q(u_k)$ and $N(rs) = \prod Q(v_i) \prod_k Q(u_k)$. Every element of $\Gamma^+(V, Q)$ is of the form $\alpha v_1 v_2 \dots v_{2j}, \alpha \in F^*$ and the claim follows from 5.1.3. □

Proposition 2. *(a) The Lie algebra of* $\Gamma^+(V, Q)$ *is the Lie algebra* L *of 4.4.*
(b) Given $ab \in L$ *we have* $N(\exp(ab)) = \exp(2(a, b))$.

Proof. (a) First, taking $ab \in L$ we claim that $\exp(t\, ab) \in \Gamma^+(V, Q), \forall t$.

Clearly $\exp(t\, ab) \in Cl^+(V)$, on the other hand, by Proposition 4.4, we have $[ab, V] \subset V$. Hence $\exp(t\, ab) V \exp(-t\, ab) \subset V$, and $\exp(t\, ab)$ is by definition in the Clifford group. To prove that L is the entire Lie algebra L' of $\Gamma^+(V, Q)$, we remark that L and L' induce the same Lie algebra $so(V)$ by acting on V. In both cases the kernel of the Lie algebra action is the scalars.

(b) $\exp(ab)^* = \exp(ba) = \exp(-ab + 2(a, b)) = \exp(-ab) \exp(2(a, b))$. □

Example. For $V = \mathbb{R}^3$ with the negative definite quadratic form, we have seen that $C(3) = \mathbb{H} \oplus \mathbb{H}$ and that $C^+(3) = \mathbb{H}$. From the preceding analysis, we may explicitly identify $C^+(2) = \mathbb{H}$ by setting $\underline{i} := e_1e_2$, $\underline{j} := e_1e_3$ and $\underline{k} := e_1e_2e_1e_3 = e_2e_3$. Let us consider $c = e_1e_2e_3$ which generates the center of $C(3)$. The map $v \mapsto cv$ is a linear map which embeds V into the subspace $\mathbb{H}^0$ of $C^+(2) = \mathbb{H}$ of quaternions q with $q = -\overline{q}$.

We claim that $\Gamma^+(\mathbb{R}^3, Q) = \mathbb{H}^*$, the group of all invertible quaternions.

In fact we have the elements cv with $v \neq 0$ which are the nonzero quaternions in $\mathbb{H}^0$, and we leave to the reader the easy verification that these elements generate the group $\mathbb{H}^*$.

Definition 2. The spin group is the subgroup:

$$\text{Spin}(V) := \{r \in \Gamma^+(V, Q) \mid N(r) = 1\}. \tag{5.1.4}$$

We assume now that F is the field of real numbers and Q is definite, or F is the complex numbers.

Theorem. *(a) The spin group is a double cover of the special orthogonal group. We have an exact sequence* $1 \to \pm 1 \to \text{Spin}(V) \xrightarrow{\pi} SO(V) \to 1$.

(b) The Lie algebra of $\text{Spin}(V)$ *is* $\bigwedge^2 V = [L, L]$ *(notations of 4.4).*

Proof. (a) Consider $r := v_1v_2 \dots v_{2j}$, and compute $N(r) = \prod_{i=1}^{2j} Q(v_i)$. If we are in the case of complex numbers we can fix an $f \in \mathbb{C}$ so that $f^2N(r) = 1$ and $N(fr) = 1$. Similarly, if $F = \mathbb{R}$ and Q is definite, we have that $N(r) > 0$ and we can fix an $f \in \mathbb{R}$ so that $f^2N(r) = 1$. In both cases we see that $\text{Spin}(V) \xrightarrow{\pi} SO(V)$ is surjective. As for the kernel, if $f \in F^*$ we have $N(f) = f^2$. So $f \in \text{Spin}(V)$ if and only if $f = \pm 1$.

(b) Since the spin group is a double cover of $SO(V)$ it has the same Lie algebra, which is $[L, L] = so(V)$ by Proposition 4.4. □

When $V = F^n$ with form $-\sum_{i=1}^n x_i^2$ we denote $\text{Spin}(V) = \text{Spin}(n, F)$.

Example. For $V = \mathbb{R}^3$ with the negative definite quadratic form, we have seen that $\Gamma^+(\mathbb{R}^3, Q) = \mathbb{H}^*$, the group of all invertible quaternions, hence:

$$\text{Spin}(3, \mathbb{R}) = \{q \in \mathbb{H} \mid q\overline{q} = 1\}, \quad q = a + bi + cj + dk,$$
$$N(q) = q\overline{q} = a^2 + b^2 + c^2 + d^2.$$

Therefore, topologically $\text{Spin}(3, \mathbb{R}) = S^3$, the 3-dimensional sphere.

As groups, $\mathbb{H}^* = R^+ \times SU(2, \mathbb{C})$, $\text{Spin}(3, \mathbb{R}) = SU(2, \mathbb{C}) = Sp(1)$. This can be seen using the formalism of 5.2 for $\mathbb{H}$.

$$q = \alpha + j\beta, \quad N(q) = \alpha\overline{\alpha} + \beta\overline{\beta}, \quad (\alpha + j\beta)j = -\overline{\beta} + j\overline{\alpha}.$$

Formula 5.2.1 expresses q as the matrix

$$q := \begin{vmatrix} \alpha & -\overline{\beta} \\ \beta & \overline{\alpha} \end{vmatrix}, \quad N(q) = \det(q) = \alpha\overline{\alpha} + \beta\overline{\beta}.$$

The previous statements follow immediately from this matrix representation.

If we take $F = \mathbb{C}$, the spin group is an algebraic group (see next chapter) so we have the *algebraic form* $\mathrm{Spin}(n, \mathbb{C})$. If we take $F = \mathbb{R}$ and the negative definite quadratic form we have the *compact form* $\mathrm{Spin}(n, \mathbb{R})$ of the spin group.

The main point is that the extension $1 \to \pm 1 \to \mathrm{Spin}(n, \mathbb{R}) \xrightarrow{\pi} SO(n, \mathbb{R}) \to 1$ $(n > 2)$ is not split, or in better words, that $\mathrm{Spin}(n, \mathbb{R})$ is a connected and simply connected group.

Let us sketch the proof using some elementary algebraic topology. First, the map $\mathrm{Spin}(n, \mathbb{R}) \xrightarrow{\pi} SO(n, \mathbb{R})$ is a locally trivial fibration (as for all surjective homomorphisms of Lie groups, cf. Chapter 4, 3.7). Since $SO(n, \mathbb{R})$ is connected it is enough to exhibit a curve connecting $\pm 1 \in \mathrm{Spin}(n, \mathbb{R})$. Since $\mathrm{Spin}(n-1, \mathbb{R}) \subset \mathrm{Spin}(n, \mathbb{R})$ it is enough to look at $\mathrm{Spin}(2, \mathbb{R})$.

In this case the Clifford algebra $C(2)$ is the quaternion algebra. The space V is spanned by $\underline{i}, \underline{j}$ and $C^+(2) = \mathbb{C} = \mathbb{R} + \mathbb{R}\underline{k}$ $\quad (\underline{k} = \underline{i}\,\underline{j})$.

$$\mathrm{Spin}(2, \mathbb{R}) = U(1) = \{\alpha \in \mathbb{C} \mid |\alpha| = 1\} = \{\cos\phi + \sin\phi\, \underline{k} = e^{\phi \underline{k}}\}$$

and we have, from $\underline{k}\,\underline{i} = -\underline{i}\,\underline{k}, \underline{k}\,\underline{j} = -\underline{j}\,\underline{k}$

$$e^{\phi \underline{k}} \underline{i} e^{-\phi \underline{k}} = e^{2\phi \underline{k}} \underline{i}, \quad e^{\phi \underline{k}} \underline{j} e^{-\phi \underline{k}} = e^{2\phi \underline{k}} \underline{j},$$

from which the double covering and the connectedness is clear.

For the simple connectedness of $\mathrm{Spin}(n, \mathbb{R})$, we need some computations in homotopy theory. Basically we need to compute the fundamental group of the special orthogonal group and prove that

$$\pi_1(SO(n, \mathbb{R})) = \mathbb{Z}/(2), \quad \forall n \geq 3.$$

This can be seen by considering the transitive action of $SO(n, \mathbb{R})$ on the $n-1$-dimensional sphere S^{n-1} by rotations. The stabilizer of a given point is $SO(n-1, \mathbb{R})$, and thus $S^{n-1} = SO(n, \mathbb{R})/SO(n-1, \mathbb{R})$. We thus have that $SO(n, \mathbb{R})$ fibers over S^{n-1} with fiber $SO(n-1, \mathbb{R})$. We have therefore an exact sequence of homotopy groups:

$$\pi_2(S^{n-1}) \to \pi_1(SO(n-1, \mathbb{R})) \to \pi_1(SO(n, \mathbb{R})) \to \pi_1(S^{n-1}).$$

If $n > 3$, $\pi_2(S^{n-1}) = \pi_1(S^{n-1}) = 0$. Hence $\pi_1(SO(n-1, \mathbb{R})) = \pi_1(SO(n, \mathbb{R}))$ and we have $\pi_1(SO(n, \mathbb{R})) = \pi_1(SO(3, \mathbb{R}))$, $\forall n \geq 3$. For $n = 3$ we have seen that we have a double covering $1 \to \mathbb{Z}/(2) \to SU(2, \mathbb{C}) \to SO(3, \mathbb{R}) \to 1$. Since $SU(2, \mathbb{C}) = S^3$, it is simply connected. The exact sequence of the fibration gives the isomorphism

$$\pi_1(S^3) = 0 \to \pi_1(SO(3, \mathbb{R})) \to \mathbb{Z}/(2) = \pi_0(\mathbb{Z}/(2)) \to 0 = \pi_0(S^3).$$

For further details we refer to standard texts in algebraic topology (cf. [Sp], [Hat]).

6 Basic Constructions on Representations

6.1 Tensor Product of Representations

Having the formalism of tensor algebra we can go back to representation theory. Here representations are assumed to be finite dimensional.

The distinctive feature of the theory of representations of a group, versus the general theory of modules, lies in the fact that we have several ways to compose representations to construct new ones. This is a feature that groups share with Lie algebras and which, once it is axiomatized, leads to the idea of Hopf algebra (cf. Chapter 8, §7).

Suppose we are given two representations V, W of a group, or of a Lie algebra.

Theorem. *There are canonical actions on* V^*, $V \otimes W$ *and* $\hom(V, W)$, *such that the natural mapping* $V^* \otimes W \to \hom(V, W)$ *is equivariant.*

First we consider the case of a group. We already have (cf. Chapter 1, 2.4.2) general definitions for the actions of a group on $\hom(V, W)$: recall that we set $(gf)(v) := g(f(g^{-1}))$. This definition applies in particular when $W = F$ with the trivial action and so defines the action on the dual (the contragredient action).

The action on $V \otimes W$ is suggested by the existence of the tensor product of operators. We set $g(v \otimes w) := gv \otimes gw$. In other words, if we denote by $\varrho_1, \varrho_2, \varrho$ the representation maps of G into: $GL(V)$, $GL(W)$, $GL(V \otimes W)$ we have $\varrho(g) = \varrho_1(g) \otimes \varrho_2(g)$. Summarizing

$$\text{for } \hom(V, W), \quad (gf)(v) := g(f(g^{-1})),$$

$$\text{for } V^*, \quad \langle g\phi | v\rangle := \langle \phi | g^{-1}v\rangle,$$

$$\text{for } V \otimes W, \quad g(v \otimes w) := gv \otimes gw.$$

We can now verify:

Proposition 1. *The natural mapping* $i : W \otimes V^* \to \hom(V, W)$ *is equivariant.*

Proof. Given $g \in G$, $a = w \otimes \varphi \in W \otimes V^*$, we have $ga = gw \otimes g\varphi$ where $\langle g\varphi | v\rangle = \langle \varphi | g^{-1}v\rangle$. Thus, $(ga)(v) = \langle g\varphi | v\rangle gw = \langle \varphi | g^{-1}v\rangle gw = g(a(g^{-1}v))$, which is the required equivariance by definition of the action of G on $\hom(V, W)$. □

Let us now consider the action at the level of Lie algebras.

First, let us assume that G is a Lie group with Lie algebra $\mathrm{Lie}(G)$, and let us consider a one-parameter subgroup $\exp(tA)$ generated by an element $A \in \mathrm{Lie}(G)$.

Given a representation ϱ of G we have the induced representation $d\varrho$ of $\mathrm{Lie}(G)$ such that $\varrho(\exp(tA)) = \exp(td\varrho(A))$. In order to understand the mapping $d\varrho$ it is enough to expand $\varrho(\exp(tA))$ in a power series up to the first term.

We do this for the representations V^*, $V \otimes W$ and $\hom(V, W)$. We denote the actions on V, W simply as gv or Aw both for the group or Lie algebra.

Since $\langle \exp(tA)\varphi | v\rangle = \langle \varphi | \exp(-tA)v\rangle$ the Lie algebra action on V^* is given by

$$\langle A\varphi | v\rangle = \langle \varphi | -Av\rangle.$$

In matrix notation the contragredient action of a Lie algebra is given by minus the transpose of a given matrix.

Similarly we have the formulas:

$$(6.1.1) \qquad A(v \otimes w) = Av \otimes w + v \otimes Aw, \ (Af)(v) = A(f(v)) - f(A(v)).$$

for the action on tensor product or on homomorphisms. As a consequence we have:

Proposition 2. *If G is a connected Lie group*

$$\hom_G(V, W) = \{f \in \hom(V, W) | Af = 0, \ A \in \mathrm{Lie}(G)\}.$$

Proof. Same as Chapter 4, Remark 1.4 on fixed points. □

One final remark, which is part of the requirements when axiomatizing Hopf algebras, is that both a group G and a Lie algebra L have a trivial 1-dimensional representation, which behaves as unit element under tensor product.

On the various algebras $T(U)$, $S(U)$, $\bigwedge U$ the group $GL(U)$ acts as automorphisms. Hence the Lie algebra $gl(U)$ acts as derivations induced by the linear action on the space U of generators. For example,

$$\begin{aligned} A(u_1 \wedge u_2 \ldots \wedge u_k) &= Au_1 \wedge u_2 \ldots \wedge u_k + u_1 \wedge Au_2 \ldots \wedge u_k + \cdots \\ (6.1.2) \qquad & + u_1 \wedge u_2 \ldots \wedge Au_k. \end{aligned}$$

On the Clifford algebra we have an action as derivations only of the Lie algebra of the orthogonal group of the quadratic form, since only this group preserves the defining ideal. We have seen in 4.4 that these derivations are inner (Chapter 4, §1.1) and induced by the elements of $\bigwedge^2 V$.

6.2 One-dimensional Representations

We complete this part with some properties of 1-dimensional representations.

A 1-dimensional representation is just a homomorphism of G into the multiplicative group F^* of the base field. Such a homomorphism is also called a *multiplicative character.*

The tensor product of two 1-dimensional spaces is 1-dimensional and so is the dual.

Moreover a linear operator on a 1-dimensional space is just a scalar, the tensor product of two scalars is their product and the inverse transpose is the inverse. Thus:

Theorem. *The product of two multiplicative characters is a multiplicative character, and so is the inverse. The multiplicative characters of a group G form a group, called the character group of G (usually denoted by $\hat{G}$).*

Notice in particular that if V is one-dimensional, $V \otimes V^*$ is canonically isomorphic to the trivial representation by the map $v \otimes \phi \to \langle \phi | v \rangle$ (the trace). Sometimes for a 1-dimensional representation it is convenient to use the notation V^{-1} instead of V^*.

Let us show a typical application of this discussion:

Proposition. *If L and U are representations of a group G such that* $\dim L = 1$, *then U is irreducible if and only if $L \otimes U$ is irreducible.*

Proof. If $W \subset U$ is a proper submodule then also $L \otimes W \subset L \otimes U$ is a proper submodule, so we have the implication in one direction. But now

$$U = (L^{-1} \otimes L) \otimes U = L^{-1} \otimes (L \otimes U),$$

and we have also the reverse implication. □

7 Universal Enveloping Algebras

7.1 Universal Enveloping Algebras

There is one further construction we want to briefly discuss since it is the natural development of the theory of Capelli of Chapter 3. Given a Lie algebra L we consider in the tensor algebra $T(L)$ the ideal I_L generated by the quadratic relations $a \otimes b - b \otimes a - [a, b]$.

Definition. The associative algebra $U(L) := T(L)/I_L$ is called the *universal enveloping algebra* of the Lie algebra L.

The meaning of these relations is that the commutator of the elements $a, b \in L \subset T(L)$ performed in the associative algebra $U(L)$ must coincide with the commutator defined in L by the Lie algebra law. In other words we impose the minimal relations which imply that the morphism $L \to T(L)/I_L$ is a Lie homomorphism (where on the associative algebra $T(L)/I_L$ the Lie algebra structure is induced by the usual commutator).

As for other universal constructions this algebra satisfies the universal property of *mapping into associative algebras*. In fact we have that:

Proposition 1. *A Lie homomorphism $i : L \to A$ where A is an associative algebra with induced Lie structure, extends uniquely to a homomorphism of algebras $U(L) \to A$.*

Proof. By the universal property of tensor algebra the linear map i extends to a homomorphism $i : T(L) \to A$. Since i is a Lie homomorphism we have

$$i(a \otimes b - b \otimes a - [a, b]) = i(a)i(b) - i(b)i(a) - i([a, b]) = 0$$

so i factors through I_L to the required homomorphism. □

The first important result on universal enveloping algebras is the Poincaré–Birkhoff–Witt theorem, which states that:

PBW Theorem. *(1) If $u_1, u_2, \dots, u_k \dots$ form a linear basis of L, the ordered monomials $u_1^{h_1} u_2^{h_2} \dots u_k^{h_k} \dots$ give a linear basis of $U(L)$.*

(2) In characteristic 0 we have a direct sum decomposition $T(L) = I_L \oplus \sum_{i=0}^{\infty} S_i(L)$ (where $S_i(L)$ denotes the space of symmetric tensors of degree i).

Proof. It is almost immediate by induction that the monomials $u_1^{h_1} u_2^{h_2} \dots u_k^{h_k} \dots$ are linear generators. In fact if we have in a product a pair $u_i u_j$ in the *wrong order*, $j < i$ we replace it be $u_j u_i - [u_i, u_j]$. Expanding $[u_i, u_j]$ in the given basis we obtain lower degree monomials and proceed by induction. The independence requires a nontrivial argument.

Consider thus the tensors $M = u_{i_1} \otimes \cdots \otimes u_{i_k}$ which as the indices i_t and k vary, give a basis of the tensor algebra.

We look at the sequence of indices $i_1, i_2, \dots, i_k$, for a tensor M, and count the number of *descents*, i.e., the positions j for which $i_j > i_{j+1}$: we call this number the *index*, $i(M)$ of M. When $i(M) = 0$, i.e., when $i_1 \leq i_2 \leq \cdots \leq i_k$ we say that M is *standard*. Let us define T_n to be the span of all tensors of degree $\leq n$ and by $T_n^s \subset T_n$ the span of the standard tensors. We need a basic lemma.

Lemma. *For each n, there is a unique linear map $\sigma : T_n \to T_n^s$ such that:*

(1) σ is the identity on T_n^s.

(2) Given a tensor $A \otimes a \otimes b \otimes B$, $a, b \in L$, $A, B \in T(L)$, we have

$$\sigma(A \otimes a \otimes b \otimes B) = \sigma(A \otimes b \otimes a \otimes B) + \sigma(A \otimes [a, b] \otimes B).$$

Proof. We define σ on the tensors $M = u_{i_1} \otimes \cdots \otimes u_{i_k}$ by induction on the degree k and on the index $i(M)$. When $i(M) = 0$ by definition we must set $\sigma(M) = M$. When $i(M) > 0$ we have an expression $A \otimes u_i \otimes u_j \otimes B$ with $i > j$ and hence $i(A \otimes u_j \otimes u_i \otimes B) = i(M) - 1$. Thus we may set recursively:

$$\sigma(A \otimes u_i \otimes u_j \otimes B) = \sigma(A \otimes u_j \otimes u_i \otimes B) + \sigma(A \otimes [u_i, u_j] \otimes B).$$

If $i(M) = 1$ this definition is well posed; otherwise, when we have at least two descents we have to prove that the definition of $\sigma(M)$ is independent of the descent we choose. We have two cases: (1) The descents are in $A \otimes b \otimes a \otimes B \otimes d \otimes c \otimes C$. (2) We have consecutive descents $A \otimes c \otimes b \otimes a \otimes B$.

In the first case we have by induction, starting from the descent in $b \otimes a$:

$$\begin{aligned}
&\sigma(A \otimes a \otimes b \otimes B \otimes d \otimes c \otimes C) + \sigma(A \otimes [b, a] \otimes B \otimes d \otimes c \otimes C) \\
&\quad = \sigma(A \otimes a \otimes b \otimes B \otimes c \otimes d \otimes C) + \sigma(A \otimes a \otimes b \otimes B \otimes [d, c] \otimes C) \\
&\qquad + \sigma(A \otimes [b, a] \otimes B \otimes c \otimes d \otimes C) + \sigma(A \otimes [b, a] \otimes B \otimes [d, c] \otimes C).
\end{aligned}$$

Clearly when we start from the other descent we obtain the same result.

For the other case, write for convenience $\tau(X) := \sigma(A \otimes X \otimes B)$. We need to compare:

$$\tau(b \otimes c \otimes a + [c, b] \otimes a), \quad \tau(c \otimes a \otimes b + c \otimes [b, a])$$

We iterate the formulas by induction, the two terms are:

$$1: \ \tau(b \otimes a \otimes c + b \otimes [c, a] + [c, b] \otimes a),$$
$$2: \ \tau(a \otimes c \otimes b + [c, a] \otimes b + [b, a] \otimes c + [c, [b, a]])$$

Applying again the same rules (notice that either the index or the degree is diminished so we can apply induction) we have:

$$1: \ \tau(a \otimes b \otimes c + [b, a] \otimes c + [c, a] \otimes b + [b, [c, a]] + [c, b] \otimes a)$$

$$2: \ \tau(a \otimes b \otimes c + [b, a] \otimes c + [c, a] \otimes b + [c, b] \otimes a + [a, [c, b]] + [c, [b, a]]).$$

From the Jacobi identity $[b, [c, a]] = [a, [c, b]] + [c, [b, a]]$ so the claim follows. □

We can now conclude the proof of the PBW theorem.

The linear map σ by definition vanishes on the ideal I_L defining $U(L)$, thus it defines a linear map $U(L) \to T^s$ which, by the previous remarks, maps the images of the standard tensors which span $U(L)$ into themselves, thus it is a linear isomorphism.

The second part follows easily since a basis of symmetric tensors is given by symmetrization of standard tensors. The image under σ of the symmetrization of a standard tensor M is M. □

There is a simple but important corollary. Let us filter $U(L)$ by setting $U(L)_i$ to be the span of all monomials in elements of L of degree $\leq i$. Then we have:

Proposition 2. *(i) The graded algebra $\oplus U(L)_i / U(L)_{i-1}$ is isomorphic to the symmetric algebra $S(L)$.*

(ii) If the characteristic is 0, for every i we have $U_{i+1}(L) = \overline{S}_i(L) \oplus U_i(L)$, where $\overline{S}_i(L)$ is the image in $U_{i+1}(L)$ of the symmetric tensors.

The importance of the second statement is this. The Lie algebra L acts on $T(L)$ by derivations and, over $\mathbb{C}$, on its associated group G by automorphisms. Both I_L and $S_i(L)$ are stable subspaces. Thus the actions factor to actions on $U(L)$. The subspaces $U_i(L)$ are subrepresentations and in characteristic 0, we have that $U_i(L)$ has the invariant complement $\overline{S}_i(L)$ in $U_{i+1}(L)$.

Exercise. If $L \subset M$ are Lie algebras, the PBW theorem for M implies the same theorem for L; we also can prove it for linear Lie algebras from Capelli's theory.[40]

[40] The PBW theorem holds for any Lie algebra, not necessarily finite dimensional. For finite-dimensional algebras there is a deep theorem stating that these algebras are indeed linear (Ado's theorem, Chapter 10). Usually this theorem is proved *after* proving the PBW theorem.

7.2 Theorem of Capelli

We want to use the previous analysis to study the center of $U(L)$, in particular to give the full details of the Theorem of Capelli, sketched in Chapter 3, §5.3 on the center of the algebra of polarizations.

From Proposition 2 of the preceding section, if G is a group of automorphisms of the Lie algebra L, G acts as automorphisms of the tensor algebra $T(L)$ and preserves the ideal I_L. Thus G extends to a group of automorphisms of $U(L)$. Moreover it clearly preserves the spaces $\overline{S}_i(L)$. In particular we can consider as in Chapter 4, §4.1 the *adjoint group* G^0 generated by the one-parameter groups $e^{t\,\mathrm{ad}(a)}$. Notice that $\mathrm{ad}(a)$ extends to the *inner derivation*: $r \to ar - ra$ of $U(L)$, preserving all the terms $U(L)_i$ of the filtration. We have:

Proposition. *The center of $U(L)$ coincides with the invariants under G^0.*

Proof. By definition an element of $U(L)$ is fixed under G^0 if and only if it is fixed by all the one-parameter subgroups $e^{t\,\mathrm{ad}(a)}$. An element is fixed by a one-parameter subgroup if and only if it is in the kernel of the generator, in our case $\mathrm{ad}(a)$, i.e., if it commutes with a. Since $U(L)$ is generated by L it is clear that its center is the set of elements which commute with L. □

Let us apply the theory to $gl(n, \mathbb{C})$, the Lie algebra of $n \times n$ matrices. In this case $gl(n, \mathbb{C})$ is also an associative algebra. Its group G of automorphisms is induced by conjugation by invertible matrices. Given a matrix A, we have that the group of linear transformations $B \to e^{tA}Be^{-tA}$ has as infinitesimal generator $B \mapsto AB - BA = \mathrm{ad}(A)(B)$.

The Lie algebra $gl(n, \mathbb{C})$ is isomorphic to the Lie algebra of polarizations of Chapter 3. The elementary matrix e_{ij} with 1 in the i, j position and 0 otherwise corresponds to the operator Δ_{ij}. The universal enveloping algebra of $gl(n, \mathbb{C})$ is isomorphic to the algebra $\mathcal{U}_n$ generated by polarizations. Formulas 5.3.3 and 5.3.6 of Chapter 3, give elements K_i in the center of $\mathcal{U}_n$, K_i is a polynomial of degree i in the generators.

Let us analyze the development of the determinant 5.3.3 and, in particular, the terms which contribute to $K_i\rho^{m-i}$. In such a term the contribution of a factor $\Delta_{ii} + m - i$ can be split into the part involving Δ_{ii} and the one involving $m - i$. This last one produces terms of strictly lower degree. Therefore in the associated grading the images of the K_i can be computed by dropping the constants on the diagonal and thus are, up to sign, the coefficients σ_i of the characteristic polynomial of a matrix with entries the classes x_{ij} of the $e_{ij} = \Delta_{ij}$. By Chapter 2, Theorem 5.1 we have that these coefficients generate the invariants under conjugation. We then get:

Theorem. *The elements K_i generate the center of U_n which is the polynomial ring in these generators.*

Proof. Let f be in the center, say $f \in (\mathcal{U}_n)_i$. Its symbol is an invariant in the graded algebra. Hence it is a polynomial in the coefficients σ_i. We have thus a polynomial g in the K_i which lies also $(\mathcal{U}_n)_i$ and has the same symbol. Therefore $f - g \in (\mathcal{U}_n)_{i-1}$, and we can finish by induction. □

This theorem has of course a natural generalization to semisimple Lie algebras. One has to replace the argument of Capelli with more general arguments of Chevalley and Harish-Chandra but the final result is quite similar. The center of the universal enveloping algebra of a semisimple Lie algebra is a ring of polynomials in generators which correspond to symmetric functions under the appropriate group, the Weyl group.

7.3 Free Lie Algebras

As usual, given a Lie algebra L and a set $X \subset L$ we say:

Definition. L is free over X if, given any Lie algebra M and any map $f : X \to M$, f extends to a unique homomorphism $f : L \to M$ of Lie algebras.

The PBW Theorem immediately tells us how to construct free Lie algebras. Let $F\langle X\rangle$ be the free associative noncommutative polynomial algebra over X (the tensor algebra on a vector space with basis X). Let L be the Lie subalgebra of $F\langle X\rangle$ generated by X.

Proposition. *L is free over X.*

Proof. Let M and $f : X \to M$ be given. Consider the universal enveloping algebra U_M of M. By PBW we have $M \subset U_M$. Since $F\langle X\rangle$ is the free associative algebra, f extends to a homomorphism $\tilde{f} : F\langle X\rangle \to U_M$. Since L is generated by the elements X as Lie algebra, $\tilde{f}$ restricted to L maps L to M extending the map f on the generators X. The extended map is also uniquely determined since L by construction is generated by X. □

The free Lie algebra is a very interesting object. It has been extensively studied ([Reu]).

6

Semisimple Algebras

1 Semisimple Algebras

1.1 Semisimple Representations

One of the main themes of our theory will be related to completely reducible representations. It is thus important to establish these notions in full detail and generality.

Definition 1. Let S be a set of operators acting on a vector space U.

(i) We say that the vector space U is *irreducible* or *simple* under the given set S of operators if the only subspaces of U which are stable under S are 0 and U.

(ii) We say that the vector space U is *completely reducible* or *semisimple* under the given set S of operators if U decomposes as a direct sum of S-stable irreducible subspaces.

(iii) We say that the vector space U is *indecomposable* under the given set S of operators if the space U cannot be decomposed in the direct sum of two nontrivial S-stable subspaces.

A space is irreducible if and only if it is completely reducible and indecomposable.

The previous notions are essentially notions of the theory of modules. In fact, let S be a set of linear operators acting on a vector space U. From S, taking linear combinations and products, we can construct an algebra $\mathcal{E}(S)$[41] of linear operators on U. U is then an $\mathcal{E}(S)$-module. It is clear that a subspace $W \subset U$ is stable under S if and only if it is stable under the algebra $\mathcal{E}(S)$, so the notions introduced for S are equivalent to the same notions for $\mathcal{E}(S)$.

A typical example of completely reducible sets of operators is the following. Let $U = \mathbb{C}^n$ and let S be a set of matrices. For a matrix A let $A^* = \overline{A}^t$ be its adjoint (Chapter 5, 3.8). For a set S of matrices we denote by S^* the set of elements A^*, $A \in S$.

Lemma 1. *If a subspace M of $\mathbb{C}^n$ is stable under A, then $M^\perp$ (the orthogonal under the Hermitian product) is stable under A^*.*

[41] This is sometimes called the *envelope* of S.

Proof. If $m \in M, u \in M^\perp$, we have $(m, A^*u) = (Am, u) = 0$ since M is A stable. Thus $A^*u \in M^\perp$. □

Proposition 1. *If $S = S^*$, then $\mathbb{C}^n$ is the orthogonal sum of irreducible submodules, in particular it is semisimple.*

Proof. Take an S-stable subspace M of $\mathbb{C}^n$ of minimal dimension. It is then necessarily irreducible. Consider its orthogonal complement $M^\perp$. By adjunction and the previous lemma we get that $M^\perp$ is S stable and $\mathbb{C}^n = M \oplus M^\perp$.

We then proceed in the same way on $M^\perp$. □

A special case is when S is a group of unitary operators. More generally, we say that S is *unitarizable* if there is a Hermitian product for which the operators of S are unitary. If we consider a matrix mapping the standard basis of $\mathbb{C}^n$ to a basis orthonormal for some given Hermitian product we see

Lemma 2. *A set of matrices is unitarizable if and only if it is conjugate to a set of unitary matrices.*

These ideas have an important consequence.

Theorem 1. *A finite group G of linear operators on a finite-dimensional complex space U is unitarizable and hence the module is semisimple.*

Proof. We fix an arbitrary positive Hermitian product (u, v) on U. Define a new Hermitian product as

$$\langle u, v \rangle := \frac{1}{|G|} \sum_{g \in G} (gu, gv). \tag{1.1.1}$$

Then $\langle hu, hv \rangle = \frac{1}{|G|} \sum_{g \in G} (ghu, ghv) = \frac{1}{|G|} \sum_{g \in G} (gu, gv) = \langle u, v \rangle$ and G is unitary for this new product. If G was already unitary the new product coincides with the initial one. □

The previous theorem has a far-reaching generalization, by replacing the average given by the sum with an integral, as we will see in Chapter 8 where we prove among other things:

Theorem 3. *A compact group G of linear operators on a finite-dimensional complex space U is unitarizable and hence the module is semisimple.*

1.2 Self-Adjoint Groups

For noncompact groups there is an important class that we have already introduced for which similar results are valid. These are the self-adjoint subgroups of $GL(n, \mathbb{C})$, i.e., the subgroups H such that $A \in H$ implies $A^* \in H$.

For a self-adjoint group on a given (finite-dimensional) Hilbert space U the orthogonal of every invariant subspace is also invariant; thus any subspace or quotient module of U is completely reducible.

Take a self-adjoint group G and consider its induced action on the tensor algebra. The tensor powers of $U = \mathbb{C}^n$ have an induced canonical Hermitian form for which

$$\langle u_1 \otimes u_2 \otimes \cdots \otimes u_n | v_1 \otimes v_2 \otimes \cdots \otimes v_n \rangle = \langle u_1 | v_1 \rangle \langle u_2 | v_2 \rangle \cdots \langle u_n | v_n \rangle.$$

It is clear that this is a positive Hermitian form for which the tensor power of an orthonormal basis is also an orthonormal basis.

The map $g \to g^{\otimes n}$ is compatible with adjunction, i.e., $(g^*)^{\otimes n} = (g^{\otimes n})^*$:

$$(g^{\otimes m}(v_1 \otimes v_2 \otimes \cdots \otimes v_m) | w_1 \otimes W_2 \otimes \cdots \otimes w_m) := \prod_{i=1}^{m} (g v_i | w_i)$$

$$= \prod_{i=1}^{m} (v_i | g^* w_i) = (v_1 \otimes v_2 \otimes \cdots \otimes v_m | (g^*)^{\otimes m} (w_1 \otimes w_2 \otimes \cdots \otimes w_m)).$$

Thus:

Proposition. *If G is self-adjoint, the action of G on $T(U)$ is self-adjoint, hence all tensor powers of U are completely reducible under G.*

Corollary. *The action of G on the polynomial ring $\mathcal{P}(U)$ is completely reducible.*

Proof. The action of G on U^* is self-adjoint, so it is also self-adjoint on $T(U^*)$, and $\mathcal{P}(U)$ is a graded quotient of $T(U^*)$. □

1.3 Centralizers

It is usually more convenient to use the language of modules since the irreducibility or complete reducibility of a space U under a set S of operators is clearly equivalent to the same property under the subalgebra of operators generated by S.

Let us recall that in Chaper 1, §3.2, given a group G, one can form its *group algebra* $F[G]$. Every linear representation of G extends by linearity to an $F[G]$-module, and conversely. A map between $F[G]$ modules is a (module) homomorphism if and only if it is G-equivariant. Thus from the point of view of representation theory it is equivalent to studying the category of G representations or that of $F[G]$ modules.

We thus consider a ring R and its modules, using the same definitions for reducible, irreducible modules. We define $R^\vee$ to be the set of (isomorphism classes) of irreducible modules of R. We may call it the *spectrum* of R.

Given an irreducible module N we will say that it is of *type* α if $\alpha \in R^\vee$ is its isomorphism class.

Given a set S of operators on U we set $S' := \{A \in \mathrm{End}(U) | As = sA, \forall s \in S\}$. S' is called the *centralizer* of S. Equivalently S' should be thought of as the set of all S-linear endomorphisms. One immediately verifies:

Proposition 1.

(i) S' is an algebra.
(ii) $S \subset S''$.
(iii) $S' = S'''$.

The centralizer of the operators induced by R in a module M is also usually indicated by $\hom_R(M, M)$ or $\mathrm{End}_R(M)$ and called the *endomorphism ring*.

Any ring R can be considered as a module on itself by left multiplication (and as a module on the opposite of R by right multiplication).

Definition 1. R considered as module over itself is called *the regular representation*.

Of course, for the regular representation, a submodule is the same as a left ideal. An irreducible submodule is also referred to as a *minimal left ideal*.

A trivial but useful fact on the regular representation is:

Proposition 2. *The ring of endomorphisms of the regular representation is the opposite of R acting by right multiplications.*

Proof. Letting $f \in \mathrm{End}_R(R)$ we have $f(a) = f(a1) = af(1)$ by linearity, thus f is the right multiplication by $f(1)$.

Given two homomorphisms f, g we have $fg(1) = f(g(1)) = g(1)f(1)$, and so the mapping $f \to f(1)$ is an isomorphism between $\mathrm{End}_R(R)$ and R^o. $\square$

Exercise. Another action of some interest is the action of $R \otimes R^o$ on R given by $a \otimes b(c) := acb$; in this case the submodules are the two-sided ideals and the centralizer is easily seen to be the *center* of R.

One can generalize the previous considerations as follows. Let R be a ring.

Definition 2. A cyclic module is a module generated by a single element.

A cyclic module should be thought of as the linear analogue of a single orbit. The structure of cyclic modules is quite simple. If M is generated by an element m, we have the map $\varphi : R \to M$ given by $\varphi(r) = rm$ (analogue of the orbit map).

By hypothesis φ is surjective, its kernel is a left ideal J, and so M is identified with R/J. Thus a module is cyclic if and only if it is a quotient of the regular representation.

Example. Consider $M_n(F)$, the full ring of $n \times n$ matrices over a field F. As a module we take F^n and in it the basis element e_1.

Its annihilator is the left ideal I_1 of matrices with the first column 0. In this case though we have a more precise picture.

Let J_1 denote the left ideal of matrices having 0 in all columns except for the first. Then $M_n(F) = J_1 \oplus I_1$ and the map $a \to ae_1$ restricted to J_1 is an isomorphism.

In fact we can define in the same way J_i (the matrices with 0 outside the ith column).

Proposition 3. $M_n(F) = \bigoplus_{i=1}^n J_i$ *is a direct sum of the algebra* $M_n(F)$ *into irreducible left ideals isomorphic, as modules, to the representation* F^n.

Remark. This proof, with small variations, applies to a division ring D in place of F.

Lemma. *The module* D^n *is irreducible. We will call it the standard module of* $M_n(D)$.

Proof. Let us consider a column vector u with its ith coordinate u_i nonzero. Acting on u with a diagonal matrix which has u_i^{-1} in the ith position, we transform u into a vector with i^{th} coordinate 1. Acting with elementary matrices we can make all the other coordinates 0. Finally, acting with a permutation matrix we can bring 1 into the first position. This shows that any submodule contains the vector of coordinates $(1, 0, 0, \dots, 0)$. This vector, in turn, generates the entire space again acting on it by elementary matrices. $\square$

Theorem. *The regular representation of* $M_n(D)$ *is the direct sum of* n *copies of the standard module.*

Proof. We decompose $M_n(D)$ as direct sum of its columns. $\square$

Remark. In order to understand $M_n(D)$ as module endomorphisms we have to take D^n as a *right* vector space over D or as a left vector space over D^o.

As done for groups in Chapter 1, §3.1, given two cyclic modules R/J, R/I we can compute $\hom_R(R/J, R/I)$ as follows.

Letting $f : R/J \to R/I$ be a homomorphism and $\overline{1} \in R/J$ is the class of 1, set $f(\overline{1}) = \overline{x}$, $x \in R$ so that $f : r\overline{1} \to r\overline{x}$. We must have then $J\overline{x} = f(J\overline{1}) = 0$, hence $Jx \subset I$. Conversely, if $Jx \subset I$ the map $f : r\overline{1} \to r\overline{x}$ is a well-defined homomorphism.

Thus if we define the set $(I : J) := \{x \in R \mid Jx \subset I\}$, we have

$$I \subset (I : J), \quad \hom_R(R/J, R/I) = (I : J)/I.$$

In particular for $J = I$ we have the *idealizer,* $\mathcal{I}(I)$ of I, $\mathcal{I}(I) := \{x \in R \mid Ix \subset I\}$.

The idealizer is the maximal subring of R in which I is a two-sided ideal, and $\mathcal{I}(I)/I$ is the ring $\hom_R(R/I, R/I) = \operatorname{End}_R(R/I)$.

1.4 Idempotents

It is convenient in the structure theory of algebras to introduce the simple idea of idempotent elements.

Definition. An idempotent in an algebra R, is an element e, such that $e^2 = e$. Two idempotents e, f are *orthogonal* if $ef = fe = 0$; in this case $e + f$ is also an idempotent.

Exercise. In a ring R consider an idempotent e and set $f := 1 - e$. We have the decomposition

$$R = eRe \oplus eRf \oplus fRe \oplus fRf$$

which presents R as matrices:

$$R = \begin{vmatrix} eRe & eRf \\ fRe & fRf \end{vmatrix}.$$

Prove that $\mathrm{End}_R(Re) = eRe$.

1.5 Semisimple Algebras

The example of matrices suggests the following:

Definition. A ring R is *semisimple* if it is semisimple as a left module on itself.

This definition is a priori not symmetric although it will be proved to be so from the structure theorem of semisimple rings.

Remark. Let us decompose a semisimple ring R as direct sum of irreducible left ideals. Since 1 generates R and 1 is in a finite sum we see:

Proposition 1. *A semisimple ring is a direct sum of finitely many minimal left ideals.*

Corollary. *If D is a division ring then $M_n(D)$ is semisimple.*

We wish to collect some examples of semisimple rings.

First, from the results in 1.1 and 1.2 we deduce:

Maschke's Theorem. *The group algebra $\mathbb{C}[G]$ of a finite group is semisimple.*

Remark. In fact it is not difficult to generalize to an arbitrary field. We have (cf. [JBA]):

The group algebra $F[G]$ of a finite group over a field F is semisimple if and only if the characteristic of F does not divide the order of G.

Next we have the obvious fact:

Proposition 2. *The direct sum of two semisimple rings is semisimple.*

In fact we let the following simple exercise to the reader.

Exercise. Decomposing a ring A in a direct sum of two rings $A = A_1 \oplus A_2$ is equivalent to giving an element $e \in A$ such that

$$e^2 = e, \; ea = ae, \; \forall a \in A,$$

$$A_1 = Ae, \; A_2 = A(1 - e) \qquad e \text{ is called a } \textit{central idempotent}.$$

Having a central idempotent e, every A module M decomposes canonically as the direct sum $eM \oplus (1 - e)M$. Where eM is an A_1 module, $(1 - e)M$ an A_2 module. Thus the module theory of $A = A_1 \oplus A_2$ reduces to the ones of A_1, A_2.

From the previous corollary, Proposition 2, and these remarks we deduce:

Theorem. *A ring $A := \bigoplus_i M_{n_i}(D_i)$, with the D_i division rings, is semisimple.*

1.6 Matrices over Division Rings

Our next task will be to show that also the converse to Theorem 1.5 is true, i.e., that every semisimple ring is a finite direct sum of rings of type $M_m(D)$, D a division ring.

For the moment we collect one further remark. Let us recall that:

Definition. A ring R is called *simple* if it does not possess any nontrivial two-sided ideals, or equivalently, if R is irreducible as a module over $R \otimes R^o$ under the left and right action $(a \otimes b)r := arb$.

This definition is slightly confusing since a simple ring is by no means semisimple, unless it satisfies further properties (the d.c.c. on left ideals [JBA]).

A classical example of an infinite-dimensional simple algebra is the algebra of differential operators $F\langle x_i, \frac{\partial}{\partial x_i}\rangle$ (F a field of characteristic 0) (cf. [Cou]).

We have:

Proposition. *If D is a division ring, $M_m(D)$ is simple.*

Proof. Let I be a nontrivial two-sided ideal, $a \in I$ a nonzero element. We write a as a linear combination of elementary matrices $a = \sum a_{ij}e_{ij}$; thus $e_{ii}ae_{jj} = a_{ij}e_{ij}$ and at least one of these elements must be nonzero. Multiplying it by a scalar matrix we can obtain an element e_{ij} in the ideal I. Then we have $e_{hk} = e_{hi}e_{ij}e_{jk}$ and we see that the ideal coincides with the full ring of matrices. □

Exercise. The same argument shows more generally that for any ring A the ideals of the ring $M_m(A)$ are all of the form $M_m(I)$ for I an ideal of A.

1.7 Schur's Lemma

We start the general theory with the following basic fact:

Theorem (Schur's lemma). *The centralizer $\Delta := \mathrm{End}_R(M)$ of an irreducible module M is a division ring.*

Proof. Let $a : M \to M$ be a nonzero R-linear endomorphism. Its kernel and image are submodules of M. Since M is irreducible and $a \neq 0$ we must have $\mathrm{Ker}(a) = 0$, $\mathrm{Im}(a) = M$, hence a is an isomorphism and so it is invertible. This means that every nonzero element in Δ is invertible. This is the definition of a division ring. □

This lemma has several variations. The same proof shows that:

Corollary. *If $a : M \to N$ is a homomorphism between two irreducible modules, then either $a = 0$ or a is an isomorphism.*

1.8 Endomorphisms

A particularly important case is when M is a finite-dimensional vector space over $\mathbb{C}$. In this case since the only division ring over $\mathbb{C}$ is $\mathbb{C}$ itself, we have that:

Theorem. *Given an irreducible set S of operators on a finite-dimensional space over $\mathbb{C}$, then its centralizer S' is formed by the scalars $\mathbb{C}$.*

Proof. Rather than applying the structure theorem of finite-dimensional division algebras one can argue that, given an element $x \in S'$ and an eigenvalue α of x, the space of eigenvectors of x for this eigenvalue is stable under S and so, by irreducibility, it is the whole space. Hence $x = \alpha$. □

Remarks. 1. If the base field is the field of real numbers, we have (according to the theorem of Frobenius (cf. [Her])) three possibilities for Δ: $\mathbb{R}$, $\mathbb{C}$, or $\mathbb{H}$, the algebra of quaternions.

2. It is not necessary to assume that M is finite dimensional: it is enough to assume that it is of countable dimension.

Sketch of proof of the remarks. In fact M is also a vector space over Δ and so Δ, being isomorphic to an $\mathbb{R}$-subspace of M, is also countably dimensional over $\mathbb{R}$.

This implies that every element of Δ is algebraic over $\mathbb{R}$. Otherwise Δ would contain a field isomorphic to the rational function field $\mathbb{R}(t)$ which is impossible, since this field contains the uncountably many linearly independent elements $1/(t-r)$, $r \in \mathbb{R}$.

Now one can prove that a division algebra[42] over $\mathbb{R}$ in which every element is algebraic is necessarily finite dimensional,[43] and thus the theorem of Frobenius applies. □

Our next task is to prove the converse of Theorem 1.5, that is, to prove that every semisimple algebra A is a finite direct sum of matrix algebras over division algebras (Theorem 1.9). In order to do this we start from a general remark about matrices.

Let $M = M_1 \oplus M_2 \oplus M_3 \oplus \cdots \oplus M_k$ be an R-module decomposed in a direct sum.

For each i, j, consider $A(j,i) := \hom_R(M_i, M_j)$. For three indices we have the composition map $A(k,j) \times A(j,i) \to A(k,i)$.

The groups $A(j,i)$ together with the composition maps allow us to recover the full endomorphism algebra of M as *block matrices*:

$$A = (a_{ji}),\ a_{ji} \in A(j,i).$$

(One can give a formal abstract construction starting from the associativity properties).

---

[42] When we want to stress the fact that a division ring Δ contains a field F in the center, we say that Δ is a division algebra over F.

[43] This depends on the fact that every element of Δ algebraic over $\mathbb{R}$ satisfies a quadratic polynomial.

In more concrete form let $e_i \in \mathrm{End}(M)$ be the projection on the summand M_i with kernel $\bigoplus_{j\neq i} M_j$. The elements e_i are *a complete set of orthogonal idempotents in* $\mathrm{End}(M)$, i.e., they satisfy the properties

$$e_i^2 = e_i, \; e_i e_j = e_j e_i = 0, \; i \neq j, \text{ and } \sum_{i=1}^{k} e_i = 1.$$

When we have in a ring S such a set of idempotents we decompose S as

$$S = \Big(\sum_{i=1}^{k} e_i\Big) S \Big(\sum_{i=1}^{k} e_i\Big) = \bigoplus_{i,j} e_i S e_j.$$

This sum is direct by the orthogonality of the idempotents.

We have $e_i S e_j e_j S e_k \subset e_i S e_k$, $e_i S e_j e_h S e_k = 0$, when $j \neq h$. In our case $S = \mathrm{End}_R(M)$ and $e_i S e_j = \hom_R(M_j, M_i)$.

In particular assume that the M_i are all isomorphic to a module N and let $A := \mathrm{End}_R(N)$. Then

(1.8.1) $$\mathrm{End}_R(N^{\oplus k}) = M_k(A).$$

Assume now that we have two modules N, P such that $\hom_R(N, P) = \hom_R(P, N) = 0$. Let $A := \mathrm{End}_R(N)$, $B := \mathrm{End}_R(P)$; then

$$\mathrm{End}_R(N^{\oplus k} \oplus P^{\oplus h}) = M_k(A) \oplus M_h(B).$$

Clearly we have a similar statement for several modules.

We can add together all these remarks in the case in which a module M is a finite direct sum of irreducibles.

Assume $N_1, N_2, \ldots, N_k$ are the distinct irreducible which appear with multiplicities $h_1, h_2, \ldots, h_k$ in M. Let $D_i = \mathrm{End}_R(N_i)$ (a division ring). Then

(1.8.2) $$\mathrm{End}_R\Big(\bigoplus_{i=1}^{k} N_i^{h_i}\Big) = \bigoplus_{i=1}^{k} M_{h_i}(D_i).$$

1.9 Structure Theorem

We are now ready to characterize semisimple rings. If R is semisimple we have that $R = \bigoplus_{i=1}^{k} N_i^{m_i}$ (as in the previous section) as a left R module; then

$$R^o = \mathrm{End}_R(R) = \mathrm{End}_R\big(\bigoplus_{i\in I} N_i^{m_i}\big) = \bigoplus_{i\in I} M_{m_i}(\Delta_i).$$

We deduce that $R = R^{oo} = \bigoplus_{i\in I} M_{m_i}(\Delta_i)^o$.

The opposite of the matrix ring over a ring A is the matrices over the opposite ring (use transposition) and so we deduce finally:

Theorem. *A semisimple ring is isomorphic to the direct sum of matrix rings R_i over division rings.*

Some comments are in order.

1. We have seen that the various blocks of this sum are simple rings. They are thus distinct irreducible representations of the ring $R \otimes R^o$ acting by the left and right action.

   We deduce that the matrix blocks are minimal two-sided ideals. From the theory of isotypic components which we will discuss presently, it follows that the only ideals of R are direct sums of these minimal ideals.
2. We have now the left-right symmetry: if R is semisimple, so is R^o.

Since any irreducible module N is cyclic, there is a surjective map $R \to N$. This map restricted to one of the N_i must be nonzero, hence:

Corollary 1. *Each irreducible R module is isomorphic to one of the N_i (appearing in the regular representation).*

Let us draw another consequence, a very weak form of a more general theorem.

Corollary 2. *For a field F every automorphism ϕ of the F-algebra $M_n(F)$ is inner.*

Proof. Recall that an inner automorphism is an automorphism of the form $X \mapsto AXA^{-1}$. Given ϕ we define a new module structure F^n_ϕ on the standard module F^n by setting $X \circ_\phi v := \phi(X)v$. Clearly F^n_ϕ is still irreducible and so it is isomorphic to F^n. We have thus an isomorphism (given by an invertible matrix A) between F^n and F^n_ϕ. By definition then, for every $X \in M_n(F)$ and every vector $v \in F^n$ we must have $\phi(X)Av = AXv$, hence $\phi(X)A = AX$ or $\phi(X) = AXA^{-1}$. □

A very general statement by Skolem–Noether is discussed in [Jac-BA2], Theorem 4.9.

2 Isotypic Components

2.1 Semisimple Modules

We will complete the theory with some general remarks:

Lemma 1. *Given a module M and two submodules A, B such that A is irreducible, either $A \subset B$ or $A \cap B = 0$.*

Proof. Trivial since $A \cap B$ is a submodule of A and A is irreducible. □

Lemma 2. *Given a module M a submodule N and an element $m \notin N$ there exists a maximal submodule $N_0 \supset N$ such that $m \notin N_0$. M/N_0 is indecomposable.*

Proof. Consider the set of all submodules containing N and which do not contain m. This has a maximal element since it satisfies the hypotheses of Zorn's lemma; call this maximal element N_0. Suppose we could decompose M/N_0. Since the class of m cannot be contained in both summands we could find a larger submodule not containing m. □

The basic fact on semisimple modules is the following:

Theorem. *For a module M the following conditions are equivalent:*

(i) M is a sum of irreducible submodules.
(ii) Every submodule N of M admits a complement, i.e., a submodule P such that $M = N \oplus P$.
(iii) M is completely reducible.

Proof. This is a rather abstract theorem and the proof is correspondingly abstract.

Clearly (iii) implies (i), so we prove (i) implies (ii) implies (iii).

(i) implies (ii). Assume (i) holds and write $M = \sum_{i \in I} N_i$ where I is some set of indices.

For a subset A of I set $N_A := \sum_{i \in A} N_i$. Let N be a given submodule and consider all subsets A such that $N \cap N_A = 0$. It is clear that these subsets satisfy the conditions of Zorn's lemma and so we can find a maximal set among them; let this be A_0. We have then the submodule $N \oplus N_{A_0}$ and claim that $M = N \oplus N_{A_0}$.

For every $i \in I$ consider $(N \oplus N_{A_0}) \cap N_i$. If $(N \oplus N_{A_0}) \cap N_i = 0$ we have that $i \notin A_0$. We can add i to A_0 getting a contradiction to the maximality.

Hence by the first lemma $N_i \subset (N \oplus N_{A_0})$, and since i is arbitrary, $M = \sum_{i \in I} N_i \subset (N \oplus N_{A_0})$ as desired.

(ii) implies (iii). Assume (ii) and consider the set J of all irreducible submodules of M (at this point we do not even know that J is not empty!).

Consider all the subsets A of J for which the modules in A form a direct sum. This clearly satisfies the hypotheses of Zorn's lemma, and so we can find a maximal set adding to a submodule N.

We must prove that $N = M$. Otherwise we can find an element $m \notin N$ and a maximal submodule $N_0 \supset N$ such that $m \notin N$. By hypothesis there is a direct summand P of N_0.

We claim that P is irreducible. Otherwise let T be a nontrivial submodule of P and consider a complement Q to $N_0 \oplus T$. We have thus that M/N_0 is isomorphic to $T \oplus Q$ and so is decomposable, against the conclusions of Lemma 2.

P irreducible is also a contradiction since P and N form a direct sum, and this contradicts the maximal choice of N as direct sum of irreducibles. □

Comment. If the reader is confused by the transfinite induction he should easily realize that all these inductions, in the case where M is a finite-dimensional vector space, can be replaced with ordinary inductions on the dimensions of the various submodules constructed.

Corollary 1. *Let $M = \sum_{i \in I} N_i$ be a semisimple module, presented as a sum of irreducible submodules. We can extract from this sum a direct sum decomposing M.*

Proof. We consider a maximal direct sum out of the given one. Then any other irreducible module N_i must be in the sum, and so this sum gives M. □

Corollary 2. *Let $M = \bigoplus_{i \in I} N_i$ be a semisimple module, presented as a direct sum of irreducible submodules. Let N be an irreducible submodule of M. Then the projection to one of the N_i, restricted to N, is an isomorphism.*

2.2 Submodules and Quotients

Proposition.

(i) Submodules and quotients of a semisimple module are semisimple, as well as direct sums of semisimple modules.

(ii) R is semisimple if and only if every R module is semisimple. In this case its spectrum is finite and consists of the irreducible modules appearing in the regular representation.

(iii) If R has a faithful semisimple module M, then R is semisimple.

Proof. (i) Since the quotient of a sum of irreducible modules is again a sum of irreducibles the statement is clear for quotients. But every submodule has a complement and so it is isomorphic to a quotient. For direct sums the statement is clear.

(ii) If every module is semisimple clearly R is also semisimple. Conversely, let R be semisimple. Since every R-module is a quotient of a free module, we get from (i) that every module is semisimple. Proposition 1.4 implies that R is a finite direct sum of irreducibles and Corollary 1, §1.9 implies that these are the only irreducibles.

(iii) For each element m of M take a copy M_m of M and form the direct sum $M := \bigoplus_{m \in M} M_m$. M is a semisimple module.

Map $R \to \bigoplus_{m \in M} M_m$ by $r \mapsto (rm)_{m \in M}$. This map is clearly injective so R, as a submodule of a semisimple module, is semisimple. □

2.3 Isotypic Components

An essential notion in the theory of semisimple modules is that of *isotypic component*.

Definition. Given an isomorphism class of irreducible representations, i.e., a point of the spectrum $\alpha \in R^\vee$ and a module M, we set M^α to be the sum of all the irreducible submodules of M of type α. This submodule is called *the isotypic component of type α*.

Let us also use the notation M_α to be the sum of all the irreducible submodules of M which are *not* of type α.

Theorem. *The isotypic components of M decompose M into a direct sum.*

Proof. We must only prove that given an isomorphism class α, $M^\alpha \cap M_\alpha = 0$.

M^α can be presented as a direct sum of irreducibles of type α, while M_α can be presented as a direct sum of irreducibles of type different from α.

Thus every irreducible submodule in their intersection must be 0; otherwise by Corollary 2 of 2.1, it is at the same time of type α and of type different from α. From Proposition 2.2 (i), any submodule is semisimple, and so this implies that the intersection is 0. □

Proposition 1. *Given any homomorphism $f : M \to N$ between semisimple modules it induces a morphism $f_\alpha : M^\alpha \to N^\alpha$ for every α and f is the direct sum of the f_α. Conversely,*

$$\hom_R(M, N) = \prod_\alpha \hom(M^\alpha, N^\alpha). \tag{2.3.1}$$

Proof. The image under a homomorphism of an irreducible module of type α is either 0 or of the same type, since the isotypic component is the sum of all the submodules of a given type the claim follows. □

Now that we have the canonical decomposition $M = \bigoplus_{\alpha \in R^\vee} M^\alpha$, we can consider the projection $\pi^\alpha : M \to M^\alpha$ with kernel M_α. We have:

Proposition 2. *Given any homomorphism $f : M \to N$ between semisimple modules we have a commutative diagram, for each $\alpha \in R^\vee$:*

$$\begin{array}{ccc} M & \xrightarrow{f} & N \\ \pi^\alpha \downarrow & & \pi^\alpha \downarrow \\ M^\alpha & \xrightarrow{f} & N^\alpha \end{array} \tag{2.3.2}$$

Proof. This is an immediate consequence of the previous proposition. □

2.4 Reynold's Operator

This rather formal analysis has an important implication. Let us assume that we have a group G acting as automorphisms on an algebra A. Let us furthermore assume that A is semisimple as a G-module. Thus the subalgebra of invariants A^G is the isotypic component of the trivial representation.

Let us denote by A_G the sum of all the other irreducible representations, so that $A = A^G \oplus A_G$.

Definition. The canonical projection $\pi^G : A \to A^G$ is usually indicated by the symbol R and called *the Reynold's operator.*

Let us now consider an element $a \in A^G$; since, by hypothesis, G acts as algebra automorphisms, both left and right multiplication by a are G equivariant. We thus have the commutative diagram 2.3.2, for $\pi_\alpha = R$ and f equal to the left or right multiplication by a, and deduce the so-called *Reynold's identities.*

Proposition.

$$R(ab) = aR(b), \quad R(ba) = R(b)a, \quad \forall b \in A, \ a \in A^G.$$

We have stated these identities since they are the main tool to develop the theory of Hilbert on invariants of forms (and its generalizations) (see [DL], [SP1]) and Chapter 14.

2.5 Double Centralizer Theorem

Although the theory could be pursued in the generality of Artinian rings (cf. [JBA]), let us revert to finite-dimensional representations.

Let $R = \bigoplus_{i \in I} R_i = \bigoplus_{i \in I} M_{m_i}(\Delta_i)$ be a finite-dimensional semisimple algebra over a field F. In particular, now all the division algebras Δ_i will be finite dimensional over F. In the case of the complex numbers (or of an algebraically closed field) they will coincide with F. For the real numbers we have the three possibilities already discussed.

If we consider any finite-dimensional module M over R we have seen that M is isomorphic to a finite sum

$$M = \bigoplus_i M_i = \bigoplus_{i \in I} N_i^{p_i},$$

where M_i is the isotypic component relative to the block R_i and $N_i = \Delta_i^{m_i}$.

We have also from 1.8.2:

$$S := \mathrm{End}_R(M) = \bigoplus_{i \in I} \mathrm{End}_R(M_i) = \bigoplus_{i \in I} \mathrm{End}_R(N_i^{p_i}) = \bigoplus_{i \in I} M_{p_i}({\Delta_i}^o)$$
$$:= \bigoplus_{i \in I} S_i.$$

The block S_i acts on $N_i^{p_i}$ as m_i copies of its standard representation, and as zero on the other isotypic components. In fact by definition $S_i = \mathrm{End}_R(N_i^{p_i})$ acts as 0 on all isotypic components different from the ith one.

As for the action on this component we may identify $N_i := \Delta_i^{m_i}$ and thus view the space $N_i^{p_i}$ as the set $M_{m_i,p_i}(\Delta_i)$ of $m_i \times p_i$ matrices. Multiplication on the right by a $p_i \times p_i$ matrix with entries in Δ_i induces a typical endomorphism in S_i. The algebra of such endomorphisms is isomorphic to $M_{p_i}(\Delta_i^o)$, acting by multiplication on the right, and the m_i subspaces of $M_{m_i,p_i}(\Delta_i)$ formed by the rows decompose this space into irreducible representations of S_i isomorphic to the standard representation $(\Delta_i^o)^{p_i}$. Summarizing:

Theorem.

(i) Given a finite-dimensional semisimple R module M the centralizer S of R is semisimple.

(ii) The isotypic components of R and S coincide.

(iii) The multiplicities and the dimensions (relative to the corresponding division ring) of the irreducibles appearing in an isotypic component are exchanged, passing from R to S.

(iv) If for a given i, $M_i \neq 0$, then the centralizer of S on M_i is R_i (or rather the ring of operators induced by R_i on M_i). In particular if R acts faithfully on M we have $R = S' = R''$ (Double Centralizer Theorem).

All the statements are implicit in our previous analysis.

We wish to restate this in case $F = \mathbb{C}$ for a semisimple algebra of operators as follows:

Given two sequences of positive integers $m_1, m_2, \ldots, m_k$ and $p_1, p_2, \ldots, p_k$ we form the two semisimple algebras $A = \bigoplus_{i=1}^k M_{m_i}(\mathbb{C})$ and $B = \bigoplus_{i=1}^k M_{p_i}(\mathbb{C})$.

We form the vector space $W = \bigoplus_{i=1}^k \mathbb{C}^{m_i} \otimes \mathbb{C}^{p_i}$ and consider A, B as commuting algebras of operators on W in the obvious way, i.e., $(a_1, a_2, \ldots, a_k) \sum u_i \otimes v_i = \sum a_i u_i \otimes v_i$ for A and $(b_1, b_2, \ldots, b_k) \sum u_i \otimes v_i = \sum u_i \otimes b_i v_i$ for B. Then:

Corollary. *Given a semisimple algebra R of operators on a finite-dimensional vector space M over $\mathbb{C}$ and calling $S = R'$ its centralizer, there exist two sequences of integers $m_1, m_2, \ldots, m_k$ and $p_1, p_2, \ldots, p_k$ and an isomorphism of M with $W = \bigoplus_{i=1}^k \mathbb{C}^{m_i} \otimes \mathbb{C}^{p_i}$ under which the algebras R, S are identified with the algebras A, B.*

This corollary gives very precise information on the nature of the two algebras since it claims that on each isotypic component we can find a basis indexed by pairs of indices such that, if we order the pairs by setting first all the pairs which *end* with 1, then all that *end* with 2 and so on, the matrices of R appear as diagonal block matrices.

We get a similar result for the matrices of S if we order the indices by setting first all the pairs which *begin* with 1 then all that *begin* with 2 and so on.

Let us continue a moment with the same hypotheses as in the previous section. Choose a semisimple algebra $A = \bigoplus_{i=1}^k M_{m_i}(\mathbb{C})$ and two representations:

$$W_1 = \bigoplus_{i=1}^k \mathbb{C}^{m_i} \otimes \mathbb{C}^{p_i}, \quad \text{and } W_2 = \bigoplus_{i=1}^k \mathbb{C}^{m_i} \otimes \mathbb{C}^{q_i}$$

which we have presented as decomposed into isotypic components. According to 1.8.1 we can compute as follows:

$$\begin{aligned} \hom_A(W_1, W_2) &= \bigoplus_{i=1}^k \hom_A(\mathbb{C}^{m_i} \otimes \mathbb{C}^{p_i}, \mathbb{C}^{m_i} \otimes \mathbb{C}^{q_i}) \\ &= \bigoplus_{i=1}^k \hom_{\mathbb{C}}(\mathbb{C}^{p_i}, \mathbb{C}^{q_i}). \end{aligned} \tag{2.5.1}$$

We will need this computation for the theory of invariants.

2.6 Products

We want to deduce an important application. Let H, K be two groups (not necessarily finite) and let us choose two finite-dimensional irreducible representations U, V of these two groups over $\mathbb{C}$.

Proposition 1. *$U \otimes V$ is an irreducible representation of $H \times K$, and any finite-dimensional irreducible representation of $H \times K$ is of this form.*

Proof. The maps $\mathbb{C}[H] \to \mathrm{End}(U)$, $\mathbb{C}[K] \to \mathrm{End}(V)$ are surjective, hence the map $\mathbb{C}[H \times K] \to \mathrm{End}(U) \otimes \mathrm{End}(V) = \mathrm{End}(U \otimes V)$ is also surjective, and so $U \otimes V$ is irreducible.

Conversely, assume that we are given an irreducible representation W of $H \times K$ so that the image of the algebra $\mathbb{C}[H \times K]$ is the whole algebra $\mathrm{End}(W)$.

Let W' be the sum of all irreducible H-submodules of a given type appearing in W. Since K commutes with H we have that W' is K stable. Since W is irreducible we have $W = W'$. So W is a semisimple $\mathbb{C}[H]$-module with a unique isotypic component.

The algebra of operators induced by H is isomorphic to the full matrix algebra $M_n(\mathbb{C})$, and its centralizer is isomorphic to $M_m(\mathbb{C})$ for some m, n. W is nm-dimensional and $M_n(\mathbb{C}) \otimes M_m(\mathbb{C})$ is isomorphic to $\mathrm{End}(W)$.

The image R of $\mathbb{C}[K]$ is contained in the centralizer $M_m(\mathbb{C})$. Since the operators from $H \times K$ span $\mathrm{End}(W)$ the algebra R must coincide with $M_m(\mathbb{C})$, and the theorem follows. □

This theorem has an important application to matrix coefficients.

Let $\rho : G \to GL(U)$ be a finite-dimensional representation of G. Then we have a mapping $i_U : U^* \otimes U \to \mathbb{C}[G]$ to the functions on G defined by (cf. Chapter 5, §1.5.4):

(2.6.1) $$i_U(\phi \otimes u)(g) := \langle \phi \mid gu \rangle = \mathrm{tr}(\rho(g) \circ u \otimes \phi).$$

Proposition 2. *i_U is $G \times G$ equivariant. The image of i_U is called the* ***space of matrix coefficients*** *of U.*

Proof. We have $i_U(h\phi \otimes ku)(g) = \langle h\phi \mid gku \rangle = \langle \phi \mid h^{-1}gku \rangle = {}^h i_U(\phi \otimes u)^k$. □

In the last part of 2.6.1 we are using the identification of $U^* \otimes U$ with $U \otimes U^* = \mathrm{End}(U)$. Under this identification the map i_U becomes $X \mapsto \mathrm{tr}(X\rho(g))$, for $X \in \mathrm{End}(U)$.

Theorem.

(i) *If U is irreducible, the map $i_U : U^* \otimes U \to \mathbb{C}[G]$ is injective.*

(ii) *Its image equals the isotypic component of type U in $\mathbb{C}[G]$ under the right action and equals the isotypic component of type U^* in $\mathbb{C}[G]$ under the left action.*

Proof. (i) $U^* \otimes U$ is an irreducible $G \times G$ module, and i_U is clearly nonzero, then i_U is injective.

(ii) We do it for the right action; the left is similar.

Let us consider a G-equivariant embedding $j : U \to \mathbb{C}[G]$ where $\mathbb{C}[G]$ is considered as a G-module under right action. We must show that its image is in $i_U(U^* \otimes U)$.

Let $\phi \in U^*$ be defined by

$$\langle \phi | u \rangle := j(u)(1).$$

Then

(2.6.2) $$j(u)(g) = j(u)(1g) = j(gu)(1) = \langle \phi | gu \rangle = i_U(\phi \otimes u)(g).$$

Thus $j(u) = i_U(\phi \otimes u)$. □

Remark. The theorem proved is completely general and refers to any group. It will be possible to apply it also to continuous representations of topological groups and to rational representations of algebraic groups.

Notice that, for a finite group G, since the group algebra is semisimple, we have:

Corollary.

(2.6.3)
$$\mathbb{C}[G] = \bigoplus_i U_i^* \otimes U_i$$

where U_i runs over all the irreducible representations.

$\mathbb{C}[G]$ is isomorphic as an algebra to $\bigoplus_i \mathrm{End}(U_i)$.

Proof. By definition, for each i we have a homomorphism of $\mathbb{C}[G]$ to $\mathrm{End}(U_i)$, given by the module structure. Since it is clear that restricted to the corresponding matrix coefficients this is a linear isomorphism, the claim follows. □

Important Remark. This basic decomposition will be the guiding principle throughout all of our presentation. It will reappear in other contexts: In Chapter 7. §3.1.1 for linearly reductive algebraic groups, in Chapter 8, §3.2 as the Peter–Weyl theorem for compact groups and as a tool to pass from compact to linearly reductive groups; finally, in Chapter 10, §6.1.1 as a statement on Lie algebras, to establish the relation between semisimple Lie algebras and semisimple simply connected algebraic groups. We will see that it is also related to Cauchy's formula of Chapter 2 and the theory of Schur's functions, in Chapter 9, §6.

2.7 Jacobson Density Theorem

We discuss now the Jacobson density theorem. This is a generalization of Wedderburn's theorem which we will discuss presently.

Theorem. *Let N be an irreducible R-module, Δ its centralizer,*

$$u_1, u_2, \dots, u_n \in N$$

elements which are linearly independent relative to Δ and

$$v_1, v_2, \dots, v_n \in N$$

arbitrary. Then there exists an element $r \in R$ such that $ru_i = v_i$, $\forall i$.

Proof. The theorem states that the module N^n is generated over R by the element $a := (u_1, u_2, \dots, u_n)$.

Since N^n is completely reducible, N^n decomposes as $Ra \oplus P$. Let $\pi \in \mathrm{End}_R(N^n)$ be the projection to P vanishing on the submodule Ra.

By 1.8.1 this operator is given by an $n \times n$ matrix d_{ij} in Δ and so we have $\sum_j d_{ij} u_j = 0$, $\forall i$ since these are the components of $\pi(a)$.

By hypothesis the elements $u_1, u_2, \dots, u_n \in N$ are linearly independent over Δ; thus the elements d_{ij} must be 0 and so $\pi = 0$ and $P = 0$ as desired. □

The term "density" comes from the fact that one can define a topology (of finite approximations) on the ring $\mathrm{End}_\Delta(N)$ so that R is dense in it (cf. [JBA]).

2.8 Wedderburn's Theorem

Again let N be an irreducible R-module, and Δ its centralizer. Assume that N is a finite-dimensional vector space over Δ of dimension n.

There are a few formal difficulties in the noncommutative case to be discussed.

If we choose a basis $u_1, u_2, \ldots, u_n$ of N we identify N with Δ^n. Given the set of n-tuples of elements of a ring A thought of as column vectors, we can act on the left with the algebra $M_n(A)$ of $n \times n$ matrices. This action clearly commutes with the multiplication *on the right* by elements of A.

If we want to think of operators as always acting on the left, then we have to think of left multiplication for the *opposite* ring A^o.

We thus have dually the general fact that the endomorphism ring of a free module of rank n on a ring A is the ring of $n \times n$ matrices over A^o. We return now to modules.

Theorem 1 (Wedderburn). *R induces on N the full ring* $\mathrm{End}_\Delta(N)$ *isomorphic to the ring of $m \times m$ matrices $M_m(\Delta^o)$.*

Proof. Immediate consequence of the density theorem, taking a basis $u_1, u_2, \ldots, u_n$ of N. □

We end our abstract discussion with another generality on characters.

Let R be an algebra over $\mathbb{C}$ (we make no assumption of finite dimensionality). Let M be a finite-dimensional semisimple representation. The homomorphism $\rho_M : R \to \mathrm{End}(M)$ allows us to define a *character* on R setting $t_M(a) := \mathrm{tr}(\rho_M(a))$.

Theorem 2. *Two finite-dimensional semisimple modules M, N are isomorphic if and only if they have the same character.*

Proof. It is clear that if the two modules are isomorphic the traces are the same. Conversely, let I_M, I_N be the kernels respectively of ρ_M, ρ_N.

By the theory of semisimple algebras we know that R/I_M is isomorphic to a direct sum $\bigoplus_i M_{n_i}(\mathbb{C})$ of matrix algebras and similarly for R/I_N.

Assume that M decomposes under $\bigoplus_{i=1}^k M_{n_i}(\mathbb{C})$ with multiplicity $p_i > 0$ for the ith isotypic component. Then the trace of an element $r = (a_1, a_2, \ldots, a_k)$ as operator on M is $\sum_{i=1}^k p_i \,\mathrm{Tr}(a_i)$ where $\mathrm{Tr}(a_i)$ is the ordinary trace as an $n_i \times n_i$ matrix.

We deduce that the bilinear form $\mathrm{tr}(ab)$ is nondegenerate on R/I_M and so I_M is the kernel of the form induced by this trace on R. Similarly for R/I_N.

If the two traces are the same we deduce that the kernel is also the same and so $I_M = I_N$, and $R/I_M = \bigoplus_i M_{n_i}(\mathbb{C}) = R/I_N$. In order to prove that the representations are the same we check that the isotypic components have the same multiplicities. This is clear since p_i is the trace of e_i, a central unit of $M_{n_i}(\mathbb{C})$. □

3 Primitive Idempotents

3.1 Primitive Idempotents

Definition. An idempotent is called *primitive* if it cannot be decomposed as a sum $e = e_1 + e_2$ of two nonzero orthogonal idempotents.

If $R = M_k(F)$ is a matrix algebra over a field F, an idempotent $e \in M_k(F)$ is a projection to some subspace $W := eF^k \subset F^k$. It is then easily verified that a primitive idempotent is a projection to a 1-dimensional subspace, i.e., an idempotent matrix of rank 1 which, in a suitable basis, can be identified with the elementary matrix $e_{1,1}$.

When $e = e_{1,1}$, the left ideal Re is formed by all matrices with 0 on the columns different from the first one. As an R-module it is irreducible and isomorphic to F^k. Finally $eRe = Fe$.

More generally the same analysis holds for matrices over a division ring D, thought of as endomorphisms of the *right* vector space D^n. In this case, again if e is primitive, one can identify $eRe = D$.

If an algebra $R = R_1 \oplus R_2$ is the direct sum of two algebras, then every idempotent in R is the sum $e_1 + e_2$ of two orthogonal idempotents $e_i \in R_i$. In particular the primitive idempotents in R are the primitive idempotents in R_1 and the ones in R_2.

Thus, if $R = \sum_i M_{n_i}(F)$ is semisimple, then a primitive idempotent $e \in R$ is just a primitive idempotent in one of the summands $M_{n_i}(F)$. Thus $M_{n_i}(F) = ReR$ and Re is irreducible as an R-module (and isomorphic to the module F^{n_i} for the summand $M_{n_i}(F)$).

We want a converse of this statement. We first need a lemma:

Lemma. *Let M be a semisimple module direct sum of a finite number of irreducible modules. Let P, Q be two submodules of M and $i : P \to Q$ a module isomorphism. Then i extends to an automorphism of the module M.*

Proof. We first decompose M, P, Q into isotypic components. Since every isotypic component under a homomorphism is mapped to the isotypic component of the same type we can reduce to a singe isotypic component. Let N be the irreducible module relative to this component. M is isomorphic to N^m for some m.

By isomorphism we must have P, Q both isomorphic to N^k for a given k. If we complete $M = P \oplus P'$, $M = Q \oplus Q'$ we must have that P', Q' are both isomorphic to N^{m-k}. Any choice of such an isomorphism will produce an extension of i. □

Theorem. *Let $R = \oplus R_i$ be a semisimple algebra over a field F with R_i simple and isomorphic to the algebra of matrices $M_{k_i}(D_i)$ over a division algebra D_i.*

(1) Given a primitive idempotent $e \in R$ we have that Re is a minimal left ideal, i.e., an irreducible module.
(2) All minimal left ideals of R are of the previous form.
(3) Two primitive idempotents $e, f \in R$ give isomorphic modules Re, Rf if and only if $eRf \neq 0$.
(4) Two primitive idempotents e, f give isomorphic modules Re, Rf if and only if they are conjugate, i.e., there is an invertible element $r \in R$ with $f = rer^{-1}$.
(5) A sufficient condition for an idempotent $e \in R$ to be primitive is that $\dim_F eRe = 1$. *In this case ReR is a matrix algebra over F.*

Proof. (1) We have seen that if e is primitive, then $e \in R_i$ for some i. Hence $Re = R_ie$, e is the projection on a 1-dimensional subspace and by change of basis, $R_i = M_{k_i}(D_i)$. R_ie is the *first column* isomorphic to the standard irreducible module $D_i^{k_i}$.

(2) We first decompose R as direct sum of the $R_i = M_{k_i}(D_i)$ and then each $M_{k_i}(D_i)$ as direct sum of the columns $R_i e_{i,i}$ ($e_{i,i}$ being the diagonal matrix units). A minimal left ideal is an irreducible module $N \subset R$, and it must be isomorphic to one of the columns, call this Re with e primitive. By the previous lemma, there is an isomorphism $\phi : R \to R$, such that $\phi(Re) = N$. By Proposition 2 of 1.3, we have that $\phi(a) = ar$ for some invertible r, hence $N = Rf$ with $f = r^{-1}er$.

(3) We have that each primitive idempotent lies in a given summand R_i. If Re is isomorphic to Rf, the two idempotents must lie in the same summand R_i. If they lie in different summands, then $eRf \subset R_i \cap R_j = 0$. Otherwise, $eR = eR_i$, $Rf = R_i f$ and, since R_i is a simple algebra we have $R_i = R_i e R_i = R_i f R_i$ and $R_i = R_i R_i = R_i e R f R_i \neq 0$.

(4) If $e = rfr^{-1}$ we have that multiplication on the right by r establishes an isomorphism between Re and Rf. Conversely, let Re, Rf be isomorphic. We may reduce to $R = M_k(D)$ and e, f are each a projection to a 1-dimensional subspace. In two suitable bases the two idempotents equal the matrix unit $e_{1,1}$, so the invertible matrix of base change conjugates one into the other.

(5) $eRe = \bigoplus_i eR_i e$. Hence if $\dim_F eRe = 1$, we must have that $eR_{i_0}e \neq 0$ for only one index i_0 and $e \in R_{i_0}$. If $e = a + b$ were a decomposition into orthogonal idempotents, we would have $a = aea$, $b = beb \in eRe$, a contradiction. Since e is primitive in $R_{i_0} = M_k(D_{i_0})$ in a suitable basis, it is a matrix unit and so the division algebra $eRe = D_{i_0}$ reduces to F since $\dim_F D_{i_0} = \dim_F eRe = 1$. □

Assume as before that $R = \oplus R_i$ is a semisimple algebra over a field F and $\dim_F eRe = 1$.

Proposition 1. *Given a module M over R the multiplicity of Re in its isotypic component in M is equal to $\dim_F eM$.*

Proof. Let $M_0 := \oplus^k Re$ be the isotypic component. We have $eM = eM_0 = \oplus^k eRe = F^k$. □

Proposition 2.

(1) Let R be a semisimple algebra and I an ideal; then $I^2 = I$.
(2) If $aRb = 0$, then $bRa = 0$. Furthermore $aRa = 0$ implies $a = 0$.

Proof. (1) An ideal of $\bigoplus_i R_i$ is a sum of some R_j and $R_j^2 = R_j$.

(2) In fact from $aRb = 0$ we deduce $(RbRaR)^2 = 0$. Since $RbRaR$ is an ideal, by (1) we must have $RbRaR = 0$. Hence $bRa = 0$.

Similarly, $aRa = 0$ implies $(RaR)^2 = 0$. Hence the claim. □

In particular we see that a semisimple algebra has no nonzero nilpotent ideals, or ideals I for which $I^k = 0$ for some k.

It can in fact be proved that for a finite-dimensional algebra this condition is equivalent to semisimplicity (cf. [JBA]).

3.2 Real Algebras

It is interesting to study also real semisimple algebras R and real representations. We state the basic facts leaving the proofs to the reader.

A real semisimple algebra is a direct sum of matrix algebras over the three basic division algebras $\mathbb{R}$, $\mathbb{C}$ or $\mathbb{H}$. The representations will decompose according to these blocks. Let us analyze one single block, $M_h(\Delta)$ in the three cases. It corresponds to the irreducible module Δ^h with centralizer Δ. When we complexify the algebra and the module we have $M_h(\Delta \otimes_{\mathbb{R}} \mathbb{C})$ acting on $(\Delta \otimes_{\mathbb{R}} \mathbb{C})^h$ with centralizer $\Delta \otimes_{\mathbb{R}} \mathbb{C}$. We have

(3.2.1) $$\mathbb{R} \otimes_{\mathbb{R}} \mathbb{C} = \mathbb{C}, \quad \mathbb{C} \otimes_{\mathbb{R}} \mathbb{C} = \mathbb{C} \oplus \mathbb{C}, \quad \mathbb{H} \otimes_{\mathbb{R}} \mathbb{C} = M_2(\mathbb{C}).$$

Exercise. Deduce that the given irreducible module for R, in the three cases, remains irreducible, splits into the sum of two non-isomorphic irreducibles, splits into the sum of two isomorphic irreducibles.

7

Algebraic Groups

Summary. In this chapter we want to have a first look into algebraic groups. We will use the necessary techniques from elementary algebraic geometry, referring to standard textbooks. Our aim is to introduce a few techniques from algebraic geometry commonly used in representation theory.

1 Algebraic Groups

1.1 Algebraic Varieties

There are several reasons to introduce algebraic geometric methods in representation theory. One which we will not pursue at all is related to the classification of finite simple groups. A standard textbook on this subject is ([Ca]). Our point of view is based instead on the fact that in a precise sense, compact Lie groups can be extended to make them algebraic, and representations should be found inside algebraic functions. We start to explain these ideas in this chapter.

We start from an algebraically closed field k, usually the complex numbers.

Recall that an *affine variety* is a subset V, of some space k^m, defined as the vanishing locus of polynomial equations (in $k[x_1, \dots, x_m]$).

The set I_V of polynomials vanishing on V is its *defining ideal.*

A *regular algebraic function* (or just algebraic function) on V is the restriction to V of a polynomial function on k^m. The set $k[V]$ of these functions is the *coordinate ring* of V; it is isomorphic to the algebra $k[x_1, \dots, x_m]/I_V = k[V]$.

Besides being a finitely generated commutative algebra over k, the ring $k[V]$, being made of functions, does not have any nilpotent elements. $k[V]$ can have zero divisors: this happens when the variety V is not *irreducible*, for instance the variety given by $xy = 0$ in k^2, which consists of two lines.

The notion of subvariety of a variety is clear. Since the subvarieties naturally form the closed sets of a topology (the Zariski topology) one often speaks of a Zariski closed subset rather than a subvariety. The Zariski topology is a rather weak topology. It is clear that a Zariski closed set is also closed in the complex topology. More important is that if V is an irreducible variety and $W \subset V$ a proper subvariety, the open set $V - W$ is dense in V, and also in the complex topology.

Finally, given two affine varieties $V \subset k^n$, $W \subset k^m$ a *regular map* or *morphism* between the two varieties, $f : V \to W$ is a map which, in coordinates, is given by regular functions $(f_1(x_1, \dots, x_n), \dots, f_m(x_1, \dots, x_n))$.

Remark 1. A morphism $f : V \to W$ induces a *comorphism* of coordinate rings

$$f^* : k[W] \to k[V], \quad f^*(g) := g \circ f.$$

Remark 2. One should free the concept of affine variety from its embedding, i.e., verify that embedding an affine variety V in a space k^m is the same as choosing m elements in $k[V]$ which generate this algebra over k.

The main fact that one uses at the beginning of the theory is the Hilbert Nullstellensatz, which implies that given an affine algebraic variety V over an algebraically closed field k, with coordinate ring $k[V]$, there is a bijective correspondence between the three sets:

1. Points of V.
2. Homomorphisms, $\phi : k[V] \to k$.
3. Maximal ideals of $k[V]$.

For the basic affine space k^n this means that the maximal ideals of $k[x_1, \dots, x_n]$ are all of the type $(x_1 - a_1, x_2 - a_2, \dots, x_n - a_n)$, $(a_1, \dots, a_n) \in k^n$.

This general fact allows one to pass from the language of affine varieties to that of *reduced affine algebras*, that is commutative algebras A, finitely generated over k and without nilpotent elements. In the end one can state that the category of affine varieties is antiisomorphic to that of reduced affine algebras. In this sense one translates from algebra to geometry and conversely.

Affine varieties do not by any means exhaust all varieties. In fact in the theory of algebraic groups one has to use systematically (see §2) at least one other class of varieties, *projective varieties*. These are defined by passing to the projective space $P^n(k)$ of lines of k^{n+1} (i.e., 1-dimensional subspaces of k^{n+1}). In this space now the coordinates $(x_0, x_1, x_2, \dots, x_n)$ must represent a line (through the given point and 0) so they are not all 0 and *homogeneous* in the sense that if $a \neq 0$, $(x_0, x_1, x_2, \dots, x_n)$ and $(ax_0, ax_1, ax_2, \dots, ax_n)$ represent the same point. Then the projective varieties are the subsets V of $P^n(k)$ defined by the vanishing of systems of homogeneous equations.

A system of homogeneous equations of course also defines an affine variety in k^{n+1}: it is the union of 0 and all the lines which correspond to points of V. This set is called the *associated cone* $C(V)$ to the projective variety V and the graded coordinate ring $k[C(V)]$ of $C(V)$ the *homogeneous coordinate ring* of V. Of course $k[C(V)]$ is not made of functions on V. If one wants to retain the functional language one has to introduce a new concept of line bundles and sections of line bundles which we do not want to discuss now.

The main feature of projective space over the complex numbers is that it has a natural topology which makes it compact. In fact, without giving too many details, the reader can understand that one can normalize the homogeneous coordinates x_i so that $\sum_{i=0}^{n} |x_i|^2 = 1$. This is a compact set, the $2n+1$-dimensional sphere S^{2n+1}. Two points in S^{2n+1} give the same point in $P^n(\mathbb{C})$ if and only if they differ by multiplication by a complex number of absolute value 1. One then gives $P^n(\mathbb{C})$ the quotient topology.

Clearly now, in order to give a regular map $f : V \to W$ between two such varieties $V \subset P^n(k)$, $W \subset P^m(k)$ one has to give a map in coordinates, given by regular functions $(f_0(x_0, x_1, \dots, x_n), f_1(x_0, x_1, \dots, x_n), \dots, f_m(x_0, x_1, \dots, x_n))$ which, to respect the homogeneity of the coordinates, must necessarily be *homogeneous of the same degree.*

It is useful also to introduce two auxiliary notions:

A *quasi-affine variety* is a (Zariski) open set of an affine variety.

A *quasi-projective variety* is a (Zariski) open set of a projective variety.

In projective geometry one has to use the ideas of affine varieties in the following way. If we fix a coordinate x_i (but we could also first make a linear change of coordinates), the points in projective space $P^n(k)$ where the coordinate $x_i \neq 0$ can be described by *dehomogenizing this coordinate* and fixing it to be $x_i = 1$. Then this open set is just an n-dimensional affine space $U_i = k^n \subset P^n(k)$. So projective space is naturally *covered* by these $n + 1$ *affine charts*. Given a projective variety $V \subset P^n(k)$, its intersection with U_i is an affine variety, obtained by setting $x_i = 1$ in the homogeneous equations of V.

The theory of projective varieties is developed by analyzing how the open affine sets $V \cap U_i$ glue together to produce the variety V.

What makes projective geometry essentially different from affine geometry is the fact that projective varieties are really *compact*. In characteristic 0 this really means that they are compact in the usual topology. In positive characteristic one rather uses the word *complete* since the usual topology does not exist.

1.2 Algebraic Groups

Consider $GL(n, k)$, the group of all invertible $n \times n$ matrices, and the *special linear group* $SL(n, k) := \{A \in M_n(k) \mid \det(A) = 1\}$ given by a single equation in the space of matrices. This latter group thus has the natural structure of an *affine variety*. Also the full linear group is an affine variety. We can identify it with the set of pairs A, c, $A \in M_n(k)$, $c \in k$ with $\det(A)c = 1$. Alternatively, we can embed $GL(n, k)$ as a closed subgroup of $SL(n + 1, k)$ as block matrices $\begin{vmatrix} A & 0 \\ 0 & \det(A)^{-1} \end{vmatrix}$.

The *regular algebraic functions* on $GL(n, k)$ are the rational functions $f(x_{i,j})d^{-p}$ where f is a polynomial in the entries of the matrix, d is the determinant and p can be taken to be a nonnegative integer; thus its coordinate ring (over the field k) is the ring of polynomials $k[x_{ij}, d^{-1}]$ in n^2 variables with the determinant d inverted.

Definition 1. A subgroup H of $GL(n,k)$ is called a *linear group*.
A Zariski closed subgroup H of $GL(n,k)$ is a *linear algebraic group*.

The coordinate ring of such a subgroup is then of the form $k[x_{ij}, d^{-1}]/I$, with I the defining ideal of H.

Examples of Linear Algebraic Groups

(1) As $GL(n+1,k)$, $SL(n+1,k)$ act linearly on the space k^{n+1}, they induce a group of *projective transformations* on the projective space $P^n(k)$ of lines in k^{n+1}.

In homogeneous coordinates these actions are still given by matrix multiplication. We remark that if one takes a scalar matrix, i.e., a scalar multiple of the identity zI_{n+1}, $z \neq 0$, this acts trivially on projective space, and conversely a matrix acts trivially if and only if it is a scalar. We identify the multiplicative group k^* of nonzero elements of k with invertible scalar matrices. Thus the group of projective transformations is

$$PGL(n+1,k) = GL(n+1,k)/k^*, \qquad \textit{projective linear group}$$

the quotient of $GL(n+1,k)$ or of $SL(n+1,k)$ by the respective centers. In the case of $GL(n+1,k)$ the center is formed by the nonzero scalars z while for $SL(n+1,k)$ we have the constraint $z^{n+1} = 1$. The fact that $PGL(n+1,k)$ is a linear algebraic group is not evident at this point; we will explain why it is so in Section 2.

(2) The *orthogonal and symplectic groups* are clearly algebraic since they are given by quadratic equations (cf. Chapter 5, §3):

$$O(n,k) := \{A \in GL(n,k) \mid A^t A = 1\},$$

$$Sp(2n,k) := \{A \in GL(2n,k) \mid AJA = J\},$$

where $J = \begin{vmatrix} 0 & 1_m \\ -1_m & 0 \end{vmatrix}$.

(3) The special orthogonal group is given by the further equation $\det(X) = 1$.

(4) The group T_n, called a *torus*, of invertible diagonal matrices, given by the equations $x_{i,j} = 0, \forall i \neq j$.

(5) The group $B_n \subset GL(n,k)$ of invertible upper triangular matrices, given by the equations $x_{i,j} = 0, \forall i > j$.

(6) The subgroup $U_n \subset B_n$ of strictly upper triangular matrices, given by the further equations $x_{i,i} = 1, \forall i$.

It may be useful to remark that the group

$$(1.2.1) \qquad U_2 = \left\{ \begin{vmatrix} 1 & a \\ 0 & 1 \end{vmatrix} \right\}, \qquad a \in k, \qquad \begin{vmatrix} 1 & a \\ 0 & 1 \end{vmatrix} \begin{vmatrix} 1 & b \\ 0 & 1 \end{vmatrix} = \begin{vmatrix} 1 & a+b \\ 0 & 1 \end{vmatrix}$$

is the *additive group* of the field k.

Exercise. Show that the Clifford and spin group (Chapter 5, §4, §5) over $\mathbb{C}$ are algebraic.

Although we will not really use it seriously, let us give the more general definition:

Definition 2. An algebraic group G is an algebraic variety with a group structure, such that the two maps of multiplication $G \times G \to G$ and inverse $G \to G$ are algebraic.

If G is an affine variety it is called an affine algebraic group.

For an elementary introduction to these concepts one can see [Sp].

1.3 Rational Actions

For algebraic groups it is important to study regular algebraic actions.

Definition 1. An action $\pi : G \times V \to V$ is algebraic if V is an algebraic variety and the map π is algebraic (i.e., it is a morphism of varieties).

If both G and V are affine, it is very important to treat all these concepts using the coordinate rings and the comorphism.

The essential theorem is that, given two affine varieties V, W, we have (cf. [Ha], [Sh]):

$$k[V \times W] = k[V] \otimes k[W].$$

Thus, to an action $\pi : G \times V \to V$ is associated a comorphism $\pi^* : k[V] \to k[G] \otimes k[V]$, satisfying conditions corresponding to the definition of action.

The first class of algebraic actions of a linear algebraic group are the induced actions on tensors (and the associated ones on symmetric, skew-symmetric tensors).

Basic example. Consider the action of $GL(n, k)$ on the linear space k^n. We have $k[k^n] = k[x_1, \dots, x_n]$, $\quad k[GL(n, k)] = k[y_{i,j}, d^{-1}]$ and the comorphism is

$$x_i \mapsto \sum_{j=1}^{n} y_{i,j} x_j.$$

Among algebraic actions there are the left and right action of G on itself:

$$(g, h) \mapsto gh, \ (g, h) \mapsto hg^{-1}.$$

Remark. The map $h \mapsto h^{-1}$ is an isomorphism between the left and right action.

For algebraic groups we will usually restrict to regular algebraic homomorphisms. A linear representation $\rho : G \to GL(n, k)$ is called *rational* if the homomorphism is algebraic.

It is useful to extend the notion to infinite-dimensional representations.

Definition 2. A linear action of an algebraic group G on a vector space V is called *rational* if V is the union of finite-dimensional subrepresentations which are algebraic.

Example. For the general linear group $GL(n, k)$ a rational representation is one in which the entries are polynomials in x_{ij} and d^{-1}.

Thus linear algebraic groups are affine. Non-affine groups belong essentially to the disjoint theory of abelian varieties.

Consider an algebraic action $\pi : G \times V \to V$ on an affine algebraic variety V. The action induces an action on functions. The regular functions $k[V]$ on V are then a representation.

Proposition. *$k[V]$ is a rational representation.*

Proof. Let $k[G]$ be the coordinate ring of G so that $k[G] \otimes k[V]$ is the coordinate ring of $G \times V$. The action π induces a map $\pi^* : k[G] \to k[G] \otimes k[V]$, where $\pi^* f(g, v) := f(gv)$.

For a given function $f(v)$ on V we have that $f(gv) = \sum_i a_i(g) b_i(v)$. Thus we see that the translated functions f^g lie in the linear span of the functions b_i.

This shows that any finite-dimensional subspace U of the space $k[V]$ is contained in a finite-dimensional G-stable subspace W. Given a basis u_i of W, we have that $u_i(g^{-1}v) = \sum_j a_{ij}(g) u_i(v)$, with the a_{ij} regular functions on G. Thus W is a rational representation. The union of these representations is clearly $k[V]$. □

The previous proposition has an important consequence.

Theorem. *(i) Given an action of an affine group G on an affine variety V there exists a linear representation W of G and a G-equivariant embedding of V in W.*

(ii) An affine group is isomorphic to a linear algebraic group.

Proof. (i) Choose a finite set of generators of the algebra $k[V]$ and then a finite-dimensional G-stable subspace $W \subset k[V]$ containing this set of generators.

W defines an embedding i of V into W^* by $\langle i(v)|w\rangle := w(v)$. This embedding is clearly equivariant if on W^* we put the dual of the action on W.

(ii) Consider the right action of G on itself. If W is as before and $u_i, i = 1, \dots, n$, is a basis of W, we have $u_i(xy) = \sum_j a_{ij}(y) u_j(x)$.

Consider the homomorphism ρ from G to matrices given by the matrix $(a_{ij}(y))$. Since $u_i(y) = \sum_j a_{ij}(y) u_j(1)$ we have that the functions a_{ij} generate the coordinate ring of G and thus ρ is an embedding of G into matrices. Thus the image of ρ is a linear algebraic group and ρ is an isomorphism from G to its image. □

1.4 Tensor Representations

We start with:

Lemma. *Any finite-dimensional rational representation U of an algebraic group G can be embedded in an equivariant way in the direct sum of finitely many copies of the coordinate ring $k[G]$ under the right (or the left) action.*

Proof. Take a basis $u^i, i = 1, \dots, n$, of the dual of U. The map

$$j : U \to k[G]^n, \quad j(u) := (\langle u^1|gu\rangle, \dots, \langle u^n|gu\rangle)$$

is clearly equivariant, with respect to the right action on $k[G]$. Computing these functions in $g = 1$ we see that this map is an embedding. □

Remark. An irreducible representation U can be embedded in a single copy, i.e., in $k[G]$.

Most of this book deals with methods of tensor algebra. Therefore it is quite useful to understand a general statement on rational representations versus tensor representations.

Theorem. *Let $G \subset GL(V)$ be a linear algebraic group. Denote by d the determinant, as a 1-dimensional representation.*

Given any finite-dimensional rational representation U of G we have that, after possibly tensoring by d^r for some r and setting $M := U \otimes d^r$, the representation M is a quotient of a subrepresentation of a direct sum of tensor powers $V^{\otimes k_i}$.

Proof. Let A, B be the coordinate rings of the space $\mathrm{End}(V)$ of all matrices, and of the group $GL(V)$, respectively. The coordinate ring $k[G]$ is a quotient of $B \supset A$.

By the previous lemma, U embeds in a direct sum $\oplus^p k[G]$. We consider the action of G by right multiplication on these spaces.

Since the algebra $B = \cup_{i=0}^{\infty} d^{-i}A$, where d is the determinant function, for some r we have that $d^r U$ is in the image of A^p.

The space of endomorphisms $\mathrm{End}(V)$ as a $G \times G$ module is isomorphic to $V \otimes V^*$, so the ring A is isomorphic to $S(V^* \otimes V) = S(V^{\oplus m})$ as a right G-module if $m = \dim V$.

As a representation $S(V^{\oplus m})$ is a quotient of the tensor algebra $\bigoplus_n (V^{\oplus m})^{\otimes n}$ which in turn is isomorphic to a direct sum of tensor powers of V. Therefore, we can construct a map from a direct sum of tensor powers $V^{\otimes m_i}$ to $k[G]^m$ so that $d^r U$ is in its image.

Since a sum of tensor powers is a rational representation we deduce that $d^r U$ is also the image of a finite-dimensional submodule of such a sum of tensor powers, as required. □

1.5 Jordan Decomposition

The previous theorem, although somewhat technical, has many corollaries. An essential tool in algebraic groups is Jordan decomposition. Given a matrix X we have seen in Chapter 4, §6.1 its *additive* Jordan decomposition, $X = X_s + X_n$ where X_s is diagonalizable, i.e., semisimple, X_n is nilpotent and $[X_s, X_n] = 0$. If X is invertible so is X_s, and it is then better to use the *multiplicative* Jordan decomposition, $X = X_s X_u$ where $X_u := 1 + X_s^{-1} X_n$ is unipotent, i.e., all of its eigenvalues are 1. We still have $[X_s, X_u] = 0$.

It is quite easy to see the following compatibility if X, Y are two invertible matrices:

$$(X \oplus Y)_s = X_s \oplus Y_s,\ (X \oplus Y)_u = X_u \oplus Y_u,$$

$$(X \otimes Y)_s = X_s \otimes Y_s,\ (X \otimes Y)_u = X_u \otimes Y_u. \tag{1.5.1}$$

Furthermore, if X acts on the space V and U is stable under X, then U is stable under X_s, X_u, and the Jordan decomposition of X restricts to the Jordan decomposition of the operator that X induces on U and on V/U.

Finally we can obviously extend this language to infinite-dimensional rational representations. This can be given a very general framework.

Theorem (Jordan–Chevalley decomposition). *Let $G \subset GL(n,k)$ be a linear algebraic group, $g \in G$ and $g = g_s g_u$ its Jordan decomposition. Then*

(i) $g_s, g_u \in G$.
(ii) For every rational representation $\rho : G \to GL(W)$ we have that $\rho(g) = \rho(g_s)\rho(g_u)$ is the Jordan decomposition of $\rho(g)$.

Proof. From Theorem 1.4, formulas 1.5.1 and the compatibility of the Jordan decomposition with direct sums, tensor products and subquotients, clearly (ii) follows from (i).

(i) is subtler. Consider the usual homomorphism $\pi : k[GL(n,k)] \to k[G]$, and the action R_g of g on functions $f(x) \mapsto f(xg)$ on $k[GL(n,k)]$ and $k[G]$, which are both rational representations.

On $k[GL(n,k)]$ we have the Jordan decomposition $R_g = R_{g_s} R_{g_u}$ and from the general properties of submodules and subquotients we deduce that the two maps R_{g_s}, R_{g_u} also induce maps on $k[G]$ which decompose the right action of g. This means in the language of algebraic varieties that the right multiplication by g_s, g_u on $GL(n,k)$ preserves the subgroup G, but this means exactly that $g_s, g_u \in G$.

(ii) From the proof of (i) it follows that $R_g = R_{g_s} R_{g_u}$ is also the Jordan decomposition on $k[G]$. We apply Lemma 1.4 and have that W embeds in $k[G]^m$ for some m. Now we apply again the fact that the Jordan decomposition is preserved when we restrict an operator to a stable subspace. □

1.6 Lie Algebras

For an algebraic group G the Lie algebra of left-invariant vector fields can be defined algebraically. For an algebraic function f and $g \in G$ we have an expression $L_g f(x) = f(gx) = \sum_i f_1^{(i)}(g) f_2^{(i)}(x)$. Applying formula 3.1.1 of Chapter 4, we have that a left-invariant vector field gives the derivation of $k[G]$ given by the formula

$$X_a f(g) := dL_g(a)(f) = a(f(gx)) = \sum_i f_1^{(i)}(g) a(f_2^{(i)}(x)).$$

The linear map a is a tangent vector at 1, i.e., a derivation of the coordinate ring of G at 1. According to basic algebraic geometry, such an a is an element of the dual of m/m^2, where m is the maximal ideal of $k[G]$ of elements vanishing at 1. Notice that this construction can be carried out over any base field.

2 Quotients

2.1 Quotients

To finish the general theory we should understand, given a linear algebraic group G and a closed subgroup H, the nature of G/H as an algebraic variety and, if H is a normal subgroup, of G/H as an algebraic group. The key results are due to Chevalley:

Theorem 1. *(a) For a linear algebraic group G and a closed subgroup H, there is a finite-dimensional rational representation V and a line $L \subset V$ such that $H = \{g \in G \mid gL = L\}$.*

(b) If H is a normal subgroup, we can assume furthermore that H acts on V by diagonal matrices (in some basis).

Proof. (a) Consider $k[G]$ as a G-module under the right action. Let I be the defining ideal of the subgroup H. Since I is a rational representation of H and finitely generated as an ideal, we can find an H-stable subspace $W \subset I$ which generates I as an ideal.

Next we can find a G-stable subspace U of $k[G]$ containing W. Thus U is a rational representation of G and we claim that $H = \{g \in G \mid gW = W\}$.

To see this let $u_1(x), \dots, u_m(x)$ be a basis of W. If $u_i(xg) \in W$ for all i we have that $u_i(xg) = \sum_{j=1}^m a_{ij}(g)u_j(x)$ is a change of basis. Compute both sides at $x = 1$ and obtain

$$u_i(g) = \sum_{j=1}^{m} a_{ij}(g)u_j(1) = 0, \quad \forall i = 1, \dots, m,$$

as the $u_i(x)$ vanish on H. As the $u_i(x)$ generate the ideal of H, we have $g \in H$, as desired.

Let $V := \bigwedge^m U$ be the exterior power and $L = \bigwedge^m W$. Given a linear transformation A of V we have that $\bigwedge^m(A)$ fixes L if and only if A fixes W,[44] and so the claim follows.

(b) Assume now that H is a normal subgroup, and let V, L be as in the previous step. Consider the sum $S \subset V$ of all the 1-dimensional H-submodules (eigenvectors). Let us show that S is a G-submodule. For this consider a vector $v \in S$ which is an eigenvector of H, $hv = \chi(h)v$.

[44] We have not proved this rather simple fact; there is a very simple proof in Chapter 13, Proposition 3.1.

We have for $g \in G$ that $hgv = gg^{-1}hgv$ and $g^{-1}hg \in H$. So $hgv = g\chi(g^{-1}hg)v = \chi(g^{-1}hg)gv \in S$. If we replace V with S we have satisfied the condition (b). □

Theorem 2. *Given a linear algebraic group G and a closed normal subgroup H, there is a linear rational representation $\rho : G \to GL(Z)$ such that H is the kernel of ρ.*

Proof. Let L, S be as in part (b) of the previous theorem. Let Z be the set of linear transformations centralizing H. If $g \in G$ and $a \in Z$ we have, for every $h \in H$,

$$h(gag^{-1})h^{-1} = g(g^{-1}hg)a(g^{-1}h^{-1}g)g^{-1}$$

$$= ga(g^{-1}hg)(g^{-1}h^{-1}g)g^{-1} = gag^{-1}. \tag{2.1.1}$$

In other words, the conjugation action of G on End(S) preserves the linear space Z.

By definition of centralizer, H acts trivially by conjugation on Z. We need only prove that if $g \in G$ acts trivially by conjugation in Z, then $g \in H$.

For this, observe that since H acts as diagonal matrices in S, there is an H-invariant complementary space to L in S and so an H-invariant projection $\pi : S \to L$. By definition $\pi \in Z$. If $g\pi g^{-1} = \pi$, we must have that $gL = L$, hence that $g \in H$. □

Example. In the case of the projective linear group $PGL(V) = GL(V, k)/k^*$ there is a very canonical representation. If we act with $GL(V)$ on the space End(V) by conjugation we see that the scalar matrices are exactly the kernel.

The purpose of these two theorems is to show that G/H can be thought of as a *quasi-projective variety*.

When H is normal we have shown that H is the kernel of a homomorphism. In Proposition 1 we will show that the image of a homomorphism is in fact a closed subgroup which will allow us to define G/H as a linear algebraic group.

Now we should make two disclaimers. First, it is not completely clear what we mean by these two statements, nor is it clear if what we have proved up to now is enough. The second point is that in characteristic $p > 0$ one has to be more precise in the theorems in order to avoid the difficult problems coming from possible inseparability. Since this discussion would really take us away from our path we refer to [Sp] for a thorough understanding of the issues involved. Here we will limit ourselves to some simple geometric remarks which, in characteristic 0, are sufficient to justify all our claims.

The basic facts we need from algebraic geometry are the following.

1. An algebraic variety decomposes uniquely as a union of irreducible varieties.

2. An irreducible variety V has a *dimension*, which can be characterized by the following inductive property: if $W \subset V$ is a maximal proper irreducible subvariety of V, we have $\dim V = \dim W + 1$. A zero-dimensional irreducible variety is a single point.

For a non-irreducible variety W one defines its dimension as the maximum of the dimensions of its irreducible components.

3. Given a map $\pi : V \to W$ of varieties, if V is irreducible, then $\overline{\pi(V)}$ is irreducible. If $\overline{\pi(V)} = W$, we say that π is *dominant*.

4. $\pi(V)$ contains a nonempty Zariski open set U of $\overline{\pi(V)}$.

For a non-irreducible variety V it is also useful to speak of the dimension of V at a point $P \in V$. By definition it is the maximum dimension of an irreducible component of V passing through P.

Usually an open set of a closed subset is called a *locally closed set*. Then what one has is that $\pi(V)$ is a finite union of locally closed sets; these types of sets are usually called *constructible*.

The second class of geometric ideas needed is related to the concept of smoothness.

Algebraic varieties, contrary to manifolds, may have *singularities*. Intuitively, a point P of an irreducible variety of dimension n is *smooth* if the variety can be described locally by n-parameters. This means that given the maximal ideal m_P of functions vanishing in P, the vector space $T_P^*(V) := m_P/m_P^2$ (which should be thought of as the space of *infinitesimals of first order*) has dimension exactly n (over the base field k).

One then defines the *tangent space* of V in P as the dual space :

$$T_P(V) := \hom(m_P/m_P^2, k).$$

This definition can be given in general. Then we say that a variety V is *smooth at a point* $P \in V$ if the dimension of $T_P(V)$ equals the dimension of P in V.

5. If V is not smooth at P, the dimension of $T_P(V)$ is strictly bigger than the dimension of P in V.

6. The set of smooth points of V is open and dense in V.

Given a map $\pi : V \to W$, $\pi(P) = Q$ and the comorphism $\pi^* : k[W] \to k[V]$ we have $\pi^*(m_Q) \subset m_P$. Thus we obtain a map $m_Q/m_Q^2 \to m_P/m_P^2$ and dually the *differential*:

$$d\pi_P : T_P(V) \to T_{\pi(P)}(W). \tag{2.1.2}$$

In characteristic 0, one has the following:

Theorem 3. *Given a dominant map $\pi : V \to W$ between irreducible varieties, there is a nonempty open set $U \subset W$ such that U, $\pi^{-1}(U)$ are smooth. If $P \in \pi^{-1}(U)$, then $d\pi_P : T_P(V) \to T_{\pi(P)}(W)$ is surjective.*

This basic theorem fails in positive characteristic due to the phenomenon of *inseparability*. This is best explained by the simplest example. We assume the characteristic $p > 0$ and take as varieties the affine line $V = W = k$. Consider the map $x \mapsto x^p$. By simple field theory it is a bijective map. Its differential can be computed with the usual rules of calculus as $dx^p = px^{p-1}dx \equiv 0$. If the differential is not identically 0, we will say that the map is *separable*, otherwise *inseparable*. Thus, in positive characteristic one usually has to take care also of these facts. Theorem 3 remains valid if we assume π to be separable. Thus we also need to add the separability condition to the properties of the orbit maps. This is discussed, for instance, in [Sp].

Apart from this delicate issue, let us see the consequences of this analysis.

Proposition 1. *(a) An orbit variety G/H is a smooth quasiprojective variety.*

(b) The image G/H under a group homomorphism $\rho : G \to GL(V)$ with kernel H is a closed subgroup.

Proof. (a) From the first lemma of this section there is a linear representation V of G and a line L such that H is the stabilizer of L. Hence if we consider the projective space $P(V)$ of lines in V, the line L becomes a point of $P(V)$. H is the stabilizer of this point and G/H its orbit. According to the previous property 4, G/H contains a nonempty set U, open in the closure $\overline{G/H}$. From property 6 we may assume that U is made of smooth points. Since G acts algebraically on G/H, it also acts on its closure and $\cup_{g\in G} gU$ is open in $\overline{G/H}$ and made of smooth points. Clearly $\bigcup_{g\in G} gU = G/H$.

(b) By (a) G/H is a group, open in $\overline{G/H} \subset GL(V)$. If we had an element $x \in \overline{G/H} - G/H$ we would have also the coset $(G/H)x \subset \overline{G/H} - G/H$. This is absurd by a dimension argument, since as varieties $(G/H)x$ and G/H are isomorphic, and thus they have the same dimension, so $(G/H)x$ cannot be contained properly in the closure of G/H. □

The reader should wonder at this point if our analysis is really satisfactory. In fact it is not, the reason being that we have never explained if G/H really has an intrinsic structure of an algebraic variety. A priori this may depend on the embeddings that we have constructed. In fact what one would like to have is a canonical structure of an algebraic variety on G/H such that:

Universal property (of orbits). If G acts on any algebraic variety X, $p \in X$ is a point fixed by H, then the map, which is defined set-theoretically by $gH \mapsto gp$, is a regular map of algebraic varieties from G/H to the orbit of p.

Similarly, when H is a normal subgroup we would like to know that:

Universal property (of quotient groups). If $\rho : G \to K$ is a homomorphism of algebraic groups and H is in the kernel of ρ, then the induced homomorphism $G/H \to K$ is algebraic.

To see what we really need in order to resolve this question we should go back to the general properties of algebraic varieties.

One major difficulty that one has to face is the following: if a map $\pi : V \to W$ is bijective it is not necessarily an isomorphism of varieties!

We have already seen the example of $x \to x^p$, but this may happen also in characteristic 0. The simplest example is the bijective *parameterization* of the cubic $C := \{x^3 - y^2 = 0\}$ given by $x = t^2, y = t^3$ (a cusp). In the associated comorphism the image of the coordinate ring $k[C]$, in the coordinate ring $k[t]$ of the line, is the proper subring $k[t^2, t^3] \subsetneq k[t]$, so the map cannot be an isomorphism.

The solution to this puzzle is in the concept of *normality*: For an affine variety V this means that its coordinate ring $k[V]$ is integrally closed in its field of fractions.

For a projective variety it means that its affine open sets are indeed normal. Normality is a weaker condition than smoothness and one has the basic fact that (cf. [Ra]):

Theorem (ZMT, Zariski's main theorem).[45] *A bijective separable morphism $V \to W$, where W is normal, is an isomorphism.*

Assume that we have found an action of G on some projective space $\mathbb{P}$, and a point $p \in \mathbb{P}$ such that H is the stabilizer. We then have the orbit map $\pi : G \to \mathbb{P}, \pi(g) = gp$ which identifies, set-theoretically, the orbit of p with G/H. The key observation is:

Proposition 2. *If π is separable, then the orbit map satisfies the universal property.*

Proof. Then let G act on some other algebraic variety X and $q \in X$ be fixed by H. Consider the product $X \times P$. Inside it the point (q, p) is clearly stabilized by exactly H. Let $A = G(x, p)$ be its orbit which set-theoretically is G/H and $\phi : g \mapsto g(q, p)$ the orbit map. The two projections p_1, p_2 on the two factors X, P give rise to maps $G \xrightarrow{\phi} A \xrightarrow{p_1} Gq, \quad G \xrightarrow{\phi} A \xrightarrow{p_2} Gp$. The second map $A \to Gp$ is bijective. Since the map from G to Gp is separable also the map from A to Gp must be separable. Hence by ZMT it is an isomorphism. Then its inverse composed with the first projection is the required map. □

Summing up we see that in characteristic 0 we do not need any further requirements for the constructions of the lemma and theorem in order to obtain the required universal properties. In positive characteristic one needs to prove (and we send the reader to [Sp]) that in fact the separability can also be granted by the construction.

One final remark, given a group homomorphism $\rho : G \to K$ with kernel H, we would like to say that ρ induces an isomorphism between G/H and the image $\rho(G)$. In fact this is *not always true* as the homomorphism $x \to x^p$ of the additive group shows. This phenomenon of course can occur only in positive characteristic and only if the morphism is not separable. Notice that in this case the notion of *kernel* has to be refined. In our basic example the kernel is defined by the equation $x^p = 0$. This equation defines the point 0 with some multiplicity, as a *scheme*. In general this can be made into the rather solid but complicated theory of *group schemes*, for which the reader can consult [DG].

3 Linearly Reductive Groups

3.1 Linearly Reductive Groups

We come to the main class of algebraic groups of our interest.

Proposition 1. *For an affine group G the following are equivalent:*

(i) Every finite-dimensional rational representation is semisimple.

---

[45] This is actually just a version of ZMT.

(ii) Every rational representation is semisimple.
(iii) The coordinate ring $k[G]$ is semisimple under the right (or the left) action.
(iv) If G is a closed subgroup of $GL(V)$ then all the tensor powers V^n are semisimple.

A group satisfying the previous properties is called a ***linearly reductive group.***

Proof. (i) implies (ii) by abstract representation theory (Chapter 6, Theorem 2.1). Clearly (ii) implies (iii) and (iv), (iii) implies (i) by Lemma 1.4 and the fact that direct sums and submodules of semisimple modules are semisimple. Assume (iv); we want to deduce (i).

Let d be a multiplicative character of G, as for instance the determinant in a linear representation. A finite-dimensional representation U is semisimple if and only if $U \otimes d^r$ is semisimple. Now we apply Theorem 1.4: since tensor powers are semisimple, so is any subrepresentation of a direct sum. Finally the quotient of a semisimple representation is also semisimple, so $U \otimes d^r$ is semisimple. □

Then let G be a linearly reductive group. For every irreducible representation U, the argument of Chapter 6, Theorem 2.6 proves that:

Lemma. *$U^* \otimes U$ appears in $k[G]$ as a $G \times G$ submodule: namely, the isotypic component of type U.*

It follows that

Theorem. *If G is a linearly reductive group we have only countably many non-isomorphic irreducible representations and*

$$k[G] = \bigoplus_i U_i^* \otimes U_i \ (\text{as} \quad G \times G \quad \text{modules}), \tag{3.1.1}$$

where U_i runs over the set of all non-isomorphic irreducible representations of G.

Proof. Since G is linearly reductive and $k[G]$ is a rational representation we must have that $k[G]$ is the direct sum of its isotypic components $U_i^* \otimes U_i$. □

Remark. Observe that this formula is the exact analogue of the decomposition formula for the group algebra of a finite group, Chapter 6, §2.6.3 (see also Chapter 8, §3.2).

Corollary. *If G, H are linearly reductive, so is $G \times H$. The irreducible representations of $G \times H$ are $U \otimes V$ where U (respectively, V) is an irreducible representation of G (respectively, H).*

Proof. We have $k[G] = \bigoplus_i U_i^* \otimes U_i$, $k[H] = \bigoplus_j V_j^* \otimes V_j$; so

$$k[G \times H] = \bigoplus_{i,j} U_i^* \otimes U_i \otimes V_j^* \otimes V_j = \bigoplus_{i,j} (U_i \otimes V_j)^* \otimes U_i \otimes V_j.$$

The theorem follows from Chapter 6, §2.6. □

Lemma. *An algebraic group G is linearly reductive if and only if, given any finite dimensional module U there is a G equivariant projection π_U to the invariants U^G, such that if $f : U_1 \to U_2$ is a $G-$equivariant map of modules we have a (functorial) commutative diagram:*

$$\begin{array}{ccc} U_1 & \xrightarrow{f} & U_2 \\ \pi_{U_1}\downarrow & & \pi_{U_2}\downarrow \\ U_1^G & \xrightarrow{f} & U_2^G \end{array}$$

Proof. If G is linearly reductive we have a canonical decomposition $U = U^G \oplus U_G$ where U_G is the sum of the non trivial irreducible submodules which induces a functorial projection. Conversely assume such a functorial projection exists, for every module U.

It is enough to prove that, given a module V and a submodule W we have a G-invariant complement P in V to W. In other words it is enough to prove that there is a G-equivariant projection of V to W.

For this let $\rho : V \to W$ be any projection, think of $\rho \in \hom(V, W)$ and $\hom(V, W)$ is a G-module. Then by hypothesis there is a G-equivariant projection π of $\hom(V, W)$ to the invariant elements which are $\hom_G(V, W)$. Let us show that $\pi(\rho)$ is the required projection. It is equivariant by construction so it is enough to show that, restricted to W, it is the identity.

Since ρ ia a projection to W under the restriction map $\hom(V, W) \to \hom(W, W)$ we have that ρ maps to the identity 1_W. Thus the functoriality of the commutative diagram implies that also $\pi(\rho)$ maps to 1_W that is $\pi(\rho)$ is a projection. □

Proposition 2. *An algebraic group G is linearly reductive if and only if its algebra of regular functions $k[G]$ has an* **integral**, *that is a $G \times G$ equivariant projection $f \mapsto \int f$ to the constant functions.*

Proof. We want to show that G satisfies the conditions of the previous lemma. Let V be any representation, $v \in V, \phi \in V^*$ consider the function $c_{\phi,v}(g) := \langle \phi \mid gv \rangle \in k[G]$, the map $\phi \mapsto c_{\phi,v}(g)$ is linear. Its integral $\int \langle \phi \mid gv \rangle$ is thus a linear function of ϕ i.e. it can be uniquely represented in the form:

$$\int \langle \phi \mid gv \rangle = \langle \phi \mid \pi(v) \rangle, \ \pi(v) \in V.$$

By right invariance we deduce that $\pi(hv) = \pi(v)$ and by left invariance that

$$\langle \phi \mid \pi(v) \rangle = \int \langle \phi \mid gv \rangle = \int \langle \phi \mid h^{-1} gv \rangle = \langle h\phi \mid \pi(v) \rangle,$$

hence $\pi(v) \in V^G$. By linearity we also have that π is linear and finally, if v is invariant $\langle \phi \mid gv \rangle$ is constant, hence $\pi(v) = v$. We have thus found an equivariant linear projection from V to V^G.

We have thus only to verify that the projection to the invariants is functorial (in the sense of the commutative diagram).

If $f : U_1 \to U_2$ is a map by construction we have

$$\langle \phi \mid \pi_{U_2}(f(v)) = \int \langle \phi \mid gf(v) \rangle = \int \langle f^*(\phi) \mid gv \rangle = \langle f^*(\phi) \mid \pi_{U_1}(v) = \langle \phi \mid f\pi_{U_1}(v)$$

Hence $\pi_{U_2}(f(v)) = f\pi_{U_1}(v)$ we can apply the previous lemma and finish the proof.

□

3.2 Self-adjoint Groups

Given a linearly reductive group, as in the case of finite groups, an explicit description of the decomposition 3.1.1 implies a knowledge of its representation theory.

We need some condition to recognize that an algebraic group is linearly reductive. There is a very simple sufficient condition which is easy to apply. This has been proved in Chapter 6, Proposition 1.2. We recall the statement.

Theorem. *Given a subgroup $G \subset GL(V) = GL(n, \mathbb{C})$, let $G^* := \{g^* = \overline{g}^t | g \in G\}$. If $G = G^*$, all tensor powers $V^{\otimes m}$ are completely reducible under G. In particular, if G is an algebraic subgroup, then it is linearly reductive.*

As a consequence one easily verifies:

Corollary. *The groups $GL(n, \mathbb{C})$, $SL(n, \mathbb{C})$, $O(n, \mathbb{C})$, $SO(n, \mathbb{C})$, $Sp(n, \mathbb{C})$, D are linearly reductive (D denotes the group of invertible diagonal matrices).*

Exercise. Prove that the spin group is self-adjoint under a suitable Hilbert structure.

We should finally remark from the theory developed:

Proposition. *If $G \subset GL(V)$ is linearly reductive, all of the irreducible representations of G, up to tensor product with powers of the determinant, can be found as subrepresentations of $V^{\otimes n}$ for some n.*

If $G \subset SL(V)$, all of the irreducible representations of G can be found as subrepresentations of $V^{\otimes n}$ for some n.

Proof. In Theorem 1.4 we have seen that a representation tensored by the determinant appears in the quotient of a direct sum of tensor powers. If it is irreducible, it must appear in one on these tensor powers.

□

We will apply this idea to classify irreducible representations of classical groups.

Remark. For a connected linear group G to be self-adjoint it is necessary and sufficient that its Lie algebra be self-adjoint.

Proof. G is generated by the exponentials $\exp(a)$, $a \in L$, and $\exp(a)^* = \exp(a^*)$. □

3.3 Tori

The simplest example of a linearly reductive group is the torus T_n, isomorphic to the product of n copies of the multiplicative group, which can be viewed as the group D of invertible diagonal matrices. Its coordinate ring is the ring of Laurent polynomials $k[T] = k[x_i, x_i^{-1}]$ in n variables. A basis of $k[T]$ is given by the monomials:

(3.3.1) $$x^{\underline{m}} = x_1^{m_1} x_2^{m_2} \dots x_n^{m_n},$$

as $\underline{m} = (m_1, m_2, \dots, m_n)$ varies in the free abelian group $\mathbb{Z}^n$.

The 1-dimensional subspace $x^{\underline{m}}$ is a subrepresentation. Under right action, if $t = (t_1, t_2, \dots, t_n)$, we have

(3.3.2) $$(x^{\underline{m}})^t = (x_1 t_1)^{m_1} (x_2 t_2)^{m_2} \dots (x_n t_n)^{m_n} = t^{\underline{m}} x^{\underline{m}}.$$

Theorem 1. *The irreducible representations of the torus T_n are the irreducible characters $t \mapsto t^{\underline{m}}$. They form a free abelian group of rank n called the **character group**.*

Proof. Apply 3.1.1. □

Proposition. *Every rational representation V of T has a basis in which the action is diagonal.*

Proof. This is the consequence of the fact that every rational representation is semisimple and that the irreducible representations are the 1-dimensional characters. □

Definition. (1) A vector generating a T-stable subspace is called a *weight vector* and the corresponding character χ or eigenvalue is called the *weight*.

(2) The set $V_\chi := \{v \in V \mid tv = \chi(t)v, \forall t \in T\}$ is called the *weight space* of V of weight χ.

The weights of the representation can of course appear with any multiplicity, and the corresponding character $\mathrm{tr}(t)$ can be identified with a Laurent polynomial $\mathrm{tr}(t) = \sum_{\underline{m}} c_{\underline{m}} t^{\underline{m}}$ with the $c_{\underline{m}}$ positive integers. One should remark that weights are a generalization of degrees of homogeneity. Let us illustrate this in the simple case of a vector space $V = U_1 \oplus U_2$.

To such a decomposition of a space corresponds a (2-dimensional) torus T, with coordinates x, y, formed by the linear transformations $(u_1, u_2) \to (xu_1, yu_2)$. The decompositions of the various spaces one constructs from V associated to the given direct sum decomposition are just weight space decompositions. For instance

$$S^n(V) = \bigoplus_{i=0}^{n} S^i(U_1) \otimes S^{n-i}(U_2), \quad \bigwedge^n(V) = \bigoplus_{i=0}^{n} \bigwedge^i(U_1) \otimes \bigwedge^{n-i}(U_2).$$

Both $S^i(U_1) \otimes S^{n-i}(U_2)$, $\bigwedge^i(U_1) \otimes \bigwedge^{n-i}(U_2)$ are weight spaces of weight $x^i y^{n-i}$.

We complete this section discussing the structure of subgroups of tori.

Let T be an n-dimensional torus, H a closed subgroup. Since T is abelian, H is normal, and hence there is a linear representation of T such that H is the kernel of the representation.

We know that such a linear representation is a direct sum of 1-dimensional characters. Thus we deduce that H is the subgroup where a set of characters χ take value 1.

The character group is $\hat{T} = \mathbb{Z}^n$ and setting $H^\perp := \{\chi \in \hat{T} \mid \chi(h) = 1, \forall h \in H\}$ we have that $H^\perp$ is a subgroup of $\hat{T} = \mathbb{Z}^n$. By the elementary theory of abelian groups we can change the character basis of $\hat{T} = \mathbb{Z}^n$ to a basis e_i so that $H^\perp = \sum_{i=1}^h \mathbb{Z} n_i e_i$ for some positive integers n_i and a suitable h. It follows that in these coordinates, H is the set

$$(3.3.3) \qquad H = \{(t_1, \dots, t_n) \mid t_i^{n_i} = 1,\ i = 1, \dots, h\} = \prod_{i=1}^h \mathbb{Z}/(n_i) \times T_{n-h}.$$

Here T_k denotes a k-dimensional torus. In particular we have:

Theorem 2. *A closed connected subgroup of a torus is a torus. A quotient of a torus is a torus.*

Proof. We have seen in 3.3.3 the description of any closed subgroup. If it is connected we have $n_i = 1$ and $H = T_{n-h}$.

As for the second part, T/H is the image of the homomorphism $(t_1, \dots, t_n) \mapsto (t_1^{n_1}, \dots, t_h^{n_h})$ with image the torus T_h. □

Remark. We have again a problem in characteristic $p > 0$. In this case the n_i should be prime with p (or else we have to deal with group schemes).

3.4 Additive and Unipotent Groups

Although we will not be using them as much as tori, one should take a look at the other algebraic abelian groups. For tori, the prototype is the multiplicative group. For the other groups as the prototype we have in mind the *additive group*. We have seen in 1.2.1 that this is made of unipotent matrices, so according to Theorem 1.4 in all representations it will be unipotent. In particular let us start from the action on its coordinate ring $k[x]$. If $a \in k$ is an element of the additive group, the right action on functions is given by $f(x + a)$, so we see that:

Proposition 1. *For every positive integer m, the subspace P_m of $k[x]$ formed by the polynomials of degree $\leq m$ is a submodule. In characteristic 0 these are the only finite-dimensional submodules of P_m.*

Proof. That these spaces P_m of polynomials are stable under the substitutions $x \mapsto x + a$ is clear. Conversely, let M be a finite-dimensional submodule. Suppose that m is the maximum degree of a polynomial contained in M, so $M \subset P_k$. Let $f(x) = x^m + u(x) \in M$ where $u(x) = bx^{m-1} + \cdots$ has degree strictly less than m. We have that for $a \in k$

$$f_a(x) := f(x+a) - f(x) = (ma+b)x^{m-1} + \cdots \in M.$$

If $ma + b \neq 0$, this is a polynomial of degree exactly $m - 1$. In characteristic 0, $m \neq 0$, and so we can find an a with $ma + b \neq 0$. By induction all the monomials $x^i, i \leq m - 1$ are in M, hence also $x^m \in M$ and $P_m \subset M$. □

Remark. In characteristic $p > 0$, we have that $k[x^p]$ is a submodule.

Unipotent elements tend to behave in a more complicated way in positive characteristic. One reason is this. Let us work in characteristic 0, i.e., $k = \mathbb{C}$. If A is a nilpotent matrix, the exponential series $\exp(A) = \sum_{i=0}^{\infty} \frac{A^i}{i!}$ terminates after finitely many steps, so it is indeed a polynomial, and $\exp(A)$ is a unipotent matrix.

Similarly, let $1 + A$ be unipotent, so that A is nilpotent. The logarithmic series terminates after finitely many steps: it is a polynomial, and $\log(A)$ is a nilpotent matrix. We have:

Proposition 2. *The variety of complex nilpotent matrices is isomorphic to the variety of unipotent matrices by the map* exp, *and its inverse* log.

Notice that both maps are equivariant with respect to the conjugation action.

In positive characteristic neither of these two series makes sense (unless the characteristic is large with respect to the size of the matrices). The previous proposition is not true in general.

We complete this discussion by studying the 1-dimensional connected algebraic groups. We assume $k = \mathbb{C}$.

Lemma. *Let $g = \exp(A)$, with $A \neq 0$ a nilpotent matrix. Let $\{g\}$ be the Zariski closure of the cyclic subgroup generated by g. Then $\{g\} = \exp(t\,A)$, $t \in \mathbb{C}$.*

Proof. Since $\{g\}$ is made of unipotent elements, it is equivalent to prove that $\log(\{g\}) = \mathbb{C}A$. Since $\exp(mA) = \exp(A)^m$ we have that $mA \in \log(\{g\})$, $\forall m \in \mathbb{Z}$. By the previous proposition it follows that $\log(\{g\})$ is a closed subvariety of the nilpotent matrices and the closure of the elements $\log(g^m) = mA$. In characteristic 0 we easily see that the closure of the integral multiples of A (in the Zariski topology) is $\mathbb{C}A$, as desired. □

Theorem. *A 1-dimensional connected algebraic group is isomorphic to the additive or the multiplicative group.*

Proof. Let G be such a group. The proof in positive characteristic is somewhat elaborate and we send the reader to ([Sp], [Bor]). Let us look in characteristic 0. First, if $g \in G$, we know that $g_s, g_u \in G$. Let us study the case in which there is a nontrivial unipotent element. By the previous lemma G contains the 1-dimensional group $\exp(\mathbb{C}\,A)$. Hence, being connected and 1-dimensional, it must coincide with $\exp(\mathbb{C}\,A)$ and the map $t \mapsto \exp(t\,A)$ is an isomorphism between $(\mathbb{C}, +)$ and G.

Consider next the case in which G does not contain unipotent elements; hence it is made of commuting semisimple elements and thus it can be put into diagonal form.

In this case it is no longer true that the closure of a cyclic group is 1-dimensional; in fact, it can be of any given dimension. But from Theorem 2 of 3.3 we know that a closed connected subgroup of a torus is a torus. Thus G is a torus, but since it is 1-dimensional it is isomorphic to the multiplicative group. □

We have seen in Chapter 4, §7 that a connected Lie group is solvable if and only if its Lie algebra is solvable. Thus Lie's theorem, Chapter 4, §6.3, implies that in characteristic 0, a connected solvable algebraic linear group is conjugate to a subgroup of the group of upper triangular matrices. In other words, in a suitable basis, it is made of upper triangular matrices. In fact this is true in any characteristic ([Sp], [Bor], [Hu2]), as we will discuss in §4. Let us assume this basic fact for the moment.

Definition. A linear group is called *unipotent* if and only if its elements are all unipotent.

Thus, from the previous discussion, we have seen in characteristic 0 that a unipotent group is conjugate to a subgroup of the group U_n of strictly upper triangular matrices. In characteristic 0 in fact we can go further. The argument of Proposition 2 shows that the two maps exp, log are also bijective algebraic isomorphisms between the variety N_n of upper triangular matrices with 0 on the diagonal, a Lie algebra, and the variety U_n of upper triangular matrices with 1 on the diagonal, an algebraic unipotent group.

Theorem. *Under the map* $\exp : N_n \to U_n$ *the image of a Lie subalgebra is an algebraic group.*

Under the map $\log : U_n \to N_n$ *the image of an algebraic group is a Lie algebra.*

Proof. Let $A \subset N_n$ be a Lie algebra. We know by the isomorphism statement that $\exp(A)$ is a closed subvariety of U_n. On the other hand, exp maps A into the analytic Lie group of Lie algebra A and it is even a local isomorphism. It follows that this analytic group is closed and coincides with $\exp(A)$. As for the converse, it is enough, by the previous statement, to prove that any algebraic subgroup H of U_n is connected, since then it will follow that it is the exponential of its Lie algebra. Now if $a = \exp(b) \neq 1$, $a \in H$ is a unipotent matrix, we have that $\exp(\mathbb{C}b) \subset H$ by the previous lemma, thus H is connected. □

3.5 Basic Structure Theory

When studying the structure of algebraic groups it is important to recognize which constructions produce algebraic groups. Let G be an algebraic group with Lie algebra L. The first important remark is that L is in a natural way a complex Lie algebra: in fact, for $GL(n, \mathbb{C})$ it is the complex matrices and for an algebraic subgroup it is clearly a complex subalgebra of matrices. Given a Lie subgroup $H \subset G$, a necessary condition for H to also be algebraic is that its Lie algebra M should be also complex. This is *not sufficient* as the following trivial example shows. Let A be a matrix which is neither semisimple nor nilpotent. The complex Lie algebra generated by A is not the Lie algebra of an algebraic group (from 1.5). On the other hand, we want to prove that the derived and lower central series as well as the solvable and nilpotent radicals of an algebraic group (considered as a Lie group) are algebraic. We thus need a criterion to ensure that a subgroup is algebraic. We use the following:

Proposition 1. *Let G be an algebraic group and $V \subset G$ an irreducible subvariety, $1 \in V$. Then the closed (in the usual complex topology) subgroup H generated by V is algebraic and connected.*

Proof. Let V^{-1} be the set of inverses of V, which is still an irreducible variety containing 1. Let $U = \overline{VV^{-1}}$. Since U is the closure of the image of an irreducible variety $V \times V^{-1}$ under an algebraic map $\pi : V \times V^{-1} \to G, \pi(a, b) = ab$, it is also an irreducible variety. Since $1 \in V \cap V^{-1}$ we have that $V, V^{-1} \subset U$ and $U = U^{-1}$. Consider for each positive integer m the closure $\overline{U^m}$ of the product $UU \dots U = U^m$. For the same reasons as before $\overline{U^m}$ is an irreducible variety, and $\overline{U^m} \subset \overline{U^{m+1}} \subset H$. An increasing sequence of irreducible varieties must at some point stop increasing.[46] Assume that $\overline{U^n} = \overline{U^k}, \forall k \geq n$. By continuity $\overline{U^n}\,\overline{U^n} \subset \overline{U^{2n}} = \overline{U^n}$. A similar argument shows $\overline{U^n} = \overline{U^n}^{-1}$, so $\overline{U^n}$ is a group and a subvariety; hence $\overline{U^n} = H$. □

Theorem. *Let G be a connected algebraic group. The terms of the derived and lower central series are all algebraic and connected.*

Proof. The proof is by induction. We do one case; the other is similar. Assume we know that G^i is algebraic and irreducible. G^{i+1} is the algebraic subgroup generated by the set X_i of elements $\{x, y\}$, $x \in G$, $y \in G^i$. The map $(x, y) \mapsto \{x, y\}$ is algebraic, so X_i is dense in an irreducible subvariety, and we can apply the previous proposition. □

Proposition 2. *The center of an algebraic group is algebraic (but in general not connected).*

The solvable radical of an algebraic group, as a Lie group, is algebraic.

Proof. The center is the kernel of the adjoint representation which is algebraic.

For the second part look at the image R in the adjoint representation of the solvable radical. Since Lie(R) is solvable, R can be be put into some basis in the form of upper triangular matrices. Then the Zariski closure in G of R is still made of upper triangular matrices and hence solvable, and clearly a normal connected subgroup. Since R is maximal closed connected normal solvable, it must be closed in the Zariski topology. □

In the same way one has, for subgroups of an algebraic group G, the following.

Proposition 3. *The Zariski closure of a connected solvable Lie subgroup of G is solvable.*

A maximal connected solvable Lie subgroup of G is algebraic.

---

[46] Here is where we use the fact that we started from an irreducible set; otherwise the number of components could go to infinity.

3.6 Reductive Groups

Definition 1. A linear algebraic group is called *reductive* if it does not contain any closed unipotent normal subgroup.

A linear algebraic group is called *semisimple* if it is connected and its solvable radical is trivial.

Notice that semisimple implies reductive (but not conversely).

Proposition 1. *(1) If G is a connected abelian linearly reductive group, it is a torus.*

(2) If G is solvable and connected, $\{G, G\}$ is unipotent.

(3) A unipotent linearly reductive group reduces to $\{1\}$.

Proof. (1) Let V be a faithful representation of G. Decompose it into irreducibles. Since G is abelian each irreducible is 1-dimensional, hence G is a subgroup of the diagonal matrices. From Theorem 2 of 3.3 it is a torus.

(2) In a linear representation G can be put into triangular form; in characteristic 0 this follows from Chapter 4. §7.1 and the analogous theorem of Lie for solvable Lie algebras. In general it is the Lie–Kolchin theorem which we discuss in Section 4.1. Then all the commutators lie in the strictly upper triangular matrices, a unipotent group.

(3) A unipotent group U, in a linear representation, in a suitable basis is made of upper triangular matrices, which are unipotent by Theorem 1.5. Hence U has a fixed vector. If U is linearly reductive, this fixed vector has an invariant complement. By induction U acts trivially on this complement. So U acts trivially on any representation, hence $U = 1$. □

Theorem 1. *Let G be an algebraic group and N a normal subgroup. Then G is linearly reductive if and only if G/N and N are linearly reductive.*

Proof. Assume G is linearly reductive. Since every representation of G/N is also a representation of G, clearly G/N is linearly reductive. We have to prove that $k[N]$ is completely reducible as an N-module. Since $k[N]$ is a quotient of $k[G]$, it is enough to prove that $k[G]$ is completely reducible as an N-module. Let M be the sum of all irreducible N-submodules of $k[G]$, we need to show that $M = k[G]$. In any case, since N is normal, it is clear that M is G-stable. Since G is linearly reductive, we have a G-stable complement P, and $k[G] = M \oplus P$. P must be 0, otherwise it contains an irreducible N-module, which by definition is in M.

Conversely, using the fact that N is linearly reductive we have an $N \times N$-equivariant projection $\pi : k[G] \to k[G]^{N\times N} = k[G/N]$. This projection is canonical since it is the only projection to the isoptypic component of invariants. Therefore it commutes with the $G \times G$ action. On $k[G]^{N\times N}$ we have an action of $G/N \times G/N$ so, since this is also linearly reductive, we have an integral projecting to the $G/N \times G/N$ invariants. Composing, we have the required projection $k[G] \to k$. We apply now Proposition 2 of §3.1. □

Proposition 2. *Let G be an algebraic group and G_0 the connected component of* 1. *Then G is linearly reductive if and only if G_0 is linearly reductive and the order of the finite group G/G_0 is not divisible by the characteristic of the base field.*

Proof. Since G_0 is normal, one direction comes from the previous theorem. Assume G_0 linearly reductive, M a rational representation of G, and N a submodule. We need to prove that N has a G-stable complement. N has a G_0-stable complement: in other words, there is a projection $\pi : M \to N$ which commutes with G_0. It follows that given $a \in G$, the element $a\pi a^{-1}$ depends only on the class of a modulo G_0. Then we can average and define $\rho = \frac{1}{|G/G_0|}\sum_{a \in G/G_0} a\pi a^{-1}$ and obtain a G-equivariant projection to N. □

The main structure theorem is (cf. [Sp]):

Theorem 2. *In characteristic* 0 *a linear algebraic group is linearly reductive if and only if it is reductive.*[47]

For a reductive group the solvable radical is a torus contained in the center.

Proof. Let G be linearly reductive. If it is not reductive it has a normal unipotent subgroup U, which by Theorem 1 would be linearly reductive, hence trivial by Proposition 1. Conversely, let G be reductive. Since the radical is a torus it is enough to prove that a semisimple algebraic group is linearly reductive. This is not easy but is a consequence of the theory of Chapter 10. □

The second statement of Theorem 2 follows from the more precise:

Lemma. *Let G be a connected algebraic group and $T \subset G$ a torus which is also a normal subgroup; then T is in the center.*

Proof. The idea is fairly simple. By assumption G induces by conjugation a group of automorphisms of T. Since the group of automorphisms of T is discrete and G is connected it must act trivially. To make this proof more formal, let M be a faithful representation of G and decompose M into eigenspaces for T for the different eigenvalues present. Clearly G permutes these spaces, and (now it should be clear) since G is connected the permutations which it induces must be the identity. Hence for any weight λ of T appearing in M and $g \in G, t \in T$ we have $\lambda(gtg^{-1}) = \lambda(t)$. Since M is faithful, the weights λ generate the character group. Hence $gtg^{-1} = t$, $\forall g \in G, \forall t \in T$. □

With these facts we can explain the program of classification of linearly reductive groups over $\mathbb{C}$.

Linearly reductive groups and their representations can be fully classified.

The steps are the following. First, decompose the adjoint representation L of a linearly reductive group G into irreducibles, $L = \bigoplus_i L_i$. Each L_i is then a simple Lie

---

[47] Linear reductiveness in characteristic $p > 0$ is a rare event and one has to generalize most of the theory in a very nontrivial way.

algebra. We separate the sum of the trivial representations, which is the Lie algebra of the center Z. By Proposition 1, the connected component Z_0 of Z is a torus.

In Chapter 10 we classify simple Lie algebras, and prove that the simply connected group associated to such a Lie algebra is a linearly reductive group. It follows that if G_i is the simply connected group associated to the nontrivial factors L_i, we have a map $Z_0 \times \prod_i G_i \to G$ which is an isomorphism of Lie algebras (and algebraic).

Then G is isomorphic to $Z_0 \times \prod_i G_i/A$ where A is a finite group contained in the center $Z_0 \times \prod_i Z_i$ of $Z_0 \times \prod_i G_i$. The center of a simply connected linearly reductive group is described explicitly in the classification and is a cyclic group or $\mathbb{Z}/(2) \times \mathbb{Z}/(2)$; hence the possible subgroups A can be made explicit. This is then a classification.

4 Borel Subgroups

4.1 Borel Subgroups

The notions of *maximal torus* and *Borel subgroup* play a special role in the theory of linear algebraic groups.

Definition 1. A subgroup of an algebraic group is called a *maximal torus* if it is a closed subgroup, a torus as an algebraic group, and maximal with respect to this property.

A subgroup of an algebraic group is called a *Borel subgroup* if it is closed, connected and solvable, and maximal with respect to this property.

The main structure theorem is:

Theorem 1. *All maximal tori are conjugate. All Borel subgroups are conjugate.*

We illustrate this theorem for classical groups giving an elementary proof of the first part (see Chapter 10, §5 for more details on this topic and a full proof).

Example 1. $GL(V)$. In the general linear group of a vector space V a maximal torus is given by the subgroup of all matrices which are diagonal for some fixed basis of V.

A Borel subgroup is the subgroup of matrices which fix a *maximal flag*, i.e., a sequence $V_1 \subset V_2 \subset \cdots V_{n-1} \subset V_n = V$ of subspaces of V with $\dim V_i = i$ (assuming $n = \dim V$).

Example 2. $SO(V)$. In the special orthogonal group of a vector space V, equipped with a nondegenerate symmetric form, a maximal torus is given as follows.

If $\dim V = 2n$ is even we take a hyperbolic basis $e_1, f_1, e_2, f_2, \ldots, e_n, f_n$. That is, the 2-dimensional subspaces V_i spanned by e_i, f_i are mutually orthogonal and the matrix of the form on the vectors e_i, f_i is $\begin{pmatrix} 0 & 1 \\ 1 & 0 \end{pmatrix}$.

For such a basis we get a maximal torus of matrices which stabilize each V_i and, restricted to V_i in the basis e_i, f_i, has matrix $\begin{pmatrix} \alpha_i & 0 \\ 0 & \alpha_i^{-1} \end{pmatrix}$.

The set of maximal tori is in 1-1 correspondence with the set of all decompositions of V as the direct sum of 1-dimensional subspaces spanned by hyperbolic bases. The set of Borel subgroups is in 1-1 correspondence with the set of *maximal isotropic flags*, i.e., the set of sequences $V_1 \subset V_2 \subset \cdots V_{n-1} \subset V_n$ of subspaces of V with $\dim V_i = i$ and such that the subspace V_n is totally isotropic for the form. To such a flag one associates the maximal flag $V_1 \subset V_2 \subset \cdots V_{n-1} \subset V_n = V_n^\perp \subset V_{n-1}^\perp \cdots \subset V_2^\perp \subset V_1^\perp \subset V$ which is clearly stable under the subgroup fixing the given isotropic flag. If $\dim V = 2n+1$ is odd, we take bases of the form $e_1, f_1, e_2, f_2, \dots, e_n, f_n, u$ with $e_1, f_1, e_2, f_2, \dots, e_n, f_n$ hyperbolic and u orthogonal to $e_1, f_1, e_2, f_2, \dots, e_n, f_n$. As a maximal torus we can take the same type of subgroup which now fixes u.

The analogue statement holds for Borel subgroups except that now a maximal flag is $V_1 \subset V_2 \subset \dots V_{n-1} \subset V_n \subset V_n^\perp \subset V_{n-1}^\perp \cdots \subset V_2^\perp \subset V_1^\perp \subset V$.

Example 3. *Sp*(V). In a symplectic group of a vector space V equipped with a nondegenerate skew-symmetric form a maximal torus is given as follows. Let $\dim V = 2n$. We say that a basis $e_1, f_1, e_2, f_2, \dots, e_n, f_n$ is symplectic if the 2-dimensional subspaces V_i spanned by e_i, f_i are mutually orthogonal and the matrix of the form on the vectors e_i, f_i is: $\begin{pmatrix} 0 & 1 \\ -1 & 0 \end{pmatrix}$. For such a basis we get a maximal torus of matrices that stabilize each V_i and, restricted to V_i in the basis e_i, f_i, has matrix $\begin{pmatrix} \alpha_i & 0 \\ 0 & \alpha_i^{-1} \end{pmatrix}$.

The set of maximal tori is in 1-1 correspondence with the set of all decompositions of V as the direct sum of 1-dimensional subspaces spanned by symplectic bases. The set of Borel subgroups is again in 1-1 correspondence with the set of *maximal isotropic flags*, i.e., the set of sequences $V_1 \subset V_2 \subset \cdots V_{n-1} \subset V_n$ of subspaces of V with $\dim V_i = i$ and such that the subspace V_n is totally isotropic for the form. To such a flag one associates the maximal flag $V_1 \subset V_2 \subset \cdots V_{n-1} \subset V_n = V_n^\perp \subset V_{n-1}^\perp \cdots \subset V_2^\perp \subset V_1^\perp \subset V$ which is clearly stable under the subgroup fixing the given isotropic flag.

Proof of previous statements. We use the fact that a torus action on a vector space decomposes as a direct sum of 1-dimensional irreducible representations (cf. §3.3). This implies immediately that any torus in the general linear group has a basis in which it is diagonal, and hence the maximal tori are the ones described.

For the other two cases we take advantage of the fact that, given two eigenspaces relative to two characters χ_1, χ_2, these subspaces are orthogonal under the given invariant form unless $\chi_1\chi_2 = 1$. For instance, assume we are in the symmetric case (the other is identical). Given two eigenvectors u_1, u_2 and an element t of the maximal torus $(u_1, u_2) = (tu_1, tu_2) = (\chi_1\chi_2)(t)(u_1, u_2)$. It follows that if $\chi_1\chi_2 \neq 1$, the two eigenvectors are orthogonal.

By the nondegenerate nature of the form when $\chi_1\chi_2 = 1$, the two eigenspaces relative to the two characters must be in perfect duality since they are orthogonal to the remaining weight spaces. We thus choose, for each pair of characters χ, χ^{-1}, a basis in the eigenspace V_χ of χ and the basis in $V_{\chi^{-1}}$ dual to the chosen basis. We complete these bases with an hyperbolic basis of the eigenspace of 1. In this way we have constructed a hyperbolic basis $\mathcal{B}$ for which the given torus is contained in the torus associated to $\mathcal{B}$.

All hyperbolic bases are conjugate under the orthogonal group. If the orthogonal transformation which conjugates two hyperbolic bases is improper, we may compose it with the exchange of e_i, f_i in order to get a special (proper) orthogonal transformation. So, up to exchanging e_i, f_i, an operation leaving the corresponding torus unchanged, two hyperbolic bases are also conjugate under the special orthogonal group. So the statement for maximal tori is complete in all cases. □

The discussion of Borel subgroups is a little subtler; here one has to use the basic fact:

Theorem (Lie–Kolchin). *A connected solvable group G of matrices is conjugate to a subgroup of upper triangular matrices.*

There are various proofs of this statement which can be found in the literature at various levels of generality. In characteristic 0 it is an immediate consequence of Lie's theorem and the fact that a connected Lie group is solvable if and only if its Lie algebra is solvable. The main step is to prove the existence of a common eigenvector for G from which the statement follows immediately by induction.

A particularly slick proof follows immediately from a stronger theorem.

Borel fixed-point theorem. *Given an action of a connected solvable group G on a projective variety X there exists a fixed point.*

Proof. Work by induction on the dimension of G. If G is trivial there is nothing to prove; otherwise G contains a proper maximal connected normal subgroup H. Since G is solvable $H \supset \{G, G\}$ and thus G/H is abelian (in fact it is easy to prove that it is 1-dimensional). Let X^H be the set of fixed points of H. It is clearly a projective subvariety. Since H is normal in G we have that X^H is G-stable and G/H acts on X^H. Take an orbit Dp of minimal dimension for D on X^H, and let E be the stabilizer of p. We claim that $E = D$, and hence p is the required fixed point. In any event D/E is closed and hence, since X is projective, it is compact. Now D/E is also a connected affine algebraic group. We thus have to use a basic fact from algebraic geometry: *an irreducible affine variety is complete if and only if it reduces to a point.* So D/E is a point, or $D = E$. □

The projective variety to which this theorem has to be applied to obtain the Lie–Kolchin Theorem is the flag variety whose points are the complete flags of linear subspaces $F := V_1 \subset V_2 \subset \cdots \subset V_n = V$ with $\dim V_i = i$. The flag variety is easily seen to be projective (cf. Chapter 10, §5.2).

Clearly a linear map fixes the flag F if and only if it is an upper triangular matrix with respect to a basis $e_1, \dots, e_n$ with the property that V_i is spanned by $e_1, \dots, e_i$ for each $i \leq n$. This shows that Borel's fixed point theorem implies the Lie–Kolchin theorem.

In fact, it is clear that these statements are more or less equivalent. When we have a linear group G acting on a vector space V, finding a vector which is an eigenvector of all the elements of G is the same as finding a line in V which is fixed by G, i.e., a fixed point for the action of G on the projective space of lines of V.

The geometric argument of Borel thus breaks down into two steps. The first step is a rather simple statement of algebraic geometry: when one has an algebraic group G acting on a variety V one can always find at least one closed orbit, for instance, choosing an orbit of minimal dimension. The next point is that for a solvable group the only possible projective closed orbits are points.

Given the theorem of Lie–Kolchin, the study of Borel subgroups is immediate. For the linear group it is clearly a restatement of this theorem. For the other groups, let G be a connected solvable group of linear transformations fixing the form. We do the symmetric case since the other is similar but simpler. We work by induction on the dimension of V. If $\dim V = 1$, then $G = 1$, and a maximal isotropic flag is empty and there is nothing to prove. Let u be an eigenvector of G and $u^\perp$ its orthogonal subspace which is necessarily G-stable. If u is isotropic, $u \in u^\perp$ and the space $u^\perp/\mathbb{C}u$ is equipped with the induced symmetric form (which is nondegenerate) for which G acts again as a group of orthogonal transformations, and we can apply induction.

In the case where u is not isotropic we have a direct sum orthogonal decomposition $V = u^\perp \oplus \mathbb{C}u$. If $g \in G$, we have $gu = \pm u$ since g is orthogonal. The induced map $G \to \pm 1$ is a homomorphism and, since G is connected, it must be identically 1. If $\dim u^\perp > 1$, by induction we can find an isotropic vector stabilized by G in $u^\perp$ and go back to the previous case. If $\dim u^\perp = 1$, the same argument as before shows that $G = 1$. In this case of course G fixes any isotropic flag.

Furthermore, for a connected algebraic group G we have the following important facts.

Theorem 4. *Every element of G is contained in a Borel subgroup.*

If G is a reductive group, then the union of all maximal tori is dense in G.

We leave as exercise to verify these statements directly for classical groups. They will be proved in general in Chapter 10.

8

Group Representations

In this chapter we want to have a first look into the representation theory of various groups with extra structure, such as algebraic or compact groups. We will use the necessary techniques from elementary algebraic geometry or functional analysis, referring to standard textbooks. One of the main points is a very tight relationship between a special class of algebraic groups, the reductive groups, and compact Lie groups. We plan to illustrate this in the classical examples, leaving the general theory to Chapter 10.

1 Characters

1.1 Characters

We want to deduce some of the basic theory of characters of finite groups and more generally, compact and reductive groups. We start from some general facts, valid for any group.

Definition. Given a linear representation $\rho : G \to GL(V)$ of a group G, where V is a finite-dimensional vector space over a field F, we define its *character* to be the following function on G:[48]

$$\chi_\rho(g) := \operatorname{tr}(\rho(g)).$$

Here tr is the usual trace. We say that a character is irreducible if it comes from an irreducible representation.

Some properties are immediate (cf. Chapter 6, §1.1).

---

[48] There is also a deep theory for infinite-dimensional representations. In this setting the trace of an operator is not always defined. With some analytic conditions a character may also be a distribution.

Proposition 1.

(1) $\chi_\rho(g) = \chi_\rho(aga^{-1}),\ \forall a, g \in G$. *The character is constant on conjugacy classes. Such a function is called a* ***class function***.

(2) Given two representations ρ_1, ρ_2 *we have*

(1.1.1)
$$\chi_{\rho_1 \oplus \rho_2} = \chi_{\rho_1} + \chi_{\rho_2}, \quad \chi_{\rho_1 \otimes \rho_2} = \chi_{\rho_1}\chi_{\rho_2}.$$

(3) If ρ *is unitarizable, the character of the dual representation* ρ^* *is the conjugate of* χ_ρ*:*

(1.1.2)
$$\chi_{\rho^*} = \chi_{\overline{\rho}}.$$

Proof. Let us prove (3) since the others are clear. If ρ is unitarizable, there is a basis in which the matrices $A(g)$ of $\rho(g)$ are unitary. In the dual representation and in the dual basis the matrix $A^*(g)$ of $\rho^*(g)$ is the transposed inverse of $A(g)$. Under our assumption $A(g)$ is unitary, hence $(A(g)^{-1})^t = \overline{A(g)}$ and $\chi_{\rho^*}(g) = \operatorname{tr}(A^*(g)) = \operatorname{tr}(\overline{A(g)}) = \overline{\operatorname{tr}(A(g))} = \overline{\chi_\rho(g)}$. □

We have just seen that characters can be added and multiplied. Sometimes it is convenient to extend the operations to include the difference $\chi_1 - \chi_2$ of two characters. Of course such a function is no longer necessarily a character but it is called a *virtual character*.

Proposition 2. *The virtual characters of a group G form a commutative ring called the character ring of G.*

Proof. This follows immediately from 1.1.1. □

Of course, if the group G has extra structure, we may want to restrict the representations, hence the characters, to be compatible with the structure. For a topological group we will restrict to continuous representations, while restricting to rational ones for algebraic groups. We will thus speak of continuous or rational characters.

In each case the class of representations is closed under direct sum and tensor product. Thus we also have a character ring, made by the virtual continuous (respectively, algebraic) characters.

Example. In Chapter 7, §3.3 we have seen that for a torus T of dimension n, the (rational) irreducible characters are the elements of a free abelian group of rank n. Thus the character ring is the ring of Laurent polynomials $\mathbb{Z}[x_1^{\pm 1}, \dots, x_n^{\pm 1}]$. Inside this ring the characters are the polynomials with nonnegative coefficients.

1.2 Haar Measure

In order to discuss representations and characters for compact groups we need some basic facts from the theory of integration on groups.

The type of measure theory needed is a special case of the classical approach to the Daniell integral (cf. [DS]).

Let X be a locally compact topological space. We use the following notation: $C_0(X, \mathbb{R})$ denotes the algebra of real-valued continuous functions with compact support, while $C_0(X)$ and $C(X)$ denote the complex-valued continuous functions with compact support, respectively, all continuous functions. If X is compact, every function has compact support, hence we drop the subscript 0.

Definition. An integral on X is a nonzero linear map $I : C_0(X, \mathbb{R}) \to \mathbb{R}$, such that if $f \in C_0(X, \mathbb{R})$ and $f(x) \geq 0, \ \forall x \in X$ (a positive function) we have $I(f) \geq 0$.[49]

If $X = G$ is a topological group, we say that an integral I is left-invariant if, for every function $f(x) \in C_0(X, \mathbb{R})$ and every $g \in G$, we have $I(f(x)) = I(f(g^{-1}x))$.

Measure theory allows us to extend a Daniell integral to larger classes of functions, in particular to the characteristic functions of *measurable sets*, and hence deduce a measure theory on X in which all closed and open sets are measurable. This measure theory is essentially equivalent to the given integral. Therefore one uses often the notation dx for the measure and $I(f) = \int f(x)dx$ for the integral.

In the case of groups the measure associated to a left-invariant integral is called a *left-invariant Haar measure*.

In our treatment we will mostly use L^2 functions on X. They form a Hilbert space $L^2(X)$, containing as a dense subspace the space $C_0(X)$; the Hermitian product is $I(f(x)\overline{g}(x))$. A basic theorem (cf. [Ho]) states that:

Theorem. *On a locally compact topological group G, there is a left-invariant measure called the Haar measure.*

The Haar measure is unique up to a scale factor.

This means that if I and J are two left-invariant integrals, there is a positive constant c with $I(f) = cJ(f)$ for all functions.

Exercise. If I is a left-invariant integral on a group G and f a nonzero positive function, we have $I(f) > 0$.

When G is compact, the Haar measure is usually normalized so that the volume of G is 1, i.e., $I(1) = 1$. Of course G also has a right-invariant Haar measure. In general the two measures are not equal.

Exercise. Compute the left and right-invariant Haar measure for the two-dimensional Lie group of affine transformations of $\mathbb{R}$, $x \mapsto ax + b$.

If $h \in G$ and we are given a left-invariant integral $\int f(x)$, it is clear that $f \mapsto \int f(xh)$ is still a left-invariant integral, so it equals some multiple $c(h) \int f(x)$. The function $c(h)$ is immediately seen to be a continuous multiplicative character with values positive numbers.

Proposition 1. *For a compact group, the left and right-invariant Haar measures are equal.*

---

[49] The axioms of the Daniell integral in this special case are simple consequences of these hypotheses.

Proof. Since G is compact, $c(G)$ is a bounded set of positive numbers. If for some $h \in G$ we had $c(h) \neq 1$, we then have $\lim_{n\to\infty} c(h^n) = \lim_{n\to\infty} c(h)^n$ is 0 or ∞, a contradiction. □

We need only the Haar measure on Lie groups. Since Lie groups are differentiable manifolds one can use the approach to integration on manifolds using differential forms (cf. [Spi]). In fact, as for vector fields, one can find $n = \dim G$ left-invariant differential linear forms ψ_i, which are a basis of the cotangent space at 1 and so too at each point.

Proposition 2. *The exterior product $\omega := \psi_1 \wedge \psi_2 \wedge \ldots \wedge \psi_n$ is a top-dimensional differential form which is left-invariant and defines the volume form for an invariant integration.*

Proof. Take a left translation L_g. By hypothesis $L_g^*(\psi_i) = \psi_i$ for all i. Since L_g^* preserves the exterior product we have that ω is a left-invariant form. Moreover, since the ψ_i are a basis at each point, ω is nowhere 0. Hence ω defines an orientation and a measure on G, which is clearly left-invariant. □

1.3 Compact Groups

The Haar measure on a compact group allows us to average functions, thus getting projections to invariants. Recall that for a representation V of G, the space of invariants is denoted by V^G.

Proposition 1. *Let $\rho : G \to GL(V)$ be a continuous complex finite-dimensional representation of a compact group G (in particular a finite group). Then (using Haar measure),*

$$\dim_{\mathbb{C}} V^G = \int_G \chi_\rho(g)dg. \tag{1.3.1}$$

Proof. Let us consider the operator $\pi := \int \rho(g)dg$. We claim that it is the projection operator on V^G. In fact, if $v \in V^G$,

$$\pi(v) = \int_G \rho(g)(v)dg = \int_G v\,dg = v.$$

Otherwise,

$$\rho(h)\pi(v) = \int_G \rho(h)\rho(g)v\,dg = \int_G \rho(hg)v\,dg = \pi(v)$$

by left invariance of the Haar integral.

We then have $\dim_{\mathbb{C}} V^G = \mathrm{tr}(\pi) = \mathrm{tr}(\int_G \rho(g)dg) = \int_G \mathrm{tr}(\rho(g))dg = \int_G \chi_{\rho(g)}dg$ by linearity of the trace and of the integral. □

The previous proposition has an important consequence.

Theorem 1 (Orthogonality of characters). *Let χ_1, χ_2 be the characters of two irreducible representations ρ_1, ρ_2 of a compact group G. Then*

$$\int_G \chi_1(g)\overline{\chi}_2(g)dg = \begin{cases} 0 & \text{if } \rho_1 \neq \rho_2 \\ 1 & \text{if } \rho_1 = \rho_2 \end{cases}. \tag{1.3.2}$$

Proof. Let V_1, V_2 be the spaces of the two representations. Consider $\hom(V_2, V_1) = V_1 \otimes V_2^*$. As a representation $V_1 \otimes V_2^*$ has character $\chi_1(g)\overline{\chi}_2(g)$, from 1.1.1 and 1.1.2.

We have seen that $\hom_G(V_2, V_1) = (V_1 \otimes V_2^*)^G$, hence, from the previous proposition, $\dim_{\mathbb{C}} \hom_G(V_2, V_1) = \int_G \chi_1(g)\overline{\chi}_2(g)dg$. Finally, by Schur's lemma and the fact that V_1 and V_2 are irreducible, $\hom_G(V_2, V_1)$ has dimension 0 if $\rho_1 \neq \rho_2$ and 1 if they are equal. The theorem follows. □

In fact a more precise theorem holds. Let us consider the Hilbert space of L^2 functions on G. Inside we consider the subspace $L^2_c(G)$ of class functions, which is clearly a closed subspace. Then:

Theorem 2. *The irreducible characters are an orthonormal basis of $L^2_c(G)$.*

Let us give the proof for finite groups. The general case requires some basic functional analysis and will be discussed in Section 2.4. For a finite group G decompose the group algebra in matrix blocks according to Chapter 6, §2.6.3 as $\mathbb{C}[G] = \bigoplus_i^m M_{h_i}(\mathbb{C})$.

The m blocks correspond to the m irreducible representations. Their irreducible characters are the composition of the projection to a factor $M_{h_i}(\mathbb{C})$ followed by the ordinary trace.

A function $f = \sum_{g \in G} f(g)g \in \mathbb{C}[G]$ is a class function if and only if $f(ga) = f(ag)$ or $f(a) = f(gag^{-1})$, for all $a, g \in G$. This means that f lies in the center of the group algebra.

The space of class functions is identified with the center of $\mathbb{C}[G]$.

The center of a matrix algebra $M_h(\mathbb{C})$ is formed by the scalar matrices. Thus the center of $\bigoplus_i^m M_{h_i}(\mathbb{C})$ equals $\mathbb{C}^{\oplus m}$.

It follows that the number of irreducible characters equals the dimension of the space of class functions. Since the irreducible characters are orthonormal they are a basis.

As a corollary we have:

Corollary. *(a) The number of irreducible representations of a finite group G equals the number of conjugacy classes in G.*

(b) If $h_1, \ldots, h_r$ are the dimensions of the distinct irreducible representations of G, we have $|G| = \sum_i h_i^2$.

Proof. (a) Since a class function is a function which is constant on conjugacy classes, a basis for class functions is given by the characteristic functions of conjugacy classes.

(b) This is just the consequence of 2.6.3 of Chapter 6. □

There is a deeper result regarding the dimensions of irreducible representations (see [CR]):

Theorem 3. *The dimension h of an irreducible representation of a finite group G divides the order of G.*

The previous theorem allows us to compute a priori the dimensions h_i in some simple cases but in general this is only a small piece of information.

We need one more general result on unitary representations which is a simple consequence of the definitions.

Proposition 2. *Let V be a Hilbert space and a unitary representation of a compact group G. If V_1, V_2 are two non-isomorphic irreducible G-submodules of V, they are orthogonal.*

Proof. The Hermitian pairing (u, v) induces a G-equivariant, antilinear map $j : V_2 \to V_1^*, j(u)(v) = (v, u)H$. Since G acts by unitary operators, $V_1^* = \overline{V_1}$. Thus j can be interpreted as a linear G-equivariant map between V_2 and V_1. Since these irreducible modules are non-isomorphic we have $j = 0$. □

1.4 Induced Characters

We now perform a computation on induced characters which will be useful when we discuss the symmetric group.

Let G be a finite group, H a subgroup and V a representation of H with character χ_V. We want to compute the character χ of $\mathrm{Ind}_H^G(V) = \bigoplus_{x \in G/H} xV$ (Chapter 1, §3.2.2, 3.2.3). An element $g \in G$ induces a transformation on $\bigoplus_{x \in G/H} xV$ which can be thought of as a matrix in block form. Its trace comes only from the contributions of the blocks xV for which $gxV = xV$, and this happens if and only if $gx \in xH$, which means that the coset xH is a fixed point under g acting on G/H. As usual, we denote by $(G/H)^g$ these fixed points. The condition that $xH \in (G/H)^g$ can also be expressed as $x^{-1}gx \in H$.

If $gxV = xV$, the map g on xV has the same trace as the map $x^{-1}gx$ on V; thus

(1.4.1) $$\chi(g) = \sum_{(G/H)^g} \chi_V(x^{-1}gx).$$

It is useful to transform the previous formula. Let $X_g := \{x \in G | x^{-1}gx \in H\}$. The next assertions are easily verified:

(i) *The set X_g is a union of right cosets $G(g)x$ where $G(g)$ is the centralizer of g in G.*

(ii) *The map $\pi : x \mapsto x^{-1}gx$ is a bijection between the set of such cosets and the intersection of the conjugacy class C_g of g with H.*

Proof. (i) is clear. As for (ii) observe that $x^{-1}gx = (ax)^{-1}gax$ if and only if $a \in C(g)$. Thus the $G(g)$ cosets of X_g are the nonempty fibers of π. The image of π is clearly the intersection of the conjugacy class C_g of g with H. □

Decompose $C_g \cap H = \cup_i O_i$ into H-conjugacy classes. Of course if $a \in C_g$ we have $|G(a)| = |G(g)|$ since these two subgroups are conjugate. Fix an element $g_i \in O_i$ in each class and let $H(g_i)$ be the centralizer of g_i in H. Then $|O_i| = |H|/|H(g_i)|$ and finally

(1.4.2)

$$\chi(g) = \frac{1}{|H|}\sum_{x\in X}\chi_V(x^{-1}gx) = \frac{1}{|H|}\sum_i\sum_{a\in O_i}|G(a)|\chi_V(a) = \sum_i \frac{|G(g)|}{|H(g_i)|}\chi_V(g_i).$$

In particular one can apply this to the case $V = 1$. This is the example of the permutation representation of G on G/H.

Proposition. *The number of fixed points of g on G/H equals the character of the permutation representation $\mathbb{C}[G/H]$ and is*

$$\chi(g) = \frac{|C_g \cap H||G(g)|}{|H|} = \sum_i \frac{|G(g)|}{|H(g_i)|}. \tag{1.4.3}$$

2 Matrix Coefficients

2.1 Representative Functions

Let G be a topological group. We have seen in Chapter 6, §2.6 the notion of matrix coefficients for G. Given a continuous representation $\rho : G \to GL(U)$ we have a linear map $i_U : \mathrm{End}(U) \to C(G)$ given by $i_U(X)(g) := \mathrm{tr}(X\rho(g))$. We want to return to this concept in a more systematic way.

We will use the following simple fact, which we leave as an exercise. X is a set, F a field.

Lemma. *The n functions $f_i(x)$ on a set X, with values in F, are linearly independent if and only if there exist n points $p_1, \dots, p_n \in X$ with the determinant of the matrix $f_i(p_j)$ nonzero.*

Lemma-Definition. *For a continuous function $f \in C(G)$ the following are equivalent:*

(1) The space spanned by the left translates $f(gx)$, $g \in G$ is finite dimensional.
(2) The space spanned by the right translates $f(xg)$, $g \in G$ is finite dimensional.
(3) The space spanned by the bitranslates $f(gxh)$, $g, h \in G$ is finite dimensional.
(4) There is a finite expansion $f(xy) := \sum_{i=1}^k u_i(x)v_i(y)$.

A function satisfying the previous conditions is called a ***representative function****.*

(5) Moreover, in the expansion (4) the functions u_i, v_i can be taken as representative functions.

Proof. Assume (1) and let $u_i(x),\ i = 1, \dots, m$, be a basis of the space spanned by the functions $f(gx),\ g \in G$.

Write $f(gx) = \sum_i v_i(g)u_i(x)$. This expression is continuous in g. By the previous lemma we can find m points p_j such that the determinant of the matrix with entries $u_i(p_j)$ is nonzero.

Thus we can solve the system of linear equations $f(gp_j) = \sum_i v_i(g)u_i(p_j)$ by Cramer's rule, so that the coefficients $v_i(g)$ are continuous functions, and (4) follows. (4) is a symmetric property and clearly implies (1) and (2).

In the expansion $f(xy) := \sum_{i=1}^k u_i(x)v_i(y)$ we can take the functions v_i to be a basis of the space spanned by the left translates of f. They are representative functions. We have that

$$f(xzy) := \sum_{i=1}^k u_i(xz)v_i(y) = \sum_{i=1}^k u_i(x)v_i(zy) = \sum_{i=1}^k u_i(x) \sum_{h=1}^k c_{i,h}(z)v_h(y)$$

implies $u_i(xz) = \sum_{h=1}^k u_h(x)c_{hi}(z)$ implying (5) and also (3). □

Proposition 1. *The set $\mathcal{T}_G$ of representative functions is an algebra spanned by the matrix coefficients of the finite-dimensional continuous representations of G.*

Proof. The fact that it is an algebra is obvious. Let us check the second statement. First, a continuous finite-dimensional representation is given by a homomorphism $\rho : G \to GL(n, \mathbb{C})$. The entries $\rho(g)_{i,j}$ by definition span the space of the corresponding matrix coefficients. We have that $\rho(xy) = \rho(x)\rho(y)$, which in matrix entries shows that the functions $\rho(g)_{i,j}$ satisfy (4).

Conversely, let $f(x)$ be a representative function. Clearly also $f(x^{-1})$ is representative. Let $u_i(x)$ be a basis of the space U of left translates, $f(g^{-1}) = \sum_{i=1}^k a_i u_i(g)$. U is a linear representation by the left action and $u_i(g^{-1}x) = \sum_j c_{ij}(g)u_j(x)$ where the functions $c_{i,j}(g)$ are the matrix coefficients of U in the given basis. We thus have $u_i(g^{-1}) = \sum_j c_{ij}(g)u_j(1)$ and $f(g) = \sum_{i=1}^k a_i \sum_j c_{ij}(g)u_j(1)$.

If G, K are two topological groups, we have that

Proposition 2. *Under multiplication $f(x)g(y)$ we have an isomorphism*

$$\mathcal{T}_G \otimes \mathcal{T}_K = \mathcal{T}_{G\times K}.$$

Proof. The multiplication map of functions on two distinct spaces to the product space is always an isomorphism of the tensor product of the spaces of functions to the image. So we only have to prove that the space of representative functions of $G \times K$ is spanned by the functions $\psi(x, y) := f(x)g(y),\ f(x) \in \mathcal{T}_G,\ g(y) \in \mathcal{T}_K$.

Using the property (4) of the definition of representative function we have that if $f(x_1x_2) = \sum_i u_i(x_1)v_i(x_2),\ g(y_1, y_2) = \sum_k w_k(y_1)z_k(y_2)$, then

$$\psi((x_1, y_1)(x_2, y_2)) = \sum_{i,k} u_i(x_1)w_k(y_1)v_i(x_2)z_k(y_2).$$

Conversely, if $\psi(x, y)$ is representative, writing $(x, y) = (x, 1)(1, y)$ one immediately sees that ψ is in the span of the product of representative functions. □

Finally, let $\rho : H \to K$ be a continuous homomorphism of topological groups.

Proposition 3. *If $f(k)$ is representative on K, then $f(\rho(k))$ is representative in H.*

Proof. We have $f(xy) = \sum_i u_i(x)v_i(y)$, hence $f(\rho(ab)) = f(\rho(a)\rho(b)) = \sum_i u_i(\rho(a))v_i(\rho(b))$. □

In terms of matrix coefficients what we are doing is to take a representation of K and deduce, by composition with ρ, a representation of H.

Particularly important for us will be the case of a compact group K, when all the finite-dimensional representations are semisimple. We then have an analogue of

Theorem. *The space $\mathcal{T}_K$ is the direct sum of the matrix coefficients $V_i^* \otimes V_i$ as $V_i \in \hat{K}$ runs on the set of different irreducible representations of K.*

(2.1.1) $$\mathcal{T}_K = \bigoplus_{V \in \hat{K}} V^* \otimes V.$$

Proof. The proof is essentially identical to that of Chapter 7, §3.1. □

2.2 Preliminaries on Functions

Before we continue our analysis we wish to collect two standard results on function theory which will be useful in the sequel. The first is the Stone–Weierstrass theorem. This theorem is a generalization of the classical theorem of Weierstrass on approximation of continuous functions by polynomials.

In its general form it says:

Stone–Weierstrass theorem. *Let A be an algebra of real-valued continuous functions on a compact space X which separates points.*[50] *Then either A is dense in $C(X, \mathbb{R})$ or it is dense in the subspace of $C(X, \mathbb{R})$ of functions vanishing at a given point a.*[51]

Proof. Let A be such an algebra. If $1 \in A$, then we cannot be in the second case, where $f(a) = 0, \forall f \in A$. Otherwise we can add 1 to A and assume that $1 \in A$. Then let S be the uniform closure of A. The theorem can thus be reformulated as follows: if S is an algebra of continuous functions, which separates points, $1 \in S$, and S is closed under uniform convergence, then $S = C_0(X, \mathbb{R})$.

We will use only one statement of the classical theorem of Weierstrass, the fact that given any interval $[-n, n]$, the function $|x|$ can be uniformly approximated by polynomials in this interval. This implies for our algebra S that if $f(x) \in S$, then $|f(x)| \in S$. From this we immediately see that if f and $g \in S$, the two functions $\min(f, g) = (f + g - |f - g|)/2$ and $\max(f, g) = (f + g + |f - g|)/2$ are in S.

Let x, y be two distinct points in X. By assumption there is a function $a \in S$ with $a(x) \neq a(y)$. Since the function $1 \in S$ takes the value 1 at x and y, we can

[50] This means that, given $a, b \in X, a \neq b$, there is an $f \in A$ with $f(a) \neq f(b)$.

[51] If $X = \{p_1, \dots, p_n\}$ is a finite set, the theorem is really a theorem of algebra, a form of the Chinese Remainder Theorem.

find a linear combination g of $a, 1$ which takes at x, y any prescribed values. Let $f \in C_0(X, \mathbb{R})$ be a function. By the previous remark we can find a function $g_{x,y} \in S$ with $g_{x,y}(x) = f(x),\ g_{x,y}(y) = f(y)$. Given any $\epsilon > 0$ we can thus find an open set U_y such that $g_{x,y}(z) > f(z) - \epsilon$ for all $z \in U_y$. By compactness of X we can find a finite number of such open sets U_{y_i} covering X. Take the corresponding functions g_{x,y_i}. We have that the function $g_x := \max(g_{x,y_i}) \in S$ has the property $g_x(x) = f(x),\ g_x(z) > f(z) - \epsilon, \forall z \in X$. Again there is a neighborhood V_x of x such that $g_x(z) < f(z) + \epsilon, \forall z \in V_x$. Cover X with a finite number of these neighborhoods V_{x_j}. Take the corresponding functions g_{x_j}. We have that the function $g := \min(g_{x_j}) \in S$ has the property $|g(z) - f(z)| < \epsilon, \forall z \in X$. Letting ϵ tend to 0, since S is closed under uniform convergence, we find that $f \in S$, as desired. □

We will often apply this theorem to an algebra A of complex functions. In this case we easily see that the statement is:

Corollary. *If $A \subset C(X)$ is an algebra of complex functions which separates points in X, $1 \in A$, A is closed under uniform convergence and A is closed under complex conjugation, then $A = C(X)$.*

For the next theorem we need to recall two simple notions. These results can be generalized but we prove them in a simple case.

Definition. A set A of continuous functions on a space X is said to be *uniformly bounded* if there is a positive constant M such that $|f(x)| < M$ for every $f \in A, x \in X$.

A set A of continuous functions on a metric space X is said to be *equicontinuous* if, for every $\epsilon > 0$, there is a $\delta > 0$ with the property that $|f(x) - f(y)| < \epsilon, \forall (x, y)$ with $\overline{xy} < \delta$ and $\forall f \in A$.

We are denoting by $\overline{xy}$ the distance between the two points x, y.

Recall that a topological space is *first countable* if it has a dense countable subset.

Theorem (Ascoli–Arzelà). *A uniformly bounded and equicontinuous set A of continuous functions on a first countable compact metric space X is relatively compact in $C(X)$, i.e., from any sequence $f_i \in A$ we may extract one which is uniformly convergent.*

Proof. Let $p_1, p_2, \dots, p_k, \dots$ be a dense sequence of points in X. Since the functions f_i are uniformly bounded, we can extract a subsequence $s_1 := f_1^1, f_2^1, \dots, f_i^1, \dots$ from the given sequence for which the sequence of numbers $f_i^1(p_1)$ is convergent. Inductively, we construct sequences s_k where s_k is extracted from s_{k-1} and the sequence of numbers $f_i^k(p_k)$ is convergent. It follows that for the *diagonal* sequence $F_i := f_i^i$, we have that the sequence of numbers $F_i(p_j)$ is convergent for each j. We want to show that F_i is uniformly convergent on X. We need to show that F_i is a Cauchy sequence. Given $\epsilon > 0$ we can find by equicontinuity a $\delta > 0$ with the property that $|f(x) - f(y)| < \epsilon, \forall (x, y)$ with $\overline{xy} < \delta$ and $\forall f \in A$. By compactness we can find a finite number of points $q_j,\ j = 1, \dots, m$, from our list p_i such that for all $x \in X$, there is one of the q_j at a distance less than δ from x. Let k be such that $|F_s(q_j) - F_t(q_j)| < \epsilon, \forall j = 1, \dots, m, \ \forall s, t > k$. For each x find a q_j at a distance less than δ, then

$$\begin{aligned}|F_s(x) - F_t(x)| &= |F_s(x) - F_s(q_j) - F_t(x) + F_t(q_j) + F_s(q_j) - F_t(q_j)| \\ &< 3\epsilon, \quad \forall s, t > k.\end{aligned}$$

□

2.3 Matrix Coefficients of Linear Groups

One possible approach to finding the representations of a compact group could be to identify the representative functions. In general this may be difficult but in a special case it is quite easy.

Theorem. *Let $G \subset U(n, \mathbb{C})$ be a compact linear group. Then the ring of representative functions of G is generated by the matrix entries and the inverse of the determinant.*

Proof. Let A be the algebra of functions generated by the matrix entries and the inverse of the determinant. Clearly $A \subset \mathcal{T}_G$ by Proposition 2.2. Moreover, by matrix multiplication it is clear that the space of matrix entries is stable under left and right G action, similarly for the inverse of the determinant and thus A is $G \times G$-stable.

Let us prove now that A is dense in the algebra of continuous functions. We want to apply the Stone–Weierstrass theorem to the algebra A which is made up of complex functions and contains 1. In this case, besides verifying that A separates points, we also need to show that A is closed under complex conjugation. Then we can apply the previous theorem to the real and imaginary parts of the functions of A and conclude that they are both dense.

In our case A separates points since two distinct matrices must have two different coordinates. A is closed under complex conjugation. In fact the conjugate of the determinant is the inverse, while the conjugates of the entries of a unitary matrix X are entries of X^{-1}. The entries of this matrix, by the usual Cramer rule, are indeed polynomials in the entries of X divided by the determinants, hence are in A.

At this point we can conclude. If $A \neq \mathcal{T}_G$, since they are both $G \times G$ representations and $\mathcal{T}_G = \bigoplus_i V_i^* \otimes V_i$ is a direct sum of irreducible $G \times G$ representations, for some i we have $V_i^* \otimes V_i \cap A = 0$. By Proposition 2 of 1.3 this implies that $V_i^* \otimes V_i$ is orthogonal to A and this contradicts thc fact that A is dense in $C(G)$. □

Given a compact Lie group G it is not restrictive to assume that $G \subset U(n, \mathbb{C})$. This will be proved in Section 4.3 as a consequence of the Peter–Weyl theorem.

3 The Peter–Weyl Theorem

3.1 Operators on a Hilbert Space

The representation theory of compact groups requires some basic functional analysis. Let us recall some simple definitions.

Definition 1. A *norm* on a complex vector space V is a map $v \mapsto \|v\| \in \mathbb{R}^+$, satisfying the properties:

$$\|v\| = 0 \iff v = 0, \quad \|av\| = |a|\|v\|, \quad \|v + w\| \le \|v\| + \|w\|.$$

A vector space with a norm is called a *normed space*.

From a norm one deduces the structure of metric space setting as distance $\overline{xy} := \|x - y\|$.

Definition 2. A Banach space is a normed space complete under the induced metric.

Most important for us are Hilbert spaces. These are the Banach spaces where the norm is deduced from a positive Hermitian form $\|v\|^2 = (v, v)$. When we talk about convergence in a Hilbert space we usually mean in this norm and also speak of *convergence in mean*.[52] All our Hilbert spaces are assumed to be countable, and to have a countable orthonormal basis.

The special properties of Hilbert spaces are the *Schwarz inequality* $|(u, v)| \le \|u\| \|v\|$, and the existence of orthonormal bases u_i with $(u_i, u_j) = \delta_i^j$. Then $v = \sum_{i=1}^{\infty}(v, u_i)u_i$ for every vector $v \in H$, from which $\|v\|^2 = \sum_{i=1}^{\infty} |(v, u_i)|^2$. This is called the *Parseval formula*.[53]

The other Banach space which we will occasionally use is the space $C(X)$ of continuous functions on a compact space X with norm the *uniform norm* $\|f\|_\infty := \max_{x \in X} |f(x)|$. Convergence in this norm is *uniform convergence*.

Definition 3. A linear operator $T : A \to B$ between normed spaces is *bounded* if there is a positive constant C such that $\|T(v)\| \le C\|v\|, \forall v \in A$.

The minimum such constant is the *operator norm* $\|T\|$ of T.

By linearity it is clear that $\|T\| = \sup_{\|v\|=1} \|T(v)\|$.

Exercise.

1. The sum and product of bounded operators are bounded.

$$\|aT\| = |a|\|T\|, \|aT_1 + bT_2\| \le |a|\|T_1\| + |b|\|T_2\|, \|T_1 \circ T_2\| \le \|T_1\|\|T_2\|.$$

2. If B is complete, bounded operators are complete under the norm $\|T\|$.

When $A = B$ bounded operators on A will be denoted by $\mathcal{B}(A)$. They form an algebra. The previous properties can be taken as the axioms of a *Banach algebra*.

Given a bounded[54] operator T on a Hilbert space, its adjoint T^* is defined by $(Tv, w) = (v, T^*w)$. We are particularly interested in bounded Hermitian operators (or self-adjoint), i.e., bounded operators T for which $(Tu, v) = (u, Tv), \forall u, v \in H$.

---

[52] There are several other notions of convergence but they do not play a role in our work.

[53] For every n we also have $\|v\|^2 \ge \sum_{i=1}^{n} |(v, u_i)|^2$, which is called *Bessel's inequality*.

[54] We are simplifying the theory drastically.

The typical example is the Hilbert space of L^2 functions on a measure space X, with Hermitian product $(f, g) = \int_X f(x)\overline{g(x)}dx$. An important class of bounded operators are the integral operators $Tf(x) := \int_X K(x, y)f(y)\,dy$. The *integral kernel* $K(x, y)$ is itself a function on $X \times X$ with some suitable restrictions (like L^2). If $K(x, y) = \overline{K(y, x)}$ we have a self-adjoint operator.

Theorem 1. *(1) If A is a self-adjoint bounded operator, $\|A\| = \sup_{\|v\|=1} |(Av, v)|$.*
(2) For any bounded operator $\|T\|^2 = \|T^*T\|$.[55]

Proof. (1) By definition if $\|v\| = 1$ we have $\|Av\| \leq \|A\|$ hence $|(Av, v)| \leq \|A\|$ by the Schwarz inequality.[56] In the self-adjoint case $(A^2v, v) = (Av, Av)$. Set $N := \sup_{\|v\|=1} |(Av, v)|$. If $\lambda > 0$, we have

$$\begin{aligned}\|Av\|^2 &= \frac{1}{4}\left[\left(A\Big(\lambda v + \frac{1}{\lambda}Av\Big), \lambda v + \frac{1}{\lambda}Av\right) - \left(A\Big(\lambda v - \frac{1}{\lambda}Av\Big), \lambda v - \frac{1}{\lambda}Av\right)\right] \\ &\leq \frac{1}{4}\left[N\left\|\lambda v + \frac{1}{\lambda}Av\right\|^2 + N\left\|\lambda v - \frac{1}{\lambda}Av\right\|^2\right] \\ &= \frac{N\|v\|^2}{2}\left[\lambda^2 + \frac{1}{\lambda^2}\frac{\|Av\|^2}{\|v\|^2}\right].\end{aligned}$$

For $Av \neq 0$ the minimum on the right-hand side is obtained when

$$\lambda^2 = \frac{\|Av\|}{\|v\|}, \quad \text{since} \quad \left(\lambda^2 + \frac{1}{\lambda^2}c^2 = \Big(\lambda - \frac{c}{\lambda}\Big)^2 + 2c \geq 2c\right).$$

Hence

$$\|Av\|^2 \leq N\|Av\|\|v\| \implies \|Av\| \leq N\|v\|.$$

Of course this holds also when $Av = 0$, hence $\|A\| \leq N$.
(2) $\|T\|^2 = \sup_{\|v\|=1}(Tv, Tv) = \sup_{\|v\|=1} |(T^*Tv, v)| = \|T^*T\|$ from 1. □

Recall that for a linear operator T an eigenvector v of eigenvalue $\lambda \in \mathbb{C}$ is a vector with $Tv = \lambda v$. If T is self-adjoint, necessarily $\lambda \in \mathbb{R}$. Eigenvalues are bounded by the operator norm. If λ is an eigenvalue, from $Tv = \lambda v$ we get $\|Tv\| = |\lambda|\|v\|$, hence $\|T\| \geq |\lambda|$.

In general, actual eigenvectors need not exist, the typical example being the operator $f(x) \mapsto g(x)f(x)$ of multiplication by a continuous function on L^2 functions on $[0, 1]$.

Lemma 1. *If A is a self-adjoint operator v, w two eigenvectors of eigenvalues $\alpha \neq \beta$, we have $(v, w) = 0$.*

---

[55] This property is taken as an axiom for C^* algebras.

[56] This does not need self-adjointness.

Proof.

$$\begin{aligned}\alpha(v, w) &= (Av, w) = (v, Aw) \\ &= \beta(v, w) \implies (\alpha - \beta)(v, w) = 0 \implies (v, w) = 0. \qquad \square\end{aligned}$$

There is a very important class of operators for which the theory better resembles the finite-dimensional theory. These are the *completely continuous operators* or compact operators.

Definition 4. A bounded operator A of a Hilbert space is *completely continuous*, or *compact* if, given any sequence v_i of vectors of norm 1, from the sequence $A(v_i)$ one can extract a convergent sequence.

In other words this means that A transforms the sphere $\|v\| = 1$ in a relatively compact set of vectors, where compactness is by convergence of sequences.

We will denote by $\mathcal{I}$ the set of completely continuous operators on H.

Proposition. *$\mathcal{I}$ is a two-sided ideal in $\mathcal{B}(H)$, closed in the operator norm.*

Proof. Suppose that $A = \lim_{i\to\infty} A_i$ is a limit of completely continuous operators A_i. Given a sequence v_i of vectors of norm 1 we can construct by hypothesis for each i, and by induction a sequence $s_i := (v_{i_1(i)}, v_{i_2(i)}, \dots, v_{i_k(i)}, \dots)$ so that s_i is a subsequence of s_{i-1} and the sequence $A_i(v_{i_1(i)}), A_i(v_{i_2(i)}), \dots, A_i(v_{i_k(i)}), \dots$ is convergent.

We then take the diagonal sequence $w_k := v_{i_k(k)}$ and see that $A(w_k)$ is a Cauchy sequence. In fact given $\epsilon > 0$ there is an N such that $\|A - A_i\| < \epsilon/3$ for all $i \geq N$, there is also an M such that $\|A_N(v_{i_k(N)}) - A_N(v_{i_h(N)})\| < \epsilon/3$ for all $h, k > M$. Thus when $h \leq k > \max(N, M)$ we have that $v_{i_k(k)} = v_{i_h(t)}$ for some $t \geq k$, and so

$$\begin{aligned}\|A(w_h) - A(w_k)\| &= \|A(v_{i_h(h)}) - A_h(v_{i_h(h)}) + A_h(v_{i_h(h)}) - A(v_{i_k(k)})\| \\ &\leq \|A(v_{i_h(h)}) - A_h(v_{i_h(h)})\| + \|A_h(v_{i_h(h)}) - A_h(v_{i_h(t)})\| \\ &\quad + \|A_h(v_{i_h(t)}) - A(v_{i_h(t)})\| < \epsilon.\end{aligned}$$

The property of being a two-sided ideal is almost trivial to verify and we leave it as exercise. $\square$

From an abstract point of view, completely continuous operators are related to the notion of the complete tensor product $H\hat{\otimes}\overline{H}$, discussed in Chapter 5, §3.8. Here $\overline{H}$ is the conjugate space. We want to associate an element $\rho(u) \in \mathcal{I}$ to an element $u \in H\hat{\otimes}\overline{H}$.

The construction is an extension of the algebraic formula 3.4.4 of Chapter 5.[57] We first define the map on the algebraic tensor product as in the formula $\rho(u \otimes v)w := u(w, v)$. Clearly the image of $\rho(u \otimes v)$ is the space generated by u; hence $\rho(H \otimes \overline{H})$ is made of operators with finite-dimensional image.

[57] We see now why we want to use the conjugate space: it is to have bilinearity of the map ρ.

Lemma 2. *The map* $\rho : H \otimes \overline{H} \to \mathcal{B}(H)$ *decreases the norms and extends to a continuous map* $\rho : H \hat{\otimes} \overline{H} \to \mathcal{I} \subset \mathcal{B}(H)$.

Proof. Let us fix an orthonormal basis u_i of H. We can write an element $v \in H \otimes \overline{H}$ as a finite sum $v = \sum_{i,j=1}^{m} c_{i,j} u_i \otimes u_j$. Its norm in $H \otimes \overline{H}$ is $\sqrt{\sum_{i,j} |c_{i,j}|^2}$.

$$\rho(v)\left(\sum_h a_h u_h\right) = \sum_h \left(\sum_i c_{i,h} a_h\right) u_i = \sum_i \left(\sum_h a_h c_{i,h}\right) u_i.$$

Given $w := \sum_h a_h u_h$ we deduce from the Schwarz inequality for $\|\rho(v)(w)\|$ that

$$\left\|\rho(v)\left(\sum_h a_h u_h\right)\right\| = \sqrt{\sum_i \left|\left(\sum_h a_h c_{i,h}\right)\right|^2}$$
$$\leq \sqrt{\sum_i \left(\sum_h |a_h|^2\right)\left(\sum_h |c_{i,h}|^2\right)} = \|w\|\|v\|.$$

Since the map decreases norms it extends by continuity. Clearly bounded operators with finite range are completely continuous. From the previous proposition, limits of these operators are also completely continuous. $\square$

In fact we will see presently that the image of ρ is dense in $\mathcal{I}$, i.e., that every completely continuous operator is a limit of operators with finite-dimensional image.

Warning. The image of ρ is not $\mathcal{I}$. For instance the operator T, which in an orthonormal basis is defined by $T(e_i) := \frac{1}{\sqrt{i}} e_i$, is not in $\mathrm{Im}(\rho)$.

The main example of the previous construction is given by taking $H = L^2(X)$ with X a space with a measure. We recall a basic fact of measure theory (cf. [Ru]). If X, Y are measure spaces, with measures $d\mu, d\nu$, one can define a product measure $d\mu \times d\nu$ on $X \times Y$. If $f(x), g(y)$ are L^1 functions on X, Y respectively, we have that $f(x)g(y)$ is L^1 on $X \times Y$ and

$$\int_{X\times Y} f(x)g(y)d\mu d\nu = \int_X f(x)d\mu \int_Y g(y)\, d\nu.$$

Lemma 3. *The map* $i : f(x) \otimes g(y) \mapsto f(x)g(y)$ *extends to a Hilbert space isomorphism* $L^2(X)\hat{\otimes}L^2(Y) = L^2(X \times Y)$.

Proof. We have clearly that the map i is well defined and preserves the Hermitian product; hence it extends to a Hilbert space isomorphism of $L^2(X)\hat{\otimes}L^2(Y)$ with some closed subspace of $L^2(X \times Y)$. To prove that it is surjective we use the fact that, given measurable sets $A \subset X, B \subset Y$ of finite measure, the characteristic function $\chi_{A\times B}$ of $A \times B$ is the tensor product of the characteristic functions of A and B. By standard measure theory, since the sets $A \times B$ generate the σ-algebra of measurable sets in $X \times Y$, the functions $\chi_{A\times B}$ span a dense subspace of $L^2(X \times Y)$. $\square$

Proposition 1. *An integral operator $Tf(x) := \int_X K(x,y)f(y)dy$, with the integral kernel in $L^2(X \times X)$, is completely continuous.*

Proof. By the previous lemma, we can write $K(x,y) = \sum_{i,j} c_{i,j}u_i(x)\overline{u}_j(y)$ with $u_i(x)$ an orthonormal basis of $L^2(X)$. We see that $Tf(x) = \sum_{i,j} c_{i,j}u_i(x) \int_X \overline{u}_j(y) f(y)dy = \sum_{i,j} c_{i,j}u_i(f,u_j)$. Now we apply Lemma 2. □

We will also need a variation of this theme. Assume now that X is a locally compact metric space with a Daniell integral. Assume further that the integral kernel $K(x,y)$ is continuous and has compact support.

Proposition 2. *The operator $Tf(x) := \int_X K(x,y)f(y)dy$ is a bounded operator from $L^2(X)$ to $C_0(X)$. It maps bounded sets of functions into uniformly bounded and equicontinuous sets of continuous functions.*[58]

Proof. Assume that the support of the kernel is contained in $A \times B$ with A, B compact. Let m be the measure of B. First, $Tf(x)$ is supported in A, and it is a continuous function. In fact if $x \in A$, by compactness and continuity of $K(x,y)$, there is a neighborhood U of x such that $|K(x,y) - K(x_0,y)| < \epsilon$, $\forall y \in B, \forall x \in U$ so that (Schwarz inequality)

(3.1.1)

$$\forall x \in U, \quad |Tf(x) - Tf(x_0)| \le \int_B |K(x,y) - K(x_0,y)|\,|f(y)|dy \le \epsilon\, m^{1/2}\, \|f\|.$$

Moreover, if $M = \max|K(x,y)|$ we have

$$\|T(f)\|_\infty = \sup\left(\left|\int_X K(x,y)f(y)dy\right|\right) \le \sup\left(\sqrt{\int_X |K(x,y)|^2dy}\right)\|f\| \le m^{1/2}M\|f\|.$$

Let us show that the functions $Tf(x)$, $\|f\| = 1$ are equicontinuous and uniformly bounded. In fact $|Tf(x)| \le m^{1/2}M$ where $M = \max|K(x,y)|$. The equicontinuity follows from the previous argument. Given $\epsilon > 0$ we can, by the compactness of $A \times B$, find $\eta > 0$ so that $|K(x_1,y) - K(x_0,y)| < \epsilon$ if $\overline{x_1x_0} < \eta$, $\forall y \in B$. Hence if $\|f\| \le M$, we have $|Tf(x_1) - Tf(x_0)| \le Mm^{1/2}\epsilon$ when $\overline{x_1x_0} < \eta$. □

Proposition 3. *If A is a self-adjoint, completely continuous operator, there is an eigenvector v with eigenvalue $\pm\|A\|$.*

Proof. By Theorem 1, there is a sequence of vectors v_i of norm 1 for which $\lim_{i\to\infty}(Av_i, v_i) = \mu = \pm\|A\|$. By hypothesis we can extract a subsequence, which

[58] If X is not compact, $C_0(X)$ is not complete, but in fact T maps into the complete subspace of functions with support in a fixed compact subset $A \subset X$.

we still call v_i, such that $\lim_{i\to\infty} A(v_i) = w$. Since $\mu := \lim_{i\to\infty}(A(v_i), v_i)$, the inequality

$$0 \le \|Av_i - \mu v_i\|^2 = \|Av_i\|^2 - 2\mu(Av_i, v_i) + \mu^2 \le 2\mu^2 - 2\mu(Av_i, v_i)$$

implies that $\lim_{i\to\infty}(A(v_i) - \mu v_i) = 0$. Thus $\lim_{i\to\infty} \mu v_i = \lim_{i\to\infty} A(v_i) = w$. In particular v_i must converge to some vector v such that $w = \mu v$, and $w = \lim_{i\to\infty} Av_i = Av$. Since $\mu = (Aw, w)$ if $A \neq 0$ we have $\mu \neq 0$, hence $v \neq 0$ is the required eigenvector. □

Given a Hilbert space H an orthogonal decomposition for H is a family of mutually orthogonal closed subspaces H_i, $i = 1, \dots, \infty$, such that every element $v \in H$ can be expressed (in a unique way) as a series $v = \sum_{i=1}^{\infty} v_i$, $v_i \in H_i$. An orthogonal decomposition is a generalization of the decomposition of H given by an orthonormal basis.

Definition 5. A self-adjoint operator is *positive* if $(Av, v) \ge 0, \forall v$.

Remark. If T is any operator T^*T is positive self-adjoint. The eigenvalues of a positive operator are all positive or 0.

Theorem 2. *Let A be a self-adjoint, completely continuous operator.*

If the image of A is not finite dimensional, there is a sequence of numbers λ_i and orthonormal vectors v_i such that

$$\|A\| = |\lambda_1| \ge |\lambda_2| \ge \cdots |\lambda_n| \ge \cdots$$

and

(1) Each λ_i is an eigenvalue of A, $Av_i = \lambda_i v_i$.

(2) The numbers λ_i are the only nonzero eigenvalues.

(3) $\lim_{i\to\infty} \lambda_i = 0$.

(4) The eigenspace H_i of each eigenvalue $\lambda \neq 0$ is finite dimensional with basis the v_i for which $\lambda_i = \lambda$.

(5) H is the orthogonal sum of the kernel of A and the subspace with basis the orthonormal vectors v_i.

If the image of A is finite dimensional, the sequence $\lambda_1, \lambda_2, \dots, \lambda_m$ is finite and $H = \mathrm{Ker}(A) \bigoplus_{i=1}^m \mathbb{C}v_i$.

Proof. Call $A = A_1$; by Proposition 3 there is an eigenvector v_1, of absolute value 1, with eigenvalue λ_1 and $|\lambda_1| = \|A_1\| > 0$. Decompose $H = \mathbb{C}v_1 \oplus v_1^\perp$, the operator A_1 induces on $v_1^\perp$ a completely continuous operator A_2 with $\|A_2\| \le \|A_1\|$. Repeating the reasoning for A_2 there is an eigenvector v_2, $|v_2| = 1$ with eigenvalue λ_2 and $|\lambda_2| = \|A_2\| > 0$.

Continuing in this way we find a sequence of orthonormal vectors $v_1, \dots, v_k, \dots$ with $Av_i = \lambda_i v_i$ and such that, setting $H_i := (\mathbb{C}v_1 + \cdots + \mathbb{C}v_{i-1})^\perp$, we have that $|\lambda_i| = \|A_i\|$ where A_i is the restriction of A to H_i.

This sequence stops after finitely many steps if for some i we have that $A_i = 0$; this implies that the image of A is spanned by $v_1, \dots, v_{i-1}$, otherwise it continues indefinitely. We must then have $\lim_{i\to\infty} \lambda_i = 0$, otherwise there is a positive constant $0 < b < |\lambda_i|$, for all i. In this case no subsequence of the sequence $Av_i = \lambda_i v_i$ can be chosen to be convergent since these vectors are orthogonal and all of absolute value $> b$, which contradicts the hypothesis that A is compact. Let $\overline{H}$ be the Hilbert subspace with basis the vectors v_i and decompose $H = \overline{H} \oplus \overline{H}^{\perp}$. On $\overline{H}^{\perp}$ the restriction of A has a norm $\leq |\lambda_i|$ for all i, hence it must be 0 and $\overline{H}^{\perp}$ is the kernel of A. Now given a vector $v = \sum_i c_i v_i + u$, $u \in \overline{H}^{\perp}$, we have $Av = \sum_i c_i \lambda_i v_i$; thus v is an eigenvector if and only if either $v = u$ or $u = 0$ and all the indices i for which $c_i \neq 0$ have the same eigenvalue λ. Since $\lim \lambda_i = 0$, this can happen only for a finite number of these indices. This proves also (2), (4), and finishes the proof. □

Exercise 2. Prove that $\mathcal{I}$ is the closure in the operator norm of the operators of finite-dimensional image. *Hint*: Use the spectral theory of T^*T and Exercise 1.

Let us now specialize to an integral operator $Tf(x) := \int_X K(x, y) f(y) dy$ with the integral kernel continuous and with compact support in $A \times B$ as before. Suppose further that T is self-adjoint, i.e., $K(x, y) = \overline{K(y, x)}$.

By Proposition 2, the eigenvectors of T corresponding to nonzero eigenvalues are continuous functions. Let us then take an element $f = \sum_{i=1}^{\infty} c_i u_i + u$ expanded, as before, in an orthonormal basis of eigenfunctions u_1, u with $Tu_i = \lambda_i u_i$, $Tu = 0$. The u_i are continuous functions with support in the compact set A.

Proposition 4. *The sequence* $g_k := T\left(\sum_{i=1}^{k} c_i u_i\right) = \sum_{i=1}^{k} c_i \lambda_i u_i$ *of continuous functions converges uniformly to* Tf.

Proof. By continuity g_k converges to Tf in the L^2-norm.

But g_k is also a Cauchy sequence in the uniform norm, as follows from the continuity of the operator from $L^2(X)$ to $C_0(X)$ (for the L^2 and uniform norm, respectively) so it converges uniformly to some function g. The inclusion $C_0(X) \subset L^2(X)$, when restricted to the functions with support in A, is continuous for the two norms ∞ and L^2 (since $\int_X \|f\|^2 d\mu \leq \mu(A) \|f\|_\infty^2$, where μ_a is the measure of A); therefore we must have $g = Tf$. □

We want to apply the theory to a locally compact group G with a left-invariant Haar measure. This measure allows us to define the *convolution* product, which is the generalization of the product of elements of the group algebra.

The convolution product is defined first of all on the space of L^1-functions by the formula

(3.1.2) $$(f * g)(x) := \int_G f(y) g(y^{-1}x) dy = \int_G f(xy) g(y^{-1}) dy.$$

When G is compact we normalize Haar measure so that the measure of G is 1. We have the continuous inclusion maps

(3.1.3) $$C_0(G) \subset L^2(G) \subset L^1(G).$$

The three spaces have respectively the uniform L^∞, L^2, L^1 norms; the inclusions decrease norms. In fact the L^1 norm of f equals the Hilbert scalar product of $|f|$ with 1, so by the Schwarz inequality, $|f|_1 \leq |f|_2$ while $|f|_2 \leq |f|_\infty$ for obvious reasons.

Proposition 5. *If G is compact, then the space of L^2 functions is also an algebra under convolution.*[59]

Both algebras $L^1(G)$, $L^2(G)$ are useful. In the next section we shall use $L^2(G)$, and we will compute its algebra structure in §3.3. On the other hand, $L^1(G)$ is also useful for representation theory.

One can pursue the algebraic relationship between group representations and modules over the group algebra in the continuous case, replacing the group algebra with the convolution algebra (cf. [Ki], [Di]).

3.2 Peter–Weyl Theorem

Theorem (Peter–Weyl).

(i) The direct sum $\bigoplus_i V_i^* \otimes V_i$ equals the space $\mathcal{T}_G$ of representative functions.
(ii) The direct sum $\bigoplus_i V_i^* \otimes V_i$ is dense in $L^2(G)$.

In other words every L^2-function f on G can be developed uniquely as a series $f = \sum_i u_i$ with $u_i \in V_i^* \otimes V_i$.

Proof. (i) We have seen (Chapter 6, Theorem 2.6) that for every continuous finite-dimensional irreducible representation V of G, the space of matrix coefficients $V^* \otimes V$ appears in the space $C(G)$ of continuous functions on G. Every finite-dimensional continuous representation of G is semisimple, and the matrix coefficients of a direct sum are the sum of the respective matrix coefficients.

(ii) For distinct irreducible representations V_1, V_2, the corresponding spaces of matrix coefficients, are irreducible non-isomorphic representations of $G \times G$. We can thus apply Proposition 2 of 1.3 to deduce that they are orthogonal.

Next we must show that the representative functions are dense in $C(G)$. For this we take a continuous function $\phi(x)$ with $\phi(x) = \phi(x^{-1})$ and consider the convolution map $R_\phi : f \mapsto f * \phi := \int_G f(y)\phi(y^{-1}x)dy$. By Proposition 2 of §3.1, R_ϕ maps $L^2(G)$ in $C(G)$ and it is compact. From Proposition 4 of §3.1 its image is in the uniform closure of the space spanned by its eigenfunctions corresponding to nonzero eigenvalues.

By construction, the convolution R_ϕ is G-equivariant for the left action, hence it follows that the eigenspaces of this operator are G-stable. Since R_ϕ is a compact operator, its eigenspaces relative to nonzero eigenvalues are finite dimensional and hence in $\mathcal{T}_G$, by the definition of representative functions. Thus the image of R_ϕ is contained in the uniform closure of $\mathcal{T}_G$.

[59] One has to be careful about the normalization. When G is a finite group the usual multiplication in the group algebra is convolution, but for the normalized measure in which G has measure $|G|$ and not 1, as we usually assume for compact groups.

The next step is to show that, given a continuous function f, as ϕ varies, one can approximate f with elements in the image of R_ϕ as close as possible.

Given $\epsilon > 0$, take an open neighborhood U of 1 such that $|f(x) - f(y)| < \epsilon$ if $xy^{-1} \in U$. Take a continuous function $\phi(x)$ with support in U, positive, with integral 1 and $\phi(x) = \phi(x^{-1})$. We claim that $|f - f * \phi| < \epsilon$:

$$\begin{aligned}|f(x) - (f * \phi)(x)| &= \left| f(x) \int_G \phi(y^{-1}x)dy - \int_G f(y)\phi(y^{-1}x)dy \right| \\ &= \left| \int_{y^{-1}x \in U} (f(x) - f(y))\phi(y^{-1}x)dy \right| \\ &\leq \int_{y^{-1}x \in U} |f(x) - f(y)| |\phi(y^{-1}x)| dy \leq \epsilon. \quad \square\end{aligned}$$

Remark. If G is separable as a topological group, for instance if G is a Lie group, the Hilbert space $L^2(G)$ is separable. It follows again that we can have only countably many spaces $V_i^* \otimes V_i$ with V_i irreducible.

We can now apply the theory developed for L^2 class functions. Recall that a class function is one that is invariant under the action of G embedded diagonally in $G \times G$, i.e., $f(x) = f(g^{-1}xg)$ for all $g \in G$.

Express $f = \sum_i f_i$ with $f_i \in V_i^* \otimes V_i$. By the invariance property and the uniqueness of the expression it follows that each f_i is invariant, i.e., a class function.

We know that in $V_i^* \otimes V_i$ the only invariant functions under the diagonal action are the multiples of the corresponding character. Hence we see that

Corollary. *The irreducible characters are an orthonormal basis of the Hilbert space of L^2 class functions.*

Example. When G is commutative, for instance if $G = S_1^k$ is a torus, all irreducible representations are 1-dimensional. Hence we have that the irreducible characters are an orthonormal basis of the space of L^2 functions.

In coordinates $G = \{(\alpha_1, \dots, \alpha_n)\} \mid |\alpha_i| = 1$, the irreducible characters are the monomials $\prod_{i=1}^n \alpha_i^{h_i}$ and we have the usual theory of Fourier series (in this case one often uses the angular coordinates $\alpha_k = e^{2\pi i \theta_k}$).

3.3 Fourier Analysis

In order to compute integrals of L^2-functions we need to know the Hilbert space structure, induced by the L^2 norm, on each space $V_i^* \otimes V_i$ into which $L^2(G)$ decomposes.

We can do this via the following simple remark. The space $V_i^* \otimes V_i = \text{End}(V_i)$ is irreducible under $G \times G$ and it has two Hilbert space structures for which $G \times G$ is unitary. One is the restriction of the L^2 structure. The other is the Hermitian product on $\text{End}(V_i)$, deduced by the Hilbert space structure on V_i and given by the form $\text{tr}(XY^*)$. Arguing as in Proposition 2 of 3.1, every invariant Hermitian product on an irreducible representation U induces an isomorphism with the conjugate dual.

By Schur's lemma it follows that any two invariant Hilbert space structures are then proportional. Therefore the Hilbert space structure on $\mathrm{End}(V_i)$ induced by the L^2 norm equals $c\ \mathrm{tr}(XY^*)$, c a scalar.

Denote by $\rho_i : G \to GL(V_i)$ the representation. By definition (Chapter 6, §2.6) an element $X \in \mathrm{End}(V_i)$ gives the matrix coefficient $\mathrm{tr}(X\rho_i(g))$. In order to compute c take $X = 1$. We have $\mathrm{tr}(1_{V_i}) = \dim V_i$. The matrix coefficient corresponding to 1_{V_i} is the irreducible character $\chi_{V_i}(g) = \mathrm{tr}(\rho_i(g))$ and its L^2 norm is 1. Thus we deduce that $c = \dim V_i^{-1}$. In other words:

Theorem 1. *If* $X, Y \in \mathrm{End}(V_i)$ *and* $c_X(g) = \mathrm{tr}(\rho_i(g)X)$, $c_Y = \mathrm{tr}(\rho_i(g)Y)$ *are the corresponding matrix coefficients, we have*

$$\int_G c_X(g)\overline{c_Y(g)}dg = \dim V_i^{-1}\,\mathrm{tr}(XY^*). \tag{3.3.1}$$

Let us finally understand convolution. We want to extend the basic isomorphism theorem for the group algebra of a finite group proved in Chapter 6, §2.6. Given a finite-dimensional representation $\rho : G \to GL(U)$ of G and a function $f \in L^2(G)$ we can define an operator T_f on U by the formula $T_f(u) := \int_G f(g)\rho(g)(u)\,dg$.

Lemma. *The map* $f \mapsto T_f$ *is a homomorphism from* $L^2(G)$ *with convolution to the algebra of endomorphisms of* U.

Proof.

$$\begin{aligned} T_{a*b}(u) &:= \int_G (a*b)(g)\rho(g)(u)dg = \int_G\int_G a(h)b(h^{-1}g)\rho(g)(u)\,dh\,dg \\ &= \int_G\int_G a(h)b(g)\rho(hg)(u)\,dh\,dg \\ &= \int_G a(h)\rho(h)\left(\int_G b(g)\ \rho(g)(u)\,dg\right)dh = T_a(T_b(u)). \end{aligned}$$ □

We have already remarked that convolution $f * g$ is G-equivariant for the left action on f; similarly it is G-equivariant for the right action on g. In particular it maps the representative functions into themselves. Moreover, since the spaces $V_i^* \otimes V_i = \mathrm{End}(V_i)$ are distinct irreducibles under the $G \times G$ action and isotypic components under the left or right action, it follows that under convolution, $\mathrm{End}(V_i) * \mathrm{End}(V_j) = 0$ if $i \neq j$ and $\mathrm{End}(V_i) * \mathrm{End}(V_i) \subset \mathrm{End}(V_i)$.

Theorem 2. *For each irreducible representation* $\rho : G \to GL(V)$ *embed* $\mathrm{End}(V)$ *in* $L^2(G)$ *by the map* $j_V : X \mapsto \dim V\ \mathrm{tr}(X\rho(g^{-1}))$. *Then on* $\mathrm{End}(V)$ *convolution coincides with multiplication of endomorphisms.*

Proof. Same proof as in Chapter 6. By the previous lemma we have a homomorphism $\pi_V : L^2(G) \to \mathrm{End}(V)$. By the previous remarks $\mathrm{End}(V) \subset L^2(G)$ is a subalgebra under convolution. Finally we have to show that $\pi_V j_V$ is the identity of $\mathrm{End}(V)$.

In fact given $X \in \mathrm{End}(V)$, we have $j_V(X)(g) = \mathrm{tr}(\rho(g^{-1})X)\dim V$. In order to prove that $\pi_V j_X(X) = (\dim V)\int_G \mathrm{tr}(\rho(g^{-1})X)\rho(g)dg = X$ it is enough to prove that for any $Y \in \mathrm{End}(V)$, we have $(\dim V)\,\mathrm{tr}(\int_G \mathrm{tr}(\rho(g^{-1})X)\rho(g)dg\, Y) = \mathrm{tr}(XY)$. We have by 3.3.1.

$$\begin{aligned}\dim V tr\left(\int_G \mathrm{tr}(\rho(g^{-1})X)\rho(g)dg\, Y\right) &= \dim V \int_G \mathrm{tr}(\rho(g^{-1})X)\,\mathrm{tr}(\rho(g)Y)dg \\ &= \dim V \int_G \mathrm{tr}(\rho(g)Y)\overline{\mathrm{tr}(\rho(g)X^*)}dg \\ &= \mathrm{tr}(YX^{**}) = \mathrm{tr}(XY).\end{aligned}$$ □

Warning. For finite groups, when we define convolution, that is, multiplication in the group algebra, we use the non-normalized Haar measure. Then for j_V we obtain the formula $j_V : X \mapsto \frac{\dim V}{|G|}\,\mathrm{tr}(X\rho(g^{-1}))$.

3.4 Compact Lie Groups

We draw some consequences of the Peter–Weyl Theorem. Let G be a compact group. Consider any continuous representation of G in a Hilbert space H.

A vector v such that the elements gv, $g \in G$, span a finite-dimensional vector space is called a *finite vector.*

Proposition 1. *The set of finite vectors is dense.*

Proof. By module theory, if $u \in H$, the set $\mathcal{T}_G u$, spanned by applying the representative functions, is made of finite vectors, but by continuity $u = 1u$ is a limit of these vectors. □

Proposition 2. *The intersection $K = \cap_i K_i$, of all the kernels K_i of all the finite-dimensional irreducible representations V_i of a compact group G is* $\{1\}$.

Proof. From the Peter–Weyl theorem we know that $\mathcal{T}_G = \bigoplus_i V_i^* \otimes V_i$ is dense in the continuous functions; since these functions do not separate the points of the intersection of kernels we must have $K = \{1\}$. □

Theorem. *A compact Lie group G has a faithful finite-dimensional representation.*

Proof. Each of the kernels K_i is a Lie subgroup with some Lie algebra L_i and we must have, from the previous proposition, that the intersection $\cap_i L_i = 0$, of all these Lie algebras is 0. This implies that there are finitely many representations $V_i, i = 1, \dots, m$, with the property that the set of elements in all the kernels K_i of the V_i is a subgroup with Lie algebra equal to 0. Thus $\cap_{i=1}^m K_i$ is discrete and hence finite since we are in a compact group. By the previous proposition we can find finitely many representations so that also the non-identity elements of this finite group are not in the kernel of all these representations. Taking the direct sum we find the required faithful representation. □

Let us make a final consideration about the Haar integral. Since the Haar integral is both left- and right-invariant it is a $G \times G$-equivariant map from $L^2(G)$ to the trivial representation. In particular if we restrict it to the representative functions $\bigoplus_i V_i^* \otimes V_i$, it must vanish on each irreducible component $V_i^* \otimes V_i$, which is different from the trivial representation which is afforded by the constant functions. Thus,

Proposition 3. *The Haar integral restricted to $\bigoplus_i V_i^* \otimes V_i$ is the projection to the constant functions, which are the isotypic component of the trivial representation, with kernel all the other nontrivial isotypic components.*

4 Representations of Linearly Reductive Groups

4.1 Characters for Linearly Reductive Groups

We have already stressed several times that we will show a very tight relationship between compact Lie groups and linearly reductive groups. We thus start to discuss characters for linearly reductive groups.

Consider the action by conjugation of G on itself. It is the restriction to G, embedded diagonally in $G \times G$, of the left and right actions.

Let $Z[G]$ denote the space of regular functions f which are invariant under conjugation.

From the decomposition of Chapter 7, §3.1.1, $F[G] = \bigoplus_i U_i^* \otimes U_i$, it follows that the space $Z[G]$ decomposes as a direct sum of the spaces $Z[U_i]$ of conjugation-invariant functions in $U_i^* \otimes U_i$. We claim that:

Lemma. *$Z[U_i]$ is 1-dimensional, generated by the character of the representation U_i.*

Proof. Since U_i is irreducible and $U_i^* \otimes U_i = \mathrm{End}(U_i)^*$ we have by Schur's lemma that $Z[U_i]$ is 1-dimensional, generated by the element corresponding to the trace on $\mathrm{End}(U_i)$.

Now we follow the identifications. An element u of $\mathrm{End}(U_i)^*$ gives the matrix coefficient $u(\rho_i(g))$ where $\rho_i : G \to GL(U_i) \subset \mathrm{End}(U_i)$ denotes the representation map.

We obtain the function $\chi_i(g) = \mathrm{tr}(\rho_i(g))$ as the desired invariant element. □

Corollary. *For a linearly reductive group, the G-irreducible characters are a basis of the conjugation-invariant functions.*

We will see in Chapter 10 that any two maximal tori are conjugate and the union of all maximal tori in a reductive group G is dense in G. One of the implications of this theorem is the fact that the character of a representation M of G is determined by its restriction to a given maximal torus T. On M the group T acts as a direct sum of irreducible 1-dimensional characters in $\hat{T}$, and thus the character of M can be expressed as a sum of these characters with nonnegative coefficients, expressing their multiplicities.

After restriction to a maximal torus T, the fact that a character is a class function implies a further symmetry. Let N_T denote the normalizer of T. N_T acts on T by conjugation, and a class function restricted to T is invariant under this action. There are many important theorems about this action, the first of which is:

Theorem 1. *T equals its centralizer and N_T/T is a finite group, called the Weyl group and denoted by W.*

Under restriction to a maximal torus T, the ring of characters of G is isomorphic to the subring of W-invariant characters of T.

Let us illustrate the first part of this theorem for classical groups, leaving the general proof to Chapter 10. We always take advantage of the same idea.

Let T be a torus contained in the linear group of a vector space V.

Decompose $V := \bigoplus_\chi V_\chi$ in weight spaces under T and let $g \in GL(V)$ be a linear transformation normalizing T. Clearly g induces by conjugation an automorphism of T, which we still denote by g, which permutes the characters of T by the formula $\chi^g(t) := \chi(g^{-1}tg)$.

We thus have, for $v \in V_\chi$, $t \in T$, $tgv = gg^{-1}tgv = \chi^g(t)gv$.

We deduce that $gV_\chi = V_{\chi^g}$. In particular g permutes the weight spaces. We thus have a homomorphism from the normalizer of the torus to the group of permutations of the weight spaces. Let us now analyze this for T a maximal torus in the general linear, orthogonal and symplectic groups. We refer to Chapter 7, §4.1 for the description of the maximal tori in these three cases. First, analyze the kernel N_T^0 of this homomorphism. We prove that $N_T^0 = T$ in the four cases.

(1) **General linear group.** Let D be the group of all diagonal matrices (in the standard basis e_i). It is exactly the full subgroup of linear transformations fixing the 1-dimensional weight spaces generated by the given basis vectors.

An element in N_D^0 by definition fixes all these subspaces and thus in this case $N_D^0 = D$.

(2) **Even orthogonal group.** Again the space decomposes into 1-dimensional eigenspaces spanned by the vectors e_i, f_i giving a hyperbolic basis. One immediately verifies that a diagonal matrix g given by $ge_i = \alpha_i e_i$, $gf_i = \beta_i f_i$ is orthogonal if and only if $\alpha_i\beta_i = 1$. The matrices form a maximal torus T. Again $N_T^0 = T$.

(3) **Odd orthogonal group.** The case is similar to the previous case except that now we have an extra non-isotropic basis vector u and g is orthogonal if furthermore $gu = \pm u$. It is special orthogonal only if $gu = u$. Again $N_T^0 = T$.

(4) **Symplectic group.** Identical to (2).

Now for the full normalizer. (1) In the case of the general linear group, N_D contains the symmetric group S_n acting as permutations on the given basis.

If $a \in N_D$ we must have that $a(e_i) \in \mathbb{C}e_{\sigma(i)}$ for some $\sigma \in S_n$. Thus $\sigma^{-1}a$ is a diagonal matrix and it follows that $N_D = D \ltimes S_n$, the semidirect product.

In the case of the special linear group, we leave it to the reader to verify that we still have an exact sequence $0 \to D \to N_D \to S_n \to 0$, but this does not split, since only the even permutations are in the special linear group.

(2) In the even orthogonal case $\dim V = 2n$, the characters come in opposite pairs and their weight spaces are spanned by the vectors $e_1, e_2, \dots, e_n;\ f_1, f_2, \dots, f_n$ of a hyperbolic basis (Chapter 7, §4.1). Clearly the normalizer permutes this set of n pairs of subspaces $\{\mathbb{C}e_i, \mathbb{C}f_i\}$.

In the same way as before, we see now that the symmetric group S_n, permuting simultaneously with the same permutation the elements $e_1, e_2, \dots, e_n$ and $f_1, f_2, \dots, f_n$ consists of special orthogonal matrices.

The kernel of the map $N_T \to S_n$ is formed by the matrices diagonal of 2×2 blocks. Each 2×2 block is the orthogonal group of the 2-dimensional space spanned by e_i, f_i and it is the semidirect product of the torus part $\begin{pmatrix} a & 0 \\ 0 & a^{-1} \end{pmatrix}$ with the permutation matrix $\begin{pmatrix} 0 & 1 \\ 1 & 0 \end{pmatrix}$.

In the special orthogonal group only an even number of permutation matrices $\begin{pmatrix} 0 & 1 \\ 1 & 0 \end{pmatrix}$ can appear. It follows that the Weyl group is the semidirect product of the symmetric group S_n with the subgroup of index 2 of $\mathbb{Z}/(2)^n$ formed by the n-tuples $a_1, \dots, a_n$ with $\sum_{i=1}^n a_i = 0, \pmod 2$.

(3) The odd special orthogonal group is slightly different. We use the notations of Chapter 5. Now one has also the possibility to act on the basis $e_1, f_1, e_2, f_2, \dots, e_n, f_n, u$ by -1 on u and this corrects the fact that the determinant of an element defined on $e_1, f_1, e_2, f_2, \dots, e_n, f_n$ may be -1.

We deduce then that the Weyl group is the semidirect product of the symmetric group S_n with $\mathbb{Z}/(2)^n$.

(4) The symplectic group. The discussion starts as in the even orthogonal group, except now the 2-dimensional symplectic group is $SL(2)$. Its torus of 2×2 diagonal matrices has index 2 in its normalizer and as representatives of the Weyl group we can choose the identity and the matrix $\begin{pmatrix} 0 & 1 \\ -1 & 0 \end{pmatrix}$.

This matrix has determinant 1 and again we deduce that the Weyl group is the semidirect product of the symmetric group S_n with $\mathbb{Z}/(2)^n$.

Now we have to discuss the action of the Weyl group on the characters of a maximal torus. In the case of the general linear group a diagonal matrix X with entries $x_1, \dots, x_n$ is conjugated by a permutation matrix σ which maps $\sigma e_i = e_{\sigma(i)}$ by $\sigma X \sigma^{-1} e_i = x_{\sigma(i)} e_i$; thus the action of S_n on the *characters* x_i is the usual permutation of variables.

For the orthogonal groups and the symplectic group one has the torus of diagonal matrices of the form $x_1, x_1^{-1}, x_2, x_2^{-1}, \dots, x_n, x_n^{-1}$. Besides the permutations of the variables we now have also the inversions $x_i \to x_i^{-1}$, except that for the even orthogonal group one has to restrict to products of only an even number of inversions.

This analysis suggests an interpretation of the characters of the classical groups as particular symmetric functions. In the case of the linear group the coordinate ring of the maximal torus can be viewed as the polynomial ring $\mathbb{C}[x_1, \dots, x_n][d^{-1}]$ with $d := \prod_{i=1}^n x_i$ inverted.

d is the nth elementary symmetric function and thus the invariant elements are the polynomial in the elementary symmetric functions $\sigma_i(x),\ i = 1, \dots, n-1$ and $\sigma_n(x)^{\pm 1}$.

In the case of the inversions we make a remark. Consider the ring $A[t, t^{-1}]$ of Laurent polynomials over a commutative ring A. An element $\sum_i a_i t^i$ is invariant under $t \to t^{-1}$ if and only if $a_i = a_{-i}$. We claim then that it is a polynomial in $u := t + t^{-1}$. In fact $t^i + t^{-i} = (t + t^{-1})^i + r(t)$ where $r(t)$ has lower degree and one can work by induction. We deduce

Theorem 2. *The ring of invariants of* $\mathbb{C}[x_1, x_1^{-1}, \dots, x_n, x_n^{-1}]$ *under* $S_n \ltimes \mathbb{Z}/(2)^n$ *is the polynomial ring in the elementary symmetric functions* $\sigma_i(u)$ *in the variables* $u_i := x_i + x_i^{-1}$.

Proof. We can compute the invariants in two steps. First we compute the invariants under $\mathbb{Z}/(2)^n$ which, by the previous argument, are the polynomials in the u_i. Then we compute the invariants under the action of S_n which permutes the u_i. The claim follows. □

For the even orthogonal group we need a different computation since now we only want the invariants under a subgroup. Let $H \subset \mathbb{Z}/(2)^n$ be the subgroup defined by $\sum_i a_i = 0$.

Start from the monomial $M := x_1 x_2 \dots x_n$, the orbit of this monomial, under the group of inversions $\mathbb{Z}/(2)^n$ consists of all the monomials $x_1^{\epsilon_1} x_2^{\epsilon_2} \dots x_n^{\epsilon_n}$ where the elements $\epsilon_i = \pm 1$. We next define

$$E := \sum_{\prod_{i=1}^n \epsilon_i = 1} x_1^{\epsilon_1} x_2^{\epsilon_2} \dots x_n^{\epsilon_n}, \quad \overline{E} := \sum_{\prod_{i=1}^n \epsilon_i = -1} x_1^{\epsilon_1} x_2^{\epsilon_2} \dots x_n^{\epsilon_n};$$

E is clearly invariant under H and $E + \overline{E}$, $E\overline{E}$ are invariant under $\mathbb{Z}/(2)^n$.

We claim that any H-invariant is of the form $a + bE$ where a, b are $\mathbb{Z}/(2)^n$-invariants.

Consider the set of all Laurent monomials which is permuted by $\mathbb{Z}/(2)^n$. A basis of invariants under $\mathbb{Z}/(2)^n$ is clearly given by the sums of the vectors in each orbit, and similarly for the H-invariants. Now let K be the stabilizer of an element of the orbit, which thus has $\frac{2^n}{|K|}$ elements. The stabilizer in H is $K \cap H$, hence a $\mathbb{Z}/(2)^n$ orbit is either an H-orbit or it splits into two orbits, according to whether $K \not\subset H$ or $K \subset H$.

We get H-invariants which are not $\mathbb{Z}/(2)^n$-invariants from the last type of orbits.

A monomial $M = \prod x_i^{h_i}$ is stabilized by all the inversions in the variables x_i that have exponent 0. Thus the only case in which the stabilizer is contained in H is when all the variables x_i appear. In this case, in the $\mathbb{Z}/(2)^n$ orbit of M there is a unique element, which by abuse of notations we still call M, for which $h_i > 0$ for all i. Let $S^i_{h_1,\dots,h_n}$, $i = 1, 2$, be the sum on the two orbits of M under H.

Since $S^1_{h_1,\dots,h_n} + S^2_{h_1,\dots,h_n}$ is invariant under $\mathbb{Z}/(2)^n$ it is only necessary to show that $S^1_{h_1,\dots,h_n}$ has the required form. The multiplication $S^1_{h_1-1,\dots,h_n-1} S^1_{1,1,\dots,1}$ gives rise to $S^1_{h_1,\dots,h_n}$ plus terms which are lower in the lexicographic ordering of the h_i's, and

$S^1_{1,1,\dots,1} = E$. By induction we assume that the lower terms are of the required form. Also by induction $S^1_{h_1-1,\dots,h_n-1} = a + bE$, and so we have derived the required form:

$$S^1_{h_1,\dots,h_n} = (a + bE)E = (a + b(E + \overline{E}))E - b(E\overline{E}).$$

We can now discuss the invariants under the Weyl group. Again, the ring of invariants under H is stabilized by S_n which acts by permuting the elements u_i, and fixing the element E. We deduce that the ring of W-invariants is formed by elements of the form $a+bE$ where a, b are polynomials in the elementary symmetric functions in the elements u_i.

It remains to understand the quadratic equation satisfied by E over the ring of symmetric functions in the u_i. E satisfies the relation $E^2 - (E + \overline{E})E + E\overline{E} = 0$ and so we must compute the symmetric functions $E + \overline{E}$, $E\overline{E}$.

We easily see that $E+\overline{E} = \prod_{i=1}^n (x_i + x_i^{-1})$ which is the nth elementary symmetric function in the u_i's. As for $E\overline{E}$, it can be easily described as a sum of monomials in which the exponents are either 2 or -2, with multiplicities expressed by binomial coefficients. We leave the details to the reader.

5 Induction and Restriction

5.1 Clifford's Theorem

We now collect some general facts about representations of groups. First, let H be a group, $\phi : H \to H$ an automorphism, and $\rho : H \to GL(V)$ a linear representation.

Composing with ϕ we get a new representation V^ϕ given by $H \xrightarrow{\phi} H \xrightarrow{\rho} GL(V)$; it is immediately verified that if ϕ is an inner automorphism, V^ϕ is equivalent to ϕ.

Let $H \subset G$ be a normal subgroup. Every element $g \in G$ induces by inner conjugation in G an automorphism $\phi_g : h \mapsto ghg^{-1}$ of H.

Let M be a representation of G and $N \subset M$ an H-submodule. Since $hg^{-1}n = g^{-1}(ghg^{-1})n$, we clearly have that $g^{-1}N \subset M$ is again an H-submodule and canonically isomorphic to N^{ϕ_g}. It depends only on the coset $g^{-1}H$.

In particular, assume that M is irreducible as a G-module and N is irreducible as an H-module. Then all the submodules gN are irreducible H-modules and $\sum_{g\in G/H} gN$ is a G-submodule, hence $\sum_{g\in G/H} gN = M$.

We want in particular to apply this when H has index 2 in $G = H \cup uH$. We shall use the canonical sign representation ϵ of $\mathbb{Z}/(2) = G/H$, $\epsilon(u) = -1$, $\epsilon(H) = 1$.

Clifford's Theorem. *(1) An irreducible representation N of H extends to a representation of G if and only if N is isomorphic to N^{ϕ_u}. In this case it extends in two ways up to the sign representation.*

(2) An irreducible representation M of G restricted to H remains irreducible if M is not isomorphic to $M \otimes \epsilon$. It splits into two irreducible representations $N \oplus N^{\phi_u}$ if M is isomorphic to $M \otimes \epsilon$.

Proof. Let $h_0 = u^2 \in H$. If N is also a G-representation, the map $u : N \to N$ is an isomorphism with N^{ϕ_u}. Conversely, let $t : N \to N = N^{\phi_u}$ be an isomorphism so that $tht^{-1} = \phi_u(h)$ as operators on N. Then $t^2ht^{-2} = h_0hh_0^{-1}$, hence $h_0^{-1}t^2$ commutes with H.

Since N is irreducible we must have $h_0^{-1}t^2 = \lambda$ is a scalar. We can substitute t with $t\sqrt{\lambda}^{-1}$ and can thus assume that $t^2 = h_0$ (on N).

It follows that mapping $u \mapsto t$ gives the required extension of the representation. It also is clear that the choice of $-t$ is the other possible choice changing the sign of the representation.

(2) From our previous discussion if $N \subset M$ is an irreducible H-submodule, then $M = N + uM$, $uM = u^{-1}N \cong N^{\phi_u}$, and we clearly have two cases: $M = N$ or $M = N \oplus uN$.

In the first case, tensoring by the sign representation changes the representation. In fact if we had an isomorphism t between N and $N \otimes \epsilon$ this would also be an isomorphism of N to N as H-modules. Since N is irreducible over H, t must be a scalar, but then the identity is an isomorphism between N and $N \otimes \epsilon$, which is clearly absurd.

In the second case, $M = N \oplus u^{-1}N$; on $n_1 + u^{-1}n_2$ the action of H is by $hn_1 + u^{-1}\phi_u(h)n_2$, while $u(n_1 + u^{-1}n_2) = n_2 + u^{-1}h_0n_1$.

On $M \otimes \epsilon$ the action of u changes to $u(n_1 + u^{-1}n_2) = -n_2 - u^{-1}h_0n_1$. Then it is immediately seen that the map $n_1 + u^{-1}n_2 \mapsto n_1 - u^{-1}n_2$ is an isomorphism between M and $M \otimes \epsilon$. □

One should compare this property of the possible splitting of irreducible representations with the similar feature for conjugacy classes.

Exercise (same notation as before). A conjugacy class C of G contained in H is either a unique conjugacy class in H or it splits into two conjugacy classes permuted by exterior conjugation by u. The second case occurs if and only if the stabilizer in G of an element in the conjugacy class is contained in H. Study $A_n \subset S_n$ (the alternating group).

5.2 Induced Characters

Now let G be a group, H a subgroup and N a representation of H (over some field k).

In Chapter 1, §3.2 we have given the notion of induced representation. Let us rephrase it in the language of modules. Consider $k[G]$ as a left $k[H]$-module by the right action. The space $\hom_{k[H]}(k[G], N)$ is a representation under G by the action of G deduced from the left action on $k[G]$.

$$\hom_{k[H]}(k[G], N) := \{f : G \to N \mid f(gh^{-1}) = hf(g)\}, \ (gf)(k) := f(g^{-1}k).$$

We recover the notion given $\mathrm{Ind}_H^G(N) = \hom_{k[H]}(k[G], N)$. We already remarked that this may be called coinduced.

To be precise, this construction is the induced representation only when H has finite index in G. Otherwise one has a different construction which we leave to the reader to compare with the one presented:

Consider $k[G] \otimes_{k[H]} N$, where now $k[G]$ is thought of as a right module under $k[H]$. It is a representation under G by the left action of G on $k[G]$.

Exercise.

(1) If $G \supset H \supset K$ are groups and N is a K-module we have

$$\mathrm{Ind}_H^G(\mathrm{Ind}_K^H N)) = \mathrm{Ind}_K^G N.$$

(2) The representation $\mathrm{Ind}_H^G N$ is in a natural way described by $\bigoplus_{g \in G/H} gN$ where, by $g \in G/H$, we mean that g runs over a choice of representatives of cosets. The action of G on such a sum is easily described.

The definition we have given of induced representation extends in a simple way to algebraic groups and rational representations. In this case $k[G]$ denotes the space of regular functions on G. If H is a closed subgroup of G, one can define $\hom_{k[H]}(k[G], N)$ as the set of regular maps $G \to N$ which are H-equivariant (for the right action on G).

The regular maps from an affine algebraic variety V to a vector space U can be identified to $A(V) \otimes U$ where $A(V)$ is the ring of regular functions on V. Hence if V has an action under an algebraic group H and U is a rational representation of H, the space of H-equivariant maps $V \to U$ is identified with the space of invariants $(A(V) \otimes U)^H$.

Assume now that G is linearly reductive and let us invoke the decomposition 3.1.1 of Chapter 7, $k[G] = \bigoplus_i U_i^* \otimes U_i$. Since by right action H acts only on the factor U_i,

$$\hom_{k[H]}(k[G], N) = \bigoplus_i U_i^* \otimes \hom_H(U_i, N).$$

Finally, if N is irreducible (under H), and H is also linearly reductive, it follows from Schur's Lemma that the dimension of $\hom_H(U_i, N)$ is the multiplicity with which N appears in U_i. We thus deduce

Theorem (Frobenius reciprocity for coinduced representations). *The multiplicity with which an irreducible representation V of G appears in* $\hom_{k[H]}(k[G], N)$ *equals the multiplicity with which N appears in V^* as a representation of H.*

5.3 Homogeneous Spaces

There are several interesting results of Fourier analysis on homogeneous spaces which are explained easily by the previous discussion. Suppose we have a finite-dimensional complex unitary or real orthogonal representation V of a compact group K. Let $v \in V$ be a vector and consider its orbit Kv, which is isomorphic to the homogeneous space K/K_v where K_v is the stabilizer of v. Under the simple condition that $\overline{v} \in Kv$ (no condition in the real orthogonal case) the polynomial functions on V, restricted to Kv, form an algebra of functions satisfying the properties of the Stone–Weierstrass theorem. The Euclidean space structure on V induces on the manifold Kv a K-invariant metric, hence also a measure and a unitary representation of K on the space of L^2 functions on Kv. Thus the same analysis as in 3.2 shows that we can decompose the restriction of the polynomial functions to Kv into an orthogonal direct sum of irreducible representations. The whole space $L^2(K/K_v)$ then decomposes in Fourier series obtained from these irreducible blocks. One method to understand which representations appear and with which multiplicity is to apply Frobenius reciprocity. Another is to apply methods of algebraic geometry to the associated action of the associated linearly reductive group, see §9. A classical example comes from the theory of *spherical harmonics* obtained restricting the polynomial functions to the unit sphere.

6 The Unitary Trick

6.1 Polar Decomposition

There are several ways in which linearly reductive groups are connected to compact Lie groups. The use of this (rather strict) connection goes under the name of the unitary trick. This is done in many different ways. Here we want to discuss it with particular reference to the examples of classical groups which we are studying.

We start from the remark that the unitary group $U(n, \mathbb{C}) := \{A | AA^* = 1\}$ is a bounded and closed set in $M_n(\mathbb{C})$, hence it is compact.

Proposition 1. *$U(n, \mathbb{C})$ is a maximal compact subgroup of $GL(n, \mathbb{C})$. Any other maximal compact subgroup of $GL(n, \mathbb{C})$ is conjugate to $U(n, \mathbb{C})$.*

Proof. Let K be a compact linear group. Since K is unitarizable there exists a matrix g such that $K \subset gU(n, \mathbb{C})g^{-1}$. If K is maximal this inclusion is an equality. □

The way in which $U(n, \mathbb{C})$ sits in $GL(n, \mathbb{C})$ is very special and common to maximal compact subgroups of linearly reductive groups. The analysis passes through the polar decomposition for matrices and the Cartan decomposition for groups.

Theorem. *(1) The map $B \to e^B$ establishes a diffeomorphism between the space of Hermitian matrices and the space of positive Hermitian matrices.*

(2) Every invertible matrix X is uniquely expressible in the form

$$X = e^B A \qquad \textbf{\textit{(polar decomposition)}} \tag{6.1.1}$$

where A is unitary and B is Hermitian.

Proof. (1) We leave it as an exercise, using the eigenvalues and eigenspaces.

(2) Consider $XX^* := X\overline{X}^t$ which is clearly a positive Hermitian matrix.

If $X = e^B A$ is decomposed as in 6.1.1, then $XX^* = e^B AA^* e^B = e^{2B}$. So B is uniquely determined. Conversely, by decomposing the space into eigenspaces, it is clear that a positive Hermitian matrix is uniquely of the form e^{2B} with B Hermitian. Hence there is a unique B with $XX^* = e^{2B}$. Setting $A := e^{-B}X$ we see that A is unitary and $X = e^B A$. □

The previous theorem has two corollaries, both of which are sometimes used as unitary tricks, the first of algebro-geometric nature and the second topological.

Corollary. *(1) $U(n, \mathbb{C})$ is Zariski dense in $GL(n, \mathbb{C})$.*

(2) $GL(n, \mathbb{C})$ is diffeomorphic to $U(n, \mathbb{C}) \times \mathbb{R}^{n^2}$ via $\phi(A, B) = e^B A$. In particular $U(n, \mathbb{C})$ is a deformation retract of $GL(n, \mathbb{C})$.

Proof. The first part follows from the fact that one has the exponential map $X \to e^X$ from complex $n \times n$ matrices to $GL(n, \mathbb{C})$. In this holomorphic map the two subspaces $i\mathcal{H}$ and $\mathcal{H}$ of anti-Hermitian and Hermitian matrices map to the two factors of the polar decomposition, i.e., unitary and positive Hermitian matrices.

Since $M_n(\mathbb{C}) = \mathcal{H} + i\mathcal{H}$, any two holomorphic functions on $M_n(\mathbb{C})$ coinciding on $i\mathcal{H}$ necessarily coincide. So by the exponential and the connectedness of $GL(n, \mathbb{C})$, the same holds in $GL(n, \mathbb{C})$: two holomorphic functions on $GL(n, \mathbb{C})$ coinciding on $U(n, \mathbb{C})$ coincide. □

There is a partial converse to this analysis.

Proposition 2. *Let $G \subset GL(n, \mathbb{C})$ be an algebraic group. Suppose that $K := G \cap U(n, \mathbb{C})$ is Zariski dense in G. Then G is self-adjoint.*

Proof. Let us consider the antilinear map $g \mapsto g^*$. Although it is not algebraic, it maps algebraic varieties to algebraic varieties (conjugating the equations). Thus G^* is an algebraic variety in which K^* is Zariski dense. Since $K^* = K$ we have $G^* = G$. □

6.2 Cartan Decomposition

The polar decomposition induces, on a self-adjoint group $G \subset GL(n, \mathbb{C})$ of matrices, a *Cartan decomposition*, under a mild topological condition.

Let $u(n, \mathbb{C})$ be the anti-Hermitian matrices, the Lie algebra of $U(n, \mathbb{C})$. Then $iu(n, \mathbb{C})$ are the Hermitian matrices. Let $\mathfrak{g} \subset gl(n, \mathbb{C})$ be the Lie algebra of G.

Theorem (Cartan decomposition). *Let $G \subset GL(n, \mathbb{C})$ be a self-adjoint Lie group with finitely many connected components, $\mathfrak{g}$ its Lie algebra.*

(i) For every element $A \in G$ in polar form $A = e^B U$, we have that $U \in G$, $B \in \mathfrak{g}$. Let $K := G \cap U(n, \mathbb{C})$ and let $\mathfrak{k}$ be the Lie algebra of K.

(ii) We have $\mathfrak{g} = \mathfrak{k} \oplus \mathfrak{p}$, $\mathfrak{p} = \mathfrak{g} \cap iu(n, \mathbb{C})$. The map $\phi : K \times \mathfrak{p} \to G$ given by $\phi : (u, p) \mapsto e^p u$ is a diffeomorphism.

(iii) If $\mathfrak{g}$ is a complex Lie algebra we have $\mathfrak{p} = i\mathfrak{k}$.

Proof. If G is a self-adjoint group, clearly (taking 1-parameter subgroups) also its Lie algebra is self-adjoint. Since $X \mapsto X^*$ is a linear map of order 2, by self-adjointness $\mathfrak{g} = \mathfrak{k} \oplus \mathfrak{p}$, with $\mathfrak{k}$ the space of anti-Hermitian and $\mathfrak{p}$ of Hermitian elements of $\mathfrak{g}$. We have that $\mathfrak{k} := \mathfrak{g} \cap u(n, \mathbb{C})$ is the Lie algebra of $K := G \cap U(n, \mathbb{C})$. $K \times \mathfrak{p}$ is a submanifold of $U(n, \mathbb{C}) \times iu(n, \mathbb{C})$. The map $\phi : K \times \mathfrak{p} \to G$, being the restriction to a submanifold of a diffeomorphism, is a diffeomorphism with its image. Thus the key to the proof is to show that its image is G. In other words that if $A = e^B U \in G$ is in polar form, we have that $U \in K$, $B \in \mathfrak{p}$.

Now $e^{2B} = AA^* \in G$ by hypothesis, so it suffices to see that if B is an Hermitian matrix with $e^B \in G$, we have $B \in \mathfrak{g}$. Since $e^{nB} \in G$, $\forall n \in \mathbb{Z}$, the hypothesis that G has finitely many connected components implies that for some n, $e^{nB} \in G_0$, where G_0 denotes the connected component of the identity. We are reduced to the case G connected. In the diffeomorphism $U(n, \mathbb{C}) \times iu(n, \mathbb{C}) \to GL(n, \mathbb{C})$, $(U, B) \mapsto e^B U$, we have that $K \times \mathfrak{p}$ maps diffeomorphically to a closed submanifold of $GL(n, \mathbb{C})$ contained in G. Since clearly this submanifold has the same dimension as G and G is connected we must have $G = K \times e^{\mathfrak{p}}$, the Cartan decomposition for G.

Finally, if $\mathfrak{g}$ is a complex Lie algebra, multiplication by i maps the Hermitian to the anti-Hermitian matrices in $\mathfrak{g}$, and conversely. □

Exercise. See that the condition on finitely many components cannot be dropped.

Corollary. *The homogeneous space G/K is diffeomorphic to $\mathfrak{p}$.*

It is useful to make explicit the action of an element of G, written in its polar decomposition, on the homogeneous space G/K. Denote by $P := e^{\mathfrak{p}}$. We have a map $\rho : G \to P$ given by $\rho(g) := gg^*$. ρ is a G-equivariant map if we act with G on G by left multiplication and on P by gpg^*. ρ is an orbit map, P is the orbit of 1, and the stabilizer of 1 is K. Thus ρ identifies G/K with P and the action of G on P is gpg^*.[60]

Theorem 2. *Let G be as before and let M be a compact subgroup of G. Then M is conjugate to a subgroup of K.*

K is maximal compact and all maximal compact subgroups are conjugate in G.

The second statement follows clearly from the first. By the fixed point principle (Chapter 1, §2.2), this is equivalent to proving that M has a fixed point on G/K. This may be achieved in several ways. The classical proof is via Riemannian geometry, showing that G/K is a Riemannian symmetric space of constant negative curvature.[61] We follow the more direct approach of [OV]. For this we need some preparation.

We need to study an auxiliary function on the space P and its closure $\overline{P}$, the set of all positive semidefinite Hermitian matrices. Let $G = GL(n, \mathbb{C})$. Consider the two-variable function $\mathrm{tr}(xy^{-1})$, $x \in \overline{P}$, $y \in P$. Since $(gxg^*)(gyg^*)^{-1} = gxy^{-1}g^{-1}$,

60 Observe that, restricted to P, the orbit map is $p \mapsto p^2$.

61 The geometry of these Riemannian manifolds is a rather fascinating part of mathematics; it is the proper setting to understand non-Euclidean geometry in general; we refer to [He].

this function is invariant under the G action on $\overline{P} \times P$. Since x, y are Hermitian, $\overline{\mathrm{tr}(xy^{-1})} = \mathrm{tr}(\overline{x}\overline{y}^{-1}) = \mathrm{tr}((\overline{y}^{-1})^t\,\overline{x}^t) = \mathrm{tr}(xy^{-1})$, so $\mathrm{tr}(xy^{-1})$ is a real function.

Let $\Omega \subset P$ be a compact set. We want to analyze the function

$$(6.2.1) \qquad \rho_\Omega(x) := \max_{a\in\Omega} \mathrm{tr}(xa^{-1}).$$

Remark. If $g \in G$, we have

$$\rho_\Omega(gxg^*) := \max_{a\in\Omega} \mathrm{tr}(gxg^*a^{-1}) = \max_{a\in\Omega} \mathrm{tr}(xg^*a^{-1}g) = \rho_{g^{-1}\Omega}(x).$$

Lemma 1. *The function $\rho_\Omega(x)$ is continuous, and there is a positive constant b such that if $x \neq 0$, $\rho_\Omega(x) > b\|x\|$, where $\|x\|$ is the operator norm.*

Proof. Since Ω is compact, $\rho_\Omega(x)$ is obviously well defined and continuous. Let us estimate $\mathrm{tr}(xa^{-1})$. Fix an orthonormal basis e_i in which x is diagonal, with eigenvalues $x_i \geq 0$. If a^{-1} has matrix a_{ij}, we have $\mathrm{tr}(xa^{-1}) = \sum_i x_i a_{ii}$. Since a is positive Hermitian, $a_{ii} > 0$ for all i and for all orthonormal bases. Since the set of orthonormal bases is compact, there is a positive constant $b > 0$, independent of a and of the basis, such that $a_{ii} > b, \forall i, \forall a \in \Omega$. Hence, if $x \neq 0$, $\mathrm{tr}(xa^{-1}) > \max_i x_i b = \|x\|b$. □

Lemma 2. *Given $C > 0$, the set P_C of matrices $x \in P$ with* $\det(x) = 1$ *and* $\|x\| \leq C$ *is compact.*

Proof. P_C is stable under conjugation by unitary matrices. Since this group is compact, it is enough to see that the set of diagonal matrices in P_C is compact. This is the set of n-tuples of numbers x_i with $\prod_i x_i = 1,\ C \geq x_i > 0$. This is the intersection of the closed set $\prod_i x_i = 1$ with the compact set $C \geq x_i \geq 0, \forall i$. □

From the previous two lemmas it follows that:

Lemma 3. *The function $\rho_\Omega(x)$ admits an absolute minimum on the set of matrices $x \in P$ with* $\det(x) = 1$.

Proof. Let $x_0 \in P$, $\det(x_0) = 1$ and let $c := \rho_\Omega(x_0)$. From Lemma 1, if $x \in P$ is such that $\|x\| > cb^{-1}$, then $\rho_\Omega(x) > c$. Thus the minimum is taken on the set of elements x such that $\|x\| \leq cb^{-1}$ which is compact by Lemma 2. Hence an absolute minimum exists. □

Recall that an element $x \in P$ is of the form $x = e^A$ for a unique Hermitian matrix A. Therefore the function of the real variable u, $x^u := e^{uA}$ is well defined. The key geometric property of our functions is:

Proposition 3. *Given $x, y \in P, x \neq 1$, the two functions of the real variable u, $\phi_{x,y}(u) :=$ $\mathrm{tr}(x^u y^{-1})$ and $\rho_\Omega(x^u)$, are strictly convex.*

Proof. One way to check convexity is to prove that the second derivative is strictly positive. If $x = e^A \neq 1$, we have that $A \neq 0$ is a Hermitian matrix. The same proof as in Lemma 1 shows that $\ddot{\phi}_{x,y}(u) = \operatorname{tr}(A^2 e^{Au} y^{-1}) > 0$, since $0 \neq A^2 e^{Au} \in \overline{P}$.

Now for $\rho_\Omega(x^u) = \max_{a\in\Omega} \operatorname{tr}(x^u a^{-1}) = \max_{a\in\Omega} \phi_{x,a}(u)$ it is enough to remark that if we have a family of strictly convex functions depending on a parameter in a compact set, the maximum is clearly a strictly convex function. □

Now revert to a self-adjoint group $G \subset GL(n, \mathbb{C}) \subset GL(2n, \mathbb{R})$, its associated P and $\Omega \subset P$ a compact set. Assume furthermore that $G \subset SL(2n, R)$.

Lemma 4. *$\rho_\Omega(x)$ has a unique minimum on P.*

Proof. First, the hypothesis that the matrices have determinant 1 implies from Lemma 3 that an absolute minimum exists. Assume by contradiction that we have two minima in A, B. By the first remark, changing Ω, since G acts transitively on P we may assume $A = 1$. Furthermore, $\lim_{u\to 0} B^u = 1$ (and it is a curve in P). By convexity and the fact that B is a minimum we have that $\rho_\Omega(B^u)$ is a strictly decreasing function for $u \in (0, 1]$, hence $\rho_\Omega(1) = \lim_{u\to 0} \rho_\Omega(B^u) > \rho_\Omega(B)$, a contradiction. □

Proof of Theorem 2. We will apply the fixed point principle of Chapter 1, §2.2, to M acting on $P = G/K$. Observe that $GL(n, \mathbb{C}) \subset GL(2n, \mathbb{R}) \subset GL^+(4n, \mathbb{R})$, the matrices of positive determinant. Thus embed $G \subset GL^+(4n, \mathbb{R})$. The determinant is then a homomorphism to $\mathbb{R}^+$. Any compact subgroup of $GL^+(m, \mathbb{R})$ is contained in the subgroup of matrices with determinant 1, and we can reduce to the case $G \subset SL(2n, \mathbb{R})$.

Let $\Omega := M1$ be the orbit of 1 in $G/K = P$. The function $\rho_{M1}(x)$ on P, by Lemma 4, has a unique minimum point p_0. We claim that $\rho_{M1}(x)$ is M-invariant. In fact, by the first remark, we have for $k \in M$ that $\rho_{M1}(kxk^*) = \rho_{k^{-1}M1}(x) = \rho_{M1}(x)$. It follows that p_0 is necessarily a fixed point of M. □

Exercise. Let G be a group with finitely many components and G_0 the connected component of 1. If G_0 is self-adjoint with respect to some positive Hermitian form, then G is also self-adjoint (under a possibly different Hermitian form).

The application of this theory to algebraic groups will be proved in Chapter 10, §6.3:

Theorem 3. *If $G \subset GL(n, \mathbb{C})$ is a self-adjoint Lie group with finitely many connected components and complex Lie algebra, then G is a linearly reductive algebraic group.*

Conversely, given a linearly reductive group G and a finite-dimensional linear representation of G on a space V, there is a Hilbert space structure on V such that G is self-adjoint.

If V is faithful, the unitary elements of G form a maximal compact subgroup K and we have a canonical polar decomposition $G = Ke^{i\mathfrak{k}}$ where $\mathfrak{k}$ is the Lie algebra of K.

All maximal compact subgroups of G are conjugate in G.

Every compact Lie group appears in this way in a canonical form.

In fact, as Hilbert structure, one takes any one for which a given maximal compact subgroup is formed of unitary elements.

6.3 Classical Groups

For the other linearly reductive groups that we know, we want to make the Cartan decomposition explicit. We are dealing with self-adjoint complex groups, hence with a complex Lie algebra $\mathfrak{g}$. In the notation of §6.2 we have $\mathfrak{p} = i\mathfrak{k}$. We leave some simple details as exercise.

1. First, the diagonal group $T = (\mathbb{C}^*)^n$ decomposes as $U(1,\mathbb{C})^n \times (\mathbb{R}^+)^n$ and the multiplicative group $(\mathbb{R}^+)^n$ is isomorphic under logarithm to the additive group of $\mathbb{R}^n$. It is easily seen that this group does not contain any nontrivial compact subgroup, hence if $K \subset T$ is compact, by projecting to $(\mathbb{R}^+)^n$ we see that $K \subset U(1,\mathbb{C})^n$.

   The *compact torus* $U(1,\mathbb{C})^n = (S^1)^n$ is the unique maximal compact subgroup of T.
2. The orthogonal group $O(n,\mathbb{C})$. We have $O(n,\mathbb{C}) \cap U(n,\mathbb{C}) = O(n,\mathbb{R})$; thus $O(n,\mathbb{R})$ is a maximal compact subgroup of $O(n,\mathbb{C})$.

Exercise. Describe the orbit map $XX^*,\ X \in O(n,\mathbb{C})$.

3. The symplectic group and quaternions: We can consider the quaternions $\mathbb{H} := \mathbb{C} + j\mathbb{C}$ with the commutation rules $j^2 = -1,\ j\alpha := \overline{\alpha}j,\ \forall \alpha \in \mathbb{C}$, and set

$$\overline{\alpha + j\beta} := \overline{\alpha} - \overline{\beta}j = \overline{\alpha} - j\beta.$$

Consider the right vector space $\mathbb{H}^n = \bigoplus_{i=1}^n e_i\mathbb{H}$ over the quaternions, with basis e_i. As a right vector space over $\mathbb{C}$ this has as basis $e_1, e_1 j, e_2, e_2 j, \dots, e_n, e_n j$. For a vector $u := (q_1, q_2, \dots, q_n) \in \mathbb{H}^n$ define $\|u\| := \sum_{i=1}^n q_i\overline{q}_i$. If $q_i = \alpha_i + j\beta_i$, we have $\sum_{i=1}^n q_i\overline{q}_i = \sum_{i=1}^n |\alpha_i|^2 + |\beta_i|^2$. Let $Sp(n,\mathbb{H})$ be the group of quaternionic linear transformations preserving this norm. It is easily seen that this group can be described as the group of $n \times n$ matrices $X := (q_{ij})$ with $X^* := \overline{X}^t = X^{-1}$ where X^* is the matrix with $\overline{q}_{ji}$ in the ij entry. This is again clearly a closed bounded group, hence compact.

$$Sp(n,\mathbb{H}) := \{A \in M_n(\mathbb{H}) \mid AA^* = 1\}.$$

On $\mathbb{H}^n = \mathbb{C}^{2n}$, right multiplication by j induces an antilinear transformation, with matrix a diagonal matrix J of 2×2 blocks of the form

$$\begin{pmatrix} 0 & -1 \\ 1 & 0 \end{pmatrix}.$$

Since a complex $2n \times 2n$ matrix is quaternionic if and only if it commutes with j, we see that the group $Sp(n,\mathbb{H})$ is the subgroup of the unitary group $U(2n,\mathbb{C})$ commuting with the operator j.

If, on a complex vector space, we have a linear operator X with matrix A and an antilinear operator Y with matrix B, it is clear that both XY and YX are antilinear with matrices AB and $B\overline{A}$, respectively. In particular the two operators commute if and only if $AB = B\overline{A}$. We apply this now to $Sp(n, \mathbb{H})$. We see that it is formed by those matrices X in $U(2n, \mathbb{C})$ such that $XJ = J\overline{X} = J(X^{-1})^t$. Its Lie algebra $\mathfrak{k}$ is formed by the anti-Hermitian matrices Y with $YJ = J\overline{Y}$.

Taking $Sp(2n, \mathbb{C})$ to be the symplectic group associated to this matrix J, we have $X \in Sp(2n, \mathbb{C})$ if and only if $X^t J = JX^{-1}$ or $XJ = J(X^{-1})^t$. Thus we have that

$$Sp(n, \mathbb{H}) = U(2n, \mathbb{C}) \cap Sp(2n, \mathbb{C}). \tag{6.3.1}$$

We deduce again that $Sp(n, \mathbb{H})$ is maximal compact in $Sp(2n, \mathbb{C})$.

Exercise. Describe the orbit XX^*, $X \in Sp(2n, \mathbb{C})$.

Although this is not the theme of this book, there are other *real forms* of the groups we studied. For instance, the orthogonal groups or the unitary groups for indefinite forms are noncompact non-algebraic but self-adjoint. We have as further examples:

Proposition. *$O(n, \mathbb{R})$ is maximal compact both in $GL(n, \mathbb{R})$ and in $O(n, \mathbb{C})$.*

7 Hopf Algebras and Tannaka–Krein Duality

7.1 Reductive and Compact Groups

We use the fact, which will be proved in Chapter 10, §7.2, that a reductive group G has a Cartan decomposition $G = Ke^{i\mathfrak{k}}$. Given two rational representations M, N of G we consider them as continuous representations of K.

Lemma. *(1)* $\hom_G(M, N) = \hom_K(M, N)$.

(2) An irreducible representation V of G remains irreducible under K.

Proof. (1) It is enough to show that $\hom_K(M, N) \subset \hom_G(M, N)$.

If $A \in \hom_K(M, N)$, the set of elements $g \in G$ commuting with A is clearly an algebraic subgroup of G containing K. Since K is Zariski dense in G, the claim follows.

(2) is clearly a consequence of (1). □

The next step is to understand:

Proposition. *Let G be a linearly reductive group and K a maximal compact subgroup of G. The restriction map, from the space of regular functions on G to the space of continuous functions on K, is an isomorphism to the space of representative functions of K.*

Proof. First, since the compact group K is Zariski dense in G, the restriction to K of the algebraic functions is injective. It is also clearly equivariant with respect to the left and right action of K.

Since $GL(n, k)$ can be embedded in $SL(n+1, k)$ we can choose a specific faithful representation of G as a self-adjoint group of matrices of determinant 1. In this representation K is the set of unitary matrices in G. The matrix coefficients of this representation as functions on G generate the algebra of regular functions. By Theorem 2.3 the same matrix coefficients generate, as functions on K, the algebra of representative functions. □

Corollary. *The category of finite-dimensional rational representations of G is equivalent to the category of continuous representations of K.*

Proof. Every irreducible representation of K appears in the space of representative functions, while every algebraic irreducible representation of G appears in the space of regular functions. Since these two spaces coincide algebraically the previous lemma (2) shows that all irreducible representations of K are obtained by restriction from irreducible representations of G. The first part of the lemma shows that the restriction is an equivalence of categories. □

In fact we can immediately see that the two canonical decompositions, $\mathcal{T}_K = \bigoplus_{V\in\hat{K}} V^* \otimes V$ (formula 2.1.1) and $k[G] = \bigoplus_i U_i^* \otimes U_i$ of Chapter 7, §3.1.1, coincide under the identification between regular functions on G and representative functions on K.

7.2 Hopf Algebras

We want now to discuss an important structure, the Hopf algebra structure, on the space of representative functions $\mathcal{T}_K$. We will deduce some important consequences for compact Lie groups. Recall that in 2.2 we have seen:

If $f_1(x), f_2(x)$ are representative functions of K, then also $f_1(x)f_2(x)$ is representative.

If $f(x)$ is representative $f(xy)$ is representative as a function on $K \times K$, and it is obvious that $f(x^{-1})$ is representative. Finally

$$\mathcal{T}_{K\times K} = \mathcal{T}_K \otimes \mathcal{T}_K.$$

In the case of a compact group,

$$\mathcal{T}_K = \bigoplus\nolimits_{i\in\hat{K}} (V_i^* \otimes V_i),$$

$$\mathcal{T}_{K\times K} = \mathcal{T}_K \otimes \mathcal{T}_K = \bigoplus\nolimits_{i,j} (V_i^* \otimes V_i) \otimes (V_j^* \otimes V_j) = \oplus (V_i \otimes V_j)^* \otimes (V_i \otimes V_j).$$

$\hat{K}$ denotes the set of isomorphism classes of irreducible representations of K.

For simplicity set $\mathcal{T}_K = A$. We want to extract, from the formal properties of the previous constructions, the notion of a (commutative) Hopf algebra.[62]

[62] Hopf algebras appear in various contexts in mathematics. In particular Hopf used them to compute the cohomology of compact Lie groups.

This structure consists of several operations on A. In the general setting A need not be commutative as an algebra.

(1) A is a (commutative and) associative algebra under multiplication with 1. We set $m : A \otimes A \to A$ to be the multiplication.
(2) The map $\Delta : f \to f(xy)$ from A to $A \otimes A$ is called a *coalgebra structure.* It is a homomorphism of algebras and *coassociative* $f((xy)z) = f(x(yz))$ or equivalently, the diagram

$$\begin{array}{ccc} A & \xrightarrow{\Delta} & A \otimes A \\ \Delta \big\downarrow & & \big\downarrow 1_A \otimes \Delta \\ A \otimes A & \xrightarrow{\Delta \otimes 1_A} & A \otimes A \otimes A \end{array}$$

is commutative. In general Δ is not cocommutative, i.e., $f(xy) \neq f(yx)$.
(3) $(fg)(xy) = f(xy)g(xy)$, that is, Δ is a morphism of algebras. Since $m(f(x) \otimes g(y)) = f(x)g(x)$ we see that also m is a morphism of coalgebras, i.e., the diagram

$$\begin{array}{ccc} A \otimes A & \xrightarrow{m} & A \\ \Delta \otimes \Delta \big\downarrow & & \big\downarrow \Delta \\ (A \otimes A) \otimes (A \otimes A) & \xrightarrow{m \otimes m \circ 1_A \otimes \tau \otimes 1_A} & A \otimes A \end{array}$$

is commutative. Here $\tau(a \otimes b) = b \otimes a$.
(4) The map $S : f(x) \to f(x^{-1})$ is called an *antipode*.
Clearly S is a homomorphism of the algebra structure. Also $f(x^{-1}y^{-1}) = f((yx)^{-1})$, hence S *is an anti-homomorphism of the coalgebra structure.* When A is not commutative the correct axiom to use is that S *is also an anti-homomorphism of the algebra structure.*
(5) It is convenient to also think of the unit element as a map $\eta : \mathbb{C} \to A$ satisfying

$$m \circ (1_A \otimes \eta) = 1_A = m \circ (\eta \otimes 1_A), \quad \epsilon\eta = 1_{\mathbb{C}}.$$

(6) We have the *counit* map $\epsilon : f \mapsto f(1)$, an algebra homomorphism $\epsilon : A \to \mathbb{C}$. With respect to the coalgebra structure, we have $f(x) = f(x1) = f(1x)$ or

$$1_A \otimes \epsilon \circ \Delta = \epsilon \otimes 1_A \circ \Delta = 1_A.$$

Also $f(xx^{-1}) = f(x^{-1}x) = f(1)$ or

$$\eta \circ \epsilon = m \circ 1_A \otimes S \circ \Delta = m \circ S \otimes 1_A \circ \Delta.$$

All the previous properties except for the axioms on commutativity or cocommutativity can be taken as the axiomatic definition of a Hopf algebra.[63]

[63] Part of the axioms are dropped by some authors. For an extensive treatment one can see [Ab], [Sw].

Example. When $A = k[x_{i,j}, d^{-1}]$ is the coordinate ring of the linear group we have

$$(7.2.1) \qquad \Delta(x_{i,j}) = \sum_h x_{i,h} \otimes x_{h,j}, \ \Delta(d) = d \otimes d, \quad \sum x_{i,h} S(x_{h,j}) = \delta_{i,j}.$$

One clearly has the notion of homomorphism of Hopf algebras, ideals, etc. We leave it to the reader to make explicit what we will use. The way we have set the definitions implies:

Theorem 1. *Given a topological group G, the algebra $\mathcal{T}_G$ is a Hopf algebra. The construction that associates to G the algebra $\mathcal{T}_G$ is a contravariant functor, from the category of topological groups, to the category of commutative Hopf algebras.*

Proof. Apart from some trivial details, this is the content of the propositions of 2.1. □

A commutative Hopf algebra A can be thought of abstractly as a *group in the opposite category of commutative algebras*, due to the following remark.

Given a commutative algebra B let $G_A(B) := \{\phi : A \to B\}$ be the set of homomorphisms.

Exercise. The operations:

$$\phi * \psi(a) := \sum_i \phi(u_i)\psi(v_i), \quad \Delta(a) = \sum_i u_i \otimes v_i,$$

$$\phi^{-1}(a) := \phi(S(a)), \quad 1(a) := \eta(a)$$

are the multiplication, inverse and unit of a group law on $G_A(B)$.

In fact, in a twisted way, these are the formulas we have used for representative functions on a group! The *twist* consists of the fact that when we consider the homomorphisms of A to B as points we should also consider the elements of A as functions. Thus we should write $a(\phi)$ instead of $\phi(a)$. If we do this, all the formulas become the same as for representative functions.

This allows us to go back from Hopf algebras to topological groups. This is best done in the abstract framework by considering Hopf algebras over the real numbers. In the case of groups we must change the point of view and take only real representative functions.

When we work over the reals, the abstract group $G_A(\mathbb{R})$ can be naturally given the *finite topology* induced from the product topology $\prod_{a \in A} \mathbb{R}$ of functions from A to $\mathbb{R}$.

The abstract theorem of Tannaka duality shows that under a further restriction, which consists of axiomatizing the notion of Haar integral for Hopf algebras, we have a duality.

Formally a Haar integral on a real Hopf algebra A is defined by mimicking the group properties $\int f(xy)dy = \int f(xy)dx = \int f(x)dx$:

$$\int : A \to \mathbb{R}, \quad \forall a \in A,$$

$$\Delta(a) = \sum_i u_i \otimes v_i \implies \int a = \sum_i a_i \int v_i = \sum_i u_i \int v_i.$$

One also imposes the further positivity condition: if $a \neq 0$, $\int a^2 > 0$. Under these conditions one has:

Theorem 2. *If A is a real Hopf algebra, with an integral satisfying the previous properties, then $G_A(\mathbb{R})$ is a compact group and A is its Hopf algebra of representative functions.*

The proof is not particularly difficult and can be found for instance in [Ho]. For our treatment we do not need it but rather, in some sense, we need a refinement. This establishes the correspondence between compact Lie groups and linearly reductive algebraic groups.

The case of interest to us is when A, as an algebra, is the coordinate ring of an affine algebraic variety V, i.e., A is finitely generated, commutative and without nilpotent elements.

Recall that giving a morphism between two affine algebraic varieties is equivalent to giving a morphism in the opposite direction between their coordinate rings. Since $A \otimes A$ is the coordinate ring of $V \times V$, it easily follows that the given axioms translate on the coordinate ring the axioms of an algebraic group structure on V.

Also the converse is true. If A is a finitely generated commutative Hopf algebra without nilpotent elements over an algebraically closed field k, then by the correspondence between affine algebraic varieties and finitely generated reduced algebras we see that A is the coordinate ring of an algebraic group. In characteristic 0 the condition to be reduced is automatically satisfied (Theorem 7.3).

Now let K be a linear compact group (K is a Lie group by Chapter 3, §3.2). We claim:

Proposition. *The ring T_K of representative functions is finitely generated. T_K is the coordinate ring of an algebraic group G, the* ***complexification*** *of K.*

Proof. In fact, by Theorem 2.2, T_K is generated by the coordinates of the matrix representation and the inverse of the determinant. Since it is obviously without nilpotent elements, the previous discussion implies the claim. □

We know (Proposition 3.4 and Chapter 4, Theorem 3.2) that linear compact groups are the same as compact Lie groups, hence:

Theorem 3. *To any compact Lie group K there is canonically associated a reductive linear algebraic group G, having the representative functions of K as regular functions.*

G is linearly reductive with the same representations as K. K is maximal compact and Zariski dense in G.

If V is a faithful representation of K, it is a faithful representation of G. For any K-invariant Hilbert structure on V, G is self-adjoint.

Proof. Let G be the algebraic group with coordinate ring $\mathcal{T}_K$. By definition its points correspond to the homomorphisms $\mathcal{T}_K \to \mathbb{C}$. In particular evaluating the functions of $\mathcal{T}_K$ in K we see that $K \subset G$ is Zariski dense. Therefore, by the argument in 7.1, every K-submodule of a rational representation of G is automatically a G-submodule. Hence the decomposition $\mathcal{T}_K = \bigoplus_i V_i^* \otimes V_i$ is in $G \times G$-modules, and G is linearly reductive with the same irreducible representations as K.

Let $K \subset H \subset G$ be a larger compact subgroup. By definition of G, the functions $\mathcal{T}_K$ separate the points of G and hence of H. $\mathcal{T}_K$ is closed under complex conjugation so it is dense in the space of continuous functions of H. The decomposition $\mathcal{T}_K = \bigoplus_i V_i^* \otimes V_i$ is composed of irreducible representations of K and G; it is also composed of irreducible representations of H. Thus the Haar integral performed on H is 0 on all the nontrivial irreducible summands. Thus if we take a function $f \in \mathcal{T}_K$ and form its Haar integral either on K or on H we obtain the same result. By density this then occurs for all continuous functions. If $H \neq K$ we can find a nonzero, nonnegative function f on H, which vanishes on K, a contradiction.

The matrix coefficients of a faithful representation of K generate the algebra $\mathcal{T}_K$. So this representation is also faithful for G. To prove that $G = G^*$ notice that although the map $g \mapsto g^*$ is not algebraic, it is an antilinear map, so it transforms affine varieties into affine varieties (conjugating the coefficients in the equations), and thus G^* is algebraic and clearly K^* is Zariski dense in G^*. Since $K^* = K$ we must have $G = G^*$. □

At this point, since G is algebraic, it has a finite number of connected components; using the Cartan decomposition of 6.1 we have:

Corollary. *(i) The Lie algebra $\mathfrak{g}$ of G is the complexification of the Lie algebra $\mathfrak{k}$ of K.*

(ii) One has the Cartan decomposition $G = K \times e^{i\mathfrak{k}}$.

7.3 Hopf Ideals

The definition of Hopf algebra is sufficiently general so that it does not need to have a base coefficient field. For instance, for the general linear group we can work over $\mathbb{Z}$, or even any commutative base ring. The corresponding Hopf algebra is $A[n] := \mathbb{Z}[x_{i,j}, d^{-1}]$, where $d = \det(X)$ and X is the *generic matrix* with entries $x_{i,j}$. The defining formulas for Δ, S, η are the same as in 7.2.1. One notices that by Cramer's rule, the elements $d\,S(x_{i,j})$ are the *cofactors*, i.e., the entries of $\bigwedge^{n-1} X$. These are all polynomials with integer coefficients.

To define a Hopf algebra corresponding to a subgroup of the linear group one can do it by constructing a *Hopf ideal.*

Definition. A *Hopf ideal* of a Hopf algebra A is an ideal I such that

$$(7.3.1) \qquad \Delta(I) \subset I \otimes A + A \otimes I, \qquad S(I) \subset I, \qquad \eta(I) = 0.$$

Clearly, if I is a Hopf ideal, A/I inherits a structure of a Hopf algebra such that the quotient map, $A \to A/I$ is a homomorphism of Hopf algebras.

As an example let us see the orthogonal and symplectic group over $\mathbb{Z}$. It is convenient to write all the equations in an intrinsic form using the generic matrix X. We do the case of the orthogonal group, the symplectic being the same. The ideal I of the orthogonal group by definition is generated by the entries of the equation $XX^t - 1 = 0$. We have

$$(7.3.2) \qquad \Delta(XX^t - 1) = XX^t \otimes XX^t - 1 \otimes 1 = (XX^t - 1) \otimes XX^t + 1 \otimes (XX^t - 1)$$

$$(7.3.3) \qquad S(XX^t - 1) = S(X)S(X^t) - 1 = d^{-2} \bigwedge^{n-1}(X) \bigwedge^{n-1}(X^t) - 1 = d^{-2} \bigwedge^{n-1}(XX^t) - 1$$

$$(7.3.4) \qquad \eta(XX^t - 1) = \eta(X)\eta(X^t) - 1 = 1 - 1 = 0.$$

Thus the first and last conditions for Hopf ideals are verified by 7.3.2 and 7.3.4. To see that $S(I) \subset I$ notice that modulo I we have $XX^t = 1$, hence $d^2 = 1$ and $\bigwedge^{n-1}(XX^t) = \bigwedge^{n-1}(1) = 1$ from which it follows that modulo I we have $S(XX^t - 1) = 0$.

Although this discussion is quite satisfactory from the point of view of Hopf algebras, it leaves open the geometric question whether the ideal we found is really the full ideal vanishing on the geometric points of the orthogonal group. By the general theory of correspondence between varieties and ideal this is equivalent to proving that $A[n]/I$ has no nilpotent elements.

If instead of working over $\mathbb{Z}$ we work over $\mathbb{Q}$ and we can use a very general fact [Sw]:

Theorem 1. *A commutative Hopf algebra A over a field of characteristic 0 has no nilpotent elements (i.e., it is reduced).*

Proof. Let us see the proof when A is finitely generated over $\mathbb{C}$. It is possible to reduce the general case to this. By standard facts of commutative algebra it is enough to see that the localization A_m has no nilpotent elements for every maximal ideal $\mathfrak{m}$. Let G be the set of points of A, i.e., the homomorphisms to $\mathbb{C}$. Since G is a group we can easily see (thinking that A is like a ring of functions) that G acts as group of automorphisms of A, transitively on the points. In fact the analogue of the formula for $f(xg)$ when $g : A \to \mathbb{C}$ is a point is the composition $R_g : A \xrightarrow{\Delta} A \otimes A \xrightarrow{1\otimes g} A \otimes \mathbb{C} = A$.

It follows from axiom (5) that $g = \epsilon \circ R_g$, as desired. Thus it suffices to see that A, localized at the maximal ideal $\mathfrak{m}$, kernel of the counit ϵ (i.e., at the point 1) has no nilpotent elements. Since the intersection of the powers of the maximal ideal is 0, this is equivalent to showing that $\bigoplus_{i=1}^{\infty} \mathfrak{m}^i/\mathfrak{m}^{i+1}$ has no nilpotent ideals.[64] If $m \in \mathfrak{m}$

[64] One thinks of this ring as the coordinate ring of the tangent cone at 1.

and $\Delta(m) = \sum_i x_i \otimes y_i$ we have $m = \sum_i \epsilon(x_i)y_i = \sum_i x_i\epsilon(y_i)$, $0 = \sum_i \epsilon(x_i)\epsilon(y_i)$. Hence

$$\Delta(m) = \sum_i x_i \otimes y_i - \sum_i \epsilon(x_i) \otimes y_i + \sum_i \epsilon(y_i) \otimes x_i - \sum_i \epsilon(x_i)\epsilon(y_i)$$

$$= \sum_i (x_i - \epsilon(x_i)) \otimes y_i + \sum_i \epsilon(y_i) \otimes (x_i - \epsilon(x_i)) \in \mathfrak{m} \otimes 1 + 1 \otimes \mathfrak{m}. \quad (7.3.5)$$

Similarly, $S(\mathfrak{m}) \subset \mathfrak{m}$. It follows easily that $B := \bigoplus_{i=1}^{\infty} \mathfrak{m}^i/\mathfrak{m}^{i+1}$ inherits the structure of a commutative graded Hopf algebra, with $B_0 = \mathbb{C}$. Graded Hopf algebras are well understood; in fact in a more general settings they were originally studied by Hopf as the cohomology algebras of Lie groups. In our case the theorem we need says that B is a polynomial ring, hence an integral domain, proving the claim. □

The theorem we need to conclude is an extremely special case of a general theorem of Milnor and Moore [MM]. Their theory generalizes the original theorem of Hopf, which was only for finite-dimensional graded Hopf algebras and treats several classes of algebras, in particular, the ones which are generated by their primitive elements (see the end of the next section).

We need a special case of the characterization of graded connected commutative and co-commutative Hopf algebras. Graded commutative means that the algebra satisfies $ab = (-1)^{|a||b|}ba$ where $|a|, |b|$ are the degrees of the two elements. The condition to be *connected* is simply $B_0 = \mathbb{C}$. In case the algebra is a cohomology algebra of a space X it reflects the condition that X is connected. The usual commutative case is obtained when we assume that all elements have even degree. In our previous case we should consider $\mathfrak{m}/\mathfrak{m}^2$ as in degree 2. In this language one unifies the notions of symmetric and exterior powers: one thinks of a usual symmetric algebra as being generated by elements of even degree and an extrerior algebra is still called by abuse a symmetric algebra, but it is generated by elements of odd degree. In more general language one can talk of the *symmetric algebra*, $S(\underline{V})$ of a graded vector space $\underline{V} = \sum V_i$, which is $S(\sum_i V_{2i}) \otimes \bigwedge(\sum_i V_{2i+1})$.

One of the theorems of Milnor and Moore. *Let B be a finitely generated*[65] *positively graded commutative and connected; then B is the symmetric algebra over the space $P := \{u \in B \mid \Delta(u) = u \otimes 1 + 1 \otimes u\}$ of* ***primitive elements****.*

We need only a special case of this theorem, so let us show only the very small part needed to finish the proof of Theorem 1.

Finishing the proof. In the theorem above $B := \bigoplus_{i=1}^{\infty} \mathfrak{m}^i/\mathfrak{m}^{i+1}$ is a graded commutative Hopf algebra generated by the elements of lowest degree $\mathfrak{m}/\mathfrak{m}^2$ (we should give to them degree 2 to be compatible with the definitions). Let $x \in \mathfrak{m}/\mathfrak{m}^2$. We have $\Delta x = a \otimes 1 + 1 \otimes b$, $a, b \in \mathfrak{m}/\mathfrak{m}^2$ by the minimality of the degree. Applying axiom (5) we see that $a = b = x$ and x is primitive. What we need to prove is thus that if

[65] This condition can be weakened.

$x_1, \dots, x_n$ constitute a basis of $\mathfrak{m}/\mathfrak{m}^2$, then the x_i are algebraically independent. Assume by contradiction that $f(x_1, \dots, x_n) = 0$ is a homogeneous polynomial relation of minimum degree h. We also have

$$0 = \Delta f(x_1, \dots, x_n) = f(x_1 \otimes 1 + 1 \otimes x_1, \dots, x_n \otimes 1 + 1 \otimes x_n) = 0.$$

Expand $\Delta f \in \sum_{i=0}^{h} B_{h-i} \otimes B_i$ and consider the term $T_{h-1,1}$ of bidegree $h-1, 1$. This is really a polarization and in fact it is $\sum_{j=1}^{n} \frac{\partial f}{\partial x_j}(x_1, \dots, x_n) \otimes x_i$. Since the x_i are linearly independent the condition $T_{h-1,1} = 0$ implies $\frac{\partial f}{\partial x_j}(x_1, \dots, x_n) = 0, \forall j$. Since we are in characteristic 0, at least one of these equations is nontrivial and of degree $h-1$, a contradiction. □

As a consequence, a Hopf ideal of the coordinate ring of an algebraic group in characteristic 0 is always the defining ideal of an algebraic subgroup.

Exercise. Let G be a linear algebraic group, $\rho : G \to GL(V)$ a linear representation and $v \in V$ a vector. Prove that the ideal of the stabilizer of v generated by the equations $\rho(g)v - v$ is a Hopf ideal.

It is still true that the algebra modulo the ideal I generated by the entries of the equations $XX^t = 1$ has no nilpotent ideals when we take as coefficients a field of characteristic $\neq 2$.

The proof requires a little commutative algebra (cf. [E]). Let k be a field of characteristic $\neq 2$. The matrix $XX^t - 1$ is a symmetric $n \times n$ matrix, so the equations $XX^t - 1 = 0$ are of dimension $\binom{n+1}{2}$, while the dimension of the orthogonal group is $\binom{n}{2}$ (this follows from Cayley's parametrization in any characteristic $\neq 2$) and $\binom{n+1}{2} + \binom{n}{2} = n^2$ the number of variables. We are thus in the case of a *complete intersection*, i.e., the number of equations equals the codimension of the variety. Since a group is a smooth variety we must then expect that the Jacobian of these equations has everywhere maximal rank. In more geometric language let $S_n(k)$ be the space of symmetric $n \times n$ matrices. Consider the mapping $\pi : M_n(k) \to S_n(k)$ given by $X \to XX^t$. In order to show that for some $A \in S_n(k)$ the equations $XX^t = A$ generate the ideal of definition of the corresponding variety, it is enough to show that the differential $d\pi$ of the map is always surjective on the points X such that $XX^t = A$. The differential can be computed by substituting for X a matrix $X + Y$ and saving only the linear terms in Y, getting the formula $YX^t + XY^t = YX^t + (YX^t)^t$.

Thus we have to show that given any symmetric matrix Z, we can solve the equation $Z = YX^t + (YX^t)^t$ if $XX^t = 1$. We set $Y := ZX/2$ and have $Z = 1/2(ZXX^t + (ZXX^t)^t)$.

In characteristic 2 the statement is simply not true since

$$\sum_j x_{i,j}^2 - 1 = \Big(\sum_j x_{i,j} - 1\Big)^2.$$

So $\sum_j x_{i,j} - 1$ vanishes on the variety but it is not in the ideal.

Exercise. Let L be a Lie algebra and U_L its universal enveloping algebra. Show that U_L is a Hopf algebra under the operations defined on L as

$$\Delta(a) = a \otimes 1 + 1 \otimes a, \quad S(a) = -a, \quad \eta(a) = 0, \qquad a \in L. \tag{7.3.6}$$

Show that $L = \{u \in U_L \mid \Delta(u) = u \otimes 1 + 1 \otimes u\}$, the set of *primitive elements*. Study the Hopf ideals of U_L.

Remark. One of the theorems of Milnor and Moore is the characterization of universal enveloping algebras of Lie algebras as suitable primitively generated Hopf algebras.

9

Tensor Symmetry

1 Symmetry in Tensor Spaces

With all the preliminary work done this will now be a short section; it serves as an introduction to the *first fundamental theorem* of invariant theory, according to the terminology of H.Weyl.

1.1 Intertwiners and Invariants

We have seen in Chapter 1, §2.4 that, given two actions of a group G, an equivariant map is just an invariant under the action of G on maps.

For linear representations the action of G preserves the space of linear maps, so if U, V are two linear representations,

$$\hom_G(U, V) = \hom(U, V)^G.$$

For finite-dimensional representations, we have identified, in a G-equivariant way,

$$\hom(U, V) = U^* \otimes V = (U \otimes V^*)^*.$$

This last space is the space of bilinear functions on $U \times V^*$.

Explicitly, a homomorphism $f : U \to V$ corresponds to the bilinear form

$$\langle f | u \otimes \varphi \rangle = \langle \varphi | f(u) \rangle.$$

We thus have a correspondence between intertwiners and invariants.

We will find it particularly useful, according to the Aronhold method, to use this correspondence when the representations are tensor powers $U = A^{\otimes m}$; $V = B^{\otimes p}$ and $\hom(U, V) = A^{*\otimes m} \otimes B^{\otimes p}$.

In particular when $A = B$; $m = p$ we have

$$\text{(1.1.1)} \qquad \mathrm{End}(A^{\otimes m}) = \mathrm{End}(A)^{\otimes m} = A^{*\otimes m} \otimes A^{\otimes m} = (A^{*\otimes m} \otimes A^{\otimes m})^*.$$

Thus in this case we have

Proposition. *We can identify, at least as vector spaces, the G-endomorphisms of $A^{\otimes m}$ with the multilinear invariant functions on m variables in A and m variables in A^*.*

Let V be an m-dimensional space. On the tensor space $V^{\otimes n}$ we consider two group actions, one given by the linear group $GL(V)$ by the formula

$$g(v_1 \otimes v_2 \otimes \cdots \otimes v_n) := gv_1 \otimes gv_2 \otimes \cdots \otimes gv_n, \tag{1.1.2}$$

and the other by the symmetric group S_n given by

$$\sigma(v_1 \otimes v_2 \otimes \cdots \otimes v_n) = v_{\sigma^{-1}1} \otimes v_{\sigma^{-1}2} \otimes \cdots \otimes v_{\sigma^{-1}n}. \tag{1.1.3}$$

We will refer to this second action as the *symmetry action* on tensors. By the definition it is clear that these two actions commute.

Before we make any further analysis of these actions, recall that in Chapter 5, §2.3 we studied symmetric tensors. Let us recall the main points of that analysis. Given a vector $v \in V$ the tensor $v^n = v \otimes v \otimes v \cdots \otimes v$ is symmetric.

Fix a basis $e_1, e_2, \ldots, e_m$ of V. The basis elements $e_{i_1} \otimes e_{i_2} \cdots \otimes e_{i_n}$ are permuted by S_n and the orbits are classified by the multiplicities $h_1, h_2, \ldots, h_m$ with which the elements $e_1, e_2, \ldots, e_m$ appear in the term $e_{i_1} \otimes e_{i_2} \ldots \otimes e_{i_n}$.

The sum of the elements of the corresponding orbit are a basis of the symmetric tensors. The multiplicities $h_1, h_2, \ldots, h_m$ are nonnegative integers, subject only to $\sum_i h_i = n$.

If $\underline{h} := h_1, h_2, \ldots, h_m$ is such a sequence, we denote by $e_{\underline{h}}$ the sum of elements in the corresponding orbit. The image of the symmetric tensor $e_{\underline{h}}$ in the symmetric algebra is

$$\binom{n}{h_1\, h_2 \cdots h_m} e_1^{h_1} e_2^{h_2} \cdots e_m^{h_m}.$$

If $v = \sum_k x_k e_k$, we have

$$v^n = \sum_{h_1+h_2+\cdots+h_m=n} x_1^{h_1} x_2^{h_2} \cdots x_m^{h_m} e_{\underline{h}}.$$

A linear function ϕ on the space of symmetric tensors is defined by $\langle \phi | e_{\underline{h}} \rangle = a_{\underline{h}}$ and computing on the tensor v^n gives

$$\left\langle \phi \left| \left(\sum_k x_k e_k \right)^n \right. \right\rangle = \sum_{h_1+h_2+\cdots+h_m=n} x_1^{h_1} x_2^{h_2} \ldots x_m^{h_m} a_{\underline{h}}.$$

This formula shows that the dual of the space of symmetric tensors of degree n is identified with the space of homogeneous polynomials of degree n.

Let us recall that a subset $X \subset V$ is Zariski dense if the only polynomial vanishing on X is 0. A typical example that we will use is: when the base field is infinite, the set of vectors where a given polynomial is nonzero (easy to verify).

Lemma. *(i) The elements* $v^{\otimes n}$, $v \in V$, *span the space of symmetric tensors.*

(ii) More generally, given a Zariski dense set $X \subset V$, *the elements* $v^{\otimes n}$, $v \in X$, *span the space of symmetric tensors.*

Proof. Given a linear form on the space of symmetric tensors we restrict it to the tensors $v^{\otimes n}$, $v \in X$, obtaining the values of a homogeneous polynomial on X. Since X is Zariski dense this polynomial vanishes if and only if the form is 0, hence the tensors $v^{\otimes n}$, $v \in X$ span the space of symmetric tensors. □

Of course the use of the word symmetric is coherent with the general idea of *invariant under the symmetric group.*

1.2 Schur–Weyl Duality

We want to apply the general theory of semisimple algebras to the two group actions introduced in the previous section. It is convenient to introduce the two algebras of linear operators spanned by these actions; thus

(1) We call A the span of the operators induced by $GL(V)$ in $\mathrm{End}(V^{\otimes n})$.
(2) We call B the span of the operators induced by S_n in $\mathrm{End}(V^{\otimes n})$.

Our aim is to prove:

Proposition. *If* V *is a finite-dimensional vector space over an infinite field of any characteristic, then* B *is the centralizer of* A.

Proof. We start by identifying

$$\mathrm{End}(V^{\otimes n}) = \mathrm{End}(V)^{\otimes n}.$$

The decomposable tensor $A_1 \otimes A_2 \otimes \cdots \otimes A_n$ corresponds to the operator:

$$A_1 \otimes A_2 \otimes \cdots \otimes A_n(v_1 \otimes v_2 \otimes \cdots \otimes v_n) = A_1v_1 \otimes A_2v_2 \otimes \cdots \otimes A_nv_n.$$

Thus, if $g \in GL(V)$, the corresponding operator in $V^{\otimes n}$ is $g \otimes g \otimes \cdots \otimes g$. From Lemma 1.1 it follows that the algebra A coincides with the symmetric tensors in $\mathrm{End}(V)^{\otimes n}$ since $GL(V)$ is Zariski dense.

It is thus sufficient to show that for an operator in $\mathrm{End}(V)^{\otimes n}$, the condition of commuting with S_n is equivalent to being symmetric as a tensor.

It is sufficient to prove that the conjugation action of the symmetric group on $\mathrm{End}(V^{\otimes n})$ coincides with the symmetry action on $\mathrm{End}(V)^{\otimes n}$.

It is enough to verify the previous statement on decomposable tensors since they span the tensor space; thus we compute:

$$\begin{aligned}
&\sigma A_1 \otimes A_2 \otimes \cdots \otimes A_n\sigma^{-1}(v_1 \otimes v_2 \otimes \cdots \otimes v_n) \\
&\quad = \sigma A_1 \otimes A_2 \otimes \cdots \otimes A_n(v_{\sigma 1} \otimes v_{\sigma 2} \otimes \cdots \otimes v_{\sigma n}) \\
&\quad = \sigma(A_1v_{\sigma 1} \otimes A_2v_{\sigma 2} \otimes \cdots \otimes A_nv_{\sigma n}) = A_{\sigma^{-1}1}v_1 \otimes A_{\sigma^{-1}2}v_2 \ldots A_{\sigma^{-1}n}v_n \\
&\quad = (A_{\sigma^{-1}1} \otimes A_{\sigma^{-1}2} \ldots A_{\sigma^{-1}n})(v_1 \otimes v_2 \otimes \cdots \otimes v_n).
\end{aligned}$$

This computation shows that the conjugation action is in fact the symmetry action and finishes the proof. □

We now draw a main conclusion:

Theorem. *If the characteristic of F is 0, the algebras A, B are semisimple and each is the centralizer of the other.*

Proof. Since B is the span of the operators of a finite group it is semisimple by Maschke's theorem (Chapter 6, §1.5); therefore, by the Double Centralizer Theorem (Chapter 6, Theorem 2.5) all statements follow from the previous theorem which states that A is the centralizer of B. □

Remark. If the characteristic of the field F is not 0, in general the algebras A, B are not semisimple. Nevertheless it is still true (at least if F is infinite or big enough) that each is the centralizer of the other (cf. Chapter 13, Theorem 7.1).

1.3 Invariants of Vectors

We formulate Theorem 1.2 in a different language.

Given two vector spaces V, W we have identified $\hom(V, W)$ with $W \otimes V^*$ and with the space of bilinear functions on $W^* \times V$ by the formulas ($A \in \hom(V, W)$, $\alpha \in W^*$, $v \in V$):

$$\langle \alpha | Av \rangle. \tag{1.3.1}$$

In case V, W are linear representations of a group G, A is in $\hom_G(V, W)$ if and only if the bilinear function $\langle \alpha | Av \rangle$ is G-invariant.

In particular we see that for a linear representation V the space of G-linear endomorphisms of $V^{\otimes n}$ is identified with the space of multilinear functions of an *n covector*[66] *and n vector variables* $f(\alpha_1, \alpha_2, \dots, \alpha_n, v_1, v_2, \dots, v_n)$ which are G-invariant.

Let us see the meaning of this for $G = GL(V)$, V an m-dimensional vector space. In this case we know that the space of G-endomorphisms of $V^{\otimes n}$ is spanned by the symmetric group S_n. We want to see which invariant function f_σ corresponds to a permutation σ. By the formula 1.3.1 evaluated on decomposable tensors we get

$$\begin{aligned} f_\sigma(\alpha_1, \alpha_2, \dots, \alpha_n, v_1, v_2, \dots, v_n) &= \langle \alpha_1 \otimes \alpha_2 \otimes \cdots \otimes \alpha_n | \sigma(v_1 \otimes v_2 \otimes \cdots \otimes v_n) \rangle \\ &= \langle \alpha_1 \otimes \alpha_2 \otimes \cdots \otimes \alpha_n | v_{\sigma^{-1}1} \\ &\quad \otimes v_{\sigma^{-1}2} \otimes \cdots \otimes v_{\sigma^{-1}n} \rangle \\ &= \prod_{i=1}^n \langle \alpha_i | v_{\sigma^{-1}i} \rangle = \prod_{i=1}^n \langle \alpha_{\sigma i} | v_i \rangle. \end{aligned}$$

We can thus deduce:

66 Covector means linear form.

Proposition. *The space of $GL(V)$ invariant multilinear functions of n covector and n vector variables is spanned by the functions*

$$f_\sigma(\alpha_1, \alpha_2, \dots, \alpha_n, v_1, v_2, \dots, v_n) := \prod_{i=1}^{n} \langle \alpha_{\sigma i} | v_i \rangle. \tag{1.3.2}$$

1.4 First Fundamental Theorem for the Linear Group (FFT)

Up to now we have made no claim on the linear dependence or independence of the operators in S_n or of the corresponding functions f_σ. This will be analyzed in Chapter 13, §8.

We want to drop now the restriction that the invariants be multilinear.

Take the space $(V^*)^p \times V^q$ of p covector and q vector variables as the representation of $GL(V)(\dim(V) = m)$. A typical element is a sequence

$$(\alpha_1, \alpha_2, \dots, \alpha_p, v_1, v_2, \dots, v_q),\ \alpha_i \in V^*,\ v_j \in V.$$

On this space consider the pq polynomial functions $\langle \alpha_i | v_j \rangle$ which are clearly $GL(V)$ invariant. We prove:[67]

Theorem (FFT First fundamental theorem for the linear group). The ring of polynomial functions on $V^{*p} \times V^q$ that are $GL(V)$-invariant is generated by the functions $\langle \alpha_i | v_j \rangle$.

Before starting to prove this theorem we want to make some remarks about its meaning.

Fix a basis of V and its dual basis in V^*. With these bases, V is identified with the set of m-dimensional column vectors and V^* with the space of m-dimensional row vectors.

The group $GL(V)$ is then identified with the group $Gl(m, \mathbb{C})$ of $m \times m$ invertible matrices. Its action on column vectors is the product $Av,\ A \in Gl(m, \mathbb{C}),\ v \in V$, while on the row vectors the action is by αA^{-1}.

The invariant function $\langle \alpha_i | v_j \rangle$ is then identified with the product of the row vector α_i with the column vector v_j. In other words identify the space $(V^*)^p$ of p-tuples of row vectors with the space of $p \times m$ matrices (in which the p rows are the coordinates of the covectors) and (V^q) with the space of $m \times q$ matrices. Thus our representation is identified with the space of pairs:

$$(X, Y) | X \in M_{p,m},\ Y \in M_{m,q}.$$

The action of the matrix group is by

$$A(X, Y) := (XA^{-1}, AY).$$

---

[67] At this moment we are in characteristic 0, but in Chapter 13 we will generalize our results to all characteristics.

Consider the multiplication map:

(1.4.1) $$f : M_{p,m} \times M_{m,q} \to M_{p,q}, \; f(X, Y) := XY.$$

The entries of the matrix XY are the basic invariants $\langle \alpha_i | v_j \rangle$; thus the theorem can also be formulated as:

Theorem. *The ring of polynomial functions on $M_{p,m} \times M_{m,q}$ that are $Gl(m, \mathbb{C})$-invariant is given by the polynomial functions on $M_{p,q}$ composed with the map f.*

Proof. We will now prove the theorem in its first form by the Aronhold method.

Let $g(\alpha_1, \alpha_2, \dots, \alpha_p, v_1, v_2, \dots, v_q)$ be a polynomial invariant. Without loss of generality we may assume that it is homogeneous in each of its variables; then we polarize it with respect to each of its variables and obtain a new multilinear invariant of the form $\overline{g}(\alpha_1, \alpha_2, \dots, \alpha_N, v_1, v_2, \dots, v_M)$ where N and M are the total degrees of g in the α, v respectively.

First we show that $N = M$. In fact, among the elements of the linear group we have scalar matrices. Given a scalar λ, by definition it transforms v to λv and α to $\lambda^{-1}\alpha$ and thus, by the multilinearity hypothesis, it transforms the function $\overline{g}$ in $\lambda^{M-N}\overline{g}$. The invariance condition implies $M = N$.

We can now apply Proposition 1.3 and deduce that $\overline{g}$ is a linear combination of functions of the form $\prod_{i=1}^{N} \langle \alpha_{\sigma i} | v_i \rangle$.

We now apply restitution to compute g from $\overline{g}$. It is clear that g has the desired form. □

The study of the relations among invariants will be the topic of the *Second Fundamental Theorem, SFT*. Here we only remark that by elementary linear algebra, the multiplication map f has, as image, the subvariety $D_{p,q}(m)$ of $p \times q$ matrices of rank $\leq m$. This is the whole space if $m \geq \min(p, q)$; otherwise, it is a proper subvariety, called a *determinantal variety* defined, at least set theoretically, by the vanishing of the determinants of the $(m+1) \times (m+1)$ minors of the matrix of coordinate functions x_{ij} on $M_{p,q}$.

The Second Fundamental Theorem will prove that these determinants generate a prime ideal which is thus the full ideal of relations among the invariants $\langle \alpha_i | v_j \rangle$.

In fact it is even better to introduce a formal language. Suppose that V is an affine algebraic variety with the action of an algebraic group G. Suppose that $p : V \to W$ is a morphism of affine varieties, inducing the comorphism $p^* : k[W] \to k[V]$.

Definition. We say that $p : V \to W$ is a quotient under G and write $W := V//G$ if p^* is an isomorphism from $k[W]$ to the ring of invariants $k[V]^G$.

Thus the FFT says in this geometric language that the determinantal variety $D_{p,q}(m)$ is the quotient under $GL(m, \mathbb{C})$ of $(V^*)^{\oplus p} \oplus V^{\oplus q}$.

2 Young Symmetrizers

2.1 Young Diagrams

We now discuss the symmetric group. The theory of cycles (cf. Chapter 1, §2.2) implies that the conjugacy classes of S_n are in one-to-one correspondence with the isomorphism classes of $\mathbb{Z}$ actions on $[1, 2, \ldots, n]$ and these are parameterized by partitions of n.

As in Chapter 1, we express that $\mu := k_1, k_2, \ldots, k_n$ is a partition of n by $\mu \vdash n$.

We shall denote by $C(\mu)$ the conjugacy class in S_n formed by the permutations decomposed in cycles of length $k_1, k_2, \ldots, k_n$, hence $S_n = \sqcup_{\mu\vdash n} C(\mu)$.

Consider the group algebra $R := \mathbb{Q}[S_n]$ of the symmetric group. We wish to work over $\mathbb{Q}$ since the theory has really this more arithmetic flavor. We will (implicitly) exhibit a decomposition as a direct sum of matrix algebras over $\mathbb{Q}$:[68]

$$R = \mathbb{Q}[S_n] := \bigoplus_{\mu\vdash n} M_{d(\mu)}(\mathbb{Q}). \tag{2.1.1}$$

The numbers $d(\mu)$ will be computed in several ways from the partition μ.

Recall, from the theory of group characters, that we know at least that

$$R_{\mathbb{C}} := \mathbb{C}[S_n] := \sum_i M_{n_i}(\mathbb{C}),$$

where the number of summands is equal to the number of conjugacy classes, hence the number of partitions of n. For every partition $\lambda \vdash n$ we will construct a primitive idempotent e_λ in R so that $R = \bigoplus_{\lambda\vdash n} \mathrm{R}e_\lambda R$ and $\dim_{\mathbb{Q}} e_\lambda \mathrm{R}e_\lambda = 1$. In this way the left ideals $\mathrm{R}e_\lambda$ will exhaust all irreducible representations. The description of 2.1.1 then follows from Chapter 6, Theorem 3.1 (5).

In fact we will construct idempotents e_λ, $\lambda \vdash n$ so that $\dim_{\mathbb{Q}} e_\lambda \mathrm{R}e_\lambda = 1$ and $e_\lambda \mathrm{R}e_\mu = 0$ if $\lambda \neq \mu$. By the previous results we have that R contains a direct summand of the form $\bigoplus_{\mu\vdash n} M_{n(\mu)}(\mathbb{Q})$, or $R = \bigoplus_{\mu\vdash n} M_{n(\mu)}(\mathbb{Q}) \oplus R'$. We claim that $R' = 0$; otherwise, once we complexify, the algebra $R_{\mathbb{C}} = \bigoplus_{\mu\vdash n} M_{n(\mu)}(\mathbb{C}) \oplus R'_{\mathbb{C}}$ would contain more simple summands than the number of partitions of n, a contradiction.

For a partition $\lambda \vdash n$ let B be the corresponding Young diagram, formed by n boxes which are partitioned in rows or in columns. The intersection between a row and a column is either empty or it reduces to a single box.

In a more formal language consider the set $\mathbb{N}^+ \times \mathbb{N}^+$ of pairs of positive integers. For a pair $(i, j) \in \mathbb{N} \times \mathbb{N}$ set $C_{i,j} := \{(h, k) | 1 \leq h \leq i,\ 1 \leq k \leq j\}$ (this is a *rectangle*).

These rectangular sets have the following simple but useful properties:

(1) $C_{i,j} \subset C_{h,k}$ if and only if $(i, j) \in C_{h,k}$.
(2) If a rectangle is contained in the union of rectangles, then it is contained in one of them.

[68] So in this case all the division algebras coincide with $\mathbb{Q}$.

Definition. A Young diagram is a subset of $\mathbb{N}^+ \times \mathbb{N}^+$ consisting of a finite union of rectangles $C_{i,j}$.

In the literature this particular way of representing a Young diagram is also called a *Ferrer diagram*. Sometimes we will use this expression when we want to stress the formal point of view.

There are two conventional ways to display a Young diagram (sometimes referred to as the French and the English way) either as points in the first quadrant or in the fourth:

Example. The partition 4311:

French 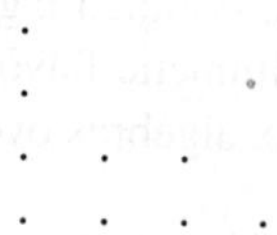 English 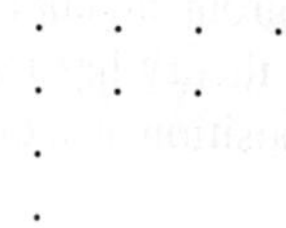

Any Young diagram can be written uniquely as a union of sets $C_{i,j}$ so that no rectangle in this union can be removed. The corresponding elements (i, j) will be called the *vertices* of the diagram.

Given a Young diagram D (in French form) the set $C_i := \{(i, j) \in D\}$, i fixed, will be called the ith *column*, the set $R_j := \{(i, j) \in D\}$, j fixed, will be called the jth *row*.

The lengths $k_1, k_2, k_3, \dots$ of the rows are a decreasing sequence of numbers which completely determine the diagrams. Thus we can identify the set of diagrams with n boxes with the set of partitions of n; this partition is called *the row shape* of the diagram.

Of course we could also have used the column lengths and the so-called *dual* partition which is *the column shape* of the diagram.

The map that to a partition associates its dual is an involutory map which geometrically can be visualized as flipping the Ferrer diagram around its diagonal.

The elements (h, k) in a diagram will be called *boxes* and displayed more pictorially as (e.g., diagrams with 6 boxes, French display):

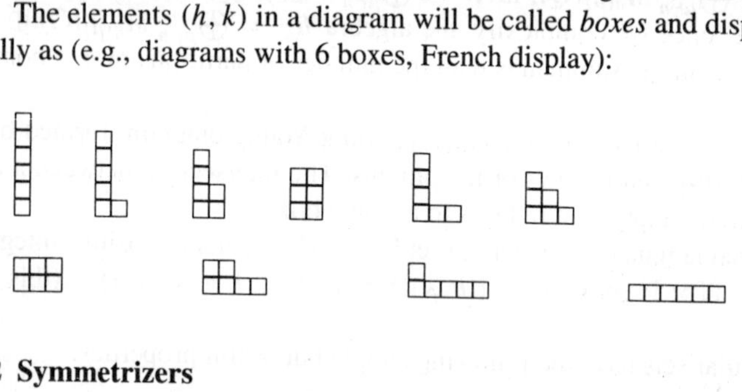

2.2 Symmetrizers

Definition 1. A bijective map from the set of boxes to the interval $(1, 2, 3, \dots, n-1, n)$ is called a *tableau*. It can be thought as a *filling* of the diagram with numbers. The given partition λ is called the *shape* of the tableau.

Example. The partition 4311:[69]

$$
\text{French}\quad
\begin{array}{llll}
3 & & & \\
1 & & & \\
5 & 2 & 7 & \\
4 & 9 & 6 & 8
\end{array}
,\qquad
\begin{array}{llll}
7 & & & \\
4 & & & \\
3 & 6 & 8 & \\
1 & 2 & 5 & 9
\end{array}
$$

The symmetric group S_n acts on the tableaux by composition:

$$\sigma T : B \xrightarrow{T} (1, 2, 3, \dots, n-1, n) \xrightarrow{\sigma} (1, 2, 3, \dots, n-1, n).$$

A tableau induces two partitions on $(1, 2, 3, \dots, n-1, n)$:

The *row partition* is defined by: i, j are in the same part if they appear in the same row of T. The *column partition* is defined similarly.

To a partition π of $(1, 2, 3, \dots, n-1, n)$[70] one associates the subgroup S_π of the symmetric group of permutations which preserve the partition. It is isomorphic to the product of the symmetric groups of all the parts of the partition. To a tableau T one associates two subgroups $\mathrm{R}_T, \mathrm{C}_T$ of S_n.

(1) R_T is the group preserving the row partition.
(2) C_T is the subgroup preserving the column partition.

It is clear that $\mathrm{R}_T \cap \mathrm{C}_T = 1$ since each box is an intersection of a row and a column.

Notice that if $s \in S_n$, the row and column partitions associated to sT are obtained by applying s to the corresponding partitions of T. Thus

$$\mathrm{R}_{sT} = s\mathrm{R}_T s^{-1}, \quad \mathrm{C}_{sT} = s\mathrm{C}_T s^{-1}. \tag{2.2.1}$$

We define two elements in $R = \mathbb{Q}[S_n]$:

$$s_T = \sum_{\sigma \in \mathrm{R}_T} \sigma \qquad \text{the \textit{symmetrizer on the rows}} \tag{2.2.2}$$

$$a_T = \sum_{\sigma \in \mathrm{C}_T} \epsilon_\sigma \sigma \qquad \text{the \textit{antisymmetrizer on the columns}.}$$

Recall that ϵ_σ denotes the sign of the permutation. The two identities are clear:

$$s_T^2 = \prod_i h_i!\, s_T, \qquad a_T^2 = \prod_i k_i!\, a_T$$

where the h_i are the lengths of the rows and k_i are the lengths of the columns.

It is better to get acquainted with these two elements from which we will build our main object of interest.

---

[69] The reader will notice the peculiar properties of the right tableau, which we will encounter over and over in the future.

[70] There is an ambiguity in the use of the word *partition*. A partition of n is just a non-increasing sequence of numbers adding to n, while a partition of a set is in fact a decomposition of the set into disjoint parts.

(2.2.3) $\qquad ps_T = s_T = s_T p, \ \forall p \in R_T; \qquad qa_T = a_T q = \epsilon_q a_T, \ \forall q \in C_T.$

Conversely $ps_T = s_T$ or $s_p = s_T$ implies $p \in R_T$. Similarly $qa_T = \epsilon_q a_T$ or $a_T q = \epsilon_q a_T$ implies $q \in C_T$. It is then an easy exercise to check the following.

Proposition. *The left ideal $\mathbb{Q}[S_n]s_T$ has as a basis the elements gs_T as g runs over a set of representatives of the cosets gR_T and it equals, as a representation, the permutation representation on such cosets.*

The left ideal $\mathbb{Q}[S_n]a_T$ has as a basis the elements ga_T as g runs over a set of representatives of the cosets gC_T and it equals, as a representation, the representation induced to S_n by the sign representation of C_T.

Now the remarkable fact comes. Consider the product

(2.2.4) $$c_T := s_T a_T = \sum_{p \in R_T,\ q \in C_T} \epsilon_q pq.$$

We will show that:

Theorem. *There exists a positive integer $p(T)$ such that the element $e_T := \frac{c_T}{p(T)}$ is a primitive idempotent.*

Definition 2. The idempotent $e_T := \frac{c_T}{p(T)}$ is called the *Young symmetrizer* relative to the given tableau.

Remark.

(2.2.5) $$c_{sT} = sc_T s^{-1}.$$

We thus have for a given $\lambda \vdash n$ several conjugate idempotents, which we will show to be primitive, associated to tableaux of row shape λ. Each will generate an irreducible module associated to λ which will be denoted by M_λ.

For the moment, let us remark that from 2.2.5 it follows that the integer $p(T)$ depends only on the shape λ of T, and thus we will denote it by $p(T) = p(\lambda)$.

2.3 The Main Lemma

The main property of the element c_T which we will explore is the following, which is clear from its definition and 2.2.3:

(2.3.1) $\qquad pc_T = c_T, \ \forall p \in R_T; \ c_T q = \epsilon_q c_T, \ \forall q \in C_T.$

We need a fundamental combinatorial lemma. Consider the partitions of n as decreasing sequences of integers (including 0) and order them lexicographically.[71]

For example, the partitions of 6 in increasing lexicographic order:

$$111111, 21111, 2211, 222, 3111, 321, 411, 42, 51, 6.$$

[71] We often drop 0 in the display.

Lemma. *Let S and T be two tableaux of row shapes:*

$$\lambda = h_1 \geq h_2 \geq \ldots \geq h_n, \ \mu = k_1 \geq k_2 \geq \ldots \geq k_n$$

with $\lambda \geq \mu$. *Then one and only one of the two following possibilities holds:*

(i) Two numbers i, j *appear in the same row in* S *and in the same column in* T.
(ii) $\lambda = \mu$ *and* $pS = qT$ *where* $p \in \mathrm{R}_S$, $q \in \mathrm{C}_T$.

Proof. We consider the first row r_1 of S. Since $h_1 \geq k_1$, by the pigeonhole principle either there are two numbers in r_1 which are in the same column in T or $h_1 = k_1$ and we can act on T with a permutation s in C_T so that S and sT have the first row filled with the same elements (possibly in a different order).

Observe that two numbers appear in the same column in T if and only if they appear in the same column in sT or $\mathrm{C}_T = \mathrm{C}_{sT}$.

We now remove the first row in both S and T and proceed as before. At the end we are either in case (i) or $\lambda = \mu$ and we have found a permutation $q \in \mathrm{C}_T$ such that S and qT have each row filled with the same elements.

In this case we can find a permutation $p \in \mathrm{R}_S$ such that $pS = qT$.

In order to complete our claim we need to show that these two cases are mutually exclusive. Thus we have to remark that if $pS = qT$ as before, then case (i) is not verified. In fact two elements are in the same row in S if and only if they are in the same row in pS, while they appear in the same column in T if and only if they appear in the same column in qT. Since $pS = qT$ two elements in the same row of pS are in different columns of qT. □

Corollary. *(i) Given* $\lambda > \mu$ *partitions, S and T tableaux of row shapes* λ, μ *respectively, and s any permutation, there exists a transposition* $u \in \mathrm{R}_S$ *and a transposition* $v \in \mathrm{C}_T$ *such that* $us = sv$.

(ii) If, for a tableau T, s *is a permutation not in* $R_T C_T$, *then there exists a transposition* $u \in R_T$ *and a transposition* $v \in C_T$ *such that* $us = sv$.

Proof. (i) From the previous lemma there are two numbers i and j in the same row for S and in the same column for sT. If $u = (i, j)$ is the corresponding transposition, we have $u \in R_S$, $u \in C_{sT}$. We set $v := s^{-1}us$ and we have $v \in s^{-1}C_{sT}s = C_T$ by 2.2.1. By definition $sv = uv$.

(ii) The proof is similar. We consider the tableau T, construct $s^{-1}T$, and apply the lemma to $s^{-1}T, T$.

If there exists a $p' \in R_{s^{-1}T}$, $q \in C_T$ with $p's^{-1}T = qT$, since $p' = s^{-1}ps$, $p \in R_T$, we would have that $s^{-1}p = q$, $s = pq^{-1}$ against the hypothesis. Hence there is a transposition $v \in C_T$ and $v \in R_{s^{-1}T}$ or $v = s^{-1}us$, $u \in R_T$, as required. □

2.4 Young Symmetrizers 2

We now draw the conclusions relative to Young symmetrizers.

Proposition. *(i) Let S and T be two tableaux of row shapes $\lambda > \mu$. If an element a in the group algebra is such that*

$$pa = a, \ \forall p \in R_S, \ \text{and } aq = \epsilon_q a, \ \forall q \in C_T,$$

then $a = 0$.

(ii) Given a tableau T and an element a in the group algebra such that

$$pa = a, \ \forall p \in R_T, \ \text{and } aq = \epsilon_q a, \ \forall q \in C_T,$$

then a is a scalar multiple of the element c_T.

Proof. (i) Let us write $a = \sum_{s \in S_n} a(s)s$; for any given s we can find u, v as in the previous lemma.

By hypothesis $ua = a, av = -a$. Then $a(s) = a(us) = a(sv) = -a(s) = 0$ and thus $a = 0$.

(ii) Using the same argument as above, we can say that if $s \notin R_T C_T$, then $a(s) = 0$. Instead, let $s = pq, \ p \in R_T, q \in C_T$. Then $a(pq) = \epsilon_q a(1)$, hence $a = a(1)c_T$. □

Before we conclude let us recall some simple facts about algebras and group algebras.

If R is a finite-dimensional algebra over a field F, we can consider any element $r \in R$ as a linear operator on R (as vector space) by right or left action. Let us define $\mathrm{tr}(r)$ to be the trace of the operator $x \mapsto xr$.[72] Clearly $\mathrm{tr}(1) = \dim_F R$. For a group algebra $F[G]$ of a finite group G, an element $g \in G, \ g \neq 1$, gives rise to a permutation $x \to xg, \ x \in G$ of the basis elements without fixed points. Hence, $\mathrm{tr}(1) = |G|, \mathrm{tr}(g) = 0$ if $g \neq 0$.

We are now ready to conclude. For $R = \mathbb{Q}[S_n]$ the theorems that we aim at are:

Theorem 1.

(i) $c_T R c_T = c_T R a_T = s_T R c_T = s_T R a_T = \mathbb{Q} c_T$.
(ii) $c_T^2 = p(\lambda) c_T$ with $p(\lambda) \neq 0$ a positive integer.
(iii) $\dim_{\mathbb{Q}} R c_T = \frac{n!}{p(\lambda)}$.
(iv) If U, V are two tableaux of shapes $\lambda > \mu$, then $s_U R a_V = a_V R s_U = 0$.
(v) If U, V are tableaux of different shapes λ, μ, we have $c_U R c_V = 0 = s_U R a_V$.

Proof. (i) We cannot have $c_T R c_T = 0$ since R is semisimple. Hence it is enough to prove $s_T R a_T = \mathbb{Q} c_T$. We apply the previous proposition and get that every element of $s_T R a_T$ satisfies (ii) of that proposition, hence $s_T R a_T = \mathbb{Q} c_T$.

(ii) In particular we have $c_T^2 = p(\lambda) c_T$. Now compute the trace of c_T for the right regular representation. From the previous discussion we have $\mathrm{tr}(c_T) = n!$, hence

---

[72] One can prove in fact that the operator $x \to rx$ has the same trace.

$c_T^2 \neq 0$. Since $c_T^2 = p(\lambda)c_T$ we have that $p(\lambda) \neq 0$. Since $p(\lambda)$ is the coefficient of 1 in the product c_T^2, it is clear that it is an integer.

(iii) $e_T := \frac{c_T}{p(\lambda)}$ is idempotent and $\frac{n!}{p(\lambda)} = \frac{\mathrm{tr}(c_T)}{p(\lambda)} = \mathrm{tr}(e_T)$. The trace of an idempotent operator is the dimension of its image. In our case $Re_T = Rc_T$, hence $\frac{n!}{p(\lambda)} = \dim_{\mathbb{Q}} Rc_T$. In particular this shows that $p(\lambda)$ is positive.

(iv) If $\lambda > \mu$ we have, by part (i), $s_U R a_V = 0$.

(v) If $\lambda > \mu$ we have, by (iv), $c_U R c_V = s_U a_U R s_V a_V \subset s_U R a_V = 0$. Otherwise $c_V R c_U = 0$, which, since R has no nilpotent ideals, implies $c_U R c_V = 0$ (Chapter 6, §3.1).

From the general discussion performed in 2.1 we finally obtain

Theorem 2. *(i) The elements $e_T := \frac{c_T}{p(\lambda)}$ are primitive idempotents in $R = \mathbb{Q}[S_n]$.*

(ii) The left ideals Re_T give all the irreducible representations of S_n explicitly indexed by partitions.

(iii) These representations are defined over $\mathbb{Q}$.

We will indicate by M_λ the irreducible representation associated to a (row) partition λ.

Remark. The Young symmetrizer a priori does not depend only on the partition λ but also on the labeling of the diagram. Two different labelings give rise to conjugate Young symmetrizers which therefore correspond to isomorphic irreducible representations.

We could have used, instead of the product $s_T a_T$, the product $a_T s_T$ in reverse order. We claim that also in this way we obtain a primitive idempotent $\frac{a_T s_T}{p(\lambda)}$, relative to the same irreducible representation.

The same proof could be applied, but we can also argue by applying the antiautomorphism $a \to \overline{a}$ of the group algebra which sends a permutation σ to σ^{-1}. Clearly,

$$\overline{a}_T = a_T, \quad \overline{s}_T = s_T, \quad \overline{s_T a_T} = a_T s_T.$$

Thus $\frac{1}{p(T)} a_T s_T = \overline{e}_T$ is a primitive idempotent.

Since clearly $c_T a_T s_T = s_T a_T a_T s_T$ is nonzero (a_T^2 is a nonzero multiple of a_T and so $(c_T a_T s_T) a_T$ is a nonzero multiple of c_T^2) we get that e_T and $\overline{e}_T$ are primitive idempotents relative to the same irreducible representation, and the claim is proved.

We will need two more remarks in the computation of the characters of the symmetric group.

Consider the two left ideals Rs_T, Ra_T. We have given a first description of their structure as representations in §2.2. They contain respectively $a_T R s_T$, $s_T R a_T$ which are both 1 dimensional. Thus we have

Lemma. *M_λ appears in its isotypic component in Rs_T (resp. Ra_T) with multiplicity 1. If M_μ appears in Rs_T, then $\mu \leq \lambda$, and if it appears in Ra_T, then $\mu \geq \lambda$.*[73]

[73] We shall prove a more precise theorem later.

Proof. To see the multiplicity with which M_μ appears in a representation V it suffices to compute the dimension of $c_T V$or of $\overline{c}_T V$ where T is a tableau of shape μ. Therefore the statement follows from the previous results. □

In particular we see that the only irreducible representation which appears in both Rs_T, Ra_T is M_λ.

The reader should apply to the idempotents that we have discussed the following fact:

Exercise. Given two idempotents e, f in a ring R we can identify

$$\hom_R(Re, Rf) = eRf.$$

2.5 Duality

There are several deeper results on the representation theory of the symmetric group which we will describe.

A first remark is about an obvious duality between diagrams. Given a tableau T relative to a partition λ, we can exchange its rows and columns obtaining a new tableau $\tilde{T}$ relative to the partition $\tilde{\lambda}$, which in general is different from λ. It is thus natural to ask in which way the two representations are tied.

Let $\mathbb{Q}(\epsilon)$ denote the sign representation.

Proposition. $M_{\tilde{\lambda}} = M_\lambda \otimes \mathbb{Q}(\epsilon)$.

Proof. Consider the automorphism τ of the group algebra defined on the group elements by $\tau(\sigma) := \epsilon_\sigma \sigma$.

Clearly, given a representation ϱ, the composition $\varrho\tau$ is equal to the tensor product with the sign representation; thus, if we apply τ to a primitive idempotent associated to M_λ, we obtain a primitive idempotent for $M_{\tilde{\lambda}}$.

Let us therefore use a tableau T of shape λ and construct the symmetrizer. We have

$$\tau(c_T) = \sum_{p \in \mathrm{R}_T,\, q \in \mathrm{C}_T} \epsilon_p \tau(pq) = \Big(\sum_{p \in \mathrm{R}_T} \epsilon_p p\Big)\Big(\sum_{q \in \mathrm{C}_T} q\Big).$$

We remark now that since $\tilde{\lambda}$ is obtained from λ by exchanging rows and columns we have

$$\mathrm{R}_T = \mathrm{C}_{\tilde{T}},\ \mathrm{C}_T = \mathrm{R}_{\tilde{T}}.$$

Thus $\tau(c_T) = a_{\tilde{T}} s_{\tilde{T}} = \overline{c}_{\tilde{T}}$, hence $\tau(e_T) = \overline{e}_{\tilde{T}}$. □

Remark. From the previous result it also follows that $p(\lambda) = p(\tilde{\lambda})$.

3 The Irreducible Representations of the Linear Group 1

3.1 Representations of the Linear Groups

We now apply the theory of symmetrizers to the linear group.

Let M be a representation of a semisimple algebra A and B its centralizer. By the structure theorem (Chapter 6) $M = \oplus N_i \otimes_{\Delta_i} P_i$ with N_i and P_i irreducible representations, respectively of A, B. If $e \in B$ is a primitive idempotent, then the subspace $eP_i \neq 0$ for a unique index i_0 and $eM = N_{i_0} \otimes eP_i \cong N_i$ is irreducible as a representation of A (associated to the irreducible representation of B relative to e).

Thus, from Theorem 2 of §2.4, to get a list of the irreducible representations of the linear group $Gl(V)$ appearing in $V^{\otimes n}$, we may apply the Young symmetrizers e_T to the tensor space and see when $e_T V^{\otimes n} \neq 0$.

Assume we have t columns of length $n_1, n_2, \ldots, n_t$, and decompose the column preserving group C_T as a product $\prod_{i=1}^t S_{n_i}$ of the symmetric groups of all columns.

By definition we get $a_T = \prod a_{n_i}$, the product of the antisymmetrizers relative to the various symmetric groups of the columns.

Let us assume, for simplicity of notation, that the first n_1 indices appear in the first column in increasing order, the next n_2 indices in the second column, and so on, so that

$$V^{\otimes n} = V^{\otimes n_1} \otimes V^{\otimes n_2} \otimes \cdots \otimes V^{\otimes n_t},$$

$$a_T V^{\otimes n} = a_{n_1} V^{\otimes n_1} \otimes a_{n_2} V^{\otimes n_2} \otimes \cdots \otimes a_{n_t} V^{\otimes n_t} = \bigwedge^{n_1} V \otimes \bigwedge^{n_2} V \otimes \cdots \otimes \bigwedge^{n_t} V.$$

Therefore we have that if there is a column of length $> \dim(V)$, then $a_T V^{\otimes n} = 0$.

Otherwise, we have $n_i \leq \dim(V)$, $\forall i$, and we prove the equivalent statement that $a_T s_T V^{\otimes n} \neq 0$. Let $e_1, e_2, \ldots, e_m$ be a basis of V and use the corresponding basis of decomposable tensors for $V^{\otimes n}$; let us consider the tensor

(3.1.1)
$$U = (e_1 \otimes e_2 \otimes \cdots \otimes e_{n_1}) \otimes (e_1 \otimes e_2 \otimes \cdots \otimes e_{n_2}) \otimes \cdots \otimes (e_1 \otimes e_2 \otimes \cdots \otimes e_{n_t}).$$

This is the decomposable tensor having e_i in the positions corresponding to the indices of the ith row. By construction it is symmetric with respect to the group R_T of row preserving permutations, hence $s_T U = pU$, $p = |R_T| \neq 0$.

Finally,

(3.1.2) $$a_T U = (e_1 \wedge e_2 \wedge \cdots \wedge e_{n_1}) \otimes (e_1 \wedge e_2 \wedge \cdots \wedge e_{n_2}) \otimes \cdots \otimes (e_1 \wedge e_2 \wedge \cdots \wedge e_{n_t}) \neq 0.$$

Recall that the length of the first column of a partition λ (equal to the number of its rows) is called the *height* of λ and indicated by $ht(\lambda)$. We have thus proved:

Proposition. *If T is a tableau of shape λ, then $e_T V^{\otimes n} = 0$ if and only if $ht(\lambda) > \dim(V)$.*

For a tableau T of shape λ, define

(3.1.3) $$S_\lambda(V) := e_T V^{\otimes n}, \qquad \text{the Schur functor associated to } \lambda.$$

We are implicitly using the fact that for two different tableaux T and T' of the same shape we have a unique permutation σ with $\sigma(T) = T'$. Hence we have a canonical ismorphism between the two spaces $e_T V^{\otimes n}, e_{T'} V^{\otimes n}$.

Remark. We shall justify the word *functor* in 7.1.

As a consequence, we thus have a description of $V^{\otimes n}$ as a representation of $S_n \times GL(V)$.

Theorem.

(3.1.4) $$V^{\otimes n} = \bigoplus_{ht(\lambda)\leq \dim(V)} M_\lambda \otimes S_\lambda(V).$$

Proof. We know that the two algebras A and B, spanned by the linear and the symmetric group, are semisimple and each the centralizer of the other. By the structure theorem we thus have $V^{\otimes n} = \bigoplus_i M_i \otimes S_i$ where the M_i are the irreducible representations of S_n which appear. We have proved that the ones which appear are the M_λ, $ht(\lambda) \leq \dim(V)$ and that $S_\lambda(V)$ is the corresponding irreducible representation of the linear group.

4 Characters of the Symmetric Group

As one can easily imagine, the character theory of the symmetric and general linear group are intimately tied together. There are basically two approaches: a combinatorial approach due to Frobenius, which first computes the characters of the symmetric group and then deduces those of the linear group, and an analytic approach based on Weyl's character formula, which proceeds in the reverse order. It is instructive to see both. There is in fact also a more recent algebraic approach to Weyl's character formula which we will not discuss (cf. [Hu1]).

4.1 Character Table

Up to now we have been able to explicitly parameterize both the conjugacy classes and the irreducible representations of S_n by partitions of n. A way to present a partition is to give the number of times that each number i appears.

If i appears k_i times in a partition μ, the partition is indicated by

(4.1.1) $$\mu := 1^{k_1}2^{k_2}3^{k_3}\dots i^{k_i}\dots .$$

Let us write

(4.1.2) $$a(\mu) := k_1!k_2!k_3!\dots k_i!\dots, \qquad b(\mu) := 1^{k_1}2^{k_2}3^{k_3}\dots i^{k_i}\dots$$

(4.1.3) $$n(\mu) = a(\mu)b(\mu) := k_1!1^{k_1}k_2!2^{k_2}k_3!3^{k_3}\dots k_i!i^{k_i}\dots$$

We need to interpret the number $n(\mu)$ in terms of the conjugacy class $C(\mu)$:

Proposition. *If $s \in C(\mu)$, then $n(\mu)$ is the order of the centralizer G_s of s and $|C(\mu)|n(\mu) = n!$.*

Proof. Let us write the permutation s as a product of a list of cycles c_i. If g centralizes s, we have that the cycles gc_ig^{-1} are a permutation of the given list of cycles.

It is clear that in this way we get all possible permutations of the cycles of equal length. Thus we have a surjective homomorphism of G_s to a product of symmetric groups $\prod S_{k_i}$; its kernel H is formed by permutations which fix each cycle.

A permutation of this type is just a product of permutations, each on the set of indices appearing in the corresponding cycle, and fixing it. For a full cycle the centralizer is the cyclic group generated by the cycle, so H is a product of cyclic groups of order the length of each cycle. The formula follows. □

The computation of the character table of S_n consists, given two partitions λ, μ, of computing the value of the character of an element of the conjugacy class $C(\mu)$ on the irreducible representation M_λ. Let us denote this value by $\chi_\lambda(\mu)$.

The final result of this analysis is expressed in compact form through symmetric functions. Recall that we denote $\psi_k(x) = \sum_{i=1}^n x_i^k$. For a partition $\mu \vdash n := k_1, k_2, \dots, k_n$, set

$$\psi_\mu(x) := \psi_{k_1}(x)\psi_{k_2}(x)\dots\psi_{k_n}(x).$$

Using the fact that the Schur functions are an integral basis of the symmetric functions there exist (unique) integers $c_\lambda(\mu)$ for which

$$\psi_\mu(x) = \sum_\lambda c_\lambda(\mu)S_\lambda(x). \tag{4.1.4}$$

We interpret these numbers as class functions c_λ on the symmetric group

$$c_\lambda(C(\mu)) := c_\lambda(\mu)$$

and we have

Theorem (Frobenius). *For all partitions $\lambda, \mu \vdash n$ we have*

$$\chi_\lambda(\mu) = c_\lambda(\mu). \tag{4.1.5}$$

The proof of this theorem is quite elaborate, and we divide it into five steps.

Step 1 First we transform the Cauchy formula into a new identity.
Step 2 Next we prove that the class functions c_λ are orthonormal.
Step 3 To each partition we associate a permutation character β_λ.
Step 4 We prove that the matrix expressing the functions β_λ in terms of the c_μ is triangular with 1 on the diagonal.
Step 5 We formulate the Theorem of Frobenius in a more precise way and prove it.

Step 1 In order to follow the Frobenius approach we go back to symmetric functions in n variables $x_1, x_2, \dots, x_n$. We shall freely use the Schur functions and the Cauchy formula for symmetric functions:

$$\prod_{i,j=1,n} \frac{1}{1 - x_i y_j} = \sum_\lambda S_\lambda(x) S_\lambda(y)$$

proved in Chapter 2, §4.1. We change its right-hand side as follows. Compute

$$\log\left(\prod_{i,j=1}^{n} \frac{1}{1 - x_i y_j}\right) = \sum_{i,j=1}^{n} \sum_{h=1}^{\infty} \frac{(x_i y_j)^h}{h} = \sum_{h=1}^{\infty} \sum_{i,j=1}^{n} \frac{(x_i y_j)^h}{h}$$

$$(4.1.6) \qquad = \sum_{h=1}^{\infty} \frac{\psi_h(x)\psi_h(y)}{h}.$$

Taking the exponential we get the following expression:

$$(4.1.7) \qquad \exp\left(\sum_{h=1}^{\infty} \frac{\psi_h(x)\psi_h(y)}{h}\right) = \sum_{k=0}^{\infty} \frac{1}{k!}\left(\sum_{h=1}^{\infty} \frac{\psi_h(x)\psi_h(y)}{h}\right)^k$$

$$= \sum_{k=0}^{\infty} \frac{1}{k!} \sum_{\sum_{i=1}^{\infty} k_i = k} \binom{k}{k_1\, k_2\, \dots} \frac{\psi_1(x)^{k_1}\psi_1(y)^{k_1}}{1} \frac{\psi_2(x)^{k_2}\psi_2(y)^{k_2}}{2^{k_2}}$$

$$(4.1.8) \qquad \times \frac{\psi_3(x)^{k_3}\psi_3(y)^{k_3}}{3^{k_3}} \dots.$$

Then from 4.1.3 we deduce

$$(4.1.9) \qquad \sum_\mu \frac{1}{n(\mu)} \psi_\mu(x)\psi_\mu(y) = \sum_\lambda S_\lambda(x) S_\lambda(y).$$

Step 2 Consider two class functions a and b as functions on partitions. Their Hermitian product is

$$\sum_{\mu \vdash n} \frac{1}{n!} \sum_{g \in C(\mu)} a(g)\bar{b}(g) = \sum_{\mu \vdash n} \frac{1}{n!} |C(\mu)| a(\mu)\bar{b}(\mu) = \sum_{\mu \vdash n} \frac{1}{n(\mu)} a(\mu)\bar{b}(\mu).$$

Let us now substitute in the identity 4.1.9 the expression $\psi_\mu = \sum_\lambda c_\lambda(\mu) S_\lambda$, and get

$$(4.1.10) \qquad \sum_{\mu \vdash n} \frac{1}{n(\mu)} c_{\lambda_1}(\mu) c_{\lambda_2}(\mu) = \begin{cases} 0 & \text{if} \quad \lambda_1 \neq \lambda_2 \\ 1 & \text{if} \quad \lambda_1 = \lambda_2. \end{cases}$$

We thus have that the class functions c_λ are an orthonormal basis, completing Step 2.

Step 3 We consider now some permutation characters.

Take a partition $\lambda := h_1, h_2, \dots, h_k$ of n. Consider the subgroup $S_\lambda := S_{h_1} \times S_{h_2} \times \dots \times S_{h_k}$ and the permutation representation on:

(4.1.11) $$S_n/S_{h_1} \times S_{h_2} \times \cdots \times S_{h_k}.$$

We will indicate the corresponding character by β_λ.

A permutation character is given by the formula $\chi(g) = \sum_i \frac{|G(g)|}{|H(g_i)|}$ (§1.4.3 of Chapter 8). Let us apply it to the case $G/H = S_n/S_{h_1} \times S_{h_2} \times \cdots \times S_{h_k}$, and for a permutation g relative to a partition $\mu := 1^{p_1}2^{p_2}3^{p_3}\dots i^{p_i}\dots n^{p_n}$.

A conjugacy class in $S_{h_1} \times S_{h_2} \times \cdots \times S_{h_k}$ is given by k partitions $\mu_i \vdash h_i$ of the numbers $h_1, h_2, \dots, h_k$. The conjugacy class of type μ, intersected with $S_{h_1} \times S_{h_2} \times \cdots \times S_{h_k}$, gives all possible k tuples of partitions $\mu_1, \mu_2, \dots, \mu_k$ of type

$$\mu_h := 1^{p_{1h}}2^{p_{2h}}3^{p_{3h}}\dots i^{p_{ih}}\dots$$

and

$$\sum_{h=1}^{k} p_{ih} = p_i.$$

In a more formal way we may define the direct sum of two partitions $\lambda = 1^{p_1}2^{p_2}3^{p_3}\dots i^{p_i}\dots, \mu = 1^{q_1}2^{q_2}3^{q_3}\dots i^{q_i}\dots$ as the partition

$$\lambda \oplus \mu := 1^{p_1+q_1}2^{p_2+q_2}3^{p_3+q_3}\dots i^{p_i+q_i}\dots$$

and remark that, with the notations of 4.1.2, $b(\lambda \oplus \mu) = b(\lambda)b(\mu)$.

When we decompose $\mu = \bigoplus_{i=1}^k \mu_i$, we have $b(\mu) = \prod b(\mu_i)$.

The cardinality $m_{\mu_1,\mu_2,\dots,\mu_k}$ of the class $\mu_1, \mu_2, \dots, \mu_k$ in $S_{h_1} \times S_{h_2} \times \cdots \times S_{h_k}$ is

$$m_{\mu_1,\mu_2,\dots,\mu_k} = \prod_{j=1}^{k} \frac{h_j!}{n(\mu_j)} = \prod_{j=1}^{k} \frac{h_j!}{a(\mu_j)} \frac{1}{b(\mu)}.$$

Now

$$\prod_{j=1}^{k} a(\mu_j) = \prod_{h=1}^{k} \left(\prod_{i=1}^{n} p_{ih}! \right).$$

So we get

$$m_{\mu_1,\mu_2,\dots,\mu_k} = \frac{1}{n(\mu)} \prod_{j=1}^{k} h_j! \prod_{i=1}^{n} \binom{p_i}{p_{i1}\,p_{i2}\dots p_{ik}}.$$

Finally for the number $\beta_\lambda(\mu)$ we have

$$\beta_\lambda(\mu) = \frac{n(\mu)}{\prod_{i=1}^k h_i!} \sum_{\mu=\bigoplus_{i=1}^k \mu_i,\ \mu_i \vdash h_i} m_{\mu_1,\mu_2,\dots,\mu_k}$$

$$= \sum_{\mu=\bigoplus_{i=1}^k \mu_i,\ \mu_i \vdash h_i} \prod_{i=1}^{n} \binom{p_i}{p_{i1}\,p_{i2}\dots p_{ik}}.$$

This sum is manifestly the coefficient of $x_1^{h_1} x_2^{h_2} \dots x_k^{h_k}$ in the symmetric function $\psi_\mu(x)$. In fact when we expand

$$\psi_\mu(x) = \psi_1(x)^{p_1} \psi_2(x)^{p_2} \dots \psi_i(x)^{p_i} \dots$$

for each factor $\psi_k(x) = \sum_{i=1}^n x_i^k$, one selects the index of the variable chosen and constructs a corresponding product monomial.

For each such monomial, denote by p_{ij} the number of choices of the term x_j^i in the p_i factors $\psi_i(x)$. We have $\prod_i \binom{p_i}{p_{i1} p_{i2} \dots p_{ik}}$ such choices and they contribute to the monomial $x_1^{h_1} x_2^{h_2} \dots x_k^{h_k}$ if and only if $\sum_i i p_{ij} = h_j$.

Step 4 If m_λ denotes the sum of all monomials in the orbit of $x_1^{h_1} x_2^{h_2} \dots x_k^{h_k}$, we get the formula

(4.1.12)
$$\psi_\mu(x) = \sum_\lambda \beta_\lambda(\mu) m_\lambda(x).$$

We wish now to expand the basis $m_\lambda(x)$ in terms of the basis $S_\lambda(x)$ and conversely:

(4.1.13)
$$m_\lambda(x) = \sum_\mu p_{\lambda,\mu} S_\mu(x), \ S_\lambda(x) = \sum_\mu k_{\lambda,\mu} m_\mu(x).$$

In order to make explicit some information about the matrices:

$$(p_{\lambda,\mu}), \ (k_{\lambda,\mu})$$

recall that the partitions are totally ordered by lexicographic ordering. We also order the monomials by the lexicographic ordering of the sequence of exponents $h_1, h_2, \dots, h_n$ of the variables $x_1, x_2, \dots, x_n$.

We remark that the ordering of monomials has the following immediate property:

If M_1, M_2, N are 3 monomials and $M_1 < M_2$, then $M_1 N < M_2 N$. For any polynomial $p(x)$, we can thus select the leading monomial $l(p)$ and for two polynomials $p(x), q(x)$ we have

$$l(pq) = l(p) l(q).$$

For a partition $\mu \vdash n := h_1 \geq h_2 \geq \dots \geq h_n$ the leading monomial of m_μ is

$$x^\mu := x_1^{h_1} x_2^{h_2} \dots x_n^{h_n}.$$

Similarly, the leading monomial of the alternating function $A_{\mu+\varrho}(x)$ is

$$x_1^{h_1+n-1} x_2^{h_2+n-2} \dots x_n^{h_n} = x^{\mu+\varrho}.$$

We now compute the leading monomial of the Schur function S_μ:

$$x^{\mu+\varrho} = l(A_{\mu+\varrho}(x)) = l(S_\mu(x) V(x)) = l(S_\mu(x)) x^\varrho.$$

We deduce that

$$l(S_\mu(x)) = x^\mu.$$

This computation has the following immediate consequence:

Corollary. *The matrices* $P := (p_{\lambda,\mu}),\ Q := (k_{\lambda,\mu})$ *are upper triangular with* 1 *on the diagonal.*

Proof. A symmetric polynomial with leading coefficient x^μ is clearly equal to m_μ plus a linear combination of the $m_\lambda,\ \lambda < \mu$. This proves the claim for the matrix Q. The matrix P is the inverse of Q and the claim follows. □

Step 5 We can now conclude a refinement of the computation of Frobenius:

Theorem 2. *(i)* $\beta_\lambda = c_\lambda + \sum_{\phi<\lambda} k_{\phi,\lambda} c_\phi,\ k_{\phi,\lambda} \in \mathbb{N}.\ c_\lambda = \sum_{\mu\geq\lambda} p_{\mu\lambda} b_\mu.$

(ii) The functions $c_\lambda(\mu)$ *are a list of the irreducible characters of the symmetric group.*

(iii) $\chi_\lambda = c_\lambda$.

Proof. From the various definitions we get

$$c_\lambda = \sum_\phi p_{\phi,\lambda} b_\phi,\ \beta_\lambda = \sum_\phi k_{\phi,\lambda} c_\phi. \tag{4.1.14}$$

Therefore the functions c_λ are virtual characters. Since they are orthonormal they are $\pm$ the irreducible characters.

From the recursive formulas it follows that $\beta_\lambda = c_\lambda + \sum_{\phi<\lambda} k_{\phi,\lambda} c_\phi,\ m_{\lambda,\phi} \in \mathbb{Z}$.

Since β_λ is a character it is a positive linear combination of the irreducible characters. It follows that each c_λ is an irreducible character and that the coefficients $k_{\phi,\lambda} \in \mathbb{N}$ represent the multiplicities of the decomposition of the permutation representation into irreducible components.[74]

(iii) Now we prove the equality $\chi_\lambda = c_\lambda$ by decreasing induction. If $\lambda = n$ is one row, then the module M_λ is the trivial representation as well as the permutation representation on S_n/S_n.

Assume $\chi_\mu = c_\mu$ for all $\mu > \lambda$. We may use Lemma 2.4 and we know that M_λ appears in its isotypic component in R_{S_T} with multiplicity 1 and does not appear in R_{S_U} for any tableau of shape $\mu > \lambda$.

We have remarked that R_{S_T} is the permutation representation of character β_λ in which, by assumption, the representation M_λ appears for the first time (with respect to the ordering of the λ). Thus the contribution of M_λ to its character must be given by the term c_λ. □

Remark. The basic formula $\psi_\mu(x) = \sum_\lambda c_\lambda(\mu) S_\lambda(x)$ can be multiplied by the Vandermonde determinant, obtaining

$$\psi_\mu(x)V(x) = \sum_\lambda c_\lambda(\mu) A_{\lambda+\varrho}(x). \tag{4.1.15}$$

Now we may apply the leading monomial theory and deduce that $c_\lambda(\mu)$ is the coefficient in $\psi_\mu(x)V(x)$ belonging to the leading monomial $x^{\lambda+\rho}$ of $A_{\lambda+\varrho}$.

This furnishes a possible algorithm; we will discuss later some features of this formula.

---

[74] The numbers $k_{\phi,\lambda}$ are called Kostka numbers. As we shall see they count some combinatorial objects called semistandard tableaux.

4.2 Frobenius Character

There is a nice interpretation of the theorem of Frobenius.

Definition. The linear isomorphism between characters of S_n and symmetric functions of degree n which assigns to χ_λ the Schur function S_λ is called the *Frobenius character.* It is denoted by $\chi \mapsto F(\chi)$.

Lemma. *The Frobenius character can be computed by the formula*

(4.2.1) $$F(\chi) = \frac{1}{n!}\sum_{\sigma\in S_n}\chi(\sigma)\psi_{\mu(\sigma)}(x) = \sum_{\mu\vdash n}\frac{\chi(\mu)}{n(\mu)}\psi_\mu(x).$$

Proof. By linearity it is enough to prove it for $\chi = \chi_\lambda$. From 4.1.4 and 4.1.10 we have

$$F(\chi_\lambda) = \sum_{\mu\vdash n}\frac{c_\lambda(\mu)}{n(\mu)}\psi_\mu(x) = \sum_{\mu\vdash n}\frac{c_\lambda(\mu)}{n(\mu)}\sum_\gamma c_\gamma(\mu)S_\gamma(x)$$
$$= \sum_\gamma\sum_\mu\frac{c_\lambda(\mu)c_\gamma(\mu)}{n(\mu)}S_\gamma(x) = S_\lambda(x). \qquad \square$$

Recall that $n(\mu)$ is the order of the centralizer of a permutation with cycle structure μ. This shows the following important multiplicative behavior of the Frobenius character.

Theorem. *Given two representations V, W of S_m, S_n, respectively, we have*

(4.2.2) $$F(\mathrm{Ind}_{S_n\times S_m}^{S_{n+m}}(V\otimes W)) = F(V)F(W).$$

Proof. Let us denote by χ the character of $\mathrm{Ind}_{S_m\times S_n}^{S_{m+n}}(V\otimes W)$. Recall the discussion of induced characters in Chapter 8. There we proved (formula 1.4.2) $\chi(g) = \sum_i\frac{|G(g)|}{|H(g_i)|}\chi_V(g_i)$. Where $|G(g)|$ is the order of the centralizer of g in G, the elements g_i run over representatives of the conjugacy classes O_i in H, decomposing the intersection of the conjugacy class of g in G with H.

In our case we deduce that $\chi(\sigma) = 0$ unless σ is conjugate to an element (a, b) of $S_n \times S_m$. In terms of partitions, the partitions $\nu \vdash n + m$ which contribute to the characters are the ones of type $\lambda\oplus\mu$. In the language of partitions the previous formula 1.4.2 becomes

$$\chi(\nu) = \sum_{\nu=\lambda+\mu}\frac{n(\lambda+\mu)}{n(\lambda)n(\mu)}\chi_V(\lambda)\chi_W(\mu);$$

since $\psi_{\lambda\oplus\mu} = \psi_\lambda\psi_\mu$ we obtain for $F(\chi)$:

$$F(\chi) = \sum_{\nu\vdash m+n}\frac{\chi(\nu)\psi_\nu}{n(\nu)} = \sum_\nu\frac{\psi_\nu}{n(\nu)}\sum_{\nu=\lambda+\mu}\frac{n(\lambda+\mu)}{n(\lambda)n(\mu)}\chi_V(\lambda)\chi_W(\mu)$$
$$= \sum_{\lambda\vdash m,\mu\vdash n}\frac{\chi_V(\lambda)\chi_W(\mu)}{n(\lambda)n(\mu)}\psi_\lambda\psi_\mu = F(\lambda)F(\mu). \qquad \square$$

4.3 Molien's Formula

We discuss a complement to the representation theory of S_n.

It will be necessary to work formally with symmetric functions in infinitely many variables, a formalism which has been justified in Chapter 2, §1.1. With this in mind we think of the identities of §4 as identities in infinitely many variables.

First, a convention. If we are given a representation of a group on a graded vector space $U := \{U_i\}_{i=0}^{\infty}$ (i.e., a representation on each U_i) its character is usually written as a power series with coefficients in the character ring in a variable q:[75]

(4.3.1)
$$\chi_U(t) := \sum_i \chi_i q^i,$$

where χ_i is the character of the representation U_i.

Definition. The expression 4.3.1 is called a graded character.

Graded characters have some formal similarities with characters. Given two graded representations $U = \{U_i\}_i$, $V = \{V_i\}_i$ we have their direct sum, and their tensor product

$$(U \oplus V)_i := U_i \oplus V_i, \quad (U \otimes V)_i := \bigoplus_{h=0}^{i} U_h \otimes V_{i-h}.$$

For the graded characters we clearly have

(4.3.2)
$$\chi_{U\oplus V}(q) = \chi_U(q) + \chi_V(q), \ \chi_{U\otimes V}(q) = \chi_U(q)\chi_V(q).$$

Let us consider a simple example.[76]

Lemma (Molien's formula). *Given a linear operator A on a vector space U its action on the symmetric algebra $S(U)$ has as graded character:*

(4.3.3)
$$\sum_{i=0}^{\infty} \operatorname{tr}(S^i(A))q^i = \frac{1}{\det(1 - qA)}.$$

Its action on the exterior algebra $\bigwedge U$ has as graded character:

(4.3.4)
$$\sum_{i=0}^{\dim U} \operatorname{tr}(\wedge^i(A))q^i = \det(1 + qA).$$

75 It is now quite customary to use q as a variable since it often appears to come from computations on finite fields where $q = p^r$ or as a quantum deformation parameter.

76 Strictly speaking we are not treating a group now, but the set of all matrices under multiplication, which is only a semigroup, for this set tensor product of representations makes sense, but not duality.

Proof. For every symmetric power $S^k(U)$ the character of the operator induced by A is a polynomial in A. Thus it is enough to prove the formula by continuity and invariance when A is diagonal.

Take a basis of eigenvectors u_i, $i = 1, \dots, n$ with eigenvalue λ_i. Then

$$S(U) = S(u_1) \otimes S(u_2) \otimes \cdots \otimes S(u_n) \quad \text{and} \quad S(u_i) = \sum_{h=0}^{\infty} F u_i^h.$$

The graded character of $S(u_i)$ is $\sum_{h=0}^{\infty} \lambda_i^h q^h = \frac{1}{1-\lambda_i q}$, hence

$$\chi_{S(U)}(q) = \prod_{i=1}^{n} \chi_{S(u_i)}(q) = \frac{1}{\prod_{i=1}^{n}(1-\lambda_i q)} = \frac{1}{\det(1-qA)}.$$

Similarly, $\bigwedge U = \wedge[u_1] \otimes \wedge[u_2] \otimes \cdots \otimes \wedge[u_n]$ and $\wedge[u_i] = F \oplus F u_i$, hence

$$\chi_{\bigwedge[U]}(q) = \prod_{i=1}^{n} \chi_{\bigwedge[u_i]}(q) = \prod_{i=1}^{n}(1+\lambda_i q) = \det(1+qA). \qquad \square$$

We apply the previous discussion to S_n acting on the space $\mathbb{C}^n$ permuting the coordinates and the representation that it induces on the polynomial ring $\mathbb{C}[x_1, x_2, \dots, x_n]$.

We denote by $\sum_{i=0}^{\infty} \chi_i q^i$ the corresponding graded character.

If σ is a permutation with cycle decomposition of lengths $\mu(\sigma) = \mu := m_1, m_2, \dots, m_k$, the standard basis of $\mathbb{C}^n$ decomposes into k-cycles each of length m_i. On the subspace relative to a cycle of length m, σ acts with eigenvalues the m-roots of 1 and

$$\det(1 - q\sigma) = \prod_{i=1}^{k} \prod_{j=1}^{m_i} \left(1 - e^{j2\pi\sqrt{-1}/m_i} q\right) = \prod_{i=1}^{k}(1 - q^{m_i}).$$

Thus the graded character of σ acting on the polynomial ring is

$$\begin{aligned}\frac{1}{\det(1-q\sigma)} &= \prod_i \sum_{j=0}^{\infty} q^{jm_i} = \prod_i \psi_{m_i}(1, q, q^2, \dots, q^k, \dots) \\ &= \psi_\mu(1, q, q^2, \dots, q^k, \dots) = \sum_{\lambda \vdash n} \chi_\lambda(\sigma) S_\lambda(1, q, q^2, \dots, q^k, \dots).\end{aligned}$$

To summarize

Theorem 1. *The graded character of S_n acting on the polynomial ring is*

(4.3.5) $$\sum_{\lambda \vdash n} \chi_\lambda S_\lambda(1, q, q^2, \dots, q^k, \dots).$$

Exercise. Prove this formula directly.

(*Hint.*) $\mathbb{C}[x_1, \dots, x_n] = \mathbb{C}[x]^{\otimes n} = \bigoplus_\lambda M_\lambda \otimes S_\lambda(\mathbb{C}[x])$.

We have a corollary of this formula. If $\lambda = h_1 \geq h_2 \ldots \geq h_n$, the term of lowest degree in q in $S_\lambda(1, q, q^2, \ldots, q^k, \ldots)$ is clearly given by the leading term $x_1^{h_1} x_2^{h_2} \ldots x_n^{h_n}$ computed in $1, q, q^2, \ldots, q^n$, and this gives $q^{h_1+2h_2+3h_3+\cdots+nh_n}$. We deduce that the representation M_λ of S_n appears for the first time in degree $h_1 + 2h_2 + 3h_3 + \cdots + nh_n$ and in this degree it appears with multiplicity 1. This particular submodule of $\mathbb{C}[x_1, x_2, \ldots, x_n]$ is called the *Specht module* and it plays an important role.[77]

Now we want to discuss another related representation.

Recall first that $\mathbb{C}[x_1, x_2, \ldots, x_n]$ is a free module over the ring of symmetric functions $\mathbb{C}[\sigma_1, \sigma_2, \ldots, \sigma_n]$ of rank $n!$. It follows that for every choice of the numbers $\underline{a} := a_1, \ldots, a_n$, the ring $R_{\underline{a}} := \mathbb{C}[x_1, x_2, \ldots, x_n]/\langle \sigma_i - a_i \rangle$ constructed from $\mathbb{C}[x_1, x_2, \ldots, x_n]$, modulo the ideal generated by the elements $\sigma_i - a_i$, is of dimension $n!$ and a representation of S_n.

We claim that it is always the regular representation.

Proof. First, we prove it in the case in which the polynomial $t^n - a_1 t^{n-1} + a_2 t^{n-2} - \cdots + (-1)^n a_n$ has distinct roots $\alpha_1, \ldots, \alpha_n$. This means that the ring $\mathbb{C}[x_1, x_2, \ldots, x_n]/\langle \sigma_i - a_i \rangle$ is the coordinate ring of the set of the $n!$ distinct points $\alpha_{\sigma(1)}, \ldots, \alpha_{\sigma(n)}$, $\sigma \in S_n$. This is clearly the regular representation.

We know that the condition for a polynomial to have distinct roots is given by the condition that the discriminant is not zero (Chapter 1). This condition defines a dense open set.

It is easily seen that the character of $R_{\underline{a}}$ is continuous in $\underline{a}$ and, since the characters of a finite group are a discrete set, this implies that the character is constant. $\square$

It is of particular interest (combinatorial and geometric) to analyze the special case $\underline{a} = 0$ and the ring $R := \mathbb{C}[x_1, x_2, \ldots, x_n]/\langle \sigma_i \rangle$ which is a graded algebra affording the regular representation. Thus the graded character $\chi_R(q)$ of R is a graded form of the regular representation. To compute it, notice that, as a graded representation, we have an isomorphism

$$\mathbb{C}[x_1, x_2, \ldots, x_n] = R \otimes \mathbb{C}[\sigma_1, \sigma_2, \ldots, \sigma_n],$$

and thus an identity of graded characters.

The ring $\mathbb{C}[\sigma_1, \sigma_2, \ldots, \sigma_n]$ has the trivial representation, by definition, and generators in degree $1, 2, \ldots, n$; so its graded character is just $\prod_{i=1}^n (1 - q^i)^{-1}$. We deduce:

Theorem 2.

$$\chi_R(q) = \sum_{\lambda \vdash n} \chi_\lambda S_\lambda(1, q, q^2, \ldots, q^k, \ldots) \prod_{i=1}^{n} (1 - q^i).$$

Notice then that the series $S_\lambda(1, q, q^2, \ldots, q^k, \ldots) \prod_{i=1}^n (1 - q^i)$ represent the multiplicities of χ_λ in the various degrees of R and thus are polynomials with positive coefficients with the sum being the dimension of χ_λ.

---

[77] It appears in the Springer representation, for instance, cf. [DP2].

Exercise. Prove that the Specht module has nonzero image in the quotient ring $R := \mathbb{C}[x_1, x_2, \dots, x_n]/\langle\sigma_i\rangle$.

The ring $R := \mathbb{Z}[x_1, x_2, \dots, x_n]/\langle\sigma_i\rangle$ has an interesting geometric interpretation as the *cohomology algebra* of the flag variety. This variety can be understood as the space of all decompositions $\mathbb{C}^n = V_1 \perp V_2 \perp \cdots \perp V_n$ into orthogonal 1-dimensional subspaces. The action of the symmetric group is induced by the topological action permuting the summands of the decomposition (Chapter 10, §6.5).

5 The Hook Formula

5.1 Dimension of M_λ

We want to now deduce a formula, due to Frobenius, for the dimension $d(\lambda)$ of the irreducible representation M_λ of the symmetric group.

From 4.1.15 applied to the partition 1^n, corresponding to the conjugacy class of the identity, we obtain

$$\left(\sum_{i=1}^{n} x_i\right)^n V(x) = \sum_{\lambda} d(\lambda) A_{\lambda+\varrho}(x). \tag{5.1.1}$$

Write the expansion of the Vandermonde determinant as

$$\sum_{\sigma\in S_n} \epsilon_\sigma \prod_{i=1}^{n} x_i^{\sigma(n-i+1)-1}.$$

Letting $\lambda+\rho = \ell_1 > \ell_2 > \cdots > \ell_n$, the number $d(\lambda)$ is the coefficient of $\prod_i x_i^{\ell_i}$ in

$$\sum_{k_1+\cdots+k_n=n} \binom{n}{k_1\, k_2 \cdots k_n} \prod_{i=1}^{n} x_i^{k_i} \sum_{\sigma\in S_n} \epsilon_\sigma \prod_{i=1}^{n} x_i^{\sigma(n-i+1)-1}.$$

Thus a term $\epsilon_\sigma \binom{n}{k_1\, k_2 \cdots k_n} \prod_{i=1}^{n} x_i^{\sigma(n-i+1)-1+k_i}$ contributes to $\prod_i x_i^{\ell_i}$ if and only if $k_i = \ell_i - \sigma(n-i+1) + 1$. We deduce

$$d(\lambda) = \sum_{\substack{\sigma\in S_n \mid \forall i \\ \ell_i - \sigma(n-i+1)+1\geq 0}} \epsilon_\sigma \frac{n!}{\prod_{i=1}^{n}(\ell_i - \sigma(n-i+1)+1)!}.$$

We change the term

$$n!\prod_{i=1}^{n} \frac{1}{(\ell_i - \sigma(n-i+1)+1)!} = \frac{n!}{\prod_{i=1}^{n} \ell_i!} \prod_{i=1}^{n} \prod_{\substack{0\leq k\leq \\ \sigma(n-i+1)-2}} (\ell_i - k)$$

and remark that this formula makes sense, and it is 0 if σ does not satisfy the restriction $\ell_i - \sigma(n-i+1) + 1 \geq 0$.

Thus

$$d(\lambda) = \frac{n!}{\prod_{i=1}^{n} \ell_i!} \prod_{i<j} (\ell_i - \ell_j) = n! \prod_{j=1}^{n} \frac{\prod_{i<j}(\ell_i - \ell_j)}{\ell_j!}.$$

$\overline{d}(\lambda)$ is the value of the determinant of a matrix with $\prod_{0 \le k \le j-2}(\ell_i - k)$ in the $n-i+1, j$ position;

$$\begin{vmatrix} 1 & \ell_n & \ell_n(\ell_n - 1) & \dots & \prod_{0\le k\le n-2}(\ell_n - k) \\ \dots & \dots & \dots & \dots & \dots \\ \dots & \dots & \dots & \dots & \dots \\ 1 & \ell_i & \ell_i(\ell_i - 1) & \dots & \prod_{0\le k\le n-2}(\ell_i - k) \\ \dots & \dots & \dots & \dots & \dots \\ \dots & \dots & \dots & \dots & \dots \\ 1 & \ell_1 & \ell_1(\ell_1 - 1) & \dots & \prod_{0\le k\le n-2}(\ell_1 - k) \end{vmatrix}.$$

This determinant, by elementary operations on the columns, reduces to the Vandermonde determinant in the ℓ_i with value $\prod_{i<j}(\ell_i - \ell_j)$. Thus we obtain the formula of Frobenius:

$$d(\lambda) = \frac{n!}{\prod_{i=1}^{n} \ell_i!} \prod_{i<j} (\ell_i - \ell_j) = n! \prod_{j=1}^{n} \frac{\prod_{i<j}(\ell_i - \ell_j)}{\ell_j!}. \tag{5.1.2}$$

5.2 Hook Formula

We want to give a combinatorial interpretation of 5.1.2. Notice that, fixing j, in $\frac{\prod_{i<j}(\ell_i - \ell_j)}{\ell_j!}$ the $j-1$ factors of the numerator cancel the corresponding factors in the denominator, leaving $\ell_j - j + 1$ factors. In all $\sum_j \ell_j - \sum_{j=1}^{n}(j-1) = n$ factors are left. These factors can be interpreted as the *hook lengths* of the boxes of the corresponding diagram.

More precisely, given a box x of a French diagram its *hook* is the set of elements of the diagram which are either on top or to the right of x, including x. For example, we mark the hooks of $1, 2;\ 2, 1;\ 2, 2$ in 4,3,1,1:

The total number of boxes in the hook of x is the *hook length* of x, denoted by h_x.

The Frobenius formula for the dimension $d(\lambda)$ can be reformulated in the settings of the *hook formula*.

Theorem. *Denote by $B(\lambda)$ the set of boxes of a diagram of shape λ. Then*

$$d(\lambda) = \frac{n!}{\prod_{x \in B(\lambda)} h_x}, \qquad \textit{hook formula}. \tag{5.2.1}$$

Proof. It is enough to show that the factors in the factorial $\ell_i!$, which are not canceled by the factors of the numerator, are the hook lengths of the boxes in the ith row. This will prove the formula.

In fact let $h_i = \ell_i + i - n$ be the length of the ith row. Given $k > i$, let us consider the $h_{k-1} - h_k$ numbers strictly between $\ell_i - \ell_{k-1} = h_i - h_{k-1} + k - i - 1$ and $\ell_i - \ell_k = h_i - h_k + k - i$.

Observe that $h_{k-1} - h_k$ is the number of cases in the ith row for which the hook ends vertically on the $k - 1$ row. It is easily seen, since the vertical *leg* of each such hook has length $k-i$ and the horizontal *arm* length goes from $h_i - h_k$ to $h_i - h_{k-1} + 1$, that the lengths of these hooks vary between $k - i + h_i - h_k - 1$ and $k - i + h_i - h_{k-1}$, the previously considered numbers. □

6 Characters of the Linear Group

6.1 Tensor Character

We plan to deducethe character theory of the linear group from previous computations. For this we need to perform another character computation. Given a permutation $s \in S_n$ and a matrix $X \in GL(V)$ consider the product sX as an operator in $V^{\otimes n}$. We want to compute its trace.

Let $\mu = h_1, h_2, \dots, h_k$ denote the cycle partition of s; introduce the obvious notation:

$$\Psi_\mu(X) = \prod_i \operatorname{tr}(X^{h_i}). \tag{6.1.1}$$

Clearly $\Psi_\mu(X) = \psi_\mu(x)$, where by x, we denote the eigenvalues of X.

Proposition. *The trace of sX as an operator in $V^{\otimes n}$ is $\Psi_\mu(X)$.*

We shall deduce this proposition as a special case of a more general formula. Given n matrices $X_1, X_2, \dots, X_n$ and $s \in S_n$ we will compute the trace of $s \circ X_1 \otimes X_2 \otimes \cdots \otimes X_n$ (an operator in $V^{\otimes n}$).

Decompose s into cycles $s = c_1 c_2 \dots c_k$ and, for a cycle $c := (i_p\ i_{p-1}\ \dots\ i_1)$, define the function of the n matrix variables $X_1, X_2, \dots, X_n$:

$$\phi_c(X) = \phi_c(X_1, X_2, \dots, X_n) := \operatorname{tr}(X_{i_1} X_{i_2} \dots X_{i_p}). \tag{6.1.2}$$

The previous proposition then follows from the following:

Theorem.

$$\operatorname{tr}(s \circ X_1 \otimes X_2 \otimes \cdots \otimes X_n) = \prod_{j=1}^{k} \phi_{c_j}(X). \tag{6.1.3}$$

Proof. We first remark that for fixed s, both sides of 6.1.3 are multilinear functions of the matrix variables X_i. Therefore in order to prove this formula it is enough to do it when $X_i = u_i \otimes \psi_i$ is decomposable.

Let us apply in this case the operator $s \circ X_1 \otimes X_2 \otimes \cdots \otimes X_n$ to a decomposable tensor $v_1 \otimes v_2 \cdots \otimes v_n$. We have

(6.1.4)
$$s \circ X_1 \otimes X_2 \otimes \cdots \otimes X_n(v_1 \otimes v_2 \cdots \otimes v_n) = \prod_{i=1}^{n} \langle \psi_i | v_i \rangle u_{s^{-1}1} \otimes u_{s^{-1}2} \ldots \otimes u_{s^{-1}n}.$$

This formula shows that

(6.1.5)
$$s \circ X_1 \otimes X_2 \otimes \cdots \otimes X_n = (u_{s^{-1}1} \otimes \psi_1) \otimes (u_{s^{-1}2} \otimes \psi_2) \ldots \otimes (u_{s^{-1}n} \otimes \psi_n),$$

so that

(6.1.6)
$$\operatorname{tr}(s \circ X_1 \otimes X_2 \otimes \cdots \otimes X_n) = \prod_{i=1}^{n} \langle \psi_i | u_{s^{-1}i} \rangle = \prod_{i=1}^{n} \langle \psi_{s(i)} | u_i \rangle.$$

Now let us compute for a cycle $c := (i_p \; i_{p-1} \; \ldots \; i_1)$ the function

$$\phi_c(X) = \operatorname{tr}(X_{i_1} X_{i_2} \ldots X_{i_p}).$$

We get

$$\begin{aligned} &\operatorname{tr}(u_{i_1} \otimes \psi_{i_1} \circ u_{i_2} \otimes \psi_{i_2} \circ \cdots \circ u_{i_p} \otimes \psi_{i_p}) \\ &= \operatorname{tr}(u_{i_1} \otimes \langle \psi_{i_1} | u_{i_2} \rangle \langle \psi_{i_2} | u_{i_3} \rangle \cdots \langle \psi_{i_{p-1}} | u_{i_p} \rangle \psi_{i_p}) \\ &= \langle \psi_{i_1} | u_{i_2} \rangle \langle \psi_{i_2} | u_{i_3} \rangle \cdots \langle \psi_{i_{p-1}} | u_{i_p} \rangle \langle \psi_{i_p} | u_{i_1} \rangle = \prod_{j=1}^{p} \langle \psi_{c(i_j)} | u_{i_j} \rangle. \end{aligned} \tag{6.1.7}$$

Formulas 6.1.6 and 6.1.7 imply the claim. □

6.2 Character of $S_\lambda(V)$

According to Theorem 3.2 of Chapter 2, the formal ring of symmetric functions in infinitely many variables has as basis all Schur functions S_λ. The restriction to symmetric functions in m-variables sets to 0 all S_λ with height $> m$.

We are ready to complete our work. Let $m = \dim V$. For a matrix $X \in GL(V)$ and a partition $\lambda \vdash n$ of height $\leq m$, let us denote by $S_\lambda(X) := S_\lambda(x)$ the Schur function evaluated at $x = (x_1, \ldots, x_m)$, the eigenvalues of X.

Theorem. *Denote $\rho_\lambda(X)$ to be the character of the representation $S_\lambda(V)$ of $GL(V)$, paired with the representation M_λ of S_n in $V^{\otimes n}$. We have $\rho_\lambda(X) = S_\lambda(X)$.*

Proof. If $s \in S_n$, $X \in GL(V)$, we have seen that the trace of $s \circ X^{\otimes n}$ on $V^{\otimes n}$ is computed by $\psi_\mu(X) = \sum_\lambda c_\lambda(\mu) S_\lambda(X)$ (definition of the c_λ).

If $m = \dim V < n$, only the partitions of height $\leq m$ contribute to the sum. On the other hand, $V^{\otimes n} = \bigoplus_{ht(\lambda) \leq \dim(V)} M_\lambda \otimes S_\lambda(V)$; thus,

$$\psi_\mu(X) = \mathrm{tr}(s \circ X^{\otimes n}) = \sum_{\lambda \vdash n, ht(\lambda) \leq m} \mathrm{tr}(s \mid M_\lambda)\, \mathrm{tr}(X^{\otimes n} \mid S_\lambda(V))$$

$$= \sum_{\lambda \vdash n, ht(\lambda) \leq m} c_\lambda(\mu) \rho_\lambda(X) = \sum_{\lambda \vdash n, ht(\lambda) \leq m} c_\lambda(\mu) S_\lambda(X).$$

If $m < n$, the $\psi_\mu(X)$ with parts of length $\leq m$ (i.e., $ht(\tilde{\mu}) \leq m$) are a basis of symmetric functions in m variables; hence we can invert the system of linear equations and get $S_\lambda(X) = \rho_\lambda(X)$. □

The eigenvalues of $X^{\otimes n}$ are monomials in the variables x_i, and thus we obtain:

Corollary. *$S_\lambda(x)$ is a sum of monomials with positive coefficients.*

We will see in Chapter 13 that one can index combinatorially the monomials which appear by *semistandard tableaux.*

We can also deduce a dimension formula for the space $S_\lambda(V)$, $\dim V = n$. Of course its value is $S_\lambda(1, 1, \ldots, 1)$ which we want to compute from the determinantal formulas giving $S_\lambda(x) = A_{\lambda+\varrho}(x)/V(x)$.

Let as usual $\lambda := h_1, h_2, \ldots, h_n$ and $l_i := h_i + n - i$. Of course we cannot substitute directly the number 1 for the x_i, or we get $0/0$. Thus we first substitute to $x_i \to x^{i-1}$ and then take the limit as $x \to 1$. Under the previous substitution we see that $A_{\lambda+\varrho}$ becomes the Vandermonde determinant of the elements x^{l_i}, hence

$$S_\lambda(1, x, x^2, \ldots, x^{n-1}) = \prod_{1 \leq i < j \leq n} \frac{(x^{l_i} - x^{l_j})}{(x^{n-i} - x^{n-j})}.$$

If $a > b$, we have $x^a - x^b = x^b(x-1)(x^{a-b-1} + x^{a-b-2} + \cdots + 1)$, hence we deduce that

$$\dim S_\lambda(V) = S_\lambda(1, 1, 1, \ldots, 1) = \prod_{1 \leq i < j \leq n} \frac{(l_i - l_j)}{(j - i)} = \prod_{1 \leq i < j \leq n} \frac{(h_i - h_j + j - i)}{(j - i)}.$$

6.3 Cauchy Formula as Representations

We want to now give an interpretation, in the language of representations, of the Cauchy formula.

Suppose we are given a vector space U over which a torus T acts with a basis of weight vectors u_i with weight χ_i.

The graded character of the action of T on the symmetric and exterior algebras are given by Molien's formula, §4.5 and are, respectively,

(6.3.1)
$$\frac{1}{\prod 1 - \chi_i q}, \qquad \prod 1 + \chi_i q.$$

As an example consider two vector spaces U, V with bases $u_1, \dots, u_m$; $v_1, \dots, v_n$, respectively. We may assume $m \leq n$.

The maximal tori of diagonal matrices have eigenvalues $x_1, \dots, x_m;\ y_1, \dots, y_n$ respectively. On the tensor product we have the action of the product torus, and the basis $u_i \otimes v_j$ has eigenvalues $x_i y_j$. Therefore the graded character on the symmetric algebra $S(U \otimes V)$ is $\prod_{i=1}^n \prod_{j=1}^m \frac{1}{1-x_i y_j q}$.

By Cauchy's formula we deduce that the character of the nth symmetric power $S^n(U \otimes V)$ equals $\sum_{\lambda \vdash n,\ ht(\lambda) \leq m} S_\lambda(x) S_\lambda(y)$.

We know that the rational representations of $GL(U) \times GL(V)$ are completely reducible and their characters can be computed by restricting to diagonal matrices. Thus we have the description:

(6.3.2)
$$S^n(U \otimes V) = \bigoplus_{\lambda \vdash n,\ ht(\lambda) \leq m} S_\lambda(U) \otimes S_\lambda(V).$$

This is also referred to as *Cauchy's formula*.

Observe that if $W \subset V$ is the subspace which is formed by the first k basis vectors, then the intersection of $S_\lambda(U) \otimes S_\lambda(V)$ with $S(U \otimes W)$ has as basis the part of the basis of weight vectors of $S_\lambda(U) \otimes S_\lambda(V)$ corresponding to weights in which the variables $y_j,\ j > k$ do not appear. Thus its character is obtained by setting to 0 these variables in $S_\lambda(y_1, y_2, \dots, y_m)$; thus we clearly get that

(6.3.3)
$$S_\lambda(U) \otimes S_\lambda(V) \cap S(U \otimes W) = S_\lambda(U) \otimes S_\lambda(W).$$

Similarly it is clear, from Definition 3.1.3: $S_\lambda(V) := e_T V^{\otimes n}$, that:

Proposition. *If $U \subset V$ is a subspace, then $S_\lambda(U) = S_\lambda(V) \cap U^{\otimes n}$.*

6.4 Multilinear Elements

Consider a rational representation $\rho : GL(n, \mathbb{C}) \to GL(W)$ for which the matrix coefficients are polynomials in the coordinates $x_{i,j}$, and thus do not contain the determinant at the denominator.

Such a representation is called a *polynomial representation*, the map ρ extends to a multiplicative map $\rho : M(n, \mathbb{C}) \to \mathrm{End}(W)$ on all matrices.

Polynomial representations are closed under taking direct sums, tensor products, subrepresentations and quotients. A typical polynomial representation of $GL(V)$ is $V^{\otimes n}$ and all its subrepresentations, for instance the $S_\lambda(V)$.

One should stress the strict connection between the two formulas, 6.3.2 and 3.1.4.

(6.3.2)
$$S^n(U \otimes V) = \bigoplus_{\lambda \vdash n,\ ht(\lambda) \leq m} S_\lambda(U) \otimes S_\lambda(V),$$

(3.1.4)
$$V^{\otimes n} = \bigoplus_{ht(\lambda) \leq \dim(V)} M_\lambda \otimes S_\lambda(V).$$

This is clearly explained when we assume that $U = \mathbb{C}^n$ with canonical basis e_i and we consider the diagonal torus T acting by matrices $Xe_i = x_i e_i$.

Let us go back to formula 6.3.2, and apply it when $\dim V = n,\ W = \mathbb{C}^n$.

Consider the subspace T_n of $S(\mathbb{C}^n \otimes V)$ formed by the elements $\prod_{i=1}^n e_i \otimes v_i$, $v_i \in V$. T_n is stable under the subgroup $S_n \times GL(V) \subset GL(n, \mathbb{C}) \times GL(V)$, where S_n is the group of permutation matrices. We have a mapping $i : V^{\otimes n} \to T_n$ defined by

(6.4.1)
$$i : v_1 \otimes v_2 \otimes \cdots \otimes v_n \mapsto \prod_{i=1}^n e_i \otimes v_i.$$

Proposition. *(i) T_n is the weight space in $S(\mathbb{C}^n \otimes V)$, of weight $\chi(X) = \prod_i x_i$ for the torus T.*

(ii) The map i is an $S_n \times GL(V)$ linear isomorphism between $V^{\otimes n}$ and T_n.

Proof. The verification is immediate and left to the reader. □

Remark. The character $\chi := \prod_{i=1}^n x_i$ is invariant under the symmetric group (and generates the group of these characters). We call it *the multilinear character.*

As usual, when we have a representation W of a torus, we denote by W^χ the weight space of character χ.

Now for every partition λ consider

(6.4.2)
$$S_\lambda(\mathbb{C}^n)^\chi := \left\{ u \in S_\lambda(\mathbb{C}^n) | Xu = \prod_i x_i u,\ \forall X \in T \right\},$$

the weight space of $S_\lambda(\mathbb{C}^n)$ formed by the elements which are formally *multilinear.*

Since the character $\prod_i x_i$ is left invariant by conjugation by permutation matrices it follows that the symmetric group $S_n \subset GL(n, \mathbb{C})$ of permutation matrices acts on $S_\lambda(\mathbb{C}^n)^\chi$. We claim that:

Proposition. *$S_\lambda(\mathbb{C}^n)^\chi = 0$ unless $\lambda \vdash n$ and in this case $S_\lambda((\mathbb{C}^n)^*)^\chi$ is identified with the irreducible representation M_λ of S_n.*

Proof. In fact assume $Xu = \prod_i x_i u$. Clearly u is in a polynomial representation of degree n. On the other hand

$$S^n(\mathbb{C}^n \otimes V) = \bigoplus_{\lambda \vdash n} S_\lambda((\mathbb{C}^n)^*) \otimes S_\lambda(V),$$

hence

(6.4.3)
$$V^{\otimes n} := S^n(\mathbb{C}^n \otimes V)^\chi = \bigoplus_{\lambda \vdash n} S_\lambda((\mathbb{C}^n)^*)^\chi \otimes S_\lambda(V) = \bigoplus_{\lambda \vdash n} M_\lambda \otimes S_\lambda(V)$$

and we get the required identification. □

Therefore, given a polynomial representation P of $GL(n, \mathbb{C})$, if it is homogeneous of degree n, in order to determine its decomposition $P = \bigoplus_{\lambda \vdash n} m_\lambda S_\lambda(\mathbb{C}^n)$ we can equivalently restrict to $M := \{p \in P \mid X \cdot p = \prod_i x_i p\}$, the multilinear weight space (for X diagonal with entries x_i) and see how it decomposes as a representation of S_n since

(6.4.4)
$$P = \bigoplus_{\lambda \vdash n} m_\lambda S_\lambda(\mathbb{C}^n) \iff M = \bigoplus_{\lambda \vdash n} m_\lambda M_\lambda.$$

7 Polynomial Functors

7.1 Schur Functors

Consider two vector spaces V, W, the space $\hom(V, W) = W \otimes V^*$, and the ring of polynomial functions on $\hom(V, W)$ decomposed as

$$\mathcal{P}[\hom(V, W)] = S(W^* \otimes V) = \bigoplus_\lambda S_\lambda(W^*) \otimes S_\lambda(V). \tag{7.1.1}$$

A way to explicitly identify the spaces $S_\lambda(W^*) \otimes S_\lambda(V)$ as spaces of functions is obtained by a variation of the method of matrix coefficients.

We start by stressing the fact that the construction of the representation $S_\lambda(V)$ from V is in a sense *natural*, in the language of categories.

Recall that a map between vector spaces is called a *polynomial map* if in coordinates it is given by polynomials.

Definition. A functor F from the category of vector spaces to itself is called a *polynomial* functor if, given two vector spaces V, W, the map $A \to F(A)$ from the vector space $\hom(V, W)$ to the vector space *hom*$(F(V), F(W))$ is a polynomial map.

We say that F is homogeneous of degree k if, for all vector spaces V, W, the map $\hom(V, W) \xrightarrow{F(-)} \hom(F(V), F(W))$ is homogeneous of degree k.

The functor $F : V \to V^{\otimes n}$ is clearly a polynomial functor, homogeneous of degree n. When $A : V \to W$ the map $F(A)$ is $A^{\otimes n}$.

We can now justify the word *Schur functor* in the Definition 3.1.3, $S_\lambda(V) := e_T V^{\otimes n}$, where e_T is a Young symmetrizer, associated to a partition λ.

As V varies, $V \mapsto S_\lambda(V)$ can be considered as a functor. In fact it is a subfunctor of the tensor power, since clearly, if $A : V \to W$ is a linear map, $A^{\otimes n}$ commutes with a_T. Thus $A^{\otimes n}(e_T V^{\otimes n}) \subset e_T W^{\otimes n}$ and we define

$$S_\lambda(A) : S_\lambda(V) \longrightarrow V^{\otimes n} \xrightarrow{A^{\otimes n}} W^{\otimes n} \xrightarrow{e_T} S_\lambda(W). \tag{7.1.2}$$

Summarizing:

Proposition 1. *Given any partition $\mu \vdash n$, $V \mapsto S_\mu(V)$ is a homogeneous polynomial functor on vector spaces of degree n, called a* ***Schur functor***.

Remark. This functor is independent of T but depends only on the partition λ. The choice of T determines an embedding of $S_\lambda(V)$ as subfunctor of $V^{\otimes n}$.

Remark. The exterior and symmetric power $\bigwedge^k V$ and $S^k(V)$ are examples of Schur functors.

Since the map $S_\mu : \hom(V, W) \to \hom(S_\mu(V), S_\mu(W))$ defined by $S_\mu : X \to S_\mu(X)$ is a homogeneous polynomial map of degree n, the dual map $S_\mu^* : \hom(S_\mu(V), S_\mu(W))^* \to \mathcal{P}[\hom(V, W)]$ defined by

$$S_\mu^*(\phi)(X) := \langle \phi | S_\mu(X) \rangle, \ \phi \in \hom(S_\mu(V), S_\mu(W))^*, \ X \in \hom(V, W)$$

is a $GL(V) \times GL(W)$-equivariant map into the homogeneous polynomials of degree n.

By the irreducibility of $\hom(S_\mu(V), S_\mu(W))^* = S_\mu(V) \otimes S_\mu(W)^*$, S_μ^* must be a linear isomorphism to an irreducible submodule of $\mathcal{P}[\hom(V, W)]$ uniquely determined by Cauchy's formula. By comparing the isotypic component of type $S_\mu(V)$ we deduce:

Proposition 2. $\mathcal{P}[\hom(V, W)] = \bigoplus_\mu \hom(S_\mu(V), S_\mu(W))^*$ *and we have the isomorphism* $S_\mu(W^*) = S_\mu(W)^*$.

Let us apply the previous discussion to $\hom(\bigwedge^i V, \bigwedge^i W)$.

Choose bases $e_i,\ i = 1, \dots, h,\ f_j,\ j = 1, \dots, k$ for V, W respectively, and identify the space $\hom(V, W)$ with the space of $k \times h$ matrices. Thus the ring $\mathcal{P}[\hom(V, W)]$ is the polynomial ring $\mathbb{C}[x_{ij}]$, $i = 1, \dots, h$, $j = 1, \dots, k$ where x_{ij} are the matrix entries.

Given a matrix X the entries of $\bigwedge^i X$ are the determinants of all the minors of order i extracted from X, and:

Corollary. $\bigwedge^i V \otimes (\bigwedge^i W)^*$ *can be identified with the space of polynomials spanned by the determinants of all the minors of order i extracted from X, which is thus irreducible as a representation of* $GL(V) \times GL(W)$.

7.2 Homogeneous Functors

We want to prove that any polynomial functor is equivalent to a direct sum of Schur functors. We start with:

Proposition 1. *A polynomial functor is a direct sum of homogeneous functors.*

Proof. The scalar multiplications by $\alpha \in \mathbb{C}^*$ on a space V induce, by functoriality, a polynomial representation of $\mathbb{C}^*$ on $F(V)$ which then decomposes as $F(V) = \bigoplus_k F_k(V)$, with $F_k(V)$ the subspace of weight α^k. Clearly $F_k(V)$ is a subfunctor and $F = \bigoplus_k F_k(V)$. Moreover $F_k(V)$ is a homogeneous functor of degree k.

We can polarize a homogeneous functor of degree k as follows. Consider, for a k-tuple $V_1, \dots, V_k$ of vector spaces, their direct sum $\bigoplus_i V_i$ together with the action of a k-dimensional torus T with the scalar multiplication x_i on each summand V_i. T acts in a polynomial way on $F(\bigoplus_i V_i)$ and we can decompose by weights

$$F(\bigoplus_i V_i) = \bigoplus_\lambda F_\lambda(V_1, \dots, V_k), \quad \lambda = x_1^{h_1} x_2^{h_2} \dots x_k^{h_k}, \ \sum h_i = k.$$

One easily verifies that the inclusion $V_i \to \oplus V_j$ induces an isomorphism between $F(V_i)$ and $F_{x_i^k}(V_1, \dots, V_k)$. $\square$

Let us now consider a polynomial functor $V \mapsto F(V)$, homogeneous of degree k. We start by performing the following constructions.

(1) First consider the functor

$$T : V \mapsto S^k(\hom(\mathbb{C}^k, V)) \otimes F(\mathbb{C}^k)$$

And the natural transformation:

$$\pi_V : S^k(\hom(\mathbb{C}^k, V)) \otimes F(\mathbb{C}^k) \to F(V)$$

defined by the formula

$$\pi_V(f^k \otimes u) := F(f)(u), \ f \in \hom(\mathbb{C}^k, V), \ u \in F(\mathbb{C}^k)$$

This formula makes sense, since $F(f)$ is a homogeneous polynomial map of degree k in f by hypothesis.

The fact that π_V is natural depends on the fact that, if $h : V \to W$ we have that $T(h)(f^k \otimes u) = (hf)^k \otimes u$ so that

$$F(h)\pi_V(f^k \otimes u) = F(hf)(u) = \pi_W((hf)^k \otimes u = \pi_W(T(h)(f^k \otimes u).$$

(2) The linear group $GL(k, \mathbb{C})$ acts by natural isomorphisms on the functor $T(V)$ by the formula

$$(f \circ g^{-1})^k \otimes F(g)u,$$

Lemma 1. *The natural transformation π_V is $GL(k, \mathbb{C})$ invariant.*

Proof. We have $F(f \circ g^{-1}) = F(f) \circ F(g)^{-1}$ and $\pi_V((f \circ g^{-1})^k \otimes F(g)u) = F(f \circ g^{-1})(F(g)u) = F(f)u$. □

These invariance properties mean that if we decompose $T(V)$, as a representation of $GL(n, \mathbb{C})$ into the invariant space $T(V)^{GL(n,\mathbb{C})}$ and the other isotypic components, the sum of the nontrivial irreducible representations $T(V)_{GL(n,\mathbb{C})}$, we have $\pi_V = 0$ on $T(V)_{GL(n,\mathbb{C})}$.

Our goal is to prove

Theorem. *The map π_V restricted to the $GL(n, \mathbb{C})$ invariants:*

$$\pi_V : [S^k(\hom(\mathbb{C}^k, V)) \otimes F(\mathbb{C}^k)]^{GL(k,\mathbb{C})} \to F(V), \quad \pi_V(f \otimes u) := F(f)(u),$$

is a functorial isomorphism.

In order to prove this theorem we need a simple general criterion:

Proposition 2. *Let $\eta : F \to G$ be a natural transformation of polynomial functors, each of degree k. Then η is an isomorphism if and only if $\eta_{\mathbb{C}^k} : F(\mathbb{C}^k) \to G(\mathbb{C}^k)$ is an isomorphism.*

Proof. Since any vector space is isomorphic to $\mathbb{C}^m$ for some m we have to prove an isomorphism for these spaces.

The diagonal torus acts on $\mathbb{C}^m$, which by functoriality acts also on $F(\mathbb{C}^m)$ and $G(\mathbb{C}^m)$. By naturality $\eta_{\mathbb{C}^m} : F(\mathbb{C}^m) \to G(\mathbb{C}^m)$ must preserve weight spaces with respect to the diagonal matrices. Now each weight involves at most k indices and so it can be deduced from the corresponding weight space for $\mathbb{C}^k$. For these weight spaces the isomorphism is guaranteed by the hypotheses. □

In order to apply this criterion to π_V we have to understand the map:

$$\pi_{\mathbb{C}^k} : [S^k(\hom(\mathbb{C}^k, \mathbb{C}^k)) \otimes F(\mathbb{C}^k)]^{GL(k,\mathbb{C})} \to F(\mathbb{C}^k).$$

The invariants are taken with respect to the diagonal action on $S^k(\hom(\mathbb{C}^k, \mathbb{C}^k))$ by acting on the source of the homomorphisms and on $F(\mathbb{C}^k)$.

Lemma 2. *$\pi_{\mathbb{C}^k}$ is an isomorphism.*

Proof. The definition of this map depends just on the fact that $F(\mathbb{C}^k)$ is a polynomial representation of $GL(k, \mathbb{C})$ which is homogeneous of degree k. It is clear that, if this map is an isomorphism for two different representations $F_1(\mathbb{C}^k)$, $F_2(\mathbb{C}^k)$ it is also an isomorphism for their direct sum. Thus we are reduced to study the case in which $F(\mathbb{C}^k) = S_\lambda(\mathbb{C}^k)$ for some partition $\lambda \vdash k$.

Identifying

$$S^k(\hom(\mathbb{C}^k, \mathbb{C}^k)) \otimes S_\lambda(\mathbb{C}^k) = \bigoplus\nolimits_{\mu\vdash k} S_\mu(\mathbb{C}^k) \otimes S_\mu(\mathbb{C}^k)^* \otimes S_\lambda(\mathbb{C}^k)$$

the invariants are by definition

$$\bigoplus\nolimits_{\mu\vdash k} S_\mu(\mathbb{C}^k) \otimes [S_\mu(\mathbb{C}^k)^* \otimes S_\lambda(\mathbb{C}^k)]^{GL(k,\mathbb{C})} = S_\lambda(\mathbb{C}^k) \otimes \mathbb{C}1_{S_\lambda(\mathbb{C}^k)}.$$

By the irreducibility of the representations $S_\lambda(\mathbb{C}^k)$.

Since clearly $\pi_{\mathbb{C}^k}(u \otimes 1_{S_\lambda(\mathbb{C}^k)}) = u$ we have proved the claim. □

By the classification of polynomial representations, we have that

$$F(\mathbb{C}^k) = \bigoplus\nolimits_{\lambda\vdash k} m_\lambda S_\lambda(\mathbb{C}^k)$$

for some nonnegative integers m_i. We deduce:

Corollary. *A polynomial functor F of degree k is of the form*

$$F(V) = \bigoplus\nolimits_{\lambda\vdash k} m_\lambda S_\lambda(V).$$

Proof.

$$\begin{aligned}[S^k(\hom(\mathbb{C}^k, V)) \otimes S_\lambda(\mathbb{C}^k)]^{GL(k,\mathbb{C})} &= \bigoplus\nolimits_{\mu\vdash k} S_\mu(V) \otimes [S_\mu(\mathbb{C}^k)^* \otimes S_\lambda(\mathbb{C}^k)]^{GL(k,\mathbb{C})} \\ &= S_\lambda(V) \otimes \mathbb{C}1_{S_\lambda(\mathbb{C}^k)}.\end{aligned}$$

□

Polynomial functors can be summed (direct sum) multiplied (tensor product) and composed. All these operations can be extended to a ring whose elements are purely formal differences of functors (a Grothendieck type of ring). In analogy with the theory of characters an element of this ring is called a *virtual functor*.

Proposition 3. *The ring of virtual functors is canonically isomorphic to the ring of infinite symmetric functions.*

Proof. We identify the functor S_λ with the symmetric function $S_\lambda(x)$. □

Exercise. Given two polynomial functors F, G of degree k prove that we have an isomorphism between the space $\mathrm{Nat}(F, G)$ of natural transformations between the two functors, and the space $\hom_{GL(k,\mathbb{C})}(F(\mathbb{C}^k), G(\mathbb{C}^k))$.

Discuss the case $F = G$ the tensor power $V^{\otimes k}$.

7.3 Plethysm

The composition of functors becomes the *Plethysm operation* on symmetric functions. In general it is quite difficult to compute such compositions, even such simple ones as $\bigwedge^i(\bigwedge^h V))$. There are formulas for $S^k(S^2(V))$, $S^k(\bigwedge^2(V))$ and some dual ones.

In general the computation $F \circ G$ should be done according to the following:

Algorithm. Apply a polynomial functor G to the space $\mathbb{C}^m$ with its standard basis.

For the corresponding linear group and diagonal torus T, $G(\mathbb{C}^m)$ is a polynomial representation of some dimension N. It then has a basis of T-weight vectors with characters a list of monomials M_i. The character of T on $G(\mathbb{C}^m)$ is $\sum_i^N M_i$, a symmetric function $S_G(x_1, \ldots, x_m)$.

If G is homogeneous of degree k, this symmetric function is determined as soon as $m \geq k$.

When we apply F to $G(\mathbb{C}^m)$ we use the basis of weight vectors to see that the symmetric function

$$S_{F\circ G}(x_1, \ldots, x_m) = S_F(M_1, \ldots, M_N).$$ □

Some simple remarks are in order. First, given a fixed functor G the map $F \mapsto F \circ G$ is clearly a ring homomorphism. Therefore it is determined by the value on a set of generators. One can choose as generators the exterior powers. In this case the operation $\bigwedge^i \circ F$ as transformations in F are called λ-operations and written λ^i.

These operations satisfy the basic law: $\lambda^i(a+b) = \sum_{h+k=i} \lambda^h(a)\lambda^k(b)$.

It is also convenient to use as generators the Newton functions $\psi_k = \sum_i x_i^k$ since then

$$\psi_k(S(x_1, \ldots, x_m)) = S(x_1^k, \ldots, x_m^k), \qquad \psi_k(\psi_h) = \psi_{kh}.$$

All of this can be formalized, giving rise to the theory of λ-rings (cf. [Knu]).

8 Representations of the Linear and Special Linear Groups

8.1 Representations of $SL(V), GL(V)$

Given an n-dimensional vector space V we want to give the complete list of irreducible representations for the general and special linear groups $GL(n) = GL(V), SL(n) = SL(V)$.

From Chapter 7, Theorem 1.4 we know that all the irreducible representations of $SL(V)$ appear in the tensor powers $V^{\otimes m}$ and all the irreducible representations of $GL(V)$ appear in the tensor powers $V^{\otimes m}$ tensored with integer powers of the determinant $\bigwedge^n(V)$. For simplicity we will denote by $D := \bigwedge^n(V)$ and by convention $D^{-1} := \bigwedge^n(V)^*$. From what we have already seen the irreducible representations of $GL(V)$ which appear in the tensor powers are the modules $S_\lambda(V)$, $ht(\lambda) \leq n$. They are all distinct since they have distinct characters. Given $S_\lambda(V) \subset V^{\otimes m}$, $\lambda \vdash m$, consider $S_\lambda(V) \otimes \bigwedge^n(V) \subset V^{\otimes m+n}$. Since $\bigwedge^n(V)$ is 1-dimensional, clearly $S_\lambda(V) \otimes \bigwedge^n(V)$ is also irreducible. Its character is $S_{\lambda+1^n}(x) = (x_1x_2\dots x_n)S_\lambda(x)$, hence (cf. Chapter 2, 6.2.1):

(8.1.1) $$S_\lambda(V) \otimes \bigwedge^n(V) = S_{\lambda+1^n}(V).$$

We now need a simple lemma. Let $\mathcal{P}_{n-1} := \{k_1 \geq k_2 \geq \ldots \geq k_{n-1} \geq 0\}$ be the set of all partitions (of any integer) of height $\leq n-1$. Consider the polynomial ring $\mathbb{Z}[x_1, \ldots, x_n, (x_1, x_2 \ldots x_n)^{-1}]$ obtained by inverting $e_n = x_1x_2\ldots x_n$.

Lemma. *(i) The ring of symmetric elements in $\mathbb{Z}[x_1, \ldots, x_n, (x_1x_2\ldots x_n)^{-1}]$ is generated by $e_1, e_2, \ldots, e_{n-1}, e_n^{\pm 1}$ and it has as basis the elements*

$$S_\lambda e_n^m, \ \lambda \in \mathcal{P}_{n-1}, \ m \in \mathbb{Z}.$$

(ii) The ring $\mathbb{Z}[e_1, e_2, \ldots, e_{n-1}, e_n]/(e_n - 1)$ has as basis the classes of the elements S_λ, $\lambda \in \mathcal{P}_{n-1}$.

Proof. (i) Since e_n is symmetric it is clear that a fraction f/e_n^k is symmetric if and only if f is symmetric, hence the first statement. Any element of $\mathbb{Z}[e_1, e_2, \ldots, e_{n-1}, e_n^{\pm 1}]$ can be written in a unique way in the form $\sum_{k\in\mathbb{Z}} a_k e_n^k$ with $a_k \in \mathbb{Z}[e_1, e_2, \ldots, e_{n-1}]$. We know that the Schur functions S_λ, $\lambda \in \mathcal{P}_{n-1}$ are a basis of $\mathbb{Z}[e_1, e_2, \ldots, e_{n-1}, e_n]/(e_n)$ and the claim follows.

(ii) follows from (i). □

Theorem. *(i) The list of irreducible representations of $SL(V)$ is*

(8.1.2) $$S_\lambda(V), \ ht(\lambda) \leq n-1.$$

(ii) The list of irreducible representations of $GL(V)$ is

(8.1.3) $$S_\lambda(V) \otimes D^k, \ ht(\lambda) \leq n-1, \ k \in \mathbb{Z}.$$

Proof. (i) The group $GL(V)$ is generated by $SL(V)$ and the scalar matrices which commute with every element. Therefore in any irreducible representation of $GL(V)$ the scalars in $GL(V)$ also act as scalars in the representation. It follows immediately that the representation remains irreducible when restricted to $SL(V)$.

Thus we have to understand when two irreducible representations $S_\lambda(V)$, $S_\mu(V)$, with $ht(\lambda) \leq n$, $ht(\mu) \leq n$, are isomorphic once restricted to $SL(V)$.

Any λ can be uniquely written in the form $(m, m, m, \dots, m) + (k_1, k_2, \dots, k_{n-1}, 0)$ or $\lambda = \mu + m\,1^n$, $ht(\mu) \leq n-1$, and so $S_\lambda(V) = S_\mu(V) \otimes D^m$. Clearly $S_\lambda(V) = S_\mu(V)$ as representations of $SL(V)$. Thus to finish we have to show that if $\lambda \neq \mu$ are two partitions of height $\leq n-1$, the two $SL(V)$ representations $S_\lambda(V)$, $S_\mu(V)$ are not isomorphic. This follows from the fact that the characters of the representations $S_\lambda(V)$, $\lambda \in \mathcal{P}_{n-1}$, are a basis of the invariant functions on $SL(V)$, by the previous lemma.

(ii) We have seen at the beginning of this section that all irreducible representations of $GL(V)$ appear in 8.1.3 above. Now if two different elements of this list in 8.1.3 were isomorphic, by multiplying by a high enough power of D we would obtain two isomorphic polynomial representations belonging to two different partitions, a contradiction. □

Remark. $\bigwedge^i V$ corresponds to the partition 1^i, made of a single column of length i. Its associated Schur function S_{1^i} is the ith elementary function e_i. Instead the symmetric power $S^i(V)$ corresponds to the partition made of a single row of length i, it corresponds to a symmetric function S_i which is often denoted by h_i and it is the sum of all the monomials of degree i.

8.2 The Coordinate Ring of the Linear Group

We can interpret the previous theory in terms of the coordinate ring $\mathbb{C}[GL(V)]$ of the general linear group.

Since $GL(V)$ is the open set of $\mathrm{End}(V) = V \otimes V^*$ where the determinant $d \neq 0$, its coordinate ring is the localization at d of the ring $S(V^* \otimes V)$ which, under the two actions of $GL(V)$, decomposes as $\bigoplus_{ht(\lambda)\leq n} S_\lambda(V^*) \otimes S_\lambda(V) = \bigoplus_{ht(\lambda)\leq n-1,\, k\geq 0} d^k S_\lambda(V^*) \otimes S_\lambda(V)$. It follows immediately then that

$$\mathbb{C}[GL(V)] = \bigoplus_{ht(\lambda)\leq n-1,\, k\in\mathbb{Z}} d^k S_\lambda(V^*) \otimes S_\lambda(V). \tag{8.2.1}$$

This of course is, for the linear group, the explicit form of formula 3.1.1 of Chapter 7. From it we deduce that

$$S_\lambda(V^*) = S_\lambda(V)^*, \qquad \forall\lambda,\ ht(\lambda) \leq n-1, \quad (d^* = d^{-1}). \tag{8.2.2}$$

Similarly,

$$\mathbb{C}[SL(V)] = \bigoplus_{ht(\lambda)\leq n-1} S_\lambda(V^*) \otimes S_\lambda(V). \tag{8.2.3}$$

8.3 Determinantal Expressions for Schur Functions

In this section we want to discuss a determinant development for Schur functions which is often used.

Recall that $V_\lambda = e_T(V^{\otimes n})$ is a quotient of

$$a_T(V^{\otimes n}) = \bigwedge^{k_1} V \otimes \bigwedge^{k_2} V \cdots \otimes \bigwedge^{k_n} V,$$

where the k_i are the columns of a tableau T, and is contained in

$$s_T(V^{\otimes n}) = S^{h_1}(V) \otimes S^{h_2}(V) \cdots \otimes S^{h_n}(V)$$

where the h_i are the rows of T. Here one has to interpret both antisymmetrization and symmetrization as occurring respectively in the columns and row indices.[78] The composition $e_T = \frac{1}{p(\lambda)} s_T a_T$ can be viewed as the result of a map

$$\bigwedge^{k_1} V \otimes \bigwedge^{k_2} V \cdots \otimes \bigwedge^{k_n} V \to s_T(V^{\otimes n}) = S^{h_1}(V) \otimes S^{h_2}(V) \cdots \otimes S^{h_n}(V).$$

As representations $\bigwedge^{k_1} V \otimes \bigwedge^{k_2} V \cdots \otimes \bigwedge^{k_n} V$ and $S^{h_1}(V) \otimes S^{h_2}(V) \cdots \otimes S^{h_n}(V)$ decompose in the direct sum of a copy of V_λ and other irreducible representations.

The character of the exterior power $\bigwedge^i(V)$ is the elementary symmetric function $e_i(x)$. The one of $S^i(V)$ is the function $h_i(x)$ sum of all monomials of degree i.

In the formal ring of symmetric functions there is a formal duality between the elements e_i and the h_j. From the definition of the e_i and from Molien's formula:

$$\sum_{i=0}^{\infty} h_i(x) q^i = \frac{1}{\prod(1 - x_i q)}, \quad \sum_{i=0}^{\infty} (-1)^i e_i(x) q^i = \prod(1 - x_i q),$$

$$1 = \left(\sum_{i=0}^{\infty} (-1)^i e_i(x) q^i\right)\left(\sum_{i=0}^{\infty} h_i(x) q^i\right).$$

Hence for $m > 0$ we have $\sum_{i+j=m} (-1)^i e_i(x) h_j(x) = 0$. These identities tell us that

$$\mathbb{Z}[e_1, e_2, \ldots, e_i, \ldots] = \mathbb{Z}[h_1, h_2, \ldots, h_i, \ldots]$$

and also that we can present the ring of infinite symmetric functions with generators e_i, h_j and the previous relations:

$$\mathbb{Z}[e_1, e_2, \ldots, e_i, \ldots; h_1, h_2, \ldots, h_i, \ldots] \Big/ \left(\sum_{i+j=m} (-1)^i e_i h_j\right). \tag{8.3.1}$$

The mapping $\tau : e_i(x) \mapsto h_i(x),\ h_i(x) \mapsto e_i(x)$ preserves these relations and gives an involutory automorphism in the ring of symmetric functions. Take the Cauchy identity

[78] It is awkward to denote symmetrization on non-consecutive indices as we did. More correctly, one should compose with the appropriate permutation which places the indices in the correct positions.

$$\sum S_\lambda(x)S_\lambda(y) = \prod_{i,j} \frac{1}{1 - x_i y_j} = \prod_j \sum_{k=0}^{\infty} h_k(x)y_j^k \tag{8.3.2}$$

and multiply it by the Vandermonde determinant $V(y)$, getting

$$\sum S_\lambda(x)A_{\lambda+\varrho}(y) = \prod_j \sum_{k=0}^{\infty} h_k(x)y_j^k V(y).$$

For a given $\lambda = a_1, \dots, a_n$ we see that $S_\lambda(x)$ is the coefficient of the monomial $y_1^{a_1+n-1} y_2^{a_2+n-2} \dots y_n^{a_n}$, and we easily see that this is

$$\sum_{\sigma \in S_n} \epsilon_\sigma \prod_{i=1}^{n} h_{\sigma(i)-i+a_i}; \tag{8.3.3}$$

thus

Proposition. *The Schur function S_λ is the determinant of the $n \times n$ matrix, which in the position i, j, has the element h_{j-i+a_i} with the convention that $h_k = 0, \ \forall k < 0$.*

8.4 Skew Cauchy Formula

We want to complete this discussion with an interesting variation of the Cauchy formula (which is used in the computation of the cohomology of the linear group, cf. [AD]).

Given two vector spaces V, W we want to describe $\bigwedge(V \otimes W)$ as a representation of $GL(V) \times GL(W)$.

Theorem.

$$\bigwedge(V \otimes W) = \sum_\lambda S_\lambda(V) \otimes S_{\tilde{\lambda}}(W). \tag{8.4.1}$$

Proof. $\tilde{\lambda}$ denotes, as in Chapter 1, §1.1, the dual partition.

We argue in the following way. For very k, $\bigwedge^k(V \otimes W)$ is a polynomial representation of degree k of both groups; hence by the general theory $\bigwedge^k(V \otimes W) = \bigoplus_{\lambda \vdash k} S_\lambda(V) \otimes P_\mu(W)$ for some representations $P_\mu(W)$ to be determined. To do it in the stable case, where $\dim W = k$, we use Proposition 6.4 and formula 6.4.4, and we compute the multilinear elements of $P_\mu(W)$ as representations of S_n.

For this, as in 6.4, identify $W = \mathbb{C}^k$ with basis e_i and

$$\bigwedge^k(V \otimes W) = \bigwedge(\bigoplus_{i=1}^k V \otimes e_i) = \otimes_{i=1}^k \bigwedge(V \otimes e_i).$$

When we restrict to the multilinear elements we have $V \otimes e_1 \wedge \dots \wedge V \otimes e_k$ which, as a representation of $GL(V)$, can be identified with $V^{\otimes n}$ except that the natural representation of the symmetric group $S_n \subset GL(n, \mathbb{C})$ is the canonical action on $V^{\otimes n}$ tensored by the sign representation.

Thus we deduce that if χ is the multilinear weight,

$$\bigoplus_{\lambda\vdash k} S_\lambda(V)\otimes M_{\tilde\lambda} = \left(\bigwedge^k V\otimes \mathbb{C}^k\right)^\chi.$$

This implies $P_\lambda(W) = M_{\tilde\lambda}$, hence $P_\lambda(W) = S_{\tilde\lambda}(W)$ from which the claim follows. □

Remark. In terms of characters, formula 8.4.1 is equivalent to the Cauchy formula Chapter 2, §4.1:

$$\prod_{i=1,\ j=1}^{n,\ m}(1+x_iy_j)=\sum_\lambda S_\lambda(x)S_{\tilde\lambda}(y). \tag{8.4.2}$$

There is a simple determinantal formula corollary of this identity, as in §8.3.

Here we remark that $\prod_{i=1}^n(1+x_iy)=\sum_{j=0}^n e_j(x)y^j$ where the e_j are the elementary symmetric functions.

The same reasoning as in §8.3 then gives the formula

$$S_\lambda(x)=\sum_{\sigma\in S_n}\epsilon_\sigma\prod_{i=1}^n e_{\sigma(i)-i+k_i} \tag{8.4.3}$$

where k_i are the columns of $\tilde\lambda$, i.e., the rows of λ.

Proposition. *The Schur function S_λ is the determinant of the $n\times n$ matrix which in the position i,j has the element e_{j-i+k_i}. The k_i are the rows of λ, with the convention that $e_k=0,\ \forall k<0$.*

From the two determinantal formulas found we deduce:

Corollary. *Under the involutive map $\tau : e_i\mapsto h_i$ we have $\tau : S_\lambda\mapsto S_{\tilde\lambda}$.*

Proof. In fact when we apply τ to the first determinantal formula for S_λ, we find the second determinantal formula for $S_{\tilde\lambda}$. □

9 Branching Rules for S_n, Standard Diagrams

9.1 Mumaghan's Rule

We wish to describe now a fairly simple recursive algorithm, due to Mumaghan, to compute the numbers $c_\lambda(\mu)$. It is based on the knowledge of the multiplication of $\psi_k S_\lambda$ in the ring of symmetric functions.

We assume the number n of variables to be more than $k+|\lambda|$, i.e., to be in a stable range for the formula.

Let h_i denote the rows of λ. We may as well compute $\psi_k(x)S_\lambda(x)V(x) = \psi_k(x)A_{\lambda+\varrho}(x)$:

$$\psi_k(x)A_{\lambda+\varrho}(x) = \left(\sum_{i=1}^{n} x_i^k\right)\left(\sum_{s\in S_n} \epsilon_s x_{s1}^{h_1+n-1} x_{s2}^{h_1+n-2} \dots x_{sn}^{h_n}\right). \tag{9.1.1}$$

Write $k_i = h_i + n - i$. We inspect the monomials appearing in the alternating function which is at the right of 9.1.1. Each term is a monomial with exponents obtained from the sequence k_i by adding to one of them, say k_j, the number k. If the resulting sequence has two equal numbers it cannot contribute a term to an alternating sum, and so it must be dropped. Otherwise, reorder it, getting a sequence:

$$k_1 > k_2 > \dots k_i > k_j + k > k_{i+1} > \dots k_{j-1} > k_{j+1} > \dots > k_n.$$

Then we see that the partition $\lambda' : h_1', \dots, h_i', \dots, h_n'$ associated to this sequence is

$$\begin{aligned} &h_t' = h_t, &&\text{if} \quad t \le i \quad \text{or} \quad t > j, \\ &h_t' = h_{t-1} + 1 &&\text{if} \quad i+2 \le t \le j, \ h_{i+1}' = h_j + k - j + i + 1. \end{aligned}$$

The coefficient of $S_{\lambda'}$ in $\psi_k(x)S_\lambda(x)$ is $(-1)^{j-1-i}$ by reordering the rows.

To understand the λ' which appear let us define the *rim* or boundary of a diagram λ as the set of points $(i, j) \in \lambda$ for which there is no point $(h, k) \in \lambda$ with $i < h$, $j < k$.

There is a simple way of visualizing the various partitions λ' which arise in this way.

Notice that we have modified $j - i$ consecutive rows, adding a total of k new boxes. Each row of this set, except for the bottom row, has been replaced by the row immediately below it plus one extra box. We add the remaining boxes to the bottom row.

This property appears to be saying that the new diagram λ' is any diagram which contains the diagram λ and such that their difference is connected, made of k boxes of the rim of λ'. Intuitively it is like a *slinky*.[79] So one has to think of a slinky made of k boxes, sliding in all possible ways down the diagram.

The sign to attribute to such a configuration is $+1$ if the number of rows occupied is odd, -1 otherwise. More formally we have:

Mumaghan's rule. *$\psi_k(x)S_\lambda(x) = \sum \pm S_{\lambda'}$, where λ' runs over all diagrams, such that by removing a connected set of k boxes of the rim of λ' we have λ.*

The sign to attribute to λ' is $+1$ if the number of rows modified from λ is odd, -1 otherwise.

For instance we can visualize $\psi_3 S_{321} = S_{321^4} - S_{32^3} - S_{3^3} - S_{4^21} + S_{621}$ as

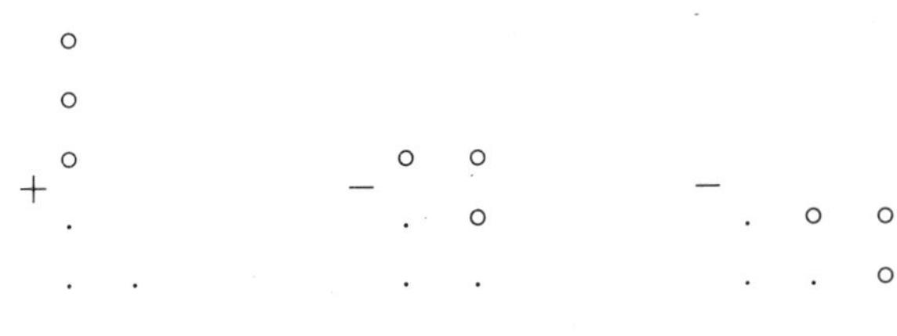

[79] This was explained to me by A. Garsia and refers to a spring toy sold in novelty shops.

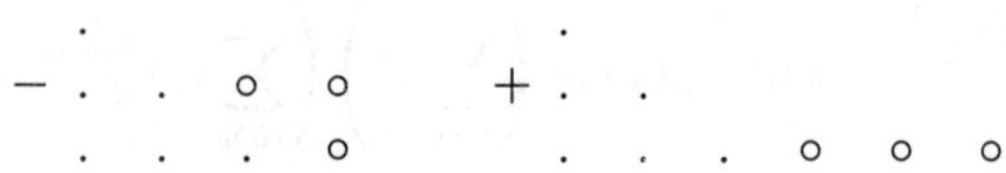

Formally one can define a k-slinky as a walk in the plane $\mathbb{N}^2$ made of k-steps, and each step is either one step down or one step to the right. The sign of the slinky is -1 if it occupies an even number of rows, and $+1$ otherwise.

Next, one defines a *striped tableau* of type $\mu := k_1, k_2, \dots, k_t$ to be a tableau filled, for each $i = 1, \dots, t$, with exactly k_i entries of the number i subject to fill a k_i-slinky. Moreover, we assume that the set of boxes filled with the numbers up to i, for each i is still a diagram. For example, a $3, 4, 2, 5, 6, 3, 4, 1$ striped diagram:

$$\begin{array}{ccccccccc} 8 \\ 4 \\ 4 & 4 & 4 & 5 \\ 3 & 3 & 4 & 5 \\ 1 & 2 & 2 & 5 & 5 & 7 & 7 & 7 & 7 \\ 1 & 1 & 2 & 2 & 5 & 5 & 6 & 6 & 6 \end{array}$$

To such a striped tableau we associate a sign: the product of the signs of all its slinkies. In our case it is the sign pattern $- - + + - + + +$ for a total $-$ sign.

Mumaghan's rule can be formulated as:

Proposition. *$c_\lambda(\mu)$ equals the number of striped tableaux of type μ and shape λ each counted with its sign.*

Notice that when $\mu = 1^n$ the slinky is one box. The condition is that the diagram is filled with all the distinct numbers $1, \dots, n$. The filling is increasing from left to right and from the bottom to the top. Let us formalize:

Definition. A *standard tableau* of shape $\lambda \vdash n$ is a filling of a Young diagram with n-boxes of shape λ, with all the distinct numbers $1, \dots, n$. The filling is strictly increasing from left to right on each row and from the bottom to the top on each column.

From the previous discussion we have:

Theorem. *$d(\lambda)$ equals the number of standard tableaux of shape λ.*

Example. Standard tableaux of shape 3, 2 (compute $d(\lambda) = 5$):

$$\begin{array}{ccc} 2 & 4 \\ 1 & 3 & 5 \end{array}, \quad \begin{array}{ccc} 2 & 5 \\ 1 & 3 & 4 \end{array}, \quad \begin{array}{ccc} 4 & 5 \\ 1 & 2 & 3 \end{array}, \quad \begin{array}{ccc} 3 & 4 \\ 1 & 2 & 5 \end{array}, \quad \begin{array}{ccc} 3 & 5 \\ 1 & 2 & 4 \end{array}.$$

9.2 Branching Rule for S_n

We want to draw another important consequence of the previous multiplication formula between Newton functions and Schur functions.

For a given partition $\mu \vdash n$, consider the module M_μ for S_n and the subgroup $S_{n-1} \subset S_n$ permuting the first $n-1$ numbers. We want to analyze M_μ as a representation of the subgroup S_{n-1}. For this we perform a character computation.

We first introduce a simple notation. Given two partitions $\mu \vdash m$ and $\lambda \vdash n$ we say that $\mu \subset \lambda$ if we have an inclusion of the corresponding Ferrer diagrams, or equivalently, if each row of μ is less than or equal to the corresponding row of λ.

If $\mu \subset \lambda$ and $n = m + 1$ we will also say that μ, λ are *adjacent*,[80] in this case clearly λ is obtained from μ by removing a box lying in a corner.

With these remarks we notice a special case of Theorem 9.1:

$$\psi_1 S_\mu = \sum_{\lambda \vdash |\mu|+1, \mu \subset \lambda} S_\lambda. \tag{9.2.1}$$

Now consider an element of S_{n-1} to which is associated a partition ν. The same element, considered as a permutation in S_n, has associated the partition $\nu 1$. Computing characters we have

$$\begin{aligned} \sum_{\lambda \vdash n} c_\lambda(\nu 1) S_\lambda = \psi_{\nu 1} = \psi_1 \psi_\nu &= \sum_{\tau \vdash (n-1)} c_\tau(\nu) \psi_1 S_\tau \\ &= \sum_{\tau \vdash (n-1)} c_\tau(\nu) \sum_{\mu \vdash n,\, \tau \subset \mu} S_\mu. \end{aligned} \tag{9.2.2}$$

In other words

$$c_\lambda(\nu 1) = \sum_{\mu \vdash (n-1),\, \mu \subset \lambda} c_\mu(\nu), \tag{9.2.3}$$

This identity between characters becomes in module notation:

Theorem (Branching rule for the symmetric group). *When restricting from S_n to S_{n-1} we have*

$$M_\lambda = \bigoplus_{\mu \vdash (n-1),\, \mu \subset \lambda} M_\mu. \tag{9.2.4}$$

A remarkable feature of this decomposition is that each irreducible S_{n-1}-module appearing in M_λ has multiplicity 1, which implies in particular that the decomposition in 9.2.4 is unique.

A convenient way to record a partition $\mu \vdash n-1$ obtained from $\lambda \vdash n$ by removing a box is given by marking this box with n. We can repeat the branching to S_{n-2} and get

[80] Adjacency is a general notion in a poset; here the order is inclusion.

(9.2.5)
$$M_\lambda = \bigoplus_{\substack{\mu_2 \vdash n-2,\\ \mu_1 \vdash n-1,\\ \mu_2 \subset \mu_1 \subset \lambda}} M_{\mu_2}.$$

Again, we mark a pair $\mu_2 \vdash (n-2)$, $\mu_1 \vdash (n-1)$, $\mu_2 \subset \mu_1 \subset \lambda$ by marking the first box removed to get μ_1 with n and the second box with $n-1$.

Example. From 4, 2, 1, 1, branching once:

. . . . 8	+	. . . 8	+	8 . . . .

and twice:

. . . . . . 7 8	+	. . . 7 . . . 8	+	7 8	+	. . . 8 . . . 7

+	7 . . 8	+	8 7	+	8 . . 7	+	8 7 . . . . . .

In general we give the following definitions: Given $\mu \subset \lambda$ two diagrams, the complement of μ in λ is called a *skew diagram* indicated by λ/μ. A *standard skew tableau* of shape λ/μ consists of filling the boxes of λ/μ with distinct numbers such that each row and each column is strictly increasing.

An example of a skew tableau of shape 6, 5, 2, 2/3, 2, 1:

6 7
. 2
. . 2 3 4
. . . 1 2 4

Notice that we have placed some dots in the position of the partition 3, 2, 1 which has been removed.

If $\mu = \emptyset$ we speak of a standard tableau. We can easily convince ourselves that if $\lambda \vdash n$, $\mu \vdash n-k$, and $\mu \subset \lambda$, there is a 1-1 correspondence between:

(1) sequences $\mu = \mu_k \subset \mu_{k-1} \subset \mu_{k-2} \ldots \subset \mu_1 \subset \lambda$ with $\mu_i \vdash n-i$;
(2) standard skew diagrams of shape λ/μ filled with the numbers

$$n-k+1, n-k+2, \ldots, n-1, n.$$

The correspondence is established by associating to a standard skew tableau T the sequence of diagrams μ_i where μ_i is obtained from λ by removing the boxes occupied by the numbers $n, n-1, \dots, n-i+1$.

When we apply the branching rule several times, passing from S_n to S_{n-k} we obtain a decomposition of M_λ into a sum of modules indexed by all possible skew standard tableaux of shape λ/μ filled with the numbers $n-k+1, n-k+2, \dots, n-1, n$.

In particular, for a given shape $\mu \vdash n-k$, the multiplicity of M_μ in M_λ equals the number of such tableaux.

Finally we may go all the way down to S_1 and obtain a canonical decomposition of M_λ into 1-dimensional spaces indexed by all the standard tableaux of shape λ. We recover in a more precise way what we discussed in the previous section.

Proposition. *The dimension of M_λ equals the number of standard tableaux of shape λ.*

It is of some interest to discuss the previous decomposition in the following way.

For every k, let S_k be the symmetric group on k elements contained in S_n, so that $\mathbb{Q}[S_k] \subset \mathbb{Q}[S_n]$ as a subalgebra.

Let Z_k be the center of $\mathbb{Q}[S_k]$. The algebras $Z_k \subset \mathbb{Q}[S_n]$ generate a commutative subalgebra C. In fact, for every k, we have that the center of $\mathbb{Q}[S_k]$ has a basis of idempotents u_λ indexed by the partitions of k. On any irreducible representation, this subalgebra, by the analysis made above, has a basis of common eigenvectors given by the decomposition into 1-dimensional spaces previously described.

Exercise. Prove that the common eigenvalues of the u_λ are distinct and so this decomposition is again unique.

Remark. The decomposition just obtained is almost equivalent to selecting a basis of M_λ indexed by standard diagrams. Fixing an invariant scalar product in M_λ, we immediately see by induction that the decomposition is orthogonal (because non-isomorphic representations are necessarily orthogonal). If we work over $\mathbb{R}$, we can thus select a vector of norm 1 in each summand. This still leaves some sign ambiguity which can be resolved by suitable conventions. The selection of a standard basis is in fact a rather fascinating topic. It can be done in several quite inequivalent ways suggested by very different considerations; we will see some in the next chapters.

A possible goal is to exhibit not only an explicit basis but also explicit matrices for the permutations of S_n, or at least for a set of generating permutations (usually one chooses the Coxeter generators $(i\ i+1),\ i = 1, \dots, n-1$). We will discuss this question when we deal in a more systematic way with standard tableaux in Chapter 13.

10 Branching Rules for the Linear Group, Semistandard Diagrams

10.1 Branching Rule

When we deal with representations of the linear group we can use the character theory which identifies the Schur functions $S_\lambda(x_1, \ldots, x_n)$ as the irreducible characters of $GL(n, \mathbb{C}) = GL(V)$. In general, the strategy is to interpret the various constructions on representations by corresponding operations on characters. There are two main ones: *branching* and *tensor product*. When we branch from $GL(n, \mathbb{C})$ to $GL(n-1, \mathbb{C})$ embedded as block $\begin{vmatrix} A & 0 \\ 0 & 1 \end{vmatrix}$ matrices, we can operate on characters by just setting $x_n = 1$ so the character of the restriction of $S_\lambda(V)$ to $GL(n-1, \mathbb{C})$ is

$$(10.1.1) \qquad S_\lambda(x_1, \ldots, x_{n-1}, 1) = \sum c_\mu S_\mu(x_1, \ldots, x_{n-1}).$$

Similarly, when we take two irreducible representations $S_\lambda(V)$, $S_\mu(V)$ and form their tensor product $S_\lambda(V) \otimes S_\mu(V)$, its character is given by the symmetric function

$$(10.1.2) \qquad S_\lambda(x_1, \ldots, x_n) S_\mu(x_1, \ldots, x_n) = \sum_\nu c^\nu_{\lambda,\mu} S_\nu(x_1, \ldots, x_n).$$

The coefficients in both formulas can be made explicit but, while in 10.1.1 the answer is fairly simple, 10.1.2 has a rather complicated answer given by the *Littlewood–Richardson rule* (discussed in Chapter 12, §5).

The reason why 10.1.1 is rather simple is that all the μ which appear actually appear with coefficient $c_\mu = 1$, so it is only necessary to explain which partitions appear. It is best to describe them geometrically by the diagrams.

WARNING For the linear group we will use *English notation*, for reasons that will be clearer in Chapter 13. Also assume that if $\lambda = h_1, h_2, \ldots, h_r$, these numbers represent the *lengths of the columns*,[81] and hence r must be at most n (we assume $h_r > 0$). In 10.3 we will show that the conditions for μ to appear are the following.

1. $\mu = k_1, \ldots, k_s$ is a diagram contained in λ, i.e., $s \le r,\ k_i \le h_i,\ \forall i \le s$.
2. $s \le n - 1$.
3. μ is obtained from λ by removing at most one box from each row.

The last condition means that we can remove only boxes at the end of each row, which form the *rim* of the diagram. It is convenient to mark the removed boxes by n.

For instance take $\lambda = 4, 2, 2, n = 5$ (we mark the rim). The possible 9 branchings are:

[81] Unfortunately the notation for Young diagrams is not coherent because in the literature they have arisen in different contexts, each having its notational needs.

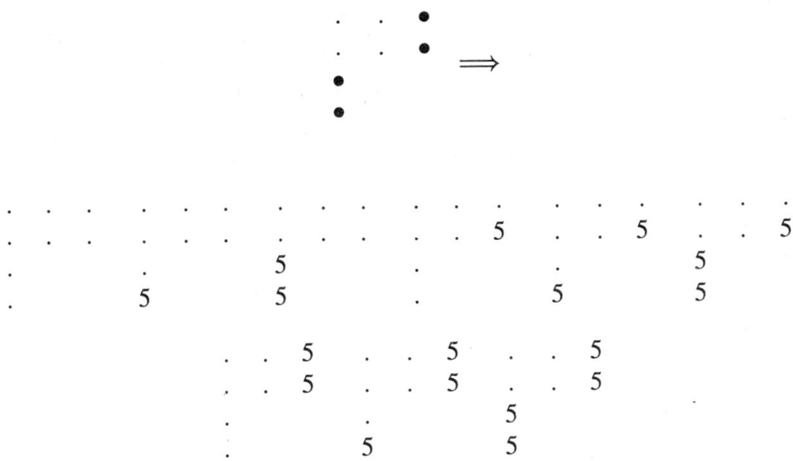

If we repeat these branchings and markings, we see that a sequence of branchings produces a *semistandard tableau* (cf. Chapter 12 §1.1 for a formal definition) like:

$$\begin{array}{ccc} 1 & 2 & 3 \\ 1 & 3 & 5 \\ 5 & & \\ 5 & & \end{array}$$

As in the case of the symmetric group we can deduce a basis of the representation indexed by semistandard tableaux. Conversely, we shall see that one can start from such a basis and deduce a stronger branching theorem which is valid over the integers (Chapter 13, 5.4).

10.2 Pieri's Formula

Although we shall discuss the general Littlewood–Richardson rule in Chapter 12, we start with an example, the study of $S_\lambda(V) \otimes \bigwedge^i(V)$. By previous analysis this can be computed by computing the product $S_\lambda(x)e_i(x)$, where $e_i(x) = S_{1^i}(x)$ is the character of $\bigwedge^i(V)$. For this set $\lambda = h_1, h_2, \dots, h_r$ and $\{\lambda\}_i := \{\mu \mid \mu \supset \lambda,\ |\mu| = |\lambda| + i$ and each column k_i of μ satisfies $h_i \leq k_i \leq h_i + 1\}$.

Theorem (Pieri's formula).

$$(10.2.1) \qquad S_\lambda(x)e_i(x) = \sum_{\mu \in \{\lambda\}_i} S_\mu(x), \quad S_\lambda(V) \otimes \bigwedge^i(V) = \bigoplus\nolimits_{\mu \in \{\lambda\}_i} S_\mu(V).$$

Proof. Let $\lambda = h_1, h_2, \dots, h_n$ where we take n sufficiently large and allow some h_i to be 0, and multiply $S_\lambda(x)e_i(x)V(x) = A_{\lambda+\rho}(x)e_i(x)$. We must decompose the alternating function $A_{\lambda+\rho}(x)e_i(x)$ in terms of functions $A_{\mu+\rho}(x)$. Let $l_i = h_i + n - i$. The only way to obtain in $A_{\lambda+\rho}(x)e_i(x)$ a monomial $x_1^{m_1}x_2^{m_2}\dots x_n^{m_n}$ with $m_1 > m_2 > \cdots > m_n$ is possibly by multiplying $x_1^{l_1}x_2^{l_2}\dots x_n^{l_n}x_{j_1}x_{j_2}\dots x_{j_i}$. This monomial has strictly decreasing exponents for the variables $x_1, \dots, x_n$ if and only if the following condition is satisfied. Set $k_a = h_a$ if a does not appear in the indices $j_1, j_2, \dots, j_i$, and $k_a = h_a + 1$ otherwise. We must have that $k_1 \geq k_2 \cdots \geq k_n$, in other words $\mu := k_1 \geq k_2 \cdots \geq k_n$ is a diagram in $\{\lambda\}_i$. The coefficient of such a monomial is 1, hence we deduce the claim

$$A_{\lambda+\rho}(x)e_i(x) = \sum_{\mu \in \{\lambda\}_i} A_{\mu+\rho}(x). \qquad \square$$

We may now deduce also by duality, using the involutory map $\tau : e_i \to h_i$ (cf. 8.3) and the fact that $h_i(x) = S_i(x)$ is the character of $S^i(V)$, the formula

$$(10.2.2) \qquad S_\lambda(x)h_i(x) = \sum_{\tilde{\mu} \in \{\tilde{\lambda}\}_i} S_\mu(x), \quad S_\lambda(V) \otimes S^i(V) = \bigoplus\nolimits_{\tilde{\mu} \in \{\tilde{\lambda}\}_i} S_\mu(V).$$

In other words, when we perform $S_\lambda(V) \otimes \bigwedge^i(V)$ we get a sum of $S_\mu(V)$ where μ runs over all diagrams obtained from λ by adding i boxes and at most one box in each column, while when we perform $S_\lambda(V) \otimes S^i(V)$ we get a sum of $S_\mu(V)$ where μ runs over all diagrams obtained from λ by adding i boxes and at most one box in each row.[82]

Recall that, for the linear group, we have exchanged rows and columns in our conventions.

10.3 Proof of the Rule

We can now discuss the branching rule from $GL(n, \mathbb{C})$ to $GL(n-1, \mathbb{C})$. From the point of view of characters it is clear that if $f(x_1, \dots, x_n)$ is the character of a representation of $GL(n, \mathbb{C})$, $f(x_1, \dots, x_{n-1}, 1)$ is the character of the restriction of the representation to $GL(n-1, \mathbb{C})$. We thus want to compute $S_\lambda(x_1, \dots, x_{n-1}, 1)$. For this we use Cauchy's formula, getting

$$\sum_\lambda S_\lambda(x_1, \dots, x_{n-1}, 1) S_\lambda(y_1, \dots, y_{n-1}, y_n)$$

$$= \prod_{i=1,\, j=1}^{n-1,\, n} \frac{1}{1 - x_i y_j} \prod_{j=1}^{n} \frac{1}{1 - y_j}$$

$$= \sum_\mu S_\mu(x_1, \dots, x_{n-1}) S_\mu(y_1, \dots, y_{n-1}, y_n) \sum_{j=0}^{\infty} h_j(y_1, \dots, y_{n-1}, y_n).$$

Use 10.2.2 to get

$$\sum_\lambda S_\lambda(x_1, \dots, x_{n-1}, 1) S_\lambda(y_1, \dots, y_{n-1}, y_n)$$

$$= \sum_\mu S_\mu(x_1, \dots, x_{n-1}) \sum_{j=0}^{\infty} \sum_{\tilde{\lambda} \in \{\tilde{\mu}\}_j} S_\lambda(y_1, \dots, y_{n-1}, y_n).$$

82 These two rules are sometimes referred to as Pieri's rule.

Comparing the coefficients of $S_\lambda(y_1, \dots, y_{n-1}, y_n)$, we obtain

$$S_\lambda(x_1, \dots, x_{n-1}, 1) = \sum_{\mu \,|\, \tilde{\lambda} \in \{\tilde{\mu}\}_j} S_\mu(x_1, \dots, x_{n-1}).$$

In other words, let $\{\lambda\}^j$ be the set of diagrams which are obtained from λ by removing j boxes and at most one box in each row. Then the branching of $S_\lambda(\mathbb{C}^n)$ to $GL(n-1)$ is $\bigoplus_{\mu\in\{\lambda\}^j} S_\mu(\mathbb{C}^{n-1})$. In particular we have the property that the irreducible representations which appear come with multiplicity 1.

Since $S_\lambda(x_1, \dots, x_{n-1}, x_n)$ is homogeneous of degree $|\lambda|$ while $\mu \in \{\lambda\}^j$ is homogeneous of degree $|\lambda| - j$, we must have

$$S_\lambda(x_1, \dots, x_{n-1}, x_n) = \sum_{j=0}^{|\lambda|} x_n^j \sum_{\mu\in\{\lambda\}^j} S_\mu(x_1, \dots, x_{n-1}). \qquad \square$$

We may iterate the branching. At each step the branching to $GL(n-i)$ is a direct sum of representations $S_\mu(\mathbb{C}^{n-1})$ with indexing a sequence of diagrams $\mu = \mu_i \subset \mu_{i-1} \subset \cdots \subset \mu_0 = \lambda$ where each μ_j is obtained from μ_{j-1} by removing u_j boxes and at most one box in each row. Furthermore we must have $ht(\mu_j) \leq n-j$. Correspondingly,

$$S_\lambda(x_1, \dots, x_{n-1}, x_n) = \sum_{\mu=\mu_i\subset\mu_{i-1}\subset\cdots\subset\mu_0=\lambda} x_n^{u_1} x_{n-1}^{u_2} \cdots x_{n-i+1}^{u_i} S_\mu(x_1, \dots, x_{n-j}).$$

If we continue the branching all the way to 1, we decompose the space $S_\lambda(V)$ into 1-dimensional subspaces which are weight vectors for the diagonal matrices. Each such weight vector is indexed by a complete *flag* of subdiagrams

$$\emptyset = \mu_n \subset \mu_1 \subset \dots \mu_i \subset \cdots \subset \mu_0 = \lambda$$

and weight $\prod_{i=1}^n x_i^{u_{n-i+1}}$.

A convenient way to encode such flags of subdiagrams is by filling the diagram λ as a semistandard tableau, placing $n-i$ in all the boxes of μ_i not in μ_{i-1}. The restriction we have placed implies that all the rows are strictly increasing, since we remove at most one box from each row, while the columns are weakly increasing, since we may remove more than one box at each step but we fill the columns with a strictly decreasing sequence of numbers. Thus we get a semistandard tableau T of (column-) shape λ filled with the numbers $1, 2, \dots, n$. Conversely, such a semistandard tableau corresponds to an allowed sequence of subdiagrams $\emptyset = \mu_n \subset \mu_1 \subset \dots \mu_i \subset \cdots \subset \mu_0 = \lambda$. Then the monomial $\prod_{i=1}^n x_i^{u_{n-i+1}}$ is deduced directly from T, since u_{n-i+1} is the number of appearances of i in the tableau.

We set $x^T := \prod_{i=1}^n x_i^{u_{n-i+1}}$ and call it the *weight of the tableau* T. Finally we have:

Theorem.

$$S_\lambda(x_1, \dots, x_{n-1}, x_n) = \sum_{T \text{ semistandard of shape } \lambda} x^T. \tag{10.3.1}$$

Of course the set of semistandard tableaux depends on the set of numbers $1, \dots, n$. Since the rows have to be filled by strictly increasing numbers we must have a restriction on height. The rows have at most n-elements.

Example. $S_{3,2,2}(x_1, x_2, x_3)$ is obtained from the tableaux:

$$\begin{matrix} 1&2&3 \\ 2&3 \\ 2&3 \end{matrix} \quad \begin{matrix} 1&2&3 \\ 1&3 \\ 2&3 \end{matrix} \quad \begin{matrix} 1&2&3 \\ 1&2 \\ 2&3 \end{matrix} \quad \begin{matrix} 1&2&3 \\ 1&3 \\ 1&3 \end{matrix} \quad \begin{matrix} 1&2&3 \\ 1&2 \\ 1&3 \end{matrix} \quad \begin{matrix} 1&2&3 \\ 1&2 \\ 1&2 \end{matrix}$$

$$S_{3,2,2}(x_1, x_2, x_3) = x_1x_2^3x_3^3 + x_1^2x_2^2x_3^3 + x_1^2x_2^3x_3^2 + x_1^3x_2x_3^3 + x_1^3x_2^2x_3^2 + x_1^3x_2^3x_3.$$

Of course if we increase the number of variables, then also the number and types of monomials will increase.

We may apply at the same time the branching rule for the symmetric and the linear group. We take an n-dimensional vector space V and consider

$$V^{\otimes m} = \bigoplus_{\lambda \vdash m} M_\lambda \otimes S_\lambda(V).$$

When we branch on both sides we decompose $V^{\otimes m}$ into a direct sum of 1-dimensional weight spaces indexed by pairs $T_1 \mid T_2$ where T_1 is a standard diagram of shape $\lambda \vdash m$ and T_2 is a semistandard diagram of shape λ filled with $1, 2, \dots, n$. We will see, in Chapter 12, §1, that this construction of a basis has a purely combinatorial counterpart, the Robinson–Schensted correspondence.

Note that from Theorem 10.3.1 it is not evident that the function $S_\lambda(x_1, \dots, x_{n-1}, x_n)$ is even symmetric. Nevertheless there is a purely combinatorial approach to Schur functions which takes Theorem 10.3.1 as definition. In this approach the proof of the symmetry of the formula is done by a simple *marriage* argument.

10

Semisimple Lie Groups and Algebras

In this chapter, unless expressly stated otherwise, by Lie algebra we mean a *complex Lie algebra*. Since every real Lie algebra can be complexified, most of our results also have immediate consequences for real Lie algebras.

1 Semisimple Lie Algebras

1.1 $sl(2, \mathbb{C})$

The first and most basic example, which in fact needs to be developed first, is $sl(2, \mathbb{C})$. For this one takes the usual basis

$$e := \begin{vmatrix} 0 & 1 \\ 0 & 0 \end{vmatrix}, \quad f := \begin{vmatrix} 0 & 0 \\ 1 & 0 \end{vmatrix}, \quad h := \begin{vmatrix} 1 & 0 \\ 0 & -1 \end{vmatrix}.$$

These elements satisfy the commutation relations

$$(1.1.1) \qquad [e, f] = h, \quad [h, e] = 2e, \quad [h, f] = -2f.$$

From the theory developed in Chapter 9, we know that the symmetric powers $S^k(V)$ of the 2-dimensional vector space V are the list of rational irreducible representations for $SL(V)$. Hence they are irreducible representations of $sl(2, \mathbb{C})$. To prove that they are also all the irreducible representations of $sl(2, \mathbb{C})$ we start with

Lemma. *Let M be a representation of $sl(2, \mathbb{C})$, $v \in M$ a vector such that $hv = kv, k \in \mathbb{C}$.*

(i) For all i we have that $he^i v = (k + 2i)e^i v,\ hf^i v = (k - 2i) f^i v$.
(ii) If furthermore $ev = 0$, then $ef^i v = i(k - i + 1) f^{i-1} v$.
(iii) Finally if $ev = 0,\ f^m v \neq 0,\ f^{m+1} v = 0$, we have $k = m$ and the elements $f^i v, i = 0, \dots, m$, are a basis of an irreducible representation of $sl(2, \mathbb{C})$.

Proof. (i) We have $2ev = (he - eh)v = hev - kev \implies hev = (k+2)ev$. Hence ev is an eigenvector for h of weight $k+2$. Similarly $hfv = (k-2)fv$. Then the first statement follows by induction.

(ii) From $[e, f] = h$ we see that $ef^i v = (k - 2i + 2) f^{i-1} v + fef^{i-1} v$ and thus, recursively, we check that $ef^i v = i(k - i + 1) f^{i-1} v$.

(iii) If we assume $f^{m+1} v = 0$ for some minimal m, then by the previous identity we have $0 = ef^{m+1} v = (m+1)(k-m) f^m v$. This implies $m = k$ and the vectors $v_i := \frac{1}{i!} f^i v, i = 0, \ldots, k$, span a submodule N with the explicit action

$$(1.1.2) \qquad hv_i = (k - 2i)v_i, \; fv_i = (i+1)v_{i+1}, \; ev_i = (k - i + 1)v_{i-1}.$$

The fact that N is irreducible is clear from these formulas. □

Theorem. *The representations $S^k(V)$ form the list of irreducible finite-dimensional representations of $sl(2, \mathbb{C})$.*

Proof. Let N be a finite-dimensional irreducible representation. Since h has at least one eigenvector, by the previous lemma, if we choose one v_0 with maximal eigenvalue, we have $ev_0 = 0$. Since N is finite dimensional, $f^{m+1} v = 0$ for some minimal m, and we have the module given by formula 1.1.1. Call this representation V_k. Notice that $V = V_1$. We identify V_k with $S^k(V)$ since, if V has basis x, y, the elements $\binom{k}{i} x^{k-i} y^i$ behave as the elements v_i under the action of the elements e, f, h. □

Remark. It is useful to distinguish among the *even* and *odd* representations,[83] according to the parity of k. In an even representation all weights for h are even, and there is a unique weight vector for h of weight 0. In the odd case, all weights are odd and there is a unique weight vector of weight 1.

It is natural to call a vector v with $ev = 0$ a *highest weight vector*. This idea carries over to all semisimple Lie algebras with the appropriate modifications (§5).

There is one expository difficulty in the theory. We have proved that rational representations of $SL(2, \mathbb{C})$ are completely reducible and we have seen that its irreducible representations correspond exactly to the irreducible representations of $sl(2, \mathbb{C})$. It is not clear though why representations of the Lie algebra $sl(2, \mathbb{C})$ are completely reducible, nor why they correspond to rational representations of $SL(2, \mathbb{C})$. There are in fact several ways to prove this which then extend to all semisimple Lie algebras and their corresponding groups.

1. One proves by algebraic means that all finite-dimensional representations of the Lie algebra $sl(2, \mathbb{C})$ are completely reducible.
2. One integrates a finite-dimensional representation of $su(2, \mathbb{C})$ to $SU(2, \mathbb{C})$. Since $SU(2, \mathbb{C})$ is compact, the representation under the group is completely reducible.
3. One integrates a finite-dimensional representation of $sl(2, \mathbb{C})$ to $SL(2, \mathbb{C})$ and then proves that it is rational.

[83] In physics it is usual to divide the highest weight by 2 and talk of integral or half-integral spin.

1.2 Complete Reducibility

Let us discuss the algebraic approach to complete reducibility.[84] First, we remark that a representation of a 1-dimensional Lie algebra is just a linear operator. Since not all linear operators are semisimple it follows that if a Lie algebra $L \supsetneq [L, L]$, then it has representations which are not completely reducible.

If $L = [L, L]$ we have that a 1-dimensional representation is necessarily trivial, and we denote it by $\mathbb{C}$.

Theorem 1. *For a Lie algebra $L = [L, L]$, the following properties are equivalent.*

(1) Every finite-dimensional representation is completely reducible.
(2) If M is a finite-dimensional representation, $N \subset M$ a submodule with M/N 1-dimensional, then $M = N \oplus \mathbb{C}$ as modules.
(3) If M is a finite-dimensional representation, $N \subset M$ an irreducible submodule, with M/N 1-dimensional, then $M = N \oplus \mathbb{C}$ as modules.

Proof. Clearly (1) $\implies$ (2) $\implies$ (3). Let us show the converse.

(3) $\implies$ (2) by a simple induction on $\dim N$. Suppose we are in the hypotheses of (2) assuming (3). If N is irreducible we can just apply (3). Otherwise N contains a nonzero irreducible submodule P and we have the new setting $M' := M/P$, $N' := N/P$ with M'/N' 1-dimensional. Thus, by induction, there is a complement $\mathbb{C}$ to N' in M'. Consider the submodule Q of M with $Q/P = \mathbb{C}$. By part (3) P has a 1-dimensional complement in Q and this is also a 1-dimensional complement of N in M.

(2) $\implies$ (1) is delicate. We check complete reducibility as in Chapter 6 and show that, given a module M and a submodule N, N has a complementary submodule P, i.e., $M = N \oplus P$.

Consider the space of linear maps $\hom(M, N)$. The formula $(l\phi)(m) := l(\phi(m)) - \phi(lm)$ makes this space a module under L. It is immediately verified that a linear map $\phi : M \to N$ is an L homomorphism if and only if $L\phi = 0$.

Since N is a submodule, the restriction $\pi : \hom(M, N) \to \hom(N, N)$ is a homomorphism of L-modules. In $\hom(N, N)$ we have the trivial 1-dimensional submodule $\mathbb{C}1_N$ formed by the multiples of the identity map. Thus take $A := \pi^{-1}(\mathbb{C}1_N)$ and let $B := \pi^{-1}(0)$. Both A, B are L modules and A/B is the trivial module. Assuming (2) we can thus find an element $\phi \in A, \phi \notin B$ with $L\phi = 0$. In other words, $\phi : M \to N$ is an L-homomorphism, which restricted to N, is a nonzero multiple of the identity. Its kernel is thus a complement to N which is a submodule. □

The previous theorem gives us a criterion for complete reducibility which can be used for semisimple algebras once we develop enough of the theory, in particular after we introduce the *Casimir element*. Let us use it immediately to prove that all finite-dimensional representations of the Lie algebra $sl(2, \mathbb{C})$ are completely reducible.

[84] We work over the complex numbers just for simplicity.

Take a finite-dimensional representation M of $sl(2,\mathbb{C})$ and identify the elements e, f, h with the operators they induce on M. We claim that the operator $C := ef + h(h-2)/4 = fe + h(h+2)/4$ commutes with e, f, h. For instance,

$$[C, e] = e[f, e] + [h, e](h-2)/4 + h[h, e]/4$$
$$= -eh + e(h-2)/2 + he/2 = [h, e]/2 - e = 0.$$

Let us show that $sl(2,\mathbb{C})$ satisfies (3) of the previous theorem. Consider $N \subset M$, $M/N = \mathbb{C}$ with N irreducible of highest weight k. On $\mathbb{C}$ the operator C acts as 0 and on N as a scalar by Schur's lemma. To compute which scalar, we find its value on the highest weight vector, getting $k(k+2)/4$. So if $k > 0$, we have a nonzero scalar. On the other hand, on the trivial module, it acts as 0. If $\dim N > 1$, we have that C has the two eigenvalues $k(k+2)/4$ on N and 0 necessarily on a complement of N. It remains to understand the case $\dim N = 1$. In this case the matrices that represent L are a priori 2×2 matrices of type $\begin{vmatrix} 0 & a \\ 0 & 0 \end{vmatrix}$. The commutators of these matrices are all 0. Since $sl(2,\mathbb{C})$ is spanned by commutators, the representation is trivial. From Theorem 1 we have proved:

Theorem 2. *All finite-dimensional representations of $sl(2,\mathbb{C})$ are completely reducible.*

Occasionally one has to deal with infinite-dimensional representations of the following type:

Definition 1. We say that a representation of $sl(2,\mathbb{C})$ is rational if it is a sum of finite-dimensional representations.

A way to study special infinite-dimensional representations is through the notion:

Definition 2. We say that an operator $T : V \to V$ is *locally nilpotent* if, given any vector $v \in V$, we have $T^k v = 0$ for some positive k.

Proposition. *The following properties are equivalent for a representation M of $sl(2,\mathbb{C})$:*

(1) M integrates to a rational representation of the group $SL(2,\mathbb{C})$.
(2) M is a direct sum of finite-dimensional irreducible representations of $sl(2,\mathbb{C})$.
(3) M has a basis of eigenvectors for h and the operators e, f are locally nilpotent.

Proof. (1) implies (2), since from Chapter 7, §3.2 $SL(2,\mathbb{C})$ is linearly reductive. (2) implies (1) and (3) from Theorem 2 of 1.2 and the fact, proved in 1.1 that the finite-dimensional irreducible representations of $sl(2,\mathbb{C})$ integrate to rational representations of the group $SL(2,\mathbb{C})$.

Assume (3). Start from a weight vector v for h. Since e is locally nilpotent, we find some nonzero power k for which $w = e^k v \neq 0$ and $ew = 0$. Since f is locally nilpotent, we have again, for some m, that $f^m w \neq 0$, $f^{m+1} w = 0$. Apply

Lemma 1.1 to get that w generates a finite-dimensional irreducible representation of $sl(2,\mathbb{C})$. Now take the sum P of all the finite-dimensional irreducible representations of $sl(2,\mathbb{C})$ contained in M; we claim that $P = M$. If not, by the same discussion we can find a finite-dimensional irreducible $sl(2,\mathbb{C})$ module in M/P. Take vectors $v_i \in M$ which, modulo P, verify the equations 1.1.2. Thus the elements $hv_i - (k-2i)v_i$, $fv_i - (i+1)v_{i+1}$, $ev_i - (k-i+1)v_{i-1}$ lie in P. Clearly, we can construct a finite-dimensional subspace V of P stable under $sl(2,\mathbb{C})$ containing these elements. Therefore, adding to V the vectors v_i we have an $sl(2,\mathbb{C})$ submodule N. Since N is finite dimensional, it is a sum of irreducibles. So $N \subset P$, a contradiction. □

1.3 Semisimple Algebras and Groups

We are now going to take a very long detour into the general theory of semisimple algebras. In particular we want to explain how one classifies irreducible representations in terms of certain objects called dominant weights. The theory we are referring to is part of the theory of representations of complex semisimple Lie algebras and we shall give a short survey of its main features and illustrate it for classical groups.[85]

Semisimple Lie algebras are closely related to linearly reductive algebraic groups and compact groups. We have already seen in Chapter 7 the definition of a semisimple algebraic group as a reductive group with finite center. For compact groups we have a similar definition.

Definition. A connected compact group is called *semisimple* if it has a finite center.

Example. $U(n,\mathbb{C})$ is not semisimple. $SU(n,\mathbb{C})$ is semisimple.

Let L be a complex semisimple Lie algebra. In this chapter we shall explain the following facts:

(a) L is the Lie algebra of a semisimple algebraic group G.
(b) L is the complexification $L = K \otimes_{\mathbb{R}} \mathbb{C}$ of a real Lie algebra K with negative definite Killing form (Chapter 4, §4.4). K is the Lie algebra of a semisimple compact group, maximal compact in G.
(c) L is a direct sum of simple Lie algebras. Simple Lie algebras are completely classified. The key to the classification of Killing–Cartan is the theory of *roots* and *finite reflection groups*.

It is a quite remarkable fact that associated to a continuous group there is a finite group of Euclidean reflections and that the theory of the continuous group can be largely analyzed in terms of the combinatorics of these reflections.

[85] There are many more general results over fields of characteristic 0 or just over the real numbers, but they do not play a specific role in the theory we shall discuss.

1.4 Casimir Element and Semisimplicity

We want to see that the method used to prove the complete reducibility of $sl(2, \mathbb{C})$ works in general for semisimple Lie algebras.

We first need some simple remarks.

Let L be a simple Lie algebra. Given any nontrivial representation $\rho : L \to gl(M)$ of L we can construct its *trace form*, $(a, b) := \operatorname{tr}(\rho(a)\rho(b))$. It is then immediate to verify that this form is *associative* in the sense that $([a, b], c) = (a, [b, c])$. It follows that the kernel of this form is an ideal of L. Hence, unless this form is identically 0, it is nondegenerate.

Remark. The form cannot be identically 0 from Cartan's criterion (Chapter 4, §6.4).

Lemma. *On a simple Lie algebra a nonzero associative form is unique up to scale.*

Proof. We use the bilinear form to establish a linear isomorphism $j : L \to L^*$, through the formula $j(a)(b) = (a, b)$. We have $j([x, a])(b) = ([x, a], b]) = -(a, [x, b]) = -j(a)([x, b])$. Thus j is an isomorphism of L-modules. Since L is irreducible as an L-module, the claim follows from Schur's lemma. □

Remark. In particular, the trace form is a multiple of the Killing form.

Let L be a semisimple Lie algebra, and consider dual bases u_i, u^i for the Killing form. Since the Killing form identifies L with L^*, by general principles the Killing form can be identified with the symmetric tensor $C_L := \sum_i u_i \otimes u^i = \sum_i u^i \otimes u_i$. The associativity property of the form translates into invariance of C_L. C_L is killed by the action of the Lie algebra, i.e.,

$$\sum_i ([x, u^i] \otimes u_i + u^i \otimes [x, u_i]) = 0, \quad \forall x \in L. \tag{1.4.1}$$

Then by the multiplication map it is best to identify C_L with its image in the enveloping algebra $U(L)$. More concretely:

Theorem 1.

(1) The element $C_L := \sum_i u_i u^i$ does not depend on the dual bases chosen.

(2) The element $C_L := \sum_i u_i u^i$ commutes with all of the elements of the Lie algebra L.

(3) If the Lie algebra L decomposes as a direct sum $L = \bigoplus_i L_i$ of simple algebras, then we have $C_L = \sum_i C_{L_i}$. Each C_{L_i} commutes with L.

(4) If M is an irreducible representation of L each element C_{L_i} acts on M by a scalar. This scalar is 0 if and only if L_i is in the kernel of the representation.

Proof. (1) Let $s_i = \sum_j d_{j,i} u_j,\ s^i = \sum_j e_{j,i} u^j$ be another pair of dual bases. We have

$$\delta_i^j = (s_i, s^j) = \left(\sum_h d_{h,i} u_h, \sum_h e_{h,j} u^h \right) = \sum_h d_{h,i} e_{h,j}.$$

If D is the matrix with entries $d_{i,j}$ and E the one with entries $e_{i,j}$, we thus have $E^t D = 1$, which implies that also $ED^t = 1$. Thus

$$\sum_i s_i s^i = \sum_i \sum_h d_{h,i} u_h \sum_k e_{k,i} u^k = \sum_{h,k} \sum_i d_{h,i} e_{k,i} u_h u^k = \sum_h u_h u^h.$$

(2) Denote by (a, b) the Killing form. If $[c, u_i] = \sum_j a_{j,i} u_j,\ [c, u^j] = \sum_i b_{i,j} u^i$, we have

$$a_{j,i} = ([c, u_i], u^j) = -(u_i, [c, u^j]) = -b_{i,j}. \tag{1.4.2}$$

$$\text{Then}\quad [c, C] = \sum_i [c, u_i] u^i + \sum_i u_i [c, u^i]$$
$$= \sum_i \sum_j a_{j,i} u_j u^i + \sum_i u_i \sum_j b_{j,i} u^j$$
$$= \sum_i \sum_j a_{j,i} u_j u^i + b_{i,j} u_j u^i = 0.$$

(3) The ideals L_i are orthogonal under the Killing form, and the Killing form of L restricts to the Killing form of L_i (Chapter 4, §6.2, Theorem 2). The statement is clear, since the L_i commute with each other.

(4) Since C_{L_i} commutes with L and M is irreducible under L, by Schur's lemma, C_{L_i} must act as a scalar. We have to see that it is 0 if and only if L_i acts by 0.

If L_i does not act as 0, the trace form is nondegenerate and is a multiple of the Killing form by a nonzero scalar λ. Then $\mathrm{tr}(\rho(C_{L_i})) = \sum_i \mathrm{tr}(\rho(u_i)\rho(u^i)) = \sum_i \lambda(u_i, u^i) = \lambda \dim L \neq 0$. □

Definition. The element $C_L \in U(L)$ is called the *Casimir element* of the semisimple Lie algebra L.

We can now prove:

Theorem 2. *A finite-dimensional representation of a semisimple Lie algebra L is completely reducible.*

Proof. Apply the method of §1.2. Since $L = [L, L]$, the only 1-dimensional representations of L are trivial. Let M be a module and N an irreducible submodule with M/N 1-dimensional, hence trivial. If N is also trivial, the argument given in 1.2 for $sl(2, \mathbb{C})$ shows that M is trivial. Otherwise, let us compute the value on M of one of the Casimir elements $C_i = C_{L_i}$ which acts on N by a nonzero scalar λ (by the previous theorem). On the quotient M/N the element C_i acts by 0. Therefore C_i on M has eigenvalue λ (with eigenspace N) and 0. There is thus a vector $v \notin N$ for which $C_i v = 0$ such that v spans the 1-dimensional eigenspace of the eigenvalue 0. Since C_i commutes with L, the space generated by v is stable under L, and $Lv = 0$ satisfying the conditions of Theorem 1 of 1.2. □

1.5 Jordan Decomposition

In a Lie algebra L an element x is called *semisimple* if $\mathrm{ad}(x)$ is a semisimple (i.e., diagonalizable) operator.

As for algebraic groups we may ask if the semisimple part of the operator $\mathrm{ad}(x)$ is still of the form $\mathrm{ad}(y)$ for some y, to be called the semisimple part of x. Not all Lie algebras have this property, as simple examples show. The ones which do are called *splittable*.

We need a simple lemma.

Lemma 1. *Let A be any finite-dimensional algebra over $\mathbb{C}$, and D a derivation. Then the semisimple part D_s of D is also a derivation.*

Proof. One can give a direct computational proof (see [Hu1]). Since we have developed some theory of algebraic groups, let us instead follow this path. The group of automorphisms of A is an algebraic group, and for these groups we have seen the Jordan–Chevalley decomposition. Hence, given an automorphism of A, its semisimple part is also an automorphism. D is a derivation if and only if $\exp(tD)$ is a one parameter group of automorphisms (Chapter 3). We can conclude noticing that if $D = D_s + D_n$ is the additive Jordan decomposition, then $\exp(tD) = \exp(tD_s)\exp(tD_n)$ is the multiplicative decomposition. We deduce that $\exp(tD_s)$ is a one parameter group of automorphisms. Hence D_s is a derivation. □

Lemma 2. *Let L be a Lie algebra, and M the Lie algebra of its derivations. The inner derivations $\mathrm{ad}(L)$ are an ideal of M and $[D, \mathrm{ad}(a)] = \mathrm{ad}(D(a))$.*

Proof.

$$[D, \mathrm{ad}(a)](b) = D(\mathrm{ad}(a)(b)) - \mathrm{ad}(a)(D(b)) = D[a, b] - [a, D(b)]$$

$$= [D(a), b] + [a, D(b)] - [a, D(b)] = [D(a), b].$$

Thus $\mathrm{ad}(L)$ is an ideal in M. □

Theorem 1. *If L is a semisimple Lie algebra and D is a derivation of L, then D is inner.*

Proof. Let M be the Lie algebra of derivations of L. It contains the inner derivations $\mathrm{ad}(L)$ as an ideal. Since L is semisimple we have a direct sum decomposition $M = \mathrm{ad}(L) \oplus P$ as L modules. Since $\mathrm{ad}(L)$ is an ideal, $[P, \mathrm{ad}(L)] \subset \mathrm{ad}(L)$. Since P is an L module, $[P, \mathrm{ad}(L)] \subset P$. Hence $[P, \mathrm{ad}(L)] = 0$. From the formula $[D, \mathrm{ad}(a)] = \mathrm{ad}(D(a))$, it follows that if $D \in P$, we have $\mathrm{ad}(D(a)) = 0$. Since the center of L is 0, $P = 0$ and $M = \mathrm{ad}(L)$. □

Corollary. *If L is a semisimple Lie algebra, $a \in L$, there exist unique elements $a_s, a_n \in L$ such that*

(1.5.1) $$a = a_s + a_n, \ [a_s, a_n] = 0, \quad \mathrm{ad}(a_s) = \mathrm{ad}(a)_s, \ \mathrm{ad}(a_n) = \mathrm{ad}(a)_n.$$

Proof. By Lemma 1, the semisimple and nilpotent parts of $\mathrm{ad}(a)$ are derivations. By the previous theorem they are inner, hence induced by elements a_s, a_n. Since the map $\mathrm{ad} : L \to \mathrm{ad}(L)$ is an isomorphism the claim follows. □

Finally we want to see that the Jordan decomposition is preserved under any representation.

Theorem 2. *If ρ is any linear representation of a semisimple Lie algebra and $a \in L$, we have $\rho(a_s) = \rho(a)_s,\ \rho(a_n) = \rho(a)_n$.*

Proof. The simplest example is when we take the Lie algebra $sl(V)$ acting on V. In this case we can apply the Lemma of §6.2 of Chapter 4. This lemma shows that the usual Jordan decomposition $a = a_s + a_n$ for a linear operator $a \in sl(V)$ on V induces, under the map $a \mapsto \mathrm{ad}(a)$, a Jordan decomposition $\mathrm{ad}(a) = \mathrm{ad}(a_s) + \mathrm{ad}(a_n)$.

In general it is clear that we can restrict our analysis to simple L, V an irreducible module and $L \subset \mathrm{End}(V)$. Let $M := \{x \in \mathrm{End}(V) \mid [x, L] \subset L\}$. As before M is a Lie algebra and L an ideal of M. Decomposing $M = L \oplus P$ with P an L-module, we must have $[L, P] = 0$. Since the module is irreducible, by Schur's lemma we must have that P reduces to the scalars. Since $L = [L, L]$, the elements of L have all trace 0, hence $L = \{u \in M \mid \mathrm{tr}(u) = 0\}$. Take an element $x \in L$ and decompose it in $\mathrm{End}(V)$ as $x = y_s + y_n$ the semisimple and nilpotent part. By the Lemma of §6.2, Chapter 4 previously recalled, $\mathrm{ad}(x) = \mathrm{ad}(y_s) + \mathrm{ad}(y_n)$ is the Jordan decomposition of operators acting on $\mathrm{End}(V)$. Since $\mathrm{ad}(x)$ preserves L, also $\mathrm{ad}(y_s), \mathrm{ad}(y_n)$ preserve L, hence we must have $y_s, y_n \in M$. Since $\mathrm{tr}(y_n) = 0$ we have $y_n \in L$, hence also $y_s \in L$. By the uniqueness of the Jordan decomposition $x_s = y_s, x_n = y_n$. □

There is an immediate connection to algebraic groups.

Theorem 3. *If L is a semisimple Lie algebra, its adjoint group is the connected component of 1 of its automorphism group. It is an algebraic group with Lie algebra L.*

Proof. The automorphism group is clearly algebraic. Its connected component of 1 is generated by the 1-parameter groups $\exp(tD)$ where D is a derivation. Since all derivations are inner, it follows that its Lie algebra is $\mathrm{ad}(L)$. Since L is semisimple, $L = \mathrm{ad}(L)$. □

1.6 Levi Decomposition

Let us first make a general construction. Given a Lie algebra L, let $\mathcal{D}(L)$ be its Lie algebra of derivations. Given a Lie homomorphism ρ of a Lie algebra M into $\mathcal{D}(L)$, we can give to $M \oplus L$ a new Lie algebra structure by the formula (check it):

$$[(m_1, a), (m_2, b)] := ([m_1, m_2], \rho(m_1)(b) - \rho(m_2)(a) + [a, b]). \tag{1.6.1}$$

Definition. $M \oplus L$ with the previous structure is called a *semidirect product* and denoted $M \ltimes L$.

Formula 1.6.1 implies immediately that in $M \ltimes L$ we have that M is a subalgebra and L an ideal. Furthermore, if $m \in M, a \in L$ we have $[m, a] = \rho(m)(a)$.

As an example, take $F \ltimes L$ with $M = F$ 1-dimensional. A homomorphism of F into $D(L)$ is given by specifying a derivation D of L corresponding to 1, so we shall write $FD \ltimes L$ to remind us of the action:

Lemma 1. *(i) If L is solvable, then $FD \ltimes L$ is solvable.*

(ii) If N is a nilpotent Lie algebra and D is a derivation of N, then $FD \ltimes N$ is nilpotent if and only if D is nilpotent.

(iii) If L is semisimple, then $F \ltimes L = L \oplus F$.

Proof. (i) The first part is obvious.

(ii) Assume that $D^m N = 0$ and $N^i = 0$. By formula 4.3.1 of Chapter 4 it is enough to prove that a long enough monomial in elements $\operatorname{ad}(a_i)$, $a_i \in F \ltimes N$ is 0. In fact it is enough to show that such an operator is 0 on N since then a 1-step longer monomial is identically 0. Consider a monomial in the operators $D, \operatorname{ad}(n_j), n_j \in N$. Assume the monomial is of degree $> mi$.

Notice that $D \operatorname{ad}(n_i) = \operatorname{ad}(D(n_i)) + \operatorname{ad}(n_i)D$. We can rewrite the monomial as a sum of terms $\operatorname{ad}(D^{h_1}n_1)\operatorname{ad}(D^{h_2}n_2)\dots\operatorname{ad}(D^{h_t}n_t)D^{h_{t+1}}$ with $\sum_{k=1}^{t+1} h_k + t > mi$. If $t \geq i-1$, this is 0 by the condition $N^i = 0$. Otherwise, $\sum_{k=1}^{t+1} h_k > (m-1)i$, and since $t < i$ at least one of the exponents h_k must be bigger than $m-1$. So again we get 0 from $D^m = 0$.

Conversely, if D is not nilpotent, then $\operatorname{ad}(D) = D$ on N, hence $\operatorname{ad}(D)^m \neq 0$ for all m, so $FD \ltimes N$ is not nilpotent.

(iii) If L is semisimple, $D = \operatorname{ad}(a)$ is inner, $D - a$ is central and $F \ltimes L = L \oplus F(D-a)$. □

We need a criterion for identifying semidirect products.[86] Given L a Lie algebra and I a Lie ideal, we have

Lemma 2. *If there is a Lie subalgebra A such that as vector spaces $L = A \oplus I$, then $L = A \ltimes I$ where A acts by derivation on I as restriction to I of the inner derivations* $\operatorname{ad}(a)$.

The proof is straightforward.

A trivial example is when L/I is 1-dimensional. Then any choice of an element $a \in L$, $a \notin I$ presents $L = Fa \oplus I$ as a semidirect product.

In the general case, the existence of such an A can be treated by the *cohomological method.*

Let us define $A := L/I$, $p : L \to L/I$ the projection. We want to find a homomorphism $f : A \to L$ with $pf = 1_A$. If such a homomorphism exists it is called a *splitting*. We proceed in two steps. First, choose any linear map $f : A \to L$ with $pf = 1_A$. The condition to be a homomorphism is that the two-variable function $\phi_f(a, b) := f([a, b]) - [f(a), f(b)]$, which takes values in I, must be 0. Given such

[86] As in group theory.

an f, if it does not satisfy the homomorphism property we can *correct it* by adding to it a linear mapping $g : A \to I$. Given such a map, the new condition is that

$$\phi_{f+g}(a,b) := f([a,b]) - [f(a), f(b)] + g([a,b]) \\ - [g(a), f(b)] - [f(a), g(b)] - [g(a), g(b)] = 0.$$

In general this is not so easy to handle, but there is a special important case. When I is abelian, then I is naturally an A module. Denoting this module structure by $a.i$, one has $[f(a), g(b)] = a.g(b)$ (independently of f) and the condition becomes: find a g with

$$f([a,b]) - [f(a), f(b)] = a.g(b) - b.g(a) - g([a,b]).$$

Notice that $\phi_f(a,b) = -\phi_f(b,a)$. Given a Lie algebra A, a skew-symmetric two-variable function $\phi(a,b)$ from $A \bigwedge A$ to an A module M of the form $a.g(b) - b.g(a) - g([a,b])$ is called a *2-coboundary*.

The method consists in stressing a property which the element

$$\phi_f(a,b) := f([a,b]) - [f(a), f(b)]$$

shares with 2-coboundaries, deduced from the Jacobi identity:

Lemma 3.

$$a.\phi_f(b,c) - b.\phi_f(a,c) + c.\phi_f(a,b) \\ - \phi_f([a,b],c) + \phi_f([a,c],b) - \phi_f([b,c],a) = 0. \tag{1.6.2}$$

Proof. From the Jacobi identity

$$a.\phi_f(b,c) - b.\phi_f(a,c) + c.\phi_f(a,b) \\ = [f(a), f([b,c])] - [f(b), f([a,c])] + [f(c), f([a,b])]$$

$$\phi_f([a,b],c) - \phi_f([a,c],b) + \phi_f([b,c],a) \\ = -[f([a,b]), f(c)] + [f([a,c]), f(b)] - [f([b,c]), f(a)].$$ □

A skew-symmetric two-variable function $\phi(a,b)$ from $A \bigwedge A$ to an A module M, satisfying 1.6.2 is called a *2-cocycle*. Then one has to understand under which conditions a 2-cocycle is a 2-coboundary.

In general this terminology comes from a cochain complex associated to Lie algebras. We will not need it but give it for reference. The k-cochains are the maps $C^k(A; M) := \hom(\bigwedge^k A, M)$, the coboundary $\delta : C^k(A; M) \to C^{k+1}(A; M)$ is defined by the formula

$$\delta\phi(a_0, \dots, a_k) = \sum_{i=0}^{k} (-1)^i a_i \phi(a_1, \dots, \check{a}_i, \dots, a_k) \\ + \sum_{i<j} (-1)^{i+j} \phi([a_i, a_j], a_0, \dots, \check{a}_i, \dots \check{a}_j, \dots, a_k).$$

By convention the 0-dimensional cochains are identified with M and $\delta(m)(a) := a.m$.

The complex property means that $\delta \circ \delta = 0$ as one can check directly. Then a cocycle is a cochain ϕ with $\delta\phi = 0$, while a coboundary is a cochain $\delta\phi$. For all k, the space of k-cocycles modulo the k-coboundaries is an interesting object called k-cohomology, denoted $H^k(A; M)$. This is part of a wide class of cohomology groups, which appear as measures of obstructions of various possible constructions.

From now on, in this section we assume that the Lie algebras are finite-dimensional over $\mathbb{C}$. In our case we will again use the Casimir element to prove that:

Proposition. *For a semisimple Lie algebra L, every 2-cocycle with values in a nontrivial irreducible module M is a 2-coboundary.*

Proof. If C is the Casimir element of L, C acts by some nonzero scalar λ on M. We compute it in a different way using 1.6.2 (with $c = u_i$):

$$u_i.\phi(a, b) = a.\phi(u_i, b) - b.\phi(u_i, a) + \phi([u_i, a], b) - \phi([u_i, b], a) + \phi([a, b], u_i) \implies$$

$$u^i.u_i.\phi(a, b) = [u^i, a].\phi(u_i, b) + a.u^i.\phi(u_i, b) - [u^i, b].\phi(u_i, a) - b.u^i.\phi(u_i, a)$$

$$+ u^i.\{\phi([u_i, a], b) - \phi([u_i, b], a) - \phi(u_i, [a, b])\}. \tag{1.6.3}$$

Now the identity 1.4.1 implies $\sum_i k([x, u^i], u_i) + k(u^i, [x, u_i]) = 0, \forall x \in L$ and any bilinear map $k(x, y)$. Apply it to the bilinear maps $k(x, y) := x.\phi(y, b)$, and $x.\phi(y, a)$ getting

$$\sum_i ([a, u^i].\phi(u_i, b) + u^i.\phi([a, u_i], b)) = \sum_i ([b, u^i].\phi(u_i, a) + u^i.\phi([b, u_i], a)) = 0.$$

Now set $h(x) := \sum_i u^i.\phi(u_i, x)$. Summing all terms of 1.6.3 one has

$$\lambda\phi(a, b) = a.h(b) - b.h(a) - h([a, b]). \tag{1.6.4}$$

Dividing by λ one has the required coboundary condition. □

Cohomology in general is a deep theory with many different applications. We mention some as (difficult) exercises:

Exercise (Theorem of Whitehead). Generalize the previous method to show that, under the hypotheses of the previous proposition we have $H^i(L; M)$ for all $i \geq 0$.[87]

[87] For a semisimple Lie algebra the interesting cohomology is the cohomology of the trivial 1-dimensional representation. This can be interpreted topologically as cohomology of the associated Lie group.

Exercise. Given a Lie algebra A and a module M for A one defines an extension as a Lie algebra L with an abelian ideal identified with M such that $L/I = A$ and the induced action of A on M is the given module action. Define the notion of equivalence of extensions and prove that equivalence classes of extensions are classified by $H^2(A; M)$. In this correspondence, the semidirect product corresponds to the 0 class.

When $M = \mathbb{C}$ is the trivial module, cohomology with coefficients in $\mathbb{C}$ is nonzero and has a deep geometric interpretation:

Exercise. Let G be a connected Lie group, with Lie algebra L. In the same way in which we have constructed the Lie algebra as left invariant vector fields we can consider also the space $\Omega^i(G)$ of left invariant differential forms of degree i for each i. Clearly a left invariant form is determined by the value that it takes at 1; thus, as a vector space $\Omega^i(G) = \bigwedge^i(T_1^*(G)) = \bigwedge^i(L^*)$, the space of i-cochains on L with values in the trivial representation.

Observe that the usual differential on forms commutes with the left G action so d maps $\Omega^i(G)$ to $\Omega^{i+1}(G)$. Prove that we obtain exactly the algebraic cochain complex previously described. Show furthermore that the condition $d^2 = 0$ is another formulation of the Jacobi identity.

By a theorem of Cartan, when G is compact: *the cohomology of this complex computes exactly the de Rham cohomology of G as a manifold.*

We return to our main theme and can now prove the:

Theorem (Levi decomposition). *Let L, A be Lie algebras, A semisimple, and $\pi : L \to A$ a surjective homomorphism. Then there is a splitting $i : A \to L$ with $\pi \circ i = 1_A$.*

Proof. Let $K = \mathrm{Ker}(\pi)$ be the kernel; we will proceed by induction on its dimension. If L is semisimple, K is a direct sum of simple algebras and its only ideals are sums of these simple ideals, so the statement is clear. If L is not semisimple, it has an abelian ideal I which is necessarily in K since A has no abelian ideals. By induction $L/I \to A$ has a splitting $j : A \to L/I$; therefore there is a subalgebra $M \supset I$ with $M/I = A$, and we are reduced to the case in which K is a minimal abelian ideal. Since K is abelian, the action of L on K vanishes on K, and K is an A-module. Since the A-submodules are ideals and K is minimal, it is an irreducible module. We have two cases: $K = \mathbb{C}$ is the trivial module or K is a nontrivial irreducible. In the first case we have $[L, K] = 0$, so A acts on L, and K is a submodule. By semisimplicity we must have a stable summand $L = B \oplus K$. B is then an ideal under L and isomorphic under projection to A.

Now, the case K nontrivial. In this case the action of A on K induces a nonzero associative form on A, nondegenerate on the direct sum of the simple components which are not in the kernel of the representation K, and a corresponding Casimir element $C = \sum_i u^i u_i$. Apply now the cohomological method and construct $f : A \to L$ and the function $\phi(a, b) := [f(a), f(b)] - f([a, b])$. We can apply Lemma 3, so f is a cocycle. Then by the previous proposition it is also a coboundary and hence we can modify f so that it is a Lie algebra homomorphism, as required. $\square$

The previous theorem is usually applied to $A := L/R$ where R is the solvable radical of a Lie algebra L (finite-dimensional over $\mathbb{C}$). In this case a splitting $L = A \ltimes R$ is called a *Levi decomposition* of L.

Lemma 4. *Let L be a finite-dimensional Lie algebra over $\mathbb{C}$, I its solvable radical, and N the nilpotent radical of I as a Lie algebra.*

(i) Then N is also the nilpotent radical of L.
(ii) $[I, I] \subset N$.
(iii) If $a \in I$, $a \notin N$, then $\mathrm{ad}(a)$ *acts on N by a linear operator with at least one nonzero eigenvalue.*

Proof. (i) By definition the nilpotent radical of the Lie algebra I is the maximal nilpotent ideal in I. It is clearly invariant under any automorphism of I. Since we are in characteristic 0, if D is a derivation, N is also invariant under $\exp(tD)$, a 1-parameter group of automorphisms, hence it is invariant under D. In particular it is invariant under the restriction to I of any inner derivation $\mathrm{ad}(a)$, $a \in L$. Thus N is a nilpotent ideal in L. Since conversely the nilpotent radical of L is contained in I, the claim follows.

(ii) From the corollary of Lie's theorem, $[I, I]$ is a nilpotent ideal.

(iii) If $\mathrm{ad}(a)$ acts in a nilpotent way on N, then $\mathbb{C}a \oplus N$ is nilpotent (Lemma 1, (ii)). From (ii) $\mathbb{C}a \oplus N$ is an ideal, and from i) it follows that $\mathbb{C}a \oplus N \subset N$, a contradiction. □

Lemma 5. *Let $L = A \oplus I$ be a Levi decomposition, $N \subset I$ the nilpotent radical of L.*

Since A is semisimple and I, N are ideals, we can decompose $I = B \oplus N$ where B is stable under $\mathrm{ad}(A)$. *Then* $\mathrm{ad}(A)$ *acts trivially on B.*

Proof. Assume that the action of $\mathrm{ad}(A)$ on B is nontrivial. Then there is a semisimple element $a \in A$ such that $\mathrm{ad}(a) \neq 0$ on B. Otherwise, by Theorem 2 of 1.5, $\mathrm{ad}(A)$ on B would act by nilpotent elements and so, by Engel's Theorem, it would be nilpotent, which is absurd since a nonzero quotient of a semisimple algebra is semisimple.

Since $\mathrm{ad}(a)$ is also semisimple (Theorem 2 of 1.5) we can find a nonzero vector $v \in B$ with $\mathrm{ad}(a)(v) = \lambda v$, $\lambda \neq 0$. Consider the solvable Lie algebra $P := \mathbb{C}a \oplus I$. Then $v \in [P, P]$ and $[P, P]$ is a nilpotent ideal of P (cf. Chapter 4, Cor. 6.3). Hence $\mathbb{C}v + [I, I]$ is a nilpotent ideal of I. From Lemma 4 we then have $v \in N$, a contradiction. □

Theorem 2. *Given a Lie algebra L with semisimple part A, we can embed it into a new Lie algebra L' with the following properties:*

(i) L' has the same semisimple part A as L.
(ii) The solvable radical of L' is decomposed as $B' \oplus N'$, where N' is the nilpotent radical of L', B' is an abelian Lie algebra acting by semisimple derivations, and $[A, B'] = 0$.
(iii) $A \oplus B'$ is a subalgebra and $L' = (A \oplus B') \ltimes N'$.

Proof. Using the Levi decomposition we can start decomposing $L = A \ltimes I$ where A is semisimple and I is the solvable radical. Let N be the nilpotent radical of I. By Lemma 4 it is also an ideal of L. By the previous Lemma 5, decompose $I = B \oplus N$ with $[A, B] = 0$. Let $m := \dim(B)$. We work by induction and construct a sequence of Lie algebras $L_i = A \oplus B_i \oplus N_i, i = 1, \dots, m$, with $L_0 = L,\ L_i \subset L_{i+1}$ and with the following properties:

(i) $B_i \oplus N_i$ is the solvable radical, N_i the nilpotent radical of L_i.

(ii) $[A, B_i] = 0$, and B_i has a basis $a_1, \dots, a_i, b_{i+1}, \dots, b_m$ with a_i inducing commuting semisimple derivations of L_i.

(iii) Finally $[a_h, B_i] = 0, h = 1, \dots, i$.

$L' = L_m$ thus satisfies the requirements of the theorem.

Given L_i as before, we construct L_{i+1} as follows. Consider the derivation $\mathrm{ad}(b_{i+1})$ of L_i and denote by a_{i+1} its semisimple part, still a derivation. By hypothesis the linear map $\mathrm{ad}(b_{i+1})$ is 0 on A. $\mathrm{ad}(b_{i+1})$ preserves the ideals I_i, N_i and, since $[B_i, B_i] \subset N_i$ it maps B_i into N_i. Therefore the same properties hold for its semisimple part:

a_{i+1} preserves I_i, N_i. $a_{i+1}(x) = 0, \forall x \in A$. $a_{i+1}(a_h) = 0, h = 1, \dots, i$ and $a_{i+1}(B_i) \subset N_i$.

Construct the Lie algebra $L_{i+1} := \mathbb{C}a_{i+1} \ltimes L_i = A \oplus (\mathbb{C}a_{i+1} \oplus I_i)$. Let $I_{i+1} := \mathbb{C}a_{i+1} \oplus I_i$. Since a_{i+1} commutes with A, I_{i+1} is a solvable ideal. Since $L_{i+1}/I_{i+1} = A$, I_{i+1} is the solvable radical of L_{i+1}.

The element $\mathrm{ad}(b_{i+1} - a_{i+1})$ acts on N_i as the nilpotent part of the derivation $\mathrm{ad}(b_{i+1})$; thus the space $N_{i+1} := \mathbb{C}(a_{i+1} - b_{i+1}) \oplus N_i$ is nilpotent by §1.6, Lemma 1, (ii).

Since $[B_i, B_i] \subset N_i$ we have that N_{i+1} is an ideal in I_{i+1}. By construction we still have $I_{i+1} = B_i \oplus N_{i+1}$. If N_{i+1} is not the nilpotent radical, we can find a nonzero element $c \in B_i$ so that $\mathbb{C} \oplus N_{i+1}$ is a nilpotent algebra. By Lemma 1, this means that c induces a nilpotent derivation in N_{i+1}. This is not possible since it would imply that c also induces a nilpotent derivation in N_i, so that $\mathbb{C} \oplus N_i$ is a nilpotent algebra and an ideal in I_i, contrary to the inductive assumption that N_i is the nilpotent radical of I_i.

The elements $a_h,\ h = 1, \dots, i + 1$, induce commuting semisimple derivations on I_{i+1} which also commute with A. Thus, under the algebra R generated by the operators $\mathrm{ad}(A)$, $\mathrm{ad}(a_h)$, the representation I_{i+1} is completely reducible. Moreover R acts trivially on I_{i+1}/N_{i+1}. Thus we can find an R-stable complement C_{i+1} to $N_{i+1} \bigoplus_{h=1}^{i+1} \mathbb{C}a_h$ in I_{i+1}. By the previous remarks C_{i+1} commutes with the semisimple elements $a_h,\ h = 1, \dots, i + 1$, and with A. Choosing a basis b_j for C_{i+1}, which is $(m - i - 1)$-dimensional, we complete the inductive step.

(iii) This is just a description, in more structural terms, of the properties of the algebra L' stated in (ii). By construction B' is an abelian subalgebra commuting with A thus $A \oplus B'$ is a subalgebra and $L' = (A \oplus B') \ltimes N'$. Furthermore, the adjoint action of $A \oplus B'$ on N' is semisimple. □

The reader should try to understand this construction as follows. First analyze the question: when is a Lie algebra over $\mathbb{C}$ the Lie algebra of an algebraic group?

By the Jordan–Chevalley decomposition this is related to the problem of when is the derivation $\mathrm{ad}(a)_s$ inner for $a \in L$. So our construction just does this: it makes L closed under Jordan decomposition.

Exercise. Prove that the new Lie algebra is the Lie algebra of an algebraic group.

Warning. In order to do this exercise one first needs to understand Ado's Theorem.

1.7 Ado's Theorem

Before we state this theorem let us make a general remark:

Lemma. *Let L be a Lie algebra and D a derivation. Then D extends to a derivation of the universal enveloping algebra U_L.*

Proof.[88] First D induces a derivation on the tensor algebra by setting

$$D(a_1 \otimes a_2 \otimes \cdots \otimes a_m) = \sum_{i=1}^{m} a_1 \otimes a_2 \otimes \cdots \otimes D(a_i) \otimes \cdots \otimes a_m.$$

Given an associative algebra R, a derivation D, and an ideal I, D factors to a derivation of R/I if and only if $D(I) \subset I$ and, to check this, it is enough to do it on a set of generators. In our case:

$$\begin{aligned} D([a, b] - a \otimes b + b \otimes a) &= [D(a), b] + [a, D(b)] - D(a) \otimes b - a \otimes D(b) \\ &\quad + D(b) \otimes a + b \otimes D(a) \\ &= [D(a), b] - D(a) \otimes b + b \otimes D(a) \\ &\quad + [a, D(b)] - a \otimes D(b) + D(b) \otimes a. \end{aligned}$$

□

Ado's Theorem. *A finite-dimensional Lie algebra L can be embedded in matrices.*

The main difficulty in this theorem is the fact that L can have a center; otherwise the adjoint representation of L on L solves the problem (Chapter 4, §4.1.1). Thus it suffices to find a finite-dimensional module M on which the center $Z(L)$ acts faithfully, since then $M \oplus L$ is a faithful module. We will construct one on which the whole nilpotent radical acts faithfully. This is sufficient to solve the problem.

We give the proof in characteristic 0 and for simplicity when the base field is $\mathbb{C}$.[89]

Proof. We split the analysis into three steps.

(1) L is nilpotent, $L^i = 0$. Let U_L be its universal enveloping algebra. By the PBW Theorem we have $L \subset U_L$. Consider U_L as an L-module by multiplication on the left. Let $J := U_L^i$ be the span of all monomials $a_1 \ldots a_k$, $k \geq i$, $a_h \in L$, $\forall h$. J is a two-sided ideal of U_L and $M := U_L/U_L^i$ is clearly finite dimensional. We claim that M is a faithful L-module. Let $d_k := \dim L^k$, $k = 1, \ldots, i-1$, and fix a basis e_i

[88] In part we follow the proof given by Neretin, cf. [Ne].

[89] This restriction can be easily removed.

for L with the property that for each $k < i$ the $e_j,\ j \leq d_k$, are a basis of L^k. For an element $e \in L$ we define its *weight* $w(e)$ as the number h such that $e \in L^h - L^{h+1}$ if $e \neq 0$ and $w(0) := \infty$. Since $[L^k, L^h] \subset L^{k+h}$ we have for any two elements $w([a,b]) \geq w(a) + w(b)$. Given any monomial $M := e_{i_1}e_{i_2}\dots e_{i_s}$ we define its *weight* as the sum of the weights of the factors: $w(M) := \sum_{j=1}^{s} w(e_{i_j})$.

Now take a monomial and rewrite it as a linear combination of monomials $e_1^{h_1}\dots e_j^{h_j}$. Each time that we have to substitute a product $e_h e_k$ with $e_k e_h + [e_h, e_k]$ we obtain in the sum a monomial of degree 1 less, but its weight does not decrease. Thus, when we write an element of J in the PBW basis, we have a linear combination of elements of weight at least i. Since no monomial of degree 1 can have weight $> i-1$, we are done.

(2) Assume L has nilpotent radical N and it is a semidirect product $L = R \ltimes N,\ R = L/N$.

Then we argue as follows. The algebra R induces an algebra of derivations on U_N and clearly the span of monomials of degree $\geq k$, for each k, is stable under these derivations.

U_N/U_N^i is thus an R-module, using the action by derivations and an N-module by left multiplication. If $n \in N$ and $D : N \to N$ is a derivation, $D(na) = D(n)a + nD(a)$. In other words, if L_n is the operator $a \mapsto na$ we have $[D, L_n] = L_{D(n)}$. Thus we have that the previously constructed module U_N/U_N^i is also an $R \ltimes N$ module. Since restricted to N this module is faithful, by the initial remark we have solved the problem.

(3) We want to reduce the general case to the previous case. For this it is enough to apply the last theorem of the previous section, embedding $L = A \ltimes I$ into some $(A \oplus B') \ltimes N'$. □

1.8 Toral Subalgebras

The strategy to unravel the structure of semisimple algebras is similar to the strategy followed by Frobenius to understand characters of groups. In each case one tries to understand how the characteristic polynomial of a *generic element* (in one case of the Lie algebra, in the other of the group algebra) decomposes into factors. In other words, once we have that a generic element is semisimple, we study its eigenvalues.

Definition. A toral subalgebra $\mathfrak{t}$ of L is a subalgebra made only of semisimple elements.

Lemma. *A toral subalgebra* $\mathfrak{t}$ *is abelian.*

Proof. If not, there is an element $x \in \mathfrak{t}$ with a nonzero eigenvalue for an eigenvector $y \in \mathfrak{t}$, or $[x, y] = ay,\ a \neq 0$. On the space spanned by x, y the element y acts as $[y, x] = -ay, [y, y] = 0$. On this space the action of y is given by a nonzero nilpotent matrix. Therefore y is not semisimple as assumed. □

It follows from the lemma that the semisimple operators $\mathrm{ad}(x), x \in \mathfrak{t}$ are simultaneously diagonalizable. When we speak of an eigenvector v for $\mathfrak{t}$ we mean a nonzero vector $v \in L$ such that $[h, v] = \alpha(h)v, \forall h \in \mathfrak{t}$. $\alpha : \mathfrak{t} \to \mathbb{C}$ is then a linear form, called the *eigenvalue*.

In a semisimple Lie algebra L a maximal toral subalgebra $\mathfrak{t}$ is called a *Cartan subalgebra*.[90]

We decompose L into the eigenspaces, or weight spaces, relative to $\mathfrak{t}$. The nonzero eigenvalues define a set $\Phi \subset \mathfrak{t}^* - \{0\}$ of nonzero linear forms on $\mathfrak{t}$ called *roots*. If $\alpha \in \Phi$, $L_\alpha := \{x \in L \mid [h, x] = \alpha(h)x, \forall h \in \mathfrak{t}\}$ is the corresponding *root space*. The nonzero elements of L_α are called *root vectors* relative to α. L_α is also called a *weight space*.

We need a simple remark that we will use in the next proposition.

Consider the following 3-dimensional Lie algebra M with basis a, b, c and multiplication:

$$[a, b] = c, \ [c, a] = [c, b] = 0.$$

The element c is in the center of this Lie algebra which is nilpotent: $[M, [M, M]] = 0$. In any finite-dimensional representation of this Lie algebra, by Lie's theorem, c acts as a nilpotent element.

Proposition. *(1) A Cartan subalgebra $\mathfrak{t}$ of a semisimple Lie algebra L is nonzero.*

(2) The Killing form (a, b), restricted to $\mathfrak{t}$, is nondegenerate. L_α, L_β are orthogonal unless $\alpha + \beta = 0$. $[L_\alpha, L_\beta] \subset L_{\alpha+\beta}$.

(3) $\mathfrak{t}$ equals its 0 weight space. $\mathfrak{t} = L_0 := \{x \in L \mid [h, x] = 0, \ \forall h \in \mathfrak{t}\}$.

(4) We have $L = \mathfrak{t} \bigoplus_{\alpha\in\Phi} L_\alpha$.

From (2) there is a unique element $t_\alpha \in \mathfrak{t}$ with $(h, t_\alpha) = \alpha(h), \forall h \in \mathfrak{t}$. Then,

(5) For each $\alpha \in \Phi$, $a \in L_\alpha$, $b \in L_{-\alpha}$, we have $[a, b] = (a, b)t_\alpha$. The subspace $[L_\alpha, L_{-\alpha}] = \mathbb{C}t_\alpha$ is 1-dimensional.

(6) $(t_\alpha, t_\alpha) = \alpha(t_\alpha) \neq 0$.

(7) There exist elements $e_\alpha \in L_\alpha$, $f_\alpha \in L_{-\alpha}$, $h_\alpha \in \mathfrak{t}$ which satisfy the standard commutation relations of $sl(2, \mathbb{C})$.

Proof. (1) Every element has a Jordan decomposition. If all elements were ad nilpotent, L would be nilpotent by Engel's theorem. Hence there are nontrivial semisimple elements.

(2) First let us decompose L into weight spaces for $\mathfrak{t}$, $L = L_0 \oplus \bigoplus_{\alpha\in\Phi} L_\alpha$.

If $a \in L_\alpha, b \in L_\beta, t \in \mathfrak{t}$ we have $\alpha(t)(a, b) = ([t, a], b) = -(a, [t, b]) = -\beta(t)(a, b)$. If $\alpha + \beta \neq 0$, this implies that $(a, b) = 0$. Since the Killing form is nondegenerate we deduce that the Killing form restricted to L_0 is nondegenerate, and the space L_α is orthogonal to all L_β for $\beta \neq -\alpha$ while L_α and $L_{-\alpha}$ are in perfect duality under the Killing form. This will prove (2) once we show that $L_0 = \mathfrak{t}$. $[L_\alpha, L_\beta] \subset L_{\alpha+\beta}$ is a simple property of derivations. If $a \in L_\alpha, b \in L_\beta, t \in \mathfrak{t}$, we have $[t, [a, b]] = [[t, a], b] + [a, [t, b]] = \alpha(t)[a, b] + \beta(t)[a, b]$.

(3) This requires a more careful proof.

---

90 There is a general notion of Cartan subalgebra for any Lie algebra which we will not use, cf. [J1].

L_0 is a Lie subalgebra (from 2). Let $a \in L_0$ and decompose $a = a_s + a_n$. Since $\mathrm{ad}(a)$ is 0 on $\mathfrak{t}$, we must have that also $a_s, a_n \in L_0$ since they commute with $\mathfrak{t}$. $\mathfrak{t}+\mathbb{C}a_s$ is still toral and by maximality $a_s \in \mathfrak{t}$, so $\mathfrak{t}$ contains all the semisimple parts of the elements of L_0.

Next let us prove that the Killing form restricted to $\mathfrak{t}$ is nondegenerate.

Assume that $a \in \mathfrak{t}$ is in the kernel of the Killing form restricted to $\mathfrak{t}$. Take $c = c_s + c_n \in L_0$. The element $\mathrm{ad}(a)\,\mathrm{ad}(c_n)$ is nilpotent (since the two elements commute), so it has trace 0. The value $(a, c_s) = \mathrm{tr}(\mathrm{ad}(a)\,\mathrm{ad}(c_s)) = 0$ since $c_s \in \mathfrak{t}$, and by assumption $a \in \mathfrak{t}$ is in the kernel of the Killing form restricted to $\mathfrak{t}$. It follows that a is also in the kernel of the Killing form restricted to L_0. This restriction is nondegenerate, so $a = 0$.

Now decompose $L_0 = \mathfrak{t} \oplus \mathfrak{t}^\perp$, where $\mathfrak{t}^\perp$ is the orthogonal complement to $\mathfrak{t}$ with respect to the Killing form. From the same discussion it follows that $\mathfrak{t}^\perp$ is made of ad nilpotent elements. $t^\perp$ is a Lie ideal of L_0, since $a \in \mathfrak{t}, c \in \mathfrak{t}^\perp, b \in L_0 \implies (a, [b, c]) = ([a, b], c) = 0$. Now we claim that $\mathfrak{t}^\perp$ is in the kernel of the Killing form restricted to L_0. In fact since $\mathrm{ad}(\mathfrak{t}^\perp)$ is a Lie algebra made of nilpotent elements, it follows by Engel's theorem that the Killing form on $\mathfrak{t}^\perp$ is 0. Since the Killing form on L_0 is nondegenerate, this implies that $\mathfrak{t}^\perp = 0$ and so $\mathfrak{t} = L_0$.

(4) follows from (2), (3) and the definitions.

Since the Killing form on $\mathfrak{t}$ is nondegenerate, we can use it to identify $\mathfrak{t}$ with its dual $\mathfrak{t}^*$. In particular, for each root α we have an element $t_\alpha \in \mathfrak{t}$ with $(h, t_\alpha) = \alpha(h), h \in \mathfrak{t}$.

(5) By (2) $[L_\alpha, L_{-\alpha}]$ is contained in $\mathfrak{t}$ and, if $h \in \mathfrak{t}, a \in L_\alpha, b \in L_{-\alpha}$ we have $(h, [a, b]) = ([h, a], b) = \alpha(h)(a, b) = (h, (a, b)t_\alpha)$. This means that $[a, b] = (a, b)t_\alpha$ lies in the 1-dimensional space generated by t_α.

(6) Since $L_\alpha, L_{-\alpha}$ are paired by the Killing form, we can find $a \in L_\alpha, b \in L_{-\alpha}$ with $(a, b) = 1$, and hence $[a, b] = t_\alpha$. We have $[t_\alpha, a] = \alpha(t_\alpha)a, [t_\alpha, b] = -\alpha(t_\alpha)b$. If $\alpha(t_\alpha) = 0$ we are in the setting of the remark preceding the proposition: a, b, t_α span a solvable Lie algebra, and in any representation t_α is nilpotent. Since $t_\alpha \in \mathfrak{t}$, it is semisimple in every representation, in particular in the adjoint representation. We deduce that $\mathrm{ad}(t_\alpha) = 0$, which is a contradiction.

(7) We are claiming that one can choose nonzero elements $e_\alpha \in L_\alpha,\ f_\alpha \in L_{-\alpha}$ such that, setting $h_\alpha := [e_\alpha, f_\alpha]$, we have the canonical commutation relations of $sl(2, \mathbb{C})$:

$$h_\alpha := [e_\alpha, f_\alpha],\ [h_\alpha, e_\alpha] = 2e_\alpha,\ [h_\alpha, f_\alpha] = -2f_\alpha.$$

In fact let $e_\alpha \in L_\alpha, f_\alpha \in L_{-\alpha}$ with $(e_\alpha, f_\alpha) = 2/(t_\alpha, t_\alpha)$ and let $h_\alpha := [e_\alpha, f_\alpha] = 2/(t_\alpha, t_\alpha)t_\alpha$. We have $[h_\alpha, e_\alpha] = \alpha(2/(t_\alpha, t_\alpha)t_\alpha)e_\alpha = 2e_\alpha$. The computation for f_α is similar. □

One usually identifies $\mathfrak{t}$ with its dual $\mathfrak{t}^*$ using the Killing form. In this identification t_α corresponds to α. One can transport the Killing form to $\mathfrak{t}^*$. In particular we have for two roots α, β that

$$(1.8.1) \qquad (\alpha, \beta) = (t_\alpha, t_\beta) = \alpha(t_\beta) = \beta(t_\alpha) = \frac{(\alpha, \alpha)}{2}\beta(h_\alpha).$$

1.9 Root Spaces

At this point we can make use of the powerful results we have about the representation theory of $sl(2, \mathbb{C})$.

Lemma 1. *Given a root α and the algebra $sl_\alpha(2, \mathbb{C})$ spanned by $e_\alpha, f_\alpha, h_\alpha$, we can decompose L as a direct sum of irreducible representations of $sl_\alpha(2, \mathbb{C})$ in which the highest weight vectors are weight vectors also for $\mathfrak{t}$.*

Proof. The space of highest weight vectors is $U := \{a \in L \,|\, [e_\alpha, a] = 0\}$. If $h \in \mathfrak{t}$ and $[e_\alpha, a] = 0$, we have $[e_\alpha, [h, a]] = [[e_\alpha, h], a] = -\alpha(h)[e_\alpha, a] = 0$, so U is stable under $\mathfrak{t}$. Since $\mathfrak{t}$ is diagonalizable we have a basis of weight vectors. □

We have two possible types of highest weight vectors, either root vectors, or elements $h \in \mathfrak{t}$ with $\alpha(h) = 0$. The latter are the trivial representations of $sl_\alpha(2, \mathbb{C})$. For the others, if e_β is a highest weight vector and a root vector relative to the root β, we have that it generates under $sl_\alpha(2, \mathbb{C})$ an irreducible representation of dimension $k + 1$ where k is the weight of e_β under h_α, i.e., $\beta(h_\alpha)$. The elements $\mathrm{ad}(f_\alpha)^i(e_\beta)$, $i = 0, \ldots, k$, are all nonzero root vectors of weights $\beta - i\alpha$. These roots by definition form an *α-string*.

One of these irreducible representations is the Lie algebra $sl_\alpha(2, \mathbb{C})$ itself.

We can next use the fact that all these representations are also representations of the group $SL(2, \mathbb{C})$. In particular, let us see how the element $s_\alpha := \begin{vmatrix} 0 & 1 \\ -1 & 0 \end{vmatrix}$ acts. If $h \in \mathfrak{t}$ is in the kernel of α, we have seen that h is killed by $sl_\alpha(2, \mathbb{C})$ and so it is fixed by $SL_\alpha(2, \mathbb{C})$. Instead $s_\alpha(h_\alpha) = -h_\alpha$. We thus see that s_α induces on $\mathfrak{t}$ the orthogonal reflection relative to the *root hyperplane* $H_\alpha := \{x \in \mathfrak{t} \,|\, \alpha(x) = 0\}$.

Lemma 2. *The group $SL_\alpha(2, \mathbb{C})$ acts by automorphisms of the Lie algebra. Given two roots α, β we have that $s_\alpha(\beta) = \beta - \frac{2(\beta,\alpha)}{(\alpha,\alpha)}\alpha$ is a root, $\frac{2(\beta,\alpha)}{(\alpha,\alpha)}$ is an integer, and $s_\alpha(L_\beta) = L_{s_\alpha(\beta)}$.*

Proof. Since $sl_\alpha(2, \mathbb{C})$ acts by derivations, its exponentials, which generate $SL_\alpha(2, \mathbb{C})$, act by automorphisms. If $a \in L_\beta$, $h \in \mathfrak{t}$, we have $[h, s_\alpha(a)] = [s_\alpha^{-1}(h), a] = \beta(s_\alpha^{-1}(h))a = s_\alpha(\beta)(h)a$. The roots come in α strings $\beta - i\alpha$ with $\beta(h_\alpha)$ a positive integer. We have $2(\beta - i\alpha, \alpha)/(\alpha, \alpha) = 2(\beta, \alpha)/(\alpha, \alpha) - 2i = 2\beta(t_\alpha)/(\alpha, \alpha) - 2i = \beta(h_\alpha) - 2i$. □

Proposition. *(1) For every root α we have* $\dim L_\alpha = 1$.

(2) If $\alpha \in \Phi$ we have $c\alpha \in \Phi$ if and only if $c = \pm 1$.

(3) If $\alpha, \beta, \alpha + \beta$ are roots, $[L_\alpha, L_\beta] = L_{\alpha+\beta}$.

Proof. We want to take advantage of the fact that in each irreducible representation of $sl_\alpha(2, \mathbb{C})$ there is a unique weight vector (up to scalars) of weight either 0 or 1 (Remark 1.1).

Let us first take the sum

$$M_\alpha = \mathfrak{t} + \bigoplus_{\beta \in \Phi,\ \beta = c\alpha} L_\beta$$

of all the irreducible representations in which the weights are multiples of α. We claim that this sum coincides with $\mathfrak{t} + sl_\alpha(2, \mathbb{C})$. In M_α the 0 weights for h_α are also 0 weights for $\mathfrak{t}$ by definition. By (3) of Proposition 1.8 the zero weight space of $\mathfrak{t}$ is $\mathfrak{t}$. Hence in M_α there are no other even representations apart from $\mathfrak{t} + sl_\alpha(2, \mathbb{C})$. This already implies that $\dim L_\alpha = 1$ and that no even multiple of a root is a root. If there were a weight vector of weight 1 for h_α, this would correspond to the weight $\alpha/2$ on $\mathfrak{t}$. This is not a root; otherwise, we contradict the previous statement since $\alpha = 2(\alpha/2)$ is a root. This proves finally that $M_\alpha = \mathfrak{t} + sl_\alpha(2, \mathbb{C})$ which implies (1) and (2).

For (3), given a root β let us consider all possible roots of type $\beta + i\alpha$ with i any integer. The sum P_β of the corresponding root spaces is again a representation of $sl_\alpha(2, \mathbb{C})$. The weight under h_α of a root vector in $L_{\beta+i\alpha}$ is $\beta(h_\alpha) + 2i$. Since these numbers are all distinct and all of the same parity, from the structure of representations of $sl_\alpha(2, \mathbb{C})$, we have that P_β is irreducible. In an irreducible representation if u is a weight vector for h_α of weight λ and $\lambda + \alpha$ is a weight, we have that $e_\alpha u$ is a nonzero weight vector of weight $\lambda + \alpha$. This proves that if $\alpha + \beta$ is a root, $[e_\alpha, L_\beta] = L_{\alpha+\beta}$. □

The integer $\frac{2(\beta,\alpha)}{(\alpha,\alpha)}$ appears over and over in the theory; it deserves a name and a symbol. It is called a *Cartan integer* and denoted by $\langle \beta \mid \alpha \rangle := \frac{2(\beta,\alpha)}{(\alpha,\alpha)}$. Formula 1.8.1 becomes

$$\beta(h_\alpha) = \frac{2(\alpha, \beta)}{(\alpha, \alpha)} = \langle \beta \mid \alpha \rangle. \tag{1.9.1}$$

Warning. The symbol $\langle \beta \mid \alpha \rangle$ is linear in β but *not* in α.

The next fact is that:

Theorem. *(1) The rational subspace* $V := \sum_{\alpha\in\Phi} \mathbb{Q}\alpha$ *is such that* $\mathfrak{t}^* = V \otimes_{\mathbb{Q}} \mathbb{C}$.

(2) For the Killing form (h, k), $h, k \in \mathfrak{t}$, *we have*

$$(h, k) = \operatorname{tr}(\operatorname{ad}(h)\operatorname{ad}(k)) = \sum_{\alpha\in\Phi} \alpha(h)\alpha(k) = 2\sum_{\alpha\in\Phi^+} \alpha(h)\alpha(k). \tag{1.9.2}$$

(3) The dual of the Killing form, restricted to V, *is rational and positive definite.*

Proof. (1) We have already seen that the numbers $2(\beta, \alpha)/(\alpha, \alpha)$ are integers. First, we have that the roots α span $\mathfrak{t}^*$; otherwise there would be a nonzero element $h \in \mathfrak{t}$ in the center of L. Let α_i, $i = 1, \dots, n$, be a basis of $\mathfrak{t}^*$ extracted from the roots. If α is any other root, write $\alpha = \sum_i a_i\alpha_i$. It is enough to show that the coefficients a_i are rationals so that the α_i are a $\mathbb{Q}$-basis for V. In order to compute the coefficients a_i we may take the scalar products and get $(\alpha, \alpha_j) = \sum_i a_i(\alpha_i, \alpha_j)$. We can then solve this using Cramer's rule. To see that the solution is rational we can multiply it by $2/(\alpha_j, \alpha_j)$ and rewrite the system with integer coefficients $\langle \alpha \mid \alpha_j \rangle = \sum_i a_i \langle \alpha_i \mid \alpha_j \rangle$.

(2) If h, k are in the Cartan subalgebra, the linear operators $\operatorname{ad}(h)$, $\operatorname{ad}(k)$ commute and are diagonal with simultaneous eigenvalues $\alpha(h), \alpha(k)$, $\alpha \in \Phi$. Therefore the formula follows.

(3) For a root β apply 1.9.2 to t_β and get

$$(\beta, \beta) = 2 \sum_{\alpha\in\Phi^+} (\alpha, \beta)^2.$$

This implies $2/(\beta, \beta) = \sum_{\alpha\in\Phi^+} \langle \alpha \mid \beta \rangle^2 \in \mathbb{N}$. From 1.8.1 and the fact that $\langle \beta \mid \alpha \rangle$ is an integer, it follows that $\beta(t_\alpha)$ is rational. Thus, if h is in the rational space W spanned by the t_α, the numbers $\alpha(h)$ are rational, so on this space the Killing form is rational and positive definite.

By duality W is identified with V since t_α corresponds to α. $\square$

Remark. As we will see, in all of the important formulas the Killing form appears only through the Cartan integers which are invariant under changes of scale.

The way in which Φ sits in the Euclidean space $V_{\mathbb{R}} := V \otimes_{\mathbb{Q}} \mathbb{R}$ can be axiomatized giving rise to the abstract notion of *root system*. This is the topic of the next section.

One has to understand that the root system is independent of the chosen toral subalgebra. One basic theorem states that all Cartan subalgebras are conjugate under the group of inner automorphisms. Thus the dimension of each of them is a well-defined number called the *rank* of L. The root systems are also isomorphic (see §2.8).

2 Root Systems

2.1 Axioms for Root Systems

Root systems can be viewed in several ways. In our setting they give an axiomatized approach to the properties of the roots of a semisimple Lie algebras, but one can also think more geometrically of their connection to *reflection groups*.

In a Euclidean space E the reflection with respect to a hyperplane H is the orthogonal transformation which fixes H and sends each vector v, orthogonal to H, to $-v$. It is explicitly computed by the formula (cf. Chapter 5, §3.9):

$$r_v : x \mapsto x - \frac{2(x, v)}{(v, v)} v. \tag{2.1.1}$$

Lemma. *If X is an orthogonal transformation and v a vector,*

$$X r_v X^{-1} = r_{X(v)}. \tag{2.1.2}$$

A finite reflection group is a finite subgroup of the orthogonal group of E, generated by reflections.[91] Among finite reflection groups a special role is played by *crystallographic groups*, the groups that preserve some lattice of integral points.

Roots are a way to construct the most important crystallographic groups, the *Weyl groups*.

---

[91] There are of course many other reflection groups, infinite and defined on non-Euclidean spaces, which produce rather interesting geometry but play no role in this book.

Definition 1. Given a Euclidean space E (with positive scalar product (u, v)) and a finite set Φ of nonzero vectors in E, we say that Φ is a *reduced root system* if:

(1) the elements of Φ span E;
(2) $\forall \alpha \in \Phi, c \in \mathbb{R}$, we have $c\alpha \in \Phi$ if and only if $c = \pm 1$;
(3) the numbers

$$\langle \alpha \mid \beta \rangle := \frac{2(\alpha, \beta)}{(\beta, \beta)}$$

are integers (called Cartan integers);
(4) For every $\alpha \in \Phi$ consider the reflection $r_\alpha : x \to x - \frac{2(x,\alpha)}{(\alpha,\alpha)}\alpha$. Then $r_\alpha(\Phi) = \Phi$.

The dimension of E is also called the *rank* of the root system.

The theory developed in the previous section implies that the roots $\Phi \in \mathfrak{t}^*$ arising from a semisimple Lie algebra form a root system in the real space which they span, considered as a Euclidean space under the restriction of the dual of the Killing form.

Axiom (4) implies that the subgroup generated by the orthogonal reflections r_α is a finite group (identified with the group of permutations that it induces on Φ). This group is usually denoted by W and called the *Weyl group.* It is a basic group of symmetries similar to the symmetric group with its canonical permutation representation.

When the root system arises from a Lie algebra, it is in fact possible to describe W also in terms of the associated algebraic or compact group as N_T/T where N_T is the normalizer of the associated maximal torus (cf. 6.7 and 7.3).

Definition 2. A root system Φ is called reducible if we can divide it as $\Phi = \Phi^1 \cup \Phi^2$, into mutually orthogonal subsets; otherwise, it is irreducible.

An isomorphism between two root systems Φ_1, Φ_2 is a 1-1 correspondence between the two sets of roots which preserves the Cartan integers.

Exercise. It can be easily verified that any isomorphism between two irreducible root systems is induced by the composition of an isometry of the ambient Euclidean spaces and a homothety (i.e., a multiplication by a nonzero scalar).

First examples In Euclidean space $\mathbb{R}^n$ the standard basis is denoted by $\{e_1, \dots, e_n\}$.

Type A_n The root system is in the subspace E of $\mathbb{R}^{n+1}$ where the sum of the coordinates is 0 and the roots are the vectors $e_i - e_j$, $i \neq j$. The Weyl group is the symmetric group S_{n+1}, which permutes the basis elements. Notice that the Weyl group permutes transitively the roots. The roots are the integral vectors of V with Euclidean norm 2.

Type B_n Euclidean space $\mathbb{R}^n$, roots:

(2.1.3) $$\pm e_i, \quad e_i - e_j,\ e_i + e_j,\ -e_i - e_j,\ i \neq j \leq n.$$

The Weyl group is the semidirect product $S_n \ltimes Z/(2)^n$ of the symmetric group S_n, which permutes the coordinates, with the *sign group* $\mathbb{Z}/(2)^n := (\pm 1, \pm 1, \dots, \pm 1)$, which changes the signs of the coordinates.

Notice that in this case we have two types of roots, with Euclidean norm 2 or 1. The Weyl group has two orbits and permutes transitively the two types of roots.

Type C_n Euclidean space $\mathbb{R}^n$, roots:

$$e_i - e_j,\ e_i + e_j,\ -e_i - e_j,\ i \neq j, \quad \pm 2e_i,\ i = 1, \dots, n. \tag{2.1.4}$$

The Weyl group is the same as for B_n. Again we have two types of roots, with Euclidean norm 2 or 4, and the Weyl group permutes transitively the two types of roots.

From the previous analysis it is in fact clear that there is a duality between roots of type B_n and of type C_n, obtained formally by passing from roots e to *coroots*, $e^\vee := \frac{2e}{(e,e)}$.

Type D_n Euclidean space $\mathbb{R}^n$, roots:

$$e_i - e_j,\ e_i + e_j,\ -e_i - e_j,\ i \neq j \leq n. \tag{2.1.5}$$

The Weyl group is a semidirect product $S_n \ltimes S$. S_n is the symmetric group permuting the coordinates. S is the subgroup (of index 2), of the *sign group* $\mathbb{Z}/(2)^n := (\pm 1, \pm 1, \dots, \pm 1)$ which changes the signs of the coordinates, formed by only even number of sign changes.[92]

The Weyl group permutes transitively the roots which all have Euclidean norm 2.

Exercise. Verify the previous statements.

2.2 Regular Vectors

When dealing with root systems one should think geometrically. The hyperplanes $H_\alpha := \{v \in E \mid (v, \alpha) = 0\}$ orthogonal to the roots are the reflection hyperplanes for the reflections s_α called *root hyperplanes*.

Definition 1. The complement of the union of the root hyperplanes H_α is called the set of *regular vectors*, denoted E^{reg}. It consists of several open connected components called *Weyl chambers*.

Examples. In type A_n the regular vectors are the ones with distinct coordinates $x_i \neq x_j$. For B_n, C_n we have the further conditions $x_i \neq \pm x_j$, $x_i \neq 0$, while for D_n we have only the condition $x_i \neq \pm x_j$.

It is convenient to fix one chamber once and for all and call it the *fundamental chamber* C. In the examples we can choose:

[92] Contrary to type B_n, not all sign changes of the coordinates are possible.

$$A_n: \qquad C := \{(x_1, x_2, \dots, x_{n+1}) \mid x_1 > x_2 \dots > x_{n+1}, \sum_i x_i = 0\},$$

$$B_n, C_n: \qquad C := \{(x_1, x_2, \dots, x_n) \mid x_1 > x_2 \dots > x_n > 0\},$$

$$D_n: \qquad C := \{(x_1, x_2, \dots, x_n) \mid x_1 > x_2 \dots > x_n,\ x_{n-1} + x_n > 0\}.$$

For every regular vector v we can decompose Φ into two parts

$$\Phi_v^+ := \{\alpha \in \Phi | (v, \alpha) > 0\}, \ \Phi_v^- := \{\alpha \in \Phi | (v, \alpha) < 0\}.$$

Weyl chambers are clearly convex cones.[93] Clearly this decomposition depends only on the chamber in which v lies. When we have chosen a fundamental chamber we drop the symbol v and write simply $\Phi^+, \ \Phi^-$.

From the definition of regular vector it follows that $\Phi^- = -\Phi^+, \ \Phi = \Phi^+ \cup \Phi^-$. One calls $\Phi^+, \ \Phi^-$ the *positive* and *negative roots*.

Notation We write $\alpha \succ 0, \alpha \prec 0$, to mean that α is a positive, resp. a negative root.

To understand root systems one should first of all understand the 2-dimensional case.

From axiom (3) and the Schwarz inequality one has

$$\langle \alpha \mid \beta \rangle \langle \beta \mid \alpha \rangle = \frac{4(\alpha, \beta)^2}{(\beta, \beta)(\alpha, \alpha)} = 4\cos(\theta)^2 \le 4, \tag{2.2.1}$$

where θ is the angle between the two roots. In 2.2.1 the equality is possible if and only if the two roots are proportional or $\alpha = \pm\beta$ by axiom 2. It follows that the only possible convex angles between two roots are $0, \ \pi/6, \ \pi/4, \ \pi/3, \ \pi/2, \ 2\pi/3, \ 3\pi/4, \ 5\pi/6, \pi$.

Exercise. At this point the reader should be able to prove, by simple arguments of Euclidean geometry, that the only possibilities are those shown in the illustration on p. 318, of which one is reducible.

From the picture, it follows that:

Lemma. *If two roots α, β form an obtuse angle, i.e., $(\alpha, \beta) < 0$, then $\alpha + \beta$ is a root.*

If two roots α, β form an acute angle, i.e., $(\alpha, \beta) > 0$, then $\alpha - \beta$ is a root.

The theory of roots is built on the two notions of decomposable and simple roots:

Definition 2. We say that a root $\alpha \in \Phi^+$ is *decomposable* if $\alpha = \beta + \gamma, \ \beta, \gamma \in \Phi^+$.

An indecomposable positive root is also called a *simple* root.

We denote by Δ the set of simple roots in Φ^+. The basic construction is given by the following:

[93] A cone is a set S of Euclidean space with the property that if $x \in S$ and $r > 0$ is a positive real number, then $rx \in S$.

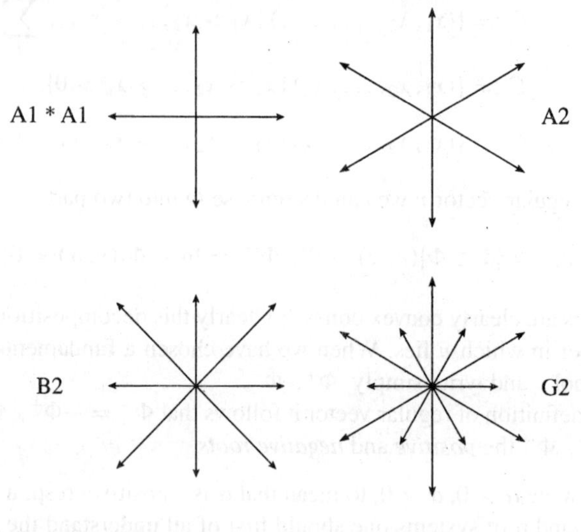

Theorem. *(i) Every element of Φ^+ is a linear combination of elements of Δ, with nonnegative integer coefficients.*

(ii) If $\alpha, \beta \in \Delta$ we have $(\alpha\,,\,\beta) \leq 0$.

*(iii) The set Δ of indecomposable roots in Φ^+ is a basis of E. $\Delta = \{\alpha_1, \dots, \alpha_n\}$ is called the set, or basis, of **simple roots** (associated to Φ^+).*

Proof. (i) We order the positive roots so that if $(v, \alpha) > (v, \beta)$ we have $\alpha > \beta$. Since $(v, \beta + \gamma) = (v, \beta) + (v, \gamma)$ is a sum of two positive numbers, by induction every positive root is a linear combination of elements of Δ with nonnegative integer coefficients.

(ii) If $\alpha, \beta \in \Delta$, we must show that $\gamma := \alpha - \beta$ is not a root, hence $(\alpha\,,\,\beta) \leq 0$ (by the previous lemma). Suppose by contradiction that γ is a positive root; then $\alpha = \gamma + \beta$ is decomposable, contrary to the definition. If γ is negative we argue in a similar way, reversing the role of the two roots.

(iii)) Since Φ^+ spans E by assumption, it remains only to see that the elements of Δ are linearly independent. Suppose not. Then, up to reordering, we have a relation $\sum_{i=1}^k a_i\alpha_i = \sum_{j=k+1}^m b_j\alpha_j$ with the $a_i, b_j \geq 0$. We deduce that

$$0 \leq \left(\sum_{i=1}^{k} a_i\alpha_i, \sum_{i=1}^{k} a_i\alpha_i\right) = \left(\sum_{i=1}^{k} a_i\alpha_i, \sum_{j=k+1}^{m} b_j\alpha_j\right) = \sum_{i,j} a_i b_j (\alpha_i, \alpha_j) \leq 0,$$

which implies $\sum_{i=1}^k a_i\alpha_i = 0$. Now $0 = (v, \sum_{i=1}^k a_i\alpha_i) = \sum_{i=1}^k a_i(v, \alpha_i)$. Since all the $(v, \alpha_i) > 0$, this finally implies all $a_i = 0$, for all i. In a similar way all $b_j = 0$. □

Now assume we are given a root system Φ of rank n. Choose a set of positive roots Φ^+ and the associated simple roots Δ.

The corresponding *fundamental chamber* is

$$C := \{x \in E \mid (x, \alpha_i) > 0, \ \forall \alpha_i \in \Delta\}.$$

In fact we could reverse the procedure, i.e., we could define a *basis* Δ as a subset of the roots with the property that each root is written in a unique way as a linear combination of elements of Δ with the coefficients either all nonnegative or all nonpositive integers.

Exercise. It is easy to realize that bases and chambers are in 1-1 correspondence.

Notice that the root hyperplanes H_{α_i} intersect the closure of C in a domain containing the open (in H_{α_i}) set $U_i := \{x \in H_{\alpha_i} \mid (x, \alpha_j) > 0, \ \forall j \neq i\}$.

The hyperplanes H_{α_i} are called the *walls* of the chamber. We write H_i instead of H_{α_i}.

2.3 Reduced Expressions

We set $s_i := s_{\alpha_i}$ and call it a *simple reflection.*

Lemma. *s_i permutes the positive roots $\Phi^+ - \{\alpha_i\}$.*

Proof. Let $\alpha \in \Phi^+$, so that $s_i(\alpha) = \alpha - \langle \alpha | \alpha_i \rangle \alpha_i$. Let $\alpha = \sum_j n_j \alpha_j$ with n_j positive coefficients. Passing to $s_i(\alpha)$, only the coefficient of α_i is modified. Hence if $\alpha \in \Phi^+ - \{\alpha_i\}$, then $s_i(\alpha)$, as a linear combination of the α_i, has at least one positive coefficient. A root can have only positive or only negative coefficients, so $s_i(\alpha)$ is positive. □

Consider an element $w \in W$, written as a product $w = s_{i_1} s_{i_2} \dots s_{i_k}$ of simple reflections.

Definition. We say that $s_{i_1} s_{i_2} \dots s_{i_k}$ is a *reduced* expression of w, if w cannot be written as a product of less than k simple reflections. In this case we write $k = \ell(w)$ and say that w has *length k*.

Remark. Assume that $s_{i_1} s_{i_2} \dots s_{i_k}$ is a reduced expression for w.

(1) $s_{i_k} s_{i_{k-1}} \dots s_{i_1}$ is a reduced expression for w^{-1}.
(2) For any $1 \leq a < b \leq k$, $s_{i_a} s_{i_{a+1}} \dots s_{i_b}$ is also a reduced expression.

Proposition 1 (Exchange property). *For $\alpha_i \in \Delta$, suppose that $s_{i_1} s_{i_2} \dots s_{i_k}(\alpha_i)$ is a negative root; then for some $h \leq k$ we have*

$$s_{i_1} s_{i_2} \dots s_{i_k} s_i = s_{i_1} s_{i_2} \dots s_{i_{h-1}} s_{i_{h+1}} s_{i_{h+2}} \dots s_{i_k}. \tag{2.3.1}$$

Proof. Consider the sequence of roots $\beta_h := s_{i_h} s_{i_{h+1}} s_{i_{h+2}} \dots s_{i_k}(\alpha_i) = s_{i_h}(\beta_{h+1})$. Since β_1 is negative and $\beta_{k+1} = \alpha_i$ is positive, there is a maximum h so that β_{h+1} is positive and β_h is negative. By the previous lemma it must be $\beta_{h+1} = \alpha_{i_h}$ and, by 2.1.2,

$$s_{i_h} = s_{i_{h+1}} s_{i_{h+2}} \dots s_{i_k} s_i (s_{i_{h+1}} s_{i_{h+2}} \dots s_{i_k})^{-1} \quad \text{or}$$

$$s_{i_h} s_{i_{h+1}} s_{i_{h+2}} \dots s_{i_k} = s_{i_{h+1}} s_{i_{h+2}} \dots s_{i_k} s_i,$$

which is equivalent to 2.3.1. □

Notice that the expression on the left-hand side of 2.3.1 is a product of $k+1$ reflections, so it is *not* reduced, since the one on the right-hand side consists of $k-1$ reflections.

Corollary. *If $s_{i_1}s_{i_2}\dots s_{i_k}$ is a reduced expression, then $s_{i_1}s_{i_2}\dots s_{i_k}(\alpha_{i_k}) \prec 0$.*

Proof. $s_{i_1}s_{i_2}\dots s_{i_k}(\alpha_{i_k}) = -s_{i_1}s_{i_2}\dots s_{i_{k-1}}(\alpha_{i_k})$. Since $s_{i_1}s_{i_2}\dots s_{i_k}$ is reduced, the previous proposition implies that $s_{i_1}s_{i_2}\dots s_{i_{k-1}}(\alpha_{i_k}) \succ 0$. □

Proposition 2. *If $w = s_{i_1}s_{i_2}\dots s_{i_k}$ is a reduced expression, then k is the number of positive roots α such that $w(\alpha) \prec 0$.*

The set of positive roots sent into negative roots by $w = s_{i_1}s_{i_2}\dots s_{i_k}$ is the set of elements $\beta_h := s_{i_k}\dots s_{i_{h+1}}(\alpha_{i_h})$, $h = 1,\dots,k$.

Proof. Since $s_{i_k}\dots s_{i_{h+1}}s_{i_h}$ is reduced, the previous proposition implies that β_h is positive. Since $s_{i_1}s_{i_2}\dots s_{i_h}$ is reduced, $w(\beta_h) = s_{i_1}s_{i_2}\dots s_{i_h}(\alpha_{i_h}) \prec 0$. Conversely, if $s_{i_1}s_{i_2}\dots s_{i_k}(\beta) \prec 0$, arguing as in the previous proposition, for some h we must have $s_{i_{h+1}}s_{i_{h+2}}\dots s_{i_k}(\beta) = \alpha_{i_h}$, i.e., $\beta = \beta_h$. □

Exercise. Let $w = s_{i_1}s_{i_2}\dots s_{i_k}$, $A_w := \{\beta \in \Phi^+ \mid w(\beta) \prec 0\}$. If $\alpha, \beta \in A_w$ and $\alpha+\beta$ is a root, then $\alpha+\beta \in A_w$. In the ordered list of the elements β_i, $\alpha+\beta$ always occurs in-between α and β. This is called a *convex ordering*.

Conversely, any convex ordering of A_w is obtained from a reduced expression of w. This is a possible combinatorial device to compute the reduced expressions of an element w.

A set $L \subset \Phi^+$ is *closed* if $\alpha, \beta \in L$ and $\alpha+\beta$ is a root implies $\alpha+\beta \in L$. Prove that $L = A_w$ for some w if and only if L, $\Phi^+ - L$ are closed.

Hint: Find a simple root $\alpha \in L$. Next consider $s_\alpha(L - \{\alpha\})$.

Remark. A permutation σ sends a positive root $\alpha_i - \alpha_j$ to a negative root if and only if $\sigma(i) > \sigma(j)$. We thus say that in the pair $i < j$, the permutation has an *inversion*. Thus, the length of a permutation counts the number of inversions. We say instead that σ has a *descent* in i if $\sigma(i) > \sigma(i+1)$. Descents clearly correspond to simple roots in A_σ.[94]

One of our goals is to prove that W is generated by the simple reflections s_i. Let us temporarily define W' to be the subgroup of W generated by the s_i. We first prove some basic statements for W' and only afterwards we prove that $W = W'$.

Lemma. *(1) W' acts in a transitive way on the chambers.*

(2) W' acts in a transitive way on the bases of simple roots.

(3) If $\alpha \in \Phi$ is a root, there is an element $w \in W'$ with $w(\alpha) \in \Delta$.

[94] There is an extensive literature on counting functions on permutations. In the literature of combinatorics these functions are usually referred to as *statistics*.

Proof. Since bases and chambers are in 1-1 correspondence, (1) and (2) are equivalent.

(1) Let $v \in C$ be in the fundamental chamber and x a regular vector in some chamber C'. Take, in the W' orbit of x, a vector $y = wx$ with the scalar product (y, v) as big as possible (so the angle between y, v is as small as possible). We claim that $y \in C$, this proves that $w(C') = C$ as required.

If $y \notin C$, for some i we have $(y, \alpha_i) < 0$. This gives a contradiction, since $(v, s_i w) = (v, y - 2(y, \alpha_i)/(\alpha_i, \alpha_i)\alpha_i) = (v, y) - 2(y, \alpha_i)/(\alpha_i, \alpha_i)(v, \alpha_i) > (v, y)$.

(3) Take a root α. In the root hyperplane H_α choose a *relatively regular* vector u, i.e., a vector u with the property that $(u, \alpha) = 0,\ (u, \beta) \neq 0,\ \forall \beta \in \Phi,\ \beta \neq \pm\alpha$. Next take a regular vector $y \in E$ with $(\alpha, y) > 0$. If $\epsilon > 0$ is sufficiently small, since the regular vectors are an open cone, $u + \epsilon y$ is regular. $(\beta, u) + \epsilon(\beta, y)$ can be made as close to (β, u) as we wish. At this point we see that α is a positive root for the regular vector $u + \epsilon y$ and, provided we take ϵ sufficiently small, the scalar product $(\alpha, u + \epsilon y)$ is strictly less than the scalar product of $u + \epsilon y$ with any other positive root. This implies readily that α is indecomposable and thus a simple root for the basis Δ' determined by the vector $u + \epsilon y$. Since we have already proved that W' acts transitively on bases we can find $w \in W'$ with $w(\Delta') = \Delta$, hence $w(\alpha) \in \Delta$. □

The next theorem collects most of the basic geometry:

Theorem. *(1) The group W is generated by the simple reflections s_i.*

(2) W acts in a simply transitive way on the chambers.

(3) W acts in a simply transitive way on the bases of simple roots.

(4) Every vector in E is in the W-orbit of a unique vector in the closure $\overline{C}$ of the fundamental chamber.

(5) The stabilizer of a vector x in $\overline{C}$ is generated by the simple reflections s_i with respect to the walls which contain x.

Proof. (1) We have to prove that $W = W'$. Since W is generated by the reflections s_α for the roots α, it is sufficient to see that $s_\alpha \in W'$ for every root α. From the previous lemma, there is a $w \in W'$ and a simple root α_i with $\alpha = w(\alpha_i)$. From 2.1.2 we have $s_\alpha = w s_i w^{-1} \in W' \implies W = W'$.

(2), (3) Suppose now that an element w fixes the fundamental chamber C and its associated basis. Write $w = s_{i_1} s_{i_2} \dots s_{i_k}$ as a reduced expression. If $w \neq 1$, we have $k \geq 1$ and $w(\alpha_{i_k}) \prec 0$ by the previous corollary, a contradiction. So the action is simply transitive.

(4), (5) Since every vector is in the closure of some chamber, from (2) it follows that every vector in Euclidean space is in the orbit of a vector in $\overline{C}$. We prove (4), (5) at the same time. Take $x, y \in \overline{C}$ and $w \in W$ with $wx = y$. We want to show that $x = y$ and w is a product of the simple reflections s_i with respect to the walls H_i containing x. Write w as a reduced product $s_{i_1} s_{i_2} \dots s_{i_k}$ and work by induction on k. If $k > 0$, by the previous propositions w^{-1} maps some simple root α_i to a negative root, hence $0 \leq (y, \alpha_i) = (wx, \alpha_i) = (x, w^{-1}\alpha_i) \leq 0$ implies that x, wx are in the wall H_i. By the exchange property 2.3.1, $\ell(ws_i) < \ell(w)$ and $x = s_i x,\ y = w s_i x$. We can now apply induction since ws_i has shorter length.

We deduce that $x = y$ and w is generated by the given reflections. □

Remark. If C is the fundamental chamber, then also $-C$ is a chamber with corresponding basis $-\Delta$. We thus have a unique element $w_0 \in W$ with the property that $w_0(\Phi^+) = \Phi^-$.

w_0 is the *longest element of* W, its length is $N = |\Phi^+|$, the number of positive roots.

Let us choose a reduced expression $w_0 = s_{i_1} s_{i_2} \dots s_{i_N}$.

Proposition. *We obtain the list of positive roots in the form*

(2.3.2) $$\beta_h := s_{i_N} \dots s_{i_{h+1}}(\alpha_{i_h}), \ h = 1, \dots, N.$$

Proof. Apply Proposition 2 of §2.3 to w_0. □

Example. In S_n the longest element is the permutation reversing the order $w_0(i) := n + 1 - i$.

The previous theorem implies that every element $w \in W$ has a reduced expression as a product of simple reflections, hence a length $l(w)$.

Exercise. Given $w \in W$ and a simple reflection s_i, prove that $l(ws_i) = l(w) \pm 1$. Moreover $l(ws_i) = l(w) + 1$ if and only if $w(\alpha_i) \succ 0$.

Given any vector $v \in E$, its stabilizer in W is generated by the reflections s_α for the roots α which satisfy $\alpha(v) = 0$.

2.4 Weights

First, let us discuss the *coroots* $\Phi^\vee$ where $e^\vee := \frac{2e}{(e,e)}$.

Proposition 1. *(1) $\Phi^\vee$ is a root system in E having the same regular vectors as Φ.*

(2) If Δ is a basis of simple roots for Φ, then $\Delta^\vee$ is a basis of simple roots for $\Phi^\vee$.

Proof. (1) is clear. For (2) we need a simple remark. Given a basis $e_1, \dots, e_n$ of a real space and the quadrant $C := \{\sum_i a_i e_i, a_i \geq 0\}$, we have that the elements $ae_i \in C$ are characterized as those vectors in C which cannot be written as sum of two linearly independent vectors in C.

If C is the fundamental Weyl chamber, a vector $v \in E$ is such that $(x, v) \geq 0$, $\forall v \in C$, if and only if x is a linear combination with positive coefficients of the elements $\Delta^\vee$ or of the elements in Δ', the simple roots for $\Phi^\vee$. The previous remark implies that the elements of $\Delta^\vee$ are multiples of those of Δ', hence $\Delta^\vee = \Delta'$ from axiom (2). □

Given the set of simple roots $\alpha_1, \dots, \alpha_n$ (of rank n), the matrix $C := (c_{ij})$ with entries the Cartan integers $c_{ij} := \langle \alpha_i \,|\, \alpha_j \rangle$ is called the *Cartan matrix* of the root system.

One can also characterize root systems by the properties of the Cartan matrix. This is an integral matrix A, which satisfies the following properties:

(1) $c_{ii} = 2,\ c_{ij} \leq 0$, if $i \neq j$. If $c_{ij} = 0$, then $c_{ji} = 0$.
(2) A is symmetrizable, i.e., there is a diagonal matrix D with positive integers entries d_i, $(= (\alpha_i, \alpha_i))$ such that $A := CD$ is symmetric.
(3) $A := CD$ is positive definite.

If the root system is irreducible, we have a corresponding irrreducibility property for C: one cannot reorder the rows and columns (with the same permutation) and make C in block form $C = \begin{pmatrix} C_1 & 0 \\ 0 & C_2 \end{pmatrix}$.

We introduce now, for a given root system, an associated lattice called the *weight lattice*. Its introduction has a full justification in the representation theory of semisimple Lie algebras (see §5). For now the *weight lattice* Λ will be introduced geometrically, as follows:

$$\Lambda := \{\lambda \in E \,|\, \langle \lambda | \alpha \rangle \in \mathbb{Z},\ \forall \alpha \in \Phi\}. \tag{2.4.1}$$

From proposition (1) it follows that we also have

$$\Lambda := \{\lambda \in E \,|\, \langle \lambda | \alpha \rangle \in \mathbb{Z},\ \forall \alpha \in \Delta\}.$$

One also sets

$$\Lambda^+ := \{\lambda \in \Lambda | \langle \lambda, \alpha \rangle \geq 0,\ \forall \alpha \in \Phi^+\}.$$

Λ^+ is called the *set of dominant weights*.

In particular we have the elements $\omega_i \in \Lambda$ with $(\omega_i \,|\, \alpha_j^\vee) = \langle \omega_i \,|\, \alpha_j \rangle = \delta_{ij}$, $\forall \alpha_j \in \Delta$, which are the basis dual to $\Delta^\vee$. Then if $\lambda = \sum_i m_i \omega_i$ we have

$$m_i = \langle \lambda \,|\, \alpha_i \rangle, \quad \lambda = \sum_i \langle \lambda \,|\, \alpha_i \rangle \omega_i, \qquad \Lambda^+ = \left\{ \sum_{i=1}^n m_i \omega_i,\ m_i \in \mathbb{N} \right\}. \tag{2.4.2}$$

$(\omega_1, \omega_2, \ldots, \omega_n)$ is called the *set of fundamental weights* (relative to Φ^+).

Exercise. Draw the picture of the simple roots and the fundamental weights for root systems of rank 2.

Let us define the *root lattice* to be the abelian group generated by the roots. We have:

Proposition 2. *The root lattice is a sublattice of the weight lattice of index the determinant of the Cartan matrix.*

Proof. By the theory developed, a basis of the root lattice is given by the simple roots α_i. From the formula 2.4.2, it follows that $\alpha_j = \sum_i \langle \alpha_j \,|\, \alpha_i \rangle \omega_i = \sum_i c_{j,i} \omega_i$. Since the $c_{j,i}$ are integers, the root lattice is a sublattice. Since the $c_{j,i}$ express a base of the root lattice in terms of a basis of the weight lattice, from the theory of elementary divisors the index is the absolute value of the determinant of the base change. In our case the determinant is positive. □

Theorem. *(a) Λ is stable under the Weyl group.*

(b) Every element of Λ is conjugate to a unique element in Λ^+.

(c) The stabilizer of a dominant weight $\sum_{i=1}^n m_i\omega_i$ is generated by the reflections s_i for the values of i such that $m_i = 0$.

Proof. Take $\lambda \in \Lambda$, $w \in W$. We have $\langle w(\lambda) \,|\, \alpha\rangle = \langle\lambda \,|\, w^{-1}(\alpha)\rangle \in \mathbb{Z}$, $\forall\alpha \in \Phi$, proving (a).

Λ^+ is the intersection of λ with the closure of the fundamental Weyl chamber, and a dominant weight $\lambda = \sum_{i=1}^n m_i\omega_i$ has $m_i = 0$ if and only if $\lambda \in H_i$, from the formula 2.4.2. Hence (b), (c) follow from (6) and (5) of Theorem 2.3. □

It is also finally useful to introduce

(2.4.3) $$\Lambda^{++} := \{\lambda \in \Lambda | \langle\lambda \,|\, \alpha\rangle > 0,\ \forall\alpha \in \Phi^+\}.$$

Λ^{++} is called the *set of regular dominant weights* or *strongly dominant*. It is the intersection of Λ with the (open) Weyl chamber. We have $\Lambda^{++} = \Lambda^+ + \rho$, where $\rho := \sum_i \omega_i$ is the *smallest* element of Λ^{++}. ρ plays a special role in the theory, as we will see when we discuss the Weyl character formula. Let us remark:

Proposition 3. *We have $\langle\rho \,|\, \alpha_i\rangle = 1$, $\forall i$. Then $\rho = 1/2\left(\sum_{\alpha\in\Phi^+}\alpha\right)$ and, for any simple reflection s_i, we have $s_i(\rho) = \rho - \alpha_i$.*

Proof. $\langle\rho \,|\, \alpha_i\rangle = 1$, $\forall i$ follows from the definition. We have $s_i(\rho) = \rho - \langle\rho \,|\, \alpha_i\rangle\alpha_i = \rho - \alpha_i$. Also the element $1/2\left(\sum_{\alpha\in\Phi^+}\alpha\right)$ satisfies the same property with respect to a simple reflection s_i, since such a reflection permutes all positive roots different from α_i sending α_i to $-\alpha_i$. Hence $\rho - 1/2\left(\sum_{\alpha\in\Phi^+}\alpha\right)$ is fixed under all simple reflections and the Weyl group. An element fixed by the Weyl group is in all the root hyperplanes, hence it is 0, and we have the claim. □

For examples see Section 5.1.

2.5 Classification

The basic result of classification is that it is equivalent to classify simple Lie algebras or irreducible root systems up to isomorphism, or irreducible Cartan matrices up to reordering rows and columns.

The final result is usually expressed by Dynkin diagrams of the types illustrated on p. 325 (see [Hu],[B2]).

The Dynkin diagram is constructed by assigning to each simple root a node ∘ and joining two nodes, corresponding to two simple roots α, β, with $\langle\alpha \,|\, \beta\rangle\langle\beta, \alpha\rangle$ edges. Finally the arrow points towards the *shorter root*. The classification is in two steps. First we see that the only possible Dynkin diagrams are the ones exhibited. Next we see that each of them corresponds to a uniquely determined root system.[95]

[95] There are by now several generalizations of this theory, first to characteristic $p > 0$, then to infinite-dimensional Lie algebras as Kac–Moody algebras. For these one considers Cartan matrices which satisfy only the first property: there is a rich theory which we shall not discuss.

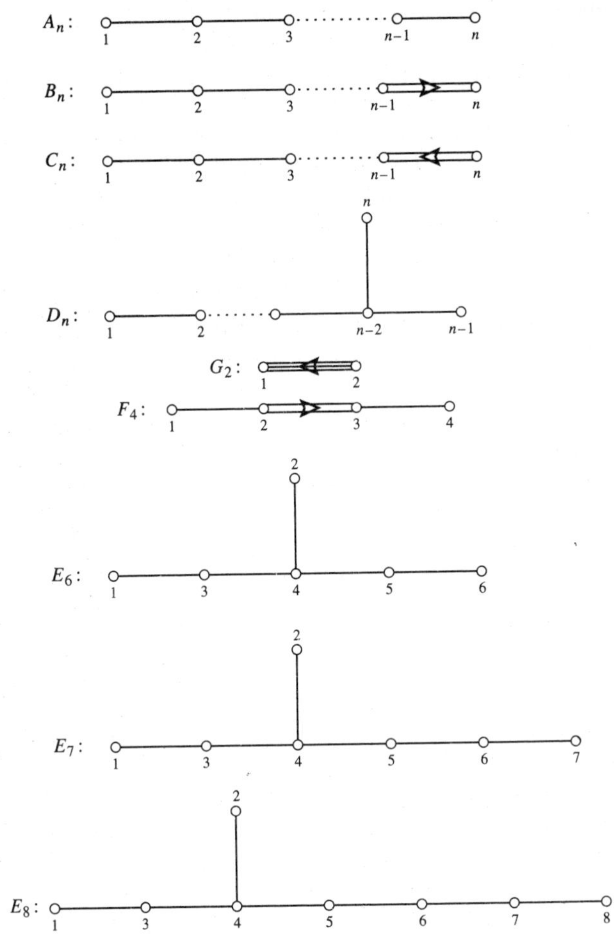

Proposition. *A connected Dynkin diagram determines the root system up to scale.*

Proof. The Dynkin diagram determines the Cartan integers. If we fix the length of one of the simple roots, the other lengths are determined for all other nodes connected to the chosen one. In fact, if α, β are connected by an edge we can use the formula $(\beta, \beta) = \frac{\langle \beta,\alpha \rangle}{\langle \alpha,\beta \rangle}(\alpha, \alpha)$. Since the scalar products of the simple roots are expressed by the Cartan integers and the lengths, the Euclidean space structure on the span of the simple roots is determined.

Next the Cartan integers determine the simple reflections, which generate the Weyl group. Hence the statement follows from Theorem 2.3. $\square$

Let us thus start formally from an irreducible Cartan matrix $C = (c_{i,j}),\ i, j = 1, \dots, n$, (we do not know yet that it is associated to a root system). We can define for C the associated Dynkin diagram as before, with n nodes i and i, j connected by $c_{i,j}c_{j,i}$ edges. If C is irreducible we have a connected diagram.

By assumption $A := CD$ is a positive symmetric matrix and set $a_{i,j} = c_{i,j}d_j$ to be the entries of A. We next construct a vector space E with basis $w_i,\ i = 1, \dots, n$, and scalar product given by A in this basis. Thus E is a Euclidean space.

Theorem. *The Dynkin diagrams associated to irreducible Cartan matrices are those of the list $A_n, B_n, C_n, D_n, E_6, E_7, E_8, F_4, G_2$ displayed as above.*

Proof. It is convenient to pass to a simplified form of the Dynkin diagram in which the lengths of the roots play no role. So we replace the vectors w_i of norm $2d_i$, by the vectors $v_i := w_i/|w_i|$ of norm 1. Thus $(v_i, v_j) = \frac{c_{i,j}d_j}{|w_i||w_j|} = \frac{c_{j,i}d_i}{|w_i||w_j|}$. Of the conditions we still have that the v_i are linearly independent, $(v_i, v_j) \leq 0,\ c_{i,j}c_{j,i} := 4(v_i, v_j)^2 = 0, 1, 2, 3$ for distinct vectors. Moreover, by assumption the quadratic form $\sum_{i,j} a_i a_j (v_i, v_j)$ is positive. Call such a list of vectors admissible. The Dynkin diagram is the same as before except that we have no arrow. In particular the diagram is connected. We deduce various restrictions on the diagram, observing that *any subset of an admissible set is admissible.*

1. *The diagram has no loops.* In fact, consider k vectors v_{i_j} out of our list. We have

$$0 < \left(\sum_{j=1}^{k} v_{i_j}, \sum_{j=1}^{k} v_{i_j} \right) = k + \sum_{h,s} 2a_{i_h, j_s}.$$

Since for all $a_{i,j} \neq 0$, we have $2a_{i,j} \leq -1$, we must have that the total number of $a_{i,j}$ present is strictly less than k. This implies that the v_i cannot be in a loop.

2. *No more than 3 edges can exit a vertex.* In fact, if $v_1, \dots, v_s$ are vertices connected to v_0, we have that they are not connected to each other since there are no loops. Since they are also linearly independent, we can find a unit vector w orthogonal to the $v_i, i = 1, \dots, s$, in the span of $v_0, v_1, \dots, v_s$. We have that $w, v_i,\ i = 1, \dots, s$, are orthonormal, so $v_0 = (v_0, w)w + \sum_{i=1}^{s}(w, v_i)v_i$ and $1 = (v_0, w)^2 + \sum_{i=1}^{s}(w, v_i)^2$. Since $(v_0, w) \neq 0$, we have $\sum_i (v_0, v_i)^2 < 1 \implies 4\sum_i (v_0, v_i)^2 < 4$, which gives the desired inequality.

3. *If we have a triple edge, then the system is G_2.* Otherwise one of the two nodes of this subgraph is connected to another node. Then out of this at least 4 edges originate.

Suppose that some vectors $v_1, \dots, v_k$ in the diagram form a *simple chain* as in type A_n; in other words $(v_i, v_{i+1}) = -1/2,\ i = 1, \dots, k-1$ (i.e., they are linked by a single edge) and no v_i, $1 < i < k$ is linked to any other node. Then:

4. *Replacing all these vectors by $v = \sum_i v_i$, creates a new admissible list.* In fact, first of all $(v, v) = k-(k-1) = 1$ is a unit vector. Next, if v_j is a vector different from the given ones, it can connect only to v_1 or v_k. In the first case $(v_j, v) = (v_j, v_1)$, and similarly for the second. The diagram associated to this new list is the one in which the simple chain has been contracted to a node. We deduce then that the diagram cannot contain any of the following subdiagrams; otherwise contracting a simple chain we obtain a node to which 4 edges are connected:

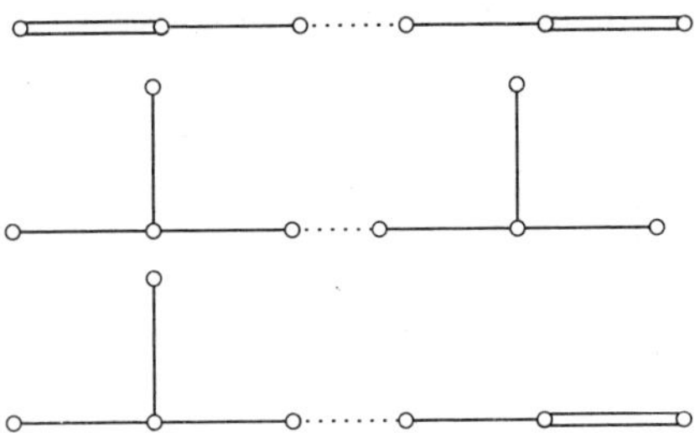

5. *We are thus left with the possible following types:*

(i) A single simple chain, this is type A_n.

(ii) Two nodes a, b connected by two edges, from each one starts a simple chain $a = a_0, a_1, \dots, a_k;\ b = b_0, b_1, \dots, b_h$.

(iii) One node from which three simple chains start.

In case (ii) we must show that it is not possible that both $h, k > 1$. In other words,

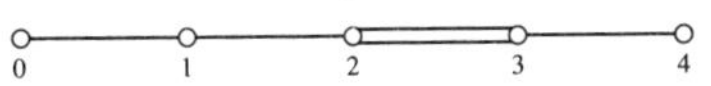

is not admissible. In fact consider $\epsilon := v_0 + 2v_1 + 3v_2 + 2\sqrt{2}v_3 + \sqrt{2}v_4$. Computing we have $(\epsilon, \epsilon) = 0$, which is impossible.

In the last case assume we have the three simple chains

$$a_1, \dots, a_{p-1}, a_p = d;\ b_1, \dots, b_{q-1}, b_q = d;\ c_1, \dots, c_{r-1}, c_r = d,$$

from the node d. Consider the three orthogonal vectors

$$x := \sum_{i=1}^{p-1} i a_i, \quad y := \sum_{i=1}^{q-1} i b_i, \quad z := \sum_{i=1}^{r-1} i c_i.$$

d is not in their span and $(x, x) = p(p-1)/2,\ (y, y) = q(q-1)/2,\ (z, z) = r(r-1)/2$.

Expanding d in an orthonormal basis of the space $\langle d, x, y, z \rangle$ we have $(d, x)^2/(x, x) + (d, y)^2/(y, y) + (d, z)^2/(z, z) < 1$. We deduce that

$$1 > \frac{(p-1)^2}{4}\frac{2}{p(p-1)} + \frac{(q-1)^2}{4}\frac{2}{q(q-1)}$$

$$+ \frac{(r-1)^2}{4}\frac{2}{r(r-1)} = \frac{1}{2}\left(3 - \left(\frac{1}{p} + \frac{1}{q} + \frac{1}{r}\right)\right). \tag{2.5.1}$$

It remains for us to discuss the basic inequality 2.5.1, which is just $\frac{1}{p} + \frac{1}{q} + \frac{1}{r} > 1$. We can assume $p \leq q \leq r$. We cannot have $p > 2$ since otherwise the three terms are $\leq 1/3$. So $p = 2$ and we have $\frac{1}{q} + \frac{1}{r} > 1/2$. We must have $q \leq 3$. If $q = 2$, r can be arbitrary and we have the diagram of type D_n. Otherwise if $q = 3$, we still have $1/r > 1/6$ or $r \leq 5$.

For $r = 3, 4, 5$, we have E_6, E_7, E_8. □

We will presently exhibit for each $A_n, B_n, C_n, D_n, E_6, E_7, E_8, F_4, G_2$ a corresponding root system. For the moment let us make explicit the corresponding Cartan matrices (see the table on p. 328).

TABLE OF CARTAN MATRICES

$$A_n := \begin{pmatrix} 2 & -1 & 0 & & \dots & & & 0 \\ -1 & 2 & -1 & 0 & \dots & & & 0 \\ 0 & -1 & 2 & -1 & & & & 0 \\ & & & \dots & \dots & & & \\ 0 & 0 & 0 & & & 0 & -1 & 2 \end{pmatrix}$$

$$B_n := \begin{pmatrix} 2 & -1 & 0 & & \dots & & & 0 \\ -1 & 2 & -1 & 0 & \dots & & & 0 \\ 0 & \dots & & & \dots & & & 0 \\ & & & \dots & \dots & -1 & 2 & -2 \\ 0 & 0 & 0 & & & 0 & -1 & 2 \end{pmatrix}$$

$$C_n := \begin{pmatrix} 2 & -1 & 0 & & \dots & & & 0 \\ -1 & 2 & -1 & 0 & \dots & & & 0 \\ 0 & \dots & & \dots & & & & 0 \\ & & & \dots & \dots & -1 & 2 & -1 \\ 0 & 0 & 0 & & & 0 & -2 & 2 \end{pmatrix}$$

$$D_n := \begin{pmatrix} 2 & -1 & 0 & & & \dots & & & 0 \\ -1 & 2 & -1 & 0 & & \dots & & & 0 \\ 0 & \dots & & \dots & & & & & 0 \\ 0 & \dots & & \dots & -1 & 2 & -1 & 0 & 0 \\ 0 & & \dots & & 0 & -1 & 2 & -1 & -1 \\ & & & \dots & & 0 & -1 & 2 & 0 \\ 0 & 0 & 0 & & & 0 & -1 & 0 & 2 \end{pmatrix}$$

$$E_6 := \begin{pmatrix} 2 & 0 & -1 & 0 & 0 & 0 \\ 0 & 2 & 0 & -1 & 0 & 0 \\ -1 & 0 & 2 & -1 & 0 & 0 \\ 0 & -1 & -1 & 2 & -1 & 0 \\ 0 & 0 & 0 & -1 & 2 & -1 \\ 0 & 0 & 0 & 0 & -1 & 2 \end{pmatrix}, \quad E_7 := \begin{pmatrix} 2 & 0 & -1 & 0 & 0 & 0 & 0 \\ 0 & 2 & 0 & -1 & 0 & 0 & 0 \\ -1 & 0 & 2 & -1 & 0 & 0 & 0 \\ 0 & -1 & -1 & 2 & -1 & 0 & 0 \\ 0 & 0 & 0 & -1 & 2 & -1 & 0 \\ 0 & 0 & 0 & 0 & -1 & 2 & -1 \\ 0 & 0 & 0 & 0 & 0 & -1 & 2 \end{pmatrix}$$

$$E_8 := \begin{pmatrix} 2 & 0 & -1 & 0 & 0 & 0 & 0 & 0 \\ 0 & 2 & 0 & -1 & 0 & 0 & 0 & 0 \\ -1 & 0 & 2 & -1 & 0 & 0 & 0 & 0 \\ 0 & -1 & -1 & 2 & -1 & 0 & 0 & 0 \\ 0 & 0 & 0 & -1 & 2 & -1 & 0 & 0 \\ 0 & 0 & 0 & 0 & -1 & 2 & -1 & 0 \\ 0 & 0 & 0 & 0 & 0 & -1 & 2 & -1 \\ 0 & 0 & 0 & 0 & 0 & 0 & -1 & 2 \end{pmatrix}, \quad F_4 = \begin{pmatrix} 2 & -1 & 0 & 0 \\ -1 & 2 & -2 & 0 \\ 0 & -1 & 2 & -1 \\ 0 & 0 & -1 & 2 \end{pmatrix}$$

$$G_2 := \begin{pmatrix} 2 & -1 \\ -3 & 2 \end{pmatrix}$$

2.6 Existence of Root Systems

The *classical* root systems of types A, B, C, D have already been exhibited as well as G_2. We leave to the reader the simple exercise:

Exercise. Given a root system Φ with Dynkin diagram Δ and a subset S of the simple roots, let E_S be the subspace spanned by S. Then $\Phi \cap E$ is a root system, with simple roots S and Dynkin diagram the subdiagram with nodes S.

Given the previous exercise, we start by exhibiting F_4, E_8 and deduce E_6, E_7 from E_8.

F_4. Consider the 4-dimensional Euclidean space with the usual basis e_1, e_2, e_3, e_4. Let $a := (e_1 + e_2 + e_3 + e_4)/2$ and let Λ be the lattice of vectors of type $\sum_{i=1}^4 n_i e_i + ma$, $n_i, m \in \mathbb{Z}$. Let $\Phi := \{u \in \Lambda \mid (u, u) = 1, \text{ or } (u, u) = 2\}$. We easily see that Φ consists of the 24 vectors $\pm e_i \pm e_j$ of norm 2 and the 24 vectors of norm 1: $\pm e_i$, $(\pm e_1 \pm e_2 \pm e_3 \pm e_4)/2$.

One verifies directly that the numbers $\langle \alpha \mid \beta \rangle$ are integers $\forall \alpha, \beta \in \Phi$. Hence Λ is stable under the reflections s_α, $\alpha \in \Phi$, so clearly Φ must also be stable under the Weyl group. The remaining properties of root systems are also clear.

It is easy to see that a choice of simple roots is

$$e_2 - e_3, e_3 - e_4, e_4, (e_1 - e_2 - e_3 - e_4)/2.$$

E_8. We start, as for F_4, from the 8-dimensional space and the vector $a = \left(\sum_{i=1}^8 e_i\right)/2$. Now it turns out that we cannot just take the lattice

$$\sum_{i=1}^8 n_i e_i + ma, \quad n_i, m \in \mathbb{Z}$$

but we have to impose a further constraint. For this, we remark that although the expression of an element of Λ as $\sum_{i=1}^8 n_i e_i + ma$ is not unique, the value of $\sum_i n_i$ is unique mod 2. In fact, $\sum_{i=1}^8 n_i e_i + ma = 0$ implies $m = 2k$ even, $n_i = -k, \forall i$ and $\sum_i n_i \equiv 0$, mod 2.

Since the map of Λ to $\mathbb{Z}/(2)$ given by $\sum_i n_i$ is a homomorphism, its kernel Λ_0 is a sublattice of Λ of index 2. We define the root system E_8 as the vectors Φ in Λ_0 of norm 2.

It is now easy to verify that the set Φ consists of the 112 vectors $\pm e_i \pm e_j$, and the 128 vectors $\sum_{i=1}^8 \pm e_i/2 = \sum_{i \in P} e_i - a$, with P the set of indices where the signs are positive. The number of positive signs must be even.

From our discussion is clear that the only point which needs to be verified is that the numbers $\langle \alpha \mid \beta \rangle$ are integers. In our case this is equivalent to proving that the scalar products between two of these vectors is an integer. The only case requiring some discussion is when both α, β are of the type $\sum_{i=1}^8 \pm e_i/2$. In this case the scalar product is of the form $(a - b)/4$, where a counts the number of contributions of 1, while b is the number of contributions of -1 in the scalar product of the two numerators. By definition $a + b = 8$, so it suffices to prove that b is even. The

8 terms ± 1 appearing in the scalar product can be divided as follows. We have t minus signs in α which pair with a plus sign in β, then u minus signs in α which pair with a minus sign in β. Finally we have r plus signs in α which pair with a minus sign in β. By the choice of Λ, the numbers $t+u, r+u$ are even, while $b = t+r$, hence $b \equiv t+u+r+u \equiv 0, \mod 2$.

For the set Δ_8, of simple roots in E_8, we take

$$1/2\Big(e_1 + e_8 - \sum_{i=2}^{7} e_i\Big),\ e_2 + e_1,\ e_2 - e_1,\ e_3 - e_2,\ e_4 - e_3,$$

(2.6.1)
$$e_5 - e_4,\ e_6 - e_5,\ e_7 - e_6.$$

E_7, E_6. Although these root systems are implicitly constructed from E_8, it is useful to extract some of their properties. We call $x_i,\ i = 1, \dots, 8$, the coordinates in $\mathbb{R}^8$. The first 7 roots in Δ_8 span the subspace E of $\mathbb{R}^8$ in which $x_7 + x_8 = 0$. Intersecting E with the roots of E_8 we see that of the 112 vectors $\pm e_i \pm e_j$, only those with $i, j \neq 7, 8$ and $\pm(e_7 - e_8)$ appear, a total of 62. Of the 128 vectors $\sum_{i=1}^{8} \pm e_i/2$ we have 64 in which the signs of e_7, e_8 do not coincide, a total of 126 roots.

For E_6 the first 6 roots in Δ_8 span the subspace F in which $x_6 = x_7 = -x_8$. Intersecting F with the roots of E_7 we find the 40 elements $\pm e_i \pm e_j, 1 \leq i < j \leq 5$ and the 32 elements $\pm 1/2(e_6 + e_7 - e_8 + \sum_{i=1}^{5} \pm e_i)$ with an even number of minus signs, a total of 72 roots.

2.7 Coxeter Groups

We have seen that the Weyl group is a reflection group, generated by the simple reflections s_i. There are further important points to this construction. First, the Dynkin diagram also determines defining relations for the generators s_i of W. Recall that if s, t are two reflections whose axes form a (convex) angle θ, then st is a rotation of angle 2θ. Apply this remark to the case of two simple reflections s_i, s_j $(i \neq j)$ in a Weyl group. Then $s_i s_j$ is a rotation of $\frac{2\pi}{m_{i,j}}$ with $m_{i,j} = 2, 3, 4, 6$ according to whether i and j are connected by $0, 1, 2, 3$ edges in the Dynkin diagram. In particular,

$$\big(s_i s_j\big)^{m_{i,j}} = 1.$$

It turns out that these relations, together with $s_i^2 = 1$, afford a presentation of W.

Theorem 1. [Hu3, 1.9] *The elements (called Coxeter relations):*

(2.7.1)
$$s_i^2, \qquad \big(s_i s_j\big)^{m_{i,j}} \qquad \textit{Coxeter relations}$$

generate the normal subgroup of defining relations for W.

In general, one defines a *Coxeter system* (W, S) as the group W with generators S and defining relations

$$(st)^{m_{s,t}} = 1, \quad s, t \in S,$$

where $m_{s,s} = 1$ and $m_{s,t} = m_{t,s} \geq 2$ for $s \neq t$. If there is no relation between s and t, we set $m_{s,t} = \infty$.

The presentation can be completely encoded by a weighted graph (the *Coxeter graph*) Γ. The vertices of Γ are the elements of S, and $s, t \in S$ are connected by an edge (with weight $m_{s,t}$) if $m_{s,t} > 2$, i.e., if they do not commute. For instance, the group corresponding to the graph consisting of two vertices and one edge labeled by ∞ is the free product of two cyclic groups of order 2 (hence it is infinite). There is a natural notion of irreducibility for Coxeter groups which corresponds exactly to the connectedness of the associated Coxeter graphs.

When drawing Coxeter graphs, it is customary to draw an edge with no label if $m_{s,t} = 3$. With this notation, to obtain the Coxeter graph from the Dynkin diagram of a Weyl group just forget about the arrows and replace a double (resp. triple) edge by an edge labeled by 4 (resp. 6).

Several questions arise naturally at this point: classify finite Coxeter groups, and single out the relationships between finite Coxeter groups, finite reflection groups and Weyl groups. In the following we give a brief outline of the answer to the previous problems, referring the reader to [Hu3, Chapters 1, 2] and [B1], [B2], [B3] for a thorough treatment of the theory.

We start from the latter problem. For any Coxeter group G, one builds up the space V generated by vectors α_s, $s \in S$ and the symmetric bilinear form

$$B_G(\alpha_s, \alpha_t) = -\cos \frac{\pi}{m_{s,t}}.$$

We can now define a "reflection", setting

$$\sigma(s)(\lambda) = \lambda - 2B_G(\lambda, \alpha_s)\alpha_s.$$

Proposition. [Ti] *The map $s \mapsto \sigma_s$ extends uniquely to a representation $\sigma : G \to GL(V)$. $\sigma(G)$ preserves the form B_G on V. The order of st in G is exactly $m(s,t)$.*

Hence Coxeter groups admit a "reflection representation" (note however that V is not in general a Euclidean space). The main result is the following:

Theorem 2. [Hu3, Theorem 6.4] *The following conditions are equivalent:*

(1) G is a finite Coxeter group.
(2) B_G is positive definite.
(3) G is a finite reflection group.

The classification problem can be reduced to determining the Coxeter graphs for which the form B_G is positive definite. Finally, the graphs of the irreducible Coxeter groups are those obtained from the Weyl groups and three more cases: the *dihedral groups* D_n, of symmetries of a regular n-gon, and two reflection groups, H_3 and H_4, which are 3-dimensional and 4-dimensional, respectively, given by the Coxeter graphs

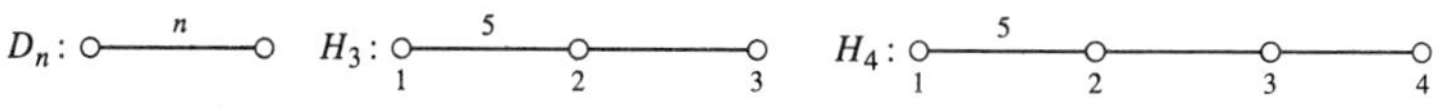

An explicit construction of the reflection group for the latter cases can be found in [Gb] or [Hu3, 2.13]. Finally, remark that Weyl groups are exactly the finite Coxeter groups for which $m_{s,t} \in \{2, 3, 4, 6\}$. This condition can be shown to be equivalent to the following: G stabilizes a lattice in V.

3 Construction of Semisimple Lie Algebras

3.1 Existence of Lie Algebras

We now pass to the applications to Lie algebras. Let L be a simple Lie algebra, $\mathfrak{t}$ a Cartan subalgebra, Φ the associated root system, and choose the simple roots $\alpha_1, \ldots, \alpha_n$. We have seen in Proposition 1.8 (7) that one can choose $e_i \in L_{\alpha_i}$, $f_i \in L_{-\alpha_i}$ so that $e_i, f_i, h_i := [e_i, f_i]$ are $sl(2, \mathbb{C})$ triples, and $h_i = 2/(t_{\alpha_i}, t_{\alpha_i})t_{\alpha_i}$. Call $sl_i(2, \mathbb{C})$ and $SL_i(2, \mathbb{C})$ the corresponding Lie algebra and group.

The previous generators are normalized so that, for each element λ of $\mathfrak{t}^*$,

$$\lambda(h_i) = \lambda(2/(t_{\alpha_i}, t_{\alpha_i})t_{\alpha_i}) = \frac{2(\lambda, \alpha_i)}{(\alpha_i, \alpha_i)} = \langle \lambda \mid \alpha_i \rangle. \tag{3.1.1}$$

Then one obtains $[h_i, e_j] := \langle \alpha_j \mid \alpha_i \rangle e_j$, $[h_i, f_j] := -\langle \alpha_i \mid \alpha_j \rangle f_j$ and one can deduce a fundamental theorem of Chevalley and Serre (using the notation $a_{ij} := \langle \alpha_j \mid \alpha_i \rangle$). Before stating it, let us make some remarks. From Proposition 1.2 it follows that, for each i, we can integrate the adjoint action of $sl_i(2, \mathbb{C})$ on L to a rational action of $SL_i(2, \mathbb{C})$. From Chapter 4, §1.5, since the adjoint action is made of derivations, these groups $SL_i(2, \mathbb{C})$ act as automorphisms of the Lie algebra. In particular, let us look at how the element $s_i := \begin{vmatrix} 0 & 1 \\ -1 & 0 \end{vmatrix} \in SL_i(2, \mathbb{C})$, acts. We intentionally use the same notation as for the simple reflections.

Lemma. *s_i preserves the Cartan subalgebra and, under the identification with the dual, acts as the simple reflection s_i.* $s_i(L_\alpha) = L_{s_i(\alpha)}$.

Proof. If $h \in \mathfrak{t}$ is such that $\alpha_i(h) = 0$, we have that h commutes with e_i, h_i, f_i, and so it is fixed by the entire group $SL_i(2, \mathbb{C})$. On the other hand, $s_i(h_i) = -h_i$, hence the first part.

Since s_i acts by automorphisms, if u is a root vector for the root α, we have

$$[h, s_i u] = [s_i^2 h, s_i u] = s_i[s_i h, u] = s_i(\alpha(s_i h)u) = s_i(\alpha)(h)s_i u. \qquad \square$$

Exercise. Consider the group $\tilde{W}$ of automorphisms of L generated by the s_i. We have a homomorphism $\pi : \tilde{W} \to W$. Its kernel is a finite group acting on each L_α with ± 1.

Theorem 1. *The Lie algebra L is generated by the $3n$ elements e_i, f_i, h_i, $i = 1, \ldots, n$, called* ***Chevalley generators***. *They satisfy the* ***Serre relations***:

$$[h_i, h_j] = 0, \quad [h_i, e_j] = a_{ij} e_j \quad [h_i, f_j] = -a_{ij} f_j \quad [e_i, f_j] = \delta_{ij} h_i \tag{3.1.2}$$

$$\mathrm{ad}(e_i)^{1-a_{ij}}(e_j) = 0, \qquad \mathrm{ad}(f_i)^{1-a_{ij}}(f_j) = 0. \tag{3.1.3}$$

Proof. Consider the Lie subalgebra L^0 generated by the given elements. It is a subrepresentation of each of the groups $SL_i(2, \mathbb{C})$; in particular, it is stable under all the s_i and the group they generate. Given any root α, there is a product w of s_i which sends α to one of the simple roots α_i; hence under the inverse w^{-1} the element e_i is mapped into L_α. This implies that $L_\alpha \subset L$ and hence $L = L^0$.

Let us see why these relations are valid. The first (3) are the definition and normalization. If $i \neq j$, the element $[e_i, f_j]$ has weight $\alpha_i - \alpha_j$. Since this is not a root $[e_i, f_j]$ must vanish.

This implies that f_j is a highest weight vector for a representation of $sl_i(2, \mathbb{C})$ of highest weight $[h_i, f_j] = -a_{i,j} f_j$. This representation thus has dimension $-a_{i,j} + 1$ and each time we apply $\operatorname{ad}(f_i)$ starting from f_j, we descend by 2 in the weight. Thus $\operatorname{ad}(f_i)^{1-a_{ij}}(f_j) = 0$. The others are similar. □

More difficult to prove is the converse. Given an irreducible root system Φ:

Theorem 2 (Serre). *Let L be the quotient of the free Lie algebra in the generators e_i, f_i, h_i modulo the Lie ideal generated by the relations* 3.1.2 *and* 3.1.3.

L is a simple finite-dimensional Lie algebra. The h_i are a basis of a Cartan subalgebra of L and the associated root system is Φ.

Proof. First, some notation: on the vector space with basis the h_i, we define α_i to be the linear form given by $\alpha_i(h_j) := a_{ij} = \langle \alpha_j \mid \alpha_i \rangle$.

We proceed in two steps. First, consider in the free Lie algebra the relations of type 3.1.2. Call L_0 the resulting Lie algebra. For L_0 we prove the following statements.

(1) In L_0 the images of the $3n$ elements e_i, f_i, h_i remain linearly independent.
(2) $L_0 = \mathfrak{u}_0^- \oplus \mathfrak{h} \oplus \mathfrak{u}_0^+$, where $\mathfrak{h}$ is the abelian Lie algebra with basis $h_1, \dots, h_n$, $\mathfrak{u}_0^-$ is the Lie subalgebra generated by the classes of the f_i, and $\mathfrak{u}_0^+$ is the Lie subalgebra generated by the classes of the e_i.
(3) $\mathfrak{u}_0^+$ (resp. $\mathfrak{u}_0^-$) has a basis of eigenvectors for the commuting operators h_i with eigenvalues $\sum_{i=1}^n m_i \alpha_i$ (resp. $-\sum_{i=1}^n m_i \alpha_i$) with m_i nonnegative integers.

The proof starts by noticing that by applying the commutation relations 3.1.2, one obtains that $L_0 = \mathfrak{u}_0^- + \mathfrak{h} + \mathfrak{u}_+$, where $\mathfrak{h}$ is abelian. Since, for a derivation of an algebra, the two formulas $D(a) = \alpha a$, $D(b) = \beta b$ imply $D(ab) = (\alpha + \beta)ab$, an easy induction proves (3). This implies (2), except for the independence of the h_i, since the three spaces of the decomposition belong to positive, 0, and negative eigenvalues for $\mathfrak{h}$. It remains to prove (1) and in particular exclude the possibility that these relations define a trivial algebra.

Consider the free associative algebra $M := \mathbb{C}\langle e_1, \dots, e_n \rangle$. We define linear operators on M which we call e_i, f_i, h_i and prove that they satisfy the commutation relations 3.1.2.

Set e_i to be left multiplication by e_i. Set h_i to be the semisimple operator which, on a tensor $u := e_{i_1} e_{i_2} \dots e_{i_k}$, has eigenvalue $\sum_{j=1}^k \alpha_{i_j}(h_i)$. Set for simplicity $\sum_{j=1}^k \alpha_{i_j} := \alpha_u$. Define f_i inductively as a map which decreases the degree of tensors by 1 and $f_i 1 = 0$, $f_i(e_j u) := e_j f_i(u) - \delta_i^j \alpha_u(h_i)u$. It is clear that these $3n$ operators are linearly independent. It suffices to prove that they satisfy the relations 3.1.2, since then they produce a representation of L_0 on M.

By definition, the elements h_i commute and

$$(h_j e_i - e_i h_j)u = (\alpha_i(h_j) + \alpha_u(h_j))e_i u - \alpha_u(h_j)e_i u = \alpha_i(h_j)e_i u.$$

For the last relations

$$(f_i e_j - e_j f_i)u = f_i e_j u - e_j f_i u = -\delta_i^j \alpha_u(h_i)u = -\delta_i^j h_i u$$

and $(f_i h_j - h_j f_i)u = \alpha_u(h_j) f_i u - \alpha_{f_i u}(h_j) f_i u$. So it suffices to remark that by the defining formula, f_i maps a vector u of weight α_u into a vector of weight $\alpha_u - \alpha_i$.

Now we can present L as the quotient of L_0 modulo the ideal I generated in L_0 by the unused relations 3.1.3. This ideal can be described as follows. Let I^+ be the ideal of $\mathfrak{u}_+$ generated by the elements $\mathrm{ad}(e_i)^{1-a_{ij}}(e_j)$, and I^- the ideal of $\mathfrak{u}_0^-$ generated by the elements $\mathrm{ad}(f_i)^{1-a_{ij}}(f_j)$. If we prove that I^+, I^- are ideals in L_0, it follows that $I = I^+ + I^-$ and that $L_0/I = \mathfrak{u}_0^+/I^+ \oplus \mathfrak{h} \oplus \mathfrak{u}_0^-/I^-$. We prove this statement in the case of the f's; the case of the e's is identical. Set $R_{i,j} := \mathrm{ad}(f_i)^{1-a_{ij}}(f_j)$. Observe first that $\mathrm{ad}(f_i)^{1-a_{ij}}(f_j)$ is a weight vector under $\mathfrak{h}$ of weight $-(\alpha_j + (1 - a_{ij})\alpha_i)$. By the commutation formulas on the elements, it is clearly enough to prove that $[e_l, \mathrm{ad}(f_i)^{1-a_{ij}}(f_j)] = \mathrm{ad}(e_l)\,\mathrm{ad}(f_i)^{1-a_{ij}}(f_j) = 0$ for all l. If l is different from i, $\mathrm{ad}(e_l)$ commutes with $\mathrm{ad}(f_i)$ and $\mathrm{ad}(f_i)^{1-a_{ij}}\,\mathrm{ad}(e_l)(f_j) = \delta_l^j\,\mathrm{ad}(f_i)^{1-a_{ij}} h_j$. If $a_{ij} = 0$, f_i commutes with h_j. If $1 - a_{ij} > 1$, we have $[f_i, [f_i, h_j]] = 0$. In either case $\mathrm{ad}(e_l)R_{ij} = 0$. We are left with the case $l = i$. In this case we use the fact that e_i, f_i, h_i generate an $sl(2, \mathbb{C})$. The element f_j is killed by e_i and is of weight $-\langle \alpha_j \,|\, \alpha_i \rangle = -a_{ij}$. Lemma 1.1 applied to $v = f_j$ implies that for all s, $\mathrm{ad}(e_i)\,\mathrm{ad}(f_i)^s f_j = s(-a_{ij} - s + 1)\,\mathrm{ad}(f_i)^{s-1} f_j$. For $s = 1 - a_{ij}$, we indeed get $\mathrm{ad}(e_i)\,\mathrm{ad}(f_i)^s f_j = 0$.

At this point of the analysis we obtain that the algebra L defined by the Chevalley generators and Serre's relations is decomposed in the form $L = \mathfrak{u}^+ \oplus \mathfrak{h} \oplus \mathfrak{u}^-$. $\mathfrak{h}$ has as a basis the elements h_i, $\mathfrak{u}^+$ (resp. $\mathfrak{u}^-$) has a basis of eigenvectors for the commuting operators h_i with eigenvalues $\sum_{i=1}^n m_i \alpha_i$ (resp. $-\sum_{i=1}^n m_i \alpha_i$), with m_i nonnegative integers.

The next step consists of proving that the elements $\mathrm{ad}(e_i)$, $\mathrm{ad}(f_i)$ are *locally nilpotent* (cf. §1.2). Observe for this that, given an algebra L and a derivation D, the set of $u \in L$ killed by a power of D is a subalgebra since

$$D^k(ab) = \sum_{i=0}^{k} \binom{k}{i} D^i(a) D^{k-i}(b)$$

Since clearly for L the elements e_i, h_i, f_i belong to this subalgebra for $\mathrm{ad}(e_j)$, $\mathrm{ad}(f_j)$, this claim is proved.

From Proposition 1.2, for each i, L is a direct sum of finite-dimensional irreducible representations of $SL_i(2, \mathbb{C})$. So we can find, for each i, an element

$s_i = \begin{vmatrix} 0 & 1 \\ -1 & 0 \end{vmatrix}$ as in §1.9 Lemma 2, which on the roots acts as the simple reflection associated to α_i. Arguing as in 1.9 we see that s_i transforms the subspace L_γ relative to a weight γ into the subspace $L_{s_i(\gamma)}$. In particular if two weights can be transformed into each other by an element of W, the corresponding weight spaces have the same dimension. Remark that, by construction, the space L_{α_i} is 1-dimensional with basis e_i, and similarly for $-\alpha_i$. We already know that the weights are of type $\sum_i m_i\alpha_i$, with the m_i all of the same sign. We deduce, using Theorem 2.3 (4), that if α is a root, $\dim L_\alpha = 1$. Suppose now α is not a root. We want to show that $L_\alpha = 0$. Let us look, for instance, at the case of positive weights, elements of type $\operatorname{ad}(e_{i_1})\operatorname{ad}(e_{i_2})\ldots\operatorname{ad}(e_{i_{k-1}})e_{i_k}$. If this monomial has degree > 1, the indices i_{k-1}, i_k must be different (or we have 0), so a multiple of a simple root never appears as a weight. By conjugation the same happens for any root. Finally, assume α is not a multiple of a root. Let us show that conjugating it with an element of W, we obtain a linear combination $\sum_i m_i\alpha_i$ in which two indices m_i have strictly opposite signs. Observe that if α is not a multiple of any root, there is a regular vector v in the hyperplane orthogonal to α. We can then find an element $w \in W$ so that wv is in the fundamental chamber. Since $(w\alpha, wv) = 0$, writing $w\alpha = \sum_i m_i\alpha_i$ we have $\sum_i m_i(\alpha_i, wv) = 0$. Since all $(\alpha_i, wv) > 0$, the claim follows.

Now we know that $w\alpha$ is not a possible weight, so also α is not a weight.

At this point we are very close to the end: We have shown that $L = \mathfrak{h} \oplus \bigoplus_{\alpha\in\Phi} L_\alpha$ and that $\dim L_\alpha = 1$. Clearly, $\mathfrak{h}$ is a maximal toral subalgebra and Φ its root system. We only have to show that L is semisimple or, assuming Φ irreducible, that L is simple. Let I be a nonzero ideal of L. Let us first show that $I \supset L_\alpha$ for some root α. Since I is stable under $\operatorname{ad}(\mathfrak{h})$, it contains a nonzero weight vector v. If v is a root vector we have achieved our first step; otherwise $v \in \mathfrak{h}$. Using a root α with $\alpha(v) \neq 0$ we see that $L_\alpha = [v, L_\alpha] \subset I$. Since I is an ideal it is stable under all the $sl(2, \mathbb{C})$ and the corresponding groups. In particular, it is stable under all the s_i, so we deduce that some $e_i, f_i, h_i \in I$. If $a_{ij} \neq 0$, we have $a_{i,j}e_j = [h_i, e_j] \in I$. We thus get all the $e_j \in I$ for all the nodes j of the Dynkin diagram connected to i. Since the Dynkin diagram is connected we obtain in this way all the elements e_j, and similarly for the f_j, h_j. Once we have all the generators, $I = L$. □

Therefore, given a root system, we have constructed an associated semisimple Lie algebra, using these generators and relations, thus proving the *existence theorem*.

This theorem, although quite satisfactory, leaves in the dark the explicit multiplication structure of the corresponding Lie algebra. In fact with some effort one can prove.

Theorem 3. *One can choose nonzero elements e_α in each of the root spaces L_α so that, if $\alpha, \beta, \alpha+\beta$ are roots, one has $[e_\alpha, e_\beta] = \pm e_{\alpha+\beta}$. These signs can be explicitly determined.*

Sketch of proof. Take the longest element of the Weyl group w_0 and write it as a reduced expression $w_0 = s_{i_1}s_{i_2}\ldots s_{i_n}$. For $\alpha = s_{i_1}s_{i_2}\cdots s_{i_{h-1}}\alpha_{i_h}$, define $e_\alpha := s_{i_1}s_{i_2}\cdots s_{i_{h-1}}e_{i_h}$. Next use Exercise 3.1 on $\tilde{W}$, which easily implies the claim. □

The determination of the explicit signs needs a long computation. We refer to Tits [Ti].

3.2 Uniqueness Theorem

We need to prove that the root system of a semisimple Lie algebra L is uniquely determined. We will use the theory of *regular elements.*

Definition. An element $h \in L$ is said to be regular semisimple if h is semisimple and the centralizer of h is a maximal toral subalgebra.

Given a maximal toral subalgebra $\mathfrak{t}$ and an element $h \in \mathfrak{t}$, we see that its centralizer is $\mathfrak{t} \bigoplus_{\alpha\in\Phi \,|\, \alpha(h)=0} L_\alpha$. So we have:

Lemma. *An element $h \in \mathfrak{t}$ is regular if and only if $h \notin \cup_{\alpha\in\Phi} H_\alpha$, $H_\alpha := \{h \in \mathfrak{t} \,|\, \alpha(h) = 0\}$.*[96]

In particular the regular elements of $\mathfrak{t}$ form an open dense set. We can now show:

Theorem. *Two maximal toral subalgebras $\mathfrak{t}_1, \mathfrak{t}_2$ are conjugate under the adjoint group.*

Proof. Let G denote the adjoint group and $\mathfrak{t}_1^{reg}$, $\mathfrak{t}_2^{reg}$ the regular elements in the two toral subalgebras. We claim that $G\mathfrak{t}_1^{reg}$, $G\mathfrak{t}_2^{reg}$ contain two Zariski open sets of L and hence have nonempty intersection. In fact, compute at the point $(1, h)$, $h \in \mathfrak{t}_1^{reg}$ the differential of the map $\pi : G \times \mathfrak{t}_1^{reg} \to L$, $\pi(g, h) := gh$. It is $L \times \mathfrak{t}_1 \to [L, h] + \mathfrak{t}_1$. Since h is regular, $[L, h] = \bigoplus_{\alpha\in\Phi} L_\alpha$. This implies that $L = [L, h] + \mathfrak{t}_1$ and the map is dominant.

Once we have found an element $g_1h_1 = g_2h_2$, $h_1 \in \mathfrak{t}_1^{reg}$, $h_2 \in \mathfrak{t}_2^{reg}$, we have that $g_2^{-1}g_1h_1$ is a regular element of $\mathfrak{t}_2$, from which it follows that for the centralizers, $\mathfrak{t}_2 = g_2^{-1}g_1(\mathfrak{t}_1)$. □

Together with the existence we now have:

Classification. *The simple Lie algebras over $\mathbb{C}$ are classified by the Dynkin diagrams.*

Proof. Serre's Theorem shows that the Lie algebra is canonically determined by the Dynkin diagram. The previous result shows that the Dynkin diagram is identified (up to isomorphism) by the Lie algebra and is independent of the toral subalgebra. □

---

96 By abuse of notation we use the symbol H_α not only for the hyperplane in the real reflection representation, but also as a hyperplane in $\mathfrak{t}$.

4 Classical Lie Algebras

4.1 Classical Lie Algebras

We want to illustrate the concepts of roots (simple and positive) and the Weyl group for classical groups. In these examples we can be very explicit, and the reader can verify all the statements directly.

We have seen that an associative form on a simple Lie algebra is unique up to scale. If we are interested in the Killing form only up to scale, we can compute the form $\mathrm{tr}(\rho(a)\rho(b))$ for any linear representation of L, not necessarily the adjoint one.

This is in particular true for the classical Lie algebras which are presented from the beginning as algebras of matrices.

In the examples we have that:

(1) $sl(n+1,\mathbb{C}) = A_n$: A Cartan algebra is formed by the space of diagonal matrices $h := \sum_{i=1}^{n+1} \alpha_i e_{ii}$, and $\sum_i \alpha_i = 0$. The spaces L_α are the 1-dimensional spaces generated by the root vectors $e_{ij},\ i \neq j$ and $[h, e_{ij}] = (\alpha_i - \alpha_j)e_{ij}$. Thus the linear forms $\sum_{i=1}^{n+1} \alpha_i e_{ii} \to \alpha_i - \alpha_j$ are the roots of $sl(n+1,\mathbb{C})$. We can consider the α_i as an orthonormal basis of a real Euclidean space $\mathbb{R}^{n+1}$. We have the root system A_n.

The positive roots are the elements $\alpha_i - \alpha_j,\ i < j$. The corresponding root vectors $e_{ij},\ i < j$ span the Lie subalgebra of strictly upper triangular matrices, and similarly for negative roots

$$(4.1.1)\qquad \mathfrak{u}^+ := \bigoplus\nolimits_{i<j} \mathbb{C}e_{ij},\ \mathfrak{u}^- := \bigoplus\nolimits_{i>j} \mathbb{C}e_{ij},\ \mathfrak{t} := \bigoplus\nolimits_{i=1}^{n} \mathbb{C}(e_{i,i} - e_{i+1,i+1}).$$

The simple roots and the root vectors associated to simple roots are

$$(4.1.2)\qquad \alpha_i - \alpha_{i+1},\ e_{i,i+1}$$

The Chevalley generators are

$$(4.1.3)\qquad e_i := e_{i,i+1}, \quad f_i := e_{i+1,i}, \quad h_i := e_{i,i} - e_{i+1,i+1}.$$

As for the Killing form let us apply 1.9.2 to a diagonal matrix with entries $x_i,\ i = 1,\dots,n+1$, $\sum x_i = 0$ to get

$$\sum_{i\neq j}(x_i - x_j)^2 = 2(n+1)\sum_{i=1}^{n+1} x_i^2.$$

Using the remarks after 1.9.2 and in §1.4, and the fact that $sl(n+1,\mathbb{C})$ is simple, we see that for any two matrices A, B the Killing form is $2(n+1)\,\mathrm{tr}(AB)$.

(2) $so(2n+1,\mathbb{C}) = B_n$: In block form a matrix $A := \begin{pmatrix} a & b & e \\ c & d & f \\ m & n & p \end{pmatrix}$ satisfies $A^t I_{2n+1} = -I_{2n+1}A$ if and only if $d = -a^t$, b, c are skew symmetric, $p = 0$, $n = -e^t$, $m = -f^t$.

A Cartan subalgebra is formed by the diagonal matrices

$$h := \sum_{i=1}^{n} \alpha_i (e_{ii} - e_{n+i,n+i}).$$

Root vectors are

$$e_{ij} - e_{n+j,n+i},\ i \neq j \leq n,\ e_{i,n+j} - e_{j,n+i},$$

$$i \neq j \leq n,\ e_{n+i,j} - e_{n+j,i}\ i \neq j \leq n$$

$$e_{i,2n+1} - e_{2n+1,i+n}, e_{n+i,2n+1} - e_{2n+1,i}, \quad i = 1, \dots, n$$

with roots

(4.1.4) $$\alpha_i - \alpha_j,\ \alpha_i + \alpha_j,\ -\alpha_i - \alpha_j,\ i \neq j \leq n, \quad \pm\alpha_i,\ i = 1, \dots, n.$$

We have the root system of type B_n. For $so(2n+1, \mathbb{C})$ we set

(4.1.5) $$\Phi^+ := \alpha_i - \alpha_j,\ \alpha_i + \alpha_j,\ i < j \leq n,\ \alpha_i.$$

The simple roots and the root vectors associated to the simple roots are

(4.1.6) $$\alpha_i - \alpha_{i+1},\ \alpha_n; \quad e_{i,i+1} - e_{n+i+1,n+i},\ e_{n,2n+1} - e_{2n+1,2n}.$$

The Chevalley generators are

(4.1.7) $$e_i := e_{i,i+1} - e_{n+i+1,n+i},\ e_n := e_{n,2n+1} - e_{2n+1,2n}.$$

(4.1.8) $$f_i := e_{i+1,i} - e_{n+i,n+i+1},\ f_n := e_{2n,2n+1} - e_{2n+1,n}.$$

$$h_i := e_{i,i} - a_{i+1,i+1} - e_{n+i,n+i} + e_{n+i+1,n+i+1},$$

(4.1.9) $$h_n := e_{n,n} - e_{2n,2n}.$$

As for the Killing form, we apply 1.9.2 to a diagonal matrix with entries x_i, $i = 1, \dots, n$, and $-x_i$, $i = n+1, \dots, 2n$ and get (using 4.1.4):

$$\sum_{i \neq j} [(x_i - x_j)^2 + (x_i + x_j)^2 + 2x_i^2] = 2(n+1) \sum_{i=1}^{n} x_i^2.$$

Using the remark after 1.9.2 and the fact that $so(2n+1, \mathbb{C})$ is simple, we see that for any two matrices A, B the Killing form is $(n+1)\operatorname{tr}(AB)$.

$$\mathfrak{u}^+ := \bigoplus_{i<j} \mathbb{C}(e_{ij} - e_{n+j,n+i}) \bigoplus_{i \neq j} \mathbb{C}(e_{i,n+j} - e_{j,n+i})$$

(4.1.10) $$\bigoplus_{i=1}^{n} \mathbb{C}(e_{i,2n+1} - e_{2n+1,i}),$$

$$\mathfrak{u}^- := \bigoplus_{i>j} \mathbb{C}(e_{ij} - e_{n+j,n+i}) \bigoplus_{i \neq j} \mathbb{C}(e_{n+i,j} - e_{n+j,i})$$

(4.1.11) $$\bigoplus_{i=n+1}^{2n} \mathbb{C}(e_{i,2n+1} - e_{2n+1,i}).$$

(3) $sp(2n, \mathbb{C}) = C_n$ In block form a matrix $A := \begin{vmatrix} a & b \\ c & d \end{vmatrix}$ satisfies $A^t J_{2n} = -J_{2n} A$ if and only if $d = -a^t$ and b, c are symmetric.
A Cartan subalgebra is formed by the diagonal matrices

$$h := \sum_{i=1}^{n} \alpha_i (e_{ii} - e_{n+i,n+i}).$$

Root vectors are the elements

$$e_{ij} - e_{n+j,n+i},\ i \neq j \leq n,\ e_{i,n+j} + e_{j,n+i},\ \ i, j \leq n,\ e_{n+i,j} + e_{n+j,i}\ i, j \leq n.$$

with roots

(4.1.12) $$\alpha_i - \alpha_j,\ \alpha_i + \alpha_j,\ -\alpha_i - \alpha_j,\ i \neq j,\ \pm 2\alpha_i,\ i = 1, \dots, n.$$

We have a root system of type C_n.
For $sp(2n, \mathbb{C})$ we set

(4.1.13) $$\Phi^+ := \alpha_i - \alpha_j,\ \alpha_i + \alpha_j,\ i < j, 2\alpha_i.$$

The simple roots and the root vectors associated to simple roots are

(4.1.14) $$\alpha_i - \alpha_{i+1}, 2\alpha_n, \quad e_{i,i+1} - e_{n+i+1,n+i},\ e_{n,2n}.$$

The Chevalley generators are

$$e_i := e_{i,i+1} - e_{n+i+1,n+i},\ i = 1, \dots, n-1,\ e_n := e_{n,2n}$$

$$f_i := e_{i+1,i} - e_{n+i,n+i+1},\ i = 1, \dots, n-1,\ f_n := e_{2n,n}$$

$$h_i := e_{i,i} - e_{i+1,i+1} + e_{n+i+1,n+i+1} - e_{n+i,n+i},\ i < n,$$

$$h_n := e_{n,n} - e_{2n,2n}.$$

As for the Killing form, we apply 1.9.2 to a diagonal matrix with entries x_i, $i = 1, \dots, n, -x_i,\ i = n+1, \dots, 2n$, and get (using 4.1.14):

$$\sum_{i \neq j} [(x_l - x_j)^2 + (x_i + x_j)^2 + 8x_i^2] = 2(n+4) \sum_{i=1}^{n} x_i^2.$$

Using the usual remarks and the fact that $sp(2n, \mathbb{C})$ is simple we see that for any two matrices A, B the Killing form is $(n+4)\ \mathrm{tr}(AB)$. We have $(i, j \leq n)$

(4.1.15) $$\mathfrak{u}^+ := \bigoplus\nolimits_{i<j} \mathbb{C}(e_{ij} - e_{n+j,n+i}) \bigoplus\nolimits_{i \neq j, \leq n} \mathbb{C}(e_{i,n+j} - e_{j,n+i}),$$

(4.1.16) $$\mathfrak{u}^+ := \bigoplus\nolimits_{i<j} \mathbb{C}(e_{ij} - e_{n+j,n+i}) \bigoplus\nolimits_{i,j \leq n} \mathbb{C}(e_{i,n+j} + e_{j,n+i}),$$

(4.1.17) $$\mathfrak{u}^- := \bigoplus\nolimits_{i>j} \mathbb{C}(e_{ij} - e_{n+j,n+i}) \bigoplus\nolimits_{i \neq j} \mathbb{C}(e_{n+i,j} + e_{n+j,i}).$$

In this case the Lie algebra $\mathfrak{u}^+$ in block matrix form is the matrices $\begin{vmatrix} a & b \\ 0 & -a^t \end{vmatrix}$ with a strictly upper triangular and b symmetric.

(4) $so(2n,\mathbb{C}) = D_n$: In block form a matrix $A := \begin{pmatrix} a & b \\ c & d \end{pmatrix}$ satisfies $A^t I_{2n} = -I_{2n} A$ if and only if $d = -a^t$ and b, c are skew symmetric.
A Cartan subalgebra is formed by the diagonal matrices

$$h := \sum_{i=1}^{n} \alpha_i (e_{ii} - e_{n+i,n+i}).$$

Root vectors are the elements

$$e_{ij} - e_{n+j,n+i},\ i \neq j \leq n,\ e_{i,n+j} - e_{j,n+i},$$
$$i \neq j \leq n,\ e_{n+i,j} - e_{n+j,i}\ i \neq j \leq n$$

with roots

(4.1.18) $$\alpha_i - \alpha_j,\ \alpha_i + \alpha_j,\ -\alpha_i - \alpha_j,\ i \neq j \leq n.$$

We have the root system D_n.
For $so(2n,\mathbb{C})$ we set

(4.1.19) $$\Phi^+ := \alpha_i - \alpha_j, i < j,\ \alpha_i + \alpha_j,\ i \neq j.$$

The simple roots and the root vectors associated to simple roots are

(4.1.20) $$\alpha_i - \alpha_{i+1},\ \alpha_{n-1} + \alpha_n,\ e_{i,i+1} - e_{n+i+1,n+i},\ e_{n-1,2n} - e_{n,2n-1}.$$

The Chevalley generators are

(4.1.21)
$$e_i := e_{i,i+1} - e_{n+i+1,n+i},\ i = 1,\dots,n-1,\ e_n := e_{n-1,2n} - e_{n,2n-1}$$
$$f_i := e_{i+1,i} - e_{n+i,n+i+1},\ i = 1,\dots,n-1,\ f_n := e_{2n,n-1} - e_{2n-1,n}$$
$$h_i := e_{i,i} - e_{i+1,i+1} + e_{n+i+1,n+i+1} - e_{n+i,n+i},\ i < n,$$
$$h_n := e_{n-1,n-1} + e_{n,n} - e_{2n-1,2n-1} - e_{2n,2n}.$$

As for the Killing form, we apply 1.9.2 to a diagonal matrix with entries $x_i,\ i = 1,\dots,n, -x_i,\ i = 1,\dots,n$, and get (using 4.1.19):

$$\sum_{i \neq j} [(x_i - x_j)^2 + (x_i + x_j)^2] = 2n \sum_{i=1}^{n} x_i^2.$$

Using Remark 1.4 and the fact that $so(2n,\mathbb{C})$ is simple (at least if $n \geq 4$), we see that for any two matrices A, B the Killing form is $n\ \mathrm{tr}(AB)$. We have

(4.1.22) $$\mathfrak{u}^+ := \bigoplus_{i<j} \mathbb{C}(e_{ij} - e_{n+j,n+i}) \bigoplus_{i \neq j, \leq n} \mathbb{C}(e_{i,n+j} - e_{j,n+i}),$$

(4.1.23) $$\mathfrak{u}^- := \bigoplus_{i>j} \mathbb{C}(e_{ij} - e_{n+j,n+i}) \bigoplus_{i \neq j, \leq n} \mathbb{C}(e_{n+i,j} - e_{n+j,i}).$$

In this case the Lie algebra $\mathfrak{u}^+$ in block matrix form is the matrices $\begin{vmatrix} a & b \\ 0 & -a^t \end{vmatrix}$ with a strictly upper triangular and b skew symmetric.

4.2 Borel Subalgebras

For all these (as well as for all semisimple) Lie algebras we have the direct sum decomposition (as vector space)

$$L = \mathfrak{u}^+ \oplus \mathfrak{t} \oplus \mathfrak{u}^-.$$

One sets

$$\mathfrak{b}^+ := \mathfrak{u}^+ \oplus \mathfrak{t}, \ \mathfrak{b}^+ := \mathfrak{u}^- \oplus \mathfrak{t};$$

these are called two opposite Borel subalgebras.

Theorem. *The fundamental property of the Borel subalgebras is that they are maximal solvable.*

Proof. To see that they are solvable we repeat a remark used previously. Suppose that a, b are two root vectors; so for $t \in \mathfrak{t}$ we have $[t, a] = \alpha(t)a, \ [t, b] = \beta(t)b$. Then $[t, [a, b]] = [[t, a], b] + [a, [t, b]] = (\alpha(t) + \beta(t))[a, b]$. In other words, $[a, b]$ is a weight vector (maybe 0) of weight $(\alpha(t) + \beta(t))$.

The next remark is that a positive root $\alpha = \sum_i n_i \alpha_i$ (the α_i simple) has a positive *height* $ht(\alpha) = \sum_i n_i$. For instance, in the case of A_n, with simple roots $\delta_i = \alpha_i - \alpha_{i+1}$, the positive root $\alpha_i - \alpha_j = \sum_{h=1}^{j-i} \delta_{i+h}$. Hence $ht(\alpha_i - \alpha_j) = j - i$.

So let $\mathfrak{b}_k$ be the subspace of $\mathfrak{b}^+$ spanned by the root vectors relative to roots of height $\geq k$ (visualize it for the classical groups). We get that $[\mathfrak{b}_k, \mathfrak{b}_h] \subset \mathfrak{b}_{k+h}$. Moreover, $\mathfrak{b}_1 = \mathfrak{u}^+$ and $[\mathfrak{b}^+, \mathfrak{b}^+] = [\mathfrak{u}^+ \oplus \mathfrak{t}, \mathfrak{u}^+ \oplus \mathfrak{t}] = [\mathfrak{u}^+, \mathfrak{u}^+] + [\mathfrak{t}, \mathfrak{u}^+] = \mathfrak{u}^+ = \mathfrak{u}_1$. From these two facts it follows inductively that the kth term of the derived series is contained in $\mathfrak{b}_k$, and so the algebra is solvable.

To see that it is maximal solvable, consider a proper subalgebra $\mathfrak{a} \supset \mathfrak{b}^+$. Since $\mathfrak{a}$ is stable under $\mathrm{ad}(t), \ t \in \mathfrak{t}$, $\mathfrak{a}$ must contain a root vector f_α for a negative root α. But then $\mathfrak{a}$ contains the subalgebra generated by f_α, e_α which is certainly not solvable, being isomorphic to $sl(2, \mathbb{C})$.

5 Highest Weight Theory

In this section we complete the work and classify the finite-dimensional irreducible representations of a semisimple Lie algebra, proving that they are in 1-1 correspondence with *dominant weights*.

5.1 Weights in Representations, Highest Weight Theory

Let L be a semisimple Lie algebra. Theorem 2, 1.4 tells us that all finite-dimensional representations of L are completely reducible.

Theorem 2, 1.5 implies that any finite-dimensional representation M of L has a basis formed by weight vectors under the Cartan subalgebra $\mathfrak{t}$.

Lemma. *The weights that may appear are exactly the weights defined abstractly for the corresponding root system.*

Proof. By the representation theory of $sl(2, \mathbb{C})$, each h_i acts in a semisimple way on M. So, since the h_i commute, M has a basis of weight vectors for $\mathfrak{t}$. Moreover, we know that if u is a weight vector for h_i, then we have $h_i u = n_i u,\ n_i \in \mathbb{Z}$. Therefore, if u is a weight vector of weight χ we have $\chi(h_i) \in \mathbb{Z},\ \forall i$. Recall (1.8.1) that $\chi(h_i) = \frac{2(\chi,\alpha_i)}{(\alpha_i,\alpha_i)} = \langle \chi \mid \alpha_i \rangle$. By Definition 2.4.1 we have $\chi \in \Lambda$. □

We should make some basic remarks in the case of the classical groups, relative to weights for the Cartan subalgebra and for the maximal torus.

Start with $GL(n, \mathbb{C})$: its maximal torus T consists of diagonal matrices with nonzero entries x_i, with Lie algebra the diagonal matrices with entries a_i. Using the exponential and setting $x_i = e^{a_i}$, we then have that given a rational representation of $GL(n, \mathbb{C})$, a vector v is a weight vector for T if and only if it is a weight vector for $\mathfrak{t}$ and the two weights are related, as they are $\prod_{i=1}^n x_i^{m_i}$, $\sum_{i=1}^n m_i a_i$.

We now treat all the classical groups as matrix groups, using the notations of 4.1, and make explicit the fundamental weights and the special weight $\rho := \sum_i \omega_i$.

For $sl(n+1)$,

$$\omega_k := \sum_{i\leq k} \alpha_i, \quad \prod_{i\leq k} x_i, \quad i \leq n, \quad \rho = \sum_{i=1}^n (n+1-i)\alpha_i, \quad \prod_{i=1}^n x_i^{n+1-i}.$$

For $so(2n)$ we again consider diagonal matrices with entries $\alpha_i, -\alpha_i$. The weights are

(5.1.1)
$$\omega_k := \sum_{i\leq k} \alpha_i,\ i \leq n-2,\ s_\pm := \frac{1}{2}\left(\sum_{i=1}^{n-1} \alpha_i \pm \alpha_n\right),$$

$$\rho = \sum_{i=1}^{n-1} (n-i-1/2)\alpha_i, \quad \prod_{i=1}^{n-1} x_i^{n-1/2-i}.$$

This shows already that the last two weights do not exponentiate to weights of the maximal torus of $SO(2n, \mathbb{C})$. The reason is that there is a double covering of this group, the spin group, which possesses these two representations called *half spin representations* which do not factor through $SO(2n, \mathbb{C})$. We study them in detail in Chapter 11, §7.2.

For $so(2n+1)$, the fundamental weights and ρ are

$$\omega_k := \sum_{i\leq k} \alpha_i,\ i \leq n-1,\ s := \frac{1}{2}\left(\sum_{i=1}^{n} \alpha_i\right),$$

(5.1.2)
$$\rho = \sum_{i=1}^{n} (n-i+1/2)\alpha_i, \quad \prod_{i=1}^{n} x_i^{n+1/2-i}.$$

The discussion of the spin group is similar (Chapter 11, §7.1) .

For $sp(2n)$ the fundamental weights and ρ are

(5.1.3) $$\omega_k := \sum_{i\leq k}\alpha_i,\ i\leq n;\quad \rho=\sum_{i=1}^n (n+1-i)\alpha_i,\quad \prod_{i=1}^n x_i^{n+1-i}.$$

5.2 Highest Weight Theory

Highest weight theory is a way of determining a *leading term* in the character of a representation. For this, it is convenient to introduce the *dominance order* of weights.

Definition 1. Given two weights λ, μ we say that $\lambda \prec \mu$ if $\lambda-\mu$ is a linear combination of simple roots with nonnegative coefficients.

Notice that $\lambda \prec \mu$, called *dominance order*, is a partial order on weights.

Proposition 1. *Given a finite-dimensional irreducible representation M of a semisimple Lie algebra L.*

(1) The space of vectors $M^+ := \{m \in M \mid \mathfrak{u}^+ m = 0\}$ is 1-dimensional and a weight space under $\mathfrak{t}$ of some weight λ. λ is called the highest weight of M. A nonzero vector $v \in M^+$ is called a highest weight vector and denoted v_λ.
(2) M^+ is the unique 1-dimensional subspace of M stable under the subalgebra $\mathfrak{b}^+$.
(3) λ is a dominant weight.
(4) M is spanned by the vectors obtained from M^+ applying elements from $\mathfrak{u}_-$.
(5) All the other weights are strictly less of λ in the dominance order.

Proof. From the theorem of Lie it follows that there is a nonzero eigenvector v, of some weight λ, for the solvable Lie algebra $\mathfrak{b}^+$. Consider the subspace of M spanned by the vectors $f_{i_1}f_{i_2}\dots f_{i_k}v$ obtained from v by acting repeatedly with the elements f_i (of weight the negative simple roots $-\alpha_i$). From the commutation relations, if $h\in\mathfrak{t}$:

$$hf_{i_1}f_{i_2}\cdots f_{i_k}v = \sum_{j=1}^k f_{i_1}f_{i_2}\dots[h, f_{i_j}]\dots f_{i_k}v + f_{i_1}f_{i_2}\dots f_{i_k}hv$$

$$= \left(\lambda-\sum_{j=1}^k \alpha_{i_j}\right)(h) f_{i_1}f_{i_2}\dots f_{i_k}v.$$

$$e_i f_{i_1}f_{i_2}\dots f_{i_k}v = \sum_{j=1}^k f_{i_1}f_{i_2}\dots[e_i, f_{i_j}]\dots f_{i_k}v + f_{i_1}f_{i_2}\dots f_{i_k}e_i v$$

$$= \sum_{j=1}^k f_{i_1}f_{i_2}\dots\delta^i_{i_j}h_i\dots f_{i_k}v.$$

We see that the vectors $f_{i_1}f_{i_2}\dots f_{i_k}v$ are weight vectors and span a stable submodule. Hence by the irreducibility of M, they span the whole of M. The weights we have computed are all strictly less than λ in the dominance order, except for the weight of v which is λ.

The set of vectors $\{u \in M \mid \mathfrak{u}^+ m = 0\}$ is clearly stable under $\mathfrak{t}$, and since M has a basis of eigenvectors for $\mathfrak{t}$, so must this subspace. If there were another vector u with this property, then there would be one which is also an eigenvector (under $\mathfrak{t}$) of some eigenvalue μ. The same argument shows that $\lambda \prec \mu$. Hence $\lambda = \mu$ and $u \in \mathbb{C}v$. This proves all points except for three. Then let v be a highest weight vector of weight λ, in particular $e_i v = 0$ for all the e_i associated to the simple roots. This means that v is a highest weight vector for each $sl(2, \mathbb{C})$ of type e_i, h_i, f_i. By the theory for $sl(2, \mathbb{C})$, §1.1, it follows that $h_i v = k_i v$ for some $k_i \in \mathbb{N}$ a nonnegative integer, or $\lambda(h_i) = k_i$ for all i. This is the condition of dominance. In fact, Formula 1.9.1 states that if α_i is the simple root corresponding to the given e_i, we have $\lambda(h_i) = \frac{2(\lambda,\alpha_i)}{(\alpha_i,\alpha_i)} = \langle \lambda \mid \alpha_i \rangle$. □

The classification of finite-dimensional irreducible modules states that each dominant weight corresponds to one and only one irreducible representation. The existence can be shown in several different ways. One can take an algebraic point of view as in [Hu1] and define the module by generators and relations in the same spirit as in the proof of existence of semisimple Lie algebras through Serre's relations.

The other approach is to relate representations of semisimple Lie algebras with that of compact groups, and use the Weyl character formula, as in [A], [Ze]. In our work on classical groups we will in fact exhibit all the finite-dimensional irreducible representations of classical groups explicitly in tensor spaces.

The uniqueness is much simpler. In general, for any representation M, a vector $m \in M$ is called a *highest weight vector* if $\mathfrak{u}^+ m = 0$ and m is a weight vector under $\mathfrak{t}$.

Proposition 2. *Let M be a finite-dimensional representation of a semisimple Lie algebra L and u a highest weight vector (of weight λ). The L-submodule generated by u is irreducible.*

Proof. We can assume without loss of generality that the L-submodule generated by u is M. The same argument given in the proof of the previous theorem shows that u is the only weight vector of weight λ. Decompose $M = \bigoplus_i N_i$ into irreducible representations. Each irreducible decomposes into weight spaces, but from the previous remark, u must be contained in one of the summands N_i. Since M is the minimal submodule containing u, we have $M = N_i$ is irreducible. □

We can now prove the uniqueness of an irreducible module with a given highest weight.

Theorem. *Two finite-dimensional irreducible representations of a semisimple Lie algebra L are isomorphic if and only if they have the same highest weight.*

Proof. Suppose we have given two such modules N_1, N_2 with highest weight λ and highest weight vectors u_1, u_2, respectively.

In $N_1 \oplus N_2$, consider the vector (u_1, u_2); it is clearly a highest weight vector, and so it generates an irreducible submodule N. Now projecting to the two summands we see that N is isomorphic to both N_1 and N_2, which are therefore isomorphic. □

It is quite important to observe that given two finite-dimensional representations M, N, we have the following:

Proposition 3. *If $u \in M$, $v \in N$ are two highest weight vectors of weight λ, μ respectively, then $u \otimes v$ is a highest weight vector of weight $\lambda + \mu$. If all other weights in M (resp. N) are strictly less than λ (resp. μ) in the dominance order, then all other weights in $M \otimes N$ are strictly less than $\lambda + \mu$ in the dominance order.*[97]

Proof. By definition we have $e(u \otimes v) = eu \otimes v + u \otimes ev$ for every element e of the Lie algebra. From this the claim follows easily. □

In particular, we will use this fact in the following forms.

(1) *Cartan multiplication* Given two dominant weights λ, μ we have $M_\lambda \otimes M_\mu = M_{\lambda+\mu} + M'$, where M' is a sum of irreducibles with highest weight strictly less than $\lambda+\mu$. In particular we have the canonical projection $\pi : M_\lambda \otimes M_\mu \to M_{\lambda+\mu}$ with kernel M'. The composition $M_\lambda \times M_\mu \to M_\lambda \otimes M_\mu \xrightarrow{\pi} M_{\lambda+\mu}$, $(m, n) \mapsto \pi(m \otimes n)$ is called *Cartan multiplication*.
(2) Take an irreducible representation V_λ with highest weight λ and highest weight vector v_λ and consider the second symmetric power $S^2(V_\lambda)$.

Corollary. *$S^2(V_\lambda)$ contains the irreducible representation $V_{2\lambda}$ with multiplicity* 1.

Proof. $v_\lambda \otimes v_\lambda$ is a highest weight vector. By the previous proposition it generates the irreducible representation $V_{2\lambda}$. It is a symmetric tensor, so $V_{2\lambda} \subset S^2(V_\lambda)$. Finally since all other weights are strictly less than 2λ, the representation $V_{2\lambda}$ appears with multiplicity 1. □

5.3 Existence of Irreducible Modules

We arrive now at the final existence theorem. It is better to use the language of associative algebras, and present irreducible L-modules as cyclic modules over the enveloping algebra U_L. The PBW theorem and the decomposition $L = \mathfrak{u}^- \oplus \mathfrak{t} \oplus \mathfrak{u}^+$ imply that U_L as a vector space is $U_L = U_{\mathfrak{u}^-} \otimes U_{\mathfrak{t}} \otimes U_{\mathfrak{u}^+}$. This allows us to perform the following construction. If $\lambda \in \mathfrak{t}^*$, consider the 1-dimensional representation $\mathbb{C}_\lambda := \mathbb{C}u_\lambda$ of $\mathfrak{t} \oplus \mathfrak{u}^+$, with basis a vector u_λ, given by $hu_\lambda = \lambda(h)u_\lambda$, $\forall h \in \mathfrak{t}$, $e_i u_\lambda = 0$. $\mathbb{C}_\lambda$ induces a module over U_L called $V(\lambda) := U_L \otimes_{U_{\mathfrak{t}} \otimes U_{\mathfrak{u}^+}} \mathbb{C}u_\lambda$, called the *Verma module*. Equivalently, it is the cyclic left U_L-module subject to the defining relations

$$hu_\lambda = \lambda(h)u_\lambda, \ \forall h \in \mathfrak{t}, \quad e_i u_\lambda = 0, \ \forall i. \tag{5.3.1}$$

We remark then that given any module M and a vector $v \in M$ subject to 5.3.1 we have a map $j : V(\lambda) \to M$ mapping u_λ to v. Such a v is also called a *singular vector*. It is easily seen from the PBW theorem that the map $a \mapsto au_\lambda$, $a \in U_{\mathfrak{u}^-}$ establishes a linear isomorphism between $U_{\mathfrak{u}^-}$ and $V(\lambda)$. Of course the extra L-module structure on $V(\lambda)$ depends on λ. The module $V(\lambda)$ shares some basic properties of irreducible modules, although in general it is not irreducible and it is always infinite dimensional.

[97] We do not assume irreducibility.

(1) It is generated by a unique vector which is a weight vector under $\mathfrak{t}$.
(2) It has a basis of weight vectors and the weights are all less than λ in the dominance order.
(3) Moreover, each weight space is finite dimensional. For a weight γ the dimension of its weight space is the number of ways we can write $\gamma = \lambda - \sum_{\alpha\in\Phi^+} m_\alpha \alpha$.

It follows in particular that the sum of all proper submodules is a proper submodule and the quotient of $V(\lambda)$ by the maximal proper submodule is an irreducible module, denoted by $L(\lambda)$. By construction, $L(\lambda)$ has a unique singular vector, the image of u_λ.

Theorem. *$L(\lambda)$ is finite dimensional if and only if λ is dominant.*

The set of highest weights coincides with the set of dominant weights.

The finite-dimensional irreducible representations of a semisimple Lie algebra L are the modules $L(\lambda)$ parameterized by the dominant weights.

When λ is dominant, and a weight μ appears in $L(\lambda)$, then μ is in the W-orbit of the finite set of dominant weights $\mu \preceq \lambda$.

Proof. First, let N_λ be a finite-dimensional irreducible module with highest weight vector v_λ and highest weight λ. By Proposition 1 of §5.2, λ is dominant. We clearly have a map of $V(\lambda)$ to N_λ mapping u_λ to v_λ. Clearly this map induces an isomorphism between $L(\lambda)$ and N_λ. Thus, the theorem is proved if we see that if λ is dominant, then $L(\lambda)$ is finite dimensional. We compute now in $L(\lambda)$. Call v_λ the class of u_λ. The first statement is that $f_i^{\lambda(h_i)+1} v_\lambda = 0$. For this it suffices to see that $f_i^{\lambda(h_i)+1} u_\lambda$ is a singular vector in $V(\lambda)$. $f_i^{\lambda(h_i)+1} u_\lambda$ is certainly a weight vector under $\mathfrak{t}$, so we need to show that $e_j f_i^{\lambda(h_i)+1} u_\lambda$ for all j. If $i \neq j$, e_j commutes with f_i and $e_j f_i^{\lambda(h_i)+1} u_\lambda = f_i^{\lambda(h_i)+1} e_j u_\lambda = 0$. Otherwise, the argument of Lemma 1.1, which we already used in Serre's existence theorem, shows that $e_i f_i^{\lambda(h_i)+1} u_\lambda = 0$. The argument of Lemma 1.1 shows that v_λ generates, under the $sl_i(2,\mathbb{C})$ given by e_i, h_i, f_i, an irreducible module of dimension $\lambda(h_i) + 1$.

The next statement we prove is:

For each i, $L(\lambda)$ is a direct sum of finite-dimensional irreducible $sl_i(2,\mathbb{C})$ modules.

To prove this let M be the sum of all such irreducibles. $M \neq 0$ from what we just proved. It is enough to see that M is an L-submodule or that $aM \subset M$, $\forall a \in L$. If $N \subset M$ is a finite-dimensional $sl_i(2,\mathbb{C})$ submodule, consider $N' := \sum_{a\in L} aN$. This is clearly a finite-dimensional subspace and we claim that it is also an $sl_i(2,\mathbb{C})$ submodule. In fact, if $u \in sl_i(2,\mathbb{C})$, we have $uaN \subset [u,a]N + auN \subset [u,a]N + aN \subset N'$. Thus M is an L-submodule.

Having established the previous statement we can integrate each $sl_i(2,\mathbb{C})$ action to an action of the group $SL_i(2,\mathbb{C})$. As usual we find an action of the elements s_i which permutes weight spaces. From our constraint on weights, it follows that the only weights which can appear are those γ such that $w(\gamma) \prec \lambda, \forall w \in W$. We know (Theorem 2.4) that each weight is W-conjugate to a dominant weight. Even if the simple roots are not a basis of the weight lattice, we can still write a weight

$\lambda = \sum_i m_i \alpha_i$ where m_i are rational numbers with denominator a fixed integer d, for instance the index of the root lattice in the weight lattice. If $\lambda = \sum_i m_i \alpha_i$, $\mu = \sum_i p_i \alpha_i$ are two dominant weights, the condition $\mu \prec \lambda$ means that $\sum_i (m_i - n_i)\alpha_i$ is a positive linear combination of positive roots. This implies that $n_i \leq m_i$ for all i. Since also $0 \leq n_i$ and dn_i are integers, we have that the set of dominant weights satisfying $\mu \prec \lambda$ is finite. We can finally deduce that $L(\lambda)$ is finite dimensional, since it is the sum of its weight spaces, each of finite dimension. The weights appearing are in the W-orbits of the finite set of dominant weights $\mu \prec \lambda$. □

Modules under a Lie algebra can be composed by tensor product and also dualized. In general, a tensor product $L(\lambda) \otimes L(\mu)$ is not irreducible. To determine its decomposition into irreducibles is a rather difficult task and the known answers are just algorithms. Nevertheless by Proposition 3 of §5.2, we know that $L(\lambda) \otimes L(\mu)$ contains the *leading term* $L(\lambda + \mu)$. Duality is a much easier issue:

Proposition. $L(\lambda)^* = L(-w_0(\lambda))$, *where* w_0 *is the longest element of* W.

Proof. The dual of an irreducible representation is also irreducible, so $L(\lambda)^* = L(\mu)$ for some dominant weight μ to be determined. Let u_i be a basis of weight vectors for $L(\lambda)$ with weights μ_i. The dual basis u^i, by the basic definition of dual action, is a basis of weight vectors for $L(\lambda)^*$ with weights $-\mu_i$. Thus the dual of the highest weight vector is a *lowest weight vector* with weight $-\lambda$ in $L(\lambda)^*$. The weights of $L(\lambda)^*$ are stable under the action of the Weyl group. The longest element w_0 of W (2.3) maps negative roots into positive roots, and hence reverses the dominance order. We deduce that $-w_0(\lambda)$ is the highest weight for $L(\lambda)^*$. □

6 Semisimple Groups

6.1 Dual Hopf Algebras

At this point there is one important fact which needs clarification. We have classified semisimple Lie algebras and their representations and we have proved that the adjoint group G_L associated to such a Lie algebra is an algebraic group.

It is not true (not even for $sl(2, \mathbb{C})$) that a representation of the Lie algebra integrates to a representation of G_L. We can see this in two ways that have to be put into correspondence. The first is by inspecting the weights. We know that in general the weight lattice is bigger than the root lattice. On the other hand, it is clear that the weights of the maximal torus of the adjoint group are generated by the roots. Thus, whenever in a representation we have a highest weight which is not in the root lattice, this representation cannot be integrated to the adjoint group. The second approach comes from the fact that in any case a representation of the Lie algebra can be integrated to the simply connected universal covering. One has to understand what this simply connected group is.

A possible construction is via the method of Hopf algebras, Chapter 8, §7.2. We can define the simply connected group as the spectrum of its *Hopf algebra of matrix coefficients*.

The axioms §7.2 of that chapter have a formal duality exchanging multiplication and comultiplication. This suggests that, given a Hopf algebra A, we can define a *dual Hopf algebra* structure on the dual A^* exchanging multiplication and comultiplication. This procedure works perfectly if A is finite dimensional, but in general we encounter the difficulty that $(A \otimes A)^*$ is much bigger than $A^* \otimes A^*$. The standard way to overcome this difficulty is to restrict the dual to:

Definition. The finite dual A^f of an algebra A, over a field F, is the space of linear forms $\phi : A \to F$ such that the kernel of ϕ contains a left ideal of finite codimension.

On A^f we can define multiplication, comultiplication, antipode, unit and counit as dual maps of comultiplication, multiplication, antipode, counit and unit in A.

Remark. If J is a left ideal and $\dim A/J < \infty$, the homomorphism $A \to \mathrm{End}(A/J)$ has as kernel a two-sided ideal I contained in J. So the condition for the finite dual could also be replaced by the condition that the kernel of ϕ contains a two-sided ideal of finite codimension.

Again we can consider the elements of the finite dual as *matrix coefficients* for finite-dimensional modules. In fact, given a left ideal J of finite codimension, one has that A/J is a finite-dimensional cyclic A-module (generated by the class $\overline{1}$ of 1), and given a linear form Φ on A vanishing on J, this induces a linear form ϕ on A/J, $\phi(a\overline{1}) = \Phi(a)$. For $a \in A$ we have the formal matrix coefficient $\Phi(a) = \langle \phi \mid a\overline{1} \rangle$.

Conversely, let M be a finite-dimensional module, $\phi \in M^*$, $u \in M$, and consider the linear form $c_{\phi,u}(a) := \langle \phi \mid au \rangle$. This form vanishes on the left ideal $J := \{a \in A \mid au = 0\}$ and we call it a *matrix coefficient*.

Exercise. The reader should verify that on the finite dual the Hopf algebra structure dualizes.[98]

Let us at least remark how one performs multiplication of matrix coefficients. By definition if $\Phi, \Psi \in A^f$, we have by duality $\Phi\Psi(a) := \langle \Phi \otimes \Psi \mid \Delta(a) \rangle$. In other words, if $\Phi = c_{\phi,u}$ is a matrix coefficient for a finite-dimensional module M and $\Psi = c_{\psi,v}$ a matrix coefficient for a finite-dimensional module N, we have

$$\Phi\Psi(a) = \langle \phi \otimes \psi \mid \Delta(a)u \otimes v \rangle.$$

The formula $\Delta(a)u \otimes v$ is the definition of the tensor product action; thus we have the basic formula

$$c_{\phi,u}c_{\psi,v} = c_{\phi\otimes\psi,u\otimes v}. \tag{6.1.1}$$

Since comultiplication is coassociative, A^f is associative as an algebra. It is commutative if Δ is also cocommutative, as for enveloping algebras of Lie algebras.

[98] Nevertheless it may not be that A is the dual of A^f.

As far as comultiplication is concerned, it is the dual of multiplication, and thus $\langle \delta(\Phi), a \otimes b \rangle = \langle \Phi, ab \rangle$. When $\Phi = c_{\phi,u}$, is a matrix coefficient for a finite-dimensional module M, choose a basis u_i of M and let u^i be the dual basis. The identity $1_M = \sum_i u_i \otimes u^i$ and

$$\langle c_{\phi,u}, ab \rangle = \langle \phi, a 1_M bu \rangle = \langle \phi, a \sum_i u_i \otimes u^i bu \rangle$$

$$= \sum_i \langle \phi, au_i \rangle \langle u^i, bu \rangle = \sum_i c_{\phi,u_i}(a) c_{u^i,u}(b).$$

In other words,

$$\Delta(c_{\phi,u}) = \sum_i c_{\phi,u_i} \otimes c_{u^i,u}. \tag{6.1.2}$$

The unit element of A^f is the counit $\eta : A \to F$ of A, which is also a matrix coefficient for the trivial representation. Finally for the antipode and counit we have $S(c_{\phi,u})(a) = (c_{\phi,u})(S(a))$, $\eta(c_{\phi,u}) = c_{\phi,u}(1) = \langle \phi \,|\, u \rangle$.

Let us apply this construction to $A = U_L$, the universal enveloping algebra of a semisimple Lie algebra.

Recall that an enveloping algebra U_L is a *cocommutative* Hopf algebra with $\Delta(a) = a \otimes 1 + 1 \otimes a$, $S(a) = -a$, $\eta(a) = 0$, $\forall a \in L$. As a consequence we have a theorem whose proof mimics the formal properties of functions on a group.

Proposition 1. *(i) We have the Peter–Weyl decomposition, indexed by the dominant weights Λ^+:*

$$U_L^f = \bigoplus_{\lambda \in \Lambda^+} \operatorname{End}(V_\lambda)^* = \bigoplus_{\lambda \in \Lambda^+} V_\lambda \otimes V_\lambda^*. \tag{6.1.3}$$

(ii) U_L^f is a finitely generated commutative Hopf algebra. One can choose as generators the matrix coefficients of $V := \bigoplus_i V_{\omega_i}$, where the ω_i are the fundamental weights.

Proof. (i) If I is a two-sided ideal of finite codimension, U_L/I is a finite-dimensional representation of L; hence it is completely reducible and it is the sum of some irreducibles V_{λ_i}, for finitely many distinct dominant weights $\lambda_i \in \Lambda^+$. In other words, U_L/I is the regular representation of a semisimple algebra. From the results of Chapter 6, §2 it follows that the mapping $U_L \to \bigoplus_i \operatorname{End}(V_{\lambda_i})$ is surjective with kernel I. Thus for any finite set of distinct dominant weights the dual $\bigoplus_i \operatorname{End}(V_{\lambda_i})^*$ maps injectively into the space of matrix coefficients and any matrix coefficient is sum of elements of $\operatorname{End}(V_\lambda)^*$ as $\lambda \in \Lambda^+$.

(ii) We can use Cartan's multiplication (5.2) and formula 6.1.1 to see that $\operatorname{End}(V_{\lambda+\mu})^* \subset \operatorname{End}(V_\lambda)^* \operatorname{End}(V_\mu)^*$. Since any dominant weight is a sum of fundamental weights, the statement follows. □

From this proposition it follows that U_L^f is the coordinate ring of an algebraic group G_s. Our next task is to identify G_s and prove that it is semisimple and simply connected with Lie algebra L. We begin with:

Lemma. *Any finite-dimensional representation V of L integrates to an action of a semisimple algebraic group G_V, whose irreducible representations appear all in the tensor powers $V^{\otimes m}$.*

Proof. In fact let us use the decomposition $L = \mathfrak{u}^- \oplus \mathfrak{t} \oplus \mathfrak{u}^+$. Both $\mathfrak{u}^-$ and $\mathfrak{u}^+$ act as Lie algebras of nilpotent elements. Therefore (Chapter 7, Theorem 3.4) the map exp establishes a polynomial isomorphism (as varieties) of $\mathfrak{u}^-$ and $\mathfrak{u}^+$ to two unipotent algebraic groups U^-, U^+, of which they are the Lie algebras, acting on V. As for $\mathfrak{t}$ we know that V is a sum of weight spaces. Set Λ' to be the lattice spanned by these weights. Thus the action of $\mathfrak{t}$ integrates to an algebraic action of the torus T' having Λ' as a group of characters. Even more explicitly, if we fix a basis e_i of V of weight vectors with the property that if λ_i, λ_j are the weights of e_i, e_j, respectively, and $\lambda_i \prec \lambda_j$, then $i > j$ we see that U^- is a closed subgroup of the group of strictly lower triangular matrices, U^+ is a closed subgroup of the group of strictly upper triangular matrices, and T' is a closed subgroup of the group of diagonal matrices. Then the multiplication map embeds $U^- \times T' \times U^+$ as a closed subvariety of the open set of matrices which are the product of a lower triangular diagonal and an upper triangular matrix. On the other hand, $U^-T'U^+$ is contained in the Lie group $G_V \subset GL(V)$ which integrates L and is open in G_V. It follows (Chapter 4, Criterion 3.3) that G_V coincides with the closure of U^-TU^+ and it is therefore algebraic. Since L is semisimple, $V^{\otimes m}$ is a semisimple representation of G_V for all m. Since $L = [L, L]$ we have that G_V is contained in $SL(V)$. We can therefore apply Theorem 1.4 of Chapter 7 to deduce all the statements of the lemma. □

Theorem. *U_L^f is the coordinate ring of a linearly reductive semisimple algebraic group $G_s(L)$, with Lie algebra L, whose irreducible representations coincide with those of the Lie algebra L.*

Proof. Consider the representation $V := \bigoplus_i V_{\omega_i}$ of L, with the sum running over all fundamental weights. We want to see that the coordinate Hopf algebra of the group $G_s(L) := G_V$ (constructed in the previous lemma) is U_L^f. Now (by Cartan's multiplication) any irreducible representation of L appears in a tensor power of V, and thus every finite-dimensional representation of L integrates to an algebraic representation of $G_s(L)$. We claim that U_L^f, as a Hopf algebra, is identified with the coordinate ring $\mathbb{C}[G_s(L)]$ of $G_s(L)$. In fact, the map is the one which identifies, for any dominant weight λ, the space $\mathrm{End}(V_\lambda)^*$ as a space of matrix coefficients either of L or of $G_s(L)$. We have only to prove that the Hopf algebra operations are the same. There are two simple ideas to follow.

First, the algebra of regular functions on $G_s(L)$ is completely determined by its restriction to the dense set of elements $\exp(a)$, $a \in L$.

Second, although the element $\exp(a)$, being given by an infinite series, is not in the algebra U_L, nevertheless, any matrix coefficient $c_{\phi,u}$ of U_L extends by continuity to a function $\langle \phi \mid \exp(a)u \rangle$ on the elements $\exp(a)$ which, being the corresponding representation of the $G_s(L)$ algebraic group, is the restriction of a regular algebraic function.

In this way we start identifying the algebra of matrix coefficients U_L^f with the coordinate ring of $G_s(L)$. At least as vector spaces they are both $\bigoplus_{\lambda\in\Lambda^+} \mathrm{End}(V_\lambda)^*$.

Next, we have to verify some identities in order to prove that we have an isomorphism as Hopf algebras. To verify an algebraic identity on functions on $G_s(L)$, it is enough to verify it on the dense set of elements $\exp(a)$, $a \in L$. By 6.1.1, 6.1.2. and the definition of matrix coefficient in the two cases, it follows that we have an isomorphism of algebras and coalgebras. Also the isomorphism respects unit and counit, as one sees easily. For the antipode, one has only to recall that in U_L, for $a \in L$ we have $S(a) = -a$ so that $S(e^a) = e^{-a}$.

Since for functions on $G_s(L)$ the antipode is $Sf(g) = f(g^{-1})$, we have the compatibility expressed by $Sf(e^a) = f((e^a)^{-1}) = f(e^{-a}) = f(S(e^a))$. We have thus an isomorphism between U_L^f and $\mathbb{C}[G_s(L)]$ as Hopf algebras. □

It is still necessary to prove that the group $G_s(L)$, the spectrum of U_L^f, is simply connected and a finite universal cover of the adjoint group of L. Before doing this we draw some consequences of what we have already proved. Nevertheless we refer to $G_s(L)$ as the simply connected group. In any case we can prove:

Proposition 2. *Let G be an algebraic group with Lie algebra a semisimple Lie algebra L. Then G is a quotient of $G_s(L)$ by a finite group.*

Proof. Any representation of L which integrates to G integrates also to a representation of $G_s(L)$; thus we have the induced homomorphism. Since this induces an isomorphism of Lie algebras its kernel is discrete, and since it is also algebraic, it is finite. □

Let us make a final remark.

Proposition 3. *Given two semisimple Lie algebras L_1, L_2, we have*

$$G_s(L_1 \oplus L_2) = G_s(L_1) \times G_s(L_2).$$

Proof. The irreducible representations of $L_1 \oplus L_2$ are the tensor products $M \otimes N$ of irreducibles for the two Lie algebras; hence $U^f_{L_1 \oplus L_2} = U^f_{L_1} \otimes U^f_{L_2}$ form which the claim follows. □

6.2 Parabolic Subgroups

Let L be a simple Lie algebra, $\mathfrak{t}$ a maximal toral subalgebra, Φ^+ a system of positive roots and $\mathfrak{b}^+$ the corresponding Borel subalgebra, and $G_s(L)$ the group constructed in the previous section. Let ω_i be the corresponding fundamental weights and V_i the corresponding fundamental representations with highest weight vectors v_i. We consider a dominant weight $\lambda = \sum_{i \in J} m_i \omega_i$, $m_i > 0$. Here J is the set of indices i which appear with nonnegative coefficient. We want now to perform a construction which in algebraic geometry is known as the *Veronese embedding*. Consider now the tensor product $M := \otimes_{i \in J} V_i^{\otimes n_i}$. It is a representation of L (not irreducible) with highest weight vector $\otimes_{i \in J} v_i^{\otimes n_i}$. This vector generates inside M the irreducible representation V_λ.

We have an induced Veronese map of projective spaces (say $k = |J|$):

$$\pi_\lambda := \prod_{i\in J} \mathbb{P}(V_i) \to \mathbb{P}(M),$$
$$\pi(\mathbb{C}a_1, \dots, \mathbb{C}a_k) := \mathbb{C}a_1^{\otimes m_1} \otimes \dots \otimes a_k^{\otimes m_k}, \quad 0 \neq a_i \in V_i.$$

This map is an embedding and it is equivariant with respect to the algebraic group $G_s(L)$ (which integrates to the Lie algebra action). In particular we deduce that:

Proposition 1. *The stabilizer in $G_s(L)$ of the line through the highest weight vector v_λ is the intersection of the stabilizers H_i of the lines through the highest weight vectors v_i for the fundamental weights. We set $H_J := \bigcap_{i\in J} H_i$.*

Proposition 2. *Let B be the Borel subgroup, with Lie algebra $\mathfrak{b}^+$. If $H \supset B$ is an algebraic subgroup, then $B = H_J$ for a subset J of the nodes of the Dynkin diagram.*

Proof. Let H be any algebraic subgroup containing B. From Theorem 1, §2.1 Chapter 7, there is a representation M of G and a line ℓ in M such that H is the stabilizer of ℓ. Since $B \subset H$, ℓ is stabilized by B. The unipotent part of B (or of its Lie algebra $\mathfrak{u}^+$) must act trivially. Hence ℓ is generated by a highest weight vector in an irreducible representation of G (Proposition 2 of §5.2). Hence by Proposition 1, we have that H must be equal to one of the groups H_J. □

To identify all these groups let us first look at the Lie algebras. Given a set $J \subset \Delta$ of nodes of the Dynkin diagram,[99] let Φ_J denote the root system generated by the simple roots not in this set, i.e., the set of roots in Φ which are linear combinations of the α_i, $i \notin J$.

Remark. The elements α_i, $i \notin J$, form a system of simple roots for the root system Φ_J.

We can easily verify that

$$\mathfrak{p}_J := \mathfrak{b}^+ \bigoplus\nolimits_{\alpha\in\Phi^+\cap\Phi_J} L_{-\alpha}, \tag{6.2.1}$$

is a Lie algebra. Moreover one easily has that $\mathfrak{p}_A \cap \mathfrak{p}_B = \mathfrak{p}_{A\cup B}$ and that $\mathfrak{p}_J$ is generated by $\mathfrak{b}^+$ and the elements f_i for $i \notin J$. We set $\mathfrak{p}_i := \mathfrak{p}_{\{i\}}$ so that $\mathfrak{p}_J = \bigcap_{i\in J} \mathfrak{p}_i$. Let B, P_J, P_i be the connected groups in $G_s(L)$ with Lie algebras $\mathfrak{b}^+$, $\mathfrak{p}_J, \mathfrak{p}_i$.

Lemma. *The Lie algebra of H_J is $\mathfrak{p}_J$.*

Proof. Since $f_i v_\lambda = 0$ if and only if $\langle \lambda, \alpha_i \rangle = 0$, we have that f_i is in the Lie algebra of the stabilizer of v_λ if and only if $m_i = 0$, i.e., $i \notin J$. Thus $\mathfrak{p}_J$ is contained in the Lie algebra of H_J. Since these Lie algebras are all distinct and the H_J exhaust the list of all groups containing B, the claim follows. □

[99] We can identify the nodes with the simple roots.

Theorem. *$P_J = H_J$, in particular H_J, is connected.*

Proof. We have that P_J stabilizes ℓ_λ and contains B, so it must coincide with one of the groups H and it can be only H_J. □

Remark. Since in Proposition 2 we have seen that the subgroups H_J exhaust all the algebraic subgroups containing B, we have in particular that all the algebraic subgroups containing B are connected.[100]

It is important to understand the Levi decomposition for these groups and algebras. Decompose $\mathfrak{t} = \mathfrak{t}_J \oplus \mathfrak{t}_J^\perp$, where $\mathfrak{t}_J$ is spanned by the elements $h_i,\ i \notin J$ and $\mathfrak{t}_J^\perp$ is the orthogonal, i.e., $\mathfrak{t}_J^\perp = \{h \in \mathfrak{t} \mid \alpha_i(h) = 0,\ \forall i \notin J\}$.

Proposition 3. *(i) The algebra $\mathfrak{l}_J := \mathfrak{t}_J \oplus_{\alpha \in \Phi_J} L_\alpha$ is the Lie algebra of the (not necessarily irreducible) root system Φ_J.*

(ii) The algebra $\mathfrak{s}_J := \mathfrak{t}_J^\perp \oplus_{\alpha \in \Phi^+ - \Phi_J} L_\alpha$ is the solvable radical of $\mathfrak{p}_J$.

(iii) $\mathfrak{p}_J = \mathfrak{l}_J \oplus \mathfrak{s}_J$ is a Levi decomposition.

Proof. (i) The elements $e_i, f_i, h_i,\ \ i \notin J$ are in $\mathfrak{l}_J$ and satisfy Serre's relations for the root system Φ_J. By Serre's theorem we thus have a homomorphism from the Lie algebra associated to Φ_J to $\mathfrak{l}_J$. This map sends a basis to a basis, so it is an isomorphism.

(ii) $\mathfrak{p}_J$ is contained in the Borel subalgebra of L, so it is solvable. It is easily seen to be an ideal of $\mathfrak{p}_J$. Since $\mathfrak{p}_J/\mathfrak{s}_J = \mathfrak{l}_J$ is semisimple, it is the solvable radical.

(iii) Follows from (i), (ii). □

We finally need to understand the weight lattice, and dominant and fundamental weights associated to Φ_J. The main remark is that if $i \notin J$ and ω_i is the corresponding fundamental weight for $\mathfrak{t}$, since the elements $h_j,\ j \notin J$ span $\mathfrak{t}_J$ we have that ω_i restricted to $\mathfrak{t}_J$ coincides with the fundamental weight dual to $\check{\alpha}_i$.

Let $L_J \subset P_J \subset G_s(L)$ be the corresponding semisimple group.

Proposition 4. *L_J is simply connected.*

Proof. To prove that a semisimple group is simply connected (according to our provisional definition), it suffices to prove that all the representations of its Lie algebra integrate to representations of the group. Now if we restrict an irreducible representation V_λ of $G_s(L)$ to L_J, it will not remain irreducible, but its highest weight vector v_λ is still a highest weight vector for the corresponding Borel subalgebra $\mathfrak{t}_J \bigoplus_{\alpha \in \Phi_J^+} L_\alpha$ of L_J. From the previous discussion, all dominant weights appear in this way. □

The subgroup $L := L_J T$ has Lie algebra $\mathfrak{l}_J \oplus \mathfrak{t}^\perp$ and it is called a *Levi factor* of P_J, L_J is the semisimple part of the Levi factor. L is a connected reductive group. If U_J denotes the unipotent radical of P_J, i.e., the unipotent group with Lie algebra $\mathfrak{u}_J := \bigoplus_{\alpha \in \Phi^+ - \Phi_J} L_\alpha$, we have that $L \cap U_J = 1$. This follows from the fact that a unipotent group with trivial Lie algebra is trivial. It gives the *Levi decomposition* for P:

100 With a more careful analysis in fact one can drop from the hypotheses the requirement to be algebraic.

Theorem (Levi decomposition). *The multiplication map $m : L \times U_J \to P$ is an isomorphism of varieties and $P = L \ltimes U_J$ is a semidirect product.*

Proof. Since L normalizes U_J and the map m is injective with invertible Jacobian, it follows that the image of m is an open subgroup, hence equal to P_J since this group is connected and clearly gives a semidirect product. □

In fact a similar argument (see [Bor], [Hu2], [OV]) shows in general that if G is any connected algebraic group with unipotent radical G_u, one can find (using the Levi decomposition for Lie algebras and the fact that the Lie algebra is algebraic) a reductive subgroup L with $G = L \ltimes G_u$ a semidirect product.

Remark. There is a certain abuse in the expression "Levi decomposition." For Lie algebras we used this term to find a presentation of a Lie algebra as a semidirect product of a semisimple and a solvable Lie algebra. Then, in order to prove Ado's theorem we corrected the decomposition so that it was really a presentation of a Lie algebra as a semidirect product of a reductive and a nilpotent Lie algebra. This is the type of decomposition which we are now stressing for P.

6.3 Borel Subgroups

We can now complete the analysis of Borel subgroups.

Let G be a connected algebraic group.

Definition. A maximal connected solvable subgroup of G is called a *Borel subgroup*.

A subgroup H with the property that G/H is projective (i.e., compact) is called a *parabolic subgroup*.

Before we start the main discussion let us make a few preliminary remarks.

Lemma 1. *(i) A Borel subgroup B of a connected algebraic group G contains the solvable radical of G.*

(ii) If $G = G_1 \times G_2$ is a product, a Borel subgroup B of G is a product $B_1 \times B_2$ of Borel subgroups in the two factors.

Proof. (i) Let R be the solvable radical, a normal subgroup. Thus BR is a subgroup; it is connected since it is the image under multiplication of $B \times R$. Finally BR is clearly solvable, hence $BR = B$.

(ii) If B is connected solvable, the two projections B_1, B_2 on the two factors are connected solvable, hence $B \subset B_1 \times B_2$. If B is maximal, we have then $B = B_1 \times B_2$. □

In this way one reduces the study of Borel subgroups to the case of semisimple groups.

Lemma 2. *Let B be connected solvable and H a parabolic subgroup of G. Then B is contained in a conjugate of H.*

Proof. By Borel's fixed point theorem, B fixes some point $aH \in G/H$, hence $B \subset aHa^{-1}$. □

Theorem. *For an algebraic subgroup $H \subset G$ the following two conditions are equivalent:*

1. *H is maximal connected solvable (a Borel subgroup).*
2. *H is minimal parabolic.*

Proof. We claim that it suffices to prove that if B is a suitably chosen Borel subgroup, then G/B is projective, hence B is parabolic. In fact, assume for a moment that this has been proved and let H be minimal parabolic. First, $B \subset aHa^{-1}$ for some a by the previous lemma. Since aHa^{-1} is minimal parabolic, we must have $B = aHa^{-1}$. Given any other Borel subgroup B', by the previous lemma, $B' \subset aBa^{-1}$. Since aBa^{-1} is solvable connected and B' maximal solvable connected, we must have $B' = aBa^{-1}$.

Let us prove now that G/B is projective for a suitable connected maximal solvable B. Since B contains the solvable radical of G, we can assume that G is semisimple. We do first the case $G = G_s(L)$, the simply connected group of a simple Lie algebra L. Consider for B the subgroup with Lie algebra $\mathfrak{b}^+$. We have proved in Theorem 6.2 that B is the stabilizer of a line ℓ generated by a highest weight vector in an irreducible representation M of $G_s(L)$ relative to a regular dominant weight. We claim that the orbit $G_s(L)/B \subset \mathbb{P}(M)$ is closed in this projective space, and hence $G_s(L)/B$ is projective. Otherwise, one could find in its closure a fixed point under B which is different from ℓ. This is impossible since it would correspond to a new highest weight vector in the irreducible representation M. Moreover, we also see that the center Z of $G_s(L)$ is contained in B. In fact, since on any irreducible representation the center of $G_s(L)$ acts as scalars, Z acts trivially on the projective space. In particular, it is contained in the stabilizer B of the line ℓ.

The general case now follows from 6.1. A semisimple group G is the quotient $G = \prod_i G_s(L_i)/Z$, of a product $\prod_i G_s(L_i)$ of simply connected groups with simple Lie algebras L_i by a finite group Z in the center. Taking the Borel subgroups B_i in $G_s(L_i)$ we have that $\prod_i B_i$ contains Z, $B := \prod_i B_i/Z$ is a Borel subgroup in G and $G/B = \prod_i G_s(L_i)/B_i$. □

Corollary of proof. *The center of G is contained in all Borel subgroups.*

All Borel subgroups are conjugate.

The normalizer of B is B.

A parabolic subgroup is conjugate to one and only one of the groups P_J, $J \subset \Delta$.

Proof. All the statements follow from the theorem and the previous lemmas. The only thing to clarify is why two groups P_J are not conjugate. Suppose $gP_Jg^{-1} = P_I$. Since gBg^{-1} is a Borel subgroup of P_I there is an $h \in P_I$ with $hgBg^{-1}h^{-1} = B$. Hence $hg \in B$ and $P_J = hgP_J(hg)^{-1} = hP_Ih^{-1} = P_I$. □

The variety G/B plays a fundamental role in the theory and, by analogy to the linear case, it is called the *(complete) flag variety.*

By the theory developed, G/B appears as the orbit of the line associated to a highest weight vector for an irreducible representation V_λ when the weight $\lambda \in \Lambda^{++}$ is strongly dominant, i.e., in the interior of the Weyl chamber C (cf. §2.4).

The other varieties G/P_J also play an important role and appear as the orbit of the line associated to a highest weight vector for an irreducible representation V_λ when the weight λ is in a given set of walls of C.

6.4 Bruhat Decomposition

Let L be a simple Lie algebra, $\mathfrak{t}$ a maximal toral subalgebra, Φ^+ a system of positive roots, $\mathfrak{b}^+X$ the corresponding Borel subalgebra, and $\mathfrak{u}^+$ its nilpotent radical. We need to make several computations with the Weyl group W and with its lift $\tilde{W}$ generated by the elements s_i in the groups $SL_i(2, \mathbb{C})$. In order to avoid unnecessary confusion let us denote by σ_i the simple reflections in W lifting to the elements $s_i \in \tilde{W}$.

For any algebraic group G with Lie algebra L we have that $G_s(L) \to G \to G_a(L)$ (G is between the simply connected and the adjoint groups). In G we have the subgroups T, B, U, i.e., the torus, Borel group, and its unipotent radical with Lie algebras $\mathfrak{t}, \mathfrak{b}^+, \mathfrak{u}^+$, respectively. Let W be the Weyl group.

Proposition 1. *The map $T \times U \to B$, $(t, u) \mapsto tu$ is an isomorphism.*

Proof. Take a faithful representation of G and a basis of eigenvectors for T, ordered such that U acts as strictly upper triangular matrices. Call D the diagonal and V the strictly upper triangular matrices in this basis. The triangular matrices form a product $D \times V$ inside which $T \times U$ is closed. Since clearly the image $TU \subset B \subset DV$ is also open in B, we must have $TU = B$ and the map is an isomorphism. □

First, a simple remark:

Definition-Proposition 2. *A maximal solvable subalgebra of L is called a Borel subalgebra. A Borel subalgebra $\mathfrak{b}$ is the Lie algebra of a Borel subgroup.*

Proof. Let B be the Lie subgroup associated to $\mathfrak{b}$. Thus B is solvable. By Proposition 3 of Chapter 7, §3.5, the Zariski closure of B is solvable and connected so B is algebraic and clearly a Borel subgroup. □

Consider the adjoint action of $G_a(L)$ on L.

Lemma 1. *The stabilizer of $\mathfrak{b}^+$ and $\mathfrak{u}^+$ under the adjoint action is B.*

Proof. Clearly B stabilizes $\mathfrak{b}^+, \mathfrak{u}^+$. If the stabilizer were larger, it would be one of the groups P_J, which is impossible. □

Remark. According to Chapter 7, we can embed G/B in a Grassmann variety. We let $N = \dim \mathfrak{u}^+$ (the number of positive roots) and consider the line $\bigwedge^N \mathfrak{u}^+ \subset \bigwedge^N L$. The orbit of this line is G/B. Now clearly a vector in $\bigwedge^N \mathfrak{u}^+$ is a highest weight vector of weight $\sum_{\alpha \in \Phi^+} \alpha = 2\rho$. It remains puzzling to understand if there is also a more geometric interpretation of the embedding of G/B associated to ρ. This in fact can be explained using a theory that we will develop in Chapter 11, §7. The adjoint group $G = G_a(L)$ preserves the Killing form and the subspace $\mathfrak{u}^+$ is totally isotropic. If we embed $\mathfrak{u}^+$ in a maximal totally isotropic subspace we can apply to it the spin formalism and associate to it a pure spinor. We leave to the reader to verify (using the theory of Chapter 7) that this spinor is a highest weight vector of weight ρ.

We want to develop some geometry of the varieties G/P. We will proceed in a geometric way which is a special case of a general theory of Bialynicki-Birula [BB].

Let us consider $\mathbb{C}^*$ acting linearly on a vector space V. Denote by $\rho : \mathbb{C}^* \to GL(V)$ the corresponding homomorphism. ρ is called a *1-parameter group*. Decompose V according to the weights $V = \sum_i V_{m_i},\ V_{m_i} = \{v \in V \mid \rho(t)v = t^{m_i}v\}$. The action induces an action on the projective space of lines $\mathbb{P}(V)$. A point $p \in \mathbb{P}(V)$ is a line $\mathbb{C}v \subset V$, and it is a fixed point under the action of $\mathbb{C}^*$ if and only if v is an eigenvector.

Given a general point in $\mathbb{P}(V)$ corresponding to the line through some vector $v = \sum_i v_i,\ v_i \in V_{m_i}$, we have $\rho(t)v = \sum_i t^{m_i}v_i$. In projective space we can dehomogenize the coordinates and choose the index i among the ones for which $v_i \neq 0$ and for which the exponent m_i is minimum. Say that this happens for $i = 1$. Choose a basis of eigenvectors among which the first is v_1, and consider the open set of projective space in which the coordinate of v_1 is nonzero and hence can be normalized to 1. In this set, in the affine coordinates chosen, we have $\rho(t)v = v_1 + \sum_{i>1} t^{m_i - m_1}v_i$. Therefore we have $\lim_{t\to 0}\rho(t)v = v_1$. We have thus proved in particular:

Lemma 2. *For a point $p \in \mathbb{P}(V)$ the limit $\lim_{t\to 0}\rho(t)p$ exists and is a fixed point of the action.*

Remark. If $W \subset \mathbb{P}(V)$ is a T-stable projective subvariety and $p \in W$, we have clearly $\lim_{t\to 0}\rho(t)p \in W$. We will apply this lemma to G/B embedded in a G-equivariant way in the projective space of a linear representation V_λ.

We want to apply this to a *regular 1-parameter subgroup* of T. By this we mean a homomorphism $\rho : \mathbb{C}^* \to T$ with the property that if $\alpha \neq \beta$ are two roots, considered as characters of T, we have that $\alpha \circ \rho \neq \beta \circ \rho$. Then we have the following simple lemma.

Lemma 3. *A subspace of L is stable under $\rho(\mathbb{C}^*)$ if and only if it is stable under T.*

Proof. A subspace of L is stable under $\rho(\mathbb{C}^*)$ if and only if it is a sum of weight spaces; since $\rho(\mathbb{C}^*)$ is regular, its weight spaces coincide with the weight spaces of T. □

We now introduce some special Borel subalgebras. For any choice Ψ of the set of positive roots, we define $\mathfrak{b}_\Psi = \mathfrak{t} \oplus \bigoplus_{\alpha\in\Psi} L_\alpha$. In particular, let $\Psi = w(\Phi^+)$ be a choice of positive roots.

We know in fact that such a Ψ corresponds to a Weyl chamber and by Theorem 2.3 3), the Weyl group acts in a simply transitive way on the chambers. Thus we have defined algebras indexed by the Weyl group and we set $\mathfrak{b}_w = \mathfrak{t} \oplus \bigoplus_{\alpha\in\Phi^+} L_{w(\alpha)}$.

Lemma 4. *Let $A \subset \Phi$ be a set of roots satisfying the two properties:*

$$\text{(S)} \qquad \alpha, \beta \in A, \ \alpha + \beta \in \Phi \implies \alpha + \beta \in A.$$

$$\text{(T)} \qquad \alpha \in A, \implies -\alpha \notin A.$$

Then $A \subset w(\Phi^+)$ for some $w \in W$.

Proof. By the theory of chambers, it suffices to find a regular vector v such that $(\alpha, v) > 0, \ \forall \alpha \in A$. In fact, since the regular vectors are dense and the previous condition is open, it suffices to find any vector v such that $(\alpha, v) > 0, \ \forall \alpha \in A$. We proceed in three steps.

(1) We prove by induction on m that given a sequence $\alpha_1, \dots, \alpha_m$ of m elements in A, we have $\alpha_1 + \alpha_2 + \cdots + \alpha_m \neq 0$. For $m = 1$ it is clear. Assume it for $m - 1$. If $-\alpha_1 = \alpha_2 + \cdots + \alpha_m$, we have $(-\alpha_1, \alpha_2 \cdots + \alpha_m) > 0$; thus, for some $j \geq 2$, we have $(\alpha_1, \alpha_j) < 0$. By Lemma 2.2, $\alpha_1 + \alpha_j$ is a root, by assumption in A, so we can rewrite the sum as a shorter sum $(\alpha_1 + \alpha_j) + \sum_{i \neq 1, j} \alpha_i = 0$, a contradiction.

(2) We find a nonzero vector v with $(\alpha, v) \geq 0, \ \forall \alpha \in A$. In fact, assume by contradiction that such a vector does not exist. In particular this implies that given $\alpha \in A$, there is a $\beta \in A$ with $(\alpha, \beta) < 0$, and hence $\alpha + \beta \in A$. Starting from any root $\alpha_0 \in A$ we find inductively an infinite sequence $\beta_i \in A$, such that $\alpha_{i+1} := \beta_i + \alpha_i \in A$. By construction $\alpha_i = \alpha_0 + \beta_1 + \beta_2 + \cdots + \beta_{i-1}, \ \forall i$. For two distinct indices $i < j$ we must have $\alpha_i = \alpha_j$, and hence $0 = \sum_{h=i+1}^{j} \beta_h$, contradicting 1.

(3) By induction on the root system induced on the hyperplane $H_v := \{x \mid (x, v) = 0\}$, we can find a vector w with $(\alpha, w) > 0, \ \forall \alpha \in A \cap H_v$. If we take w sufficiently close to 0, we can still have that $(\beta, v + w) > 0, \ \forall \beta \in A - H_v$. The vector $v + w$ solves our problem. □

Lemma 5. *Let $A \subset \Phi$ be a set of roots. $\mathfrak{h} := \mathfrak{t} \oplus \bigoplus_{\alpha \in A} L_\alpha$ is a Lie algebra if and only if A satisfies the property (S) of Lemma 4.*

Furthermore $\mathfrak{h}$ is solvable if and only if A satisfies the further property (T).

Proof. The first part follows from the formula $[L_\alpha, L_\beta] = L_{\alpha+\beta}$. For the second, if $\alpha, -\alpha \in A$, we have inside $\mathfrak{h}$ a copy of $sl(2, \mathbb{C})$ which is not solvable. Otherwise, we can apply the previous lemma and see that $\mathfrak{h} \subset \mathfrak{b}_w$ for some w. □

Proposition 3. *(i) A Borel subalgebra $\mathfrak{h}$ is stable under the adjoint action of T if and only if $\mathfrak{t} \subset \mathfrak{h}$.*

(ii) The Borel subalgebras containing $\mathfrak{t}$ are the algebras $\mathfrak{b}_w, \ w \in W$.

Proof. (i) If a subalgebra $\mathfrak{h}$ is stable under T, it is stable under the adjoint action of $\mathfrak{t}$. Hence $\mathfrak{h}' := \mathfrak{h} + \mathfrak{t}$ is a subalgebra and $\mathfrak{h}$ is an ideal in $\mathfrak{h}'$. So, if $\mathfrak{h}$ is maximal solvable, we have that $\mathfrak{t} \subset \mathfrak{h}$. The converse is clear.

(ii) Since $\mathfrak{h}$ is T-stable we must have $\mathfrak{h} = \mathfrak{t} \oplus \bigoplus_{\alpha \in A} L_\alpha$ satisfies the hypotheses of Lemma 5, and hence it is contained in a $\mathfrak{b}_w$. Since $\mathfrak{h}$ is maximal solvable, it must coincide with $\mathfrak{b}_w$. □

From Theorem 6.3 and all the results of this section we have:

Theorem 1. *The set of Borel subalgebras of L is identified under the adjoint action with the orbit of* $\mathfrak{b}$ *and with the projective homogeneous variety* G/B. *The fixed points under* T *of* G/B *are the algebras* $\mathfrak{b}_w$ *indexed by* W.

We want to decompose G/B according to the theory of 1-parameters subgroups now. We choose a regular 1-parameter subgroup of T with the further property that if α is a positive root, $\alpha(\rho(t)) = t^{m_\alpha}, \ m_\alpha < 0$. From Lemma 3, it follows that the fixed points of $\rho(\mathbb{C}^*)$ on G/B coincide with the T fixed points $\mathfrak{b}_w$. We thus define

$$C_w^- := \{p \in G/B \mid \lim_{t\to 0} \rho(t)p = \mathfrak{b}_w\}. \tag{6.4.1}$$

From Lemma 2 we deduce:

Proposition 4. *G/B is the disjoint union of the sets $C_w^-, \ w \in W$.*

We need to understand the nature of these sets C_w^-. We use Theorem 2 of §3.5 in Chapter 4. First, we study G/B in a neighborhood of $B = \mathfrak{b}$, taking as a model the variety of Borel subalgebras. We have that $\mathfrak{u}^-$, the Lie algebra of the unipotent group U^-, is a complement of $\mathfrak{b}^+$ in L, so from the cited theorem we have that the orbit map restricted to U^- gives a map $i : U^- \to G/B, \ u \mapsto \mathrm{Ad}(u)(\mathfrak{b})$ which is an open immersion at 1. Since U^- is a group and i is equivariant with respect to the actions of U^-, we must have that i is an open map with an invertible differential at each point. Moreover, i is an isomorphism onto its image, since otherwise an element of U^- would stabilize $\mathfrak{b}_+$, which is manifestly absurd. In fact, $U^- \cap B = 1$ since it is a subgroup of U^- with trivial Lie algebra, hence a finite group. In a unipotent group the only finite subgroup is the trivial group. We have thus found an open set isomorphic to U^- in G/B and we claim that this set is in fact C_1. To see this, notice that U^- is T-stable, so $G/B - U^-$ is closed and T-stable. Hence necessarily $C_1 \subset U^-$. To see that it coincides, notice that the T-action on U^- is isomorphic under the exponential map to the T-action on $\mathfrak{u}^-$. By the definition of the 1-parameter subgroup ρ, all the eigenvalues of the group on $\mathfrak{u}^-$ are strictly positive, so for every vector $u \in \mathfrak{u}^-$ we have $\lim_{t\to 0} \rho(t)u = 0$, as desired.

Let us look now instead at another fixed point $\mathfrak{b}_w$. Choose a reduced expression of w and correspondingly an element $s_w = s_{i_1}s_{i_2}\dots s_{i_k} \in G$ as in Section 3.1, $s_i \in SL_i(2,\mathbb{C})$ the matrix $\begin{vmatrix} 0 & 1 \\ -1 & 0 \end{vmatrix}$. We have that $\mathfrak{b}_w = s_w(\mathfrak{b})$ has as an open neighborhood the orbit of the group $s_w(U^-)s_w^{-1}$, which by the exponential map is isomorphic to the Lie algebra $\mathfrak{u}_w^- = \sum_{\alpha\in\Phi^-} L_{w(\alpha)}$. This neighborhood is thus isomorphic, in a T-equivariant way, to $\mathfrak{u}_w^-$ with the adjoint action of T. On this space the 1-parameter subgroup ρ has positive eigenvalues on the root spaces $L_{w(\alpha)}$ for the roots $\alpha \prec 0$ such that $w(\alpha) \prec 0$ and negative eigenvalues for the roots $\alpha \prec 0$ such that $w(\alpha) \succ 0$. Clearly the Lie algebra of the unipotent group $U_w^- := U^- \cap s_w(U^-)s_w^{-1}$ is the sum of the root spaces L_β, where $\beta \prec 0, \ w^{-1}(\beta) \prec 0$. We have:

Lemma 6. *C_w^- is the closed set $U_w^-\mathfrak{b}_w$ of the open set $s_w(U^-)s_w^{-1}\mathfrak{b}_w = s_w(U^-\mathfrak{b})$.*

The orbit map from U_w^- to C_w^- is an isomorphism.

We need one final lemma:

Lemma 7. *Decompose* $\mathfrak{u} = \mathfrak{u}_1 \oplus \mathfrak{u}_2$ *as the direct sum of two Lie subalgebras, say* $\mathfrak{u}_i = Lie(U_i)$. *The map* $i : U_1 \times U_2 \to U, i : (x, y) \mapsto xy$ *and the map* $j : \mathfrak{u}_1 \oplus \mathfrak{u}_2 \to U,\ j(a, b) := \exp(a)\exp(b)$ *are isomorphisms of varieties.*

Proof. Since U_i is isomorphic to $\mathfrak{u}_i$ under the exponential map, the two statements are equivalent. The Jacobian of j at 0 is the identity, so the same is true for i. i is equivariant with respect to the right and left actions of U_2, U_1, so the differential of i is an isomorphism at each point.[101] Moreover, i is injective since otherwise, if $x_1y_1 = x_2y_2$ we have $x_2^{-1}x_1 = y_2y_1^{-1} \in U_1 \cap U_2$. This group is a subgroup of a unipotent group with trivial Lie algebra, hence it is trivial. To conclude we need a basic fact from affine geometry. Both $U_1 \times U_2$ and U are isomorphic to some affine space $\mathbb{C}^m$. We have embedded, via i, $U_1 \times U_2$ into U. Suppose then that we have an open set A of $\mathbb{C}^m$ which is an affine variety isomorphic to $\mathbb{C}^m$. Then $A = \mathbb{C}^m$. To see this, observe that in a smooth affine variety the complement of a proper affine open set is a hypersurface, which in the case of $\mathbb{C}^m$ has an equation $f(x) = 0$. We would have then that the function $f(x)$ restricted to $A = \mathbb{C}^m$ is a nonconstant invertible function. Since on $\mathbb{C}^n$ the functions are the polynomials, this is impossible. □

Given a $w \in W$ we know by 2.3 that if $w = s_{i_1} \dots s_{i_k}$ is a reduced expression, the set $B_w := \{\beta \in \Phi^+ \mid w^{-1}(\beta) \prec 0\}$ of positive roots sent into negative roots by w^{-1} is the set of elements $\beta_h := s_{i_1}s_{i_2} \dots s_{i_{h-1}}(\alpha_{i_h}),\ h = 1, \dots, k$. Let us define the unipotent group U_w as having Lie algebra $\bigoplus_{\beta \in B_w} L_\beta$. For a root α let U_α be the additive group with Lie algebra L_α.

Corollary. *Let* $w = s_{i_1} \dots s_{i_k}$ *be a reduced expression. Then the group* U_w *is the product* $U_{\beta_1}U_{\beta_2} \dots U_{\beta_k} = U_{\beta_k}U_{\beta_{k-1}} \dots U_{\beta_1}$.

Proof. We apply induction and the fact which follows by the previous lemma that U_w is the product of $U_{ws_{i_k}}$ with U_{β_k}. □

In particular it is useful to write the unipotent group U as a product of the root subgroups for the positive roots, ordered by a convex ordering. We can then complete our analysis.

Theorem (Bruhat decomposition). *The sets* C_w^- *are the orbits of* B^- *acting on* G/B.

Each $C_w^- = U_w^-\mathfrak{b}_w$ *is a locally closed subset isomorphic to an affine space of dimension* $\ell(ww_0)$ *where* w_0 *is the longest element of the Weyl group.*

The stabilizer in B^- *of* $\mathfrak{b}_w$ *is* $B^- \cap s_w(B)s_w^{-1} = TU'_w$, *where* $U'_w = U^- \cap s_w(U)s_w^{-1}$ *has Lie algebra* $\bigoplus_{\alpha \in \Phi^-,\ w^{-1}(\alpha) \in \Phi^+} L_\alpha$.

---

[101] It is not known if this is enough. There is a famous open problem, the Jacobian conjecture, stating that if we have a polynomial map $\mathbb{C}^n \to \mathbb{C}^n$ with everywhere nonzero Jacobian, it is an isomorphism.

Proof. Most of the statements have been proved. It is clear that $s_w(B)s_w^{-1}$ is the stabilizer of $\mathfrak{b}_w = \mathrm{Ad}(s_w)(\mathfrak{b})$ in G. Hence $B^- \cap s_w(B)s_w^{-1}$ is the stabilizer in B^- of $\mathfrak{b}_w$. This is a subgroup with Lie algebra

$$\mathfrak{b}^- \cap \mathfrak{b}_w = \mathfrak{t} \oplus \bigoplus_{\substack{\alpha\in\Phi^- \\ w^{-1}(\alpha)\in\Phi^+}} L_\alpha .$$

We have a decomposition

$$\mathfrak{b}^- = \mathfrak{t} \oplus \bigoplus_{\substack{\alpha\in\Phi^- \\ w^{-1}(\alpha)\in\Phi^+}} L_\alpha \oplus \bigoplus_{\substack{\alpha\in\Phi^- \\ w^{-1}(\alpha)\in\Phi^-}} L_\alpha ,$$

which translates in groups as $B^- = TU_w^- U_w = U_w^- T U_w'$ (the products giving isomorphisms of varieties). Hence $B^-\mathfrak{b}_w = U_w^- T U_w' \mathfrak{b}_w = U_w^- \mathfrak{b}_w = C_w^-$. □

Some remarks are in order. In a similar way one can decompose G/B into B-orbits. Since $B = s_{w_0} B^- s_{w_0}^{-1}$, we have that

$$G/B = \bigcup_{w\in W} s_{w_0} C_w^- = \bigcup_{w\in W} B s_{w_0} \mathfrak{b}_w = \bigcup_{w\in W} B \mathfrak{b}_{w_0 w}.$$

Set $U_w := s_{w_0} U_{w_0 w}^- s_{w_0}^{-1}$. The cell

$$C_w^+ = s_{w_0} C_{w_0 w}^- = s_{w_0} U_{w_0 w}^- \mathfrak{b}_{w_0 w} = U_w \mathfrak{b}_w ,$$

with center $\mathfrak{b}_w$ is an orbit under B and has dimension $\ell(w)$.

U_w is the unipotent group with Lie algebra $\bigoplus_{\alpha\in B_w} L_\alpha$ where

$$B_w := \{\beta \in \Phi^+ \mid w^{-1}(\beta) \prec 0\}.$$

Finally $C_w^+ = \{p \in G/B \mid \lim_{t\to\infty} \rho(t)p = \mathfrak{b}_w\}$.

We have a decomposition of the open set $U_w^- = C_w^- \times C_{w_0 w}^+$. The cells are called *Bruhat cells* and form two *opposite* cell decompositions.

Second, if we now choose G to be any algebraic group with Lie algebra L and $G_a(L) = G/Z$ where Z is the finite center of G, we have that:

Proposition 5. *The preimage of a Borel subgroup B_a of $G_a(L)$ is a Borel subgroup B of G. Moreover there is a canonical* 1-1 *correspondence between the B-orbits on G/B and the double cosets of B in G, hence the decomposition:*

$$G = \bigsqcup_{w\in W} B s_w B = \bigsqcup_{w\in W} U_w s_w B, \qquad \textit{(Bruhat decomposition)}. \tag{6.4.2}$$

The obvious map $U_w \times B \to U_w s_w B$ is an isomorphism.

Proof. Essentially everything has already been proved. Observe only that G/B embeds in the projective space of an irreducible representation. On this space Z acts as scalars, hence Z acts trivially on projective space. It follows that $Z \subset B$ and that $G/B = G_a(L)/B_a$. □

The fact that the Bruhat cells are orbits (under B^- or B^+) implies immediately that:

Proposition 6. *The closure $S_w := \overline{C_w}$ of a cell C_w is a union of cells.*

Definition. S_w is called a *Schubert variety.*

The previous proposition has an important consequence. It defines a partial order in W given by $x \prec y \iff C_x^+ \subset \overline{C_y^+}$. This order is called the *Bruhat order*. It can be understood combinatorially as follows.

Theorem (on Bruhat order). *Given $x, y \in W$ we have $x \prec y$ in the Bruhat order if and only if there is a reduced expression $y = s_{i_1} s_{i_2} \dots s_{i+k}$ such that x can be written as a product of some of the s_{i_j} in the same order as they appear in y.*

We postpone the proof of this theorem to the next section.

6.5 Bruhat Order

Let us recall the definitions and results of 6.2. Given a subset J of the nodes of the Dynkin diagram, we have the corresponding root system Φ_J with simple roots as those corresponding to the nodes not in J. Its Weyl group W_J is the subgroup generated by the corresponding simple reflections $s_i,\ i \notin J$ (the stabilizer of a point in a suitable stratum of the closure of the Weyl chamber). We also have defined a parabolic subalgebra $\mathfrak{p}_J$ and a corresponding parabolic subgroup $P = P_J$.

Reasoning as in 6.4 we have:

Lemma. *P is the stabilizer of $\mathfrak{p}_J$ under the adjoint action.*

G/P can be identified with the set of parabolic subalgebras conjugate to $\mathfrak{p}_J$.

The parabolic subalgebras conjugated to $\mathfrak{p}_J$ and fixed by T are the ones containing $\mathfrak{t}$ are in 1-1 correspondence with the cosets W/W_J.

Proof. Let us show the last statement. If $\mathfrak{q}$ is a parabolic subalgebra containing $\mathfrak{t}$, a maximal solvable subalgebra of $\mathfrak{q}$ is a Borel subalgebra of L containing $\mathfrak{t}$, hence is equal to $\mathfrak{b}_w$. This implies that for some $w \in W$ we have that $\mathfrak{q} = s_w(\mathfrak{p}_J)$. The elements s_w are in the group $\tilde{W}$ such that $\tilde{W}/\tilde{W} \cap T = W$. One verifies immediately that the stabilizer in $\tilde{W}$ of $\mathfrak{p}_J$ is the preimage of W_J and the claim follows. $\square$

Theorem. *(i) We have a decomposition $P = \bigsqcup_{w \in W_J} U_w^+ s_w B$.*

(ii) The variety G/P has also a cell decomposition. Its cells are indexed by elements of W/W_J, and in the fibration $\pi : G/N \to G/P$, we have that $\pi^{-1} C_{xW_J} = \bigsqcup_{w \in xW_J} C_w$.

(iii) The coset xW_J has a unique minimal element in the Bruhat order whose length is the dimension of the cell C_{xW_J}.

(iv) Finally, the fiber P/B over the point P of this fibration is the complete flag variety associated to the Levi factor L_J.

Proof. All the statements follow easily from the Levi decomposition. Let L be the Levi factor, B the standard Borel subgroup, and R the solvable radical of P. We have that $L \cap B$ is a Borel subgroup in L and we have a canonical identification $L/L \cap B = P/B$. Moreover, L differs from L_J by a torus part in B. Hence we have also $P/B = L_J/B_J$, with $B_J = B \cap L_J$ a Borel subgroup in L_J. The Weyl group of L_J is W_J so the Bruhat decomposition for L_J induces a Bruhat decomposition for P given by i).

By the previous lemma, the fixed points $(G/P)^W$ are in 1-1 correspondence with W/W_J. We thus have a decomposition into locally closed subsets C_a, $a \in W/W_J$ as for the full flag variety. □

A particularly important case of the previous analysis for us is that of a *minimal parabolic*. By this we mean a parabolic P associated to the set J of all nodes except one i. In this case the root system Φ_J is a system of type A_1 and thus the semisimple Levi factor is the group $SL(2, \mathbb{C})$. The Bruhat decomposition for P, which we denote $P(i)$, reduces to only two double cosets $P(i) = B \sqcup Bs_iB$ and $P(i)/B = \mathbb{P}^1(\mathbb{C})$ is the projective line, a sphere. The two cells are the affine line and the point at infinity.

We pass now to the crucial combinatorial lemma.

Bruhat lemma. *Let* $w \in W$ *and* $s_w = s_{i_1}s_{i_2}\dots s_{i+k}$ *be associated to a reduced expression of* $w = \sigma_{i_1}\sigma_{i_2}\dots\sigma_{i+k}$. *Let us consider the element* s_i *associated to a simple reflection* σ_i. *Then*

$$Bs_iBBs_wB = \begin{cases} Bs_is_wB & \text{if } \ell(\sigma_i w) = \ell(w) + 1 \\ Bs_is_wB \cup Bs_wB & \text{if } \ell(\sigma_i w) = \ell(w) - 1. \end{cases}$$

Proof. For the first part it is enough to see that $s_iBs_wB \subset Bs_is_wB$. Since $s_i^2 \in T$ this is equivalent to proving that $s_iBs_is_is_wB \subset Bs_is_wB$. We have (by the Corollary of the previous section), that $s_iBs_i \subset BU_{-\alpha_i}$ and

$$U_{-\alpha_i}s_is_w = s_is_w(s_is_w)^{-1}U_{-\alpha_i}s_is_w = U_{-w^{-1}\sigma_i(\alpha_i)}s_is_w = U_{w^{-1}(\alpha_i)}s_is_w.$$

Since $\ell(\sigma_iw) = \ell(w) + 1$ we have $w^{-1}(\alpha_i) \succ 0$, and so $U_{w^{-1}(\alpha_i)} \subset B$.

In the other case $w^{-1}(\alpha_i) \prec 0$. Set $w = \sigma_iu$. By the previous case $Bs_wB = Bs_iBs_uB$. Let us thus compute Bs_iBBs_iB. We claim that $Bs_iBBs_iB = Bs_iB \cup B$. This will prove the claim. Clearly this is true if $G = SL(2, \mathbb{C})$ and $s_i = s = \begin{vmatrix} 0 & 1 \\ -1 & 0 \end{vmatrix}$, since in this case $BsB \cup B = SL(2, \mathbb{C})$, and $1 = (-1)s^2 \in BsBBsB$ and $s \in BsBBsB$ since

$$\begin{vmatrix} 0 & 1 \\ -1 & 0 \end{vmatrix} = \begin{vmatrix} 1 & 1 \\ 0 & 1 \end{vmatrix}\begin{vmatrix} 0 & 1 \\ -1 & 0 \end{vmatrix}\begin{vmatrix} -1 & -1 \\ 0 & -1 \end{vmatrix}\begin{vmatrix} 0 & 1 \\ -1 & 0 \end{vmatrix}\begin{vmatrix} 1 & 1 \\ 0 & 1 \end{vmatrix}.$$

For the general case we notice that this is a computation in the minimal parabolic associated to i. We have that all the double cosets contain the solvable radical and thus we can perform the computation modulo the radical, reducing to the previous special case. □

Geometric lemma. *Let $w = \sigma_i u$, $\ell(w) = \ell(u) + 1$. Then $S_y = PS_u$ where $P \supset B$ is the minimal parabolic associated to the simple reflection s_i.*

Proof. Since $P = Bs_iB \cup B$, from Bruhat's lemma we have $PC_u = C_w \cup C_u$. Moreover, from the proof of the lemma, it follows that C_u is in the closure of C_w; thus by continuity $C_w \subset PS_u \subset S_w$. Thus it is sufficient to prove that PS_u is closed. By the Levi decomposition for P using $SL(2,\mathbb{C}) \subset P$, we have that $PS_u = SL(2,\mathbb{C})S_u$. In $SL(2,\mathbb{C})$ every element can be written as product ab with $a \in SU(2,\mathbb{C})$, and b upper triangular, hence in $B \cap SL(2,\mathbb{C})$. Thus $PS_u = SL(2,\mathbb{C})S_u = SU(2,\mathbb{C})P$. Finally since $SU(2,\mathbb{C})$ is compact, the action map $SU(2,\mathbb{C}) \times S_u \to G/B$ is proper and so its image is closed. □

Remark. With a little more algebraic geometry one can give a proof which is valid in all characteristics. One forms the algebraic variety $P \times_B S_u$ whose points are pairs (p, v), $p \in P$, $v \in S_u$, modulo the identification $(pb, v) \equiv (p, bv)$, $\forall v \in S_u$.

The action map factors through a map $P \times_B S_u \to S_w$ which one proves to be proper. This in fact is the beginning of an interesting construction, the Bott–Samelson resolution of singularities for S_w.

Proof (of the theorem on Bruhat order stated in §6.4*).* Let $y \in W$ and $T_y := \cup_{x \prec y} C_x$, where $\prec$ is the Bruhat order. We have to prove that $T_y = S_y$. We work by induction on the length of y. Let $y = s_{i_1}u$, $\ell(y) = \ell(u) + 1$. By induction S_u is the union of the cells C_x where x is obtained from the reduced expressions of u, deleting some factors. Given this we have by the Bruhat decomposition of P that $PS_u = (B \cup Bs_{i_1}B)S_u = S_u \cup Bs_{i_1}S_u$. Now if $x \prec u$ in the Bruhat order, we have that also $x, s_{i_1}x$ precede y. This shows that $S_w \subset T_w$. Conversely, let $x \prec y$. Then we have a reduced expression $y = s_{j_1}s_{j_2}\dots s_{j_k} = s_{j_1}w$ and x is obtained by dropping some of the factors; hence either $x \prec w$ or $x = s_{j_1}x'$ and $x' \prec w$. The same argument as before shows that $C_x \subset S_y$. □

Remark. The theory we have discussed holds in any characteristic, and in fact also over finite fields, where it gives the basic ingredients for the representation theory of the finite Chevalley groups. For instance, in the case of a finite field F with q elements, one takes the flag variety as basis of a permutation representation $\mathbb{C}[G/B]$. One applies next the discussion of Chapter 1, §3.2, where we showed that the endomorphism algebra of $\mathbb{C}[G/B]$ is the Hecke algebra of double cosets. The theory of Bruhat implies that this algebra has a basis T_w indexed by $w \in W$ and that $T_uT_v = T_{uv}$, if $\ell(uv) = \ell(u)+\ell(v)$, while $T_{s_i}^2 = (q-1)T_{s_i} + q$ for a simple reflection. This is the beginning of a rather deep theory.

Remark. Given a representation V_λ, $\lambda = \sum_i m_i\omega_i$ such that the stabilizer of the highest weight vector v_λ is P_J, $J = \{i \mid m_i \neq 0\}$, the set of T fixed points in the orbit $G\mathbb{C}v_\lambda = G/P_J \subset \mathbb{P}(V_\lambda)$ is the set of lines of the highest weight vectors for the various algebras $\mathfrak{b}_w$. These vectors are called the *extremal weight vectors* and their weights (the W orbit of λ) the *extremal weights.* There are (very few) representations which are particularly simple, and have the extremal weight vectors as basis; these are called *minuscule*. Among them we find the exterior powers $\bigwedge^k V$ for type A_n and, as we will see, the spin representations.

Examples. In classical groups we can represent the variety G/B in a more concrete way as a *flag variety*. We have already seen the definition of flags in Chapter 7, §4.1 where we described Borel subgroups of classical groups as stabilizers of totally isotropic flags.

Examples of fixed points Let us understand in this language which points are the T-fixed points. A flag of subspaces V_i of the defining representation is fixed under T if and only if each V_i is fixed, i.e., it is a sum of eigenspaces. For $SL(n, \mathbb{C})$, where T consists of the diagonal matrices, the standard basis $e_1, \dots, e_n$ is a basis of distinct eigenvalues, so a T-stable space has as basis a subset of the e_i. Thus a stable flag is constructed from a permutation σ as the sequence of subspaces $V_i := \langle e_{\sigma(1)}, \dots, e_{\sigma(i)} \rangle$. We see concretely how the fixed points are indexed by S_n.

Exercise. Prove directly the Bruhat decomposition for $SL(n, \mathbb{C})$, using the method of *putting a matrix into canonical row echelon form.*

For the other classical groups, the argument is similar. Consider for instance the symplectic group, with basis e_i, f_i of eigenvectors with distinct eigenvalues. Again a T-stable space has as basis a subset of the e_i, f_i. The condition for such a space to be totally isotropic is that, for each i, it should contain at most one and not both of the elements e_i, f_i. This information can be encoded with a permutation plus a sequence of ± 1, setting $+1$ if e_i appears, -1 if f_i appears. It is easily seen that we are again encoding the fixed flags by the Weyl group. The even orthogonal group requires a better analysis. The problem is that in this case, the set of complete isotropic flags is no longer an orbit under $SO(2n, \mathbb{C})$. This is explained by the familiar example of the *two rulings* of lines in a quadric in projective 3-space (which correspond to totally isotropic planes in 4 space). In group theoretical terms this means that the set of totally isotropic spaces of dimension n form two orbits under $SO(2n, \mathbb{C})$. This can be seen by induction as follows. One proves:

1. If $m = \dim V > 2$, the special orthogonal group acts transitively on the set of nonzero isotropic vectors.
2. If e is such a vector the orthogonal $e^\perp$ is an $m-1$-dimensional space on which the symmetric form is degenerate with kernel generated by e. Modulo e, we have an $m-1$-dimensional space $U := e^\perp/\mathbb{C}e$. The stabilizer of e in $SO(V)$ induces on U the full special orthogonal group.
3. By induction, two k-dimensional totally isotropic spaces are in the same orbit for $k < m/2$.
4. Finally $SO(2, \mathbb{C})$ is the subgroup of $SL(2, \mathbb{C})$ stabilizing the degenerate conic $\{xe+yf \mid xy = 0\}$. $SO(2, \mathbb{C})$ consists of diagonal matrices $e \mapsto te,\ f \mapsto t^{-1}f$. There are only two isotropic lines $x = 0,\ y = 0$. This analysis explains why the fixed flags correspond to the Weyl group which has $n!2^{n-1}$ elements.

Now let us understand the *partial flag varieties* G/P with P a parabolic subgroup. We leave the details to the reader (see also Chapter 13).

For type A_n, with group $SL(V)$, a parabolic subgroup is the stabilizer of a *partial flag* $V_1 \subset V_2 \cdots \subset V_k$ with $\dim V_i = h_i$. The dimensions h_i correspond to the positions, in the Dynkin diagram A_n, of the set J of nodes that we remove.

In particular, a maximal parabolic is the stabilizer of a single subspace, for instance, in a basis $e_1, \dots, e_m$, the k-dimensional subspace spanned by $e_1, \dots, e_k$. In fact, the stabilizer of this subspace is the stabilizer of the line through the vector $e_1 \wedge e_2 \wedge \cdots \wedge e_k \in \bigwedge^k V$. The irreducible representation $\bigwedge^k V$ is a fundamental representation with weight ω_k. The orbit of the highest weight vector $e_1 \wedge \cdots \wedge e_k$ is the set of decomposable exterior vectors. It corresponds in projective space to the set of k-dimensional subspaces, which is the *Grassmann variety*, isomorphic to $G/P_{\{k\}}$ where the parabolic subgroup is the group of block matrices $\begin{vmatrix} A & B \\ 0 & C \end{vmatrix}$ with A a $k \times k$ matrix. We plan to return to these ideas in Chapter 13, where we will take a more combinatorial approach which will free us from the characteristic 0 constraint.

Exercise. Show that the corresponding group $W_{\{k\}}$ is $S_k \times S_{n+1-k}$. The fixed points correspond to the decomposable vectors $e_{i_1} \wedge e_{i_2} \wedge \cdots \wedge e_{i_k}$, $i_1 < i_2 < \cdots < i_k$.

For the other classical groups one has to impose the restriction that the subspaces of the flags be totally isotropic. One can then develop a parallel theory of isotropic Grassmannians. We have again a more explicit geometric interpretation of the fundamental representations. These representations are discussed in the next chapter.

Exercise. Visualize, using the matrix description of classical groups, the parabolic subgroups in block matrix form.

For the symplectic and orthogonal group we will use the Theorems of Chapter 11.

For the symplectic group of a space V, the fundamental weight ω_k corresponds to the irreducible representation $\bigwedge_0^k(V) \subset \bigwedge^k(V)$ consisting of traceless tensors. Its highest weight vector is $e_1 \wedge e_2 \wedge \ldots \wedge e_k$ (cf. Chapter 11, §6.7).

In $\bigwedge^k V$, the orbit under $Sp(V)$ of the highest weight vector $e_1 \wedge \cdots \wedge e_k$ is the set of decomposable exterior vectors which are traceless. The condition to be traceless corresponds to the constraint, on the corresponding k-dimensional subspace, to be totally isotropic. This is the *isotropic Grassmann variety*. It is then not hard to see that to a set $J = \{j_1, j_2, \dots, j_k\}$ of nodes corresponds the variety of *partial isotropic flags* $V_{j_1} \subset V_{j_2} \subset \cdots \subset V_{j_k}$, $\dim V_{j_t} = j_t$.

Exercise. Prove that the intersection of the usual Grassmann variety, with the projective subspace $\mathbb{P}(\bigwedge_0^k(V))$ is the isotropic Grassmann variety.

For the orthogonal groups we have, besides the exterior powers $\bigwedge^k(V)$ which remain irreducible, also the spin representations.

On each of the exterior powers $\bigwedge^k(V)$ we have a quadratic form induced by the quadratic form on V.

Exercise. Prove that a decomposable vector $u := v_1 \wedge v_2 \wedge \ldots \wedge v_k$ corresponds to an isotropic subspace if and only if it is isotropic. The condition that the vector u

corresponds to a totally isotropic subspace is more complicated although it is given by quadratic equations. In fact one can prove that it is: $u \otimes u$ belongs to the irreducible representation in $\bigwedge^k V^{\otimes 2}$ of the orthogonal group, generated by the highest weight vector.

Finally for the maximal totally isotropic subspaces we should use, as fundamental representations, the spin representations, although from §6.2 we know we can also use twice the fundamental weights and work with exterior powers (cf. Chapter 11, §6.6). The analogues of the decomposable vectors are called in this case *pure spinors*.

6.6 Quadratic Equations

E. Cartan discovered that pure spinors, as well as the usual decomposable vectors in exterior powers, can be detected by a system of quadratic equations. This phenomenon is quite general, as we will see in this section. We have seen that the parabolic subgroups P give rise to compact homogeneous spaces G/P. Such a variety is the orbit, in some projective space $\mathbb{P}(V_\lambda)$, of the highest weight vector. In this section we want to prove a theorem due to Kostant, showing that, as a subvariety of $\mathbb{P}(V_\lambda)$, the ideal of functions vanishing on G/P is given by explicit quadratic equations.

Let v_λ be the highest weight vector of V_λ. We know by §5.2 Proposition 3 that $v_\lambda \otimes v_\lambda$ is a highest weight vector of $V_{2\lambda} \subset V_\lambda \otimes V_\lambda$ and $V_\lambda \otimes V_\lambda = V_{2\lambda} \bigoplus_{\mu \prec 2\lambda} V_\mu$.

If $v \in V_\lambda$ is in the orbit of v_λ we must have that $v \otimes v \in V_{2\lambda}$.

The theorem we have in mind is proved in two steps.

1. One proves that the Casimir element C has a scalar value on $V_{2\lambda}$ which is different from the values it takes on the other V_μ of the decomposition and interprets this as quadratic equations on G/P.
2. One proves that these equations generate the desired ideal.

As a first step we compute the value of C on an irreducible representation V_λ, or equivalently on v_λ. Take as basis for L the elements $(\alpha, \alpha)e_\alpha/2$, f_α for all positive roots, and an orthonormal basis k_i of $\mathfrak{h}$. From 1.8.1 (computing the dual basis $f_\alpha, (\alpha, \alpha)e_\alpha/2, k_i$)

$$C = \sum_{\alpha \in \Phi^+} (\alpha, \alpha)(e_\alpha f_\alpha + f_\alpha e_\alpha)/2 + \sum_i k_i^2.$$

We have

$$f_\alpha e_\alpha v_\lambda = 0, \quad e_\alpha f_\alpha v_\lambda = [e_\alpha, f_\alpha]v_\lambda = h_\alpha v_\lambda = \lambda(h_\alpha)v_\lambda = \langle \lambda \,|\, \alpha \rangle v_\lambda,$$

(cf. 1.8.1). Finally, $\sum_i k_i^2 v_\lambda = \sum_i \lambda(k_i)^2 v_\lambda = (\lambda, \lambda)v_\lambda$ by duality and:

Lemma 1. *C acts on V_λ by the scalar*

$$C(\lambda) := \sum_{\alpha \in \Phi^+} (\lambda, \alpha) + (\lambda, \lambda) = (\lambda, 2\rho) + (\lambda, \lambda) = (\lambda + \rho, \lambda + \rho) - (\rho, \rho).$$

We can now complete step 2.

Lemma 2. *If $\mu \prec \lambda$ is a dominant weight, we have $C(\mu) < C(\lambda)$.*

Proof. $C(\lambda) - C(\mu) = (\lambda + \rho, \lambda + \rho) - (\mu + \rho, \mu + \rho)$. Write $\mu = \lambda - \gamma$ with γ a positive combination of positive roots. Then $(\lambda + \rho, \lambda + \rho) - (\mu + \rho, \mu + \rho) = (\lambda + \mu + 2\rho, \gamma)$. Since $\lambda + \mu + 2\rho$ is a regular dominant weight and γ a nonzero sum of positive roots, we have $(\lambda + \mu + 2\rho, \gamma) > 0$. □

Corollary. $V_{2\lambda} = \{a \in V_\lambda \otimes V_\lambda \mid Ca = C(2\lambda)a\}$.

Proof. The result follows from the lemma and the decomposition $V_\lambda \otimes V_\lambda = V_{2\lambda} \bigoplus_{\mu \prec 2\lambda} V_\mu$, which shows that the summands V_μ are eigenspaces for C with eigenvalue strictly less than the one obtained on $V_{2\lambda}$. □

Let us now establish some notations. Let R denote the polynomial ring on the vector space V_λ. It equals the symmetric algebra on the dual space which is V_μ, with $\mu = -w_0(\lambda)$, by Proposition 5.3. Notice that $-w_0(\rho) = \rho$ and $(\mu, \mu) = (\lambda, \lambda)$, so $C(\lambda) = C(\mu)$.

The space R_2 of polynomials of degree 2 consists (always, by §5.3) of $V_{2\mu}$ over which the Casimir element has value $C(2\mu)$ and lower terms. Let X denote the affine cone corresponding to $G/P \subset \mathbb{P}(V_\lambda)$. X consists of the vectors v which are in the G-orbit of the multiples of v_λ. Let A be the coordinate ring of X which is R/I, where I is the ideal of R vanishing on X. Notice that since X is stable under G, I is also stable under G, and $A = \bigoplus_k A_k$ is a graded representation. Since $v_\lambda \otimes v_\lambda \in V_{2\lambda}$, $x \otimes x \in V_{2\lambda}$ for each element $x \in X$. In particular let us look at the restriction to X of the homogeneous polynomials R_2 of degree 2 with image A_2. Let $Q \subset R_2$ be the kernel of the map $R_2 \to A_2$. Q is a set of quadratic equations for X. From the corollary it follows that Q equals the sum of all the irreducible representations different from $V_{2\mu}$, in the decomposition of R_2 into irreducibles. Since $V_{2\lambda}$ is irreducible, A_2 is dual to $V_{2\lambda}$, and so it is isomorphic to $V_{2\mu}$. The Casimir element $C = \sum_i a_i b_i$ acts as a second order differential operator on R and on A, where the elements of the Lie algebra a_i, b_i act as derivations.

Theorem (Kostant). *Let J be the ideal generated by the quadratic equations Q in R.*

(i) $R/J = A$, J is the ideal of the definition of X.
(ii) The coordinate ring of X, as a representation of G, is $\bigoplus_{k=0}^{\infty} V_{k\mu}$.

Proof. Let $R/J = B$. The Lie algebra L acts on R by derivations and, since J is generated by the subrepresentation Q, L preserves J and induces an action on B. The corresponding simply connected group acts as automorphisms. The Casimir element $C = \sum_i a_i b_i$ acts as a second order differential operator on R and B. Finally, $B_1 = V_\mu$, $B_2 = V_{2\mu}$.

Let $x, y \in B_1$. Then, $xy \in B_2 = V_{2\mu}$, hence $C(x) = C(\lambda)x$, $C(xy) = C(2\lambda)xy$. On the other hand, $C(xy) = C(x)y + xC(y) + \sum_i a_i(x)b_i(y) + b_i(x)a_i(y)$. We have hence $\sum_i a_i(x)b_i(y) + b_i(x)a_i(y) = [C(2\lambda) - 2C(\lambda)]xy = (2\lambda, 2\rho) + 4(\lambda, \lambda) - 2[(\lambda, 2\rho) + (\lambda, \lambda)] = 2(\lambda, \lambda)$. On an element of degree k we have

$$C(x_1x_2\dots x_k) = \sum_{i=1}^{k} x_1x_2\dots C(x_i)\dots x_k + \sum_{i<j}\sum_{h} x_1x_2\dots a_hx_i\dots b_hx_j\dots x_k$$
$$+ x_1x_2\dots b_hx_i\dots a_hx_j\dots x_k = [kC(\lambda) + 2\binom{k}{2}(\lambda, \lambda)]x_1x_2\dots x_k.$$

Now $[kC(\lambda) + 2\binom{k}{2}(\lambda, \lambda)] = C(k\lambda)$. We now apply Lemma 2. B_k is a quotient of $V_\mu^{\otimes k}$, and on B_k we have that C acts with the unique eigenvalue $C(k\lambda)$; therefore we must have that $B_k = V_{k\mu}$ is irreducible. We can now finish. The map $\pi : B \to A$ is surjective by definition. If it were not injective, being L-equivariant, we would have that some B_k maps to 0. This is not possible since if on a variety all polynomials of degree k vanish, this variety must be 0. Thus J is the defining ideal of X and $B = \bigoplus_{k=0}^{\infty} V_{k\mu}$ is the coordinate ring of X. □

Corollary. *A vector $v \in V_\lambda$ is such that $v \otimes v \in V_{2\lambda}$ if and only if a scalar multiple of v is in the orbit of v_λ.*

Proof. In fact $v \otimes v \in V_{2\lambda}$ if and only if v satisfies the quadratic relations Q. □

Remark. On $V_\lambda \otimes V_\lambda$ the Casimir operator is $C_\lambda \otimes 1 + 1 \otimes C_\lambda + 2D$, where

$$D = \sum_{\alpha\in\Phi^+} (\alpha, \alpha)(e_\alpha \otimes f_\alpha + f_\alpha \otimes e_\alpha)/2 + \sum_i k_i \otimes k_i.$$

Thus a vector $a \in V_\lambda \otimes V_\lambda$ is in $V_{2\lambda}$ if and only if $Da = (\lambda, \lambda)a$.

The equation $D(v \otimes v) = (\lambda, \lambda)v \otimes v$ expands in a basis to a system of quadratic equations, defining in projective space the variety G/P.

Examples. We give here a few examples of fundamental weights for classical groups. More examples can be obtained from the theory to be developed in the next chapters.

1. The defining representation.

For the special linear group acting on $\mathbb{C}^n = V_{\omega_1}$ there is a unique orbit of nonzero vectors and so $X = \mathbb{C}^n$. $V_{2\omega_1} = S^2(\mathbb{C}^n)$, the symmetric tensors, and the condition $u \otimes u \in S^2(\mathbb{C}^n)$ is always satisfied.

For the symplectic group the analysis is the same.

For the special orthogonal group it is easily seen that X is the variety of isotropic vectors. In this case the quadratic equations reduce to $(u, u) = 0$, the canonical *quadric*.

2. For the other fundamental weights in $\bigwedge^k \mathbb{C}^n$, the variety X is the set of decomposable vectors, and its associated projective variety is the Grassmann variety. In this case the quadratic equations give rise to the theory of standard diagrams. We refer to Chapter 13 for a more detailed discussion.

For the other classical groups the Grassmann variety has to be replaced by the variety of totally isotropic subspaces. We discuss in Chapter 11, §6.9 the theory of maximal totally isotropic spaces in the orthogonal case using the theory of *pure spinors*.

6.7 The Weyl Group and Characters

We now deduce the internal description of the Weyl group and its consequences for characters.

Theorem 1. *The normalizer N_T of the maximal torus T is the union $\bigcup_{w\in W} s_w T$. We have an isomorphism $W = N_T/T$.*

Proof. Clearly $N_T \supset \bigcup_{w\in W} s_w T$. Conversely, let $a \in N_T$. Since a normalizes T it permutes its fixed points in G/B. In particular we must have that for some $w \in W$ the element $s_w^{-1}a$ fixes $\mathfrak{b}^+$. By Lemma 1 of 6.4, this implies that $s_w^{-1}a \in B$. If we have an element $tu \in B,\ t \in T,\ u \in U$ in the normalizer of T we also have $u \in N_T$. We claim that $N_T \cap U = 1$. This will prove the claim. Otherwise $N_T \cap U$ is a subgroup of U which is a unipotent group normalizing T. Recall that a unipotent group is necessarily connected. The same argument of Lemma 3.6 of Chapter 7 shows that $N_T \cap U$ must commute with T. This is not possible for a nontrivial subgroup of U, since the Lie algebra of U is a sum of nontrivial eigenspaces for T. □

We collect another result which is useful for the next section.

Proposition. *If G is a semisimple algebraic group, then the center of G is contained in all maximal tori.*

Proof. From the corollary of 6.3 we have $Z \subset B = TU$. If we had $z \in Z, z = tu$, we would have that u commutes with T. We have already remarked that the normalizer of T in U is trivial so $u = 1$ and $z \in T$. Since maximal tori are conjugate, Z is in every one of them. □

We can now complete the proof of Theorem 1, Chapter 8, §4.1 in a very precise form. Recall that an element $t \in T$ is *regular* if it is not in the kernel of any root character. We denote by T^{reg} this set of regular elements. Observe first that a generic element of G is a regular element in some maximal torus T.

Lemma. *The map $c : G \times T^{\text{reg}} \to G,\ c : (g, t) \mapsto gtg^{-1}$ has surjective differential at every point. Its image is a dense open set of G.*

Proof. Since the map c is G-equivariant, with respect to the left action on $G \times T^{\text{reg}}$ and conjugation in G, it is enough to compute it at some element $(1, t_0)$ where the tangent space is identified with $L \oplus \mathfrak{t}$. To compute the differential we can compose with $L_{t_0^{-1}}$ and consider separately the two maps $g \mapsto t_0^{-1}gt_0g^{-1}$ and $t \mapsto t$. The first map is the composition $g \mapsto (t_0^{-1}gt_0, g^{-1}) \mapsto t_0^{-1}gt_0g^{-1}$, and so it has differential $a \mapsto \mathrm{Ad}(t_0^{-1})(a) - a$; the second is the identity of $\mathfrak{t}$. Thus we have to prove that the map $(a, u) \mapsto [\mathrm{Ad}(t_0^{-1}) - 1]a + u$ is surjective. Since by hypothesis $\mathrm{Ad}(t_0^{-1})$ does not possess any eigenvalue 1 on the root subspaces, the image of L under $\mathrm{Ad}(t_0^{-1}) - 1$ is $\bigoplus_{\alpha\in\Phi} L_\alpha$. We can then conclude that the image of c is open; since it is algebraic it is also dense. □

Theorem 2. *Let G be a simply connected semisimple group.*

(i) The ring of regular functions, invariant under conjugation by G, is the polynomial ring $\mathbb{C}[\chi_{\omega_i}]$ in the characters χ_{ω_i} of the fundamental representations.
(ii) The restriction to a maximal torus T of the irreducible characters χ_λ, $\lambda \in \Lambda^+$ forms an integral basis of the ring $\mathbb{Z}[\hat{T}]^W$ of W-invariant characters of T.
(iii) $\mathbb{Z}[\hat{T}]^W = \mathbb{Z}[\chi_{\omega_1}, \dots, \chi_{\omega_r}]$ is a polynomial ring over $\mathbb{Z}$ generated by the restrictions of the characters χ_{ω_i} of the fundamental representations.

Proof. From the previous theorem the restriction to T of a function on G, invariant under conjugation, is W-invariant. Since the union of all maximal tori in G is dense in G we have that this restriction is an injective map. The rings $\mathbb{Z}[\Lambda]$, $\mathbb{C}[\Lambda]$ are both permutation representations under W. From Theorem 2.4, every element of Λ is W-conjugate to a unique element $\lambda \in \Lambda^+$. Therefore if we set S_λ, $\lambda \in \Lambda^+$, to be the sum of all the conjugates under W of $\lambda \in \Lambda^+$, we have that

(6.7.1) $$\mathbb{Z}[\Lambda] = \bigoplus\nolimits_{\lambda \in \Lambda^+} \mathbb{Z}S_\lambda.$$

From the highest weight theory it follows that the restriction to a maximal torus T of the irreducible character χ_λ, $\lambda \in \Lambda^+$, which by abuse of notation we still denote by χ_λ, is of the form

$$\chi_\lambda = S_\lambda + \sum_{\mu \prec \lambda} c_{\mu,\lambda} S_\mu$$

for suitable positive integers $c_{\mu,\lambda}$ (which express the multiplicity of the space of weight μ in V_λ). In particular, we deduce that the irreducible characters χ_λ, $\lambda \in \Lambda^+$, form an integral basis of the ring $\mathbb{Z}[\hat{T}]^W$ of W-invariant characters of T. Writing a dominant character $\lambda = \sum_{i=1}^r n_i\omega_i$, $n_i \in \mathbb{N}$ we see that χ_λ and $\prod_{i=1}^r \chi_{\omega_i}^{n_i}$ have the same leading term S_λ (in the dominance order) and thus $\mathbb{Z}[\hat{T}]^W$ is a polynomial ring over $\mathbb{Z}$ generated by the restrictions of the characters χ_{ω_i} of the fundamental representations.

The statement for regular functions over $\mathbb{C}$ follows from this more precise analysis. □

The reader will note the strong connection between this general theorem and various theorems on symmetric functions and conjugation invariant functions on matrices.

6.8 The Fundamental Group

We have constructed the group $G_s(L)$ with the same representations as a semisimple Lie algebra L. We do not yet know that $G_s(L)$ is simply connected. The difficulty comes from the fact that we cannot say a priori that the simply connected group associated to L is a linear group, and so it is obtained by integrating a finite-dimensional representation. The next theorem answers this question. In it we will use some basic facts of algebraic topology for which we refer to standard books, such as [Sp], [Ha]. We need to know that if G is a Lie group and H a closed subgroup, we have a locally trivial fibration $H \to G \to G/H$. To any such fibration one has an associated long exact sequence of homotopy groups. This will allow us to compute $\pi_1(G)$ for G an adjoint semisimple group. The fibration we consider is $B \to G \to G/B$. In order to compute the long exact sequence of this fibration we need to develop some topology of B and of G/B.

First, let us analyze B. Let T_c be the compact torus in T.

Proposition. *The inclusion of $T_c \subset T \subset B$ is a homotopy equivalence.* $\pi_1(T_c) = \hom_{\mathbb{Z}}(\hat{T}, \mathbb{Z})$.

Proof. We remark that $B = TU$ is homotopic to the maximal torus T since U is homeomorphic to a vector space. $T = (\mathbb{C}^*)^n$ is homotopic to $(S^1)^n = \mathbb{R}^n/\mathbb{Z}^n$. The homotopy groups are $\pi_i(\mathbb{R}^n/\mathbb{Z}^n) = 0, \forall i > 1,\ \pi_1(\mathbb{R}^n/\mathbb{Z}^n) = \mathbb{Z}^n$. The homotopy group of $(S^1)^n$ is the free abelian group generated by the canonical inclusions of S^1 in the n factors. In precise terms, in each homotopy class we have the loop induced by a 1-parameter subgroup μ:

$$\begin{array}{ccc} S^1 & \xrightarrow{\mu} & T_c \\ {\scriptstyle i}\downarrow & & {\scriptstyle i}\downarrow \\ \mathbb{C}^* & \xrightarrow{\mu} & T \end{array}$$

More intrinsically $\pi_1(T)$ is identified with the group $\hom_{\mathbb{Z}}(\hat{T}, \mathbb{Z})$ of 1-parameter subgroups. □

By the Bruhat decomposition G/B is a union of even-dimensional cells. In order to apply the standard theory of CW-complexes we need more precise information. Let $(G/B)_h$ be the union of all the Schubert cells of complex dimension $\leq h$. We need to show that $(G/B)_h$ is the $2h$-dimensional skeleton of a CW complex and that every Schubert cell of complex dimension h is the interior of a ball of real dimension $2h$ with its boundary attached to $(G/B)_{h-1}$. If we can prove these statements, we will deduce by standard theory that $\pi_1(G/B) = 1,\ \pi_2(G/B) = H_2(G/B, \mathbb{Z})$ having as basis the orientation classes of the complex 1-dimensional Schubert varieties, which correspond to the simple reflections s_i and are each homeomorphic to a 2-dimensional sphere.

Let us first analyze the basic case of $SL(2, \mathbb{C})$. We have the action of $SL(2, \mathbb{C})$ on $\mathbb{P}^1$, and in homogeneous coordinates the two cells of $\mathbb{P}^1$ are $p_0 := \{(1, 0)\}$, $C := \{a, 1\},\ a \in \mathbb{C}$. The map

$$\begin{vmatrix} 1 & a \\ 0 & 1 \end{vmatrix} \begin{vmatrix} 0 \\ 1 \end{vmatrix} = \begin{vmatrix} a \\ 1 \end{vmatrix}$$

is the parametrization we have used for the open cell.

Consider the set D of unitary matrices:

$$D := \left\{ \begin{vmatrix} se^{i\theta} & r \\ -r & se^{-i\theta} \end{vmatrix} : r^2 + s^2 = 1,\ r, s \geq 0,\ \theta \in [0, 2\pi] \right\}.$$

Setting $se^{i\theta} = x + iy$ we see that this is in fact the 2-cell $x^2 + y^2 + r^2 = 1,\ r \geq 0$, with boundary the circle with $r = 0$. When we apply these matrices to p_0, we see that the boundary ∂D fixes p_0 and the interior $\mathring{D}$ of the cell maps isomorphically to the open cell of $\mathbb{P}^1$.

If B_0 denotes the subgroup of upper triangular matrices in $SL(2, \mathbb{C})$, we have, comparing the actions on $\mathbb{P}^1$, that

$$\mathring{D} B_0 = B_0 s B_0, \quad \partial D \subset B_0.$$

We can now use this attaching map to recursively define the attaching maps for the Bruhat cells. For each node i of the Dynkin diagram, we define D_i to be the copy of D contained in the corresponding group $SU_i(2, \mathbb{C})$.

Proposition. *Given $w = \sigma_{i_1}\sigma_{i_2}\dots\sigma_{i_k} \in W$ a reduced expression, consider the $2k$-dimensional cell $D_w = D_{i_1} \times D_{i_2} \times \cdots \times D_{i_k}$. The multiplication map:*

$$D_{i_1} \times D_{i_2} \times \cdots \times D_{i_k} \to D_{i_1} D_{i_2} \dots D_{i_k} B$$

has image S_w. The interior of D_w maps homeomorphically to the Bruhat cell C_w, while the boundary maps to $S_w - C_w$.

Proof. Let $u = \sigma_{i_1} w$. By induction we can assume the statement for u. Then by induction $D_{i_1}(S_u - C_u) \subset (S_w - C_w)$ and $\partial D_{i_1} S_u \subset S_u$ since $\partial D_{i_1} \subset B$. It remains to prove only that we have a homeomorphism $\mathring{D} \times C_u \to C_w$. By the description of the cells, every element $x \in C_w$ has a unique expression as $x = a s_{i_1} c$ where a is in the root subgroup $U_{\alpha_{i_1}}$, and $c \in C_u$. We have that $a s_{i_1} = db$, for a unique element $d \in \mathring{D}$ and $b \in SL_{i_1}(2, \mathbb{C})$ upper triangular. The claim follows. $\square$

We thus have:

Corollary. *G/B has the structure of a CW complex, with only even-dimensional cells D_w of dimension $2\ell(w)$ indexed by the elements of W.*

Each Schubert cell is a subcomplex.

If $w = s_i u$, $\ell(w) = \ell(u) + 1$, then S_w is obtained from S_u by attaching the cell D_w.

$\pi_1(G/B) = 0$, $H_i(G/B, \mathbb{Z}) = 0$ if i is odd, while $H_{2k}(G/B, \mathbb{Z}) = \bigoplus_{\ell(w)=k} \mathbb{Z}[D_w]$, where D_w is the homology class induced by the cell D_w.

$\pi_2(G/B) = H_2(G/B, \mathbb{Z})$.

Proof. These statements are all standard consequences of the CW complex structure. The main remark is that in the cellular complex which computes homology, the odd terms are 0, and thus the even terms of the complex coincide with the homology.

Given that $\pi_1(G/B) = 0$, $\pi_2(G/B) = H_2(G/B, \mathbb{Z})$ is the Hurewicz isomorphism. $\square$

For us, the two important facts are that $\pi_1(G/B) = 0$ and $\pi_2(G/B, \mathbb{Z}) = \bigoplus_{s_i} \mathbb{Z}[D_{s_i}]$, a sum on the simple reflections.

Theorem 1. *Given a root system Φ, if $G_s(L)$ is as in §6.1 then $G_s(L)$ is simply connected.*

Moreover, its center Z is isomorphic to $\hom_{\mathbb{Z}}(\Lambda/\Lambda_r, \mathbb{Q}/\mathbb{Z})$ where Λ is the weight lattice and Λ_r is the root lattice.

Finally, $Z = \pi_1(G)$, where $G := G_a(L)$ is the associated adjoint group.

Proof. Given any dominant weight λ and the corresponding irreducible representation V_λ, by Schur's lemma, Z acts as some scalars which are elements ϕ_λ of $\hat{Z} = \hom(Z, \mathbb{C}^*)$. Since Z is a finite group, any such homomorphism takes values in the roots of 1, which can be identified with $\mathbb{Q}/\mathbb{Z}$. By §5.2 Proposition 3, we have that $\phi_{\lambda+\mu} = \phi_\lambda \phi_\mu$. Hence we get a map from Λ to $\hat{Z}$. If this mapping were not surjective, we would have an element $a \in Z, a \neq 1$ in the kernel of all the ϕ_λ. This is impossible since by definition $G_s(L)$ has a faithful representation which is the sum of the V_{ω_i}. Since $Z \subset T$ and Z acts trivially on the adjoint representation, we have that the homomorphism factors to a homomorphism of Λ/Λ_r to $\hat{Z}$.

Apply the previous results to $G = G_a(L)$. We obtain that the long exact sequence of homotopy groups of the fibration $B \to G \to G/B$ gives the exact sequence:

$$(6.8.1) \qquad 0 \to \pi_2(G) \to H_2(G/B, \mathbb{Z}) \to \hom_{\mathbb{Z}}(\hat{T}, \mathbb{Z}) \to \pi_1(G) \to 0.$$

It is thus necessary to understand the mapping $H_2(G/B, \mathbb{Z}) \to \hom_{\mathbb{Z}}(\hat{T}, \mathbb{Z})$.

Next we treat $SL(2, \mathbb{C})$. In the diagram

$$\begin{array}{ccccc} U(1,\mathbb{C}) & \longrightarrow & SU(2,\mathbb{C}) & \longrightarrow & \mathbb{P}^1 \\ \downarrow & & \downarrow & & \downarrow \\ B & \longrightarrow & SL(2,\mathbb{C}) & \longrightarrow & \mathbb{P}^1 \end{array}$$

the vertical arrows are homotopy equivalences. We can thus replace $SL(2, \mathbb{C})$ by $SU(2, \mathbb{C})$.

We have the homeomorphisms $SU(2, \mathbb{C}) = S^3$, (Chapter 5, §5.1) $U(1, \mathbb{C}) = S^1$, $\mathbb{P}^1 = S^2$. The fibration $S^1 \to S^3 \to S^2$ is called the *Hopf fibration.*

Since $\pi_1(S^3) = \pi_2(S^3) = 0$ we get the isomorphism $\pi_1(S^1) = H_2(S^2, \mathbb{Z})$. A more precise analysis shows that this isomorphism preserves the standard orientations of S^1, S^2.

The way to achieve the general case, for each node i of the Dynkin diagram we embed $SU(2, \mathbb{C})$ in $SL_i(2, \mathbb{C}) \subset G_s(L) \to G = G_a(L)$ and we have a diagram

$$\begin{array}{ccccc} U(1,\mathbb{C}) & \longrightarrow & SU(2,\mathbb{C}) & \longrightarrow & \mathbb{P}^1 \\ {\scriptstyle i}\downarrow & & \downarrow & & {\scriptstyle j}\downarrow \\ B & \longrightarrow & G & \longrightarrow & G/B \end{array}$$

The mapping i of $S^1 = U(1, \mathbb{C})$ into the maximal torus T of G is given by $e^{\phi\sqrt{-1}} \mapsto e^{\phi h_i \sqrt{-1}}$. As a homotopy class in $\hom(\hat{T}, \mathbb{Z}) = \hom(\Lambda_r, \mathbb{Z})$ it is the element which is the evaluation of $\beta \in \Lambda_r$ at h_i. From 1.8.1 this value is $\langle \beta \mid \alpha_i \rangle$.

Next j maps $\mathbb{P}^1$ to the cell D_{s_i}.

We see that the homology class $[D_{s_i}]$ maps to the linear function $\tau_i \in \hom(\Lambda_r, \mathbb{Z})$, $\tau_i : \beta \mapsto \langle \beta \mid \alpha_i \rangle$. By 2.4.2 these linear functions are indeed a basis of the dual of the weight lattice and this completes the proof that

$$\hom_{\mathbb{Z}}(\Lambda/\Lambda_r, \mathbb{Q}/\mathbb{Z}) = \hom_{\mathbb{Z}}(\Lambda_r, \mathbb{Z})/\hom_{\mathbb{Z}}(\Lambda, \mathbb{Z}) = \pi_1(G). \tag{6.8.2}$$

Now we have a mapping on the universal covering group $\tilde{G}$ of G to $G_s(L)$, which maps surjectively the center of $\tilde{G}$, identified to $\hom_{\mathbb{Z}}(\Lambda/\Lambda_r, \mathbb{Q}/\mathbb{Z})$, to the center Z of $G_s(L)$. Since we have seen that Z has a surjective homomorphism to $\hom_{\mathbb{Z}}(\Lambda/\Lambda_r, \mathbb{Q}/\mathbb{Z})$ and these are all finite groups, the map from $\tilde{G}$ to $G_s(L)$ is an isomorphism. □

By inspecting the Cartan matrices and computing the determinants we have the following table for Λ/Λ_r:

A_n : $\Lambda/\Lambda_r = \mathbb{Z}/(n+1)$. In fact the determinant of the Cartan matrix is $n+1$ but $SL(n+1, \mathbb{C})$ has as center the group of $(n+1)^{\text{th}}$ roots of 1.

For G_2, F_4, E_8 the determinant is 1. Hence $\Lambda/\Lambda_r = 0$, and the adjoint groups are simply connected.

For E_7, D_n, B_n we have $\Lambda/\Lambda_r = \mathbb{Z}/(2)$. For E_6, $\Lambda/\Lambda_r = \mathbb{Z}/(3)$, by the computation of the determinant.

For type D_n the determinant is 4. There are two groups of order 4, $\mathbb{Z}/(4)$ and $\mathbb{Z}/(2) \times \mathbb{Z}/(2)$. A closer inspection of the elementary divisors of the Cartan matrix shows that we have $\mathbb{Z}/(2) \times \mathbb{Z}/(2)$ when n is even and $\mathbb{Z}/(4)$ when n is odd.

6.9 Reductive Groups

We have seen the definition of reductive groups in Chapter 7 where we proved that a reductive group is linearly reductive, modulo the same theorem for semisimple groups. We have now proved this from the representation theory of semisimple Lie algebras. From all the work done, we have now proved that if an algebraic group is semisimple, that is, if its Lie algebra $\mathfrak{g}$ is semisimple, then it is the quotient of the simply connected semisimple group of $\mathfrak{g}$ modulo a finite subgroup of its center. The simply connected semisimple group of Lie algebra $\mathfrak{g}$ is the product of the simply connected groups of the simple Lie algebras $\mathfrak{g}_i$ which decompose $\mathfrak{g}$.

Lemma. *Let G be a simply connected semisimple algebraic group with Lie algebra $\mathfrak{g}$ and H any algebraic group with Lie algebra $\mathfrak{h}$. If $\phi : \mathfrak{g} \to \mathfrak{h}$ is a complex linear homomorphism of Lie algebras, ϕ integrates to an algebraic homomorphism of algebraic groups.*

Proof. Consider a faithful linear representation of $H \subset GL(n, \mathbb{C})$. When we integrate the homomorphism ϕ, we are in fact integrating a linear representation of $\mathfrak{g}$. We know that these representations integrate to rational representations of G. □

Given a connected reductive group G, let Z be the connected component of its center. We know that Z is a torus. Decompose the Lie algebra of G as $\oplus \mathfrak{g}_i \oplus \mathfrak{z}$ where $\mathfrak{z}$ is the Lie algebra of Z and the algebras $\mathfrak{g}_i$ are simple. Let G_i be the simply connected algebraic group with Lie algebra $\mathfrak{g}_i$. The previous lemma implies that for each i, there is an algebraic homomorphism $\phi_i : G_i \to G$ inducing the inclusion of the Lie algebra. Thus we deduce a map $\phi : \prod_i G_i \times Z \to G$ which is the identity on the Lie algebras. This is thus a surjective algebraic homomorphism with finite kernel contained in the product $\prod_i Z_i \times Z$, where Z_i is the finite center of G_i. Conversely:

Theorem. *Given simply connected algebraic groups G_i with simple Lie algebras and centers Z_i, a torus Z and a finite subgroup $A \subset \prod_i Z_i \times Z$ with $A \cap Z = 1$, the group $\prod_i G_i \times Z/A$ is reductive with Z as its connected center.*

In this way all reductive groups are obtained and classified.

The irreducible representations of $\prod_i G_i \times Z/A$ are the tensor products $\otimes_i V_{\lambda_i} \otimes \chi$ with χ a character of Z with the restriction that A acts trivially.

6.10 Automorphisms

The construction of Serre (cf. §3.1) allows us to also determine the entire group of automorphisms of a simple Lie algebra L. Recall that since all derivations are inner, the adjoint group is the connected component of the automorphism group. Now let $\phi : L \to L$ be any automorphism. We use the usual notations $\mathfrak{t}$, Φ, Φ^+ for a maximal toral subalgebra, roots and positive roots. Since maximal toral subalgebras are conjugate under the adjoint group, there is an element g of the adjoint group such that $g(\mathfrak{t}) = \phi(\mathfrak{t})$. Thus setting $\psi := g^{-1}\phi$, we have $\psi(\mathfrak{t}) = \mathfrak{t}$. From Proposition 3 of §5.4 we have that $\psi(\mathfrak{b}^+) = \mathfrak{b}_w$ for some $w \in W$. Hence $s_w^{-1}\psi(\mathfrak{b}^+) = \mathfrak{b}^+$. The outcome of this discussion is that we can restrict our study to those automorphisms ϕ for which $\phi(\mathfrak{t}^+) = \mathfrak{t}^+$ and $\phi(\mathfrak{b}^+) = \mathfrak{b}^+$.

One such automorphism permutes the roots preserving the positive roots, and hence it induces a permutation of the simple roots, hence a symmetry of the Dynkin diagram. On the other hand, we see immediately that the group of symmetries of the Dynkin diagram is $\mathbb{Z}/(2)$ for type $A_n, n > 1$ (reversing the orientation), $D_n, n > 4$ (exchanging the two last nodes $n-1, n$), E_6. It is the identity in cases $B_n, C_n, G_2, F_4, E_7, E_8$. Finally, D_4 has as a symmetry group the symmetric group S_3 (see the triality in the next Chapter 7.3). Given a permutation σ of the nodes of the Dynkin diagram we have that we can define an automorphism ϕ_σ of the Lie algebra by $\phi_\sigma(h_i) = h_{\sigma(i)}$, $\phi_\sigma(e_i) = e_{\sigma(i)}$, $\phi_\sigma(f_i) = f_{\sigma(i)}$. This is well defined since the Serre relations are preserved. We finally have to understand the nature of an automorphism fixing the roots. Thus $\phi(h_i) = h_i$, $\phi(e_i) = \alpha_i e_i$, for some numbers α_i. It follows that $\phi(f_i) = \alpha_i^{-1} f_i$ and that ϕ is conjugation by an element of the maximal torus, of coordinates α_i.

Theorem. *The full group* Aut(L) *of automorphisms of the Lie algebra L is the semidirect product of the adjoint group and the group of symmetries of the Dynkin diagram.*

Proof. We have seen that we can explicitly realize the group of symmetries of the Dynkin diagram as a group S of automorphisms of L and that every element of $\mathrm{Aut}(L)$ is a product of an inner automorphism in $G_a(L)$ and an element of S. It suffices to see that $S \cap G_a(L) = 1$. For this, notice that an element of $S \cap G_a(L)$ normalizes the Borel subgroup. But we have proved that in $G_a(L)$ the normalizer of B is B. It is clear that $B \cap S = 1$. □

Examples. In A_n, as an outer automorphism we can take $x \mapsto (x^{-1})^t$.

In D_n, as an outer automorphism we can take conjugation by any improper orthogonal transformation.

7 Compact Lie Groups

7.1 Compact Lie Groups

At this point we can complete the classification of compact Lie groups. Let K be a compact Lie group and $\mathfrak{k}$ its Lie algebra. By complete reducibility we can decompose $\mathfrak{k}$ as a direct sum of irreducible modules, hence simple Lie algebras. Among simple Lie algebras we distinguish between the 1-dimensional ones, which are abelian, and the nonabelian. The abelian summands of $\mathfrak{k}$ add to the center $\mathfrak{z}$ of $\mathfrak{k}$.

The adjoint group is a compact group with Lie algebra the sum of the nonabelian simple summands of $\mathfrak{k}$. First, we study the case $\mathfrak{z} = 0$ and K is adjoint. On $\mathfrak{k}$ there is a K-invariant (positive real) scalar product for which the elements $\mathrm{ad}(a)$ are skew symmetric. For a skew-symmetric real matrix A we see that A^2 is a negative semidefinite matrix, since $(A^2v, v) = -(Av, Av) \leq 0$. For a negative semidefinite nonzero matrix, the trace is negative and we deduce

Proposition 1. *The Killing form for the Lie algebra of a compact group is negative semidefinite with kernel the Lie algebra of the center.*

Definition. A real Lie algebra with negative definite Killing form is called a *compact* Lie algebra.

Before we continue, let L be a real simple Lie algebra; complexify L to $L \otimes \mathbb{C}$. By Chapter 6, §3.2 applied to L as a module on the algebra generated by the elements $\mathrm{ad}(a),\ a \in L$, we may have that either $L \otimes \mathbb{C}$ remains simple or it decomposes as the sum of two irreducible modules.

Lemma. *Let L be a compact simple real Lie algebra. $L \otimes \mathbb{C}$ is still simple.*

Proof. Otherwise, in the same chapter the elements of $\mathrm{ad}(L)$ can be thought of as complex or quaternionic matrices (hence also complex).

If a real Lie algebra has also a complex structure we can compute the Killing form in two ways, taking either the real or the complex trace. A complex $n \times n$ matrix A is a real $2n \times 2n$ matrix. The real trace $\mathrm{tr}_{\mathbb{R}}(A)$ is obtained from the complex trace as $2\,\mathrm{Re}(\mathrm{tr}_{\mathbb{C}} A)$ twice its real part. Given a complex quadratic form, in some basis it is a sum of squares $\sum_h (x_h + iy_h)^2$, its real part is $\sum_h x_h^2 - y_h^2$. This is indefinite, contradicting the hypotheses made on L. □

Proposition 2. *Conversely, if for a group G the Killing form on the Lie algebra is negative definite, the adjoint group is a product of compact simple Lie groups.*

Proof. If the Killing form (a, a) is negative definite, the Lie algebra, endowed with $-(a, a)$ is a Euclidean space. The adjoint group G acts as a group of orthogonal transformations. We can therefore decompose $L = \bigoplus_i L_i$ as a direct sum of orthogonal irreducible subspaces. These are necessarily ideals and simple Lie algebras. Since the center of L is trivial, each L_i is noncommutative and, by the previous proposition, $L_i \otimes \mathbb{C}$ is a complex simple Lie algebra. We claim that G is a closed subgroup of the orthogonal group. Otherwise its closure $\overline{G}$ has a Lie algebra bigger than $\mathrm{ad}(L)$. Since clearly $\overline{G}$ acts as automorphisms of the Lie algebra L, this implies that there is a derivation D of L which is not inner. Since $L \otimes \mathbb{C}$ is a complex semisimple Lie algebra, D is inner in $L \otimes \mathbb{C}$, and being real, it is indeed in L. Therefore the group G is closed and the product of the adjoint groups of the simple Lie algebras L_i. Each G_i is a simple group. We can quickly prove at least that G_i is simple as a Lie group, although a finer analysis shows that it is also simple as an abstract group. Since G_i is adjoint, it has no center, hence no discrete normal subgroups. A proper connected normal subgroup would correspond to a proper two-sided ideal of L_i. This is not possible, since L_i is simple. □

To complete the first step in the classification we have to see, given a complex simple Lie algebra L, of which compact Lie algebras it is the complexification. We use the theory of Chapter 8, §6.2 and §7.1. For this we look first to the Cartan involution.

7.2 The Compact Form

We prove that the semisimple groups which we found are complexifications of compact groups. For this we need to define a suitable adjunction on the semisimple Lie algebras. This is achieved by the *Cartan involution*, which can be defined using the Serre relations. Let L be presented as in §3.1 from a root system.

Proposition 1. *There is an antilinear involution ω, called the* **Cartan involution**, *on a semisimple Lie algebra which, on the Chevalley generators, acts as*

$$\omega(e_i) = f_i, \ \omega(h_i) = h_i. \tag{7.2.1}$$

Proof. To define ω means to define a homomorphism to the conjugate opposite algebra. Since all relations are defined over $\mathbb{Q}$ the only thing to check is that the relations are preserved. This is immediate. For instance $\delta_{ij} h_i = \omega([e_i, f_j]) = [\omega(f_j), \omega(e_i)] = [e_j, f_i]$. The fact that ω is involutory is clear since it is so on the generators. □

We now need to show that:

Theorem 1. *The real subalgebra $\mathfrak{k} := \{a \in L \mid \omega(a) = -a\}$ gives a compact form for L.*

Clearly for each $a \in L$ we have $a = (a + \omega(a))/2 + (a - \omega(a))/2$, $(a - \omega(a))/2 \in \mathfrak{k}$, $(a + \omega(a))/2 \in \sqrt{-1}\mathfrak{k}$. Since ω is an antilinear involution, we can easily verify that for the Killing form we have $(\omega(a), \omega(b)) = \overline{(a, b)}$. This gives the Hermitian form $\langle a, b \rangle := (a, \omega(b))$. We claim it is positive definite. Let us compute $\langle a, a \rangle = (a, \omega(a))$ using the orthogonal decomposition for the Killing form $L = \mathfrak{t} \oplus \bigoplus_{\alpha \in \Phi^+} (L_\alpha \oplus L_{-\alpha})$. On $\mathfrak{t} = E_{\mathbb{C}}$, with E the real space generated by the elements h_i, for $a \in E, \alpha \in \mathbb{C}$ we have $(a \otimes \alpha, a \otimes \alpha) = (a, a)|\alpha|^2 > 0$.

For $L_\alpha \oplus L_{-\alpha}$ one should first remark that the elements s_i which lift the simple reflections preserve the Hermitian form. Next, one can restrict to $\mathbb{C}e_i \oplus \mathbb{C}f_i$ and compute

$$\begin{aligned}(ae_i + bf_i, \omega(ae_i + bf_i)) &= (ae_i + bf_i, \overline{a} f_i + \overline{b} e_i) \\ &= (a\overline{a} + b\overline{b})(e_i, f_i) = 2(a\overline{a} + b\overline{b})/(\alpha_i, \alpha_i) > 0.\end{aligned}$$

In conclusion we have a self-adjoint group and a compact form:

Proposition 2. *$\langle a, b \rangle$ is a Hilbert scalar product for which the adjoint of* $\mathrm{ad}(x)$, $x \in L$ *is given by* $\mathrm{ad}(\omega(x))$.

$\mathfrak{k}$ is the Lie algebra of the unitary elements in the adjoint group of L.

Proof. We have just checked positivity. For the second statement, notice that since $[x, \omega(b)] = -\omega[\omega(x), b]$, we have

$$\begin{aligned}\langle \mathrm{ad}(x)(a), b \rangle = (a, -\mathrm{ad}(x)(\omega(b))) &= (a, \omega(\mathrm{ad}(\omega(x))(b))) \\ &= \langle a, \mathrm{ad}(\omega(x))(b) \rangle.\end{aligned} \tag{7.2.2}$$

The last statement follows from the previous ones. □

We have at this point proved that the adjoint group of a semisimple algebraic group is self-adjoint for the Hilbert structure given by the Cartan involution. In particular, it has a Cartan decomposition $G = KP$ with K a maximal compact subgroup. If the Lie algebra of G is simple, G is a simple algebraic group and K a simple compact group. Let us pass now to the simply connected cover $G_s(L)$. Let K_s be the preimage of K in $G_s(L)$.

Proposition 3. *K_s is connected maximal compact and is the universal cover of K. K_s is Zariski dense in $G_s(L)$.*

Proof. Since the map $\pi : G_s(L) \to G$ is a finite covering, the map $K_s \to K$ is also a finite covering. The inclusion of K in G is a homotopy equivalence. In particular, it induces an isomorphism of fundamental groups. Thus K_s is connected compact and the universal cover of K. If it were not maximal compact, we would have a larger compact group with image a compact group strictly larger than K. The first claim follows.

The Zariski closure of K_s is an algebraic subgroup H containing K_s. Its image in G contains K, so it must coincide with G. Since the kernel of π is in H we must have $G_s(L) = H$. □

From Proposition 2, Chapter 8, §6.1 we have:

Theorem 2. *Given any rational representation M of $G_s(L)$ choose a Hilbert space structure on M invariant under K_s. Then $G_s(L)$ is self-adjoint and K_s is the subgroup of unitary elements.*

In the correspondence between a compact K and a self-adjoint algebraic group G, we have seen that the algebraic group is topologically $G = K \times V$ with V affine space. Thus G is simply connected if and only if K is simply connected.

Remark. At this point we can complete the analysis by establishing the full classification of compact connected Lie groups and their algebraic analogues, the linearly reductive groups. Summarizing all our work we have proved:

Theorem 3. *There is a correspondence between connected compact Lie groups and reductive algebraic groups, which to a compact group K associates its algebraic envelope defined in Chapter 8, §7.2.*

Conversely, to a reductive group G we associate a maximal compact subgroup K unique up to conjugacy.

In any linear representation of G, a Hilbert metric invariant under K makes G self-adjoint.

G has a Cartan decomposition relative to K.

Then Theorem 6.9 becomes the classification theorem for connected compact Lie groups:

Theorem 4. *Given simply connected compact groups K_i with simple Lie algebras and centers Z_i, a compact torus T and a finite subgroup $A \subset \prod_i Z_i \times T$ with $A \cap T = 1$, the group $\prod_i K_i \times T/A$ is compact with Z as its connected center.*

In this way all connected compact groups are obtained and classified.

Proof. The compact group $\prod_i K_i \times T/A$ is the one associated to the reductive group $\prod_i G_i \times Z/A$, where G_i is the complexification of K_i and Z the complexification of T. $\square$

From these theorems we can also deduce the classification of irreducible representations of reductive or compact Lie groups. For $G = (\prod_i G_i \times Z)/A$, we must give for each i an irreducible representation V_{λ_i} of G_i and also a character χ of Z. The representations $\otimes_i V_{\lambda_i} \otimes \chi$ are the list of irreducible representations of $\prod_i G_i \times Z$. Such a representation factors through G if and only if A acts trivially on it. For each i, λ_i induces a character on Z_i which we still call λ_i. Thus the condition is that the character $\prod_i \lambda_i \chi$ should be trivial on A.

7.3 Final Comparisons

We have now established several correspondences. One is between reductive groups and compact groups, the other between Lie algebras and groups. In particular we have associated to a complex simple Lie algebra two canonical algebraic groups, the adjoint group and the simply connected group, their compact forms and the compact Lie algebra. Several other auxiliary objects have appeared in the classification, and we should compare them all.

First, let us look at tori. Let L be a simple Lie algebra, $\mathfrak{t}$ a Cartan subalgebra, $G_a(L)$, $G_s(L)$ the adjoint and simply connected groups. $G_a(L) = G_s(L)/Z$, where Z is the center of $G_s(L)$. Consider the maximal tori T_a, T_s associated to $\mathfrak{t}$ in $G_a(L)$, $G_s(L)$, respectively. From §7.8, it follows that $Z \subset T_s$. Since T_s and T_a have the same Lie algebra it follows that $T_a = T_s/Z$. Since the exponential from the nilpotent elements to the unipotents is an isomorphism of varieties, the unipotent elements of $G_s(L)$ are mapped isomorphically to those of $G_a(L)$ under the quotient map. For the Borel subgroup associated to positive roots we thus have T_aU^+ in $G_a(L)$ and T_sU^+ in $G_s(L)$; for the Bruhat decomposition we have

$$G_a(L) = \bigsqcup_{w \in W} U_w^+ s_w T_a U^+, \qquad G_s(L) = \bigsqcup_{w \in W} U_w^+ s_w T_s U^+.$$

A similar argument shows that the normalizer of T_s in $G_s(L)$ is $N_{T_s} = \bigsqcup_{w \in W} s_w T_s$ and $N_{T_s}/Z = N_{T_a}$. In particular $N_{T_s}/T_s = N_{T_a}/T_a = W$. Another simple argument, which we leave to the reader, shows that there is a 1-1 correspondence between maximal tori in any group with Lie algebra L and maximal toral subalgebras of L. In particular maximal tori of G are all conjugate (Theorem 3.2).

More interesting is the comparison with compact groups. In this case, the second main tool, besides the Cartan decomposition, is the *Iwasawa decomposition*. We explain a special case of this theorem. Let us start with a very simple remark. The Cartan involution, by definition, maps $\mathfrak{u}^+$ to $\mathfrak{u}^-$ and $\mathfrak{t}$ into itself.

Let $\mathfrak{k}$ be the compact form associated to the Cartan involution. Let us look first at the Cartan involution on $\mathfrak{t}$. From formulas 7.2.1 we see that $\mathfrak{t}_c := \mathfrak{t} \cap \mathfrak{k}$ is the real space with basis $i\,h_j$. It is clearly the Lie algebra of the maximal compact torus T_c in T, and T has a Cartan decomposition.

Proposition 1. *(i)* $\mathfrak{k} = \mathfrak{t}_c \oplus \mathfrak{m}$ *where* $\mathfrak{m} := \{a - \omega(a),\ a \in \mathfrak{u}^-\}$.

(ii) $L = \mathfrak{k} + \mathfrak{b}^+,\ \mathfrak{t}_c = \mathfrak{k} \cap \mathfrak{b}^+$.

(iii) If K *is the compact group* $B \cap K = T_c$.

Proof. (i) The first statement follows directly from the formula 7.2.1 defining ω which shows in particular that $\omega(\mathfrak{u}^-) = \mathfrak{u}^+$, $\omega(\mathfrak{t}) = \mathfrak{t}$. It follows that every element of $\mathfrak{k}$ is of the form $-\omega(a) + t + a,\ t \in \mathfrak{t}_c,\ a \in \mathfrak{u}^-$.

(ii) Since $\mathfrak{b}^+ = \mathfrak{t} \oplus \mathfrak{u}^+$ any element $x = a + t + b,\ a \in \mathfrak{u}^+, t \in \mathfrak{t}, b \in \mathfrak{u}^-$ in L equals $a + \omega(b) + t + b - \omega(b)$, $a + \omega(b) + t \in \mathfrak{b}^+$, $b - \omega(b) \in \mathfrak{k}$ showing that $L = \mathfrak{k} + \mathfrak{b}^+$.

Consider now an element $-\omega(a) + t + a \in \mathfrak{k}$ with $t \in \mathfrak{t}_c, a \in \mathfrak{u}^-$. If we have $t + (a - \omega(a)) \in \mathfrak{b}^+$ since $t \in \mathfrak{b}^+$ we have $a - \omega(a) \in \mathfrak{b}^+$. This clearly implies that $a = 0$.

For the second part, we have from the first part that $B \cap K$ is a compact Lie group with Lie algebra $\mathfrak{t}_c$. Clearly $B \cap K \supset T_c$. Thus it suffices to remark that T_c is maximal compact in B. Let $H \supset T_c$ be maximal compact. Since unitary elements are semisimple we have that $H \cap U^+ = 1$. Hence in the quotient, H maps injectively into the maximal compact group of T. This is T_c. Hence $H = T_c$. □

The previous simple proposition has a very important geometric implication. Let K be the associated compact group. We can assume we are working in the adjoint case. As the reader will see, the other cases follow. Restrict the orbit map of G to G/B to the compact group K. The stabilizer of $[B] \in G/B$ in K is then, by Lemma 1 of 6.4 and the previous proposition, $B \cap K = T_c$. The tangent space of G/B in B is $L/\mathfrak{b}^+$. Hence from the same proposition the Lie algebra of $\mathfrak{k}$ maps surjectively to this tangent space. By the implicit function theorem this means that the image of K under the orbit map contains an open neighborhood of B. By equivariance, the image of K is open. Since K is compact the image of K is also closed. It follows that $K[B] = G/B$ and:

Theorem 1. *$KB = G$ and the homogeneous space G/B can also be described in compact form as K/T_c.*

It is interesting to see concretely what this means at least in one classical group. For $SL(n, \mathbb{C})$, the flag variety is the set of flags $V_1 \subset V_2 \subset \cdots \subset V_n = V$. Fixing a maximal compact subgroup is like fixing a Hilbert structure on V. When we do this, each V_i has an orthogonal complement L_i in V_{i+1}. The flag is equivalent to the sequence $L_1, L_2, \dots, L_n$ of mutually orthogonal lines. The group $SU(n, \mathbb{C})$ acts transitively on this set and the stabilizer of the set of lines generated by the standard orthonormal basis e_i in $SU(n, \mathbb{C})$ is the compact torus of special unitary diagonal matrices.

Consider next the normalizer N_{T_c} of the compact torus in K. First, let us recall that for each i in the Dynkin diagram, the elements s_i inducing the simple reflections belong to the corresponding $SU_i(2, \mathbb{C}) \subset SL_i(2, \mathbb{C})$. In particular all the elements s_i belong to K. We have.

Proposition 2. *$N_{T_c} = K \cap N_T$. Moreover $N_{T_c}/T_c = N_T/T = W$.*

Proof. If $a \in N_{T_c}$ since T_c is Zariski dense in T we have $a \in N_T$, hence the first statement. Since the classes of the elements $s_i \in N_{T_c}$ generate $W = N_T/T$, the second statement follows. □

We have thus proved that the Weyl group can also be recovered from the compact group. When we are dealing with compact Lie groups the notion of maximal torus is obviously that of a maximal compact abelian connected subgroup (Chapter 4, §7.1). Let us see then:

Theorem 2. *In a compact connected Lie group K all maximal tori are conjugate. Every element of K is contained in a maximal torus.*

Proof. Let Z be the connected part of the center of K. If A is a torus, it is clear that AZ is also a compact connected abelian group. It follows that all maximal tori contain Z. Hence we can pass to K/Z and assume that K is semisimple. Then K is a maximal compact subgroup of a semisimple algebraic group G. We use now the identification $G/B = K/T_c$, where T_c is the compact part of a maximal torus in G. Let A be any torus in K. Since A is abelian connected, it is contained in a maximal connected solvable subgroup P of G, that is a Borel subgroup of G. From the theory developed, we have that P has a fixed point in G/B, and hence A has a fixed point in K/T_c. By the fixed point principle A is conjugate to a subgroup of T_c. If A is a maximal torus, we must have A conjugate to T_c.

For the second part, every element of G is contained in a Borel subgroup, but a Borel subgroup intersects K in a maximal torus. □

One should also see [A] for a more direct proof based on the notion of degree of a map between manifolds.

Remark. In the algebraic case it is not true that every semisimple element of G is contained in a maximal torus!

In the description K/T_c we lose the information about the B-action and the algebraic structure, but we gain a very interesting topological insight.

Proposition 3. *Let $n \in N_{T_c}$. Given a coset $kT_c, k \in K$ the coset $kn^{-1}T_c$ is well defined and depends only on the class of n in W. In this way we define an action of W on K/T_c.*

Proof. If $t \in T_c$ we must show that $kn^{-1}T_c = ktn^{-1}T_c$. Now $ktn^{-1}T_c = kn^{-1}ntn^{-1}T_c = kn^{-1}T_c$ since $ntn^{-1} \in T_c$. It is clear that the formula, since it is well defined, defines an action of N_{T_c} on K/T_c. We have to verify that T_c acts trivially, but this is clear. □

Example. In the case of $SU(n, \mathbb{C})$ where K/T_c is the set of sequence $L_1, L_2, \dots, L_n$ of mutually orthogonal lines and W is the symmetric group, the action of a permutation σ on a sequence is just the sequence $L_{\sigma(1)}, L_{\sigma(2)}, \dots, L_{\sigma(n)}$.

Exercise. Calculate explicitly the action of $S_2 = \mathbb{Z}/(2)$ on the flag variety of $SL(2, \mathbb{C})$ which is just the projective line. Verify that it is not algebraic.

The Bruhat decomposition and the topological action of W on the flag variety arc the beginning of a very deep theory which links geometry and representations but goes beyond the limits of this book.

11

Invariants

1 Applications to Invariant Theory

In this chapter we shall make several computations of invariants of a group G acting linearly on a space V. In our examples we have that G is algebraic and often we are working over $\mathbb{C}$.

In the general case, V is a vector space over a general field F, and $G \subset GL(V) = GL(n, F)$ is a linear group. One should consider the following basic principles:

(1) Let $\overline{F}$ be the algebraic closure of F and let $\overline{G}$ be the Zariski closure of G in $GL(n, \overline{F})$. Then a polynomial on $V \otimes_F \overline{F}$ is invariant under $\overline{G}$ if and only if it is invariant under G.
(2) Suppose F is infinite. If we find a set of generating polynomial invariants f_i for the action of $\overline{G}$ on $V \otimes_F \overline{F}$, and we suppose that the f_i have coefficients in F, then the f_i form a set of generators for the polynomial invariants under the action of G on V.
(3) If $F = \mathbb{R}$, $\overline{F} = \mathbb{C}$, then the real and the imaginary part of a polynomial invariant with complex coefficients are invariant.

These remarks justify the fact that we often work geometrically over $\mathbb{C}$.

1.1 Cauchy Formulas

Assume now that we have an action of a group G on a space U of dimension m and we want to compute the invariants of n copies of U. Assume first $n \geq m$.

We think of n copies of U as $U \otimes \mathbb{C}^n$ and the linear group $GL(n, \mathbb{C})$ acts on this vector space by tensor action on the second factor, commuting with the G action on U. As we have seen in Chapter 3, the Lie algebra of $GL(n, \mathbb{C})$ acts by polarization operators. The ring of G-invariants is stable under these actions.

From Chapter 9, §6.3.2, the ring $\mathcal{P}(U^n)$ of polynomial functions on $U \otimes \mathbb{C}^n$ equals

$$S(U^* \otimes \mathbb{C}^n) = \bigoplus\nolimits_{\lambda,\, ht(\lambda) \leq m} S_\lambda(U^*) \otimes S_\lambda(\mathbb{C}^n).$$

For the invariants under G we clearly have that

$$S(U^* \otimes \mathbb{C}^n)^G = \bigoplus_{\lambda,\, ht(\lambda)\leq m} S_\lambda(U^*)^G \otimes S_\lambda(\mathbb{C}^n). \tag{1.1.1}$$

This formula describes the ring of invariants of G acting on the polynomial ring of U^n as a representation of $GL(n, \mathbb{C})$. In particular we see that the multiplicity of $S_\lambda(\mathbb{C}^n)$ in the ring $\mathcal{P}(U^n)^G$ equals the dimension of the space of invariants $S_\lambda(U^*)^G$.

The restriction on the height implies that when we restrict to $m \leq n$ copies we again have the same formula, and hence we deduce the following by comparing isotypic components and by Chapter 9, §6.3.3:

$$S_\lambda(U^*)^G \otimes S_\lambda(\mathbb{C}^n) \cap S(U^* \otimes \mathbb{C}^m) = S_\lambda(U^*)^G \otimes S_\lambda(\mathbb{C}^m).$$

We deduce:

Theorem 1. *If* $\dim U = m$, *the ring of invariants of* $S(U^* \otimes \mathbb{C}^n)$ *is generated, under polarization, by the invariants of* m *copies of* U.

Proof. Each isotypic component (under $GL(n, \mathbb{C})$) $S_\lambda(U^*)^G \otimes S_\lambda(\mathbb{C}^n)$ is generated under this group by $S_\lambda(U^*)^G \otimes S_\lambda(\mathbb{C}^m)$ since $ht(\lambda) \leq m$. □

There is a useful refinement of this theorem.

If λ is a partition of height exactly m, we can write it as $\lambda = \mu + k\,1^m$ with $ht(\mu) \leq m-1$. Then $S_\lambda(U) \otimes S_\lambda(\mathbb{C}^m) = (\bigwedge^m U \otimes \bigwedge^m \mathbb{C}^m)^k S_\mu(U) \otimes S_\mu(\mathbb{C}^m)$. If the group G is contained in the special linear group the *determinant* of m-vectors, i.e., a generator of the 1-dimensional space $\bigwedge^m U \otimes \bigwedge^m \mathbb{C}^m$ is invariant and we have

$$S_\lambda(U)^G \otimes S_\lambda(\mathbb{C}^m) = \left(\bigwedge^m U \otimes \bigwedge^m \mathbb{C}^m\right)^k S_\mu(U)^G \otimes S_\mu(\mathbb{C}^m).$$

Thus we obtain by the same reasoning:

Theorem 2. *If* $\dim U = m$ *and* $G \subset SL(U)$, *the ring of* G*-invariants of* $S(U^* \otimes \mathbb{C}^n)$ *is generated, under polarization, by the determinant and invariants of* $m-1$ *copies of* U.

Alternatively, we could use the $G \times U^+$ invariants $\oplus S_\lambda(U^*)^G \otimes S_\lambda(\mathbb{C}^n)^{U^+}$. They are contained in the polynomial ring $S(U^* \otimes C^{m-1})[d]$, where d is the determinant of the first m vectors u_i. By the theory of the highest weight, they generate, under polarization, all the invariants.[102]

[102] Every subspace $W \subset S(U^* \otimes \mathbb{C}^n)$ which is stable under polarization is generated by $W \cap S(U^* \otimes \mathbb{C}^m)$. A subspace $W \subset S(U^* \otimes \mathbb{C}^n)$ stable under polarization and multiplication by the determinant d is generated (under polarizations and multiplication by d) by $W \cap S(U^* \otimes \mathbb{C}^{m-1})$.

1.2 FFT for $SL(n, \mathbb{C})$

We shall now discuss the first fundamental theorem of the special linear group $SL(V)$, $V = \mathbb{C}^n$, acting on m copies of the fundamental representation.

In the computation of invariants we will often use two simple ideas.

(1) If two invariants, under a group G coincide on a set X, then they coincide on the set $GX := \{gx \mid g \in G, x \in X\}$.
(2) If two polynomial invariants, under a group G coincide on a set X dense in Y, then they coincide on Y.

We identify V^m with the space of $n \times m$ matrices, with $SL(n, \mathbb{C})$ acting by left multiplication.

We consider the polynomial ring on V^m as the ring $\mathbb{C}[x_{ij}]$, $i = 1, \dots, n;\ j = 1, \dots, m$, of polynomials in the entries of $n \times m$ matrices. Set $X := (x_{ij})$ the matrix whose entries are the indeterminates.

Given n indices $i_1, \dots, i_n$ between $1, \dots, m$, we shall denote by $[i_1, \dots, i_n]$ the determinant of the maximal minor of X extracted from the corresponding columns.

Theorem 1. *The ring of invariants* $\mathbb{C}[x_{ij}]^{SL(n,\mathbb{C})}$ *coincides with the ring* $A := \mathbb{C}[[i_1, \dots, i_n]]$ *generated by the* $\binom{m}{n}$ *elements* $[i_1, \dots, i_n]$, $1 \le i_1 < \cdots < i_n \le m$.

Proof. We can apply the previous theorem using the fact that the proposed ring A is certainly made of invariants, closed under polarization, and by definition it contains the determinant on the first n copies which is $[1, 2, \dots, n]$. Thus it suffices to show that A coincides with the ring of invariants for $n - 1$ copies of V.

Now it is clear that, given any $n - 1$ linearly independent vectors of V, they can be completed to a basis with determinant 1. Hence this set of $n - 1$ tuples, which is open, forms a unique orbit under $SL(V)$. Therefore the invariants of $n - 1$ copies are just the constants, and the theorem is proved. □

It is often convenient to formulate statements on invariants in a geometric way, that is, in the language of quotients (cf. Chapter 14, §3 for details).

Definition. Given an affine variety V with the action of a group G and a map $\pi : V \to W$, we say that π is a quotient under G, and denote $W = V//G$ if the comorphism $\pi^* : k[W] \to k[V]$ is an isomorphism of $k[W]$ to $k[V]^G$.

We consider the space of m-tuples of vectors of dimension n as the space of n-tuples $u_1, \dots, u_n$ of m-dimensional vectors, then the FFT of $SL(n)$ becomes:

Theorem 2. *The map* $(u_1, \dots, u_n) \mapsto u_1 \wedge \dots \wedge u_n$ *with image the decomposable vectors of* $\bigwedge^n \mathbb{C}^m$ *is the quotient of* $(\mathbb{C}^m)^n$ *under* $SL(n, \mathbb{C})$.[103]

Let us understand this ring of invariants as a representation of $GL(m, \mathbb{C})$. Notice that we recover a special case of Theorem 6.6 of Chapter 10.

[103] We have not really proved everything; the reader should check that the decomposable vectors are a subvariety.

Corollary. *The space of polynomials of degree k in the elements* $[i_1, \ldots, i_n]$ *as a representation of* $GL(m, \mathbb{C})$ *equals* $S_{k^n}(\mathbb{C}^m)$.

Proof. In the general formula $S(U^* \otimes \mathbb{C}^m)^{SL(U)} = \bigoplus_\lambda S_\lambda(U^*)^{SL(U)} \otimes S_\lambda(\mathbb{C}^m)$. We see that $S_\lambda(U^*)$ is always an irreducible representation of $SL(U)$, and hence it contains invariants only if it is the trivial representation. This happens if and only if $\lambda := k^n$ and $S_{k^n}(\mathbb{C}^n) = \mathbb{C}$. Thus we have the result that

$$S(U^* \otimes \mathbb{C}^m)^{SL(U)} = \bigoplus_k S_{k^n}(U^*) \otimes S_{k^n}(\mathbb{C}^m) = \bigoplus_k S_{k^n}(\mathbb{C}^m).$$

By comparing degrees we have the result. □

Using the results of Chapter 10, §6.6 we could also describe the quadratic equations satisfied by the invariants. We prefer to leave this task to Chapter 13, where it will be approached in a combinatorial way.

One can in fact combine this theorem with the first fundamental theorem of Chapter 9 §1.4 to get a theorem of invariants for $SL(U)$ acting on $U^m \oplus (U^*)^p$. We describe this space as pairs of matrices $X \in M_{m,n}(\mathbb{C})$, $Y \in M_{n,p}(\mathbb{C})$, X of rows $\phi_1, \ldots, \phi_m$ and Y of columns $u_1, \ldots, u_p$. A matrix $A \in SL(n, \mathbb{C})$ acts as (XA^{-1}, AY) and we have invariants

$$\langle \phi_i | u_j \rangle, \ [u_{i_1}, \ldots, u_{i_n}], \ [\phi_{j_1}, \ldots, \phi_{j_n}].$$

The $\langle \phi_i | u_j \rangle$ are the entries of XY, and the $[u_{i_1}, \ldots, u_{i_n}]$ are the determinants of the maximal minors of Y, while $[\phi_{j_1}, \ldots, \phi_{j_n}]$ are the determinants of the maximal minors of X.

Theorem 3. *The ring of invariants* $\mathcal{P}[U^m \oplus (U^*)^p]^{SL(U)}$ *is the ring generated by*

$$\langle \phi_i | u_j \rangle, \ [u_{i_1}, \ldots, u_{i_n}], \ [\phi_{j_1}, \ldots, \phi_{j_n}].$$

Proof. We apply the methods of §1.1 to U, U^* and reduce to $n-1$ copies of both U and U^*.

In this case, of the invariants described we only have the $\langle \phi_i | u_j \rangle$. Consider the open set in which the $n-1$ vectors $u_1, \ldots, u_{n-1}$ are linearly independent. Acting with $SL(U)$ we can transform these vectors to be $e_1, \ldots, e_{n-1}$. Let us thus consider the subset

$$W := \{(\phi_1, \ldots, \phi_{n-1}; e_1, \ldots, e_{n-1})\}.$$

We can still think of the linear forms $\phi_1, \ldots, \phi_{n-1}$ as rows of an $n-1 \times n$ matrix X. This set W is stable under the subgroup of $SL(n, \mathbb{C})$:

$$H := \left\{ \begin{pmatrix} 1 & 0 & 0 & \ldots & 0 & a_1 \\ 0 & 1 & 0 & \ldots & 0 & a_2 \\ & & \ldots & \ldots & & \\ 0 & 0 & 0 & \ldots & 1 & a_{n-1} \\ 0 & 0 & 0 & \ldots & 0 & 1 \end{pmatrix} \right\}.$$

A nonzero $SL(n, \mathbb{C})$ invariant function restricts to W to a nonzero H-invariant function.

The action of an element of H on a matrix X consists of subtracting, from the last column of X, the linear combination with the coefficients a_i of the first $n-1$ columns.

Thus, on the open set where the first $n-1$ columns are linearly independent, we can make the last column 0 by acting with H. Hence we see that an H-invariant function is independent of the coordinates of the last column. Now notice that the $(n-1)^2$ functions $\langle \phi_i | u_j \rangle$ restrict on W to the functions $\langle \phi_i | e_j \rangle$, $j \leq n-1$, which give all the coordinates of the first $n-1$ columns of X. It follows then that, given any invariant function f, there is a polynomial $g(\langle \phi_i | u_j \rangle)$ which coincides with f on W. By the initial discussion we must have that $f = g(\langle \phi_i | u_j \rangle)$. □

2 The Classical Groups

2.1 FFT for Classical Groups

We start here the description of the representation theory of other classical groups, in particular the orthogonal and the symplectic group; again we relate invariant theory with representation theory by the same methods used for the linear group.

The proofs work over any field F of characteristic 0. In practice we think of $F = \mathbb{R}, \mathbb{C}$.

We fix a vector space V (over F) with a nondegenerate invariant bilinear form. Let us denote by (u, v) a symmetric bilinear form, and by $O(V)$ the orthogonal group fixing it.

Let us denote by $[u, v]$ an antisymmetric form and by $Sp(V)$ the symplectic group fixing it.

We start by remarking that for the fundamental representation of either one of these two groups we have a nondegenerate invariant bilinear form which identifies this representation with its dual. Thus for the first fundamental theorem it suffices to analyze the invariants of several copies of the fundamental representation. When convenient we identify vectors, or exterior products of vectors, with functions.

The symplectic group $Sp(V)$ on a $2n$-dimensional vector space is formed by unimodular matrices (Chapter 5, §3.6.2). Fix a symplectic basis for which the matrix of the skew form is $J = \begin{pmatrix} 0 & 1_n \\ -1_n & 0 \end{pmatrix}$. Given $2n$ vectors u_i which we write as column vectors of a $2n \times 2n$ matrix A, the determinant of A equals the Pfaffian of the matrix $A^t J A$ which has as entries the skew products $[u_i, u_j]$, (Chapter 5, §3.6.2). Hence

Lemma. *The determinant of $A = [u_1, \dots, u_{2n}]$ equals the Pfaffian of the matrix $A^t J A$.*

The orthogonal group instead contains a subgroup of index 2, the special linear group $SO(V)$ formed by the orthogonal matrices of determinant 1.

When we discuss either the symplectic or the special orthogonal group we assume we have chosen a trivialization $\bigwedge^{\dim V} V = \mathbb{C}$ of the top exterior power of V.

If $m = \dim V$ and $v_1, \ldots, v_m$ are m vector variables, the element $v_1 \wedge \ldots \wedge v_m$ is to be understood as a function on $V^{\oplus m}$ invariant under the special linear group.

Given a direct sum $V^{\oplus k}$ of copies of the fundamental representation, we denote by $u_1, \ldots, u_k$ a typical element of this space.

Theorem. *Let* $\dim V = n$*:*

(i) The ring of invariants of several copies of the fundamental representation of $SO(V)$ is generated by the scalar products (u_i, u_j) and by the determinants $u_{i_1} \wedge u_{i_2} \wedge \ldots \wedge u_{i_n}$.

(ii) The ring of invariants of several copies of the fundamental representation of $O(V)$ is generated by the scalar products (u_i, u_j).

(iii) The ring of invariants of several copies of the fundamental representation of $Sp(V)$ is generated by the functions $[u_i, u_j]$.

Before proving this theorem we formulate it in the language of matrices and quotients.

Consider the group $O(n, \mathbb{C})$ of $n \times n$ matrices X with $X^t X = XX^t = 1$ and consider the space of $n \times m$ matrices Y with the action of $O(n, \mathbb{C})$ given by multiplication XY. Then:

The mapping $Y \mapsto Y^t Y$ from the variety of $n \times m$ matrices to the symmetric $m \times m$ matrices of rank $\leq n$ is a quotient under the orthogonal group.

Similarly, let

$$J_n := \begin{pmatrix} 0 & 1_n \\ -1_n & 0 \end{pmatrix}$$

be the standard $2n \times 2n$ skew-symmetric matrix and $Sp(2n, \mathbb{C})$ the standard symplectic group of the matrices X such that $X^t J_n X = J_n$. Then consider the space of $2n \times m$ matrices Y with the action of $Sp(2n, \mathbb{C})$ given by multiplication XY. Then:

The mapping $Y \to Y^t J_n Y$ from the space of $2n \times m$ matrices to the variety of antisymmetric $m \times m$ matrices of rank $\leq 2n$ is a quotient under the symplectic group.

Proof of the Theorem for $SO(V)$, $O(V)$. We prove first that the theorem for $SO(V)$ implies the theorem for $O(V)$.

One should remark that since $SO(V)$ is a normal subgroup of index 2 in $O(V)$, we have a natural action of the group $O(V)/SO(V) \cong \mathbb{Z}/(2)$ on the ring of $SO(V)$ invariants.

Let τ be the element of $\mathbb{Z}/(2)$ corresponding to the orthogonal transformations of determinant -1 (*improper transformations*). The elements (u_i, u_j) are invariants of this action while $\tau(u_{i_1} \wedge u_{i_2} \wedge \ldots u_{i_m}) = -u_{i_1} \wedge u_{i_2} \wedge \ldots u_{i_m}$. It follows that the orthogonal invariants are polynomials in the special orthogonal invariants in which every monomial contains a product of an even number of elements of type $u_{i_1} \wedge u_{i_2} \wedge \ldots u_{i_n}$.

Thus it is enough to verify the following identity:

$$(u_{i_1} \wedge u_{i_2} \wedge \ldots u_{i_n})(u_{j_1} \wedge u_{j_2} \wedge \ldots u_{j_n})$$
$$= \det \begin{pmatrix} (u_{i_1}, u_{j_1}) & (u_{i_1}, u_{j_2}) & \ldots & (u_{i_1}, u_{j_n}) \\ (u_{i_2}, u_{j_1}) & (u_{i_2}, u_{j_2}) & \ldots & (u_{i_2}, u_{j_n}) \\ \vdots & \vdots & \vdots & \vdots \\ (u_{i_n}, u_{j_1}) & (u_{i_n}, u_{j_2}) & \ldots & (u_{i_n}, u_{j_n}) \end{pmatrix}.$$

This is easily verified, since in an orthonormal basis the matrix having as rows the coordinates of the vectors u_i times the matrix having as columns the coordinates of the vectors u_j yields the matrix of scalar products.

Now we discuss $SO(V)$. Let A be the proposed ring of invariants. From the definition, this ring contains the determinants and it is closed under polarization operators.

From §1.1 we deduce that it is enough to prove the Theorem for $n-1$ copies of the fundamental representation. We work by induction on n and can assume $n > 1$.

We have to use one of the two possible reductions.

First We first prove the theorem for the case of real invariants on a real vector space $V := \mathbb{R}^n$ with the standard Euclidean norm.

This method is justified by the following analysis that we leave to the reader to justify. Suppose that we have an algebraic group G acting on a complex space $V_{\mathbb{C}}$ which is the complexification of a real space V. Assume also that we have a real subgroup H of G which acts on V and which is Zariski dense in G. Given a polynomial f on $V_{\mathbb{C}}$, we have that f is invariant under G if and only if it is invariant under H. Such a polynomial can be uniquely decomposed into $f_0(v) + i f_1(v)$ where both f_0 and f_1 have real coefficients. Moreover, f is H-invariant if and only if both f_0 and f_1 are H-invariant. Finally, a polynomial with real coefficients is H-invariant as a function on $V_{\mathbb{C}}$ if and only if it is invariant as a function on V. The previous setting applies to $SO(n, \mathbb{R}) \subset SO(n, \mathbb{C})$ and $(\mathbb{R}^n)^m \subset (\mathbb{C}^n)^m$.

Let e_i denote the canonical basis of $\mathbb{R}^n$ and consider $\overline{V} := \mathbb{R}^{n-1}$ formed by the vectors with the last coordinate 0 (and spanned by e_i, $i < n$).

We claim that any special orthogonal invariant E on V^{n-1} restricted to $\overline{V}^{n-1}$ is an invariant under the orthogonal group of $\overline{V}$: in fact, it is clear that every orthogonal transformation of $\overline{V}$ can be extended to a special orthogonal transformation of V.

By induction, therefore, we have a polynomial $F((u_i, u_j))$ which, restricted to $\overline{V}^{n-1}$, coincides with E. We claim that F, E coincide for every choice of $n-1$ vectors $u_1, \ldots, u_{n-1}$.

For any such choice there is a vector u of norm 1 and orthogonal to these vectors. There is a special orthogonal transformation which brings this vector u into e_n and thus the vectors u_i into the space $\mathbb{R}^{n-1}$. Since both F, E are invariant and they coincide on $\mathbb{R}^{n-1}$, the claim follows.

Second If one does not like the reduction to the real case, one can argue as follows. Prove first that the set of $n-1$ tuples of vectors which span a nondegenerate subspace in V are a dense open set of V^{n-1}, and then argue as before. □

Proof of the Theorem for $Sp(V)$. Again let R be the proposed ring of invariants for $Sp(V)$, $\dim V = 2m$. From the remark on the Pfaffian (Lemma 2.1) we see that R contains the determinants, and it is closed under polarization operators.

From §1.1 we deduce that it is enough to prove the theorem for $2m - 1$ copies of the fundamental representation. We work by induction on m. For $m = 1$, $Sp(2, \mathbb{C}) = SL(2, \mathbb{C})$ and the theorem is clear. Assume we have chosen a symplectic basis $e_i, f_i, i = 1, \dots, m$, and consider the space of vectors $\overline{V}$ having coordinate 0 in e_1.

On this space the symplectic form is degenerate with kernel spanned by f_1 and it is again nondegenerate on the subspace W where both the coordinates of e_1, f_1 vanish.

Claim A symplectic invariant $F(u_1, \dots, u_{2m-1})$, when computed on elements $u_i \in \overline{V}$, is a function which depends only on the coordinates in $e_i, f_i, i > 1$.

To prove the claim, consider the symplectic transformations $e_1 \mapsto te_1, f_1 \mapsto t^{-1} f_1$ and identity on W. These transformations preserve $\overline{V}$, induce multiplication by t on the coordinate of f_1, and fix the other coordinates. If a polynomial is invariant under this group of transformations, it must be independent of the coordinate of f_1, hence the claim.

Since $F(u_1, \dots, u_{2m-1})$ restricted to W^{2m-1} is invariant under the symplectic group of W, by induction there exists a polynomial $G([u_i, u_j])$ which coincides with F on W^{2m-1} and by the previous claim, also on $\overline{V}^{2m-1}$. We claim that $G([u_i, u_j]) = F(u_1, \dots, u_{2m-1})$ everywhere.

It is enough by continuity to show this on the set of $2m - 1$ vectors which are linearly independent. In this case such a set of vectors generates a subspace where the symplectic form is degenerate with a 1-dimensional kernel. Hence, by the theory of symplectic forms, there is a symplectic transformation which brings this subspace to $\overline{V}$ and the claim follows. □

We can now apply this theory to representation theory.

3 The Classical Groups (Representations)

3.1 Traceless Tensors

We start from a vector space with a nondegenerate symmetric or skew form, denoted $(\, , \,)$, $[\, , \,]$ respectively. Before we do any further computations we need to establish a basic dictionary deduced from the identification $V = V^*$ induced by the form.

We first want to study the identification $\mathrm{End}(V) = V \otimes V^* = V \otimes V$ and treat $V \otimes V$ as operators. We make explicit some formulas in the two cases (easy verification):

$$(u \otimes v)(w) = u(v, w),\ (u \otimes v) \circ (w \otimes z) = u \otimes (v, w)z,$$

$$\operatorname{tr}(u \otimes v) = (u, v), \tag{3.1.1}$$

$$(u \otimes v)(w) = u[v, w],\ (u \otimes v) \circ (w \otimes z) = u \otimes [v, w]z,$$

$$\operatorname{tr}(u \otimes v) = -[u, v]. \tag{3.1.2}$$

Furthermore, for the adjoint case we have $(u \otimes v)^* = v \otimes u$ in the orthogonal case and $(u \otimes v)^* = -v \otimes u$ in the symplectic case. Now we enter a more interesting area: we want to study the tensor powers $V^{\otimes n}$ under the action of $O(V)$ or of $Sp(V)$.

We already know that these groups are linearly reductive (Chapter 7, §3.2). In particular all the tensor powers are completely reducible and we want to study these decompositions.

Let us denote by G one of the two previous groups. We use the notation of the symmetric case but the discussion is completely formal and it applies also to the skew case.

First, we have to study $\hom_G(V^{\otimes h}, V^{\otimes k})$. From the basic principle of Chapter 9, §1.1 we identify

$$\hom_G(V^{\otimes h}, V^{\otimes k}) = [V^{\otimes h} \otimes (V^*)^{\otimes k}]^G = [(V^{\otimes h+k})^*]^G. \tag{3.1.3}$$

Thus the space of intertwiners between $V^{\otimes h}$, $V^{\otimes k}$ can be identified with the space of multilinear invariants in $h + k$ vector variables.

Explicitly, on each $V^{\otimes p}$ we have the scalar product

$$(w_1 \otimes \cdots \otimes w_p, z_1 \otimes \cdots \otimes z_p) := \prod_i (w_i, z_i)$$

and we identify $A \in \hom_G(V^{\otimes h}, V^{\otimes k})$ with the invariant function

$$\psi_A(X \otimes Y) := (A(X), Y),\ X \in V^{\otimes h},\ Y \in V^{\otimes k}. \tag{3.1.4}$$

It is convenient to denote the variables as $(u_1, \dots, u_h, v_1, \dots, v_k)$. Theorem 2.1 implies that these invariants are spanned by suitable monomials (the multilinear ones) in the scalar or skew products between these vectors. In particular there are nontrivial intertwiners if and only if $h + k = 2n$ is even.

It is necessary to identify some special intertwiners.

Contraction The map $V \otimes V \to \mathbb{C}$, given by $u \otimes v \to (u, v)$, is called an *elementary contraction.*

Extension By duality in the space $V \otimes V$, the space of G-invariants is one-dimensional; a generator can be exhibited by choosing a pair of dual bases $(e_i, f_j) = \delta_{ij}$ and setting $I := \sum_i e_i \otimes f_i$. The map $\mathbb{C} \to V \otimes V$ given by $a \to aI$ is an *elementary extension.*

We remark that since $u := \sum_i (u, f_i)e_i = \sum_i (u, e_i)f_i$, we have[104]

$$(I, u_1 \otimes u_2) = (u_1, u_2). \tag{3.1.5}$$

[104] In the skew case $u := \sum_i (u, f_i)e_i = -\sum_i (u, e_i)f_i$.

So I is identified with the given bilinear form. One can easily extend these maps to general tensor powers and consider the contractions and extensions

$$c_{ij} : V^{\otimes k} \to V^{\otimes k-2}, \; e_{ij} : V^{\otimes k-2} \to V^{\otimes k} \tag{3.1.6}$$

given by contracting in the indices i, j or inserting in the indices i, j (e.g., $e_{13} : V \to V^{\otimes 3}$ is $v \mapsto \sum_i e_i \otimes v \otimes f_i$).

Remark. Notice in particular that c_{ij} is surjective and e_{ij} is injective.

We have $c_{ij} = c_{ji}, \; e_{ij} = e_{ji}$ in the symmetric case, while $c_{ij} = -c_{ji}, \; e_{ij} = -e_{ji}$ in the skew-symmetric case.

In order to keep some order in these maps it is useful to consider the two symmetric groups S_h and S_k which act on the two tensor powers commuting with the group G and as orthogonal (or symplectic) transformations.

Thus $S_h \times S_k$ acts on $\hom_G(V^{\otimes h}, V^{\otimes k})$ with $(\sigma, \tau)A := \tau A \sigma^{-1}$. We also have an action of $S_h \times S_k$ on the space $[(V^{\otimes h+k})^*]^G$ of multilinear invariants by the inclusion $S_h \times S_k \subset S_{h+k}$.

We need to show that the identification $\psi : \hom_G(V^{\otimes h}, V^{\otimes k}) = [(V^{\otimes h+k})^*]^G$ is $S_h \times S_k$-equivariant. The formula $\psi_A(X \otimes Y) := (A(X), Y)$ gives

$$\begin{aligned} ((\sigma, \tau)\psi_A)(X \otimes Y) &:= \psi_A(\sigma^{-1}X \otimes \tau^{-1}Y) = (A(\sigma^{-1}X), \tau^{-1}Y) \\ &= (\tau A(\sigma^{-1}X), Y), \end{aligned} \tag{3.1.7}$$

as required.

Consider now a multilinear monomial in the elements $(u_i, u_j), \; (v_h, v_k), \; (u_p, v_q)$. In this monomial the $h+k = 2n$ elements $(u_1, \dots, u_h, v_1, \dots, v_k)$ each appear once and the monomial itself is described by the combinatorics of the n pairings.

Suppose we have exactly a pairings of type (u_i, v_j).[105] Then $h - a, \; k - a$ are both even and the remaining pairings are all *homosexual.*

It is clear that under the action of the group $S_h \times S_k$, this invariant can be brought to the following canonical form:

(3.1.8)

$$J_a := (u_1, v_1)(u_2, v_2) \dots (u_a, v_a) \prod_{i=1}^{(h-a)/2} (u_{a+2i-1}, u_{a+2i}) \prod_{j=1}^{(k-a)/2} (v_{a+2j-1}, v_{a+2j}).$$

Lemma. *The invariant J_a corresponds to the intertwiner*

(3.1.9)

$$C_a : u_1 \otimes u_2 \otimes \cdots \otimes u_h \mapsto \prod_{i=1}^{(h-a)/2} (u_{a+2i-1}, u_{a+2i}) u_1 \otimes \cdots \otimes u_a \otimes I^{\otimes (h-a)/2}.$$

105 Weyl calls these pairings heterosexual and the others homosexual.

Proof. We compute explicitly

$$\left(\prod_{i=1}^{(h-a)/2}(u_{a+2i-1}, u_{a+2i})u_1 \otimes \cdots \otimes u_a \otimes I^{\otimes(h-a)/2}, v_1 \otimes v_2 \otimes \cdots \otimes v_k\right) = J_a$$

from iteration of 3.1.5. □

The main consequence of the lemma is:

Theorem 1. *Any intertwiner between $V^{\oplus p}$ and $V^{\oplus q}$ is a composition of symmetries, contractions and extensions.*

Remark. It is convenient, in order to normalize this composition, to perform first all the contractions and after all the extensions.

We have seen that there are no intertwiners between $V^{\otimes h}$, $V^{\otimes k}$ if $h+k$ is odd but there are injective G-equivariant maps $V^{\otimes h-2a} \to V^h$ for all $a \leq h/2$.

In particular we can define the subspace $T^0(V^h)$ as the sum of all the irreducible representations of G which do not appear in the lower tensor powers $V^{\otimes h-2a}$. We claim:

Theorem 2. *The space $T^0(V^h)$ is the intersection of all the kernels of the maps c_{ij}. It is called the space of* ***traceless tensors****.*

Proof. If an irreducible representation M of G appears both in $V^{\otimes h-2a}$, $a > 0$, and in V^h, by semisimplicity an isomorphism between these two submodules can be extended to a nonzero intertwiner between V^h, $V^{\otimes h-2a}$.

From the previous theorem, these intertwiners vanish on $T^0(V^h)$, and so all the irreducible representations in $T^0(V^h)$ do not appear in $V^{\otimes h-2a}$, $a > 0$. The converse is also clear: if a contraction does not vanish on an irreducible submodule N of $V^{\otimes h}$, then the image of N is an isomorphic submodule of $V^{\otimes h-2}$. Thus we may say that $T^0(V^h)$ contains all the *new* representations of G in $V^{\otimes h}$. □

In particular, $T^0(V^h)$ is a sum of isotypic components and we may study the restriction of the centralizer of G in $V^{\otimes h}$ to $T^0(V^h)$.

Proposition. *$T^0(V^h)$ is stable under the action of the symmetric group S_h, which spans the centralizer of G in $T^0(V^h)$.*

Proof. Since clearly $\sigma c_{ij} \sigma^{-1} = c_{\sigma(i)\sigma(j)}$, $\forall i, j, \sigma \in S_h$, the first claim is clear.

Since the group is linearly reductive, any element of the centralizer of G acting on $T^0(V^h)$ is the restriction of an element of the centralizer of G acting on $V^{\otimes h}$.

From Theorem 1, these elements are products of symmetries, contractions and extensions. As soon as at least one contraction appears, the operator vanishes on $T^0(V^h)$. Hence only the symmetries are left. □

Thus, we have again a situation similar to the one for the linear group, except that the space $T^0(V^h)$ is a more complicated object to describe. Our next task is to decompose

$$T^0(V^h) = \bigoplus_{\lambda \vdash h} U_\lambda \otimes M_\lambda,$$

where the M_λ are the irreducible representations of S_h, which are given by the theory of Young symmetrizers, and the U_λ are the corresponding new representations of G. We thus have to discover which λ appear. In order to do this we will have to work out the second fundamental theorem.

3.2 The Envelope of $O(V)$

Let us complete the analysis with a simple presentation of the algebra U_p spanned by the orthogonal group acting on tensor space $V^{\otimes p}$, for some p.

By definition this algebra is contained in the algebra A_p spanned by the linear group acting on $V^{\otimes p}$ and this, in turn, is formed by the symmetric elements of $\mathrm{End}(V)^{\otimes p}$.

We have already, by complete reducibility, that U_p is the centralizer of $\mathrm{End}_{O(V)}(V^{\otimes p})$, and by the analysis of the previous section one sees immediately that this centralizer is generated by the symmetric group S_p and a single operator:

(3.2.1) $$C : v_1 \otimes v_2 \otimes \cdots \otimes v_p \to (v_1, v_2) I \otimes v_3 \otimes \cdots \otimes v_p.$$

Thus we want to understand the condition for an element of A_p to commute with C. We claim that this is equivalent to the following system of linear equations. Define the following linear map of $\mathrm{End}(V)^{\otimes p}$ to $\mathrm{End}(V)^{\otimes (p-1)}$:

$$\pi : A_1 \otimes A_2 \otimes A_3 \otimes \cdots \otimes A_p \to A_1 A_2^t \otimes A_3 \otimes \cdots \otimes A_p.$$

Theorem. *An element $X \in A_p$ is also in U_p if and only if $\pi(X)$ is of the form $\alpha 1_V \otimes Y$, $Y \in \mathrm{End}(V)^{\otimes(p-2)}$, $\alpha \in F$ a scalar.*

Proof. First, it is clear that the set of elements satisfying the given condition forms a linear space containing the elements $A^{\otimes p}$, $A \in O(V)$. Let us verify indeed that the condition is just the condition to commute with C. As usual we can work on decomposable elements $A_1 \otimes A_2 \otimes Y$. The condition to commute with C is that $(A_1 u, A_2 v) I = (u, v) A_1 \otimes A_2 I$, that is, taking vectors a, b, that

$$(a, b)(A_1 u, A_2 v) = (a \otimes b, (A_1 u, A_2 v) I) = (u, v)(a \otimes b, A_1 \otimes A_2 I)$$
$$= (u, v)(A_1^t a, A_2^t b)$$

implies that $B := A_1 A_2^t$ satisfies $(a, b)(u, B^t v) = (u, v)(a, Bb)$. Taking the a, b, u, v from an orthonormal basis, we deduce the identities for the entries x_{ij} of B:

$$x_{h,k}\delta_i^j = x_{i,j}\delta_h^k \implies x_{i,i} = x_{h,h}, \quad x_{i,j} = 0 \ \text{ if } \ i \neq j.$$

That is, B is a scalar matrix. □

4 Highest Weights for Classical Groups

4.1 Example: Representations of $SL(V)$

Let V be a vector space with a given basis $e_1, \ldots, e_n$. To V we associate the torus of diagonal matrices and the Borel subgroup B of upper triangular matrices.

Proposition. *The vector $e_1 \wedge e_2 \wedge \ldots \wedge e_k$ is a highest weight vector, for B, of the exterior power $\bigwedge^k V$, as an $SL(V)$ module, of weight ω_k.*

Proof. Apply

$$e_{i,j}(e_1 \wedge e_2 \wedge \ldots \wedge e_k) = \sum_{h=1}^{k} e_1 \wedge e_2 \wedge \ldots e_{i,j}(e_h) \wedge \ldots \wedge e_k$$

$$= \sum_{h=1}^{k} e_1 \wedge e_2 \wedge \ldots \delta_{j,h} e_i \wedge \ldots \wedge e_k,$$

which is 0, since, if $\delta_{j,h} \neq 0$ we obtain two factors e_i in the product. When we apply a diagonal matrix $\sum \alpha_i e_{i,i}$ to the vector we obtain the weight $\omega_k = \sum_{i=1}^{k} \alpha_i$. □

For $sl(n, \mathbb{C})$ consider the representation on a tensor power associated to a partition $\lambda := h_1, h_2, \ldots, h_n$ and dual partition $n_1, n_2, \ldots, n_t = 1^{a_1} 2^{a_2} \ldots n^{a_n}$. In Chapter 9, §3.1, the formulas 3.1.1, 3.1.2 produce the tensor

$$a_T U = (e_1 \wedge e_2 \wedge \ldots \wedge e_{n_1}) \otimes (e_1 \wedge e_2 \wedge \ldots \wedge e_{n_2}) \otimes \cdots \otimes (e_1 \wedge e_2 \wedge \ldots \wedge e_{n_t})$$

in this representation. One easily verifies that $a_T U$ is a highest weight vector of weight

$$\omega_\lambda := \sum_{j=1}^{n-1} a_j \omega_j. \tag{4.1.1}$$

Set $V = \mathbb{C}^n$. The previous statement is a consequence of the previous proposition and Chapter 10, §5.2 Proposition 3. In fact by construction $a_T U$ is the tensor product of the highest weight vectors in the factors of $V^{\otimes a_1} \otimes (\bigwedge^2 V)^{\otimes a_2} \otimes \ldots (\bigwedge^n V)^{\otimes a_n}$. The factors $\bigwedge^n V$ do not intervene since they are the trivial representation of $sl(n, \mathbb{C})$.

Remark. Notice that $h_i = \sum_{j=i}^{n} a_j$, so that the sum in the lattice of weights correspond to the sum of the sequences $\lambda = (h_1, h_2, \ldots, h_n)$ as vectors with n coordinates. This should not be confused with our use of the direct sum of partitions in Chapter 8, §4.2.

Remark. If we think of the weights of the Cartan subalgebra also as weights for the maximal torus, we can see that the highest weight gives the leading term in the character, which for $SL(n, \mathbb{C})$ is the Schur function.

One should remark that when we take a $2n$-dimensional symplectic space with basis $e_1, \ldots, e_n, f_1, \ldots, f_n$, the elements $e_1 \wedge e_2 \wedge \ldots \wedge e_k$, $k \leq n$ are still highest weight vectors of weight ω_i but, as we shall see in §6.6, the exterior powers are no longer irreducible. Moreover, for the orthogonal Lie algebras, besides the exterior powers, we need to discuss the spin representations.

4.2 Highest Weights and U-Invariants

We now pass to groups. Let G be a semisimple group. Fix a Cartan subalgebra $\mathfrak{t}$ a set of positive roots and the corresponding Lie algebras $\mathfrak{u}^+$, $\mathfrak{b}^+$. The three corresponding algebraic subgroups of G are denoted by T, U^+, B^+. One has that $B^+ = U^+T$ as a semidirect product (in particular $T = B^+/U^+$) (Chapter 10, §6.4).

As an example in $SL(n, \mathbb{C})$ we have that T is the subgroup of diagonal matrices, B^+ the subgroup of upper triangular matrices and U^+ the subgroup of strictly upper triangular matrices, that is, the upper triangular matrices with 1 on the diagonal or, equivalently, with all eigenvalues 1 (unipotent elements).

For an irreducible representation of G, the highest weight vector relative to B^+ is the unique (up to scalars) vector v invariant under U^+; it is also an eigenvector under B^+ and hence T.

Thus v determines a multiplicative character on B^+.[106] Notice that any multiplicative algebraic character of B^+ is trivial on U^+ (by Chapter 7, §1.5) and it is just induced by an (algebraic) character of $T = B^+/U^+$.

In the following, when we use the word character, we mean an algebraic multiplicative character.

The geometric interpretation of highest weights is obtained as follows.

In Chapter 10, §6.9 we extended the notion of highest weight to reductive groups and described the representations. Consider an action of a reductive group G on an affine algebraic variety V. Let $A[V]$ be the coordinate ring of V which is a rational representation under the induced action of G. Thus $A[V]$ can be decomposed into a direct sum of irreducible representations.

If $f \in A[V]$ is a highest weight vector of some weight λ, we have for every $b \in B^+$ that $bf = \lambda(b)f$, and conversely, a function f with this property is a highest weight vector.

Notice that if f_1, f_2 are highest weight vectors of weight λ_1, λ_2, then f_1f_2 is a highest weight vector of weight $\lambda_1 + \lambda_2$. Now unless f is invertible, the set

$$S_f := \{x \in V \mid f(x) = 0\}$$

is a hypersurface of V and it is clearly stable under B^+. Conversely, if V satisfies the property that every hypersurface is defined by an equation (for instance if V is an affine space) we have that to a B^+-stable hypersurface is associated a highest weight vector.

Of course, as usual in algebraic geometry, this correspondence is not bijective, but we have to take into consideration multiplicities.

Lemma. *If $A[V]$ is a unique factorization domain, then we have a 1–1 correspondence between irreducible B^+-stable hypersurfaces and irreducible (as polynomials) highest weight vectors (up to a multiplicative scalar factor).*

106 Recall that we define multiplicative character of a group G to be any homomorphism of G to the multiplicative group $\mathbb{C}^*$.

Proof. It is enough to show that if $f = \prod_i g_i$ is a highest weight vector, say of weight χ, factored into irreducible polynomials, then the g_i are also highest weight vectors.

For this take an element $b \in B^+$. We have $\chi(b)f = (bf) = \prod_i (bg_i)$. Since B^+ acts as a group of automorphisms, the bg_i are irreducible, and thus the elements bg_i must equal the g_j up to some possible permutation and scalar multiplication. Since the action of B^+ on $A[V]$ is rational there is a B^+-stable subspace $U \subset A[V]$ containing the elements g_i.

Consider the induced action of B^+ on the projective space of lines of U. By assumption the lines through the elements g_i are permuted by B^+. Since B^+ is connected, the only possible algebraic actions of B^+ on a finite set are trivial. It follows that the g_i are eigenvectors under B^+. One deduces, from the previous remarks, that they are U^+-invariant and $bg_i = \chi_i(b)g_i$, where χ_i are characters of B^+ and in fact, for the semisimple part of G, are dominant weights. □

4.3 Determinantal Loci

Let us analyze now, as a first elementary example, the orbit structure of some basic representations.

We start with $\hom(V, W)$ thought of as a representation of $GL(V) \times GL(W)$. It is convenient to introduce bases and use the usual matrix notation.

Let n, m be the dimensions of V and W. Using bases we identify $\hom(V, W)$ with the space $M_{m\,n}$ of rectangular matrices. The group $GL(V) \times GL(W)$ is also identified to $GL(n) \times GL(m)$ and the action on $M_{m\,n}$ is $(A, B)X = BXA^{-1}$.

The notion of the rank of an operator is an invariant notion. Furthermore we have:

Proposition. *Two elements of* $\hom(V, W)$ *are in the same* $GL(V) \times GL(W)$ *orbit if and only if they have the same rank.*

Proof. This is an elementary fact. One can give an abstract proof as follows.

Given a matrix C of rank k, choose a basis of V such that the last $n - k$ vectors are a basis of its kernel. Then the image of the first k vectors are linearly independent and we can complete them to a basis of W. In these bases the operator has matrix (in block form):

$$\begin{pmatrix} 1_k & 0 \\ 0 & 0 \end{pmatrix},$$

where 1_k is the identity matrix of size k. This matrix is obtained from C by the action of the group, and so it is in the same orbit. We have a canonical representative for matrices of rank k. In practice this abstract proof can be made into an effective algorithm, for instance, by using Gaussian elimination on rows and columns. □

As a consequence we also have: Consider $V \otimes W$ as a representation of $GL(V) \times GL(W)$. Then there are exactly $\min(n, m) + 1$ orbits, formed by the tensors which can be expressed as sum of k decomposable tensors (and not less), $k = 0, \dots, \min(m, n)$.

This is left to the reader using the identification $V \otimes W = \hom(V^*, W)$. We remark that these results are quite general and make no particular assumptions on the field F.

We suggest to the reader a harder exercise which is in fact quite interesting and has far-reaching generalizations.

Exercise. Consider again the space of $m \times n$ matrices but restrict the action to $B^+(m) \times B^-(n)$ where $B^+(m)$ (resp. $B^-(n)$) is the group of upper (resp. of lower) triangular matrices, and prove that also in this case there are finitely many orbits. This is a small generalization of the Bruhat decomposition (Chapter 10, §6.4):

The orbits of $B^+(n) \times B^-(n)$ acting on $GL(n)$ by $(A, B)X := AXB^{-1}$ are in 1–1 correspondence with the symmetric group S_n.

4.4 Orbits of Matrices

Let us now consider the action on bilinear forms, restricting to ϵ symmetric forms on a vector space U over a field F. Representing them as matrices, the action of the linear group is $(A, X) \mapsto AXA^t$.

For antisymmetric forms on U, the only invariant is again the rank, which is necessarily even. The rank classifies symmetric forms if F is algebraically closed, otherwise there are deeper arithmetical invariants. For instance, in the special case of the real numbers, the signature is a sufficient invariant.[107]

The proof is by induction. If an antisymmetric form is nonzero we can find a pair of vectors e_1, f_1 on which the matrix of the form is $\begin{vmatrix} 0 & 1 \\ -1 & 0 \end{vmatrix}$. If V is the span of these vectors, the space U decomposes into the orthogonal sum $V \oplus V^\perp$. Then one proceeds by induction on $V^\perp$ until one reaches a complement where the form is 0 and one chooses an arbitrary basis v_j of it, getting a basis $e_i, f_i, i = 1, \dots, k, v_j$. In the dual basis we have a canonical representative of the form as $\sum_{i=1}^k e^i \wedge f^i$.

For a nonzero symmetric form (over an algebraically closed field) instead, one can choose a vector e_1 of norm 1 and proceed with the orthogonal decomposition until one has a space where the form is 0. With obvious notation the form is $\sum_{i=1}^k (e^i)^2$.

Summarizing, we have seen in particular the orbit structure for the following representations:

1. The space $M_{n,m}(\mathbb{C})$ of $n \times m$ matrices X with the action of $GL(n, \mathbb{C}) \times GL(m, \mathbb{C})$ given by $(A, B)X := AXB^{-1}$.
2. The space $M_n^+(\mathbb{C})$ of symmetric $n \times n$ matrices X with the action of $GL(n, \mathbb{C})$ given by $A.X := AXA^t$.
3. The space $M_n^-(\mathbb{C})$ of skew-symmetric $n \times n$ matrices X with the action of $GL(n, \mathbb{C})$ given by $A.X := AXA^t$.

In each case there are only finitely many orbits under the given group action, and two matrices are in the same orbit if and only if they have the same rank.

107 The theory of quadratic forms over $\mathbb{Q}$ or $\mathbb{Z}$ is a rather deep part of arithmetic.

Exercise. (i) If V_k denotes the set of matrices of rank k (where k must be even in case 3), we have that the closure $\overline{V}_k$ is

(4.4.1) $$\overline{V}_k := \cup_{j \le k} V_j.$$

(ii) The varieties $\overline{V}_k$ are the only varieties invariant under the given group action.
(iii) The varieties $\overline{V}_k$ are irreducible.

From the correspondence between varieties and ideals we deduce that the ideals defining them are the only invariant ideals equal to their radical and they are all prime ideals.

We shall deduce from this the second fundamental theorem of invariant theory in §6.

More interesting is the fact that there are also finitely many orbits under the action of a Borel subgroup. We will not compute all the orbits, but we will restrict ourselves to analyzing the invariant hypersurfaces. We discuss the 3 cases.

1. To distinguish between $GL(n, \mathbb{C})$ and $GL(m, \mathbb{C})$ we let $T(n), T(m), U^+(n), U^+(m), \dots$, etc., denote the torus, unipotent, etc. of the two groups.

We take as a Borel subgroup of $GL(n, \mathbb{C}) \times GL(m, \mathbb{C})$ the subgroup $B(n)^- \times B(m)^+$ of pairs (A, B) where A is a lower and B is an upper triangular matrix. We may assume $n \le m$ (or we transpose).

If $(A, B) \in B(n)^- \times B(m)^+$, $X \in M_{n,m}(\mathbb{C})$, the matrix AXB^{-1} is obtained from X by elementary row and column operations of the following types:

(a) multiply a row or a column by a nonzero scalar,
(b) add to the ith row the jth row, with $j < i$, multiplied by some number,
(c) add to the ith column the jth column, with $j < i$, multiplied by some number.

This is the usual Gaussian elimination on rows and columns of X without performing any exchanges.

The usual remark about these operations is that, for every $k \le n$, they do not change the rank of the $k \times k$ minor X_k of X extracted from the first k rows and the first k columns. Moreover, if we start from a matrix X with the property that for every $k \le n$ we have $\det(X_k) \neq 0$, then the standard algorithm of Gaussian elimination proves that, under the action of $B(n)^- \times B(m)^+$, this matrix is equivalent to the matrix I with entries 1 on the diagonal and 0 elsewhere. We deduce

Theorem 1. *The open set of matrices $X \in M_{n,m}(\mathbb{C})$ with $\det(X_k) \neq 0$, $k = 1, \dots, n$, is a unique orbit under the group $B(n)^- \times B(m)^+$.*

The only $B(n)^- \times B(m)^+$ stable hypersurfaces of $M_{n,m}(\mathbb{C})$ are the ones defined by the equations $\det(X_k) = 0$, which are irreducible.

Proof. We have already remarked that the first part of the theorem is a consequence of Gaussian elimination. As for the second, since the complement of this open orbit is the union of the hypersurfaces of equations $\det(X_k) = 0$, it is clearly enough to prove that these equations are irreducible. The main property of the functions $\det(X_k)$ is the fact that they are highest weight vectors for $B(n)^- \times B(m)^+$.

Thus we compute the weight of $\det(X_k)$ directly. Given a pair $D_1, D_2 \in T(n) \times T(m)$ of diagonal matrices with entries x_i, y_j respectively, we have

$$(4.4.2) \qquad (D_1, D_2)d_k(X) := d_k(D_1^{-1} X D_2) = \prod_{i=1}^{k} x_i^{-1} y_i d_k(X).$$

Hence the weight of d_k is

$$\prod_{i=1}^{k} x_i^{-1} \prod_{i=1}^{k} y_i$$

which is the highest weight of $(\bigwedge^k \mathbb{C}^n)^* \otimes \bigwedge^k \mathbb{C}^m$, a fundamental weight.

We can now prove that the functions $\det(X_k)$ are irreducible. If $\det(X_k)$ is not irreducible, it is a product of highest weight vectors g_i and its weight is the sum of the weights of the g_i which are dominant weights. We have seen that $d_k := \det(X_k)$ is a fundamental weight, hence we are done once we remark that the only polynomials which belong to the 0 weight are constant. □

2. and 3. are treated as follows. One thinks of a symmetric or skew-symmetric matrix as the matrix of a form.

Again we choose as Borel subgroup $B(n)^-$ and Gaussian elimination is the algorithm of putting the matrix of the form in normal form by a triangular change of basis.

The generic orbit is obtained when the given form has maximal rank on all the subspaces spanned by the first k basis vectors $k \leq n$, which, in the symmetric case, means that the form is nondegenerate on these subspaces, while in the skew case it means that the form is nondegenerate on the even-dimensional subspaces.

2. On symmetric matrices this is essentially the Gram–Schmidt algorithm. We get that

Theorem 2. *The open set of symmetric matrices $X \in M_n^+(\mathbb{C})$ with $\det(X_k) \neq 0$, $\forall k$ is a unique orbit under the group $B(n)^-$.*

The only $B(n)^-$ stable hypersurfaces of $M_n^+(\mathbb{C})$ are the ones defined by the equations $s_k := \det(X_k) = 0$, which are irreducible.

Proof. Here we can proceed as in the linear case except at one point, when we arrive at the computation of the character of s_k, we discover that it is $\prod_{i=1}^k x_i^{-2}$, which is twice a fundamental weight.

Hence a priori we could have $s_k = ab$ with a, b with weight $\prod_{i=1}^k x_i^{-1}$. To see that this is not possible, set the variables $x_{ij} = 0$, $i \neq j$ getting $s_k = \prod_{i=1}^k x_{ii}$; hence a, b should specialize to two factors of $\prod_{i=1}^k x_{ii}$, but clearly these factors never have as weight $\prod_{i=1}^k x_i^{-1}$. Hence s_k is irreducible. □

3. Choose as a Borel subgroup $B(n)^-$ and perform Gaussian elimination on skew-symmetric matrices. For every k with $2k \leq n$, consider the minor X_{2k}. The condition that this skew-symmetric matrix be nonsingular is that the Pfaffian $p_k := Pf(X_{2k})$ is nonzero.

Theorem 3. *The open set of skew-symmetric matrices $X \in M_n^-(\mathbb{C})$ with $Pf(X_{2k}) \neq 0$, $\forall k$ is a unique orbit under the group $B(n)^-$.*

The only $B(n)^-$-stable hypersurfaces of $M_n^-(\mathbb{C})$ are the ones defined by the equations $p_k := Pf(X_{2k}) = 0$, which are irreducible.

Proof. If $Pf(X_{2k}) \neq 0$ for all k, we easily see that we can construct in a triangular form a symplectic basis for the matrix X, hence the first part. For the rest, again it suffices to prove that the polynomials p_k are irreducible. In fact we can compute their weight which, by the formula 3.6.2 in Chapter 5, is $\prod_{i=1}^{2k} x_i^{-1}$, a fundamental weight. □

From Lemma 4.2 we can describe, in the three previous examples, the highest weight vectors as monomials in the irreducible ones. This gives an implicit description of the polynomial ring as representation. We shall discuss this in more detail in the next sections.

4.5 Cauchy Formulas

We can deduce now the Cauchy formulas that are behind this theory. We do it in the symmetric and skew-symmetric cases, which are the new formulas, using the notations of the previous section.

In the symmetric case we have that the full list of highest weight vectors is the set of monomials $\prod_{k=1}^n s_k^{m_k}$ with weight $\prod_{k=1}^n \prod_{i=1}^k x_i^{-2m_k}$.

If we denote by V the fundamental representation of $GL(n, \mathbb{C})$, we have that

$$\prod_{k=1}^{n} \prod_{i=1}^{k} x_i^{-2m_k} = \prod_{i=1}^{n} x_i^{-\sum_{i \leq k} 2m_k}$$

is the highest weight of $S_\lambda(V)^*$, where λ is the partition $\lambda_k = 2\sum_{i \leq k} m_k$.

In the skew case we obtain the monomials $\prod_{2k \leq n} p_k^{m_k}$ with weight $\prod_{i=1}^n x_i^{-\sum_{i \leq 2k} m_k}$.

We deduce the *special plethystic formulas* (Chapter 9, §7.3).[108]

Theorem. *As a representation of $GL(V)$ the ring $S(S^2(V))$ decomposes as*

$$S(S^2(V)) := \bigoplus\nolimits_\lambda S_{2\lambda}(V). \tag{4.5.1}$$

As a representation of $GL(V)$ the ring $S(\bigwedge^2(V))$ decomposes as

$$S\left(\bigwedge\nolimits^2(V)\right) := \bigoplus\nolimits_\lambda S_{\widetilde{2\lambda}}(V). \tag{4.5.2}$$

Proof. We have that the space of symmetric forms on V is, as a representation, $S^2(V)^*$ and the action is, in matrix notation, given by $(A, X) \to (A^{-1})^t X A^{-1}$, so $S(S^2(V))$ is the ring of polynomials on this space. We can apply the previous theorem and the analysis following it. The considerations for the skew case are similar. □

[108] Recall these are formulas which describe the composition $S_\lambda(S_\mu(V))$ of Schur functors.

In more pictorial language, thinking of λ as the shape of a diagram we can say that the diagrams appearing in $S(S^2(V))$ have even rows, while the ones appearing in $S(\bigwedge^2(V))$ have even columns.[109] It is convenient to have a short notation for these concepts and write

$$\lambda \vdash^{er} n, \ \lambda \vdash^{ec} n$$

to express the fact that the diagram λ has even rows, resp., even columns.

We should remark that the previous theorem corresponds to identities of characters.

According to Molien's formula (Chapter 9, §4.3.3), given a linear operator A on a vector space U, its action on the symmetric algebra has as graded character $\frac{1}{\det(1-tA)}$.

If $e_1, \dots, e_n$ is a basis of V and X is the matrix $Xe_i = x_i e_i$, we have that $e_i e_j,\ i \leq j$, is a basis of $S^2(V)$ and $e_i \wedge e_j,\ i < j$ a basis of $\bigwedge^2(V)$. Thus from Molien's formula and the previous theorem we therefore deduce

$$\frac{1}{\prod_{i \leq j}(1 - x_i x_j)} = \sum_m \sum_{\lambda \vdash^{er} m} S_\lambda(x_1, \dots, x_n) \tag{4.5.3}$$

$$\frac{1}{\prod_{i < j}(1 - x_i x_j)} = \sum_m \sum_{\lambda \vdash^{ec} m} S_\lambda(x_1, \dots, x_n). \tag{4.5.4}$$

These are the formulas $(C2)$ and $(C3)$ stated in Chapter 2, §4.1.

4.6 Bruhat Cells

We point out also the following fact, which generalizes the discussion for the linear group and whose details we leave to the reader. If G is a semisimple simply connected algebraic group, we know that its coordinate ring $\mathbb{C}[G] = \bigoplus_{\lambda \in \Lambda^+} V_\lambda \otimes V_\lambda^*$. Given a Borel subgroup B of G, the highest weight vectors (for the $B \times B$) action are the elements $f_\lambda := v_\lambda \otimes \phi_\lambda$ where v_λ resp. ϕ_λ are the highest weight vectors of V_λ, resp. V_λ^*. If $\lambda = \sum_i m_i \omega_i$ with the ω_i fundamental, we have $f_\lambda = \prod_i f_{\omega_i}^{m_i} \cdot f_{\omega_i} = 0$ is a hypersurface of G which is stable under $B \times B$. It is not difficult to see that it is the closure of the double coset $Bs_i w_0 B$. Here s_i denotes the simple reflection by the simple root of the same index i as the fundamental weight.

5 The Second Fundamental Theorem (SFT)

5.1 Determinantal Ideals

We need to discuss now the relations among invariants. We shall take a geometric approach reserving the combinatorial approach for the section on tableaux. The

109 Unfortunately the row or column encoding is not really canonical. In Chapter 13 we will be obliged to switch the convention.

study of the relations among invariants proceeds as follows. We express the first fundamental theorem in matrix form and deduce a description of the invariant prime ideals, which is equivalent to the SFT, from the highest weight theory.

For the general linear group we have described invariant theory through the multiplication map $f : M_{p,m} \times M_{m,q} \to M_{p,q}$, $f(X, Y) := XY$ (Chapter 9, §1.4.1).

The ring of polynomial functions on $M_{p,m} \times M_{m,q}$ which are $Gl(m, \mathbb{C})$-invariant is given by the polynomial functions on $M_{p,q}$ composed with the map f. We have remarked that by elementary linear algebra, the multiplication map f has as its image the subvariety of $p \times q$ matrices of rank $\leq m$. This is the whole space if $m \geq \min(p, q)$. Otherwise, it is a proper subvariety defined, at least set theoretically, by the vanishing of the determinants of the $(m + 1) \times (m + 1)$ minors of the matrix of coordinate functions x_{ij} on $M_{p,q}$.

For the group $O(n, \mathbb{C})$ we have considered the space of $n \times m$ matrices Y with the action of $O(n, \mathbb{C})$ given by multiplication XY. Then the mapping $Y \to Y^tY$ from the space of $n \times m$ matrices to the symmetric $m \times m$ matrices of rank $\leq n$ is a quotient under the orthogonal group. Again, the determinants of the $(m+1) \times (m+1)$ minors of these matrices define this subvariety set-theoretically.

Similarly, for the symplectic group we have considered the space of $2n \times m$ matrices Y with the action of $Sp(2n, \mathbb{C})$ given by multiplication XY. Then the mapping $Y \to Y^t J_n Y$ (with J_n the standard $2n \times 2n$ skew-symmetric matrix), from the space of $2n \times m$ matrices to the antisymmetric $m \times m$ matrices of rank $\leq 2n$ is a quotient under the symplectic group. In this case the correct relations are not the determinants of the minors but rather the Pfaffians of the principal minors of order $2(n + 1)$.

We have thus identified three types of *determinantal varieties* for each of which we want to determine the ideal of relations. We will make use of the plethystic formulas developed in the previous section.

According to §4.3 we know that the determinantal varieties are the only varieties that are invariant under the appropriate group action. According to the matrix formulation of the first fundamental theorem they are also the varieties which have rings of invariants as coordinate rings. We want to describe the ideals of definition and their coordinate rings as representations.

In Chapter 9, §7.1 we have seen that, given two vector spaces V and W, we have the decomposition

$$\mathcal{P}[\hom(V, W)] = \bigoplus_\lambda S_\lambda(W^*) \otimes S_\lambda(V) = \bigoplus_\lambda \hom(S_\lambda(V), S_\lambda(W))^*.$$

Moreover if we think of $\mathcal{P}[\hom(V, W)]$ as the polynomial ring $\mathbb{C}[x_{ij}]$, its subspace D_k, spanned by the determinants of the minors of order k of the matrix $X := (x_{ij})$, is identified with the subrepresentation $D_k = \bigwedge^k W^* \otimes \bigwedge^k V$. We define

$$I_k := \mathcal{P}[\hom(V, W)] D_k \tag{5.1.1}$$

to be the determinantal ideal generated by the determinants of all the $k \times k$ minors of X.

Consider now $S(S^2(V)) = \mathcal{P}[S^2(V)^*]$ as the polynomial ring $\mathbb{C}[x_{ij}]$, $x_{ij} = x_{ji}$. Let $X := (x_{ij})$ be a symmetric matrix of variables.

We want to see how to identify, along the same lines as Chapter 9, §7.1, the subspace D_k^s of $\mathcal{P}[S^2(V)^*]$ spanned by the determinants of the minors of order k of the matrix X.

Let $\lambda \vdash m$ be a partition. Given a symmetric form A on V, it induces a symmetric form $A^{\otimes m}$ on $V^{\otimes m}$ by

$$A^{\otimes m}(u_1 \otimes \cdots \otimes u_m, v_1 \otimes \cdots \otimes v_m) := \prod_{i=1}^{m} A(u_i, v_i).$$

Thus by restriction it induces a symmetric form, which we will denote by $S_\lambda(A)$, on $S_\lambda(V)$, or equivalently, an element of $S^2(S_\lambda(V)^*)$ which we identify with $S^2(S_\lambda(V))^*$.

In other words we can identify $S_\lambda(A)$ with a linear form on $S^2(S_\lambda(V))$.

According to the Corollary in Chapter 10, §5.2, $S_{2\lambda}(V)$ appears with multiplicity 1 in $S^2(S_\lambda(V))$. So we deduce that $S_\lambda(A)$ induces a linear function $\langle S_\lambda(A) \mid v\rangle$ on $S_{2\lambda}(V)$.

As a function of A, v this function $\langle S_\lambda(A) \mid v\rangle$ is $GL(V)$-invariant, linear in v, and a homogeneous polynomial of degree $2m$ in A. Thus we have a dual map

$$S_{2\lambda}(V) \to \mathcal{P}[S^2(V)^*].$$

We claim that this map is nonzero. Since $S_{2\lambda}(V)$ is irreducible, it then identifies $S_{2\lambda}(V)$ with its corresponding factor in the decomposition 4.5.1.

To see this, we compute the linear form $S_\lambda(A)$ on the highest weight vector $U \otimes U$ of $S_{2\lambda}(V)$, $U = (e_1 \wedge e_2 \wedge \ldots \wedge e_{n_1}) \otimes (e_1 \wedge e_2 \wedge \ldots \wedge e_{n_2}) \otimes \cdots \otimes (e_1 \wedge e_2 \wedge \ldots \wedge e_{n_t})$. By definition we get a product of determinants of suitable minors of A, so in general a nonzero element.

In particular, we apply this to $\lambda = 1^k$, $S_\lambda(V) = \bigwedge^k(V)$, $S_\lambda(A) = \wedge^k A$. We see that if $A = (a_{ij})$ is the matrix of the given form in a basis $e_1, \ldots, e_n$, the matrix of $\wedge^k A$ in the basis $e_{i_1} \wedge e_{i_2} \ldots \wedge e_{i_k}$ is given by the formula

(5.1.2)

$$\wedge^k A(e_{i_1} \wedge e_{i_2} \wedge \ldots \wedge e_{i_k}, e_{j_1} \wedge e_{j_2} \wedge \ldots \wedge e_{j_k}) = \det(a_{i_r, j_s}),\ r, s = 1, \ldots, k.$$

The determinant 5.1.2 is the determinant of the minor extracted from the rows $i_1, \ldots, i_k$ and the columns $j_1, \ldots, j_k$ of A.

We have thus naturally defined a map $j_k : S^2(\bigwedge^k V) \to \mathcal{P}[S^2(V)^*]$ with the image the space D_k^s (spanned by the determinants of the minors of order k).

We need to prove that j_k, restricted to $S_{2^k}(V)$, gives an isomorphism to D_k^s, which is therefore irreducible with highest weight vector $2\omega_k$.

To see this we analyze the decomposition of $S^2(\bigwedge^k V)$ into irreducible representations. We have $S^2(\bigwedge^k V) \subset \bigwedge^k V \otimes \bigwedge^k V$; the decomposition of $\bigwedge^k V \otimes \bigwedge^k V$ is given by Pieri's formula. Since $\bigwedge^k V$ consists of just one column of length k, $\bigwedge^k V \otimes \bigwedge^k V = \bigoplus_{i=0}^k S_{1^{2i}2^{k-i}}(V)$, the sum of the $S_\mu(V)$ where μ has at most two rows, of lengths $2k - i, i$ respectively ($i \leq k$). Of these partitions, the only one with even rows is $S_{2^k}(V)$. Hence all the other irreducible representation appearing in

$\bigwedge^k V \otimes \bigwedge^k V$ do not appear in $\mathcal{P}[S^2(V)^*]$ and so the map j_k on them must be 0, proving the claim.

Consider now $\mathcal{P}[\bigwedge^2(V)^*]$ as the polynomial ring $\mathbb{C}[x_{ij}]$, $x_{ij} = -x_{ji}$. Let $X := (x_{ij})$ be a skew-symmetric matrix of variables. We carry out a similar analysis for the subspace P_k of $\mathcal{P}[\bigwedge^2(V)^*]$ spanned by the Pfaffians of the principal minors of order $2k$ of X.

In this case the analysis is simpler. The exterior power map $\bigwedge^2 V^* \to \bigwedge^{2k} V^*$, $A \to A^k := \wedge^k A$ gives for $A = \sum_{i<j} a_{ij} e_i \wedge e_j$ that

$$A^k = k! \sum_{i_1<i_2<\dots i_{2k}} [i_1, i_2, \dots, i_{2k}] e_{i_1} \wedge e_{i_2} \wedge \dots \wedge e_{i_{2k}}, \tag{5.1.3}$$

where $[i_1, i_2, \dots, i_{2k}]$ denotes the Pfaffian of the principal minor of A extracted from the row and column indices $i_1 < i_2 < \dots i_{2k}$. One has immediately by duality the required map with image the space P_k:

$$\bigwedge^{2k} V \to S\left(\bigwedge^2(V)\right).$$

Of course $\bigwedge^{2k} V$ corresponds to a single even column of length $2k$.

Using the results of the previous section and the previous discussion we have the

Second Fundamental Theorem.

(1) The only $GL(V) \times GL(W)$-invariant prime ideals in $\mathcal{P}[\hom(V, W)]$ are the ideals I_k. As representations we have that

$$I_k = \bigoplus_{\lambda, ht(\lambda) \geq k} S_\lambda(W^*) \otimes S_\lambda(V),$$

$$\mathcal{P}[\hom(V, W)]/I_k = \bigoplus_{\lambda, ht(\lambda) < k} S_\lambda(W^*) \otimes S_\lambda(V). \tag{5.1.4}$$

(2) The only $GL(V)$-invariant prime ideals in $\mathcal{P}[S^2(V)^*]$ are the determinantal ideals

$$I_k^+ := \mathcal{P}[S^2(V)^*] J_k$$

generated by the determinants of the $k \times k$ minors of the symmetric matrix X. As representations we have that

$$I_k^+ = \bigoplus_{\lambda \vdash^{er}, ht(\lambda) \geq k} S_\lambda(V), \quad \mathcal{P}[S^2(V)^*]/I_k^+ = \bigoplus_{\lambda \vdash^{er}, ht(\lambda) < k} S_\lambda(V). \tag{5.1.5}$$

(3) The only $GL(V)$-invariant prime ideals in $\mathcal{P}[\bigwedge^2(V)^*]$ are the Pfaffian ideals

$$I_k^- := \mathcal{P}\left[\bigwedge^2(V)^*\right] P_k$$

generated by the Pfaffians of the principal $2k \times 2k$ minors. As representations we have that

(5.1.6) $I_k^- = \bigoplus_{\lambda \vdash^{ec}, ht(\lambda) \geq 2k} S_\lambda(V), \ \mathcal{P}\left[\bigwedge^2(V)^*\right]/I_k^- = \bigoplus_{\lambda \vdash^{ec}, ht(\lambda) < 2k} S_\lambda(V).$

Proof. In each of the 3 cases we know that the set of highest weights is the set of monomials $\prod_{i=1}^n d_i^{h_i}$ in a certain number of elements d_i (determinants or Pfaffians) highest weights of certain representations N_i (the determinantal or Pfaffian spaces described before). Specifically, in case 1 we have $n = \min(\dim V, \dim W)$ and d_i is the determinant of the principal $i \times i$ minor of the matrix $x_{i,j}$. In case 2, again $n = \dim V$ with d_i the same type of determinants, while in type 3 we have $\dim V = 2n$ and the d_i is the Pfaffian of the principal $2i \times 2i$ submatrix.

Every invariant subspace M is identified by the set I of highest weight vectors that it contains.

If M is an ideal, and $\prod_{i=1}^n d_i^{h_i} \in I$, then certainly $\prod_{i=1}^n d_i^{k_i} \in I$ if $k_i \geq h_i$ for each i.

If M is a prime ideal it follows that, if a monomial $\prod d_i^{m_i} \in I$, then at least one d_i appearing with nonzero exponent must be in I.

It follows, for a prime ideal M, that M is necessarily generated by the subspaces N_i contained in it. To conclude, it is enough to remark, in the case of determinantal ideals, that d_i is in the ideal generated by d_k as soon as $i \geq k$. It clearly suffices to show this for d_{k+1}, i.e., to show that

(5.1.7) $$D_{k+1} \subset I_k, \ D_{k+1}^s \subset I_k^+, \ P_{k+1} \subset I_k^-.$$

The first two statements follow immediately by a row or column expansion of a determinant of a $(k+1) \times (k+1)$ minor in terms of determinants of $k \times k$ minors. As for the Pfaffians, recall the defining formula 3.6.1 of Chapter 5. $A = \sum_{i<j} x_{ij} e_i \wedge e_j$:

$$A^{k+1} = (k+1)! \sum_{i_1 < i_2 < \ldots i_{2(k+1)}} [i_1, i_2, \ldots, i_{2(k+1)}] e_{i_1} \wedge e_{i_2} \wedge \ldots \wedge e_{i_{2(k+1)}}$$

$$= A \wedge k! \sum_{i_1 < i_2 < \ldots < i_{2k}} [i_1, i_2, \ldots, i_{2k}] e_{i_1} \wedge e_{i_2} \wedge \ldots \wedge e_{i_{2k}}.$$

From the above we deduce the typical Laplace expansion for a Pfaffian:

$$(k+1)[1, 2, \ldots, 2(k+1)]$$
$$= \sum_{i<j} (-1)^{i+j-1} x_{ij} [1, 2, \ldots, i-1, i+1, \ldots, j-1, j+1, \ldots, 2(k+1)].$$

Therefore in each case we have that for a given prime invariant ideal M there is an integer k such that $\prod_{i=1}^n d_i^{m_i} \in M$ if and only if $m_i > 0$ for at least one $i \geq k$. Thus the ideal is generated by D_k, J_k^+, P_k^- for the minimal index k for which this space is contained in the ideal. The remaining statements are just a reinterpretation of what we said. □

We called the previous theorem the Second Fundamental Theorem since it describes the ideals of matrices (in the three types) of rank $< k$ (or $< 2k$ in the skew-symmetric case). We have already seen that these varieties are the varieties corresponding to the three algebras of invariants we considered, so this theorem is a description of the ideal of relations among invariants.

5.2 Spherical Weights

There is a rather general theory of spherical subgroups H of a given group G and the orthogonal and symplectic group are spherical in the corresponding group $GL(V)$.

The technical definition of a spherical subgroup H of a reductive group G is the following: We assume that a Borel subgroup B has an open orbit in G/H, or equivalently, that there is a Borel subgroup B with BH dense in G. There is a deep theory of these pairs which we will not treat.

One of the first properties of spherical subgroups is that given an irreducible representation of G, the space of invariants under H is at most 1-dimensional. When it is exactly 1, the corresponding dominant weight is then called a *spherical weight* relative to H. We want to illustrate this phenomenon for the orthogonal and symplectic groups.

1. Orthogonal group. Take an n-dimensional orthogonal space V and consider $n-1$ copies of V which we display as $V \otimes W$, $\dim W = n-1$.

We know that the orthogonal or special orthogonal invariants of $n-1$ copies of V are the scalar products (u_i, u_j) and that they are algebraically independent. As a representation of $GL(n-1, \mathbb{C}) = GL(W)$, these basic invariants transform as $S^2(W)$ and so the ring they generate is $S(S^2(W)) = \bigoplus_\mu S_{2\mu}(W)$. On the other hand, by Cauchy's formula,

$$S(V \otimes W)^{O(V)} = \bigoplus_\lambda S_\lambda(V)^{O(V)} \otimes S_\lambda(W).$$

We deduce an isomorphism as $GL(W)$-representations:

$$\bigoplus_\lambda S_\lambda(V)^{O(V)} \otimes S_\lambda(W) = \bigoplus_\mu S_{2\mu}(W). \tag{5.2.1}$$

Corollary of 5.2.1. $\dim S_\lambda(V)^{O(V)} = 1$ *if* $\lambda = 2\mu$ *and* 0 *otherwise.*

There is a simple explanation for this invariant which we leave to the reader.

The scalar product induces a nondegenerate scalar product on $V^{\otimes p}$ for all p and thus a nondegenerate scalar product on $S_\lambda(V)$ for each V. This induces a canonical invariant element in $S^2(S_\lambda(V))$, and its projection to $S_{2\lambda}(V)$ is the required invariant (cf. §5.1).

Since $SO(V) \subset SL(V)$ the restriction to dimension $n-1$ is not harmful since all representations of $SL(V)$ appear. For $O(V)$ we can take n-copies and leave the details to the reader.

2. Symplectic group. Take a $2n$-dimensional symplectic space V and consider $2n$ copies of V which we display as $V \otimes W$, $\dim W = 2n$.

We know that the symplectic invariants of $2n$ copies of V are the scalar products $[u_i, u_j]$ and that they are algebraically independent. As a representation of $GL(2n, \mathbb{C}) = GL(W)$ these basic invariants transform as $\bigwedge^2(W)$ and so the ring they generate is $S(\bigwedge^2(W)) = \bigoplus_\mu S_{\widetilde{2\mu}}(W)$. On the other hand by Cauchy's formula,

$$S(V \otimes W)^{Sp(V)} = \bigoplus_\lambda S_\lambda(V)^{Sp(V)} \otimes S_\lambda(W),$$

we deduce an isomorphism as $GL(W)$ representations:

(5.2.2) $$\bigoplus_\lambda S_\lambda(V)^{Sp(V)} \otimes S_\lambda(W) = \bigoplus_\mu S_{\widetilde{2\mu}}(W).$$

Corollary of 5.2.2. $\dim S_\lambda(V)^{Sp(V)} = 1$ *if* $\lambda = \widetilde{2\mu}$ *and* 0 *otherwise.*

There is a similar explanation for this invariant. Start from the basic invariant $J \in \bigwedge^2 V$. We get by exterior multiplication $J^k \in \bigwedge^{2k} V$, and then by tensor product, the invariant $J^{k_1} \otimes J^{k_2} \otimes \cdots \otimes J^{k_m} \in \bigwedge^{2k_1} V \otimes \cdots \otimes \bigwedge^{2k_r} V$, which projects to the invariant in $S_\lambda(V)$, where $\tilde{\lambda} = 2k_1, 2k_2, \ldots, 2k_r$.

6 The Second Fundamental Theorem for Intertwiners

6.1 Symmetric Group

We want to apply the results of the previous section to intertwiners. We need to recall the discussion of multilinear characters in §6.4 of Chapter 9. Start from $\mathcal{P}[\hom(V, W)] = S(W^* \otimes V) = \bigoplus_\lambda S_\lambda(W^*) \otimes S_\lambda(V)$, with $\dim V = n,\ W = \mathbb{C}^n$.

Using the standard basis e_i of $\mathbb{C}^n$, we have identified $V^{\otimes n}$ with a subspace of $S(\mathbb{C}^n \otimes V)$ by encoding the element $\prod_{i=1}^n e_i \otimes v_i$ as $v_1 \otimes v_2 \otimes \cdots \otimes v_n$.

The previous mapping identifies $V^{\otimes n} \subset S(\mathbb{C}^n \otimes V)$ with a weight space under the torus of diagonal matrices X with entries x_i. We have $Xe_i = x_i e_i$ and

$$X \prod_{i=1}^n e_i \otimes v_i = \prod_i x_i \prod_{i=1}^n e_i \otimes v_i.$$

Now we extend this idea to various examples. First, we identify the group algebra $\mathbb{C}[S_n]$ with the space P^n of polynomials in the variables x_{ij} which are multilinear in right and left indices, that is, we consider the span P^n of those monomials $x_{i_1,j_1} x_{i_2,j_2} \cdots x_{i_n,j_n}$ such that both $i_1, i_2, \ldots, i_n$ and $j_1, j_2, \ldots, j_n$ are permutations of $1, 2, \ldots, n$.

Of course a monomial of this type can be uniquely displayed as

$$x_{1,\sigma^{-1}(1)} x_{2,\sigma^{-1}(2)} \cdots x_{n,\sigma^{-1}(n)} = x_{\sigma(1),1} x_{\sigma(2),2} \cdots x_{\sigma(n),n}$$

which defines the required map

(6.1.1) $$\Phi : \sigma \mapsto x_{\sigma(1),1} x_{\sigma(2),2} \cdots x_{\sigma(n),n}.$$

Now remark that this space of polynomials is a weight space with respect to the product $T \times T$ of two maximal tori of diagonal matrices under the induced $GL(n, \mathbb{C}) \times GL(n, \mathbb{C})$ action on $\mathbb{C}[x_{ij}]$. Let us denote by $\chi_1; \chi_2$ the two weights.

We remark also that this mapping is equivariant under the left and right action of S_n on $\mathbb{C}[S_n]$ which correspond, respectively, to

(6.1.2) $$x_{i,j} \to x_{\sigma(i),j},\ x_{i,j} \to x_{i,\sigma(j)}.$$

Fix an m-dimensional vector space V and recall the basic symmetry homomorphism

$$i_n : \mathbb{C}[S_n] \to \mathrm{End}(V^{\otimes n}). \tag{6.1.3}$$

We have a homomorphism given by the FFT. With the notations of Chapter 9, §1.4 it is $f : \mathbb{C}[x_{ij}] \to \mathbb{C}[\langle \alpha_i | v_j \rangle]$. It maps P^n onto the space of multilinear invariants of n covector and n vector variables.

Let us denote by $\mathcal{I}_n$ this space, which is spanned by the elements $\prod_{i=1}^n \langle \alpha_{\sigma(i)} | v_i \rangle$.

Finally, we have the canonical isomorphism $j : \mathrm{End}_{GL(V)} V^{\otimes n} \xrightarrow{j} \mathcal{I}_n$. It maps the endomorphism induced by a permutation σ to $\prod_{i=1}^n \langle \alpha_{\sigma(i)} | v_i \rangle$ (cf. Chapter 9, §1.3.2).

We have a commutative diagram

$$\begin{array}{ccc} \mathbb{C}[S_n] & \xrightarrow{\Phi} & P_n \\ {\scriptstyle i_n}\downarrow & & \downarrow{\scriptstyle f} \\ \mathrm{End}_{GL(V)} V^{\otimes n} & \xrightarrow{j} & \mathcal{I}_n \end{array} \tag{6.1.4}$$

from which we deduce that the kernel of i_n can be identified, via the linear isomorphism Φ, with the intersection of P_n with the determinantal ideal I_{m+1} in $\mathbb{C}[x_{ij}]$.

Given a matrix X we denote by $(i_1, i_2, \ldots, i_{m+1} | j_1, j_2, \ldots, j_{m+1})(X)$ or, if no confusion arises, by $(i_1, i_2, \ldots, i_{m+1} | j_1, j_2, \ldots, j_{m+1})$ the determinant of the minor of X extracted from the rows of indices i and the columns of indices j.

Clearly $(i_1, i_2, \ldots, i_{m+1} | j_1, j_2, \ldots, j_{m+1})$, multiplied by any monomial in the x_{ij}, is a weight vector for $T \times T$.

In order to get a weight vector in P_n we must consider the products

$$(i_1, i_2, \ldots, i_{m+1} | j_1, j_2, \ldots, j_{m+1}) x_{i_{m+2}, j_{m+2}} \cdots x_{i_n, j_n} \tag{6.1.5}$$

with $i_1, i_2, \ldots, i_n$ and $j_1, j_2, \ldots, j_n$, both permutations of $1, 2, \ldots, n$.

Theorem. *Under the isomorphism Φ the space $P_n \cap I_{m+1}$ is 0 if $m \geq n$. If $m < n$ it corresponds to the two-sided ideal of $\mathbb{C}[S_n]$ generated by the element*

$$A_{m+1} := \sum_{\sigma \in S_{m+1}} \epsilon_\sigma \sigma, \tag{6.1.6}$$

the antisymmetrizer on $m+1$ elements (chosen arbitrarily from the given n elements).

Proof. From 6.1.2 it follows that the element corresponding to a typical element of type 6.1.5 is of the form $\sigma A \tau^{-1}$ where

$$\begin{aligned} A &= (1, 2, \ldots, m+1 | 1, 2, \ldots, m+1) x_{m+2,m+2} \cdots x_{n,n} \\ &= \Big(\sum_{\sigma \in S_{m+1}} \epsilon_\sigma x_{\sigma(1),1} x_{\sigma(2),2} \cdots x_{\sigma(m+1),m+1} \Big) x_{m+2,m+2} \cdots x_{n,n}. \end{aligned}$$

This element clearly corresponds to A_{m+1}. □

Now we want to recover in a new form the result of Chapter 9, §3.1.3. First, recall that, as any group algebra, $\mathbb{C}[S_n]$ decomposes into the sum of its minimal ideals corresponding to irreducible representations

$$\mathbb{C}[S_n] = \bigoplus_{\lambda \vdash n} M_\lambda^* \otimes M_\lambda.$$

We first decompose

$$\mathbb{C}[x_{ij}] = \bigoplus_\lambda S_\lambda(\mathbb{C}^n)^* \otimes S_\lambda(\mathbb{C}^n), \quad I_{m+1} = \bigoplus_{ht\lambda \geq m+1} S_\lambda(\mathbb{C}^n)^* \otimes S_\lambda(\mathbb{C}^n).$$

Then we pass to the weight space

$$P^n = \mathbb{C}[x_{ij}]^{\chi_1, \chi_2} = \bigoplus_\lambda S_\lambda(\mathbb{C}^n)^{*\chi_1} \otimes S_\lambda(\mathbb{C}^n)^{\chi_2},$$

$$P^n \cap I_{m+1} = \bigoplus_{ht\lambda \geq m+1} S_\lambda(\mathbb{C}^n)^{*\chi_1} \otimes S_\lambda(\mathbb{C}^n)^{\chi_2}$$

$$(6.1.8) \qquad = \bigoplus_{\lambda \vdash n,\, ht\lambda \geq m+1} M_\lambda^* \otimes M_\lambda.$$

As a consequence the image of $\mathbb{C}[S_n]$ in $\mathrm{End}(V^{\otimes n})$ is $\bigoplus_{\lambda \vdash n,\, ht\lambda \leq m} M_\lambda^* \otimes M_\lambda$.

6.2 Multilinear Spaces

Now we go on to the orthogonal and symplectic group. First, we formulate some analogue of P^n. In the symmetric or skew-symmetric cases, we take a (symmetric or skew-symmetric) matrix $Y = (y_{ij})$ of even size $2n$, and consider the span of the *multilinear monomials*

$$y_{i_1,j_1} y_{i_2,j_2} \dots y_{i_n,j_n}$$

such that $i_1, i_2, \dots, i_n, j_1, j_2, \dots, j_n$, is a permutation of $1, 2, \dots, 2n$. This is again the weight space of the weight χ of the diagonal group T of $GL(2n, \mathbb{C})$.

Although at first sight the space looks the same in the two cases, it is not so, and we will call these two cases P_+^{2n} and P_-^{2n}. In fact, the symmetric group S_{2n} acts on the polynomial ring $\mathbb{C}[y_{ij}]$ (as a subgroup of $GL(2n, \mathbb{C})$) by $\sigma(y_{ij}) = y_{\sigma(i)\sigma(j)}$.

It is clear that S_{2n} acts transitively on the given set of multilinear monomials.

First, we want to understand it as a representation on $P_\pm^n$ and for this we consider the special monomial

$$M_0 := y_{1,2} y_{3,4} \dots y_{2n-1,2n}.$$

Let H_n be the subgroup of S_{2n} which fixes the partition of $\{1, 2, \dots, 2n\}$ formed by the n sets with 2 elements $\{2i-1, 2i\}$, $i = 1, \dots, n$. Clearly, $H_n = S_n \ltimes \mathbb{Z}/(2)^n$ is the obvious semidirect product, where S_n acts in the diagonal way on odd and even numbers $\sigma(2i-1) = 2\sigma(i) - 1$, $\sigma(2i) = 2\sigma(i)$ and $\mathbb{Z}/(2)^n$ is generated by the transpositions $(2i-1, 2i)$.

In either case H_n is the stabilizer of the line through M_0. In the symmetric case H_n fixes M_0, while in the skew-symmetric case it induces on this line the sign of the permutation (remark that $S_n \subset S_{2n}$ is made of even permutations). We deduce:

Proposition. *P^n_+, as a representation of S_{2n}, coincides with the permutation representation $\mathrm{Ind}_{H_n}^{S_{2n}} \mathbb{C}$ associated to S_{2n}/H_n.*

P^n_-, as a representation of S_{2n}, coincides with the representation $\mathrm{Ind}_{H_n}^{S_{2n}} \mathbb{C}(\epsilon)$ induced to S_{2n} from the sign representation $\mathbb{C}(\epsilon)$ of H_n.

Next, one can describe these representations in terms of irreducible representations of S_{2n}. Using the formulas 4.5.1,2 and Chapter 9, §6.4 we get

Theorem. *As a representation of S_{2n} the space P^n_+ decomposes as*

(6.2.1) $$P^n_+ = S(S^2(V))^\chi := \bigoplus\nolimits_{\lambda \vdash^{er} 2n} S_\lambda(V)^\chi = \bigoplus\nolimits_{\lambda \vdash^{er} 2n} M_\lambda.$$

P^n_- decomposes as

(6.2.2) $$P^n_- = S\Big(\bigwedge^2(V)\Big)^\chi := \bigoplus\nolimits_{\lambda \vdash^{ec} 2n} S_\lambda(V)^\chi = \bigoplus\nolimits_{\lambda \vdash^{ec} 2n} M_\lambda.$$

Remark. By Frobenius reciprocity (Chapter 8, §5.2), this theorem computes the multiplicity of the trivial and the sign representation of H_n in any irreducible representation of S_{2n}. Both appear with multiplicity at most 1, and in fact the trivial representation appears when λ has even columns and the sign representation when it has even rows.

One can think of H_n as a *spherical subgroup* of S_{2n}.

Next we apply the ideas of 6.1 to intertwiners. We do it only in the simplest case.

6.3 Orthogonal and Symplectic Group

1. Orthogonal case. Let V be an orthogonal space of dimension m, and consider the space $\mathcal{I}^+_{2n}$ of multilinear invariants in $2n$ variables $u_i \in V,\ i = 1, \dots, 2n$. $\mathcal{I}^+_{2n}$ is spanned by the monomials $(u_{i_1}, u_{i_2})(u_{i_3}, u_{i_4}) \dots (u_{i_{2n-1}}, u_{i_{2n}})$, where $i_1, i_2, \dots, i_{2n}$ is a permutation of $1, 2, \dots, 2n$.

Let $y_{ij} = y_{ji}$ be symmetric variables. Under the map

$$\mathbb{C}[y_{ij}] \to \mathbb{C}[(u_i, u_j)],\ \ y_{ij} \mapsto (u_i, u_j),$$

the space P^n_+ maps surjectively onto $\mathcal{I}^+_{2n}$ with kernel $P^n_+ \cap I^+_{m+1}$.

The same proof as in 6.1 shows that

Theorem 1. *As a representation of S_{2n} we have*

$$P^n_+ \cap I^+_{m+1} = \bigoplus\nolimits_{\lambda \vdash^{er} 2n,\ ht(\lambda) \geq m+1} M_\lambda, \quad \mathcal{I}^+_{2n} = \bigoplus\nolimits_{\lambda \vdash^{er} 2n,\ ht(\lambda) \leq m} M_\lambda.$$

The interpretation of the relations in the algebras of intertwiners $\mathrm{End}_{O(V)} V^{\otimes n}$ is more complicated and we shall not describe it in full detail.

2. Symplectic case. Let V be a symplectic space of dimension $2m$ and consider the space $\mathcal{I}_{2n}^-$ of multilinear invariants in $2n$ variables $u_i \in V$, $i = 1, \dots, 2n$. $\mathcal{I}_{2n}^-$ is spanned by the monomials $[u_{i_1}, u_{i_2}][u_{i_3}, u_{i_4}]\dots[u_{i_{2n-1}}, u_{i_{2n}}]$, where $i_1, i_2, \dots, i_{2n}$ is a permutation of $1, 2, \dots, 2n$.

Let $y_{ij} = -y_{ji}$ be antisymmetric variables. Under the map

$$\mathbb{C}[y_{ij}] \to \mathbb{C}[[u_i, u_j]], \; y_{ij} \mapsto [u_i, u_j],$$

the space P_-^n maps surjectively onto $\mathcal{I}_{2n}^-$ with kernel $P_-^n \cap I_{m+1}^-$.

The same proof as in 6.1 shows that

Theorem 2. *As a representation of S_{2n}, we have*

$$P_-^n \cap I_{m+1}^- = \bigoplus\nolimits_{\lambda \vdash^{ec} 2n,\; ht(\lambda) \geq 2m+2} M_\lambda, \; \mathcal{I}_{2n}^- = \bigoplus\nolimits_{\lambda \vdash^{ec} 2n,\; ht(\lambda) \leq 2m} M_\lambda.$$

The interpretation of the relations in the algebras of intertwiners $\mathrm{End}_{Sp(V)} V^{\otimes n}$ is again more complicated and we shall not describe it in full detail. We want nevertheless to describe part of the structure of the algebra of intertwiners, the one relative to traceless tensors $T^0(V^{\otimes n})$. In both the orthogonal and symplectic cases the idea is similar. Let us first develop the symplectic case which is simpler, and leave the discussion of the orthogonal case to §6.5.

Consider a monomial $M := y_{i_1,j_1} y_{i_2,j_2} \dots y_{i_n,j_n}$. We can assume by symmetry $i_k < j_k$ for all k. We see by the formulas 3.1.8, 3.1.9 that the operator $\phi_M : V^{\otimes n} \to V^{\otimes n}$ involves at least one contraction unless all the indices i_k are $\leq n$ and $j_k > n$.

Let us denote by $\overline{\phi}_M$ the restriction of the operator ϕ_M to $T^0(V^{\otimes n})$. We have seen that $\overline{\phi}_M = 0$ if the monomial contains a variable y_{ij}, $y_{n+i,n+j}$, $i, j \leq n$. Thus the map $M \to \overline{\phi}_M$ factors through the space $\overline{P}_-^n$ of monomials obtained, setting to zero one of the two previous types of variables.

The only monomials that remain are of type $M_\sigma := \prod_{i=1}^n y_{i,n+\sigma(i)}$, $\sigma \in S_n$, and M_σ corresponds to the invariant

$$\prod [u_i, u_{n+\sigma(i)}] = [u_{\sigma^{-1}(1)} \otimes u_{\sigma^{-1}(2)} \otimes \dots \otimes u_{\sigma^{-1}(n)}, u_{n+1} \otimes u_{n+2} \otimes \dots \otimes u_{2n}]$$

which corresponds to the map induced by the permutation σ on $V^{\otimes n}$.

We have just identified $\overline{P}_-^n$ with the group algebra $\mathbb{C}[S_n]$ and the map

$$\rho : \overline{P}_-^n = \mathbb{C}[S_n] \to \mathrm{End}_{Sp(V)}(T^0(V^{\otimes n})), \quad \rho(M) := \overline{\phi}_M,$$

with the canonical map to the algebra of operators induced by the symmetric group. Since $T^0(V^{\otimes m})$ is a sum of isotypic components in $V^{\otimes m}$, the map ρ is surjective.

The image of $P_-^n \cap I_{m+1}^-$ in $\overline{P}_-^n = \mathbb{C}[S_n]$ is contained in the kernel of ρ. To identify this image, take the Pfaffian of the principal minor of Y of indices $i_1, i_2, \dots, i_{2m+2}$ and evaluate after setting $y_{ij} = y_{n+i,n+j} = 0$, $i, j \leq n$. Let us say that h of these indices are $\leq n$ and $2m + 2 - h$ are $> n$.

The specialized matrix has block form $\begin{pmatrix} 0 & Z \\ -Z^t & 0 \end{pmatrix}$ and the minor extracted from the indices $i_1, i_2, \dots, i_{2m+2}$ contains a square block matrix, made of 0's, whose size is the larger of the two numbers h and $2m + 2 - h$.

Since the maximum dimension of a totally isotropic space, for a nondegenerate symplectic form on a $2m+2$-dimensional space, is $m+1$, we deduce that the only case in which this Pfaffian can be nonzero is when $h = 2m+2-h = m+1$. In this case Z is a square $(m+1)\times(m+1)$ matrix, and the Pfaffian equals $\det(Z)$.

Thus, arguing as in the linear case, we see that the image of $P^n_- \cap I^-_{m+1}$ in $\overline{P}^n_- = \mathbb{C}[S_n]$ is the ideal generated by the antisymmetrizer on $m+1$ elements.

Theorem 3. *The algebra* $\mathrm{End}_{Sp(V)}(T^0(V^{\otimes n}))$ *equals the algebra* $\mathbb{C}[S_n]$ *modulo the ideal generated by the antisymmetrizer on* $m+1$ *elements.*

Proof. We have already seen that the given algebra is a homomorphic image of the group algebra of S_n modulo the given ideal. In order to prove that there are no further relations, we observe that if $U \subset V$ is the subspace spanned by $e_1,\dots,e_m$, it is a (maximal) totally isotropic subspace and thus $U^{\otimes n} \subset T^0(V^{\otimes n})$. On the other hand, by the linear theory, the kernel of the action of $\mathbb{C}[S_n]$ on $U^{\otimes n}$ coincides with the ideal generated by the antisymmetrizer on $m+1$ elements, and the claim follows. □

6.4 Irreducible Representations of $Sp(V)$

We are now ready to exhibit the list of irreducible rational representations of $Sp(V)$. First, using the double centralizer theorem, we have a decomposition

$$T^0(V^{\otimes n}) = \bigoplus_{\lambda\vdash n,\ ht(\lambda)\le m} M_\lambda \otimes T_\lambda(V), \tag{6.4.1}$$

where we have indicated by $T_\lambda(V)$ the irreducible representation of $Sp(V)$ paired to M_λ. We should note then that we can construct, as in 4.1, the tensor

(6.4.2)

$$(e_1\wedge e_2\wedge\dots\wedge e_{n_1})\otimes(e_1\wedge e_2\wedge\dots\wedge e_{n_2})\otimes\cdots\otimes(e_1\wedge e_2\wedge\dots\wedge e_{n_t}) \in T_\lambda(V)$$

where the partition λ has columns $n_1, n_2, \dots, n_t$. We ask the reader to verify that it is a highest weight vector for $T_\lambda(V)$ with highest weight $\sum_{j=1}^t \omega_{n_j}$.

Since $Sp(V)$ is contained in the special linear group, from Proposition 1.4 of Chapter 7, all irreducible representations appear in tensor powers of V. Since $T^0(V^{\otimes m})$ contains all the irreducible representations appearing in $V^{\otimes n}$ and not in $V^{\otimes k}$, $k<n$, we deduce:

Theorem. *The irreducible representations* $T_\lambda(V)$, $ht(\lambda)\le m$, *constitute a complete list of inequivalent irreducible representations of* $Sp(V)$.

Since $Sp(V)$ is simply connected (Chapter 5, §3.10), from Theorem 6.1 of Chapter 10 it follows that this is also the list of irreducible representations of the Lie algebra $sp(2m,\mathbb{C})$. Of course, we are recovering in a more explicit form the classification by highest weights.

6.5 Orthogonal Intertwiners

We want to describe also in the orthogonal case part of the structure of the algebra of intertwiners, the one relative to traceless tensors $T^0(V^{\otimes n})$.

We let V be an m-dimensional orthogonal space. Consider a monomial $M := y_{i_1,j_1}y_{i_2,j_2}\dots y_{i_n,j_n}$. We can assume by symmetry that $i_k < j_k$ for all k. We see by the formulas 3.1.8 and 3.1.9 that the operator $\phi_M : V^{\otimes n} \to V^{\otimes n}$ involves at least one contraction unless all the indices i_k are $\leq n$ and $j_k > n$.

Let us denote by $\overline{\phi}_M$ the restriction of the operator ϕ_M to $T^0(V^{\otimes n})$. We have seen that $\overline{\phi}_M = 0$ if the monomial contains a variable $y_{ij},\ y_{n+i,n+j},\ i, j \leq n$. Thus the map $M \to \overline{\phi}_M$ factors through the space $\overline{P}^n_+$ of monomials obtained by setting to 0 one of the two previous types of variables. The only monomials that remain are of type

$$M_\sigma := \prod_{i=1}^{n} y_{i,n+\sigma(i)}, \quad \sigma \in S_n,$$

and M_σ corresponds to the invariant

$$\prod (u_i, u_{n+\sigma(i)}) = (u_{\sigma^{-1}(1)} \otimes u_{\sigma^{-1}(2)} \otimes \cdots \otimes u_{\sigma^{-1}(n)}, u_{n+1} \otimes u_{n+2} \otimes \cdots \otimes u_{2n}),$$

which corresponds to the map induced by the permutation σ on $V^{\otimes n}$.

We have just identified $\overline{P}^n_+$ with the group algebra $\mathbb{C}[S_n]$ and the map

$$\rho : \overline{P}^n_+ = \mathbb{C}[S_n] \to \mathrm{End}_{Sp(V)}(T^0(V^{\otimes n})), \quad \rho(M) := \overline{\phi}_M,$$

with the canonical map to the algebra of operators induced by the symmetric group. Since $T^0(V^{\otimes n})$ is a sum of isotypic components, the map ρ is surjective.

Let us identify the image of $P^n_+ \cap I^+_{m+1}$ in $\overline{P}^n_+ = \mathbb{C}[S_n]$ which is in the kernel of ρ.

Take the determinant D of an $(m+1) \times (m+1)$ minor of Y extracted from the row indices $i_1, i_2, \dots, i_{m+1}$, the column indices $j_1, j_2, \dots, j_{m+1}$, and evaluate after setting $y_{ij} = y_{n+i,n+j} = 0,\ i, j \leq n$. Let us say that h of the row indices are $\leq n$ and $m+1-h$ are $> n$ and also k of the column indices are $\leq n$ and $m+1-k$ are $> n$.

The specialized matrix has block form $\begin{pmatrix} 0 & Z \\ W & 0 \end{pmatrix}$ where Z is an $h \times m+1-k$ and W an $m+1-h \times k$ matrix. If this matrix has nonzero determinant, the image of the first k basis vectors must be linearly independent. Hence $m+1-h \geq k$, and similarly $h \leq m+1-k$. Hence $h+k = m+1$, i.e., Z is a square $h \times h$ and W is a square $k \times k$ matrix.

Up to sign the determinant of this matrix is $D = \det(W)\det(Z)$.

This determinant is again a weight vector which, multiplied by a monomial M in the y_{ij}, can give rise to an element of $DM \in P^n_+$ if and only if the indices $i_1, i_2, \dots, i_{m+1}$ and $j_1, j_2, \dots, j_{m+1}$ are all distinct.

Up to a permutation in $S_n \times S_n$ we may assume then that these two sets of indices are $1, 2, \dots, h, n+h+1, n+h+2, \dots, n+h+k$ and $h+1, h+2, \dots, h+k$, $n+1, n+2, \dots, n+h$ so that using the symmetry $y_{n+h+i,h+j} = y_{h+j,n+h+i}$, we obtain that $\det(W)\det(Z)$ equals

$$\sum_{\sigma\in S_h}\epsilon_\sigma y_{\sigma(1),n+1}y_{\sigma(2),n+2}\cdots y_{\sigma(h),n+h}$$
$$\times\sum_{\sigma\in S_k}\epsilon_\sigma y_{h+\sigma(1),n+h+1}y_{h+\sigma(2),n+h+2}\cdots y_{h+\sigma(k),n+h+k}.$$

We multiply this by $M := \prod_{t=1}^{n-h-k} y_{h+k+t,n+h+k+t}$.

This element corresponds in $\mathbb{C}[S_n]$ to the antisymmetrizer relative to the partition consisting of the parts $(1, 2, \ldots, h)$, $(h+1, \ldots, h+k)$, 1^{n-h-k}. This is the element $\sum_{\sigma\in S_h\times S_k}\epsilon_\sigma\sigma$. A determinant of a minor of order $> m+1$ can be developed into a linear combination of monomials times determinants of order $m+1$; thus the ideal of $\mathbb{C}[S_n]$ corresponding to $P_+^n \cap I_{m+1}^+$ contains the ideal generated by the products of two antisymmetrizers on two disjoint sets whose union has $\geq m+1$ elements.

From the description of Young symmetrizers, it follows that each Young symmetrizer, relative to a partition with the first two columns adding to a number $\geq m+1$ is in the ideal generated by such products. Thus we see that:

The image of $P_+^n \cap I_{m+1}^+$ in $\overline{P}_+^n = \mathbb{C}[S_n]$ contains the ideal generated by all the Young symmetrizers relative to diagrams with the first two columns adding to a number $\geq m+1$.

Theorem 1. *The algebra $\mathrm{End}_{O(V)}(T^0(V^{\otimes n}))$ equals the algebra $\mathbb{C}[S_n]$ modulo the ideal generated by all the Young symmetrizers relative to diagrams with the first two columns adding to a number $\geq m+1$.*

Proof. We have already seen that the given algebra is a homomorphic image of the group algebra of S_n modulo the given ideal. In order to prove that there are no further relations it is enough to show that if λ is a partition with the first two columns adding up to at most m, then we can find a nonzero tensor u with $c_T u$ traceless and $c_T u \neq 0$, where c_T is a Young symmetrizer of type λ.

For this we cannot argue as simply as in the symplectic case: we need a variation of the theme. First, consider the diagram of λ filled in increasing order from up to down and right to left with the numbers $1, 2, \ldots, n$, e.g.,

$$\begin{matrix} 1 & 5 & 8 & 11 \\ 2 & 6 & 9 & \\ 3 & 7 & 10 & \\ 4 & & & \end{matrix}$$

Next suppose we fill it as a tableau with some of the basis vectors e_i, f_j, e.g.,

$$\begin{matrix} e_2 & f_3 & e_1 & e_1 \\ f_2 & e_4 & f_4 & \\ e_3 & e_3 & f_1 & \\ f_4 & & & \end{matrix}$$

To this display we associate a tensor u in which we place in the ith position the vector placed in the tableau in the case labeled with i, e.g., in our previous example:

$$u := e_2 \otimes f_2 \otimes e_3 \otimes f_4 \otimes f_3 \otimes e_4 \otimes e_3 \otimes e_1 \otimes f_4 \otimes f_1 \otimes e_1.$$

If the vectors that we place on the columns are all distinct, the antisymmetrization $a_T u \neq 0$.

Assume first $m = 2p$ is even. If the first column has $\leq p$ elements, we can work in the totally isotropic subspace $U := \langle e_1, \dots, e_p \rangle$. As for the symplectic group, $U^{\otimes n}$ is made of traceless tensors. On $U^{\otimes n}$ the group algebra $\mathbb{C}S_n$ acts with the kernel generated by the antisymmetrizer on $p+1$ elements. We construct the display with e_i on the ith row. The associated tensor u is a highest weight vector as in 6.4.2.

Otherwise, let $p+s,\ p-t,\ s \leq t$ be the lengths of the first two columns.

We first fill the diagram with e_i in the ith row, $i \leq p$, and we are left with s rows with just 1 element, which we fill with $f_p, f_{p-1}, \dots, f_{p-s+1}$. This tensor is symmetric in the row positions, so when we apply to it the corresponding Young symmetrizer we only have to antisymmetrize the columns.

When we perform a contraction on this tensor there are s possible contractions that are nonzero (in fact have value 1) and that correspond to the indices of the first column occupied by the pairs e_{p-i}, f_{p-i}. Notice that if we exchange these two positions, the contraction does not change.

It follows that when we antisymmetrize this element, we get a required nonzero traceless tensor in M_λ. Explicitly, up to a permutation we have the tensor:

$$(e_1 \wedge \dots \wedge e_p \wedge f_p \wedge f_{p-1} \wedge \dots \wedge f_{p-s+1}) \otimes (e_1 \wedge \dots \wedge e_{p-t})$$

$$\otimes \cdots \otimes (e_1 \wedge \dots \wedge e_{n_t}) \in T_\lambda(V). \tag{6.5.1}$$

The odd case is similar. □

Proposition. *The tensor* 6.5.1 *is a highest weight vector of weight* $\omega_{p-s} + \omega_{p-t} + \sum \omega_{n_i}$.

To prove the proposition, one applies the elements $\underline{e_i}$[110] defined in Chapter 10, §4.1.7 and 4.1.21 and one uses the formula 5.1.1 of the same chapter.

We are now ready to exhibit the list of irreducible rational representations of $O(V)$. First, using the double centralizer theorem, we have a decomposition

$$T^0(V^{\otimes n}) = \bigoplus_{\lambda \vdash n,\ h_1+h_2 \leq m} M_\lambda \otimes T_\lambda(V),$$

where we have indicated by $T_\lambda(V)$ the irreducible representation of $O(V)$ paired to M_λ and h_1, h_2 the first two columns of λ.

The determinant representation of $O(V)$ is contained in $V^{\otimes m}$ and it is equal to its inverse. Hence, from Theorem 1.4 of Chapter 7, all irreducible representations appear in tensor powers of V. Since $T^0(V^{\otimes n})$ contains all the irreducible representations appearing in $V^{\otimes n}$ and not in $V^{\otimes k},\ k < n$, we deduce

Theorem 2. *The irreducible representations* $T_\lambda(V),\ h_1 + h_2 \leq m$, *form a complete list of inequivalent irreducible representations of* $O(V)$.

[110] We underline to indicate the Chevalley generators, in order not to confuse them with the basis elements.

6.6 Irreducible Representations of $SO(V)$

We can now pass to $SO(V)$. In this case there is one more invariant $[v_1, \ldots, v_m]$ which gives rise to new intertwiners.

Moreover, since $SO(V)$ has index 2 in $O(V)$, we can apply Clifford's Theorem (Chapter 8, §5.1), and deduce that each irreducible representation M of $O(V)$ remains irreducible under $SO(V)$ or splits into two irreducibles according whether it is or it is not isomorphic to $M \otimes \epsilon$. The sign representation is the one induced by the determinant.

The theory is nontrivial only when $\dim V = 2n$ is even. In fact in the case of $\dim V$ odd, we have that $O(V) = SO(V) \times \mathbb{Z}/(2)$ where $\mathbb{Z}/(2)$ is ± 1 (plus or minus the identity matrix). Therefore an irreducible representation of $O(V)$ remains irreducible when restricted to $SO(V)$.

To put together these two facts we start with:

Lemma 1. *Let $T : V^{\otimes p} \to V^{\otimes q}$ be an $SO(V)$-invariant linear map.*

We can decompose $T = T_1 + T_2$ so that:

T_1 is $O(V)$-linear, and T_2 is $O(V)$-linear provided we twist $V^{\otimes q}$ by the determinant representation.

Given any irreducible representation $N \subset V^{\otimes p}$, if $T_2(N) \neq 0$, then $T_2(N)$ is isomorphic to $N \otimes \epsilon$ (ϵ is the determinant or sign representation of $O(V)$).

Proof. $\mathbb{Z}/(2) = O(V)/SO(V)$ acts on the space of $SO(V)$-invariant linear maps. Any such map is canonically decomposed as a sum $T = T_1 + T_2$ for the two eigenvalues ± 1 of $\mathbb{Z}/(2)$. Then T_1 is $O(V)$-invariant while for $X \in O(V)$, we have $T_2(Xa) = \det(X)X(T_2(a))$, the required condition.

The second part is an immediate consequence of the first. □

Let us use the notation $[v_1, \ldots, v_m] := v_1 \wedge v_2 \wedge \ldots \wedge v_m$. First, let us analyze, for $k \leq m = \dim(V)$, the operator

$$* : \otimes^k V \to \otimes^{m-k} V,$$

defined by the implicit formula (using the induced scalar product on tensor):

$$(*(v_1 \otimes \cdots \otimes v_k), v_{k+1} \otimes \cdots \otimes v_m) = [v_1, \ldots, v_m]. \tag{6.6.1}$$

Remark that if $\sigma \in S_{m-k}$ we have, by symmetry of the scalar product,

$$\begin{aligned}
&(\sigma * (v_1 \otimes \cdots \otimes v_k), v_{k+1} \otimes \cdots \otimes v_m) \\
&= (*(v_1 \otimes \cdots \otimes v_k), \sigma^{-1}(v_{k+1} \otimes \cdots \otimes v_m)) \\
&= [v_1, \ldots, v_k, v_{\sigma(k+1)}, \ldots, v_{\sigma(m)}] \\
&= \epsilon_\sigma [v_1, \ldots, v_m].
\end{aligned}$$

This implies that $*(v_1 \otimes \cdots \otimes v_k) \in \bigwedge^{m-k} V$. Similarly

$$*(\sigma(v_1 \otimes \cdots \otimes v_k)) = \epsilon_\sigma * (v_1 \otimes \cdots \otimes v_k)$$

implies that $*$ factors through a map

$$* : \bigwedge^k V \to \bigwedge^{m-k} V$$

still defined by the implicit formula

(6.6.2) $$(*(v_1 \wedge \ldots \wedge v_k), v_{k+1} \wedge \ldots \wedge v_m) = [v_1, \ldots, v_m].$$

In particular exterior powers are pairwise isomorphic under $SO(V)$.

In an orthonormal oriented basis u_i we have $*(u_{i_1} \wedge \ldots \wedge u_{i_k}) = \epsilon u_{j_1} \wedge \ldots \wedge u_{j_{m-k}}$, where $i_1, \ldots, i_k, j_1, \ldots, j_{m-k}$ is a permutation of $1, \ldots, m$, and ϵ is the sign of this permutation.

It is better to show one case explicitly in a hyperbolic basis, which explains in a simple way some of the formulas we are using. We do the even case in detail. Let V be of even dimension $2n$ with the usual hyperbolic basis e_i, f_i. Then,

(6.6.3) $$*(e_1 \wedge e_2 \wedge \ldots \wedge e_k) = f_{k+1} \wedge f_{k+2} \wedge \ldots \wedge f_n \wedge e_1 \wedge e_2 \wedge \ldots \wedge e_n.$$

Proof. By definition, taking the exterior products of the hyperbolic basis as the basis of $\bigwedge^{2n-k}(V)$ we see that the only nonzero scalar product of $*(e_1 \wedge e_2 \wedge \ldots \wedge e_k)$ with the basis elements is $(*(e_1 \wedge e_2 \wedge \ldots \wedge e_k), e_{k+1} \wedge e_{k+2} \wedge \ldots \wedge e_n \wedge f_1 \wedge \ldots \wedge f_n) = e_1 \wedge e_2 \wedge \ldots \wedge \ldots \wedge e_n \wedge f_1 \wedge \ldots \wedge f_n = 1$. $\square$

In particular, notice that this implies that $f_{k+1} \wedge f_{k+2} \wedge \ldots \wedge f_n \wedge e_1 \wedge e_2 \wedge \ldots \wedge e_n$ is a highest weight vector of $\bigwedge^{2n-k} V$.

In general let us understand what type of intertwiners we obtain on traceless tensors using this new invariant. By the same principle of studying only *new invariants* we start by looking at $SO(V)$ equivariant maps $\gamma : V^{\otimes p} \to V^{\otimes q}$ which induce nonzero maps $\gamma : T^0(V^{\otimes p}) \to T^0(V^{\otimes q})$.

We may restrict to a map γ corresponding to an $SO(V)$ but not to $O(V)$-invariant $(\gamma(u_1 \otimes \cdots \otimes u_p), v_1 \otimes \cdots \otimes v_q)$, which is the product of one single determinant and several scalar products. We want to restrict to the ones that give rise to operators $T : V^{\otimes p} \to V^{\otimes q}$ which do not vanish on traceless tensors and also which cannot be factored through an elementary extension. This implies that the invariant $(T(u_1 \otimes \cdots \otimes u_p), v_1 \otimes \cdots \otimes v_q)$ should not contain any *homosexual contraction.* Thus, up to permuting the u's and v's separately, we are reduced to studying the maps γ_k from $T^0(V^{k+t})$ to $T^0(V^{2n-k+t})$ induced by the invariants:

$$(\gamma_k(u_1 \otimes \cdots \otimes u_{k+t}), v_1 \otimes \cdots \otimes v_{2n+t-k})$$

(6.6.4) $$= \prod_{i=1}^{t} (u_{k+i}, v_{2n-k+i})[u_1, \ldots, u_k, v_1, \ldots, v_{2n-k}].$$

Lemma 2. *If k is the length of the first column of λ, then γ_k maps $T_\lambda(V)$ to $T_{\lambda'}(V)$, where λ' is obtained from λ substituting the first column k with $2n - k$.*

Thus $T_\lambda(V) \otimes \epsilon = T_{\lambda'}(V)$.

Proof. First, it is clear, from 6.6.1, that we have

$$\gamma(u_1 \otimes u_2 \otimes \cdots \otimes u_k \otimes \cdots \otimes u_{k+t})$$
$$= *(u_1 \otimes u_2 \otimes \cdots \otimes u_k) \otimes u_{k+1} \otimes \cdots \otimes u_{k+t}.$$

In the lemma we have two *dual* cases, $k \leq 2n - k$ or $k \geq 2n - k$. For instance in the second case, let $k = n + s,\ s \leq n$. Let us compute using 6.6.3:

$$\gamma[(e_1 \wedge e_2 \wedge \ldots \wedge e_n \wedge f_n \wedge f_{n-1} \wedge \ldots \wedge f_{n-s+1}) \otimes (e_1 \wedge e_2 \wedge \ldots \wedge e_{n-t})$$
$$\otimes \cdots \otimes (e_1 \wedge e_2 \wedge \ldots \wedge e_{h_r})]$$
$$= [*(e_1 \wedge e_2 \wedge \ldots \wedge e_n \wedge f_n \wedge f_{n-1} \wedge \ldots \wedge f_{n-s+1})]$$
$$\otimes (e_1 \wedge e_2 \wedge \ldots \wedge e_{n-t}) \otimes \ldots \otimes (e_1 \wedge e_2 \wedge \ldots \wedge e_{h_r})$$
$$= (e_1 \wedge e_2 \wedge \ldots \wedge e_s) \otimes (e_1 \wedge e_2 \wedge \ldots \wedge e_{n-t}) \otimes \ldots \otimes (e_1 \wedge e_2 \wedge \ldots \wedge e_{h_r}).$$

Thus γ maps a highest weight vector of $T_\lambda(V)$ to one of $T_{\lambda'}(V)$. Since these two representations are irreducible as $O(V)$-modules, the claim follows. □

We will say that the two partitions λ, λ' are *associated.* By definition $\lambda = \lambda'$ if and only only if the first column has length n. If the two associated partitions are distinct, one of them has the first column of length $< n$ and the other $> n$. Thus we have:

Proposition 1. *$T_\lambda(V)$ is isomorphic (as an $O(V)$ module) to $T_\lambda(V) \otimes \epsilon$ if and only if λ is self-associated, i.e., $\lambda = \lambda'$.*

Proof. We have already seen that $T_\lambda(V) \otimes \epsilon = T_{\lambda'}(V)$. □

In the case $\lambda = \lambda'$, from Clifford's theorem $T_\lambda(V)$ decomposes as the direct sum of two irreducible representations under $SO(V)$. It is interesting to write explicitly the two highest weight vectors of $SO(V)$.

Proposition 2. *In $\bigwedge^n V$, $\dim V = 2n$ we have the two highest weight vectors:*

$$e_1 \wedge e_2 \wedge \ldots \wedge e_n, \qquad e_1 \wedge e_2 \wedge \ldots \wedge e_{n-1} \wedge f_n,$$

with weights: $2s_\pm = \sum_{i=1}^{n-1} \alpha_i \pm \alpha_n$. (cf. Chapter 10, §5.1.1)

If $\dim V = 2n+1$, choose a hyperbolic basis $e_1, \ldots, e_n, f_1, \ldots, f_n, u$. $\bigwedge^n V \equiv \bigwedge^{n+1} V$ are irreducible under $SO(V)$ with highest weight vectors $e_1 \wedge e_2 \wedge \ldots \wedge e_n$, $e_1 \wedge e_2 \wedge \ldots \wedge e_n \wedge u$ and weight $2s = \sum_{i=1}^{n} \alpha_i$. (cf. Chapter 10, §5.1.2).

Proof. By definition we have to check that the two vectors are killed by the Chevalley generators of $SO(V)$. We then use the formulas in 4.1.21 of Chapter 10 (unfortunately we have a certain overlap of symbols). The analysis in the odd case is similar. □

For general $\lambda = \lambda'$ we construct the highest weight vectors as follows.

We build a tensor as in the pattern of 6.5. We insert e_i in the ith row for $i \leq n-1$, then the nth row; in one case we insert e_n, in the other f_n.

Summarizing we have:

Theorem. *If* $\dim V = 2n+1$ *is odd, the irreducible representations of* $SO(V)$ *are the* $T_\lambda(V)$ *indexed by partitions* λ *of height* $\leq n$.

If $\dim V = 2n$ *is even, the irreducible representations of* $SO(V)$ *are of two types:*

(1) The restriction to $SO(V)$ *of the irreducible representations* $T_\lambda(V)$ *of* $O(V)$ *indexed by partitions* λ *of height* $\leq n$ *which are not self-associated.*

(2) For each self-associated partition $\lambda = \lambda'$, *the two irreducible representations in which* $T_\lambda(V)$ *splits.*

In both cases their highest weights are linear combinations of the fundamental weights where the spin weights appear with even coefficients.

Remark. From the formulas of Chapter 10, §4.1.3 and our discussion it follows that, in the odd case $2n+1$, the exterior powers $\bigwedge^i V$, $i < n$, are fundamental representations. For $SO(2n+1)$, $\bigwedge^n V$ is irreducible corresponding to twice the spin representation which is fundamental.

On the other hand, when the dimension is even $2n$, the exterior powers $\bigwedge^i V$, $i < n-1$, are fundamental representations. The other two fundamental weights $\pm s$ belong to spin representations which do not appear in tensor powers. The exterior power $\bigwedge^n V$ decomposes as a direct sum of two irreducible representations corresponding to twice the half-spin representations. The exterior power $\bigwedge^{n-1} V$ appears as the leading term of the tensor product of the two half-spin representations (cf. §7.1).

The explanation is that in tensor powers we find the Lie algebra representations which, integrated to the spin group, factor through the special orthogonal group.

6.7 Fundamental Representations

We give here a complement to the previous theory by analyzing the action of the symplectic group of a space V on the exterior algebra in order to describe the *fundamental* representations.

We start with the symplectic case which in some way is more interesting. Assume $\dim V = 2n$.

First, by Theorem 6.4 we have that the traceless tensors $T^0(V^{\otimes m})$ contain a representation associated to the full antisymmetrizer (the sign representation of S_m) if and only if $m \leq n$. This is a new representation in $V^{\otimes m}$, hence it appears only in the traceless tensors and, by 6.4.1, with multiplicity 1.

Let us denote by $\bigwedge_0^m(V)$ this representation which, by what we have just seen, appears with multiplicity 1 also in $\bigwedge^m(V)$. The element $e_1 \wedge e_2 \wedge \ldots \wedge e_m$ is its highest weight vector. Its weight is the fundamental weight ω_m (cf. Chapter 10, §5.1.3).

Remark next that, by definition, $\bigwedge^2 V$ contains a canonical bivector

$$\psi := \sum_{i=1}^{n} e_i \wedge f_i$$

invariant under $Sp(V)$.

We want to compute now the $Sp(V)$ equivariant maps between $\bigwedge^k V$ and $\bigwedge^h V$.

Since the skew-symmetric tensors are direct summands in tensor space, any $Sp(V)$ equivariant map between $\bigwedge^k V$ and $\bigwedge^h V$ can be decomposed as $\bigwedge^k V \xrightarrow{i} V^{\otimes k} \xrightarrow{p} V^{\otimes h} \xrightarrow{A} \bigwedge^h V$ where i is the canonical inclusion, p is some equivariant map and A is the antisymmetrizer.

We have seen, in 3.1, how to describe equivariant maps $V^{\otimes k} \xrightarrow{p} V^{\otimes h}$ up to the symmetric group action.

If we apply the symmetric groups to i or to A, it changes at most the sign. In particular we see that the insertion maps $\bigwedge^k V \to \bigwedge^{k+2} V$ can, up to sign, be identified with $u \to \psi \wedge u$, which we shall also call ψ. For the contraction, we have a unique map (up to constant) which can be normalized to

$$(6.7.1)\quad c : v_1 \wedge v_2 \wedge \ldots \wedge v_k \to \sum_{i<j} (-1)^{i+j-1} \langle v_i, v_j \rangle v_1 \wedge v_2 \wedge \check{v}_i \ldots \check{v}_j \ldots \wedge v_k.$$

The general formula of 3.1 gives in this case that

Lemma. *All the maps between $\bigwedge^k V$ and $\bigwedge^h V$ are linear combinations of $\psi^i c^j$.*

Definition. The elements

$$\bigwedge\nolimits_0(V) := \left\{ a \in \bigwedge V \mid c(a) = 0 \right\}$$

are called *primitive*.

In order to understand the commutation relations between these two maps, let us set $h := [c, \psi]$. Go back to the spin formalism of Chapter 5, §4.1 and recall the formulas of the action of the Clifford algebra on the exterior power:

$$i(v)(u) := v \wedge u, \; j(\varphi)(v_1 \wedge v_2 \ldots \wedge v_k)$$
$$:= \sum_{t=1}^{k} (-1)^{t-1} \langle \varphi | v_t \rangle v_1 \wedge v_2 \ldots \check{v}_t \ldots \wedge v_k$$

together with the identity

$$i(v)^2 = j(\varphi)^2 = 0, \; i(v)j(\varphi) + j(\varphi)i(v) = \langle \varphi | v \rangle.$$

Now clearly as an operator we have $\psi := \sum_i i(e_i)i(f_i)$. Using the dual basis let us show that $c = \sum_i j(f^i)j(e^i)$. Let us drop the symbols i, j and compute directly in the Clifford algebra of $V \oplus V^*$ with the standard hyperbolic form:

$$\sum_i f^i e^i v_1 \wedge v_2 \wedge \ldots \wedge v_k = \sum_i \sum_{s<t} (-1)^{t+s} ((e^i|v_t)\langle f^i|v_s\rangle - \langle f^i|v_t\rangle\langle e^i|v_s\rangle)$$
$$\times v_1 \wedge v_2 \wedge \check{v}_s \ldots \check{v}_t \ldots \wedge v_k$$
$$= \sum_{s<t} (-1)^{t+s-1} \langle v_s|v_t\rangle v_1 \wedge v_2 \wedge \check{v}_s \ldots \check{v}_t \ldots \wedge v_k.$$

Now we can use the commutation relations

$$e_i e^j + e^j e_i = 0,\ i \neq j,\ e_i f^j + f^j e_i = 0,\ f_i e^j + e^j f_i = 0,$$
$$f_i f^j + f^j f_i = 0,\ i \neq j,\ e_i e^i + e^i e_i = 1,\ f_i f^i + f^i f_i = 1$$

to deduce that

$$h = [c, \psi] = \sum_{i,j} [f^j e^j, e_i f_i] = \sum_i [f^i e^i, e_i f_i] = \sum_i (f^i e^i e_i f_i - e_i f_i f^i e^i)$$
$$= \sum_i (-f^i e_i e^i f_i + f^i f_i - e_i e^i + e_i f^i f_i e^i)$$
$$= \sum_i (-f^i e_i e^i f_i + f^i f_i + e^i e_i + f^i e_i e^i f_i) \sum_i (f^i f_i + e^i e_i).$$

Now we claim that on $\bigwedge^k V$ the operator $\sum_i (f^i f_i + e^i e_i)$ acts as $2n - k$. In fact when we consider a vector $u := v_1 \wedge v_2 \ldots \wedge v_k$ with the v_i from the symplectic basis, the operators $f^i f_i, e^i e_i$ annihilate u if e_i, f_i is one of the vectors $v_1, \ldots, v_k$. Otherwise they map u into u itself.

Proposition. *The elements c, ψ, h satisfy the commutation relations of the standard generators e, f, h of $sl(2, \mathbb{C})$.*

Proof. We need only show that $[h, c] = 2c,\ [c, \psi] = -2\psi$. This follows immediately from the fact that c maps $\bigwedge^k V$ to $\bigwedge^{k-2} V$, ψ maps $\bigwedge^k V$ to $\bigwedge^{k+2} V$, while h has eigenvalue $2n - k$ on $\bigwedge^k V$. □

We can apply now the representation theory of $sl(2, \mathbb{C})$ to deduce

Theorem. *We have the direct sum decomposition:*

(1) $$\bigwedge V = \bigwedge\nolimits_0(V) \oplus \psi \wedge \left(\bigwedge V\right),$$

(2) $$\bigwedge\nolimits_0(V) = \bigoplus\nolimits_{m \leq n} \bigwedge\nolimits_0^m(V),$$

(3) $$\bigwedge^k V = \bigoplus\nolimits_{2i \geq k-n} \bigwedge\nolimits_0^{k-2i}(V) \wedge \psi^i.$$

Proof. For every finite-dimensional representation M of $sl(2, \mathbb{C})$, we have a decomposition $M = M^e \bigoplus_{i>0} f^i M,\ M^e := \{m \in M \mid em = 0\}$, by highest weight theory. Let $M = \bigwedge V$. Since all the contractions reduce to c on skew-symmetric tensors, the traceless skew-symmetric tensors $\bigoplus_{m \leq n} \bigwedge_0^m(V)$ are the kernel of c or $\bigwedge_0(V) = \bigoplus_{m \leq n} \bigwedge_0^m(V)$. Thus, $\bigwedge^k V = \bigoplus_i \bigwedge_0(V) \wedge \psi^i$. Comparing degrees in this formula finally gives all the statements. □

We can interpret the first part of the previous theorem as saying that:

Corollary. *The quotient algebra* $\bigwedge V/\psi \wedge (\bigwedge V) = \bigoplus_{m\leq n} \bigwedge_0^m(V)$ *is the direct sum of the trivial and the fundamental representations of* $Sp(V)$ *in the same way as the exterior powers* $\bigwedge^i V$, $1 \leq i < \dim V$, *are the fundamental representations of* $SL(V)$.

6.8 Invariants of Tensor Representations

When one tries to study invariants of representations different from sums of the defining representation, one finds quickly that these rings tend to become extremely complicated. The classical theory of $SL(2, \mathbb{C})$, which is essentially the theory of binary forms, (Chapter 15) shows this clearly. Nevertheless, at least in a rather theoretical sense, one could from the theory developed compute invariants of general representations for classical groups.

One starts from the general remark that for a linearly reductive group G and an equivariant surjective map $U \to V$, we have a surjective map of invariants $U^G \to V^G$. Thus if $M \supset N$ are two linear representations, it follows that the invariants of degree m on N are the restriction to N of the invariants of degree m on M.

For classical groups, up to the problem of the determinant, one can embed representations into a sum $\bigoplus_i V^{\otimes h_i}$ of tensor powers of the defining vector space V. An invariant of degree m on $\bigoplus_i V^{\otimes h_i}$ is thus an invariant linear form on $S^m(\bigoplus_i V^{\otimes h_i})$, which we may think of as being embedded into $(\bigoplus_i V^{\otimes h_i})^{\otimes m}$. We are thus led to study linear invariants on such a tensor power. This last space is clearly a (possibly very large) direct sum of tensor powers and we know, for $SL(V)$, $O(V)$, $Sp(V)$ all the linear invariants, given by various kinds of contractions. These contractions are usually expressed in symbolic form on decomosable tensors, giving rise to the symbolic method (Chapter 15). For $GL(V)$ one will have to work with direct sums of tensor powers $V^{\otimes p} \otimes (V^*)^{\otimes q}$ and the relative contractions.

In principle, these expressions give formulas of invariants which span the searched-for space of invariants for the given representation. In fact, it is almost impossible to make computations in this way due to several obstacles. First, to control the linear dependencies among the invariants given by different patterns of contraction is extremely hard. It is even worse to understand the multiplicative relations and in particular which symbolic invariants generate the whole ring of invariants, what the relations among them are, and so on. In fact, in order to give a theoretical answer to these questions Hilbert, in an impressive series of papers, laid down the foundations of modern commutative algebra (cf. Chapter 14).

7 Spinors

In this section we discuss some aspects of the theory of spinors, both from the point of view of representations as well as from the view of the theory of pure spinors.

7.1 Spin Representations

We now discuss the spin group, the spin representations and the conjugation action on the Clifford algebra of some orthogonal space W. We use the notations of Chapter 5, §4 and §5. Since $\bigwedge^2 W$ is the Lie algebra of the spin group and clearly $\bigwedge^2 W$ generates the even Clifford algebra $C^+(W)$, we have that any irreducible representation of $C^+(W)$ remains irreducible on its restriction to $\mathrm{Spin}(W)$.

We treat the case over $\mathbb{C}$, and use the symbol $C(n)$ to denote the Clifford algebra of an n-dimensional complex space. Recall that

$$C^+(n) \equiv C(n-1), \dim C(n) = 2^n, \ C(2m) = \mathrm{End}(\mathbb{C}^{2^m}),$$

$$C(2m+1) = \mathrm{End}(\mathbb{C}^{2^m}) \oplus \mathrm{End}(\mathbb{C}^{2^m}).$$

In the even-dimensional case we identify

$$W = V \oplus V^*, \quad C^+(W) = \mathrm{End}\big(\mathbb{C}^{2^{m-1}}\big) \oplus \mathrm{End}\big(\mathbb{C}^{2^{m-1}}\big).$$

Since $C^+(V \oplus V^*) \subset \mathrm{End}(\bigwedge V)$ and it has only two nonisomorphic irreducible representations each of dimension 2^{m-1}, the only possibility is that $\bigwedge_{\text{ev}} V, \bigwedge_{\text{odd}} V$, are exactly these two nonisomorphic representations.

Definition 1. These two representations are called the two *half-spin representations* of the even spin group, denoted S_0, S_1.

The odd case is similar; $C^+(2m+1) = C(2m) = \mathrm{End}(\mathbb{C}^{2^m})$ has a unique irreducible module of dimension 2^m. To analyze it let us follow the spin formalism for $C(2m)$ followed by the identification of $C^+(2m+1) = \mathbb{C}(2m)$ as in Chapter 5, §4.4. From that discussion, starting from a hyperbolic basis $e_1, \dots, e_n, f_1, \dots, f_n, u$, we have that $C^+(2m+1)$ is the Clifford algebra over the elements $a_i := e_i u, \ b_i := u f_i$. Now apply the spin formalism and, if V is the vector space spanned by the a_i, we consider $\bigwedge V$ as a representation of $\mathbb{C}^+(2m+1)$ and of the spin group.

Definition 2. This representation is called the *spin representation* of the odd spin group.

In the even case, we have the two half-spin representations.

Recall that the formalism works with $1/2$ of the standard hyperbolic form on $W = V \oplus V^*$. We take $e_1, \dots, e_n, f_1, \dots, f_n$ to be the standard basis with $(e_i, e_j) = (f_i, f_j) = 0$, $(e_i, f_j) = \delta_i^j/2$.

Let us now compute the highest weight vectors. We have to interpret the Chevalley generators:

$$\text{(7.1.1)} \qquad \underline{e}_i := e_{i,i+1} - e_{n+i+1,n+i}, \ i = 1, \dots, n-1, \ \underline{e}_n := e_{n-1,2n} - e_{n,2n-1}$$

$$\underline{f}_i := e_{i+1,i} - e_{n+i,n+i+1}, \ i = 1, \dots, n-1, \ \underline{f}_n := e_{2n,n-1} - e_{2n-1,n}$$

in terms of the spin formalism and formulas 4.4.1, 4.4.2 of Chapter 5:

$$[a \wedge b, c] = (b, c)a - (a, c)b,$$

$$[c \wedge d, a \wedge b] = (b, d)a \wedge c + (a, d)c \wedge b - (a, c)d \wedge b - (b, c)a \wedge d.$$

We deduce (recall we are using $1/2$ the standard form):

(7.1.2)

$$e_i \wedge e_j = e_i e_j = e_{i,n+j} - e_{j,n+i}, \quad e_i \wedge f_j = e_i f_j = e_{i,j} - e_{n+j,n+i}, \; i \neq j,$$

(7.1.3)

$$f_i \wedge f_j = f_i f_j = e_{n+i,j} - e_{n+j,i}, \quad e_i \wedge f_i = e_i f_i - 1/2 = e_{i,i} - e_{n+i,n+i}.$$

Hence

$$\underline{e}_i = e_i f_{i+1}, \qquad \underline{f}_i = e_{i+1} f_i, \qquad \underline{e}_n = e_{n-1} e_n, \qquad \underline{f}_n = f_n f_{n-1}.$$

We deduce that

$$e_1 \wedge e_2 \wedge \ldots \wedge e_n, \quad e_1 \wedge e_2 \wedge \ldots \wedge e_{n-1}$$

are highest weight vectors. In fact it is clear that they are both killed by $\underline{e}_n$ but also by the other $\underline{e}_i$, $i < n$, by a simple computation.

For the weights, apply

$$\sum_i^n x_i e_i \wedge f_i = \sum_i x_i (e_{i,i} - e_{n+i,n+i}) \text{ to } e_1 \wedge e_2 \wedge \ldots \wedge e_n \text{ and } e_1 \wedge e_2 \wedge \ldots \wedge e_{n-1}$$

to obtain the weights $1/2 \sum_i^n x_i$, $1/2(\sum_i^{n-1} x_i - x_n)$ respectively. We have the two half-spin weights $s_\pm$ determined in Chapter 10, §5.1.1.

In the odd case consider the Chevalley generators in the form

$$\begin{aligned} \underline{e}_i &:= e_{i,i+1} - e_{n+i+1,n+i} = e_i f_{i+1} = a_i b_{i+1}, \underline{e}_n = e_{n,2n+1} \\ &- e_{2n+1,2n} = e_n u = a_n, \end{aligned} \tag{7.1.4}$$

from which it follows that $a_1 \wedge a_2 \wedge \ldots \wedge a_n$ is a highest weight vector.

As for its weight, apply, as in the even case,

$$\sum_i^n x_i \frac{[a_i, b_i]}{2} = \sum_i^n x_i e_i \wedge f_i = \sum_i x_i (e_{i,i} - e_{n+i,n+i})$$

and obtain the weight $1/2 \sum_{i=1}^n x_i$, which is the fundamental spin weight for the odd orthogonal group.

7.2 Pure Spinors

It may be slightly more convenient to work with the lowest weight vector 1 killed by $e_i \wedge f_j, \forall i \neq j, \; f_i \wedge f_j$, with $(e_1 \wedge f_1)1 = -1/2$.

Definition. A vector in one of the two half-spin representations is called a *pure spinor* if it is in the orbit, under the spin group, of the line through the highest weight vector.

This is a special case of the theory discussed in Chapter 10, §6.2. If G is the spin group and P is the parabolic subgroup fixing the line through 1, in projective space a pure spinor corresponds to a point of the variety G/P.

We now do our computations in the exterior algebra on the generators e_i, which is the sum of the two half-spin representations.

Theorem 1. *The two varieties G/P for the two half-spin representations correspond to the two varieties of maximal totally isotropic subspaces. If a is a nonzero pure spinor, the maximal isotropic subspace to which it corresponds is $N_a := \{x \in W \mid xa = 0\}$.*

Proof. Let us look for instance at the even half-spin representation. Here 1 is a lowest weight vector and $x1 = 0$ if and only if x is in the span of $f_1, f_2, \dots, f_n$, so that we have $N_1 = \langle f_1, f_2, \dots, f_n\rangle$. If a is any nonzero pure spinor, $a = g\lambda 1$, $g \in \mathrm{Spin}(2n)$, $\lambda \in \mathbb{C}$, and hence the set of $x \in W$ with $xa = 0$ is $g(N_1)$; the claim follows. □

The parabolic subalgebra $\mathfrak{p}$ stabilizing 1 is the span of the elements $f_i \wedge f_j$, $e_i \wedge f_j$, which in matrix notation, from 7.1.2 and 7.1.3 coincides with the set of block matrices $\begin{vmatrix} A & 0 \\ B & -A^t \end{vmatrix}$, $B^t = -B$. It follows that the parabolic subgroup fixing 1 coincides with the stabilizer of the maximal isotropic space N_1.

Thus the unipotent group, opposite to $\mathfrak{p}$, has its Lie algebra spanned by the elements $e_i \wedge e_j$. Applying the exponential of an element $\omega_X := \sum_{i<j} x_{i,j} e_i \wedge e_j$ to 1, one obtains the open cell in the corresponding variety of pure spinors. This is described by formula 3.6.3 of Chapter 5, involving the Pfaffians $[i_1, i_2, \dots, i_{2k}]$:

$$\exp(\omega_X) = \sum_k \sum_{i_1<i_2<\cdots<i_{2k}} [i_1, i_2, \dots, i_{2k}] e_{i_1} \wedge e_{i_2} \wedge \dots \wedge e_{i_{2k-1}} \wedge e_{i_{2k}}. \tag{7.2.1}$$

Remark that exponentiating the toral subalgebra $\sum_i^n x_i e_i \wedge f_i$ we have a torus, whose coordinates we may call t_i such that the vector $e_{i_1} \wedge e_{i_2} \wedge \dots \wedge e_{i_k}$ has weight $t_{i_1} t_{i_2} \dots t_{i_k}$.

To obtain all the pure spinors we may apply the Weyl group. To understand it take one of the $sl_i(2, \mathbb{C})$, $i < n$, generated by $\underline{e}_i = e_i f_{i+1}$, $\underline{f}_i = e_{i+1} f_i$, $\underline{h}_i = e_i f_i - e_{i+1} f_{i+1}$.

We see by direct computation that all the vectors are killed by this Lie algebra except for the pairs $e_i \wedge u$, $e_{i+1} \wedge u$, with u a product not containing e_i, e_{i+1}. These pair of vectors transform as the standard basis of the basic representation of $sl(2, \mathbb{C})$. It follows that the reflection s_i maps $e_i \wedge u$ to $-e_{i+1} \wedge u$, and $e_{i+1} \wedge u$, to $e_i \wedge u$.

In the special linear group $SL(n, \mathbb{C})$ which acts on $\bigwedge(\sum_i \mathbb{C} e_i)$, the elements s_i defined by $s_i(e_j) = e_j$, $j \neq i, j \neq i+1$, $s_i(e_i) = -e_{i+1}$, $s_i(e_{i+1}) = e_i$ are the usual elements we choose to generate a group $\tilde{W}$ which induces, by adjoint action, the symmetric group S_n.

The last case is $sl_n(2, \mathbb{C})$ generated by $\underline{e}_n = e_{n-1}e_n$, $\underline{f}_n = f_n f_{n-1}$, $\underline{h}_n = e_{n-1}f_{n-1} - f_n e_n$. Now the vectors killed by $\underline{e}_n$ are all the vectors that contain either e_{n-1} or e_n or both.

We are left with two types of nontrivial representations: $e_{n-1} \wedge u, e_n \wedge u$ or $u, e_{n-1} \wedge e_n \wedge u$. The simple reflection s_n sends (u does not contain n):

$$e_{n-1} \wedge u \mapsto -e_n \wedge u,\ e_n \wedge u \mapsto e_{n-1} \wedge u, u \mapsto -e_{n-1} \wedge e_n \wedge u,\ e_{n-1} \wedge e_n \wedge u \mapsto u.$$

At this point we can easily check:

Proposition.

(i) The exterior products of the elements e_i are extremal vectors (for the two representations).
(ii) The stabilizer in the Weyl group of the weight of 1 is S_n and has index 2^{n-1}.
(iii) The half-spin representations are minuscule.

Proof. Let us denote by $\tilde{W}_n$ the group generated by the elements s_i, $i = 1, \dots, n$. It contains the subgroup W generated by the s_i, $i = 1, \dots, n-1$, which acts on weights as the symmetric group. Let us compute the orbit of 1 under $\tilde{W}$. We claim that it is formed by all the vectors $\pm e_{i_1} \wedge e_{i_2} \wedge \dots \wedge e_{i+2k}$. In fact, this set of vectors is stable under the action of $\tilde{W}$. The elements s_i^2, $i = 1, \dots, n-1$, change the sign of the elements e_i, e_{i+1}, while s_n^2 changes the sign of e_{n-1}, e_n. Thus the subgroup generated by the s_i^2 is the group of diagonal transformations which changes an even number of signs. The group W permutes transitively the vectors $\pm e_{i_1} \wedge e_{i_2} \wedge \dots \wedge e_{i+2k}$ for a given k up to sign. Finally, s_n allows us to pass from a suitable product with k factors to one with $k-2$ or $k+2$ factors. This shows that the 2^{n-1} vectors $e_{i_1} \wedge e_{i_2} \wedge \dots \wedge e_{i+2k}$ are extremal.

For the other half-spin representation we may start from e_1 instead of 1. Thus (i) and (iii) follow. Also (ii) follows from the fact that the Weyl group of Spin$(2n)$ is a quotient of $\tilde{W}$ and has order $n!2^{n-1}$. □

It is useful to do some computations directly inside the Clifford algebra. We consider the even case and the usual basis e_i, f_i. Set $f := f_n f_{n-1} \dots f_1$, $e := e_1 e_2 \dots e_n$. Set furthermore $A = \bigoplus_{k=0}^n A_k$ to be the (graded) subalgebra generated by the e_i's, isomorphic to the exterior algebra $\bigwedge V$, $V = \langle e_1, \dots, e_n \rangle$.

Proposition 1. *The left ideal in $C(2n)$ generated by f is minimal and equals Af. The mapping $a \mapsto af, a \in \bigwedge V$ is an isomorphism of representations of $C(2n)$.*

Proof. Since $f_i f = 0$, for all i we see that $C(2n)f = Af$. Since A is irreducible and the map $a \mapsto af$ nonzero, we must have that A maps isomorphically to Af. □

Notice in particular that $fe_{i_1}e_{i_2} \dots e_{i_k} f = 0$ unless $k = n$ and $fef = fe_1e_2 \dots e_n f = f$. This allows us to define a nondegenerate bilinear form on $\bigwedge V = Af$ invariant under the spin group. We use the principal involution which is the identity on the generators e_i, f_i and consider, given $a, b \in A$, the element $(af)^*bf = f^*a^*bf$. We can assume both elements a and b are homogeneous. Since

$f^* = (-1)^{n(n-1)/2} f$ we deduce that $(af)^*bf = 0$ unless a^*b has degree n, hence it is a multiple, $\beta(a,b)e$ of e. Then $(af)^*bf = \beta(a,b)f$. The number $\beta(a,b)$ is a bilinear form on $\bigwedge V$.

Theorem 2.

(1) The form β is nondegenerate. It pairs in a nondegenerate way $\bigwedge^k V$ and $\bigwedge^{n-k} V$.
(2) β satisfies the symmetry condition $\beta(a,b) = (-1)^{n(n-1)/2}\beta(b,a)$.
(3) β is stable under the spin group.
(4) The principal involution is the adjunction with respect to the form β.

Proof. (1) On the *decomposable* elements $e_{i_1}e_{i_2}\dots e_{i_k}$ it is clear that β pairs, with some sign, elements of complementary indices, proving 1.

(2) If a^*b has degree n, then $b^*a = (a^*b)^* = (-1)^{n(n-1)/2}a^*b$.

(3) If $s \in \mathrm{Spin}(2n)$, we have $\beta(sa, sb)$ is computed by

$$(saf)^*sbf = f^*a^*s^*sbf = f^*a^*bf.$$

(4) If $c \in C(2n)$ we see that $\beta(ca,b)f = (caf)^*bf = \beta(a, c^*b)f$. So the principal involution is the adjunction with respect to the form β. □

There are several interesting consequences of the previous theorem.

Corollary.

(1) Under the form β, the even and odd spinors are maximal totally isotropic if n is odd and instead orthogonal to each other if n is even.
(2) The two half-spin representations are self-dual if n is even, and each is the dual of the other if n is odd.
(3) If $n = 2k$ the form β is a nondegenerate form on each of the two half-spin representations. It is symmetric if k is even, skew symmetric if k is odd.

Proof. If a, b are homogeneous, and a^*b has degree n, we must have that a, b have different parity if n is odd, and the same if n is even, proving (1). (2) follows from (1). (3) follows from (2) and part (2) of Theorem 2. □

Remark. We could have worked also with $B = \bigoplus_{k=0}^{n} B_k$, the algebra generated by the f_i, and the minimal left ideal Be.

Now that we have seen that the spin representation $\bigwedge V = Af$ is self-dual we can identify $\bigwedge V \otimes \bigwedge V$ with $\mathrm{End}(\bigwedge V) = C(2n)$, thought of as a representation of the spin group under conjugation. This is best done using the internal map:

Theorem. *The map $i : Af \otimes Be \to C(2n)$, $af \otimes be \mapsto af(be)^*$, is an isomorphism as representations of the spin group.*

Proof. Since Af, Be are minimal left ideals, $(Be)^* = eB$ is a minimal right ideal. $Af(Be)^* = AfeB$, a two-sided ideal. Since $C(2n)$ is a simple algebra and the two spaces have the same dimension, i is an isomorphism. If $s \in \mathrm{Spin}(2n)$, we have $s(ae(bf)^*)s^{-1} = s(ae(bf)^*)s^* = (sae)(sbf)^*$. □

Remark. Notice that $Be = Afe$ so we can use the isomorphism i in the form $a_1 \otimes a_2 \mapsto a_1 fefa_2^* = a_1 fa_2^*$.

For a set $J := \{i_1 < i_2 \dots < i_k\}$, set $e_J := e_{i_1} e_{i_2} \dots e_{i_k}$, $f_J := e_{i_k} e_{i_{k-1}} \dots e_{i_1}$.

Lemma 1. *We have*

$$fe = \sum_{J \subset [1,2,\dots,n]} (-1)^{|J|} e_J f_J.$$

Proof. By induction

$$fe = f_n \Big(\sum_{J \subset [1,2,\dots,n-1]} (-1)^{|J|} e_J f_J \Big) e_n = \sum_{J \subset [1,2,\dots,n-1]} (-1)^{|J|} e_J f_n e_n f_J$$
$$= \sum_{J \subset [1,2,\dots,n-1]} (-1)^{|J|} e_J (-e_n f_n + 1) f_J = \sum_{J \subset [1,2,\dots,n]} (-1)^{|J|} e_J f_J. \qquad \square$$

We use two bases for $C(2n)$, $e_J f_K, e_J f e_K$, $J, K \subset [1, 2, \dots, n]$. The first basis $e_J f_K$ is adapted to the filtration by the degree of the monomials in the e_i, f_j.

Lemma 2. *The spaces $C_i := \bigoplus_{|J|+|K| \le i} \mathbb{C} e_J f_K$ are representations under the conjugation action of the spin group. C_i / C_{i-1} is isomorphic to $\bigwedge^i (V \oplus V^*)$.*

Proof. Since the spin group conjugates the space spanned by all the e_i, f_i, $i = 1, \dots, n$, into itself inducing the standard representation of the orthogonal group, the lemma follows by the commutation relations. $\square$

The second basis $e_J f e_K$ is adapted to the tensor product decomposition.

Under the tensor product the two summands $S_0 \otimes S_0 \oplus S_1 \otimes S_1$, $S_0 \otimes S_1 \oplus S_1 \otimes S_0$, correspond to $Cl^+(2n)$, $Cl^-(2n)$ when n is even and the other way if n is odd.

It is useful to describe the base change explicitly. Let J^c define the complement of J in $[1, 2, \dots, n]$ and ϵ_J the sign of the permutation such that $f = \epsilon_J f_J f_{J^c}$. Thus:

$$e_J f e_K = \epsilon_K e_J f_K f_{K^c} e_K = (-1)^{n(n-|K|)} \epsilon_K e_J f_K e_K f_{K^c}$$
$$= (-1)^{n(n-|K|)} \epsilon_K e_J \Big(\sum_{A \subset K} (-1)^{|A|} e_A f_A \Big) f_{K^c}$$
$$(7.2.2) \qquad = (-1)^{n(n-|K|)} \epsilon_K \sum_{A \subset K \cap J^c} (-1)^{|A|} e_J e_A f_A f_{K^c}.$$

Notice that the leading term of this sum is

$$(-1)^{n(n-|K|)} \epsilon_K (-1)^{|K \cap J^c|} e_J e_{K \cap J^c} f_{K \cap J^c} f_{K^c}.$$

Proposition 2. *The conjugation action of the spin group on $Cl(W)$ factors through an action of the special orthogonal group and it is isomorphic to $\bigoplus_{k=0}^{\dim W} \bigwedge^k W$.*

Proof. We have a homomorphism $T(W) \to Cl(W)$ which, restricted to the direct sum $\bigoplus_k \bigwedge^k(W)$ of antisymmetric tensors, is a linear isomorphism of representations of the orthogonal group. □

If $\dim W = 2n, 2n+1$, the even Clifford algebra $Cl^+(W)$ decomposes as $\bigoplus_{k=0}^n \bigwedge^{2k} W$. If $v_1, v_2, \dots, v_k \in W$, we have $v_1 \wedge v_2 \wedge \dots \wedge v_k = \frac{1}{k!}\sum_{\sigma \in S_k} \epsilon_\sigma v_{\sigma(1)} v_{\sigma(2)} \dots v_{\sigma(k)}$.

Lemma 3. *If $v_1, v_2, \dots, v_k \in W$ are orthogonal, we have $v_1 \wedge v_2 \wedge \dots \wedge v_k = v_1 v_2 \dots v_k$.*

Proof. From the commutation relations $v_{\sigma(1)} v_{\sigma(2)} \dots v_{\sigma(k)} = \epsilon_\sigma v_1 v_2 \dots v_k$. □

If the elements are not orthogonal, this is only the leading term in the filtration.

In particular if $v_1, v_2, \dots, v_{2n}$ form an orthogonal basis of W the element $v_1 v_2 \dots v_{2n}$ is a generator of the 1-dimensional vector space $\bigwedge^{2n} W$ and is, under the conjugation action, invariant under $SO(W)$.

Choose as basis $u_{2i-1} := e_i + f_i$, $u_{2i} := e_i - f_i$, $i = 1, \dots, n$. These vectors are an orthogonal basis of W, and so anticommute. Moreover $u_{2i-1}^2 = 1$, $u_{2i}^2 = -1$.

The element $z := u_1 u_2 \dots u_{2n}$ anticommutes with each element of W, commutes with the even Clifford algebra, and generates $\bigwedge^{2n} W$.

Theorem 3. $z^2 = 1$, $fz = f$ *and* $\bigwedge^k W z = \bigwedge^{2n-k} W$.

Proof. The permutation $2n, 2n-1, \dots, 2, 1$ has sign $(-1)^n$, so

$$z^2 = (-1)^n u_1 u_2 \dots u_{2n} u_{2n} u_{2n-1} \dots u_2 u_1 = (-1)^n (-1)^n = 1.$$

For fz let us see the first multiplications. From $f_i(e_i + f_i)(e_i - f_i) = -f_i$ we have

$$fz = (-1)^n f u_{2n} u_{2n-1} \dots u_2 u_1 = (-1)^{n-1} f_1 \dots f_{n-1} f_n u_{2n-2} u_{2n-3} \dots u_2 u_1$$
$$= (-1)^{n-1} f_1 \dots f_{n-1} u_{2n-2} u_{2n-3} \dots u_2 u_1 f_n.$$

Then work by induction. Since z is an invariant, the map $u \mapsto uz$ is an $SO(W)$-equivariant map, so it maps $\bigwedge^k W$, which is irreducible if $k < n$, to an isomorphic irreducible representation. Thus to prove that $z \bigwedge^k W = \bigwedge^{2n-k} W$, it is enough to see that there is an element $v \in \bigwedge^k W$ with $vz \in \bigwedge^{2n-k} W$. We can take $v = u_1 u_2 \dots u_k$ for which $vz = (-1)^k u_{k+1} u_{k+2} \dots u_{2n}$. We know, by §6.6, that $\bigwedge^n W$ decomposes as the direct sum of two irreducible representations of highest weights twice the two half-spin representations which are not isomorphic to the representations $\bigwedge^k W$ for $k < n$. It follows that z induces on $\bigwedge^n W$ a linear map with eigenvalues $+1$ and -1, preserving these two irreducibles. □

We will see presently that the two summands $\bigwedge^n W_+$, $\bigwedge^n W_-$ of $\bigwedge^n W$ relative to the eigenvalues $+1, -1$ coincide with the irreducibles of weights $2s_+, 2s_-$ respectively.

For $k < n$ we also decompose the direct sum $\bigwedge^k W \oplus \bigwedge^{2n-k} W = L_k^+ \oplus L_k^-$ as a sum of the two eigenspaces of eigenvalues ± 1 for z. Notice that, as representations, both L_k^+, L_k^- are isomorphic to $\bigwedge^k W \equiv \bigwedge^{2n-k} W$.

Notice that we also have $\bigwedge^k W \oplus \bigwedge^{2n-k} W = L_k^+ \oplus \bigwedge^k W = L_k^- \oplus \bigwedge^k W$.

We can now analyze $S_0 \otimes S_0$, $S_1 \otimes S_1$ by analyzing their images in $C(2n)$.

Take an element $afb^*, a, b \in S$. If $b \in S_0$, we have $afb^*z = afzb^* = afb^*$. If $b \in S_1$ is odd, we have $afb^*z = -afzb^* = -afb^*$. It follows that for each k we do have $\bigwedge^k W \not\subset SfS_0$ and $\bigwedge^k W \not\subset SfS_1$. Otherwise multiplication by z on $\bigwedge^k W$ would equal ± 1, while $\bigwedge^k Wz = \bigwedge^{2n-k} W$.

Theorem 4.

(i) The image of $S_0 \otimes S_0$ *in* $C(2n)$ *is* $\bigwedge^n W_+ \bigoplus_k L_k^+$, $k < n$ *and* $k \equiv n$, $\mod 2$.
(ii) $\bigwedge^n W_+$ *is the irreducible representation with highest weight* $2s_+$.
(iii) The image of $S_1 \otimes S_1$ *in* $C(2n)$ *is* $\bigwedge^n W_- \bigoplus_k L_k^-$, $k < n$ *and* $k \equiv n$, $\mod 2$.
(iv) $\bigwedge^n W_-$ *is the irreducible representation with highest weight* $2s_-$.

Proof. Under the equivariant map $i : S \otimes S \to C(2n)$, $i(a \otimes b) = afb^*$ we have that $S_0 \otimes S_0 \oplus S_1 \otimes S_1$ maps to the even or odd part of the Clifford algebra, depending on whether n is even or odd.

Conversely, $S_0 \otimes S_1 \oplus S_1 \otimes S_0$ maps to the even or odd part of the Clifford algebra, depending on whether n is odd or even.

In both cases, the odd or even part of the image belonging to $S \otimes S_0$ is the set of elements u with $uz = u$, and the image belonging to $S \otimes S_1$ is the set of elements u with $uz = -u$. The first claim follows now from the definitions. Next we know that $S_0 \otimes S_0$ contains (as leading term) the irreducible representation of highest weight $2s_+$ which appears in $\bigwedge^n W$ so it must coincide with $\bigwedge^n W_+$. A similar argument on $S_1 \otimes S_1$ shows that $\bigwedge^n W_-$ must be the irreducible representation of highest weight $2s_-$ which appears in $\bigwedge^n W$. □

We have that the image of $S_0 \otimes S_0$, $S_1 \otimes S_1$ is in the even or odd Clifford algebra according to whether n is even or odd. Let us assume n is even to simplify the notation; the other case is similar. Let us denote by C_{n-1}^+ the part of $C(2n)^+$ of filtration degree $\leq n-1$. Then:

Proposition 3. *We have a direct sum decomposition*

$$C(2n)^+ = S_0 \otimes S_0 \oplus C_{n-1}^+ \oplus \bigwedge^n W_-$$

and also

$$C(2n)^+ = S_1 \otimes S_1 \oplus C_{n-1}^+ \oplus \bigwedge^n W_+.$$

Proof. Let us prove one of the two statements, the other being similar.

We have $S_0 \otimes S_0 = \bigwedge^n W_+ \bigoplus_{2i<n} L_{2i}^+$.

$$C_{n-1}^{+} = \bigoplus_{2i<n} \bigwedge^{2i} W,$$
$$C(2n)^{+} = \bigoplus_{i\leq n} \bigwedge^{2i} W = \bigwedge^{n} W \bigoplus_{2i<n} \Big(\bigwedge^{2i} W \oplus \bigwedge^{2n-2i} W\Big)$$
$$= \bigwedge^{n} W_{+} \oplus \bigwedge^{n} W_{-} \bigoplus_{2i<n} \Big(\bigwedge^{2i} W \oplus L_{2i}^{+}\Big).$$

The claim follows. □

From this corollary we have a more explicit way of computing the quadratic equations defining pure spinors. We do it for S_0; for S_1 it is similar. Given $u = \sum_J x_J e_J \in S_0$ we know, by Chapter 10, §6.6, that the property of being a pure spinor is equivalent to the fact that $u \otimes u \in \bigwedge^n W_+$. This means, using the previous proposition, that when we project ufu^* to $C(2n)/C_n$ the image must be 0. This is quite an explicit condition since projecting to $C(2n)/C_n$ is the same as imposing the condition that in the expansion of ufu^*, all terms of degree $> n$ must vanish. Now

$$ufu^* = \Big(\sum_J x_J e_J\Big) f \Big(\sum_J (-1)^{|J|} x_J e_J\Big).$$

This sum can be computed using formulas 7.2.3. Notice that when applied to a pure spinor given by formula 7.2.1, one obtains some identical quadratic equations satisfied by Pfaffians. The theme of quadratic equations will be revised in Chapter 13 when we use such equations to develop standard monomial theory for tableaux.

Remarks.

(1) All the representations of $\mathrm{Spin}(V)$ which do not factor through $SO(V)$ appear in the spaces $S \otimes V^{\otimes p}$.

(2) In order to study intertwiners between these representations one must use the structure of $\mathrm{End}(S)$.

7.3 Triality

There is one special case to be noticed: Spin(8). In this case we have three fundamental representations of dimension 8. One is the defining representation, the other two are the half-spin representations. Since in each of the half-spin representations we have a nondegenerate quadratic form β preserved by the spin group, and since the Lie algebra of the spin group is simple, we get:

Proposition. *In each of the 3 fundamental 8-dimensional representations* Spin(8) *induces the full special orthogonal group.*

There is a simple explanation of this phenomenon which is called *triality* and it is specific to dimension 8. This comes from the external automorphism group of the spin group or its associated Lie algebra. We have that the symmetric group S_3 of permutations of 3 elements acts as symmetry group of the Dynkin diagram. Thus (Chapter 10, §6.10) it induces a group of external automorphisms which permutes transitively the 3 outer vertices, which correspond to the 3 fundamental representations described.

Notice the simple:

Corollary. *The set of pure spinors in a half-spin representation equals the set of isotropic vectors for the form β.*

8 Invariants of Matrices

In this section we will deduce the invariant theory of matrices from our previous work.

8.1 FFT for Matrices

We are interested now in the following problem: describe the ring of invariants of the action of the general linear group $GL(n, \mathbb{C})$ acting by simultaneous conjugation on m copies of the space $M_n(\mathbb{C})$ of square matrices.

In intrinsic language, we have an n-dimensional vector space V and $GL(V)$ acts on $\mathrm{End}(V)^{\oplus m}$.

We will denote by $(X_1, X_2, \ldots, X_m)$ an m-tuple of $n \times n$ matrices.

Before we formulate and prove the main theorem let us recall the results of Chapter 9, §1. The theorem proved there can be reformulated as follows. Suppose we are interested in the multilinear invariants of m matrices, i.e., the invariant elements of the dual of $\mathrm{End}(V)^{\otimes m}$.

First, remark that the dual of $\mathrm{End}(V)^{\otimes m}$ can be identified, in a $GL(V)$-equivariant way, with $\mathrm{End}(V)^{\otimes m}$ by the pairing formula:

$$\langle A_1 \otimes A_2 \ldots \otimes A_m | B_1 \otimes B_2 \ldots \otimes B_m \rangle$$
$$:= \mathrm{tr}(A_1 \otimes A_2 \ldots \otimes A_m \circ B_1 \otimes B_2 \ldots \otimes B_m) = \prod \mathrm{tr}(A_i B_i).$$

Therefore under this isomorphism the multilinear invariants of matrices are identified with the $GL(V)$-invariants of $\mathrm{End}(V)^{\otimes m}$ which in turn are spanned by the elements of the symmetric group. We deduce from the theory of Chapter 9 and formula 6.1.3 there:

Lemma. *The multilinear invariants of m matrices are linearly spanned by the functions:*

$$\phi_\sigma(X_1, X_2, \ldots, X_m) := \mathrm{tr}(\sigma^{-1} \circ X_1 \otimes X_2 \otimes \cdots \otimes X_m), \ \sigma \in S_m.$$

Recall that if $\sigma = (i_1 i_2 \ldots i_h)(j_1 j_2 \ldots j_k) \ldots (s_1 s_2 \ldots s_m)$ is the cycle decomposition of σ, then from Chapter 9, Theorem 6.1.3, we have that

$$\phi_\sigma(X_1, X_2, \ldots, X_m) = \mathrm{tr}(X_{i_1} X_{i_2} \ldots X_{i_h}) \,\mathrm{tr}(X_{j_1} X_{j_2} \ldots X_{j_k}) \ldots \mathrm{tr}(X_{s_1} X_{s_2} \ldots X_{s_m}).$$

This explains our convention in defining ϕ_σ.

Theorem (FFT for Matrices). *The ring of invariants of matrices under simultaneous conjugation is generated by the elements*

$$\mathrm{tr}(X_{i_1}X_{i_2}\dots X_{i_{k-1}}X_{i_k}),$$

where the formula means that we take all possible noncommutative monomials in the X_i *and form their traces.*

Proof. The proof of this theorem is an immediate consequence of the previous lemma and the Aronhold method.

The proposed ring of invariants is in fact clearly closed under polarizations and coincides with the invariants, by the previous lemma, for the multilinear elements. The claim follows. □

Exercise. It is easy to generalize this theorem when one considers as a representation $\mathrm{End}(V)^{\otimes p} \oplus V^q \oplus (V^*)^s$.

One gets as generators of invariants besides the elements $\mathrm{tr}(M)$ also the elements $\langle \phi \mid Mv\rangle = \langle M^t\phi \mid v\rangle$, $\phi \in V^*$, $v \in V$ and M a monomial.

One can compute also the $SL(V)$-invariants and then add the invariants

$$M_1^t\phi_1 \wedge \dots \wedge M_n^t\phi_n, \quad M_1v_1 \wedge \dots \wedge M_nv_n, \quad \phi_i \in V^*, v_i \in V$$

and M a monomial.

8.2 FFT for Matrices with Involution

There is a similar theorem for the orthogonal and symplectic groups.

Assume that V is equipped with a nondegenerate form $\langle u, v\rangle$ (symmetric or skew symmetric). Then we can identify $\mathrm{End}(V) = V \otimes V^* = V \otimes V$ by

$$u \otimes v(w) := \langle v, w\rangle u\,.$$

Let $\epsilon = \pm 1$ according to the symmetry, i.e., $\langle a, b\rangle = \epsilon\langle b, a\rangle$. We then get the formulas

$$(a \otimes b) \circ (c \otimes d) = a \otimes \langle b, c\rangle d, \ \mathrm{tr}(a \otimes b) = \langle b, a\rangle$$

$$\langle (a \otimes b)c, d\rangle = \langle b, c\rangle\langle a, d\rangle = \epsilon\langle c, (b \otimes a)d\rangle.$$

In particular, using the notion of adjoint X^* of an operator, $\langle Xa, b\rangle := \langle a, X^*b\rangle$ we see that $(a \otimes b)^* = \epsilon(b \otimes a)$.

We can now analyze first the multilinear invariants of m matrices under the group G (orthogonal or symplectic) fixing the given form, recalling the FFT of invariant theory for such groups.

Compute such an invariant function on $\mathrm{End}(V)^{\otimes m} = V^{\otimes 2m}$. Write a decomposable element in this space as

$$X_1 \otimes X_2 \dots \otimes X_m = u_1 \otimes v_1 \otimes u_2 \otimes v_2 \otimes \dots u_m \otimes v_m, \quad X_i = u_i \otimes v_i.$$

Then the invariants are spanned by products of m scalar products $\langle x_i, y_i\rangle$ such that the $2m$ elements x_i, y_i exhaust the list of the u_i, v_j.

Of course in these scalar products a vector u can be paired with another u or a v (homosexual or heterosexual pairings, according to Weyl).

The previous formulas show that, up to a sign, such an invariant can be expressed in the form

$$\phi_{\overline{\sigma}}(X_1, \dots, X_m) := \operatorname{tr}(Y_{i_1}Y_{i_2}\dots Y_{i_h})\operatorname{tr}(Y_{j_1}Y_{j_2}\dots Y_{j_k})\dots\operatorname{tr}(Y_{s_1}Y_{s_2}\dots Y_{s_m}),$$

where, for each i, the element Y_i is either X_i or X_i^*.

Combinatorially this can be pictured by a *marked permutation* $\overline{\sigma}$, e.g.,

$$\overline{\sigma} = \{3, 2, \overline{1}, \overline{5}, 4\}, \quad \phi_{\overline{\sigma}}(X_1, \dots, X_5) := \operatorname{tr}(X_3X_1^*X_2)\operatorname{tr}(X_4X_5^*).$$

We deduce then the:

Theorem (FFT for Matrices). *The ring of invariants of matrices under simultaneous conjugation by the group G (orthogonal or symplectic) is generated by the elements*

$$\operatorname{tr}(Y_{i_1}Y_{i_2}\dots Y_{i_{k-1}}Y_{i_k}),\ Y_{i_h} = X_{i_h} \text{ or } X_{i_h}^*.$$

Proof. Same as before. □

Exercise. Generalize this theorem when one considers as a representation $\operatorname{End}(V)^{\otimes p} \oplus V^q$.

One gets as generators of invariants, besides the elements $\operatorname{tr}(M)$, also the elements $(u\,,\ Mv) = (M^*u\,,\ v)$, $v \in V$ and M a monomial in the X_i, X_j^*.

The computation of the $SO(V)$-invariants is more complicated but can also be performed. In this case one should observe that the only new case is when $\dim V$ is even. In fact, when $\dim V$ is odd, $-1 \in SO(V)$ is an improper orthogonal transformation which acts as 1 on matrices. Thus, in this case we have no further invariants when we restrict to $SO(V)$.

Instead, when $\dim(V) = 2n$, given a skew-symmetric matrix Y we have that $Pf(Y)$ is invariant under $SO(V)$, but for $A \in O(V)$ we have $Pf(AYA^{-1}) = Pf(AYA^t) = \det(A)Pf(Y)$, from Chapter 5, §3.6.2.

Of course we may think of $Pf\left(\frac{X-X^t}{2}\right)$ as an invariant of matrices of degree n.

When we polarize it we obtain a symmetric multilinear invariant $Q(X_1, \dots, X_n)$ which under an improper orthogonal transformation is multiplied by -1. When we specialize the matrices $X_i := u_i \otimes v_i$ we get a special orthogonal, but not an orthogonal invariant of the $2n$ vectors u_i, v_j. We know that, up to a scalar, this invariant must be equal to $u_1 \wedge v_1 \wedge u_2 \wedge v_2 \wedge \dots \wedge u_n \wedge v_n$. If we set $u_i := e_i,\ v_i := f_i$ we have that the constant is 1. Now we have more general invariants, given by the functions $Q(M_1, M_2, \dots, M_n)$ where the M_i's are monomials in the X_i, X_i^t, and we see this, by the same symbolic method.

Theorem (FFT for Matrices). *The ring of invariants of matrices under simultaneous conjugation by the group* $SO(V)$, $\dim V = 2n$ *is generated by the elements*

$$\operatorname{tr}(M),\ Q(M_1, M_2, \dots, M_n),\ M, M_i \text{ are monomials in } X_i, X_i^t.$$

8.3 Algebras with Trace

It is useful to formalize the previous analysis, in the spirit of universal algebra, as follows.

Definition. An algebra with trace is an associative algebra R equipped with a unary (linear) map $\mathrm{tr}: R \to R$ satisfying the following axioms:

(1) $\mathrm{tr}(a)b = b\,\mathrm{tr}(a), \ \forall a, b \in R.$
(2) $\mathrm{tr}(\mathrm{tr}(a)b) = \mathrm{tr}(a)\,\mathrm{tr}(b), \ \forall a, b \in R.$
(3) $\mathrm{tr}(ab) = \mathrm{tr}(ba), \ \forall a, b \in R.$

An algebra with involution is an associative algebra R equipped with a unary (linear) map $* : R \to R, \ x \to x^*$ satisfying the following axioms:

$$(x^*)^* = x, \ (xy)^* = y^*x^*, \ \forall x, y \in R.$$

For an algebra with involution and a trace we shall assume the compatibility condition $\mathrm{tr}(x^*) = \mathrm{tr}(x), \forall x$.

As always happens in universal algebra, for these structures one can construct free algebras on a set S (or on the vector space spanned by S).

Explicitly, in the category of associative algebras, the free algebra on S is obtained by introducing variables $x_s, s \in S$ and considering the algebra having as basis all the *words* or monomials in the x_s. For algebras with involution we have to add also the *adjoints* x_s^* as an independent set of variables.

When we introduce a trace we have to add to these free algebras a set of commuting indeterminates $t(M)$ as M runs over the set of monomials.

Here, in order to preserve the axioms, we also need to impose the identifications given by *cyclic equivalence*: $t(AB) = t(BA)$, and *adjoint symmetry*: $t(A^*) = t(A)$, for all monomials A, B.

In all cases the free algebra with trace is the tensor product of the free algebra (without trace) and the polynomial ring in the elements $t(A)$. This polynomial ring will be called the *free trace ring* (with or without involution).

The free algebras F_S, by definition, have the universal property that given elements $f_s \in F_S, \ s \in S$, there is a unique homomorphism $F_S \to F_S$ (compatible with the structures of trace, involution) which maps x_s to f_s for all s.

In particular we can rescale independently the x_s and thus speak about multihomogeneous, in particular multilinear, elements in F_S.

Let $S = \{1, 2, \ldots, m\}$. We describe the multilinear elements in the various cases.

1. For the free associative algebra in m variables, the multilinear monomials (in all the variables) correspond to permutations in S_m as $x_{i_1}x_{i_2}\ldots x_{i_m}$.
2. For the free algebra with involution the multilinear monomials (in all the variables) correspond to marked permutations in S_m as: $y_{i_1}y_{i_2}\ldots y_{i_m}$ where $y_i = x_i$ if i is unmarked, while $y_i = x_i^*$ if i is marked.

3. For the free associative algebra with trace in m variables the multilinear monomials (in all the variables) correspond to permutations in S_{m+1} according to the following rule. Take such a permutation and decompose it into cycles isolating the cycle containing $m+1$ as follows: if $\sigma = (i_1 i_2 \dots i_h)(j_1 j_2 \dots j_k)\dots(r_1 \dots r_p)$ $(s_1 s_2 \dots s_q m+1)$, set

$$\psi_\sigma(x_1, x_2, \dots, x_m) = \operatorname{tr}(x_{i_1} x_{i_2} \dots x_{i_h}) \operatorname{tr}(x_{j_1} x_{j_2} \dots x_{j_k}) \dots \\ \times \operatorname{tr}(x_{r_1} x_{r_2} \dots x_{r_p}) x_{s_1} x_{s_2} \dots x_{s_q}.$$

4. For the free associative algebra with trace and involution in m variables the multilinear elements (in all the variables) correspond to marked permutations in S_{m+1}, but there are some equivalences due to the symmetry of trace under involution.

In fact it may be interesting to isolate, in the case of trace algebras, the part T_{m+1} of the multilinear elements in $m+1$ variables lying in the trace ring and compare it with the full set A_m of multilinear elements in m variables using the map $c_m : A_m \to T_{m+1}$, given by $c_m(M) := t(Mx_{m+1})$. We leave it to the reader to verify that this map is a linear isomorphism.

Example. The correspondence between A_2 and T_3, in the case of involutions:

$$x_1x_2, t(x_1x_2x_3);\ x_2x_1, t(x_2x_1x_3);\ x_1x_2, t(x_1x_2x_3);\ x_2x_1, t(x_2x_1x_3);$$

$$x_1^*x_2, t(x_1^*x_2x_3);\ x_2^*x_1, t(x_2^*x_1x_3);\ x_1x_2^*, t(x_1x_2^*x_3);\ x_2x_1^*, t(x_2x_1^*x_3);$$

$$t(x_1)x_2, t(x_1)t(x_2x_3);\ t(x_2)x_1, t(x_2)t(x_1x_3);\ t(x_1)x_2^*, t(x_1)t(x_2^*x_3);$$

$$t(x_2)x_1^*, t(x_2)t(x_1^*x_3);\ t(x_1)t(x_2), t(x_1)t(x_2)t(x_3);\ t(x_1x_2), t(x_1x_2)t(x_3);$$

$$t(x_1x_2^*), t(x_1x_2^*)t(x_3).$$

Let us also denote for convenience by R the free algebra with trace in infinitely many variables and by Tr the polynomial ring of traces in R. The formal trace in R is denoted by $t : R \to \mathrm{Tr}$.

We formalize these remarks as follows. We define maps

$$\Psi : \mathbb{C}[S_{m+1}] \to R, \quad \Psi(\sigma) := \psi_\sigma; \qquad \Phi : \mathbb{C}[S_m] \to \mathrm{Tr}, \quad \Phi(\sigma) := \phi_\sigma.$$

We have a simple relation between these maps. If $\sigma \in S_{m+1}$, then

$$\Phi(\sigma) = t(\Psi(\sigma)x_{m+1}).$$

We remark finally that for a permutation $\sigma \in S_m$ and an element $a \in \mathbb{C}[S_m]$ we have

$$\Phi(\sigma a \sigma^{-1}) = \Phi(a)(x_{\sigma(1)}, x_{\sigma(2)}, \dots, x_{\sigma(m)}).$$

8.4 Generic Matrices

We shall now take advantage of this formalism and study more general algebras. We want to study the set of G-equivariant maps

$$f : \mathrm{End}(V)^{\oplus m} \to \mathrm{End}(V).$$

Here $n = \dim V$ and G is either $GL(V)$ or the orthogonal or symplectic group of a form.

We shall denote this space by $R_m(n)$ in the $GL(V)$ case, and by $R_m^o(n)$, $R_m^s(n)$, respectively, in the orthogonal and symplectic cases.

First, observe that the scalar multiples of the identity $\mathbb{C}1_V \subset \mathrm{End}(V)$ form the trivial representation, and the equivariant maps with values in $\mathbb{C}1_V$ can be canonically identified with the ring of invariants of matrices. We shall denote this ring by $T_m(n)$, $T_m^o(n)$, $T_m^s(n)$ in the 3 corresponding cases.

Next remark that $\mathrm{End}(V)$ has an algebra structure and a trace, both compatible with the group action. We deduce that under pointwise multiplication of the values the space $R_m(n)$ is a (possibly noncommutative) algebra, and moreover, applying the trace function, we deduce that:

$R_m(n)$ is an algebra with trace and the trace takes values in $T_m(n)$.

For the orthogonal and symplectic case $\mathrm{End}(V)$ has also a canonical involution, so $R_m^o(n)$ and $R_m^s(n)$ are algebras with trace and involution.

Finally, observe that the *coordinate maps* $(X_1, X_2, \dots, X_m) \to X_i$ are clearly equivariant. We shall denote them (by the usual abuse of notations) by X_i. We have:

Theorem. *In the case of $GL(V)$, $R_m(n)$ is generated as an algebra over $T_m(n)$ by the variables X_i.*

$R_m^o(n)$ and $R_m^s(n)$ are generated as algebras over $T_m^o(n)$, $T_m^s(n)$ by the variables X_i, X_i^*.

Proof. Let us give the proof of the first statement; the others are similar.

Given an equivariant map $f(X_1, \dots, X_m)$ in m variables, we construct the invariant function of $m+1$ variables $g(X_1, \dots, X_m, X_{m+1}) := \mathrm{tr}(f(X_1, \dots, X_m)X_{m+1})$.

By the structure theorem of invariants, g can be expressed as a linear combination of elements of the form $\mathrm{tr}(M_1)\,\mathrm{tr}(M_2)\dots\mathrm{tr}(M_k)$, where the M_i's are monomials in the variables X_i, $i = 1, \dots, m+1$.

By construction, g is linear in X_{m+1}; thus we can assume that each term

$$\mathrm{tr}(M_1)\,\mathrm{tr}(M_2)\dots\mathrm{tr}(M_k)$$

is linear in X_{m+1} and in particular (using the cyclic equivalence of trace) we may assume that X_{m+1} appears only in M_k and $M_k = N_k X_{m+1}$. Then N_k does not contain X_{m+1} and

$$\mathrm{tr}(M_1)\,\mathrm{tr}(M_2)\dots\mathrm{tr}(M_{k-1})\,\mathrm{tr}(M_k) = \mathrm{tr}((\mathrm{tr}(M_1)\,\mathrm{tr}(M_2)\dots\mathrm{tr}(M_{k-1})N_k)X_{m+1}).$$

It follows that we can construct a polynomial $h(X_1, \dots, X_m)$ (noncommutative) in the variables X_i, $i \leq m$, with invariant coefficients and such that

$$\mathrm{tr}(h(X_1, \dots, X_m)X_{m+1}) = \mathrm{tr}(f(X_1, \dots, X_m)X_{m+1}).$$

Now we can use the fact that the trace is a nondegenerate bilinear form and that X_{m+1} is an independent variable to deduce that $h(X_1, \dots, X_m) = f(X_1, \dots, X_m)$, as desired. □

Definition. The algebra $R_m(n)$ is called the *algebra of m-generic $n \times n$ matrices with trace.*

The algebra $R_m^o(n)$ (resp. $R_m^s(n)$) is called the *algebra of m-generic $n\times n$ matrices with trace and orthogonal* (resp. *symplectic*) *involution.*

8.5 Trace Identities

Our next task is to use the second fundamental theorem to understand the relations between the invariants that we have constructed. For this we will again use the language of universal algebra. We have to view the algebras constructed as quotients of the corresponding free algebras, and we have to deduce some information on the kernel of this map.

Recall that in an algebra R with some extra operations an ideal I must be stable under these operations, so that R/I can inherit the structure. In an algebra with trace or involution, we require that I be stable under the trace or the involution.

Let us call by F_m, F_m^i the free algebra with trace in m-variables and the free algebra with trace and involution in m-variables. We have the canonical maps (compatible with trace and, when it applies, with the involution)

$$\pi : F_m \to R_m(n),\ \pi^o : F_m^i \to R_m^o(n),\ \pi^s : F_m^i \to R_m^s(n).$$

We have already seen that in the free algebras we have the operation of substituting the variables x_s by any elements f_s. Then one has the following:

Definition. An ideal of a free algebra, stable under all the substitutions of the variables, is called a *T-ideal* (or an *ideal of polynomial identities*).

The reason for this notation is the following. Given an algebra R, a morphism of the free algebra in R consists of evaluating the variables x_s in some elements r_s. The intersection of all the kernels of all possible morphisms are those expressions of the free algebra which vanish identically when evaluated in R, and it is clear that they form a T-ideal, the ideal of polynomial identities of R. Conversely, if $I \subset F_S$ is a T-ideal, it is easily seen that it is the ideal of polynomial identities of F_S/I.

Of course an intersection of T-ideals is again a T-ideal and thus we can speak of the T-ideal generated by a set of elements (polynomial identities, or trace identities).

Going back to the algebra $R_m(n)$ (or $R_m^o(n)$, $R_m^s(n)$) we also see that we can compose any equivariant map $f(X_1, \dots, X_m)$ with any m maps $g_i(X_1, \dots, X_m)$ getting a new map $f(g_1(X_1, \dots, X_m), \dots, g_m(X_1, \dots, X_m))$. Also in $R_m(n)$, we have the morphisms given by substitutions of variables.

Clearly substitution in the free algebra is compatible with substitution in the algebras $R_m(n)$, $R^o_m(n)$, $R^s_m(n)$, and thus the kernels of the maps π, π^o, π^s are all T-ideals. They are called respectively:

The ideal of trace identities of matrices.

The ideal of trace identities of matrices with orthogonal involution.

The ideal of trace identities of matrices with symplectic involution.

We can apply the language of substitutions in the free algebra and thus define polarization and restitution operators, and we see immediately (working with infinitely many variables) that

Lemma. *Two T-ideals which contain the same multilinear elements coincide.*

Thus we can deduce that the kernels of π, π^o, π^s are generated as T-ideals by their multilinear elements. We are going to prove in fact that they are generated as T-ideals by some special identities. So let us first analyze the case of $R(n) := \cup_m R_m(n)$, $T(n) := \cup_m T_m(n)$.

The multilinear elements of degree m (in the first m variables) are contained in $R_m(n)$ and are of the form

$$\sum_{\sigma \in S_{m+1}} a_\sigma \psi_\sigma(X_1, X_2, \ldots, X_m).$$

From the theory developed we know that the map

$$\Psi : \mathbb{C}[S_{m+1}] \to R_m(n), \ \sum_{\sigma \in S_{m+1}} a_\sigma \sigma \mapsto \sum_{\sigma \in S_{m+1}} a_\sigma \psi_\sigma(X_1, X_2, \ldots, X_m)$$

has as its kernel the ideal generated by the antisymmetrizer in $n+1$ elements. Thus, the multilinear identities appear only for $m \geq n$ and for $m = n$ there is a unique identity (up to scalars). So the first step consists of identifying the identity $A_n(x_1, \ldots, x_n)$ corresponding to the antisymmetrizer

$$A_n(x_1, \ldots, x_n) := \sum_{\sigma \in S_{n+1}} \epsilon_\sigma \psi_\sigma(x_1, x_2, \ldots, x_n)$$

with ϵ_σ the sign of the permutation.

For this, recall that there is a canonical identity, homogeneous of degree n in 1 variable, called *the Cayley–Hamilton identity*. Consider the characteristic polynomial of X, $\chi_X(t) := \det(t-X) = t^n - \mathrm{tr}(X)t^{n-1} + \cdots + (-1)^n \det(X)$, we have $\chi_X(X) = 0$.

We want to interpret this as a trace identity. Remark that $\mathrm{tr}(X^i)$ is the ith Newton function in the eigenvalues of X. Hence, by Chapter 2, §1.1.3 we can interpret each coefficient of the characteristic polynomial as a well-defined polynomial in the elements $\mathrm{tr}(X^i)$.

Thus we can consider $CH_n(x) := x^n - t(x)x^{n-1} + \cdots + (-1)^n \det(x)$ as a formal element of the free algebra. If we fully polarize this element we get a multilinear trace identity $CH(x_1, \ldots, x_n)$ for $n \times n$ matrices, whose terms not containing traces arise from the polarization of x^n and are thus of the form $\sum_{\tau \in S_n} x_{\tau(1)} x_{\tau(2)} \ldots x_{\tau(n)}$.

By the uniqueness of the identities in degree n we must have that the polarized Cayley–Hamilton identity is a multiple of the identity corresponding to the antisymmetrizer. To compute the scalar we may look at the terms not containing a trace in the two identities.

Clearly $x_1x_2\dots x_n = \psi_{(1\,2\dots n\,n+1)}$ and $\epsilon_{(1\,2\dots n\,n+1)} = (-1)^n$, and thus we have finally:

Proposition. $A_n(x_1,\dots,x_n) = (-1)^n CH(x_1,\dots,x_n)$.

Example. $n = 2$ *(polarize* $P_2(x)$*)*

$$A_2 = x_1x_2 + x_2x_1 - t(x_1)x_2 - t(x_1)x_2 - t(x_1x_2) + t(x_1)t(x_2);$$

$$P_2(x) = x^2 - t(x)x + \det(x) = x^2 - t(x)x + \frac{1}{2}(t(x)^2 - t(x^2)).$$

Exercise. Using this formal description (and restitution) write combinatorial expressions for the coefficients of the characteristic polynomial of X in terms of $\mathrm{tr}(X^i)$ (i.e., expressions for elementary symmetric functions in terms of Newton power sums).

It is particularly interesting to see what the polarized form is of the determinant. The terms corresponding to the determinant correspond exactly to the sum over all the permutations which fix $n+1$. Therefore we deduce:

Corollary. *The polarized form of* $\det(X)$ *is the expression*

$$\sum_{\sigma\in S_n} \epsilon_\sigma \phi_\sigma(x_1,\dots,x_n). \tag{8.5.1}$$

Let us look at some implications for relations among traces.

According to the general principle of correspondence between elements in R and T, we deduce the trace relation

$$T_{n+1}(x_1,\dots,x_n,x_{n+1}) := \sum_{\sigma\in S_{n+1}} \epsilon_\sigma \phi_\sigma(x_1,x_2,\dots,x_n,x_{n+1})$$
$$= t(A_n(x_1,\dots,x_n)x_{n+1}).$$

Recall that

$$\phi_\sigma(x_1,x_2,\dots,x_m) = \mathrm{tr}(x_{i_1}x_{i_2}\dots x_{i_h})\,\mathrm{tr}(x_{j_1}x_{j_2}\dots x_{j_k})\dots\mathrm{tr}(x_{s_1}x_{s_2}\dots x_{s_m}).$$

Example. $(n = 2)$

$$T_3 := t(x_1x_2x_3) + t(x_2x_1x_3) - t(x_1)t(x_2x_3) - t(x_1)t(x_2x_3) - t(x_1x_2)t(x_3)$$
$$+ t(x_1)t(x_2)t(x_3).$$

Observe also that when we apply the operator of restitution to T_{n+1} we get $n!t(P_n(x)x) = T_{n+1}(x,x,\dots,x)$; the vanishing of this expression for $n\times n$ matrices is precisely the identity which expresses the $n+1$ power sum of n indeterminates as a polynomial in the lower power sums, e.g., $n = 2$:

$$t(x^3) = \tfrac{3}{2}t(x)t(x^2) - \tfrac{1}{2}t(x)^3.$$

Before we pass to the general case we should remark that any substitution map in R sends the trace ring Tr into itself. Thus it also makes sense to speak of a T-ideal in Tr. In particular we have the T-ideal of trace relations, the kernel of the evaluation map of Tr into $T(n)$. We wish to prove:

Theorem. *The T-ideal of trace relations is generated (as a T-ideal) by the trace relation T_{n+1}.*

The T-ideal of (trace) identities of $n \times n$ matrices is generated (as a T-ideal) by the Cayley–Hamilton identity.

Proof. From all the remarks made it is sufficient to prove that a multilinear trace relation (resp. identity) (in the first m variables) is in the T-ideal generated by T_{n+1} (resp. A_n).

Let us first look at trace relations. By the description of trace identities it is enough to look at the relations of the type $\Phi(\tau(\sum_{\sigma\in S_{n+1}} \epsilon_\sigma \sigma)\gamma),\ \tau, \gamma \in S_{m+1}$.

We write such a relation as $\Phi(\gamma^{-1}(\gamma\tau(\sum_{\sigma\in S_{n+1}} \epsilon_\sigma\sigma))\gamma)$. We have seen that conjugation corresponds to permutation of variables, an operation allowed in the T-ideal; thus we can assume that $\gamma = 1$.

We start with the remark ($m \geq n$):

Splitting the cycles. Every permutation τ in S_{m+1} can be written as a product $\tau = \alpha \circ \beta$ where $\beta \in S_{n+1}$ and, in each cycle of α, there is at most 1 element in $1, 2, \dots, n+1$.

This is an exercise on permutations. It is based on the observation that

$$(xxxxx\ a\ yyyyy\ b\ zzzz\ c\ \dots) = (xxxxx\ a)(yyyyy\ b)(zzzz\ c)(\dots)(abc\dots).$$

From the remark it follows that (up to a sign) we can assume that τ has the property that in each cycle of τ there is at most 1 element in $1, 2, \dots, n+1$.

Assume that τ satisfies the previous property and let $\sigma \in S_{n+1}$. The cycle decomposition of the permutation $\tau\sigma$ is obtained by formally substituting in the cycles of σ, for every element $a \in [1, \dots, n+1]$, the cycle of τ containing a as a word (written formally with a at the end) and then adding all the cycles of τ not containing elements $\leq n+1$.

If we interpret this operation in terms of the corresponding trace element $\phi_\sigma(x_1, \dots, x_{m+1})$ we see that the resulting element is the product of a trace element corresponding to the cycles of τ not containing elements $\leq n+1$ and an element obtained by substituting monomials in $\phi_\sigma(x_1, \dots, x_{m+1})$ for the variables x_i (which one reads off from the cycle decomposition of τ).

As a result we have proved that a trace relation is in the T-ideal generated by T_{n+1}.

Let us pass now to trace identities. First we remark that, by the definition of an ideal in a trace ring, the relation T_{n+1} is a consequence of Cayley–Hamilton.

A trace polynomial $f(x_1, x_2, \dots, x_m)$ is a trace identity for matrices if and only if $t(f(x_1, x_2, \dots, x_m)x_{m+1})$ is a trace relation. Thus from the previous proof we have that $t(f(x_1, x_2, \dots, x_m)x_{m+1})$ is a linear combination of elements of type

$$A\, T_{n+1}(M_1, \dots, M_{n+1}) = A\, t(A_n(M_1, \dots, M_n)M_{n+1}).$$

Now we have to consider two cases: the variable x_{m+1} appears either in A or in one of the M_i. In the first case we have

$$A\, T_{n+1}(M_1, \dots, M_{n+1}) = t(T_{n+1}(M_1, \dots, M_{n+1})Bx_{m+1}),$$

and $T_{n+1}(M_1, \dots, M_{n+1})B$ is a consequence of Cayley–Hamilton. In the second, due to the antisymmetry of T_{n+1}, we can assume that x_{m+1} appears in $M_{n+1} = Bx_{m+1}C$. Hence

$$\begin{aligned} A\, t(A_n(M_1, \dots, M_n)M_{n+1}) &= A\, t(A_n(M_1, \dots, M_n)Bx_{m+1}C) \\ &= t(CAA_n(M_1, \dots, M_n)Bx_{m+1}). \end{aligned}$$

$CAA_n(M_1, \dots, M_n)B$ is also clearly a consequence of Cayley–Hamilton. □

8.6 Polynomial Identities

We now want to discuss polynomial identities of matrices. A polynomial identity is a special type of trace identity in which no traces appear. In fact these identities were studied before the trace identities, although they turn out to be harder to describe. Under the map $\Psi : \mathbb{C}[S_{m+1}] \to R(m)$ the elements that correspond to polynomials without traces are the full cycles $(i_1, i_2, \dots, i_m, m+1)$ and

$$\Psi((i_1, i_2, \dots, i_m, m+1)) = x_{i_1}x_{i_2}\dots x_{i_m}.$$

Thus:

Theorem. *The space $M_n(m)$ of multilinear polynomial identities of $n \times n$ matrices of degree m can be identified with $P_{m+1} \cap I_n$, where P_{m+1} is the subspace of $\mathbb{C}[S_{m+1}]$ spanned by the full cycles, and I_n is the ideal of $\mathbb{C}[S_{m+1}]$ generated by an antisymmetrizer on $n+1$ elements.*

A more precise description of this space is not known. Indeed we have only asymptotic information about its dimension [RA]. One simple remark is useful. Consider the automorphism τ of the group algebra defined on the group elements by $\tau(\sigma) := \epsilon_\sigma \sigma$ (Chapter 9, §2.5). τ is multiplication by $(-1)^m$ on P_{m+1} and transforms I_n into the ideal J_n of $\mathbb{C}[S_{m+1}]$ generated by a symmetrizer on $n+1$ elements. It follows that

$$M_n(m) = P_m \cap I_n \cap J_n\,.$$

The ideal I_n is the sum of all blocks corresponding to partitions λ with height $\geq n+1$, while J_n is the sum of all blocks corresponding to partitions λ with height of $\tilde{\lambda}$

$\geq n+1$. Thus $I_n \cap J_n$ is the sum of blocks relative to partitions which contain the hook $n+1, 1^n$. It follows that $M_n(m) = 0$ if $m < 2n$.[111] As for degree $2n$, we have the famous Amitsur–Levitski identity, given by the vanishing of the *standard polynomial* S_{2n}:

$$S_{2n}(x_1, \dots, x_{2n}) = \sum_{\sigma \in S_{2n}} \epsilon_\sigma x_{\sigma(1)} \dots x_{\sigma(2n)}.$$

This can be deduced from the Cayley–Hamilton identity as follows: consider

$$\sum_{\sigma \in S_{2n}} \epsilon_\sigma CH(x_{\sigma(1)} x_{\sigma(2)}, \dots, x_{\sigma(2n-1)} x_{\sigma(2n)}).$$

The terms of CH which do not contain trace expressions give rise to a multiple of the Amitsur–Levitski identity, but the terms which contain trace expressions can be grouped so as to contain each a factor of the form $\mathrm{tr}(S_{2n}(x_1, \dots, x_{2m}))$. Now we claim that $\mathrm{tr}(S_{2n}(x_1, \dots, x_{2m})) = 0$ identically as a formal expression. This can be seen, for instance, by the fact that $\mathrm{tr}(x_{i_1} x_{i_2} \dots x_{i_{2m}}) = \mathrm{tr}(x_{i_2} \dots x_{i_{2m}} x_{i_1})$, and these two permutations have opposite sign and so cancel in the expression.

It can be actually easily seen that $M_n(2n)$ is 1-dimensional, generated by S_{2n}.

8.7 Trace Identities with Involutions

One can also deduce a second fundamental theorem for invariants of matrices and trace identities in the case of involutions. To do so, it is just a question of translating the basic determinants or Pfaffians into trace relations.

As usual we will describe the multilinear relations and deduce the general relations through polarization and restitution. In order to study the multilinear relations involving $m-1$ matrix variables X_i we may restrict to decomposable matrices $X_i := u_i \otimes v_i$. A monomial of degree $m-1$ in these variables in the formal trace algebra is of the type $\mathrm{tr}(M_1)\,\mathrm{tr}(M_2) \dots \mathrm{tr}(M_{k-1}) N_k$, where M_i, N_k are monomials in X_i, X_i^* and each index $i = 1, \dots, m-1$, appears exactly once. To this monomial we can, as usual, associate the trace monomial $\mathrm{tr}(M_1)\,\mathrm{tr}(M_2) \dots \mathrm{tr}(M_{k-1})\,\mathrm{tr}(N_k X_m)$. Extending from monomials to linear combinations, a multilinear trace identity in $m-1$-variables corresponds to a multilinear trace relation in m variables.

Let us substitute in a given monomial, for the matrices X_i, $i = 1, \dots, m$, their values $u_i \otimes v_i$ and expand the monomial according to the rules in 8.2.

To each trace monomial we associate a product $\prod_{i=1}^m \langle x_i, y_i \rangle$, where the list of variables $x_1, \dots, x_m, y_1, \dots, y_m$ is just a reordering of $u_1, \dots, u_m, v_1, \dots, v_m$. Conversely, given such a product, it is easily seen that it comes from a unique trace monomial (in this correspondence one takes care of the formal identities between expressions).

For us it may be convenient to consider the scalar products between the u_i, v_j as evaluations of variables $x_{ij} = (i|j)$ which satisfy the symmetry condition $(i|j) =$

[111] This fact can be proved directly.

$\epsilon(j|i)$, with $\epsilon = 1$ in the orthogonal case and -1 in the symplectic case. We consider the variables $(i|j)$ as entries of an ϵ-symmetric $2m \times 2m$ matrix.

Consider next the space M_m spanned by the *multilinear monomials* $x_{i_1 i_2} x_{i_3 i_4} \dots x_{i_{2m} i_{2m}} = (i_1|i_2)(i_3|i_4)\dots(i_{2m}|i_{2m})$, where $i_1, i_2, i_3, i_4, \dots, i_{2m}, i_{2m}$ is a permutation of $1, 2, \dots, 2m$.

With a different notation[112] this space has been studied in §6.2 as a representation of S_{2m}.

We have a *coding map* from M_m to the vector space of multilinear pure trace expressions $f(X_1, \dots, X_{m+1})$ of degree $m+1$ in the variables $X_i,\ X_i^*$ or the noncommutative trace expressions $g(x_1, \dots, X_m)$ of degree m, with $f(X_1, \dots, X_{m+1}) = \operatorname{tr}(g(X_1, \dots, X_m)X_{m+1})$.

It is given formally by considering the numbers i as vector symbols, the basis of a free module F over the polynomial ring in the variables $(i|j)$, $X_i := (2i-1) \otimes 2i \in F \otimes F \subset \operatorname{End}(F)$, and $(i|j)$ as the scalar product $\langle i|j \rangle$, with the multiplication, trace and adjoint as in 3.1.1., deduced by the formulas:

$$i \otimes j \circ h \otimes k = i \otimes (j|h)k, \quad \operatorname{tr}(i \otimes j) = \epsilon\,(i|j), \quad (i \otimes j)^* = \epsilon\, j \otimes i.$$

For instance, for $m = 2$ we have 15 monomials. We list a few of them for $\epsilon = -1$:

$$(1|2)(3|4)(5|6) = -\operatorname{tr}(X_1)\operatorname{tr}(X_2)\operatorname{tr}(X_3),$$

$$(1|2)(3|5)(4|6) = -\operatorname{tr}(X_1)\operatorname{tr}(X_2^* X_3),$$

$$(1|2)(3|6)(4|5) = \operatorname{tr}(X_1)\operatorname{tr}(X_2 X_3),$$

$$(1|3)(2|5)(4|6) = -\operatorname{tr}(X_1^* X_2 X_3^*) = -\operatorname{tr}(X_3 X_2^* X_1)$$

and in noncommutative form:

$$(1|2)(3|4)(5|6) = -\operatorname{tr}(X_1)\operatorname{tr}(X_2), \qquad (12)(35)(46) = -\operatorname{tr}(X_1)X_2^*,$$

$$(1|2)(3|6)(4|5) = \operatorname{tr}(X_1)X_2, \qquad (1|3)(2|5)(4|6) = -X_2^* X_1.$$

Under this coding map the subspace of relations of invariants corresponds to the trace identities. Thus one can reinterpret the second fundamental theorem as giving basic identities from which all the others can be deduced.

The reinterpretation is particularly simple for the symplectic group. In this case the analogue of the Cayley–Hamilton identity is a *characteristic Pfaffian* equation.

Thus consider M_{2n} the space of $2n \times 2n$ matrices, $J := \begin{vmatrix} 0 & 1_n \\ -1_n & 0 \end{vmatrix}$ the standard matrix of the symplectic involution and $A^* := -JA^tJ$ the symplectic involution.

Let M_{2n}^+ denote the space of symmetric matrices under this involution. Under the map $A \to B := AJ$ we identify M_{2n}^+ with the space of skew-symmetric matrices, and we can define the *characteristic Pfaffian* as

112 We called it $P_+^{2n},\ P_-^{2n}$.

$$(8.7.1) \qquad Pf_A(\lambda) := Pf((\lambda I - A)J).$$

We leave it to the reader to verify that $Pf_A(A) = 0$.

Hint. A generic symmetric matrix is semisimple and one can decompose the $2n$-dimensional symplectic space into an orthogonal sum of stable 2-dimensional subspaces. On each such subspace A is a scalar.

If we polarize the identity $Pf_A(A) = 0$, we obtain a multilinear identity $P(X_1, \dots, X_n)$ satisfied by symmetric symplectic matrices. In another form we may write this identity as an identity $P(X_1 + X_1^*, \dots, X_n + X_n^*)$ satisfied by all matrices. This is an identity in degree n. On the other hand, there is a unique canonical pure trace identity of degree $n + 1$ which, under the coding map, comes from the Pfaffian of a $(2n + 2) \times (2n + 2)$ matrix of skew indeterminates; thus, up to a scalar, the polarized characteristic Pfaffian must be a multiple of the noncommutative identity coded by that Pfaffian. For instance, in the trivial case $n = 1$ we get the identity $\mathrm{tr}(X_1) = X_1 + X_1^*$. As for the Cayley–Hamilton one may prove:

Theorem. *All identities for matrices with symplectic involution are a consequence of $P(X_1 + X_1^*, \dots, X_n + X_n^*)$.*

Proof. We will give the proof for the corresponding trace relations. Consider the matrix of (skew) scalar products:

$$S := \begin{vmatrix} 0 & [u_1, u_2] & \dots & [u_1, u_m] & [u_1, v_1] & [u_1, v_2] & \dots & [u_1, v_m] \\ [u_2, u_1] & 0 & \dots & [u_2, u_m] & [u_2, v_1] & [u_1, v_2] & \dots & [u_2, v_m] \\ \dots & \dots & \dots & \dots & \dots & \dots & \dots & \dots \\ \dots & \dots & \dots & \dots & \dots & \dots & \dots & \dots \\ [u_m, u_1] & [u_m, u_2] & \dots & 0 & [u_m, v_1] & [u_m, v_2] & \dots & [u_m, v_m] \\ [v_1, u_1] & [v_1, u_2] & \dots & [v_1, u_m] & 0 & [v_1, v_2] & \dots & [v_1, v_m] \\ [v_2, u_1] & [v_2, u_2] & \dots & [v_2, u_m] & [v_2, v_1] & 0 & \dots & [v_2, v_m] \\ \dots & \dots & \dots & \dots & \dots & \dots & \dots & \dots \\ \dots & \dots & \dots & \dots & \dots & \dots & \dots & \dots \\ [v_m, u_1] & [v_m, u_2] & \dots & [v_m, u_m] & [v_m, v_1] & [v_m, v_2] & \dots & 0 \end{vmatrix}.$$

A typical identity in $m - 1$ variables corresponds to a polynomial

$$[w_1, w_2, \dots, w_{2n+2}][x_1, y_1] \dots [x_s, y_s]$$

of the following type. $m = n+1+s$, the elements $w_1, w_2, \dots, w_{2n+2}, x_1, y_1, \dots, x_s, y_s$ are given by a permutation of $u_1, \dots, u_m, v_1, \dots, v_m$ and $[w_1, w_2, \dots, w_{2n+2}]$ denotes the Pfaffian of the principal minor of S whose row and column indices correspond to the symbols $w_1, w_2, \dots, w_{2n+2}$. To understand the corresponding trace relation we argue as follows:

1. Permuting the variables X_i corresponds to permuting simultaneously the u_i and v_i with the same permutation.
2. Exchanging X_i with X_i^* corresponds to exchanging u_i with v_i. Performing these operations we can reduce our expression to one of type

$$[u_1, \dots, u_k, v_1, \dots, v_k, v_{k+1}, \dots, v_{2n+2-k}][x_1, y_1] \dots [x_s, y_s].$$

If $k = n+1$, the Pfaffian $[u_1, \dots, u_{n+1}, v_1, \dots, v_{n+1}]$ corresponds to the identity of the characteristic Pfaffian. The given expression is a product of this relation by trace monomials. Otherwise we look at the symbol u_{k+1} which does not appear in the Pfaffian but appears in one of the terms $[x_i, y_i]$. We have that u_{k+1} is paired with either a term u_i or a term v_j. If u_{k+1} is paired with v_j, let us use the formula

$$X_j X_{k+1} = u_j \otimes [v_j, u_{k+1}] v_{k+1}.$$

If u_{k+1} is paired with u_j, we use the formula

$$X_j^* X_{k+1} = -v_j \otimes [u_j, u_{k+1}] v_{k+1}.$$

In the first case we introduce a new variable $\overline{X}_j := X_j X_{k+1}$; in the second $\overline{X}_j := X_j^* X_{k+1}$

$$\overline{X}_j := \overline{u}_j \otimes \overline{v}_j, \quad \overline{u}_j := u_j, \ \overline{v}_j := [v_j, u_{k+1}] v_{k+1},$$

or

$$\overline{u}_j := -v_j, \quad \overline{v}_j := [u_j, u_{k+1}] v_{k+1}.$$

Let us analyze for instance the case $[u_j, u_{k+1}]$, the other being similar. We substitute $[v_j, u_{k+1}]$ inside the Pfaffian:

$$\begin{aligned}
&[u_1, \dots, u_k, v_1, \dots, v_k, v_{k+1}, \dots, v_{2n+2-k}][u_j, u_{k+1}] \\
&= [u_1, \dots, u_k, v_1, \dots, v_k, [u_j, u_{k+1}] v_{k+1}, \dots, v_{2n+2-k}] \\
&= [u_1, \dots, u_k, v_1, \dots, v_k, \overline{v}_j, \dots, v_{2n+2-k}].
\end{aligned}$$

We obtain a new formal expression in $m-1$-variables. The variable X_{k+1} has been suppressed, while the variable X_j has been substituted with a new variable $\overline{X}_j$. In order to recover the old m-variable expression from the new one, one has to substitute $\overline{X}_j$ with $X_j X_{k+1}$ or $X_j^* X_{k+1}$. By induction the theorem is proved. □

8.8 The Orthogonal Case

The orthogonal case is more complicated, due to the more complicated form of the second fundamental theorem for orthogonal invariants.

Consider the matrix of (symmetric) scalar products:

$$S := \begin{vmatrix} (u_1,u_1) & (u_1,u_2) & \dots & (u_1,u_m) & (u_1,v_1) & (u_1,v_2) & \dots & (u_1,v_m) \\ (u_2,u_1) & (u_2,u_2) & \dots & (u_2,u_m) & (u_2,v_1) & (u_1,v_2) & \dots & (u_2,v_m) \\ \dots & \dots & \dots & \dots & \dots & \dots & \dots & \dots \\ \dots & \dots & \dots & \dots & \dots & \dots & \dots & \dots \\ (u_m,u_1) & (u_m,u_2) & \dots & (u_m,u_m) & (u_m,v_1) & (u_m,v_2) & \dots & (u_m,v_m) \\ (v_1,u_1) & (v_1,u_2) & \dots & (v_1,u_m) & (v_1,v_1) & (v_1,v_2) & \dots & (v_1,v_m) \\ (v_2,u_1) & (v_2,u_2) & \dots & (v_2,u_m) & (v_2,v_1) & (v_2,v_2) & \dots & (v_2,v_m) \\ \dots & \dots & \dots & \dots & \dots & \dots & \dots & \dots \\ \dots & \dots & \dots & \dots & \dots & \dots & \dots & \dots \\ (v_m,u_1) & (v_m,u_2) & \dots & (v_m,u_m) & (v_m,v_1) & (v_m,v_2) & \dots & (v_m,v_m) \end{vmatrix}$$

A relation may be given by the determinant of any $(n+1) \times (n+1)$ minor times a monomial. Again we have that:

1. Permuting the variables X_i corresponds to permuting simultaneously the u_i and v_i with the same permutation.
2. Exchanging X_i with X_i^* corresponds to exchanging u_i with v_i.

In this case the basic identities to consider correspond to the determinants of scalar products, which we may denote by

$$(u_{i_1}u_{i_2}\dots u_{i_k}v_{j_1}v_{j_2}\dots v_{j_{n+1-k}} \mid u_{i_{k+1}}u_{i_{k+2}}\dots u_{i_{n+1}}v_{j_{n+2-k}}v_{j_{n+3-k}}\dots v_{j_{n+1}}),$$

where the condition that they represent a multilinear relation in the matrix variables $X_1, \dots, X_{n+1}$ means that the indices $i_1, \dots, i_{n+1}$ and also the indices $j_1, \dots, j_{n+1}$ are permutations of $1, 2, \dots, n+1$.

Using the previous symmetry laws we may first exchange all u_i, v_i for all the indices for which u_i appears on the left but v_i does not appear. After reordering the indices we end up with a determinant

$$(u_1u_2\dots u_kv_1v_2\dots v_{n+1-k} \mid u_{k+1}u_{k+2}\dots u_{n+1}v_{n+2-k}v_{n+3-k}\dots v_{n+1}),$$
$$n+1-k \geq k.$$

From the coding map, the above determinant corresponds to some trace relation which we denote by $F_k(X_1, \dots, X_{n+1})$. Unlike the symplectic case these basic identities do not have enough symmetry to consider them as polarizations of 1-variable identities.

We have, exchanging u_i, v_i,

$$F_k(X_1, \dots, X_i, \dots, X_{n+1})$$
$$\text{(a)} \qquad = -F_k(X_1, \dots, X_i^t, \dots, X_{n+1}), \text{ if } \quad i > n+k-1, \text{ or } i \leq k.$$

If σ is a permutation of $1, 2, \dots, n+1$ which preserves the subsets $(1, 2, \dots, k)$; $(k+1, \dots, n+1-k)$; $(n_1-k+1, \dots, n+1)$ we have

$$\text{(b)} \qquad F_k(X_1, \dots, X_i, \dots, X_{n+1}) = F_k(X_{\sigma(1)}, \dots, X_{\sigma(i)}, \dots, X_{\sigma(n+1)}).$$

This means that F_k can be viewed as polarization of an identity in 3 variables $F_k(X, Y, Z)$. Moreover, from the first relation, the first and third variable can be taken to be skew-symmetric variables $\frac{X-X^t}{2}, \frac{Z-Z^t}{2}$.

Finally, exchanging all u_i, v_i and transposing the determinant we get

$$\begin{aligned}
&(v_{k+1}v_{k+2}\dots v_{n+1}u_{n+2-k}u_{n+3-k}\dots u_{n+1} \mid v_1v_2\dots v_ku_1u_2\dots u_{n+1-k}) \\
&= (u_{n+1}u_{n+3-k}\dots u_{n+2-k}v_{n+1}\dots v_{k+2}v_{k+1} \mid u_{n+1-k}\dots u_2u_1v_k\dots v_2v_1) \\
\text{(c)}\quad &= F_k(X_1^t, \dots, X_i^t, \dots, X_{n+1}^t) = F_k(X_{n+1}, \dots, X_{n+1-i}, \dots, X_1).
\end{aligned}$$

We need to make a remark when we pass from the trace relation to the noncommutative trace relation. We pick a variable X_i and write

$$F_k(X_1, \dots, X_i, \dots, X_{n+1}) = \operatorname{tr}(G_k^i(X_1, \dots, \check{X}_i, \dots, X_{n+1})X_i).$$

It is clear that G_k^i depends on i only up to the symmetry embedded in the previous statements (a), (b), (c).

From the symmetry (b) it appears that we obtain 3 inequivalent identities G_k^1, G_k^2, G_k^3 from the 3 subsets. From (c) we obtain a different deduction formula from the identity in the first and the third subset.

$$F_k(X_1, \dots, X_{n+1}) = \operatorname{tr}(G_k^1(X_2, \dots, X_{n+1})X_1),$$

gives

$$\begin{aligned}
F_k(X_1^t, \dots, X_{n+1}^t) &= \operatorname{tr}(G_k^1(X_2^t, \dots, X_{n+1}^t)X_1^t) \\
&= \operatorname{tr}(G_k^1(X_2^t, \dots, X_{n+1}^t)^tX_1) = \operatorname{tr}(G_k^3(X_{n+1}, \dots, X_2)X_1).
\end{aligned}$$

Hence

$$G_k^1(X_2^t, \dots, X_{n+1}^t)^t = G_k^3(X_{n+1}, \dots, X_2).$$

In this sense only G_k^1 and G_k^2 should be taken as generating identities.

8.9 Some Estimates

Both the first and second fundamental theorem for matrices are not as precise as the corresponding theorems for vectors and forms. Several things are lacking; the first is a description of a minimal set of generators for the trace invariants. At the moment the best result is given by the following:

Theorem (Razmyslov). *The invariants of $n \times n$ matrices are generated by the elements* $\operatorname{tr}(M)$ *where M is a monomial of degree $\leq n^2$.*

Proof. The precise question is: for which values of m, can we express the trace monomial $\operatorname{tr}(X_1X_2\dots X_{m+1})$ as a product of shorter trace monomials? This in turn is equivalent to asking for which values of m, in the free algebra with trace $R(n)$, the monomial $X_1X_2\dots X_m$ lies in the ideal $T(n)^+R(n)$, where as usual $T(n)^+$ denotes the trace invariants without constant term.

If F denotes the free algebra with trace and T the trace expressions, by the general correspondence between the symmetric group and the trace expressions, we have that the elements of S_{m+1} which map to T^+F are not full cycles. Let Q_{m+1} denote the subspace that they span in $\mathbb{C}[S_{m+1}]$. Let P_{m+1} be the subspace of $\mathbb{C}[S_{m+1}]$ spanned by the full cycles so that $Q_{m+1} \oplus P_{m+1} = \mathbb{C}[S_{m+1}]$. By symmetry, the requirement that modulo the identities of $n \times n$ matrices we have $X_1X_2\dots X_m \in T(n)^+R(n)$ is equivalent to requiring that $P_{m+1} \subset Q_{m+1}+I_n$; in other words, that $\mathbb{C}[S_{m+1}] = Q_{m+1}+I_n$.

So let us prove that this equality is true for $m = n^2$. Consider the automorphism $\tau(\sigma) = \epsilon_\sigma \sigma$. τ is $(-1)^m$ on P_{m+1}. It preserves Q_{m+1} and maps I_n to J_n the ideal generated by a symmetrizer on $n+1$ elements. Therefore $\tau(Q_{m+1}+I_n) = Q_{m+1}+J_n$ so that $Q_{m+1} + I_n = Q_{m+1} + I_n + J_n$. Finally, if $m \geq n^2$, every diagram with $m+1$ cases belongs either to I_n or to J_n so that $S_{n^2+1} = I_n + J_n$, and the theorem is proved. □

If one computes explicitly for small values of n, one realizes that n^2 does not appear to be the best possible estimate. A better possible estimate should be $\binom{n+1}{2}$. It is interesting that these estimates are related to a theorem in noncommutative algebra known as the Dubnov–Ivanov–Nagata–Higman Theorem.

8.10 Free Nil Algebras

To explain the noncommutative algebra we start with a definition:

Definition. (1) An algebra R (without 1) is said to be nil of exponent n if for every $r \in R$ we have $r^n = 0$

(2) An algebra R (without 1) is said to be nilpotent of exponent m if $R^m = 0$.

To say that R is nilpotent of exponent m means that every product $r_1r_2\dots r_m = 0$, $r_i \in R$. Of course an algebra R which is nilpotent of exponent m is also nil of exponent m. The converse is not true.

Theorem 1.

(i) *The algebra $R(n)^+/T(n)^+R(n)$ is the free nil algebra of exponent n. In other words, it is the free algebra without* 1, *modulo the T-ideal generated by the polynomial identity $z^n = 0$.*

(ii) *In characteristic* 0 *an algebra R, nil of exponent n, is nilpotent of exponent $\leq n^2$.*

Proof. (i) In the free algebra with trace without 1 we impose the identity $t(z) = 0$, that is the trace of every element is 0. We obtain the free algebra without trace. Thus the free algebra without 1 modulo the polynomial identity $z^n = 0$ can be seen as the free algebra with trace without 1 modulo the two identities $t(z) = z^n = 0$. Now the Cayley–Hamilton identity plus the identity $t(z) = 0$ becomes $z^n = 0$. Hence the free algebra without one modulo the polynomial identity $z^n = 0$ can be seen also as the free algebra with trace without 1 modulo the two identities $t(z) = CH_n(z) = 0$.

But clearly this last algebra is $R(n)^+/T(n)^+R(n)$.

(ii) To say that $X_1X_2\dots X_m$ lies in the ideal $T(n)^+R(n)$ is the same as saying that $X_1X_2\dots X_m = 0$ in $R(n)^+/T(n)^+R(n)$. Since this is the free algebra modulo $z^n = 0$, to say that $X_1X_2\dots X_m = 0$ is equivalent to saying that $X_1X_2\dots X_m = 0$ is an identity in $R(n)^+/T(n)^+R(n)$ or that $R(n)^+/T(n)^+R(n)$ is nilpotent of exponent $\leq m$.

If the free algebra modulo $z^n = 0$ is nilpotent of exponent m, then so is every nil algebra of exponent n, since it is a quotient of the free algebra. Finally, from the previous theorem we know that $m = n^2$ satisfies the condition that $X_1X_2\dots X_m$ lies in the ideal $T(n)^+R(n)$. □

Remark. We have in fact proved, for fixed n, (over fields of characteristic 0) that for an integer m the following conditions are equivalent:

(i) $\mathbb{C}[S_{m+1}] = Q_{m+1} + I_n$.
(ii) The invariants of $n \times n$ matrices are generated by the elements $\mathrm{tr}(M)$ where M is a monomial of degree $\leq m$.
(iii) An algebra R that is nil of exponent n is nilpotent of exponent $\leq m$.

We have shown that there is a minimum value m_0 for which these three conditions are satisfied and $m_0 \leq n^2$.

It is known that the minimum value of m is $\geq \binom{n+1}{2}$. One may conjecture that it is exactly $\binom{n+1}{2}$. This is true by explicit computations for $n \leq 5$.

One can also interpret Theorem 8.7 on symplectic involution in a similar way and obtain:

Theorem 2.

(i) The algebra $R^s(2n)^+/T^s(2n)^+R^s(2n)$ is the free nil algebra with involution in which every symmetric element is nilpotent of exponent n; in other words, it is the free algebra without 1 *with involution modulo the polynomial identity* $(z + z^*)^n = 0$.

(ii) In characteristic 0 *if an algebra R with involution satisfies $(z + z^*)^n = 0$, then it satisfies $z^{2n} = 0$.*

Proof. The first part is like the case of nil algebras. For the second part, it is enough to remark that $2n \times 2n$ symplectic matrices satisfy the Cayley–Hamilton theorem in degree $2n$, which then can be deduced from the Pfaffian identity. □

8.11 Cohomology

We can apply the theory developed to the computation of the cohomology of the classical Lie algebras. Given a semisimple Lie algebra L, we apply Cartan's theorem that the cohomology space $H^i(L, \mathbb{C})$ with constant coefficients can be identified to the space of invariant multilinear and alternating functions of k variables in L. These spaces are the summands of a graded algebra under $\wedge$ product.

Let us start thus from $sl(n, \mathbb{C})$, and even start computing the space of invariant multilinear and alternating functions of k matrix variables X_i (we restrict to trace 0 only later).

By definition one can obtain such a function by alternating any given multilinear function $f(x_1, \dots, x_k)$. Let us call $Af(x_1, \dots, x_k) := \frac{1}{k!}\sum_{\sigma \in S_k} \epsilon_\sigma f(x_{\sigma(1)}, \dots, x_{\sigma(k)})$. Moreover, if a multilinear function is a product

$$h(x_1, \dots, x_{a+b}) = f(x_1, \dots, x_a) g(x_{a+1}, \dots, x_{a+b}),$$

one has by definition that $Ah = Af \wedge Ag$.

We want to apply these to invariant functions of matrices. A multilinear invariant function of matrices is up to permuting the variables a product of functions

$$\mathrm{tr}(X_1 X_2 \dots X_a)\,\mathrm{tr}(X_{a+1} \dots X_{a+b}) \dots tr(X_s X_s{+}1 \dots X_{s+t}).$$

Therefore we see:

Lemma. *The algebra of multilinear and alternating functions of $n \times n$ matrices is generated (under $\wedge$) by the elements* $\mathrm{tr}(S_{2h+1}(X_1, \dots, X_{2h+1}))$, $h = 1, \dots, n-1$. S_{2h+1} *is the standard polynomial.*

Proof. When we alternate one of the factors $\mathrm{tr}(X_1 X_2 \dots X_a)$ we then obtain $\mathrm{tr}(S_a(X_1, \dots, X_a))$. We have noticed in 8.6 that this function is 0 if a is even, and it vanishes on $n \times n$ matrices if $a \geq 2n$ by the Amitsur–Levitski theorem. The claim follows. □

Theorem. *The algebra $A(n)$ of multilinear and alternating functions of $n \times n$ matrices is the exterior algebra $\wedge(c_1, \dots, c_{2n-1})$ generated by the elements*

$$c_h := \mathrm{tr}(S_{2h+1}(X_1, \dots, X_{2h+1})), \ h = 0, \dots, n-1, \ \textit{of degree } 2h+1.$$

Proof. We have seen that these elements generate the given algebra. Moreover, since they are all of odd degree they satisfy the basic relations $c_h \wedge c_k = -c_k \wedge c_h$. Thus we have a surjective homomorphism $\pi : \bigwedge(c_1, \dots, c_{2n-1}) \to A(n)$. In order to prove the claim, since every ideal of $\bigwedge(c_1, \dots, c_{2n-1})$ contains the element $c_1 \wedge \dots \wedge c_{2n-1}$, it is enough to see that this element does not map to 0 under π. If not, we would have that in $A(n)$ there are no nonzero elements of degree $\sum_{h=1}^{n}(2h-1) = n^2$. Now the determinant $X_1 \wedge X_2 \wedge \dots \wedge X_{n^2}$ of n^2 matrix variables is clearly invariant alternating, of degree n^2, hence the claim. □

In fact the elements c_i have an important property which comes from the fact that the cohomology $H^*(G, \mathbb{R})$ of a group G has extra structure. From the group properties and the properties of cohomology it follows that $H^*(G, \mathbb{R})$ is also a Hopf algebra. The theorem of Hopf, of which we mentioned the generalization by Milnor–Moore in Chapter 8, implies that $H^*(G, \mathbb{R})$ is an exterior algebra generated by the primitive elements (cf. Chapter 8, §7). The comultiplication Δ in our case can be viewed as follows, we map a matrix X_i to $X_i \otimes 1 + 1 \otimes X_i$, and then

$$\Delta \operatorname{tr}(S_{2h+1}(X_1, \dots, X_{2h+1}))$$
$$= \operatorname{tr}(S_{2h+1}(X_1 \otimes 1 + 1 \otimes X_1, \dots, X_{2h+1} \otimes 1 + 1 \otimes X_{2h+1})).$$

When we expand the expression

$$\operatorname{tr}(S_{2h+1}(X_1 \otimes 1 + 1 \otimes X_1, \dots, X_{2h+1} \otimes 1 + 1 \otimes X_{2h+1}))$$

we obtain a sum:

$$\operatorname{tr}(S_{2h+1}(X_1, \dots, X_{2h+1})) \otimes 1 + 1 \otimes \operatorname{tr}(S_{2h+1}(X_1, \dots, X_{2h+1}))$$
$$+ \sum_{a+b=2h-1} c_a \operatorname{tr}(S_a(X_{i_1}, \dots, X_{i_a})) \otimes \operatorname{tr}(S_b(X_{j_1}, \dots, X_{j_b})).$$

The first two terms give $c_h \otimes 1 + 1 \otimes c_h$; the other terms with $a, b > 0$ vanish since either a or b is even. We thus see that the c_i are primitive.

In a similar way one can treat the symplectic and odd orthogonal Lie algebras. We only remark that in the odd case, the element -1 acts trivially on the Lie algebra, hence instead of $SO(2n+1, \mathbb{C})$-invariants we can work with $O(2n+1, \mathbb{C})$-invariants, for which we have the formulas $\operatorname{tr}(M)$ with M a monomial in X_i, X_i^*. Since the variables are antisymmetric the variables X_i^* disappear and we have to consider the same expressions c_h. In this case we have a further constraint. Applying the involution we have

$$\begin{aligned}\operatorname{tr}(S_{2h+1}(X_1, \dots, X_{2h+1})) &= \operatorname{tr}(S_{2h+1}(X_1, \dots, X_{2h+1})^*) \\ &= \operatorname{tr}(-S_{2h+1}(X_{2h+1}, \dots, X_1)) \\ &= \operatorname{tr}(-(-1)^h S_{2h+1}(X_1, \dots, X_{2h+1})).\end{aligned}$$

Therefore $c_h = 0$ unless h is odd. We deduce that for $Sp(2n, \mathbb{C})$, $SO(2n+1, \mathbb{C})$ the corresponding cohomology is an exterior algebra in the primitive elements

$$d_h := \operatorname{tr}(S_{4h+3}(X_1, \dots, X_{4h+3})), \ h = 0, \dots, n-1, \ \text{ of degree } 4h+3.$$

The even case $SO(2n, \mathbb{C})$ is more complicated, since in this case we really need to compute with $SO(2n, \mathbb{C})$-invariants.

It can be proved, and we leave it to the reader, that besides the elements $d_h := \operatorname{tr}(S_{4h+3}(X_1, \dots, X_{4h+3}))$, $h = 0, \dots, n-2$ of degree $4h+3$, we also have an element of degree $2n-1$ which can be obtained by antisymmetrizing the invariant $Q(X_1, X_2X_3, X_4X_5, \dots, X_{2n-2}X_{2n-1})$ (cf. 8.2). In order to understand this new invariant, and why we construct it in this way, let us recall the discussion of §8.7. Let $J = \begin{vmatrix} 0 & 1_n \\ -1_n & 0 \end{vmatrix}$. We have seen that if X is skew-symmetric, XJ is symplectic-symmetric. Generically XJ has n-distinct eigenvalues $\lambda_1, \dots, \lambda_n$, each counted with multiplicity 2. Then $Pf(X) = \prod_i \lambda_i$. To see this in a concrete way, let Λ be the diagonal $n \times n$ matrix with the λ_i as eigenvalues. Consider the matrix $X_\Lambda := \begin{vmatrix} 0 & \Lambda \\ -\Lambda & 0 \end{vmatrix}$.

Then $X_\Lambda J = \begin{vmatrix} -\Lambda & 0 \\ 0 & -\Lambda \end{vmatrix}$. Finally $Pf(X) = \det(X) = \prod_i \lambda_i$.

Next observe that

$$\operatorname{tr}((XJ)^h) = 2(-1)^h \sum_{i=1}^{n} \lambda_i^h.$$

We deduce that if $P_n(t_1, \dots, t_n)$ is the polynomial expressing the nth elementary symmetric function in terms of the Newton functions, we have

$$Pf(X) = P_n(\operatorname{Tr}(-XJ)/2, tr((-XJ)^2/2, \dots, \operatorname{tr}((-XJ)^n/2).$$

It is important to polarize this identity. Recall that in Corollary 8.5.1 we have seen that the polarization of the determinant is the expression

$$\sum_{\sigma \in S_n} \epsilon_\sigma \phi_\sigma(x_1, \dots, x_n).$$

We deduce that the polarization of $Pf(X)$ is the expression

$$(8.11.1) \qquad Q(X_1, \dots, X_n) = \sum_{\sigma \in S_n} \epsilon_\sigma 2^{-|\sigma|} \phi_\sigma(-X_1 J, \dots, -X_n J),$$

where we denote by $|\sigma|$ the number of cycles into which σ decomposes. As in the proof of the Amitsur–Levitski theorem we see that when we perform the antisymmetrization of $Q(X_1, X_2 X_3, \dots, X_{2n-2} X_{2n})$ only the terms corresponding to full cycles survive, and we obtain $\operatorname{tr}(S_{2n-1}(-X_1 J, \dots, -X_{2n-1} J))$. It is easy to see, as we did before, that $\operatorname{tr}(S_{2n-1}(-X_1 J, \dots, -X_{2n-1} J))$ is a primitive element. One can now take advantage of the following easy fact about Hopf algebras:

Proposition. *In a graded Hopf algebra, linearly independent primitive elements generate an exterior algebra.*

We have thus found primitive elements for the cohomology of $so(2n, \mathbb{C})$ of degrees $2n - 1$, $4h + 3$, $h = 0, \dots, n-2$. Their product is a nonzero element of degree $2n^2 - n = \dim so(2n, \mathbb{C})$. Therefore there cannot be any other primitive elements since there are no skew-symmetric invariants in degree $> \dim so(2n, \mathbb{C})$. We have thus completed a description of cohomology and primitive generators.

Remark. A not-so-explicit description of cohomology can be given for any simple Lie algebra (Borel). We have generators in degrees $2h_i - 1$ where the numbers h_i, called *exponents*, are the degrees of the generators of the polynomials, invariant under W in the reflection representation.

9 The Analytic Approach to Weyl's Character Formula

9.1 Weyl's Integration Formula

The analytic approach is based on the idea of applying the orthogonality relations for class functions in the compact case and obtaining directly a formula for the characters.

We illustrate this for the unitary group $U(n, \mathbb{C})$, leaving some details to the reader for the general case of a compact Lie group (see [A]).

The method is based on Weyl's integration formula for class functions.

Let $d\mu$, $d\tau$, respectively, denote the normalized Haar measures on $U(n,\mathbb{C})$ and on T, the diagonal unitary matrices. We have

Theorem (Weyl's Integration Formula). *For a class function f on $U(n,\mathbb{C})$ we have*

$$\int_{U(n,\mathbb{C})} f(g)d\mu = \frac{1}{n!}\int_T f(t)V(t)\overline{V}(t)d\tau, \tag{9.1.1}$$

where $V(t_1,\dots,t_n) := \prod_{i<j}(t_i - t_j)$ is the Vandermonde determinant.

Assuming this formula for a moment we have:

Corollary. *The Schur functions $S_\lambda(y)$ are irreducible characters.*

Proof. By Weyl's formula, we compute

$$\frac{1}{n!}\int_T S_\lambda(y)\overline{S}_\mu(y)V(y)\overline{V}(y)d\tau = \frac{1}{n!}\int_T A_\lambda(y)\overline{A}_\mu(y)d\tau = \delta_{\lambda,\mu},$$

where the last equality follows from the usual orthogonality of characters for the torus. It follows that the class functions on $U(n,\mathbb{C})$, which restricted to T give the Schur functions, are orthonormal with respect to Haar measure.

The irreducible characters restricted to T are symmetric functions which are sums of monomials with positive integer coefficients, and we have proved (Chapter 10, §6.7) that they span the symmetric functions. It follows that the Schur functions are $\pm$ irreducible characters. The sign must be plus since their leading coefficient is 1.

By the description of the ring of symmetric functions with the function $e_n = \prod y_i$ inverted, it follows immediately that the characters $S_\lambda e_n^k$, $ht(\lambda) < n$, $k \in \mathbb{Z}$, are a basis of this ring, and so they exhaust all possible irreducible characters. □

Let us explain the proof of 9.1.1. The idea is roughly the following. Decompose the unitary group into its conjugacy classes. Two unitary matrices are conjugate if and only if they have the same eigenvalues; hence each conjugacy class intersects the torus in an orbit under S_n. Generically, the eigenvalues are distinct and a conjugacy class intersects the torus in $n!$ points. Therefore if we use the torus T as a parameter space for conjugacy classes we are counting each class $n!$ times. Now perform the integral by first integrating on each conjugacy class the function $f(g)$, which now is a constant function, then on the set of conjugacy classes, or rather on T, dividing by $n!$.

If we keep track of this procedure correctly we see that the various conjugacy classes do not all have the same volume and the factor $V(t)\overline{V}(t)$ arises in this way.

Remark that given an element $t \in T$, its conjugacy class can be identified with the set $U(n,\mathbb{C})/Z_t$ where $Z_t = \{g \in U(n,\mathbb{C}) \mid gt = tg\}$. If t has distinct eigenvalues we have $Z_t = T$, and so the generic conjugacy classes can be identified with

$U(n,\mathbb{C})/T$. In any case we have a global mapping $\pi : U(n,\mathbb{C})/T \times T \to U(n,\mathbb{C})$ given by $\pi(gT, y) := gyg^{-1}$.

Under this mapping, given a class function $f(g)$ on $U(n,\mathbb{C})$, we have that $f(\pi(gT, y))$ depends only on the coordinate y and not on gT.

Now the steps that we previously described are precisely the following:

1. $U(n,\mathbb{C})/T$ is a compact manifold over which $U(n,\mathbb{C})$ acts differentiably and transitively, that is, it is a homogeneous space. $U(n,\mathbb{C})/T$ has a measure invariant under the action of $U(n,\mathbb{C})$, which we normalize to have volume 1 and still call a Haar measure.
2. Consider the open sets in T^0, $U(n,\mathbb{C})^0$ made of elements with distinct eigenvalues in T, $U(n,\mathbb{C})$, respectively.
   (i) The complements of these sets have measure 0.
   (ii) π induces a mapping $\pi^0 : U(n,\mathbb{C})/T \times T^0 \to U(n,\mathbb{C})^0$ which is an unramified covering with exactly $n!$ sheets.
3. Let $d\mu$ denote the Haar measure on $U(n,\mathbb{C})$ and dv, $d\tau$ the normalized Haar measures on $U(n,\mathbb{C})/T$ and T, respectively. Let $\pi^*(d\mu)$ denote the measure induced on $U(n,\mathbb{C})/T \times T^0$ by the covering π^0. Then

$$\pi^*(d\mu) = V(t)\overline{V}(t)dv \times d\tau. \tag{9.1.2}$$

From these 3 steps Weyl's integration formula follows immediately. From 1 and 2:

$$\int_{U(n,\mathbb{C})} f(g)d\mu = \int_{U(n,\mathbb{C})^0} f(g)d\mu = \frac{1}{n!}\int_{U(n,\mathbb{C})/T\times T^0} f(gtg^{-1})\pi^*(d\mu).$$

From 3 and Fubini's theorem we have

$$\begin{aligned}\int_{U(n,\mathbb{C})/T\times T^0} & f(gtg^{-1})V(t)\overline{V}(t)dv \times d\tau \\ &= \int_{T^0} V(t)\overline{V}(t) \int_{U(n,\mathbb{C})/T} f(gtg^{-1})dv d\tau \\ &= \int_{T^0} V(t)\overline{V}(t)f(t)d\tau.\end{aligned}$$

Let us now explain the previous 3 statements.

1. We shall use some simple facts of differential geometry. Given a Lie group G and a closed Lie subgroup H, the coset space G/H has a natural structure of differentiable manifold over which G acts in an analytic way (Chapter 4, §3.7).

On an oriented manifold M of dimension n a measure can be defined by a differential form of degree n, which in local coordinates is $f(x)dx_1 \wedge dx_2 \wedge \ldots \wedge dx_n$.

In order to define on the manifold G/H a G-invariant metric in this way we need a G-invariant top differential form. This form is determined by the nonzero value it takes at a given point, which can be the coset H. Given such a value ψ_0 the form exists if and only if ψ_0 is invariant under the action of H on the tangent space of G/H at H itself. This means that H acts on this space by matrices of determinant 1. If H is compact, then it will act by orthogonal matrices, hence of determinant ± 1. If, furthermore, H is connected, the determinant must be 1. This is our case.[113]

This tangent space of G/H at H is $\mathfrak{g}/\mathfrak{h}$ where $\mathfrak{g}, \mathfrak{h}$ are the Lie algebras of G, H.[114]

In our case the Lie algebra of $U(n, \mathbb{C})$ consists of the matrices iA with A Hermitian, the Lie algebra of T corresponds to the subspace where A is diagonal, and we can take as the complement the set H_0 of anti-Hermitian matrices with 0 in the diagonal.

H_0 has basis the elements $i(e_{hj} + e_{jh}),\ e_{hj} - e_{jh}$ with $h < j$.

2. (i) Since $T, U(n, \mathbb{C})$ are differentiable manifolds and the Haar measures are given by differential forms, to verify that a set S has measure 0, it is sufficient to work in local coordinates and see that S has measure 0 for ordinary Lebesgue measure. For T, using angular coordinates, the condition on our set S is that two coordinates coincide, clearly a set of measure 0. For $U(n, \mathbb{C})$ it is a bit more complicated but similar, and we leave it as an exercise.

(ii) We have $U(n, \mathbb{C})/T \times T^0 = \pi^{-1}U(n, \mathbb{C})^0$ and since clearly π is proper, also π^0 is proper. The centralizer of a given matrix X in T^0 is T itself and the conjugacy class of X intersects T at exactly $n!$ elements obtained by permuting the diagonal entries. This implies that the preimage of any element in $U(n, \mathbb{C})^0$ under π^0 has $n!$ elements. Finally in the proof of 3 we show that the Jacobian of π^0 is nonzero at every element. Hence π^0 is a local diffeomorphism. These facts are enough to prove (i).

3. This requires a somewhat careful analysis. The principle is that when we compare two measures on two manifolds given by differential forms, under a local diffeomorphism, we are bound to compute a Jacobian. Let us denote $\pi^*(d\mu) = F(g, t) d\nu\, d\tau$ and let $\omega_1, \omega_2, \omega_3$ represent, respectively, the value of the form defining the normalized measure in the class of 1 for $U(n, \mathbb{C})/T, T, U(n, \mathbb{C})$, respectively.

Locally $U(n, \mathbb{C})/T$ can be parameterized by the classes $e^A T,\ A \in H_0$.

Given $g \in U(n, \mathbb{C}), y \in T$ consider the map $\ell_{g,y} : U(n, \mathbb{C})/T \times T \to U(n, \mathbb{C})/T \times T$ defined by multiplication $\ell_{g,y} : (hT, z) \mapsto (ghT, yz)$. It maps $(T, 1)$ to (gT, y). Its differential induces a linear map of the tangent spaces at these points. Denote this map $d_{g,y}$.

By construction $\omega_1 \wedge \omega_2$ is the pullback under the map $d_{g,y}$ of the value of $d\nu\, d\tau$ at the point (gT, y); similarly for $U(n, \mathbb{C})$, the value of $d\mu$ at an element h is the pullback of ω_3 under the map $r_h : x \to xh^{-1}$.

On the other hand $d\pi_{(gT,y)}$ induces a linear map (which we will see is an isomorphism) between the tangent space of $U(n, \mathbb{C})/T \times T$ at (gT, y) and the tan-

[113] The orientability follows.

[114] There is also a more concrete realization of $U(n, \mathbb{C})/T$ for which one can verify all the given statements. It is enough to take the set of $n \times n$ Hermitian matrices with n prescribed distinct eigenvalues.

gent space of $U(n, \mathbb{C})$ at gyg^{-1}. By definition $\pi^*(d\mu)$ at the point (gT, y) is $d\pi^*_{(gT,y)}\mu$.

Consider the composition of maps

$$\psi : U(n, \mathbb{C})/T \times T \xrightarrow{\ell_{g,y}} U(n, \mathbb{C})/T \times T \xrightarrow{\pi} U(n, \mathbb{C})$$
$$\xrightarrow{r_{gyg^{-1}}} U(n, \mathbb{C}), \quad \psi(T, 1) = 1.$$

We get that the pullback of ω_3, (at 1) under this map ψ is $F(g, t)\omega_1 \wedge \omega_2$.

In order to compute the function $F(g, t)$, we fix a basis of the tangent spaces of $U(n, \mathbb{C})/T, T, U(n, \mathbb{C})$ at 1 and compute the Jacobian of $r_{gyg^{-1}}\pi\ell_{g,y}$ in this basis. This Jacobian is $F(g, t)$ up to some constant, independent of g, t, which measures the difference between the determinants of the given bases and the normalized invariant form.

At the end we will compute the constant by comparing the integrals of the constant function 1. Take as local coordinates in $U(n, \mathbb{C})/T \times T$ the parameters $\exp(A)T, \exp(D)$ where $A \in H_0$ and D is diagonal. Since we need to compute the linear term (differential) of a map, we can compute everything in power series, saving at each step only the linear terms (set $\mathcal{O}$ to be the higher order terms):

$$\psi((1 + A)T, 1 + D) = g(1 + A)y(1 + D)(g(1 + A))^{-1}gy^{-1}g^{-1}$$
$$= 1 + g[D + A - yAy^{-1}]g^{-1} + \mathcal{O}.$$

The required Jacobian is the determinant of $(D, A) \mapsto g[D + A - yAy^{-1}]g^{-1}$. Since conjugating by g has determinant 1 we are reduced to the map $(D, A) \to (D, A - yAy^{-1})$. This is a block matrix with one block the identity. We are reduced to $\gamma : A \mapsto A - yAy^{-1}$.

To compute this determinant we complexify the space, obtaining as basis the elements e_{ij}, $i \neq j$. We have $\gamma(e_{i,j}) = (1 - y_i y_j^{-1})e_{i,j}$. In this basis the determinant is

$$\prod_{i \neq j}(1 - y_i y_j^{-1}) = \prod_{i<j}(1 - y_i y_j^{-1})(1 - y_j y_i^{-1})$$
$$= \prod_{i<j}(y_j - y_i)(y_j^{-1} - y_i^{-1}) = V(y)\overline{V}(y)$$

since y is unitary.[115]

At this point formula 9.1.1 is true, possibly up to some multiplicative constant.

The constant is 1 since if we take $f = 1$ the left-hand side of 9.1.1 is 1. As for the right-hand side, remember that the monomials in the y_i (coordinates of T) are the irreducible characters and so they are orthonormal. It follows that $\frac{1}{n!}\int_T V(y)\overline{V}(y)d\tau = 1$, since $V(y) = \sum_{\sigma \in S_n} \epsilon_\sigma y_{\sigma(1)}^{n-1} y_{\sigma(2)}^{n-2} \cdots y_{\sigma(n-1)}$, and the proof is complete.

115 Note that this formula shows that $U(n, \mathbb{C})/T \times T^0$ is the set of points of $U(n, \mathbb{C})/T \times T$ where $d\pi$ is invertible.

We have used the group $U(n, \mathbb{C})$ to illustrate this theorem, but the proof we gave is actually quite general. Let us see why. If K is a connected compact Lie group and T a maximal torus, we know from Chapter 10, §7.3 that $K = \cup_{g\in K} gTg^{-1}$ and the normalizer N_T modulo T is the Weyl group W. The nonregular elements are the union of a finite number of submanifolds, thus a set of measure 0. The quotient K/T is the flag variety (Chapter 10, §7.3) and, when we complexify the tangent space of K/T in the class of T we have the direct sum of all the root spaces. Therefore, the determinant of the corresponding Jacobian is $\prod_{\alpha\in\Phi}(1-t^\alpha)$. When t is in the compact torus, $t^{-\alpha}$ is conjugate to t^α, so setting

$$V(t) := \prod_{\alpha\in\Phi^+}(1-t^\alpha),$$

one has the general form of Weyl's integration formula:

Theorem (Weyl's Integration Formula). *For a class function f on K we have*

$$\int_K f(g)d\mu = \frac{1}{|W|}\int_T f(t)V(t)\overline{V}(t)d\tau \tag{9.1.3}$$

where $V(t) := \prod_{\alpha\in\Phi^+}(1-t^\alpha)$.

10 Characters of Classical Groups

10.1 The Symplectic Group

From the theory developed in Chapter 10 we easily see the following facts for $Sp(2n, \mathbb{C})$ or of its compact form, which can be proved directly in an elementary way.

Fix a symplectic basis e_i, f_i, $i = 1, \dots, n$. A maximal torus in $Sp(2n, \mathbb{C})$ is formed by the diagonal matrices such that if x_i is the eigenvalue of e_i, x_i^{-1} is the eigenvalue of f_i.

Besides the standard torus there are other symplectic matrices which preserve the basis by permuting it or changing sign. In particular

(i) The permutations S_n where $\sigma(e_i) = e_{\sigma(i)}$, $\sigma(f_i) = f_{\sigma(i)}$.

(ii) For each i, the exchange ϵ_i which fixes all the elements except $\epsilon_i(e_i) = f_i$, $\epsilon_i(f_i) = -e_i$.

These elements form a semidirect product (or rather a wreath product) $S_n \ltimes \mathbb{Z}/(2)^n$.

The normalizer N_T of the standard torus T is the semidirect product of T with the Weyl group $S_n \ltimes \mathbb{Z}/(2)^n$.

Proof. If a symplectic transformation normalizes the standard torus it must permute its eigenspaces. Moreover, if it maps the eigenspace of α to that of β, it maps the eigenspace of α^{-1} to that of β^{-1}. Therefore the group of permutations on the set of the $2n$ eigenspaces $\mathbb{C}e_i, \mathbb{C}f_i$ induced by the elements of N_T is the group $S_n \ltimes \mathbb{Z}/(2)^n$, hence every element of N_T can be written in a unique way as the product of an element of $S_n \ltimes \mathbb{Z}/(2)^n$ and a diagonal matrix that is an element of T. □

It is important to remark that the same analysis applies to the compact form of the symplectic group $Sp(n, \mathbb{H}) = Sp(2n, \mathbb{C}) \cap U(2n, \mathbb{C})$. The compact maximal torus T_c is formed by the unitary matrices of T which are just the diagonal matrices with $|\alpha_i| = 1$. The matrices in $S_n \ltimes \mathbb{Z}/(2)^n$ are unitary and we have the analogues of the previous facts for the compact group $Sp(n, \mathbb{H})$.

The normalizer of T acts on T by conjugation. In particular the Weyl group acts as $\sigma(x_i) = x_{\sigma(i)}$ and $\epsilon_i(x_j) = x_j,\ i \neq j,\ \epsilon_i(x_i) = x_i^{-1}$. One can thus look at the action on the coordinate ring of T. As in the theory of symmetric functions we can even work in an arithmetic way, considering the ring of Laurent polynomials $A := \mathbb{Z}[x_i, x_i^{-1}],\ i = 1, \dots, n$, with the action of $S_n \ltimes \mathbb{Z}/(2)^n$.

When we look at the invariant functions we can proceed in two steps. First, look at the functions invariant under $\mathbb{Z}/(2)^n$. We claim that an element of A is invariant under $\mathbb{Z}/(2)^n$ if and only if it is a polynomial in the elements $x_i + x_i^{-1}$. This follows by simple induction. If R is any commutative ring, consider $R[x, x^{-1}]$. Clearly an element in $R[x, x^{-1}]$ is invariant under $x \mapsto x^{-1}$ if and only if it is of the form $\sum_j r_j(x^j + x^{-j})$. Now it suffices to prove by simple induction that, for every j, the element $x^j + x^{-j}$ is a polynomial with integer coefficients in $x + x^{-1}$ (*hint*: $x^j + x^{-j} = (x + x^{-1})^j +$ *lower terms*). The next step is to apply the symmetric group to the elements $x_i + x_i^{-1}$ and obtain:

Proposition 1. *The invariants of $\mathbb{Z}[x_i, x_i^{-1}],\ i = 1, \dots, n$, under the action of $S_n \ltimes \mathbb{Z}/(2)^n$ form the ring of polynomials $\mathbb{Z}[e_1, e_2, \dots, e_n]$ where the e_i are the elementary symmetric functions in the variables $x_i + x_i^{-1}$.*

Let us consider now invariant functions on $Sp(2n, \mathbb{C})$ under conjugation. Let us first comment on the characteristic polynomial $p(t) = t^{2n} + \sum_{i=1}^{2n}(-1)^i s_i t^{2n-i}$ of a symplectic matrix X with eigenvalues $x_1, \dots, x_n, x_1^{-1}, \dots, x_n^{-1}$.

It is $\prod_{i=1}^n (t - x_i)(t - x_i^{-1}) = \prod_{i=1}^n (t^2 - (x_i + x_i^{-1})t + 1)$. Thus it is a *reciprocal* polynomial, $t^{2n} p(t^{-1}) = p(t)$. It is easy to give explicit relations between the first n coefficients s_i of $t^{2n-i},\ i = 1, \dots, n$, and the elementary symmetric functions e_i of the previous proposition, showing that the ring of invariants is also generated by the elements s_i. We deduce the following.

Proposition 2. *The ring of invariant polynomials under conjugation on $Sp(2n, \mathbb{C})$ is generated by the coefficients $s_i(X),\ i = 1, \dots, n$, of the characteristic polynomial of X.*

Proof. One can give two simple proofs. One is essentially the same as in Chapter 2, §5.1. Another is as follows. Using the FFT of §8.2, an invariant of any matrix X under conjugation of the symplectic group is a polynomial in traces of monomials in X and X^s. If the matrix is in the symplectic group, these monomials reduce to $X^k,\ k \in \mathbb{Z}$, and then one uses the theory of symmetric functions we have developed. □

We come now to the more interesting computation of characters. First, let us see how the Weyl integration formula of 9.1 takes shape in our case for the compact group $Sp(n, \mathbb{H})$ and its compact torus T_c. The method of the proof carries over with the following change. The covering now has $2^n n!$ sheets, and the Jacobian J expressing the change of measure is always given by the same method. One takes the Lie algebra $\mathfrak{k}$ of $Sp(n, \mathbb{H})$ which one decomposes as the direct sum of the Lie algebra of T_c and the T_c-invariant complement $\mathfrak{p}$ (under adjoint action). Then J is the determinant of $1 - \text{Ad}(t)$, where $\text{Ad}(t)$ is the adjoint action of T_c on $\mathfrak{p}$. Again, in order to compute this determinant, it is convenient to complexify $\mathfrak{p}$, obtaining the direct sum of all root spaces. Hence the Jacobian is the product $\prod(1 - \alpha(t))$ where $\alpha(t)$ runs over all the roots. As usual, now we are replacing the additive roots for the Cartan subalgebra with the multiplicative roots for the torus. Looking at the formulas 4.1.12 of Chapter 10 we see that the multiplicative roots are

$$(10.1.1) \qquad x_i x_j^{-1}, \quad (x_i x_j)^{\pm 1}, \ i \neq j, \qquad x_i^{\pm 2}.$$

$$J(x_1, \dots, x_n) = \prod_{i \neq j}(1 - x_i x_j^{-1})(1 - x_i x_j)(1 - (x_i x_j)^{-1}) \prod_{i=1}^{n}(1 - x_i^2)(1 - x_i^{-2}).$$

We collect the terms from the positive roots and obtain

$$\tilde{\Delta}(x_1, \dots, x_n) = \prod_{i<j}(1 - x_i x_j^{-1})(1 - x_i x_j) \prod_{i=1}^{n}(1 - x_i^2)$$

$$= \prod_{i<j} x_i (x_i + x_i^{-1} - (x_j + x_j^{-1})) \prod_{i=1}^{n}(-x_i)(x_i - x_i^{-1}).$$

When we compute J for the compact torus where $|x_i| = 1$, we have that the factors x_i cancel and $J = \Delta_C \overline{\Delta}_C$ with

$$\Delta_C(x_1, \dots, x_n) = \prod_{i<j}(x_i + x_i^{-1} - (x_j + x_j^{-1})) \prod_{i=1}^{n}(x_i - x_i^{-1}).$$

We have:

Theorem (Weyl's Integration Formula). *For a class function f on $Sp(n, \mathbb{H})$ we have*

$$(10.1.2) \qquad \int_{Sp(n,\mathbb{H})} f(g) d\mu = \frac{1}{2^n n!} \int_{T_c} f(x) \Delta_C(x) \overline{\Delta}_C(x) d\tau$$

where $\Delta_C(x_1, \dots, x_n) = \prod_{i<j}(x_i + x_i^{-1} - (x_j + x_j^{-1})) \prod_{i=1}^{n}(x_i - x_i^{-1})$.

Let us denote for simplicity by Q_n the group $S_n \ltimes \mathbb{Z}/(2)^n$. As for ordinary symmetric functions, we also have a sign character *sign* for the group Q_n, which is the ordinary sign on S_n, and $sgn(\epsilon_i) = -1$. It is the usual determinant when we consider the group in its natural reflection representation. We can consider again polynomials which are antisymmetric, i.e., which transform as this sign representation. Clearly $\Delta_C(x_1, \dots, x_n)$ is antisymmetric. The same argument given in Chapter 2, 3.1 shows that:

Proposition 3.

(1) A Q_n-antisymmetric Laurent polynomial $f(x_1, \dots, x_n, x_1^{-1}, \dots, x_n^{-1})$ is of the form $g(x)\Delta_C(x)$ with $g(x)$ symmetric.

(2) A basis of antisymmetric polynomials is given by the polynomials:

$$A^C_{\ell_1,\ell_2,\dots,\ell_n} := \sum_{\sigma \in Q_n} sign(\sigma)\sigma(x_1^{\ell_1} x_2^{\ell_2} \dots x_n^{\ell_n}), \quad \ell_1 > \ell_2 > \dots > \ell_n > 0.$$

(3) $A^C_{n,n-1,n-2,\dots,1}(x) = \Delta_C(x)$.

(4) $A^C_{\ell_1,\ell_2,\dots,\ell_n}$ is the determinant of the matrix with $x_i^{\ell_j} - x_i^{-\ell_j}$ in the i, j position.

Proof. (1) Let $f(x_1, \dots, x_n, x_1^{-1}, \dots, x_n^{-1})$ be antisymmetric, and expand it as a sum $\sum_k a_k x_i^k + b_k x_i^{-k}$ where a_k, b_k do not depend on x_i. We then see that $a_0 = b_0 = 0$ and $a_k = -b_k$. This shows that f is divisible by $\prod_i (x_i - x_i^{-1})$. If we write $f(x) = \tilde{f}(x) \prod_i (x_i - x_i^{-1})$ we have that $\tilde{f}(x)$ is symmetric with respect to the group of exchanges $x_i \to x_i^{-1}$. Hence, by a previous argument, $\tilde{f}(x)$ is a polynomial in the variables $x_i + x_i^{-1}$. Looking at the action of S_n we notice that $\prod_i (x_i - x_i^{-1})$ is symmetric, hence $\tilde{f}(x)$ is still antisymmetric for this action. Then, by the usual theory, it is divisible by the Vandermonde determinant of the variables $x_i + x_i^{-1}$ which is the remaining factor of $\Delta_C(x)$.

(2) Since the group Q_n permutes the Laurent monomials, the usual argument shows that in order to give a basis of antisymmetric polynomials, we have to check for each orbit if the alternating sum on the orbit is 0.

In each orbit there is a unique monomial $M = x_1^{k_1} x_2^{k_2} \dots x_n^{k_n}$, $k_1 \geq k_2 \geq \dots \geq k_n \geq 0$. Unless all the inequalities are strict there is an odd element[116] of Q_n which fixes M. In this case M and its entire orbit cannot appear in an antisymmetric polynomial. The other orbits correspond to the elements $A^C_{\ell_1,\ell_2,\dots,\ell_n}$.

(3) By part (1) each $A^C_{\ell_1,\ell_2,\dots,\ell_n}$ is divisible by Δ_C, so δ, up to some constant, must equal the one of lowest degree. To see that the constant factor is 1 it is enough to remark that the leading monomial of Δ_C is indeed $x_1^n x_2^{n-1} \dots x_n$.

(4)
$$A^C_{\ell_1,\ell_2,\dots,\ell_n} = \sum_{\sigma \in Q_n} sign(\sigma)\sigma(x_1^{\ell_1} x_2^{\ell_2} \dots x_n^{\ell_n})$$

$$= \sum_{\sigma \in S_n} sign(\sigma)\sigma\left(\sum_{\tau \in \mathbb{Z}/(2)^n} sign(\tau)\tau(x_1^{\ell_1} x_2^{\ell_2} \dots x_n^{\ell_n})\right)$$

$$= \sum_{\sigma \in S_n} sign(\sigma) \prod_{i=1}^n (x_{\sigma(i)}^{\ell_i} - x_{\sigma(i)}^{-\ell_i}).$$

□

116 We use "odd" to mean that its sign is −1.

We can now define the analogues of the Schur functions for the symplectic group:

Proposition 4. *The elements*

$$S_\lambda^C(x) := A^C_{\ell_1,\ell_2,\ldots,\ell_n}/\Delta_C(x), \quad \lambda := \ell_1 - n, \ell_2 - n + 1, \ldots, \ell_n - 1$$

are an integral basis of the ring of Q_n-invariants of $\mathbb{Z}[x_i, x_i^{-1}]$.

The leading monomial of $S_\lambda^C(x)$ is $x_1^{\ell_1-n}x_2^{\ell_2-n+1}\ldots x_n^{\ell_n-1}$.

Theorem 2. *The functions $S_\lambda^C(x)$ (as functions of the eigenvalues) are the irreducible characters of the symplectic group. $S_\lambda^C(x)$ is the character of the representation $T_\lambda(V)$ of §6.4.*

Proof. We proceed as for the unitary group. First, we prove that these characters are orthonormal. This follows immediately by the Weyl integration formula. Next we know that they are an integral basis of the invariant functions, hence we deduce that they must be $\pm$ the irreducible characters. Finally, a character is a positive sum of monomials, so by the computation of the leading term we can finally conclude that they must coincide with the characters.

Part (2) follows from the highest weight theory. In §6.4 the partition λ indexing the representation is the sequence of exponents of the character of the highest weight. □

10.2 Determinantal Formula

Let us continue this discussion developing an analogue for the functions $S_\lambda^C(x)$ of the determinantal formulas for Schur functions described in Chapter 9, §8.3.

Let us use the following notation. Given n-Laurent polynomials $f_i(x)$, in a variable x we denote by

$$|f_1(x), f_2(x), \ldots, f_n(x)| := \det(f_i(x_j))$$
$$= \sum_{\sigma\in S_n} sgn(\sigma) f_1(x_{\sigma(1)}), f_2(x_{\sigma(2)}), \ldots, f_n(x_{\sigma(n)}).$$

The symbol introduced is antisymmetric and multilinear in the obvious sense. Moreover if $g(x)$ is another polynomial

$$\prod_{i=1}^n g(x_i)|f_1(x), f_2(x), \ldots, f_n(x)| = |g(x)f_1(x), g(x)f_2(x), \ldots, g(x)f_n(x)|.$$

Let us start from the step of the Cauchy formula transformed as

$$\frac{V(x)V(y)}{\prod_{i,j=1}^{n}(x_i - y_j)} = \det\left(\frac{1}{x_i - y_j}\right).$$

Substitute for x_i the element $x_i + x_i^{-1}$, and for y_i the element $y_i + y_i^{-1}$. Observe that

$$x_i + x_i^{-1} - y_j - y_j^{-1} = x_i^{-1}(1 - x_i y_j)(1 - x_i y_j^{-1})$$

and that the Vandermonde determinant in the elements $x_i + x_i^{-1}$ equals the determinant $|x^{n-1} + x^{-n+1}, x^{n-1} + x^{-n+2}, \dots, 1|$ to deduce the identity:

$$\prod_i x_i^{n-1} \frac{|x^{n-1} + x^{-n+1}, x^{n-2} + x^{-n+2}, \dots, 1|\,|y^{n-1} + y^{-n+1}, y^{n-2} + y^{-n+2}, \dots, 1|}{\prod_{i,j}(1 - y_j x_i)(1 - y_j^{-1} x_i)}$$

$$= \det\left(\frac{1}{(1 - y_j x_i)(1 - y_j^{-1} x_i)}\right)$$

$$= \frac{|x^{2n-2} + 1, x^{2n-3} + x, \dots, x^{n-1}|\,|y^{n-1} + y^{-n+1}, y^{n-1} + y^{-n+2}, \dots, 1|}{\prod_{i,j}(1 - y_j x_i)(1 - y_j^{-1} x_i)}.$$

Notice that

$$\frac{1}{(1 - yx)(1 - y^{-1}x)} = \sum_{k=0}^{\infty}(y^k + y^{k-2} + \dots + y^{-k})x^k = \sum_{k=0}^{\infty} \frac{y^{k+1} - y^{-k-1}}{y - y^{-1}} x^k.$$

Hence the determinant $\det\left|\frac{1}{(1-y_j x_i)(1-y_j^{-1}x_i)}\right|$ expands as

$$\sum_{k_1 > k_2 > \dots > k_n \geq 0} A_{k_1,k_2,\dots,k_n}(x) A^C_{k_1+1,k_2+1,\dots,k_n+1}(y) \prod_{i=1}^{n}(y_i - y_i^{-1})^{-1}. \tag{10.2.1}$$

Next, consider the polynomial $\prod_{j=1}^{n}(1 - y_j z)(1 - y_j^{-1} z) = \det(1 - zA)$, where A is a symplectic matrix with eigenvalues y_i, y_i^{-1}, $i = 1, \dots, n$.

Let us define the symmetric functions $p_f(A) = p_f(y)$ by the formula

$$\frac{1}{\det(1 - zA)} = 1 + \sum_{k=1}^{\infty} p_k(A) z^k. \tag{10.2.2}$$

Dividing both terms of the identity by $|y^{n-1} + y^{-n+1}, y^{n-2} + y^{-n+2}, \dots, 1|$ we get

$$\sum_{\lambda} A_{\lambda+\rho}(x) S^C_{\lambda}(y) = |x^{2n-2}+1, x^{2n-3}+x, \ldots, x^{n-1}| \prod_{i=1}^{n}\left(1+\sum_{k=1}^{\infty} p_k(y) x_i^k\right)$$

$$= \left|\left(1+\sum_{k=1}^{\infty} p_k(y)x^k\right)(x^{2n-2}+1), \left(1+\sum_{k=1}^{\infty} p_k(y)x^k\right)\right.$$

$$\left. \times (x^{2n-3}+x), \ldots, \left(1+\sum_{k=1}^{\infty} p_k(y)x^k\right)x^{n-1}\right|$$

$$= \sum_{k_1,k_2,,..,k_n} p_{k_1}p_{k_2}\cdots p_{k_n}|x^{k_1}(x^{2n-2}+1), x^{k_2}(x^{2n-3}+x), \ldots, x^{k_n}x^{n-1}|$$

$$= \sum_{k_1,k_2,\ldots,k_n} p_{k_1}p_{k_2}\cdots p_{k_n}|x^{k_1+2n-2}+x^{k_1}, x^{k_2+2n-3}+x^{k_2+1}, \ldots, x^{k_n+n-1}|$$

$$= \sum_{k_1,k_2,\ldots,k_n} p_{k_1-2n+2}p_{k_2}\cdots p_{k_n}|x^{k_1}, x^{k_2+2n-3}+x^{k_2+1}, \ldots, x^{k_n+n-1}|$$

$$+ \sum_{k_1,k_2,\ldots,k_n} p_{k_1}p_{k_2}\cdots p_{k_n}|x^{k_1}, x^{k_2+2n-3}+x^{k_2+1}, \ldots, x^{k_n+n-1}|$$

$$\times \sum_{k_1,k_2,\ldots,k_n} (p_{k_1-2n+2}+p_{k_1})p_{k_2}\cdots p_{k_n}|x^{k_1}, x^{k_2+2n-3}+x^{k_2+1}, \ldots, x^{k_n+n-1}|.$$

Iterating the procedure,

$$\sum_{k_1,k_2,\ldots,k_n} (p_{k_1-2n+2}+p_{k_1})(p_{k_2-2n+3}+p_{k_2-1})\ldots$$

$$(p_{k_i-2n+i}+p_{k_i-i+2})\ldots p_{k_n-n+1}|x^{k_1}, \ldots, x^{k_n}|$$

$$= \sum_{k_1>k_2>\ldots>k_n} |p_{k_1-2n+2}+p_{k_1}, p_{k_2-2n+3}+p_{k_2-1}, \ldots, p_{k_i-2n+i}$$

$$+ p_{k_i-i+2}, \ldots, p_{k_n-n+1}||x^{k_1}, \ldots, x^{k_n}|,$$

where the symbol $|p_{k_1-2n+2}+p_{k_1}, p_{k_2-2n+3}+p_{k_2-1}, \ldots, p_{k_i-2n+i+1}+p_{k_i-i+1}, \ldots, p_{k_n-n+1}|$ equals the determinant of the matrix in which the element in the ith row and j^{th} column is $p_{k_i-2n+j}+p_{k_i-j+2}$.

Thus, with the notations of partitions, the function $S^C_{\lambda}(y)$ with $\lambda = h_1, h_2, \ldots, h_n$ corresponds to the coefficient where $k_1 = h_1+n-1, k_2 = h_2+n-2, \ldots k_n = h_n$ and so it is

$$S^C_{\lambda}(y) =$$

$$|p_{h_1-n+1}+p_{h_1+n-1}, p_{h_2-n+1}+p_{h_2+n-3}, \ldots, p_{h_i-n+1}+p_{h_i+n-2i+1}, \ldots, p_{h_n-n+1}|.$$

Notice that when we evaluate for $y_i = 1$:

(10.2.3)

$$1+\sum_{k=1}^{\infty} p_k(1)z^k = \frac{1}{\det(1-zI_{2n})} = \frac{1}{(1-z)^{2n}} = 1+\sum_{k=1}^{\infty}\binom{2n+k-1}{k}z^k$$

gives a determinantal expression in terms of binomial coefficients for $S_\lambda^C(1) = \dim T_\lambda(V)$.

10.3 The Spin Groups: Odd Case

Let Spin$(2n+1)$ be the compact spin group.

Let us work with the usual hyperbolic basis e_i, f_i, u. Consider the improper orthogonal transformation $J : J(e_i) = e_i, J(f_i) = f_i, J(u) = -u$, so that $O(V) = SO(V) \cap SO(V)J$.

Its roots are $\Phi^+ := \alpha_i - \alpha_j, i < j,\ \alpha_i + \alpha_j,\ i \neq j, \alpha_i$, cf. Chapter 10, §4.1.5. Its Weyl group is a semidirect product $S_n \ltimes \mathbb{Z}/(2)^n$ as for the symplectic group.

The character group X of the corresponding maximal torus T of the special orthogonal group is the free group in the variables x_i. X must be of index 2 in the character group X' of the corresponding torus T' for the spin group, since T' is the preimage of T in the spin group, hence it is a double covering of T. X' contains the fundamental weight of the spin representation which we may call s with $s^2 = \prod_i x_i$, so it is convenient to consider X as the subgroup of the free group in variables $x_i^{1/2}$ with the usual action of the Weyl group, generated by the elements x_i and $s = (\prod_{i=1}^n x_i^{1/2})$. It is of course itself free on the variables $x_i,\ i < n,\ s$.

For the Weyl integration formula we obtain

(10.3.1)

$$\int_{O(V)} f(g)d\mu = \frac{1}{2^n n!}\left(\int_{T_c} f(x)\Delta_B(x)\overline{\Delta}_B(x)d\tau + \int_{T_cJ} f(x)\Delta_B(x)\overline{\Delta}_B(x)d\tau\right),$$

where $\Delta_B(x_1,\dots,x_n) = \prod_{i<j}(x_i + x_i^{-1} - (x_j + x_j^{-1}))\prod_i(x_i^{1/2} - x_i^{-1/2})$. A function

$$A^B_{\ell_1,\ell_2,\dots,\ell_n} := \sum_{\sigma\in Q_n} sign(\sigma)\sigma(x_1^{\ell_1}x_2^{\ell_2}\dots x_n^{\ell_n}), \qquad \ell_1 > \ell_2 > \dots > \ell_n > 0.$$

is in the character group X' if the ℓ_i are either all integers or all half integers.

$\Delta_B(x_1,\dots,x_n) = \prod_{i<j}(x_i + x_i^{-1} - (x_j + x_j^{-1}))\prod_i(x_i^{1/2} - x_i^{-1/2})$ is skew-symmetric with leading term $x_1^{n-1/2}\dots x_n^{1/2}$ so $\Delta_B = A^B_{n-1/2,n-1-1/2,\dots,1/2}$. Notice that

$$\Delta_B\prod_i(x_i^{1/2} + x_i^{-1/2}) = \prod_{i<j}(x_i + x_i^{-1} - (x_j + x_j^{-1}))\prod_i(x_i - x_i^{-1}),$$

which is the formula we found for the symplectic group (10.1). If the ℓ_i are all integers, then $A^B_{\ell_1,\ell_2,\dots,\ell_n}$ is divisible by $\Delta_B\prod_i(x_i^{1/2} + x_i^{-1/2})$ with the quotient being a linear combination of characters all with integer exponents. Otherwise, if all the ℓ_i are half integers, we multiply by $\prod_i x_i^{1/2}$ and still see that all the A^B are multiples of Δ_B. We get a theorem similar to the one for the symplectic group.

Proposition. *The elements*

$$S_\lambda^B(x) := A^B_{\ell_1,\ell_2,\dots,\ell_n}/\Delta_B(x), \quad \lambda := \ell_1 - n, \ell_2 - n + 1, \dots, \ell_n - 1$$

are an integral basis of the ring of Q_n invariants of $\mathbb{Z}[x_i, x_i^{-1}, \prod_i x_i^{1/2}]$.

The leading monomial of $S_\lambda^B(x)$ is $x_1^{\ell_1-n+1/2} x_2^{\ell_2-n+3/2} \dots x_n^{\ell_n-1/2}$.

Theorem. *The functions $S_\lambda^B(x)$ (as functions of the eigenvalues) are the irreducible characters of the spin group. The functions $S_\lambda^B(x)$ with integral weight are the irreducible characters of the special orthogonal group.*

If λ is integral, $S_\lambda^B(x)$ is the character of the representation $T_\lambda(V)$ of §6.6.

Proof. Same as for the symplectic group. □

10.4 The Spin Groups: Even Case

Let Spin$(2n)$ be the compact spin group.

Its positive roots are $\Phi^+ := \alpha_i - \alpha_j, i < j, \ \alpha_i + \alpha_j, \ i \neq j$, Chapter 10, §4.1.19. Its Weyl group is a semidirect product $S_n \ltimes S$. S_n is the symmetric group permuting the coordinates. S is the subgroup of the *sign group* $\mathbb{Z}/(2)^n := (\pm 1, \pm 1, \dots, \pm 1)$ which changes the signs of the coordinates, formed by only even number of sign changes.

For the Weyl integration formula we obtain

$$\text{(10.4.1)} \qquad \int_{\mathrm{Spin}(2n)} f(g) d\mu = \frac{1}{2^{n-1} n!} \int_{T_c} f(x) \Delta^D(x) \overline{\Delta}^D(x) d\tau,$$

where $\Delta^D(x_1, \dots, x_n) = \prod_{i<j} (x_i + x_i^{-1} - (x_j + x_j^{-1}))$, with leading monomial $x_1^{n-1} x_2^{n-2} \dots x_{n-1}$.

The character ring of the maximal torus of the special orthogonal group is the polynomial ring $\mathbb{Z}[x_i^{\pm 1}]$, while that of the spin group has also the element $\prod_i x_i^{1/2}$.

Now it is not always true in the orbit of a monomial under the action of the Weyl group W there is a monomial with all exponents positive. Thus we have a different notion of leading monomial. In fact it is easy to verify that, given $h_1 > h_2 > \dots > h_n \geq 0$, if $h_n > 0$ the two monomials $x_1^{h_1} x_1^{h_2} \dots x_n^{h_n}$ and $x_1^{h_1} x_1^{h_2} \dots x_{n-1}^{h_{n-1}} x_n^{-h_n}$ are in two different orbits and leading in each of them. In the language of weights, an element is leading in an orbit if it is in the fundamental chamber (cf. Chapter 10, Theorem 2.4). Therefore each of these two elements will be the leading term of an antisymmetric function. Dividing these functions by $\Delta^D(x_1, \dots, x_n)$ we get the list of the irreducible characters. We can understand to which representations they correspond using the theory of highest weights.

10.5 Weyl's Character Formula

As should be clear from the examples, Weyl's character formula applies to all simply connected semisimple groups, as well as the integration formula for the corresponding compact groups. In the language of roots and weights let us use the notation that if α is a weight of the Cartan subalgebra, e^α is the corresponding weight for the torus according to the formula $e^\alpha(e^t) = e^{\alpha(t)}$, so that $e^{\alpha+\beta} = e^\alpha e^\beta$.

The analogue of Δ used in the integration formula is

$$(10.5.1) \qquad \rho := \sum_{\alpha\in\Phi^+} \alpha/2, \qquad \Delta := e^\rho \prod_{\alpha\in\Phi^+} (1 - e^{-\alpha}) = \prod_{\alpha\in\Phi^+} (e^{\alpha/2} - e^{-\alpha/2}).$$

We have seen (Chapter 10, §2.4) the notions of weight lattice, dominant, regular dominant and fundamental weight. We denote by Λ the weight lattice. Λ is identified with the character group of the maximal torus T of the associated simply connected group. We also denote by Λ^+ the *dominant weights*, Λ^{++} the *regular dominant weights* and ω_i the *fundamental weights.*

The character ring for T is the group ring $\mathbb{Z}[\Lambda]$. Choosing as generators the fundamental weights ω_i this is also a Laurent polynomial ring in the variables e^{ω_i}.

From Chapter 10, Theorems 2.4 and 2.3 (6) it follows that there is one and only one dominant weight in the orbit of any weight. The stabilizer in W of a dominant weight $\sum_{i=1}^n a_i\omega_i$ is generated by the simple reflections s_i for which $a_i = 0$. Let $\rho = \sum_i \omega_i$.

It follows that the skew-symmetric elements of the character ring have a basis obtained by antisymmetrizing the dominant weights for which all $a_i > 0$, i.e., the regular dominant weights. Observe that ρ is the minimal regular dominant weight and we have the 1–1 correspondence $\lambda \mapsto \lambda + \rho$ between dominant and regular dominant weights.

We have thus a generalization of the ordinary theory of skew-symmetric polynomials:

Proposition 1. *The group $\mathbb{Z}[\Lambda]^-$ of skew-symmetric elements of the character ring has as basis the elements*

$$A_{\lambda+\rho} := \sum_{w\in W} sign(w) e^{w(\lambda+\rho)}, \qquad \lambda \in \Lambda^+.$$

First, we want to claim that $A_{\lambda+\rho}$ has the *leading term* $e^{\lambda+\rho}$.

Lemma 1. *Let λ be a dominant weight and w an element of the Weyl group. Then $\lambda - w(\lambda)$ is a linear combination with positive coefficients of the positive roots $s_{i_1}s_{i_2}\dots s_{i_{h-1}}(\alpha_{i_h})$ sent to negative roots by w^{-1}. If $w \neq 1$ and λ is regular dominant, this linear combination is nonzero and $w(\lambda) \prec \lambda$ in the dominance order.*

Proof. We will use the formula of Chapter 10, §2.3 with $w = s_{i_1}s_{i_2}\dots s_{i_k}$ a reduced expression, $u := s_{i_2}\dots s_{i_k}$, $\gamma_h := s_{i_2}\dots s_{i_{h-1}}(\alpha_{i_h})$, $\beta_h := s_{i_1}s_{i_2}\dots s_{i_{h-1}}(\alpha_{i_h})$. We have by induction

$$\begin{aligned}\lambda - w(\lambda) &= \lambda - s_{i_1}u(\lambda) = \lambda + s_{i_1}(\lambda - u(\lambda)) - s_{i_1}\lambda \\ &= \lambda + s_{i_1}\left(\sum_h n_h\gamma_h\right) - \lambda + \langle\lambda \,|\, \alpha_{i_1}\rangle\alpha_{i_1} \\ &= \sum_h n_h\beta_h + \langle\lambda \,|\, \alpha_i\rangle\alpha_i.\end{aligned}$$

Since λ is dominant, $\langle\lambda \,|\, \alpha_i\rangle \geq 0$. If, moreover, λ is regular $\langle\lambda \,|\, \alpha_i\rangle > 0$. □

In order to complete the analysis we need to be able to work as we did for Schur functions. Notice that choosing a basis of the weight lattice, the ring $\mathbb{Z}[\Lambda] = \mathbb{Z}[x_i^{\pm 1}]$, $i = 1, \dots, n$, is a unique factorization domain. Its invertible elements are the elements $\pm e^\lambda$, $\lambda \in \Lambda$. It is immediate that the elements $1 - x_i$, $1 - x_i^{-1}$ are irreducible elements.

Lemma 2. *For any root α the element $1 - e^{-\alpha}$ is irreducible in $\mathbb{Z}[\Lambda]$.*

Proof. Under the action of the Weyl group every root α is equivalent to a simple root; thus we can reduce to the case in which $\alpha = \alpha_1$ is a simple root.

We claim that α_1 is part of an integral basis of the weight lattice. This suffices to prove the lemma. In fact let ω_i be the fundamental weights (a basis). Consider $\alpha_1 = \sum_j \langle \alpha_1, \alpha_i \rangle \omega_j$ (Chapter 10, §2.4.2). By inspection, for all Cartan matrices we have that each column is formed by a set of relatively prime integers. This implies that α_1 can be completed to an integral basis of the weight lattice. □

One clearly has $s_i(\Delta) = -\Delta$ which shows the antisymmetry of Δ. Moreover one has:

Proposition 2. $\mathbb{Z}[\Lambda]^- = \mathbb{Z}[\Lambda]^W \Delta$ *is the free module over the ring of invariants* $\mathbb{Z}[\Lambda]^W$ *with generator* Δ.

$$\Delta = \sum_{w \in W} sign(w) e^{w(\rho)} = A_\rho. \tag{10.5.2}$$

The elements $S_\lambda := A_{\lambda+\rho}/\Delta$ form an integral basis of $\mathbb{Z}[\Lambda]^W$.

Proof. Given a positive root α let s_α be the corresponding reflection. By decomposing W into cosets with respect to the subgroup with two elements 1 and s_α, we see that the alternating function $A_{\lambda+\rho}$ is a sum of terms of type $e^\mu - s_\alpha(e^\mu) = e^\mu - e^{\mu - \langle \mu | \alpha \rangle \alpha} = e^\mu(1 - e^{-\langle \mu | \alpha \rangle \alpha})$. Since $\langle \mu | \alpha \rangle$ is an integer, it follows that $(1 - e^{-\langle \mu | \alpha \rangle \alpha})$ is divisible by $1 - e^{-\alpha}$; since all terms of $A_{\lambda+\rho}$ are divisible, $A_{\lambda+\rho}$ is divisible by $1 - e^{-\alpha}$ as well. Now if $\alpha \neq \beta$ are positive roots, $1 - e^{-\alpha}$ and $1 - e^{-\beta}$ are not associate irreducibles, otherwise $1 - e^{-\alpha} = \pm e^\lambda(1 - e^{-\beta})$ implies $\alpha = \pm\beta$. It follows that each function $A_{\lambda+\rho}$ is divisible by Δ and the first part follows as for symmetric functions.

The second formula $\Delta = A_\rho$ can be proved as follows. In any skew-symmetric element there must appear at least one term of type e^λ with λ strongly dominant. On the other hand, ρ is minimal in the dominance order among strongly dominant weights. The alternating function $\Delta - A_\rho$ does not contain any such term, hence it is 0. The last statement follows from the first. □

Now let us extend the leading term theory also to the S_λ.

Proposition 3. $S_\lambda = e^\lambda$ *plus a linear combination of elements e^μ where μ is less than λ in the dominance order.*

Proof. We know the statement for $A_{\lambda+\rho}$, so it follows for

$$S_\lambda = A_{\lambda+\rho}/\Delta = A_{\lambda+\rho}e^{-\rho}\prod_{\alpha\in\Phi^+}(1-e^{-\alpha})^{-1} = A_{\lambda+\rho}e^{-\rho}\prod_{\alpha\in\Phi^+}\left(\sum_{k=0}^{\infty}e^{-k\alpha}\right). \quad \square$$

The analysis culminates with:

Weyl's Character Formula. *The element S_λ is the irreducible character of the representation with highest weight λ.*

Proof. The argument is similar to the one developed for $U(n, \mathbb{C})$. First, we have the orthogonality of these functions. Let G be the corresponding compact group, T a maximal torus of G. Apply Weyl's integration formula:

$$\int_G S_\lambda \overline{S}_\mu d\mu = \frac{1}{|W|}\int_T S_\lambda \overline{S}_\mu \Delta\overline{\Delta} d\tau = \frac{1}{|W|}\int_T A_{\lambda+\rho}\overline{A}_{\mu+\rho} d\tau = \delta_\lambda^\mu.$$

The last equality comes from the fact that if $\lambda \neq \mu$, the two W-orbits of $\lambda+\rho$, $\mu+\rho$ are disjoint, and then we apply the usual orthogonality of characters for the torus.

Next, we know also that irreducible characters are symmetric functions of norm 1. When we write them in terms of the S_λ, we must have that an irreducible character must be $\pm S_\lambda$. At this point we can use the highest weight theory and the fact that e^λ with coefficient 1 is the leading term of S_λ and also the leading term of the irreducible character of the representation with highest weight λ. To finish, since all the dominant weights are highest weights the theorem is proved. $\square$

We finally deduce also:

Weyl's Dimension Formula. *The value $S_\lambda(1)$, dimension of the irreducible representation with highest weight λ, is*

$$\prod_{\alpha\in\Phi^+}\frac{(\lambda+\rho,\alpha)}{(\rho,\alpha)} = \prod_{\alpha\in\Phi^+}\left(1+\frac{(\lambda,\alpha)}{(\rho,\alpha)}\right). \tag{10.5.3}$$

Proof. Remark first that if $\alpha = \sum_i n_i\alpha_i \in \Phi^+$, we have $(\rho,\alpha) = \sum_i n_i(\rho,\alpha_i) > 0$, so the formula makes sense. One cannot evaluate the fraction of the Weyl Character formula directly at 1. In fact the denominator vanishes exactly on the nonregular elements. So we compute by the usual method of calculus. We take a regular vector $v \in \mathfrak{t}$ and compute the formula on $\exp(v)$, then take the limit for $v \to 0$ by looking at the linear terms in the numerator and denominator. By definition, $e^\beta(e^v) = e^{\beta(v)}$, so we analyze the quotient of (use 10.5.1):

$$\sum_{w\in W} sign(w)e^{w(\lambda+\rho)(v)} \text{ and } \sum_{w\in W} sign(w)e^{w(\rho)(v)} = e^{\rho(v)}\prod_{\alpha\in\Phi^+}(1-e^{-\alpha(v)}).$$

By duality, we work directly on the root space and analyze equivalently the quotient of

$$\sum_{w\in W} sign(w)e^{(w(\lambda+\rho),\beta)} \text{ and } \sum_{w\in W} sign(w)e^{(w(\rho),\beta)} = e^{(\rho,\beta)}\prod_{\alpha\in\Phi^+}(1-e^{-(\alpha,\beta)}).$$

Take $\beta = s\rho$. We have $\sum_{w\in W} sign(w)e^{(w(\lambda+\rho),s\rho)} = \sum_{w\in W} sign(w)e^{s(\lambda+\rho,w(\rho))}$. Now substituting for $\beta = s(\lambda+\rho)$ in the second identity we get

$$\sum_{w\in W} sign(w)e^{s(\lambda+\rho,w(\rho))} = \sum_{w\in W} sign(w)e^{(w(\rho),s(\lambda+\rho))}$$
$$= e^{s(\rho,\lambda+\rho)}\prod_{\alpha\in\Phi^+}(1-e^{-s(\alpha,\lambda+\rho)}).$$

The character computed in $s\rho$ is thus

$$\frac{e^{s(\rho,\lambda+\rho)}}{e^{(\rho,s\rho)}}\prod_{\alpha\in\Phi^+}\frac{(1-e^{-s(\alpha,\lambda+\rho)})}{(1-e^{-(\alpha,s\rho)})}.$$

Its limit when $s \to 0$ is clearly $\prod_{\alpha\in\Phi^+}\frac{(\lambda+\rho,\alpha)}{(\rho,\alpha)}$. □

A final remark. The computation of the character on a maximal torus determines the character on the entire group. This is obvious for the compact form since every element is conjugate to one in the maximal torus. For the algebraic group we have seen in Chapter 10, §6.7 that the set of elements conjugate to one in the maximal torus is dense and we have proved that the ring of regular functions on a linearly reductive group, invariant under conjugation, has as basis the irreducible characters.

12

Tableaux

1 The Robinson–Schensted Correspondence

We start by explaining some combinatorial aspects of representation theory by giving a beautiful combinatorial analogue of the decomposition of tensor space $V^{\otimes n} = \bigoplus_{\lambda \vdash n} M_\lambda \otimes V_\lambda$.

Recall that in Chapter 9, §10.3 we have shown that this decomposition can be refined to a basis of weight vectors with indices pairs of tableaux. In this chapter tableaux are taken as the main objects of study, and from a careful combinatorial study we will get the applications to symmetric functions and the Littlewood–Richardson rule.

1.1 Insertion

We thus start from a totally ordered set A which in combinatorics is called an *alphabet*. We consider A as an ordered basis of a vector space V.

Consider the set of *words* of length n in this alphabet, i.e., sequences $a_1 a_2 \dots a_n$, $a_i \in A$. If $|A| = m$, this is a set with m^n elements in correspondence with the induced basis of $V^{\otimes n}$.

Next we shall construct from A certain combinatorial objects called *column and row tableaux*. Let us use pictorial language, and illustrate with examples. We shall use as an alphabet either the usual alphabet or the integers.

A *standard column* of length k consists in placing k distinct elements of A in a column (i.e., one on top of the other) so that they decrease from top to bottom:

Example.

$$\begin{array}{c} t \\ g \\ e \\ b \\ a \end{array} \qquad \begin{array}{c} s \\ p \\ g \\ e \\ d \\ c \end{array} \qquad \begin{array}{c} 10 \\ 9 \\ 6 \\ 5 \\ 1 \end{array}$$

A sequence of columns of non-increasing length, placed one next to the other, identify a tableau and its *rows*. If the columns are standard and the elements in the rows going from left to right are weakly increasing (i.e., they can be also equal) the tableau is *semistandard*:

Definition 1. A semistandard tableau is a filling of a diagram with the letters from the alphabet so that the columns are strictly increasing from bottom to top while the rows are weakly increasing from left to right (cf. Chapter 9, §10.1).[117]

Example.

$$\begin{array}{llll} t & & & \\ g & g & j & \\ e & f & f & \\ b & c & d & u \\ a & b & b & f \end{array} \qquad \begin{array}{lll} s & & \\ p & & \\ g & h & p \\ e & f & g \\ d & e & f \\ c & d & e \end{array} \qquad \begin{array}{lllll} 10 & & & & \\ 9 & & & & \\ 6 & & & & \\ 5 & 5 & 5 & & \\ 1 & 1 & 1 & 1 & 1 \end{array}$$

The main algorithm which we need is that of *inserting a letter in a standard column.*

Assume we have a letter x and a column c; we begin by placing x on top of the column. If the resulting column is standard, this is the result of inserting x in c. Otherwise we start going down the column attempting to replace the entry that we encounter with x and we stop at the first step in which this produces a standard column. We thus replace the corresponding letter y with x, obtaining a new column c' and an expelled letter y.

Example.

$$\text{Inserting } h \text{ in } \begin{array}{c} t \\ g \\ e, \\ b \\ a \end{array} \text{ we obtain } \begin{array}{c} t \\ h \\ e, \\ b \\ a \end{array} \text{ expelling } g.$$

It is possible that the entering and exiting letters are the same. For instance, in the previous case if we wanted to insert g we would also extract g. A special case is when c is empty, and then inserting x just creates a column consisting of only x.

The first remark is that, from the new column c' and, if present, the expelled letter y one can reconstruct c and x. In fact we try backwards to insert y in c' from bottom upwards, stopping at the first position that makes the new column standard and expelling the relative entry. This is the reconstruction of c, x.

The second point is that we can now insert a letter x in a semistandard tableau T as follows. T is a sequence of columns $c_1, c_2, \ldots, c_i$. We first insert x in c_1; if we get an expelled element x_1 we insert it in c_2; if we get an expelled element x_2 we insert it in c_3, etc.

---

[117] Notice that we changed the display of a tableau! We are using the French notation.

Example.

1) Insert d in

$$\begin{array}{llll} t & & & \\ g & g & j & \\ e & f & f & \\ b & c & d & u \\ a & b & b & f \end{array}$$

get

$$\begin{array}{lllll} t & & & & \\ g & g & j & & \\ d & e & f & & \\ b & c & d & u & \\ a & b & b & f & f. \end{array}$$

2) Insert d in

$$\begin{array}{lll} s & & \\ p & & \\ g & h & p \\ e & f & g \\ d & e & f \\ c & d & e \end{array}$$

and get

$$\begin{array}{llll} s & & & \\ p & & & \\ g & h & p & \\ e & f & g & \\ d & e & f & \\ c & d & d & e. \end{array}$$

In any case, inserting a letter in a semistandard tableau, we always get a new semistandard tableau (verify it) with one extra box occupied by some letter. By the previous remark, the knowledge of this box allows us to recursively reconstruct the original tableau and the inserted letter.

All this can be made into a recursive construction. Starting from a word $w = a_1a_2\dots a_k = a_1w_1$ of length k, we construct two tableau $T(w),\ D(w)$ of the same shape, with k entries. The first, called *the insertion tableau,* is obtained recursively from the empty tableau by inserting a_1 in the tableau $T(w_1)$ (constructed by recursion). This tableau is semistandard and contains as entries exactly the letters appearing in w.

The tableau $D(w)$ is *the recording tableau,* and is constructed as follows. At the ith step of the insertion a new box is produced, and in this box we insert i. Thus $D(w)$ records the way in which the tableau $T(w)$ has been recursively constructed. It is filled with all the numbers from 1 to k. Each number appears only once and we will refer to this property as standard.[118]

$D(w)$ is constructed from $D(w_1)$, which by inductive hypothesis has the same shape as $T(w_1)$, by placing in the position of the new box (occupied by the procedure of inserting a_1), the number k.

An example should illustrate the construction. We take the word *standard* and construct the sequence of associated insertion tableaux $T(w_i)$, inserting its letters starting from the right. We get:

$$
\begin{array}{l} d \end{array}\quad
\begin{array}{l} r \\ d \end{array}\quad
\begin{array}{ll} r & \\ a & d \end{array}\quad
\begin{array}{ll} d & r \\ a & d \end{array}\quad
\begin{array}{ll} n & \\ d & r \\ a & d \end{array}\quad
\begin{array}{lll} n & & \\ d & r & \\ a & a & d \end{array}\quad
\begin{array}{lll} t & & \\ n & & \\ d & r & \\ a & a & d \end{array}\quad
\begin{array}{lll} s & & \\ n & t & \\ d & r & \\ a & a & d \end{array};
$$

the sequence of recording tableaux is

$$
\begin{array}{l} 1 \end{array}\quad
\begin{array}{l} 2 \\ 1 \end{array}\quad
\begin{array}{ll} 2 & \\ 1 & 3 \end{array}\quad
\begin{array}{ll} 2 & 4 \\ 1 & 3 \end{array}\quad
\begin{array}{ll} 5 & \\ 2 & 4 \\ 1 & 3 \end{array}\quad
\begin{array}{lll} 5 & & \\ 2 & 4 & \\ 1 & 3 & 6 \end{array}\quad
\begin{array}{lll} 7 & & \\ 5 & & \\ 2 & 4 & \\ 1 & 3 & 6 \end{array}\quad
\begin{array}{lll} 7 & & \\ 5 & 8 & \\ 2 & 4 & \\ 1 & 3 & 6 \end{array}.
$$

Theorem (Robinson–Schensted correspondence). *The map $w \to (D(w), T(w))$ is a bijection between the set of words of length k and pairs of tableaux of the same shape of which $D(w)$ is standard and $T(w)$ semistandard.*

Proof. The proof follows from the sequence of previous remarks about the reversibility of the operation of inserting a letter. The diagram $D(w)$ allows one to determine which box has been filled at each step and thus to reconstruct the insertion procedure and the original word. □

Definition 2. We call the *shape* of a word w the common shape of the two tableaux $(D(w), T(w))$.

Given a semistandard tableau T we call its *content* the set of elements appearing in it with the respective multiplicity. Similarly, we speak of the content of a given

[118] Some authors prefer doubly standard for this restricted type, and standard in place of semistandard.

word, denoted by $c(w)$. It is convenient to think of the content as the commutative monomial associated to the word, e.g., $w = abacbaa$ gives $c(w) = a^4b^2c$. The Robinson–Schensted correspondence preserves contents.

There is a special case to be observed. Assume that the recording tableau $D(w)$ is such that if we read it starting from left to right and then from the bottom to the top, we find the numbers $1, 2, \ldots, k$ in increasing order, e.g.,

$$\begin{matrix} 9 & & & \\ 7 & 8 & & \\ 5 & 6 & & \\ 1 & 2 & 3 & 4 \end{matrix}$$

Then the word w can be very quickly read off from $T(w)$. It is obtained by reading $T(w)$ from top to bottom and from left to right (as in the English language), e.g.,[119]

$$spuntatu \implies \begin{matrix} s & & \\ p & u & \\ n & t & \\ a & t & u \end{matrix} \qquad \begin{matrix} 8 & & \\ 6 & 7 & \\ 4 & 5 & \\ 1 & 2 & 3 \end{matrix}$$

Such a word will be called a *semistandard word.*

1.2 Knuth Equivalence

A natural construction from the R-S correspondence is the *Knuth equivalence*.

Let us ask the question of when two words w_1, w_2 have the same insertion tableau, i.e., when $T(w_1) = T(w_2)$. As the starting point let us see the case of words of length 3.

For the 6 words w in a, b, c with the 3 letters appearing we have the simple table of corresponding tableaux $T(w)$:

$$abc \to \begin{matrix} a & b & c \end{matrix};\ acb \to \begin{matrix} c & \\ a & b \end{matrix};\ bac \to \begin{matrix} b & \\ a & c \end{matrix};\ bca \to \begin{matrix} b & \\ a & c \end{matrix};\ cab \to \begin{matrix} c & \\ a & b \end{matrix};\ cba \to \begin{matrix} c \\ b \\ a \end{matrix}.$$

We write $acb \overset{K}{\cong} cab$, $bac \overset{K}{\cong} bca$ to recall that they have the same insertion tableau. For words of length 3 with repetition of letters we further have $aba \overset{K}{\cong} baa$ and $bab \overset{K}{\cong} bba$.

At this point we have two possible approaches to Knuth equivalence that we will prove are equivalent. One would be to declare equivalent two words if they have the same insertion tableau. The other, which we take as definition, is:

[119] I tried to find a real word; this is maybe dialect.

Definition. Knuth equivalence on words is the minimal equivalence generated by the previous equivalence on all possible subwords of length 3. We will write $w_1 \overset{K}{\cong} w_2$.

In other words we pass from one word to another in the same Knuth equivalence class, by substituting a string of 3 consecutive letters with an equivalent one according to the previous table.

Proposition. *Knuth equivalence is compatible with multiplication of words. It is the minimal compatible equivalence generated by the previous equivalence on words of length 3.*

Proof. This is clear by definition. □

The main result on Knuth equivalence is the following:

Theorem. *Two words w_1, w_2 are Knuth equivalent if and only if they have the same insertion tableau, i.e., when $T(w_1) = T(w_2)$.*

Proof. We start by proving that if w_1, w_2 are Knuth equivalent, then $T(w_1) = T(w_2)$. The reverse implication will be proved after we develop, in the next section, the *jeu de taquin*.

By the construction of the insertion tableau it is clear that we only need to show that if w_1, w_2 are Knuth equivalent words of length 3, and z is a word, then $T(w_1 z) = T(w_2 z)$. In other words, when we insert the word w_1 in the tableau $T(z)$ we obtain the same result as when we insert w_2.

The proof is done by case analysis. For instance, let us do the case $w_1 = wuv \overset{K}{\cong} uwv = w_2$ for 3 arbitrary letters $u < v < w$. We have to show that inserting these letters in the 2 given orderings in a semistandard tableau T produces the same result.

Let c be the first column of T, and T' the tableau obtained from T by removing the first column.

Suppose first that inserting in succession uwv in c, we place these letters in 3 distinct boxes, expelling successively some letters f, g, e. From the analysis of the positions in which these letters were, it is easily seen that $e < f < g$ and that, inserting wuv, we expel f, e, g. Thus in both cases the first column is obtained from c replacing e, f, g with u, v, w.

The tableau T' now is modified by inserting the word egf or gef, which are elementary Knuth equivalent. Thus we are in a case similar to the one in which we started for a smaller tableau and induction applies.

Some other cases are possible and are similarly analyzed.

If the top element of c is $< u$, the result of insertion is, in both cases, to place u, w on top of c and insert v in T'. The analysis is similar if w or u expels v. □

The set of words modulo Knuth equivalence is thus a monoid under multiplication, called by Schützenberger *le monoide plactique*. We will see how to use it.

There is a very powerful method to understand these operations which was invented by Schützenberger and it is called *jeu de taquin*[120] [Sch]. In order to explain it we must first of all discuss skew diagrams and tableaux.

2 Jeu de Taquin

2.1 Slides

In our convention (the French convention) we draw our tableaux in the *quadrant* $Q = \{(i, j) \mid i, j \in \mathbb{N}^+\}$. We refer to the elements of Q as *boxes*. In Q we have the partial order $(i, j) \leq (h, k) \iff i \leq h, \text{ and } j \leq k$. Given $c \leq d,\ c, d \in Q$ the set $R(c, d) := \{a \in Q \mid c \leq a \leq d\}$ is a *rectangle*, the set $R_c := \{a \in Q \mid a \leq c\}$ is the rectangle $R((1, 1), c)$ while the set $Q_c := \{a \in Q \mid c \leq a\}$ is itself a quadrant.

Definition 1. A finite set $S \subset Q$ is called:

1) A diagram, if for every $c \in S$ we have $R_c \subset S$.
2) A skew diagram if given $c \leq d$, both in S, we have $R(c, d) \subset S$.
3) A box c is called an outer (resp. inner) box of S if $S \cap Q_c = \emptyset$ (resp. $S \cap R_c = \emptyset$).

Observe that, in our definition, it is possible that a box is at the same time inner and outer.

We will usually denote diagrams with greek letters. If $\mu \subset \lambda$ are diagrams we see that $\lambda - \mu$ is a skew diagram, indicated by λ/μ. Each skew diagram can be expressed in this way although not uniquely. A skew diagram is also called a *shape* and one refers to a diagram as a *normal shape*.

Definition 2. A standard (resp. semistandard) skew tableau T is a filling of a skew diagram λ/μ satisfying the restrictions as for diagrams (strictly increasing upwards on columns and nondecreasing from left to right on rows). We call $\lambda/\mu = sh(T)$ the shape of the tableau.

If $\mu = \emptyset$ and $sh(T) = \lambda$ we say that the tableau is in normal shape.

Example. Let $\lambda = 4, 4, 3, 2,\ \mu = 2, 2, 2$. We show λ/μ and a semistandard diagram of this shape:

$$\begin{array}{cccccccc}
\cdot & \cdot & & & a & a & & \\
 & & \cdot & & & & f & \\
 & & \cdot & \cdot & & & d & u \\
 & & \cdot & \cdot & & & b & f
\end{array}$$

Given a semistandard skew tableau, its *row word* or *reading word* is the word one obtains by successively reading the rows of the tableau, starting from the top one, and proceeding downwards.

[120] *Il gioco del 15.* Sam Loyd (1841–1911) was the creator of famous mathematical puzzles and recreations. One of his most famous was the "15 Puzzle," which consisted of a 4×4 square of tiles numbered 1 to 15, with one empty space. The challenge was, starting from an arbitrary ordering and sliding tiles one by one to an adjacent empty space, to rearrange them in numerical order.

In the previous example, the row word is $aafdubf$.

Conversely, given any word w there is a unique way of decomposing it as a product of maximal standard rows $w = w_1 w_2 \dots w_r$. For instance,[121]

$$pr \mid e \mid cip \mid it \mid ev \mid o \mid liss \mid im \mid ev \mid o \mid lm \mid ent \mid e$$

we can immediately use this decomposition to present the word as a semistandard skew tableau, stringing the standard rows together in the obvious trivial way:

$$\begin{array}{cccccccccccccccccccccccccc}
p & r \\
&& e \\
&&& c & i & p \\
&&&&&& i & t \\
&&&&&&&& e & v \\
&&&&&&&&&& o \\
&&&&&&&&&&& l \\
&&&&&&&&&&&& i & s & s \\
&&&&&&&&&&&&&&& i & m \\
&&&&&&&&&&&&&&&&& e & v \\
&&&&&&&&&&&&&&&&&&& o \\
&&&&&&&&&&&&&&&&&&&& l & m \\
&&&&&&&&&&&&&&&&&&&&&& e & n & t \\
&&&&&&&&&&&&&&&&&&&&&&&&& e
\end{array}$$

There is also another trivial way that consists of sliding each row as far to the left as is possible, maintaining the structure of semistandard tableau:

$$\begin{array}{cccccccccccc}
p & r \\
& e \\
& c & i & p \\
&&& i & t \\
&&&& e & v \\
&&&&& o \\
&&&&& l \\
&&&&& i & s & s \\
&&&&&& i & m \\
&&&&&&& e & v \\
&&&&&&&& o \\
&&&&&&&& l & m \\
&&&&&&&&& e & n & t \\
&&&&&&&&&&& e
\end{array}$$

[121] This is the longest word in Italian.

The word we started from is a standard word if and only if, after this trivial slide, it gives a tableau (not a skew one). In general this trivial slide is in fact part of a more subtle slide game, the *jeu de taquin*, which eventually leads to the standard insertion tableau of the word. The game is applied in general to a semistandard skew tableau with one *empty box,* which we visualize by a dot, as in

$$\begin{array}{cccc} g & g & & \\ e & . & f & \\ & b & d & u \\ & & b & f \end{array}$$

When we have such a tableau we can perform either a forward or a backward slide. For a backward slide we move one of the adjacent letters into the empty box from right to left or from top to bottom and empty the corresponding box, provided we maintain the standard structure of the tableau. One uses the opposite procedure for a forward slide:

$$\begin{array}{cccc} g & g & & \\ e & . & f & \\ & b & d & u \\ & & b & f \end{array} \implies \begin{array}{cccc} g & g & & \\ e & f & . & \\ & b & d & u \\ & & b & f \end{array} \qquad \text{backward.}$$

It is clear that at each step we can perform one and only one backward slide until the empty box is expelled from the tableau. Similarly for forward slides.

A typical sequence of slides comes from Knuth equivalence. Consider for instance the equivalences $bca \overset{K}{\cong} bac$ and $acb \overset{K}{\cong} cab$. We obtain them through the backward slides:

$$\begin{array}{cc} b & c \\ . & a \end{array} \implies \begin{array}{cc} b & c \\ a & . \end{array} \implies \begin{array}{cc} b & . \\ a & c \end{array} \; ; \quad \begin{array}{cc} a & c \\ . & b \end{array} \implies \begin{array}{cc} . & c \\ a & b \end{array} \implies \begin{array}{cc} c & . \\ a & b \end{array}$$

Definition 3. An inner corner (resp. an outer corner) of a skew diagram S is an inner (resp. outer) box c such that $S \cup \{c\}$ is still a diagram.

A typical step of *jeu de taquin* consists in picking one of the *inner corners* of a skew diagram and perform the backward slides until the empty box (the hole) exits the diagram and becomes an outer corner. We will call this a *complete backward slide*, similarly for *complete forward slides*.

Example. We have 2 inner corners and show sequences of complete backward slides:

$$\begin{array}{ccc} i & t & \\ \bullet & e & v \\ . & . & o \end{array} \Rightarrow \begin{array}{ccc} i & t & \\ e & \bullet & v \\ . & . & o \end{array} \Rightarrow \begin{array}{ccc} i & \bullet & \\ e & t & v \\ . & . & o \end{array} \Rightarrow \begin{array}{ccc} i & & \\ e & t & v \\ . & o & \bullet \end{array} \Rightarrow \begin{array}{ccc} i & & \\ e & t & \\ \bullet & o & v \end{array} \Rightarrow \begin{array}{ccc} i & & \\ \bullet & t & \\ e & o & v \end{array} \Rightarrow \begin{array}{ccc} i & t & \\ e & o & v \end{array}$$

We could have started from the other inner corner:

$$\begin{array}{ccc} i & t & \\ . & e & v \\ . & . & o \end{array} \Rightarrow \begin{array}{ccc} i & t & \\ . & e & v \\ . & o & . \end{array} \Rightarrow \begin{array}{ccc} i & t & \\ . & e & \\ . & o & v \end{array} \Rightarrow \begin{array}{ccc} i & t & \\ e & . & \\ . & o & v \end{array} \Rightarrow \begin{array}{ccc} i & & \\ e & t & \\ . & o & v \end{array} \Rightarrow \begin{array}{ccc} i & & \\ . & t & \\ e & o & v \end{array} \Rightarrow \begin{array}{ccc} & & \\ i & t & \\ e & o & v \end{array}$$

In this example we discover that, at the end, we always obtain the same tableau.

One should also take care of the trivial case in which an inner corner is also outer and nothing happens:

$$\begin{array}{cc} i & \\ . & \bullet \\ . & o \end{array} \qquad \begin{array}{cc} i & \\ . & . \\ . & o \end{array}$$

We want to introduce a notation for this procedure. Starting from an inner corner c of a semistandard tableau T, proceed with the corresponding backward slides on T. As the final step we vacate a cell d of T obtaining a new semistandard tableau T'. We set

$$T' := j_c(T), \quad d := v_c(T).$$

Then d is an outer corner of T' and if we make the forward slides starting from it, we just invert the previous sequence, restore T and vacate c. We set

$$T := j^d(T'), \quad c := v^d(T').$$

We thus have the identity $T = j^d j_c(T)$ if $d = v_c(T)$. We have a similar identity if we start from forward slides.

We see in this example the fundamental results of Schützenberger:

Theorem 1. *Two skew semistandard tableaux of row words w_1, w_2 can be transformed one into the other, performing in some order complete backward slides starting from inner corners or complete forward slides from outer corners, if and only if the two row words are Knuth equivalent.*

Corollary. *Starting from a skew semistandard tableau of row word w, if we perform in any order, complete backward slides starting from inner corners, at the end we always arrive at the semistandard insertion tableau of w.*

In order to prove this theorem we first need a:

Lemma. *Let $a_1 a_2 \dots a_n$ be a standard row (i.e., $a_i \leq a_{i+1}, \forall i$).*

(i) If $y \leq a_1$ we have $a_1 a_2 \dots a_n y \overset{K}{\cong} a_1 y a_2 \dots a_n$.
(ii) If $y \geq a_n$ we have $y a_1 a_2 \dots a_n \overset{K}{\cong} a_1 a_2 \dots a_{n-1} y a_n$.

Proof. By induction on n. If $n = 2$ we are just using the definition of Knuth equivalence on words of length 3. Otherwise, in the first case if $y < a_1$ (or $y = a_1$) by induction $a_1 a_2 \dots a_n y \overset{K}{\cong} a_1 a_2 y \dots a_n$. Next we use the equivalence $a_1 y a_2 \overset{K}{\cong} a_1 a_2 y$.

In the second case $y > a_n$ (or $y = a_n$). By induction $y a_1 a_2 \dots a_n \overset{K}{\cong} a_1 a_2 \dots y a_{n-1} a_n$ and then we use the equivalence $y a_{n-1} a_n \overset{K}{\cong} a_{n-1} y a_n$. □

The reader should remark that this inductive proof corresponds in fact to a sequence of backward, respectively forward, slides for the tableaux:

$$y < a_1 : \quad \begin{matrix} a_1 & a_2 & a_3 & \dots & a_n & \\ \cdot & \cdot & \cdot & \dots & \cdot & y \end{matrix} \implies \begin{matrix} a_1 & \cdot & \cdot & \dots & \cdot \\ y & a_2 & a_3 & \dots & a_n \end{matrix}$$

$$y > a_n : \quad \begin{matrix} & y & \cdot & \cdot & \dots & \\ a_1 & a_2 & a_3 & \dots & a_n \end{matrix} \implies \begin{matrix} a_1 & a_2 & a_3 & \dots & y \\ \cdot & \cdot & \cdot & \dots & a_n \end{matrix}$$

Proof of the Theorem. We must show that the row words we obtain through the slides are always Knuth equivalent. The slides along a row do not change the row word, so we only need to analyze slides from top to bottom or conversely. Let us look at one such move:

$$\begin{matrix} \dots u & c_1 & c_2 & \dots & c_n & x & d_1 & d_2 & \dots & d_m & \\ & a_1 & a_2 & \dots & a_n & \cdot & b_1 & b_2 & \dots & b_m & \dots v \end{matrix}$$

to

$$\begin{matrix} \dots u & c_1 & c_2 & \dots & c_n & \cdot & d_1 & d_2 & \dots & d_m & \\ & a_1 & a_2 & \dots & a_n & x & b_1 & b_2 & \dots & b_m & \dots v \end{matrix}$$

Since Knuth equivalence is compatible with multiplication, we are reduced to analyze:

$$\begin{matrix} c_1 & c_2 & \dots & c_n & x & d_1 & d_2 & \dots & d_m \\ a_1 & a_2 & \dots & a_n & \cdot & b_1 & b_2 & \dots & b_m \end{matrix}$$

to

$$\begin{matrix} c_1 & c_2 & \dots & c_n & \cdot & d_1 & d_2 & \dots & d_m \\ a_1 & a_2 & \dots & a_n & x & b_1 & b_2 & \dots & b_m \end{matrix}$$

Here $a_n < c_n \le x \le b_1$. If $n = m = 0$ there is nothing to prove. Let $n > 0$. Perform the slide which preserves the row word:

$$\begin{matrix} c_1 & & & & & & & & \\ \cdot & c_2 & \dots & c_n & x & d_1 & d_2 & \dots & d_m \\ a_1 & a_2 & \dots & a_n & \cdot & b_1 & b_2 & \dots & b_m \end{matrix}$$

Next the slide:

$$\begin{array}{ccccccccc} c_1 & & & & & & & & \\ a_1 & c_2 & \dots & c_n & x & d_1 & d_2 & \dots & d_m \\ . & a_2 & \dots & a_n & . & b_1 & b_2 & \dots & b_m \end{array}$$

is compatible with Knuth equivalence by the previous Lemma (i) applied to $y = a_1 < c_1$ and the standard row $c_1c_2 \dots c_nxd_1d_2 \dots d_m$. Now we can apply induction and get Knuth equivalence for the slide:

$$\begin{array}{ccccccccc} c_1 & & & & & & & & \\ a_1 & c_2 & \dots & c_n & . & d_1 & d_2 & \dots & d_m \\ . & a_2 & \dots & a_n & x & b_1 & b_2 & \dots & b_m \end{array}$$

Next we slide

$$\begin{array}{ccccccccc} c_1 & & & & & & & & \\ . & c_2 & \dots & c_n & . & d_1 & d_2 & \dots & d_m \\ a_1 & a_2 & \dots & a_n & x & b_1 & b_2 & \dots & b_m \end{array} \qquad \begin{array}{ccccccccc} . & & & & & & & & \\ c_1 & c_2 & \dots & c_n & . & d_1 & d_2 & \dots & d_m \\ a_1 & a_2 & \dots & a_n & x & b_1 & b_2 & \dots & b_m \end{array}$$

To justify that the first slide preserves Knuth equivalence, notice that we can slide to the left:

$$\begin{array}{ccccccccc} . & c_2 & \dots & c_n & d_1 & d_2 & \dots & d_m & \\ a_1 & a_2 & \dots & a_n & x & b_1 & b_2 & \dots & b_m \end{array}$$

The same argument as before allows us to take away b_m and analyze the same type of slide as before but in fewer elements, and apply induction.

In the induction we performed we were reducing the number n, thus we still have to justify the slide in the case $n = 0,\ m > 0$. In this case we have to use the second part of the lemma. Slide:

$$\begin{array}{ccccc} x & d_1 & d_2 & \dots & d_m \\ . & b_1 & b_2 & \dots & b_m \end{array} \Rightarrow \begin{array}{cccccc} x & d_1 & d_2 & \dots & d_{m-1} & d_m \\ . & b_1 & b_2 & \dots & b_{m-1} & . \\ & & & & & b_m \end{array} \Rightarrow \begin{array}{cccccc} x & d_1 & d_2 & \dots & d_{m-1} & . \\ . & b_1 & b_2 & \dots & b_{m-1} & d_m \\ & & & & & b_m \end{array}$$

The last slide preserves Knuth equivalence by the previous case. Now by induction we pass to the $\stackrel{K}{\cong}$ equivalent:

$$\begin{array}{cccccc} . & d_1 & d_2 & \dots & d_{m-1} & . \\ x & b_1 & b_2 & \dots & b_{m-1} & d_m \\ & & & & & b_m \end{array}$$

Apply next Lemma (ii) to $y = d_m > b_m$ and the standard row $xb_1b_2 \dots b_{m-1}b_m$, obtaining

$$\begin{array}{cccccc} . & d_1 & d_2 & \dots & d_{m-1} & d_m \\ x & b_1 & b_2 & \dots & b_{m-1} & . \\ & & & & & b_m \end{array} \Rightarrow \begin{array}{cccccc} . & d_1 & d_2 & \dots & d_{m-1} & d_m \\ x & b_1 & b_2 & \dots & b_{m-1} & b_m \end{array}$$

completing the required slide and preserving Knuth equivalence. □

We can now complete the proof of:

Theorem 2. *Two words w_1, w_2 are Knuth equivalent if and only if $T(w_1) = T(w_2)$.*

Proof. We have already proved one part of this statement. We are left to show that if $T(w_1) = T(w_2)$, then $w_1 \overset{K}{\cong} w_2$. The algorithm given by *jeu de taquin* shows that, starting from w_1, we can construct first a semistandard tableau of row word w_1 then, by a sequence of slides, a standard tableau P, of semistandard word s_1, which is Knuth equivalent to w_1. Then $P = T(s_1) = T(w_1)$. Similarly for w_2, we construct a semistandard word s_2. Since $T(s_1) = T(w_1) = T(w_2) = T(s_2)$ we must have $s_1 = s_2$. Hence $w_1 \overset{K}{\cong} w_2$ by transitivity of equivalence. □

We can finally draw the important consequences for the monoid plactique. We start by:

Remark. In each Knuth equivalence class of words we can choose as canonical representative the unique semistandard word (defined in 1.1).

This means that we can formally identify the monoid plactique with the set of semistandard words. Of course the product by juxtaposition of two semistandard words is not in general semistandard, so one has to perform the Robinson–Schensted algorithm to transform it into its equivalent semistandard word and compute in this way the multiplication of the monoide plactique.

2.2 Vacating a Box

Suppose we apply a sequence of complete backward slides to a given skew tableau T and in this way n boxes of T are vacated. We can mark in succession with decreasing integers, starting from n, the boxes of T which are emptied by the procedure. We call the resulting standard skew tableau the *vacated tableau.*

We can proceed similarly for forward slides, but with increasing markings.

One explicit way of specifying a sequence of backward (resp. forward) slides for a semistandard skew tableau T of shape λ/μ is to construct a standard skew tableau U of shape μ/ν (resp. ν/λ). The sequence is determined in the following way. Start from the box of U occupied by the maximum element, say n. It is clearly an inner corner of T, so we can perform a complete backward slide from this corner. The result is a new tableau T' which occupies the same boxes of T, except that it has a new box which previously was occupied by n in U. At the same time it has vacated a box in the rim, which becomes the first box to build the vacated tableau. Then proceed with U_1, which is U once we have removed n. Let us give an example-exercise in which we write the elements of U with numbers and of T with letters. In order not to cause confusion, we draw a square around the boxes, the boxes vacated at each step.

$$\begin{array}{llll} g & g & & \\ 4 & 6 & f & \\ 2 & 5 & d & u \\ 1 & 3 & b & f \end{array}, \quad \begin{array}{llll} g & g & & \\ 4 & f & \boxed{6} & \\ 2 & 5 & d & u \\ 1 & 3 & b & f \end{array}, \quad \begin{array}{llll} g & g & & \\ 4 & f & \boxed{6} & \\ 2 & d & u & \boxed{5} \\ 1 & 3 & b & f \end{array}, \quad \begin{array}{llll} g & \boxed{4} & & \\ f & g & \boxed{6} & \\ 2 & d & u & \boxed{5} \\ 1 & 3 & b & f \end{array}$$

$$\begin{array}{llll} g & \boxed{4} & & \\ f & g & \boxed{6} & \\ 2 & d & u & \boxed{5} \\ 1 & b & f & \boxed{3} \end{array}, \quad \begin{array}{llll} g & \boxed{4} & & \\ f & \boxed{2} & \boxed{6} & \\ d & g & u & \boxed{5} \\ 1 & b & f & \boxed{3} \end{array}, \quad \begin{array}{llll} g & \boxed{4} & & \\ f & \boxed{2} & \boxed{6} & \\ d & g & \boxed{1} & \boxed{5} \\ b & f & u & \boxed{3} \end{array}$$

Let us give some formal definitions. Let T be a semistandard skew tableau of shape λ/μ, and U, V standard of shapes μ/ν, ν/λ. We set

$$j_U(T),\ v_U(T);\quad j^V(T),\ v^V(T)$$

to be the tableaux we obtain by performing backward slides with U or forward slides with V on T. If $Q := v_U(T)$; $R := v^V(T)$ we have seen:

Proposition. $T = j^Q j_U(T),\ T = j_R j^V(T)$

From the previous theorem it follows, in particular, that a skew tableau T can be transformed by *jeu de taquin* into a tableau of normal shape, and this normal shape is uniquely determined. We may sometimes refer to it as *the normal shape of T*.

3 Dual Knuth Equivalence

3.1 Dual equivalence

A fundamental discovery of Schützenberger has been the following (cf. [Sch2]).

Consider a permutation σ of $1, 2, \dots, n$, as a word $\sigma(1), \sigma(2), \dots, \sigma(n)$. We associate to it, by the Robinson–Schensted correspondence, a pair of standard tableaux P, Q.

Theorem 1. *If a permutation σ corresponds, by R-S, to the pair of tableaux A, B, then σ^{-1} corresponds, by R-S, to the pair of tableaux B, A.*

We do not prove this here since we will not need it (see [Sa]); what we shall use instead are the ideas on duality that are introduced in this proof.

Let us start by defining and studying dual Knuth equivalence for words.

Definition 1. We say that two words are *dually Knuth equivalent* if they have the same recording tableau.

For our purposes it will be more useful to study a general dual equivalence of semistandard tableaux. Let us consider two semistandard skew tableaux T, T' of shape λ/μ, having the same content.

Definition 2. A sequence of slides for a tableau T is a sequence of boxes $c_1, c_2, \ldots, c_m$, satisfying the following recursive property:

c_1 is either an inner or an outer corner of T. Set $T_1 := j_{c_1}(T)$ if c_1 is inner or $T_1 := j^{c_1}(T)$ if c_1 is outer. We then have that $c_2, \ldots, c_m$ is a sequence of slides for T_1.

The definition is so organized that it defines a corresponding sequence of tableaux T_i with $T_{i+1} = j_{c_{i+1}} T_i$ or $T_{i+1} = j^{c_{i+1}} T_i$.

Now we give the notion of dual equivalence for semistandard tableaux.

Definition 3. Two tableaux T, T' are said to be *dually equivalent,* denoted $T \overset{*}{\cong} T'$ if and only if any given sequence of slide operations that can be done on T can also be performed on T', thus producing tableaux of the same shape.[122]

We analyze dual equivalence, following closely Haiman (see [Hai]).

Let us take a sequence of diagrams $\lambda \supset \mu \supset \nu$. Let P be a standard tableau of shape λ/ν. We say that P decomposes as $P_1 \cup P_2$ with P_1 of shape μ/ν and P_2 of shape λ/μ if the entries of P_2 are all larger than the entries of P_1.

Let us observe how decompositions arise. If $n = |P|$ is the number of boxes of P and $k = |P_2|$, then P_2 is formed by the boxes in which the numbers $1, \ldots, k$ appear. Conversely, any number k with $1 \leq k < n$ determines such a decomposition.

We have the converse, starting from diagrams $\nu \subset \mu \subset \lambda$. Consider two standard tableaux P_1, P_2 with shapes λ/μ (a diagram with h boxes) and μ/ν (a diagram with k boxes). By definition, P_1, P_2 have respectively entries $1, 2, \ldots, h$ and $1, 2, \ldots, k$. We then can form the tableau $P_1 \cup P_2$ by placing the tableau P_2 in μ/ν and the tableau P_1 in λ/μ, but with each entry shifted by k.

Remark. Since all the numbers appearing in P_1 are strictly larger than those appearing in P_2, one easily verifies the following: when we perform a slide, for instance a complete backward slide from some cell c on $P_1 \cup P_2$, we first have to apply the complete slide to P_2 which leave some cell d of P_2 vacant, and then we apply the slide determined by d on P_1. In other words

$$j_c(P_1 \cup P_2) = j_d(P_1) \cup j_c(P_2).$$

For forward slides we have with similar notation the following:

$$j^c(P_1 \cup P_2) = j^c(P_1) \cup j^d(P_2).$$

[122] In fact, requiring that any sequence performed on T can also be performed on T' implies that the two shapes must be the same.

Lemma 1. *Consider two standard skew tableaux of the same shape and decomposed as $P = X \cup S \cup Y,\ Q = X \cup T \cup Y$.*

If $S \stackrel{*}{\cong} T$, then also $P \stackrel{*}{\cong} Q$.

Proof. This follows from the *jeu de taquin* and the previous remark. Start from a slide, for instance a complete backward slide from some cell c. In both cases (of standard skew tableaux) we first have to apply the slide to Y, which leaves some cell d of Y vacant. Then, in the first case we have to apply the backward slide from d to S, and in the second case we have to apply it to T.

By hypothesis of equivalence, this will leave the same cell vacant in both cases. Next we have to apply the slide to X. So we see that under this slide, the two tableaux are transformed into tableaux $P' = X' \cup S' \cup Y',\ Q' = X' \cup T' \cup Y'$, with $S' \stackrel{*}{\cong} T'$. We can thus repeat the argument for any number of backward slides; for forward slides the proof is similar. □

We want to start the analysis with the first special case.

We say that a shape λ/μ is *miniature* if it has exactly three boxes. So the first result to prove is:

Proposition 1. *Two miniature tableaux of the same shape are dual equivalent if and only if they produce an insertion tableau of the same shape and with the same recording tableau.*

Proof. Assume the two tableaux dual equivalent: If we apply the *jeu de taquin* to the first tableau, in order to construct the insertion tableau, we have to apply a sequence of backward slides, which, by assumption, can also be applied to the second tableau, which then results in an insertion tableau of the same shape and with the same recording tableau.

As for the converse, we need first a reduction to some basic cases by applying translations; then we must do a case analysis on the reading word. We leave the details to the reader.

For the basic example for anti-chains we have the following equivalence:

(3.0.1)
$$\begin{array}{ccc} 1 & & \\ & 3 & \\ & & 2 \end{array} \quad \stackrel{*}{\cong} \quad \begin{array}{ccc} 2 & & \\ & 3 & \\ & & 1 \end{array}$$

with recording tableau

$$\begin{array}{cc} 2 & \\ 1 & 3 \end{array}$$

(3.0.2)
$$\begin{array}{ccc} 2 & & \\ & 1 & \\ & & 3 \end{array} \quad \stackrel{*}{\cong} \quad \begin{array}{ccc} 3 & & \\ & 1 & \\ & & 2 \end{array}$$

with recording tableau

$$\begin{array}{cc} 3 & \\ 1 & 2 \end{array}$$

□

The other two form single dual equivalence classes

$$\begin{matrix} 1 & & \\ & 2 & \\ & & 3 \end{matrix}\ ,\quad \begin{matrix} 3 & & \\ & 2 & \\ & & 1 \end{matrix}$$

with recording tableaux

$$1\ 2\ 3, \quad \begin{matrix} 3 \\ 2 \\ 1 \end{matrix}$$

For the main reduction we use:

Definition 4. An *elementary dual equivalence* is one of the form $X \cup S \cup Y \overset{*}{\cong} X \cup T \cup Y$ in which S and T are miniature.

Lemma 2. *Let $U \overset{*}{\cong} V$ be an elementary dual equivalence. Applying any slide to U and V respectively yields U' and V' so that $U' \overset{*}{\cong} V'$ is elementary.*

Proof. This follows from the description of the slide given in Lemma 1. □

Proposition 2. *If S, T are two standard tableaux having the same normal shape λ, then S and T are connected by a chain of elementary dual equivalences.*

Proof. The proof is by induction on $n = |\lambda|$. We may assume $n \geq 3$, otherwise the two tableaux would be identical and there would be nothing to prove.

Consider the box in which n is placed. If it is the same in both tableaux, we remove it, thus obtaining tableaux of the same shape to which we now can apply induction. Otherwise, n appears in two distinct boxes, c and c': they are both corners of λ, and we may assume that c lies in a row higher than c'. We can then find another box c'' in the next row which is lower than the one where c is and as far to the right as possible. We now place $n-2$ in c'' and $n, n-1$ in c, c' in two possible ways. We fill the remaining boxes with the numbers $1, 2, \ldots, n-3$ so as to make the tableaux standard. We obtain two standard tableaux of shape λ, say S' and T', which are elementary dual equivalent by construction. The first S' has n in box c, and thus by induction is connected by a chain of elementary dual equivalences to S; the second, by the same reasoning is connected to T, and the claim follows. □

Corollary. *Two tableaux of the same normal shape are dual equivalent.*

The main theorem which we will need about dual equivalence is the following.

Theorem 2. *Two standard tableaux S, T of the same shape are dual equivalent if and only if they are connected by a chain of elementary dual equivalences.*

Proof. In one direction the theorem is obvious. Let us prove the converse. Let λ/μ be the shape of S, T, and choose a standard tableau U of normal shape μ which we will use to define the sequence of slides to put both S and T in normal shape.

By the dual equivalence of S, T we obtain two standard tableaux of the same normal shape $S' = j_U(S), T' = j_U(T)$, and in so doing we obtain the same tableau that was vacated, i.e., $V = v_U(T) = v_U(S)$. We know that S', T', which can be connected by a chain of elementary dual equivalences, yields $S' \overset{*}{\cong} R_1 \overset{*}{\cong} R_2 \overset{*}{\cong} \dots \overset{*}{\cong} R_k \overset{*}{\cong} T'$. Now we can apply Lemma 2; using the forward slides j^V, we have $T = j^V(T') \cdot S = j^V(S')$, and thus

$$S = j^V(S') \overset{*}{\cong} j^V(R_1) \overset{*}{\cong} j^V(R_2) \overset{*}{\cong} \dots \overset{*}{\cong} j^V(R_k) \overset{*}{\cong} j^V(T') = T$$

is the required chain of elementary equivalences. □

Let us now apply the theory to permutations.

First, write a permutation as a word, and then as a skew tableau as a *diagonal* as in

$$(3,4,1,5,2) = \begin{matrix} 3 & & & & \\ & 4 & & & \\ & & 1 & & \\ & & & 5 & \\ & & & & 2 \end{matrix}$$

We call such a standard tableau a *permutation tableau.*

Theorem 3. *Two permutation tableaux S, T are dual equivalent, if and only if the recording tableaux of their words are the same.*

Proof. If the two tableaux are dual equivalent, when we apply the Robinson–Schensted algorithm to both we follow the same sequence of slides and thus produce the same recording tableaux. Conversely, let U be a standard tableau, which we add to S or T to obtain a triangular tableau, which we use to define a sequence of slides of *jeu de taquin* that will produce in both cases the corresponding insertion tableaux $j_X(S), j_X(T)$. Let λ be the shape of $j_X(S)$.

Consider the tableau $Y = v_X(S)$, which was vacated. We know that j_X and j^Y establish a 1–1 correspondence between dual equivalent permutation tableaux and tableaux of normal shape λ. By the Schensted correspondence, j_X establishes a 1–1 correspondence between tableaux of shape λ and the set $R(S)$ of permutation tableaux whose recording tableau is the same as that of S.

Now we claim that by the first part, $R(S)$ contains the class of permutation tableaux dually equivalent to S. Since by the previous remark j^Y produces as many dually equivalent tableaux as the number of standard tableaux of shape λ, we must have that $R(S)$ coincides with the dual equivalence class of S. □

But now let us understand directly the elementary dual equivalences on words; we see that the two elementary equivalences given by formulas 3.0.1, 3.0.2 are

$$\dots k \dots k+2 \dots k+1 \dots \overset{*}{\cong} \dots k+1 \dots k+2 \dots k \dots$$

$$\dots k+1 \dots k \dots k+2 \dots \overset{*}{\cong} \dots k+2 \dots k \dots k+1 \dots$$

Now we have the remarkable fact that the two previous basic elementary dual Knuth relations for a permutation σ correspond to the usual elementary Knuth relation for σ^{-1}. We deduce that:

Theorem 4. *Two permutations σ, τ are* ***dually Knuth equivalent if and only if*** *σ^{-1}, τ^{-1}* ***are Knuth equivalent.***

We have up to now been restricted to standard tableaux or words without repeated letters or permutations. We need to extend the theory to semistandard tableaux or words with repeated letters. Fortunately there is a simple reduction to the previous case.

Definition 5. Given a standard tableau T with entries $1, 2, \dots, n$, call i a *descent* if $i + 1$ is in a row higher than i.

We say that a word $a_1 \leq a_2 \leq \cdots \leq a_n$ is *compatible with the descent set of T* if it is such that $a_i < a_{i+1}$ if and only if i is a descent of T.

Proposition 3. *Replacing the entries of T with $a_1, \dots, a_n$ gives a semistandard tableau. This is a bijection between semistandard tableaux S and pairs (standard tableau T, word a compatible with the descent set of T).*

Proof. In one direction, given a standard tableau filled with $1, 2, \dots, n$ we replace each i with a_i. In the reverse direction we read the semistandard tableau as follows.

We start by finding the positions of a_1, then of a_2, and so on, for each letter a_j reading from left to right. At the ith step of this procedure, we place i in the corresponding case. For instance, for

$$\begin{matrix} 5 & 5 \\ 4 & 4 & 4 \\ 3 & 3 & 3 & 5 \\ 2 & 2 & 2 & 2 & 4 \\ 1 & 1 & 1 & 1 & 2 & 3 \end{matrix} \qquad \Longrightarrow T = \begin{matrix} 18 & 19 \\ 14 & 15 & 16 \\ 10 & 11 & 12 & 20 \\ 5 & 6 & 7 & 8 & 17 \\ 1 & 2 & 3 & 4 & 9 & 13 \end{matrix}\,.$$

The a sequence is of course in this case

$$1, 1, 1, 1, 2, 2, 2, 2, 2, 3, 3, 3, 3, 4, 4, 4, 4, 5, 5, 5.$$

The descent set of T is in fact 4, 9, 13, 17. □

One checks that the *jeu-de-taquin* operations and elementary dual equivalences do not change the descent set. Now define *jeu-de-taquin* operations, dual equivalence, etc., on semistandard tableaux S to operate on T while keeping a fixed.

It follows immediately that the semistandard tableaux T and U of the same normal shape are dual equivalent by the corresponding result for standard tableaux. Also, their reading words are dual Knuth equivalent, because elementary dual equivalences on the underlying standard tableaux induce dual Knuth relations on the reading words of the corresponding semistandard tableaux.

Example: If we switch $a+1, a, a+2 <-> a+1, a+2, a$ in a standard tableau, then both before and after the switch, a is a descent and $a + 1$ is not. So this translates into either $bac \overset{K}{\cong} bca$ (with $a < b < c$ consecutive) or $yxy \overset{K}{\cong} yyx$ (with $x < y$ consecutive) in the semistandard tableau.

4 Formal Schur Functions

4.1 Schur Functions

First, let us better understand the map which associates to a skew semistandard tableau T of shape λ/μ its associated semistandard tableau (insertion tableau of the row word w of T). To study this take a fixed standard tableau P of shape μ. The *jeu de taquin* shows that the map we are studying is $T \to U := j_P(T)$. Let $\nu = sh(U)$ be the normal shape of T. If $Q := v_P(T)$ we also have that $sh(Q) = \lambda/\nu$.

The set $S_{\lambda/\mu}$ of tableaux of shape λ/μ decomposes as the union:

$$S_{\lambda/\mu} = \bigcup\nolimits_{\nu\subset\lambda} S^{\nu}_{\lambda/\mu}, \quad S^{\nu}_{\lambda/\mu} := \{T \in S_{\lambda/\mu},\ sh(j_P(T)) = \nu\}.$$

So let now U be a fixed semistandard tableau of shape ν and consider the set

$$S_{\lambda/\mu}(U) := \{T \in S^{\nu}_{\lambda/\mu},\ j_P(T) = U\}. \tag{4.1.3}$$

We have not put the symbol P in the definition since P is just auxiliary. The result is independent of P by the basic theorem on *jeu de taquin*.

For $T \in S_{\lambda/\mu}(U)$ consider the vacated tableau $Q = v_P(T)$ (of shape λ/ν). Given another semistandard tableau U' with the same shape as U, consider the tableau $T' := j^Q(U')$. We claim that:

Lemma. *(i)* $sh(T') = \lambda/\mu$.

(ii) The map $\rho_P^{U'} : T \to T' := j^{v_P(T)}(U')$ *is a bijection between* $S_{\lambda/\mu}(U)$ *and* $S_{\lambda/\mu}(U')$.

Proof. (i) Since U, U' are semistandard tableaux of the same normal shape they are dually equivalent. Hence $T = j^Q(U)$ and $T' = j^Q(U')$ have the same shape. Moreover, the dual equivalence implies that the shapes are the same at each step of the operations leading to $j^Q(U), j^Q(U')$. Hence $v^Q(U) = v^Q(U')$.

(ii) We shall show now that the inverse of the map $\rho_P^{U'} : S_{\lambda/\mu}(U) \to S_{\lambda/\mu}(U')$ is the map $\rho_P^U : T' \to j^{v_P(T')}(U)$.

Since $T = j^{v_P(T)}(U)$, $T' = j^{v_P(T)}(U')$ we have that $T \overset{*}{\cong} T'$ (§3.1 Corollary to Theorem 2). Thus we must have $Q = v_P(T) = v_P(T')$ since these tableaux record the changes of shape under the operations j_P. So

$$\rho_P^U \rho_P^{U'}(T) = \rho_P^U(T') = j^{v_P(T')}(U) = j^Q(U) = T, \tag{4.1.4}$$

and we have inverse correspondences between $S_{\lambda/\mu}(U)$ and $S_{\lambda/\mu}(U')$:

$$T \to T' := j^{v_P(T)}(U'), \quad T' \to T := j^{v_P(T')}(U). \qquad \square$$

Let us then define $d^{\nu}_{\lambda,\mu} := |S_{\lambda/\mu}(U)|$ for any semistandard tableau U of shape ν.

We arrive now at the construction of Schützenberger. Consider the monoide plactique $\mathcal{M}$ in an alphabet and define a *formal Schur function* $\mathcal{S}_\lambda \in \mathbb{Z}[\mathcal{M}]$ defined by $\mathcal{S}_\lambda = \sum w$ where w runs over all semistandard words of shape λ. Similarly define $\mathcal{S}_{\lambda/\mu}$ to be the sum of all row words which correspond to all semistandard skew tableaux of shape λ/μ. We have in the algebra of the monoide plactique:

Theorem.

$$S_{\lambda/\mu} = \sum_{\nu \subset \lambda} d^{\nu}_{\lambda,\mu} S_{\nu}, \quad S_{\lambda} S_{\mu} = \sum_{\nu} c^{\nu}_{\lambda,\mu} S_{\nu}. \tag{4.1.5}$$

for some nonnegative integers $d^{\nu}_{\lambda,\mu}$, $c^{\nu}_{\lambda,\mu}$.

Proof. The first statement is just a consequence of the previous lemma. As for the second it is enough to remark that $\mathcal{S}_{\lambda}\mathcal{S}_{\mu} = \mathcal{S}_{\gamma/\rho}$ where ρ is a rectangular diagram and γ is obtained from ρ by placing λ on its top and μ on its right, as in

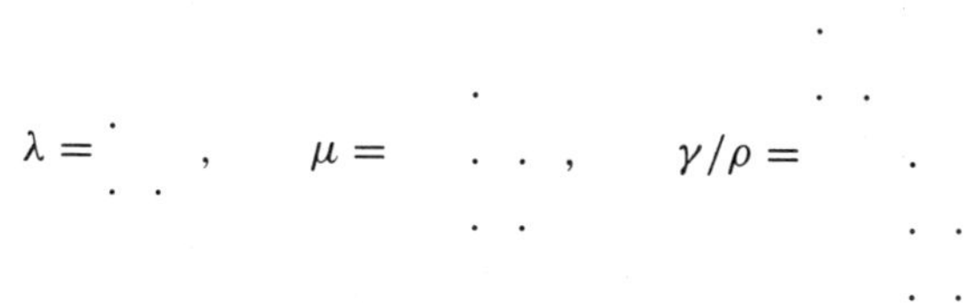

□

Now assume that we are using as an alphabet $1, 2, \dots, m$. Then we consider the content of a word as a monomial in the variables x_i. Since $c(ab) = c(a)c(b)$ and content is compatible with Knuth equivalence, we get a morphism: $c : \mathbb{Z}[\mathcal{M}] \to \mathbb{Z}[x_1, \dots, x_m]$.

Proposition. $c(\mathcal{S}_{\lambda}) = S_{\lambda}(x_1, x_2, \dots, x_m)$.

Proof. This is a consequence of Chapter 9, Theorem 10.3.1, stating that

$$S_{\lambda}(x_1, x_2, \dots, x_m) = \sum_{T} x^T,$$

where the sum is indexed by semistandard tableaux T of shape λ, filled with $1, \dots, m$. x^T is, in monomial form, the content $c(T)$. □

It follows that $S_{\lambda}(x)S_{\mu}(x) = \sum_{\nu} c^{\nu}_{\lambda,\mu} S_{\nu}(x)$. The interpretation of the symmetric function associated to $\mathcal{S}_{\lambda/\mu}$ will be given in the next section.

5 The Littlewood–Richardson Rule

5.1 Skew Schur Functions

The Littlewood–Richardson rule describes in a combinatorial way the multiplicities of the irreducible representations of $GL(V)$ that decompose a tensor product $S_{\lambda}(V) \otimes S_{\mu}(V) = \bigoplus_{\nu} c^{\nu}_{\lambda,\mu} S_{\nu}(V)$. Using characters, this is equivalent to finding the multiplication between symmetric Schur functions.[123]

[123] These types of formulas are usually called Clebsch–Gordan formulas, since for $SL(2, \mathbb{C})$ they are really the ones discussed in Chapter 3.

$$(5.1.1) \qquad S_\lambda(x)S_\mu(x) = \sum_\nu c^\nu_{\lambda,\mu} S_\nu(x).$$

We revert first to symmetric functions. Let us consider $\lambda \vdash n$ and the Schur function $S_\lambda(x_1, x_2, \dots, x_n, z_1, z_2, \dots, z_n)$ in a double set of variables. Since this is also symmetric separately in the x and z, we can expand it as a sum in the $S_\mu(x)$:

$$S_\lambda(x_1, x_2, \dots, x_n, z_1, z_2, \dots, z_n) = \sum_\mu S_\mu(x_1, x_2, \dots, x_n) S_\lambda/\mu(z_1, z_2, \dots, z_n).$$

$S_{\lambda/\mu}(z)$ is defined by this formula and it is symmetric in the z's. Now take Cauchy's formula for the variables $x_1, x_2, \dots, x_n, z_1, z_2, \dots, z_n$ and $y_1, \dots, y_n$:

$$\prod_{i,j} \frac{1}{1 - x_i y_i} \prod_{i,j} \frac{1}{1 - z_i y_i} = \sum_\lambda S_\lambda(x, z) S_\lambda(y)$$

$$= \Big(\sum_\mu S_\mu(x) S_\mu(y)\Big)\Big(\sum_\nu S_\nu(z) S_\nu(y)\Big).$$

Expand $S_\lambda(x, z) = \sum_\mu S_\mu(x) S_{\lambda/\mu}(z)$ to get

$$\sum_\lambda \sum_\mu S_\mu(x) S_{\lambda/\mu}(z) S_\lambda(y) = \sum_{\nu,\mu} S_\mu(x) S_\nu(z) \sum_\lambda c^\lambda_{\nu,\mu} S_\lambda(y),$$

hence

$$(5.1.2) \qquad S_{\lambda/\mu}(z) = \sum_\nu c^\lambda_{\nu,\mu} S_\nu(z).$$

5.2 Clebsch–Gordan Coefficients

The last formula allows one to develop a different approach to compute the numbers $c^\lambda_{\nu,\mu}$, relating these numbers to semistandard skew tableaux.

In fact, if we consider the variables x_i as less than the z_j in the given order, $S_\lambda(x, z)$ is the sum of the contents of all the semistandard tableaux of shape λ filled with the x_i and z_j (or indices corresponding to them). In each such semistandard tableau the letters x must fill a subdiagram μ and the remaining letters fill a skew diagram of shape λ/μ. We can thus separate the sum according to the μ. Since, given μ, we can fill independently with x's the diagram μ and with z's the skew diagram λ/μ, we have that this contribution to the sum $S_\lambda(x, z)$ is the product of $S_\mu(x, z)$ with a function $S'_{\lambda/\mu}(z)$ sum of the contents of all the skew tableaux filled with the z's of shape λ/μ. We deduce finally that the skew Schur function $S_{\lambda/\mu}(z)$ equals to the content $c(\mathcal{S}_{\lambda/\mu})$ of the skew formal Schur function $\mathcal{S}_{\lambda/\mu}$ defined in 4.1, and so

$$(5.2.1) \qquad S_{\lambda/\mu}(z) = c(\mathcal{S}_{\lambda/\mu}), \quad \sum_\nu c^\lambda_{\nu,\mu} S_\nu(z) = \sum_{\nu \subset \lambda} d^\nu_{\lambda,\mu} S_\nu(z), \quad c^\lambda_{\nu,\mu} = d^\nu_{\lambda,\mu}.$$

We deduce:

Proposition. $c^{\lambda}_{\nu,\mu} = d^{\nu}_{\lambda,\mu} := |S_{\lambda/\mu}(U)|$ *for any semistandard tableau U of shape ν.*

It is convenient to choose as U the semistandard tableau with i on the i^{th}row as

$$\begin{array}{lllll} 4 & 4 \\ 3 & 3 \\ 2 & 2 & 2 & 2 \\ 1 & 1 & 1 & 1 & 1 \end{array}$$

There is a unique such tableau for each shape λ; this tableau is called *the supercanonical tableau of shape* λ and denoted by C_λ. It allows us to interpret the combinatorics in terms of *lattice permutations.*

Definition. A word w in the numbers $1, \dots, r$ is called a lattice permutation if, for each initial subword (prefix) a, i.e., such that $w = ab$, setting k_i to be the number of occurrences of i in a, we have $k_1 \geq k_2 \geq \cdots \geq k_r$. A reverse lattice permutation is a word w such that the opposite word[124] w^o is a lattice permutation.

Of course the word of C_λ has this property (e.g., for $w = 4433222211111$, we have $w^o = 1111122223344$). Conversely:

Lemma. *The row word of a semistandard tableau U of shape λ is a reverse lattice permutation if and only if $U = C_\lambda$ is supercanonical.*

Proof. When we read in reverse the first row, we have by definition of semistandardness a decreasing sequence. By the lattice permutation condition this sequence must start with 1 and so it is the constant sequence 1. Now for the next row we must start with some $i > 1$ by standardness, but then $i = 2$ by the lattice permutation property, and so on. □

5.3 Reverse Lattice Permutations

Notice that for a supercanonical tableau the shape is determined by the content. To use these facts we must prove:

Proposition. *The property for the row word w of a tableau T to be a reverse lattice permutation is invariant under* jeu de taquin.

Proof. We must prove it for just one elementary vertical slide.

Let us look at one such move (the other is similar) from the word w of:

$$\begin{array}{cccccccccc} \dots u & c_1 & c_2 & \dots & c_n & x & d_1 & d_2 & \dots & d_m \\ & a_1 & a_2 & \dots & a_n & . & b_1 & b_2 & \dots & b_m & \dots v \end{array}$$

with $a_n < c_n \leq x \leq b_1 < d_1$, to:

[124] The opposite word is just the word read from right to left.

$$\begin{array}{ccccccccccccc} \dots u & c_1 & c_2 & \dots & c_n & . & d_1 & d_2 & \dots & d_m & \\ & a_1 & a_2 & \dots & a_n & x & b_1 & b_2 & \dots & b_m & \dots v \end{array}$$

The opposite of the row word is read from right to left and from bottom to top. The only changes in the contents of the prefix words of w^o (or suffixes of w) can occur in the prefixes ending with one of the a's or d's. Here the number of x's has increased by 1, so we must only show that it is still $\leq$ than the number of $x - 1$. Let us see where these elements may occur. By semistandardness, no $x - 1$ occurs among the d's and no x among the a's or the d's. Clearly if $x - 1$ appears in the a's, there is a minimum j such that $x - 1 = a_j = a_{j+1} = \cdots = a_n$. By semistandardness $x \geq c_j > a_j = x - 1$ implies that $x = c_j = c_{j+1} = \cdots = c_n$. Let $A \geq B$ (resp $A' \geq B'$) be the number of occurrences of $x - 1, x$ in the suffix of w starting with b_1 (resp with c_j). We have $A' = A + n - j + 1$ while $B' = B + n - j + 2$. Hence $A > B$. This is sufficient to conclude since in the new word for the changed suffixes the number of occurrences of x is always $B + 1 \leq A$. □

Let us say that a word has content ν if it has the same content as C_ν. We can now conclude (cf. 4.1.1):

Theorem (The Littlewood–Richardson rule).

(i) $S_{\lambda/\mu}(C_\nu)$ equals the set of skew semistandard tableaux of shape λ/μ and content ν whose associated row word is a reverse lattice permutation.

(ii) The multiplicity $c^\lambda_{\nu,\mu}$ of the representation $S_\lambda(V)$ in the tensor product $S_\mu(V) \otimes S_\nu(V)$, equals the number of skew semistandard tableaux of shape λ/μ and content ν whose associated row word is a reverse lattice permutation.

Proof. (i) $\implies$ (ii) By Proposition 5.2, $c^\lambda_{\nu,\mu} = d^\nu_{\lambda,\mu} := |S_{\lambda/\mu}(U)|$ for any tableau U of shape ν. If we choose $U = C_\nu$ supercanonical, i.e., of content ν, clearly we get the claim.

(i) By the previous proposition $S_{\lambda/\mu}(C_\nu)$ is formed by skew semistandard tableaux of shape λ/μ and content ν whose associated row word is a reverse lattice permutation. Conversely, given such a tableau T, from the previous proposition and Lemma 5.2, its associated semistandard tableau is supercanonical of content ν. Hence we have $T \in S_{\lambda/\mu}(C_\nu)$. □

13

Standard Monomials

While in the previous chapter standard tableaux had a purely combinatorial meaning, in the present chapter they will acquire a more algebro-geometric interpretation. This allows one to develop some invariant theory and representation theory in a characteristic-free way. The theory is at the same time a special case and a generalization of the results of Chapter 10, §6. In fact there are full generalizations of this theory to all semisimple groups in all characteristics, which we do not discuss (cf. [L-S], [Lit], [Lit2]).

1 Standard Monomials

1.1 Standard Monomials

We start with a somewhat axiomatic approach. Suppose that we are given: a commutative algebra R over a ring A, and a set $S := \{s_1, \dots, s_N\}$ of elements of R together with a partial ordering of S.

Definition.

(1) An ordered product $s_{i_1}s_{i_2}\dots s_{i_k}$ of elements of S is said to be standard (or to be a *standard monomial*) if the elements appear in nondecreasing order (with respect to the given partial ordering).

(2) We say that R has a standard monomial theory for S if the standard monomials form a basis of R over A.

Suppose that R has a standard monomial theory for S; given $s, t \in S$ which are not comparable in the given partial order. By axiom (2) we have a unique expression, called a *straightening law*:

(1.1.1) $$st = \sum_i \alpha_i M_i, \ \alpha_i \in A, \ M_i \text{ standard.}$$

We devise now a possible algorithm to replace any monomial $s_{i_1}s_{i_2}\dots s_{i_k}$ with a linear combination of standard ones. If, in the monomial, we find a product st with $s > t$, we replace st with ts. If instead s, t are not comparable we replace st with the right-hand side of 1.1.1.

(3) We say that R *has a straightening algorithm* if the previous replacement algorithm always stops after finitely many steps (giving the expression of the given product in terms of standard monomials).

Our prime example will be the following:

We let $A = \mathbb{Z},\ R := \mathbb{Z}[x_{ij}],\ i = 1, \dots, n;\ j = 1 \dots m$, the polynomial ring in nm variables, and S be the set of determinants of all square minors of the $m \times n$ matrix with entries the x_{ij}.

Combinatorially it is useful to describe a determinant of a $k \times k$ minor as two sequences

(1.1.2) $$(i_k\, i_{k-1}\, \dots i_1 | j_1\, j_2 \dots j_k), \quad \text{determinant of a minor}$$

where the i_t are the indices of the rows and the j_s the indices of the columns. It is customary to write the i in decreasing order and the j in increasing order.

In this notation a variable $x_{i,j}$ is denoted by $(i|j)$, e.g.,

$$(2|3) = x_{23}, \quad (2\,1|1\,3) := x_{11}x_{23} - x_{21}x_{13}.$$

The partial ordering will be defined as follows:

$$(i_h\, i_{h-1} \dots i_1 | j_1\, j_2 \dots j_h) \le (u_k\, u_{k-1} \dots u_1 | v_1\, v_2 \dots v_k) \text{ iff } h \le k,$$

$$i_s \ge u_s;\ j_t \ge v_t,\ \forall s, t \le h.$$

In other words, if we display the two determinants as rows of a bi-tableau, the left- and right-hand parts of the bi-tableau are semistandard tableaux. It is customary to call such a bi-tableau *standard*.[125]

$$\begin{array}{c} u_k \dots u_h\, u_{h-1} \dots u_1 | v_1\, v_2 \dots v_h \dots v_k \\ i_h\, i_{h-1}\, \dots\, i_1\, | j_1\, j_2 \dots\, j_h. \end{array}$$

Let us give the full partially ordered set of the 9 minors of a 2×3 matrix; see the figure on p. 501

In the next sections we will show that $\mathbb{Z}[x_{ij}]$ has a standard monomial theory with respect to this partially ordered set of minors and we will give explicitly the straightening algorithm.

---

125 Now we are using the English notation. The left tableau is in fact a mirror of a semistandard tableau.

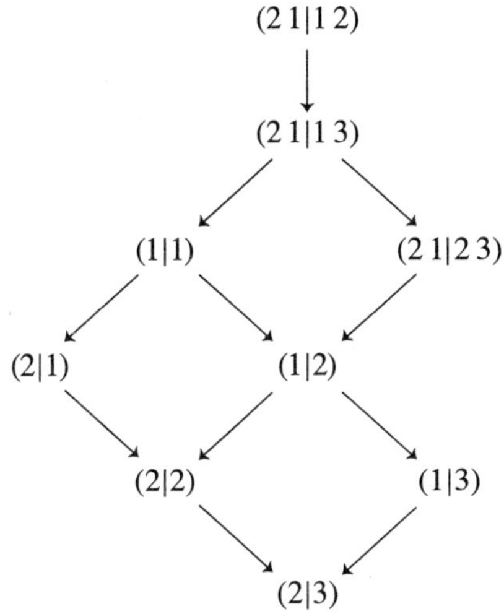

2 Plücker Coordinates

2.1 Combinatorial Approach

We start with a very simple combinatorial approach to which we will soon give a deeper geometrical meaning.

Denote by $M_{n,m}$ the space of $n \times m$ matrices. Assume $n \leq m$. We denote by $x_1, x_2, \ldots, x_m$ the columns of a matrix in $M_{n,m}$. Let $A := \mathbb{Z}[x_{ij}]$ be the ring of polynomial functions on $M_{n,m}$ with integer coefficients. We may wish to consider an element in A as a function of the columns and then we will write it as $f(x_1, x_2, \ldots, x_m)$. Consider the *generic matrix* $X := (x_{ij}),\ i = 1, \ldots, n;\ j = 1, \ldots, m$ of indeterminates. We use the following notation: Given n integers $i_1, i_2, \ldots, i_n$ chosen from the numbers $1, 2, \ldots, m$ we use the symbol:

(2.1.1) $$[i_1, i_2, \ldots, i_n] \qquad \text{Plücker coordinate}$$

to denote the determinant of the maximal minor of X which has as columns the columns of indices $i_1, i_2, \ldots, i_n$. We call such a polynomial a *Plücker coordinate*.

The first properties of these symbols are:

S1) $[i_1, i_2, \ldots, i_n] = 0$ if and only if 2 indices coincide.
S2) $[i_1, i_2, \ldots, i_n]$ is *antisymmetric* (under permutation of the indices).
S3) $[i_1, i_2, \ldots, i_n]$ is multilinear as a function of the vector variables.

We are now going to show that the Plücker coordinates satisfy some basic quadratic equations. Assume $m \geq 2n$ and consider the product:

(2.1.2) $$f(x_1, x_2, \ldots, x_{2n}) := [1, 2, \ldots, n][n+1, n+2, \ldots, 2n].$$

Select now an index $k \leq n$ and the $n+1$ variables $x_k, x_{k+1}, \dots, x_n, x_{n+1}, x_{n+2}, \dots, x_{n+k}$.

Next alternate the function f in these variables:[126]

$$\sum_{\sigma \in S_{n+1}} \epsilon_\sigma f(x_1, \dots, x_{k-1}, x_{\sigma(k)}, x_{\sigma(k+1)}, \dots, x_{\sigma(n)},$$

$$x_{\sigma(n+1)}, \dots, x_{\sigma(n+k)}, x_{n+k+1}, \dots, x_{2n}).$$

The result is a multilinear and alternating expression in the $n+1$ vector variables

$$x_k, x_{k+1}, \dots, x_n, x_{n+1}, x_{n+2}, \dots, x_{n+k}.$$

This is necessarily 0 since the vector variables are n-dimensional.

We have thus already found a *quadratic relation* among Plücker coordinates. We need to simplify it and expand it.

The symmetric group S_{2n} acts on the space of $2n$-tuples of vectors x_i by permuting the indices. Then we have an induced action on functions by

$$(\sigma g)(x_1, x_2, \dots, x_{2n}) := g(x_{\sigma(1)}, x_{\sigma(2)}, \dots, x_{\sigma(2n)}).$$

The function $[1, 2, \dots, n][n+1, n+2, \dots, 2n]$ is alternating with respect to the subgroup $S_n \times S_n$ acting separately on the first n and last n indices.

Given $k \leq n$, consider the symmetric group S_{n+1} (subgroup of S_{2n}), permuting only the indices $k, k+1, \dots, n+k$. With respect to the action of this subgroup, the function $[1, 2, \dots, n][n+1, n+2, \dots, 2n]$ is alternating with respect to the subgroup $S_{n-k+1} \times S_k$ of the permutations which permute separately the variables $k, k+1, \dots, n$ and $n+1, n+2, \dots, n+k$.

Thus if $g \in S_{n+1}, h \in S_{n-k+1} \times S_k$, we have

$$ghf(x_1, x_2, \dots, x_{2n}) = \epsilon_h g f(x_1, x_2, \dots, x_{2n}).$$

We deduce that, if $g_1, g_2, \dots, g_N$ are representatives of the left cosets $g(S_{n-k+1} \times S_k)$,

$$0 = \sum_{i=1}^{N} \epsilon_{g_i} g_i f(x_1, x_2, \dots, x_{2n}). \tag{2.1.3}$$

As representatives of the cosets we may choose some canonical elements. Remark that two elements $a, b \in S_{n+1}$ are in the same left coset with respect to $S_{n-k+1} \times S_k$ if and only if they transform the numbers $k, k+1, \dots, n$ and $n+1, n+2, \dots, n+k$ into the same sets of elements. Therefore we can choose as representatives for right cosets the following permutations:

(i) Choose a number h and select h elements out of $k, k+1, \dots, n$ and another h out of $n+1, n+2, \dots, n+k$. Then exchange in order the first set of h elements with the second. Call this permutation an exchange. Its sign is $(-1)^h$.

---

[126] In this chapter ϵ_σ denotes the sign of a permutation.

(ii) A better choice may be the permutation obtained by composing such an exchange with a reordering of the indices in each Plücker coordinate. This is an *inverse shuffle* (inverse of the operation performed on a deck of cards by a single shuffle).

The inverse of a shuffle is a permutation σ such that

$$\sigma(k) < \sigma(k+1) < \cdots < \sigma(n) \quad \text{and} \quad \sigma(n+1) < \sigma(n+2) < \cdots < \sigma(n+k).$$

Thus the basic relation is: the sum (with signs) of all exchanges, or inverse shuffles, in the polynomial $f(x_1, x_2, \ldots, x_{2n})$, of the variables $x_k, x_{k+1}, \ldots, x_n$, with the variables $x_{n+1}, x_{n+2}, \ldots, x_{n+k}$ equal to 0.

The simplest example is the *Klein quadric*, where $n = 2, m = 4$. It is the equation of the set of lines in 3-dimensional projective space. We start to use the combinatorial display of a product of Plücker coordinates as a tableau. In this case we write

$$\begin{vmatrix} a & b \\ c & d \end{vmatrix} := [a, b][c, d], \qquad \text{product of two Plücker coordinates.}$$

$$0 = \begin{vmatrix} 1 & 4 \\ 2 & 3 \end{vmatrix} - \begin{vmatrix} 1 & 2 \\ 4 & 3 \end{vmatrix} - \begin{vmatrix} 1 & 3 \\ 2 & 4 \end{vmatrix} = \begin{vmatrix} 1 & 4 \\ 2 & 3 \end{vmatrix} + \begin{vmatrix} 1 & 2 \\ 3 & 4 \end{vmatrix} - \begin{vmatrix} 1 & 3 \\ 2 & 4 \end{vmatrix}$$

which expresses the fact that the variety of lines in $\mathbb{P}^3$ is a quadric in $\mathbb{P}^5$.

We can now choose any indices $i_1, i_2, \ldots, i_n; j_1, j_2, \ldots, j_n$ and substitute in the basic relation 2.1.3 for the vector variables $x_h,\ h = 1, \ldots, n$, the variable x_{i_h} and for $x_{n+h}, h = 1, \ldots, n$ the variables x_{j_h}. The resulting relation will be denoted symbolically by

$$\text{(2.1.4)} \qquad \sum \epsilon \begin{vmatrix} i_1, i_2, \ldots, \underline{i_k, \ldots, i_n} \\ \underline{j_1, j_2, \ldots, j_k}, \ldots, j_n \end{vmatrix} \cong 0$$

where the symbol should remind us that we should sum over all exchanges of the underlined indices with the sign of the exchange, and the two-line tableau represents the product of the two corresponding Plücker coordinates.

We want to work in a formal way and consider the polynomial ring in the symbols $|i_1, i_2, \ldots, i_n|$ as independent variables only subject to the symmetry conditions S1, S2. The expressions 2.1.4 are to be thought of as quadratic polynomials in this polynomial ring.

When we substitute for the symbol $|i_1, i_2, \ldots, i_n|$ the corresponding Plücker coordinate $[i_1, i_2, \ldots, i_n]$, the quadratic polynomials 2.1.4 vanish, i.e., they are *quadratic equations*.

Remark. If some of the indices i coincide with indices j, it is possible that several terms of the quadratic relation vanish or cancel each other.

Let us thus define a ring A as the polynomial ring $\mathbb{Z}[\,|i_1, i_2, \ldots, i_n|\,]$ modulo the ideal J generated by the quadratic polynomials 2.1.4. The previous discussion shows that we have a homomorphism:

$$\text{(2.1.5)} \qquad j : A = \mathbb{Z}[\,|i_1, i_2, \ldots, i_n|\,]/J \to \mathbb{Z}[\,[i_1, i_2, \ldots, i_n]\,].$$

One of our goals is to prove:

Theorem. *The map j is an isomorphism.*

2.2 Straightening Algorithm

Before we can prove Theorem 2.1, we need to draw a first consequence of the quadratic relations. For the moment when we speak of a Plücker coordinate $[i_1, i_2, \dots, i_n]$ we will mean only the class of $|i_1, i_2, \dots, i_n|$ in A. Of course with Theorem 2.1 this use will be consistent with our previous one.

Consider a product of m Plücker coordinates

$$[i_{11}, i_{12}, \dots, i_{1k}, \dots, i_{1n}][i_{21}, i_{22}, \dots, i_{2k}, \dots, i_{2n}] \dots \times [i_{m1}, i_{m2}, \dots, i_{mk}, \dots, i_{mn}]$$

and display it as an m-row tableau:

$$(2.2.1) \qquad \begin{vmatrix} i_{11} \, i_{12} \dots \, i_{1k} \, \dots \, i_{1n} \\ i_{21} \, i_{22} \dots \, i_{2k} \, \dots \, i_{2n} \\ \dots \\ \dots \\ i_{m1} \, i_{m2} \dots \, i_{mk} \dots \, i_{mn} \end{vmatrix}$$

Due to the antisymmetry properties of the coordinates let us assume that the indices in each row are strictly increasing; otherwise the product is either 0, or up to sign, equals the one in which each row has been reordered.

Definition. We say that a rectangular tableau is *standard* if its rows are strictly increasing and its columns are non-decreasing (i.e., $i_{hk} < i_{h\,k+1}$ and $i_{hk} \le i_{h+1\,k}$). The corresponding monomial is then called a *standard monomial.*

It is convenient, for what follows, to associate to a tableau the word obtained by sequentially reading the numbers on each row:

$$(2.2.2) \qquad i_{11}\, i_{12} \dots i_{1k} \dots i_{1n}, i_{21}\, i_{22} \dots i_{2k} \dots \, i_{2n} \dots \dots i_{m1}\, i_{m2} \dots i_{mk} \dots \, i_{mn}$$

and order these words lexicographically. It is then clear that if the rows of a tableaux T are not strictly increasing, the tableaux T' obtained from T by reordering the rows in an increasing way is strictly smaller than T in the lexicographic order.

The main algorithm is given by:

Lemma. *A product T of two Plücker coordinates*

$$T := \begin{vmatrix} i_1, i_2, \dots, i_k, \dots, i_n \\ j_1, j_2, \dots, j_k, \dots, j_n \end{vmatrix}$$

can be expressed through the quadratic relations 2.1.4 as a linear combination with integer coefficients of standard tableaux with 2 rows, preceding T in the lexicographic order and filled with the same indices $i_1, i_2, \dots, i_k, \dots, i_n, j_1, j_2, \dots, j_k, \dots, j_n$.

Proof. We may assume first that the 2 rows are strictly increasing. Next, if the tableau is not standard, there is a position k for which $i_k > j_k$, and hence

$$j_1 < j_2 < \cdots < j_k < i_k < \cdots < i_n.$$

We call such a position a *violation* of the standard form. We then apply the corresponding quadratic equation. In this equation, every inverse shuffle different from the identity, replaces some of the indices $i_k < \cdots < i_n$ with indices from $j_1 < j_2 < \cdots < j_k$. It produces thus a tableau which is strictly lower lexicographically than T. Thus, if T is not standard it can be expressed, via the relations 2.1.4, as a linear combination of lexicographically smaller tableaux. We say that we have applied a step of a *straightening algorithm.*

Take the resulting expression, if it is a linear combination of standard tableaux we stop. Otherwise we repeat the algorithm to all the non-standard tableaux which appear. Each non-standard tableau is replaced with a linear combination of strictly smaller tableaux. Since the two-line tableaux filled with the indices $i_1, i_2, \ldots, i_k, \ldots, i_n, j_1, j_2, \ldots, j_k, \ldots, j_n$ are a finite number, totally ordered lexicographically, the straightening algorithm must terminate after a finite number of steps, giving an expression with only standard two-row tableaux. □

We can now pass to the general case:

Theorem. *Any rectangular tableau with m rows is a linear combination with integer coefficients of standard tableaux. The standard form can be obtained by a repeated application of the straightening algorithm to pairs of consecutive rows.*

Proof. The proof is essentially obvious. We first reorder each row, and then inspect the tableau for a possible violation in two consecutive rows. If there is no violation, the tableau is standard. Otherwise we replace the two given rows with a sum of two-row tableaux which are strictly lower than these two rows, and then we repeat the algorithm. The same reasoning of the lemma shows that the algorithm stops after a finite number of steps. □

2.3 Remarks

Some remarks on the previous algorithm are in order. First, we can express the same ideas in the language of §1.1. On the set S of $\binom{m}{n}$ symbols $|i_1\ i_2\ \ldots\ i_n|$ where $1 \leq i_1 < i_2 < \cdots < i_n \leq m$ we consider the partial ordering (the *Bruhat order*) given by

(2.3.1) $$|i_1\ i_2\ \ldots\ i_n| \leq |j_1\ j_2\ \ldots\ j_n| \text{ if and only if } i_k \leq j_k,\ \forall k = 1, \ldots, n.$$

Observe that $|i_1\ i_2\ \ldots\ i_n| \leq |j_1\ j_2\ \ldots\ j_n|$ if and only if the tableau:

$$\begin{vmatrix} i_1\ i_2\ \ldots\ i_n \\ j_1\ j_2\ \ldots\ j_n \end{vmatrix}$$

is standard. In this language, a standard monomial is a product

$$[i_{11}, i_{12}, \ldots, i_{1k}, \ldots, i_{1n}][i_{21}, i_{22}, \ldots, i_{2k}, \ldots, i_{2n}] \cdots [i_{m1}, i_{m2}, \ldots, i_{mk}, \ldots, i_{mn}]$$

in which the coordinates appearing are increasing from left to right in the order 2.3.1. This means that the associated tableau of 2.2 is standard.

If $a = |i_1\, i_2\, \ldots\, i_n|$, $b = |j_1\, j_2\, \ldots\, j_n|$ and the product ab is not standard, then we can apply a quadratic equation and obtain $ab = \sum_i \lambda_i a_i b_i$ with λ_i coefficients and a_i, b_i obtained from a, b by the shuffle procedure of Lemma 2.2. The proof we have given shows that this is indeed a straightening algorithm in the sense of 1.1. The proof of that lemma shows in fact that $a < a_i,\ b > b_i$. It is useful to axiomatize the setting.

Definition. Suppose we have a commutative algebra R over a commutative ring A, a finite set $S \subset R$, and a partial ordering in S for which R has a standard monomial theory and a straightening algorithm.

We say that R is a *quadratic Hodge algebra* over S if wherever $a, b \in S$ are not comparable,

$$ab = \sum_i \lambda_i a_i b_i \tag{2.3.2}$$

with $\lambda_i \in A$ and $a < a_i,\ b > b_i$.

Notice that the quadratic relations 2.3.2 give the straightening law for R. The fact that the straightening algorithm terminates after a finite number of steps is clear from the condition $a < a_i,\ b > b_i$.

Our main goal is a theorem which includes Theorem 2.1:

Theorem. *The standard tableaux form a $\mathbb{Z}$-basis of A and A is a quadratic Hodge algebra isomorphic under the map j (cf. 2.1.5), to the ring $\mathbb{Z}[[i_1, i_2, \ldots, i_n]] \subset \mathbb{Z}[x_{ij}]$.*

Proof. Since the standard monomials span A linearly and since by construction j is clearly surjective, it suffices to show that the standard monomials are linearly independent in the ring $\mathbb{Z}[[i_1, i_2, \ldots, i_n]]$. This point can be achieved in several different ways, we will follow first a combinatorial and then, in §3.6, a more complete geometric approach through Schubert cells.

The algebraic combinatorial proof starts as follows:

Remark that, in a standard tableau, each index i can appear only in the first i columns.

Let us define a tableau to be k-canonical if, for each $i \leq k$, the indices i which appear are all in the ith column. Of course a tableau (with n columns and h rows) is n-canonical if and only if the ith column is filled with i for each i, i.e., it is of type $|1\ 2\ 3\ \ldots\ n-1\ n|^h$.

Suppose we are given a standard tableau T which is k-canonical. Let $p = p(T)$ be the minimum index (greater than k) which appears in T in a column $j < p$. Set

$m_p(T)$ to be the minimum of such column indices. The entries to the left of p, in the corresponding row, are then the indices $1\ 2\ 3\ \ldots\ j-1$.

Given an index j, let us consider the set $\mathcal{T}^k_{p,j,h}$ of k-canonical standard tableaux for which p is the minimum index (greater than k) which appears in T in a column $j < p$. $m_p(T) = j$ and in the jth column p occurs exactly h times (necessarily in h consecutive rows). In other words, reading the jth column from top to bottom, one finds first a sequence of j's and then h occurrences of p. What comes after is not relevant for the discussion.

The main combinatorial remark we make is that if we substitute p with j in all these positions, we see that we have a map which to distinct tableaux associates distinct k-canonical tableaux T', with either $p(T') > p(T)$ or $p(T') = p(T)$ and $m_p(T') > m_p(T)$.

To prove the injectivity of this map it is enough to observe that if a tableau T is transformed into a tableau T', then the tableau T is obtained from T' by substituting p for the last h occurrences of j (which are in the jth column).

The next remark is that if we substitute the variable x_i with $x_i + \lambda x_j$, $(i \neq j)$ in a Plücker coordinate $[i_1, i_2, \ldots, i_n]$, then the result of the substitution is $[i_1, i_2, \ldots, i_n]$ if i does not appear among the indices $i_1, i_2, \ldots, i_n$ or if both indices i, j appear.

If instead $i = i_k$, the result of the substitution is

$$[i_1, i_2, \ldots, i_n] + \lambda[i_1, i_2, \ldots, i_{k-1}, j, i_{k+1}, \ldots, i_n].$$

Suppose we make the same substitution in a tableau, i.e., in a product of Plücker coordinates; then by expanding the product of the transformed coordinates we obtain a polynomial in λ of degree equal to the number of entries i which appear in rows of T where j does not appear. The leading coefficient of this polynomial is the tableau obtained from T by substituting j for all the entries i which appear in rows of T where j does not appear.

After these preliminary remarks we can give a proof of the linear independence of the standard monomials in the Plücker coordinates.

Let us assume by contradiction that

(2.3.3) $$0 = f(x_1, \ldots, x_m) = \sum_i c_i T_i$$

is a dependence relation among (distinct) standard tableaux. We may assume it is homogeneous of some degree k.

At least one of the T_i must be different from a power $|1\ 2\ 3\ \ldots\ n-1\ n|^h$, since such a relation is not valid.

Then let p be the minimum index which appears in one of the T_i in a column $j < p$, and let j be the minimum of these column indices. Also let h be the maximum number of such occurrences of p and assume that the tableaux T_i, $i \leq k$ are the ones for which this happens. This implies that if in the relation 2.3.3 we substitute x_p with $x_p + \lambda x_j$, where λ is a parameter, we get a new relation which can be written as a polynomial in λ of degree h. Since this is identically 0, each coefficient must be zero. Its leading coefficient is

(2.3.4)
$$\sum_{i=1}^{k} c_i T_i'$$

where T_i' is obtained from T_i replacing the h indices p appearing in the j column with j.

According to our previous combinatorial remark the tableaux T_i' are distinct and thus 2.3.4 is a new relation. We are thus in an inductive procedure which terminates with a relation of type

$$0 = |1\ 2\ 3\ \dots\ n-1\ n|^k,$$

which is a contradiction. □

3 The Grassmann Variety and Its Schubert Cells

In this section we discuss in a very concrete way what we have already done quickly but in general in Chapter 10 on parabolic subgroups. The reader should compare the two.

3.1 Grassmann Varieties

The theory of Schubert cells has several interesting features. We start now with an elementary treatment. Let us start with an m-dimensional vector space V over a field F and consider $\bigwedge^n V$ for some $n \leq m$.

Proposition.

(1) Given n vectors $v_1, v_2, \dots, v_n \in V$, the decomposable vector

$$v_1 \wedge v_2 \wedge \cdots \wedge v_n \neq 0$$

if and only if the vectors are linearly independent.

(2) Given n linearly independent vectors $v_1, v_2, \dots, v_n \in V$ and a vector v:

$$v \wedge v_1 \wedge v_2 \wedge \cdots \wedge v_n = 0$$

if and only if v lies in the subspace spanned by the vectors v_i.

(3) If $v_1, v_2, \dots, v_n$ and $w_1, w_2, \dots, w_n$ are both linearly independent sets of vectors, then

$$w_1 \wedge w_2 \wedge \cdots \wedge w_n = \alpha v_1 \wedge v_2 \wedge \cdots \wedge v_n, \quad 0 \neq \alpha \in F$$

if and only if the two sets span the same n-dimensional subspace W of V.

Proof. Clearly (2) is a consequence of (1). As for (1), if the $v_i's$ are linearly independent they may be completed to a basis, and then the statement follows from the fact that $v_1 \wedge v_2 \wedge \cdots \wedge v_n$ is one of the basis elements of $\bigwedge^n V$.

If, conversely, one of the v_i is a linear combination of the others, we replace this expression in the product and have a sum of products with a repeated vector, which is then 0.

For (3), assume first that both sets span the same subspace. By hypothesis $w_i = \sum_j c_{ij} v_j$ with $C = (c_{ij})$ an invertible matrix, hence

$$w_1 \wedge w_2 \wedge \cdots \wedge w_n = \det(C) v_1 \wedge v_2 \wedge \cdots \wedge v_n.$$

Conversely by (2) we see that

$$W := \{v \in V \mid v \wedge w_1 \wedge w_2 \wedge \cdots \wedge w_n = 0\}. \qquad \square$$

We have an immediate geometric corollary. Given an n-dimensional subspace $W \subset V$ with basis $v_1, v_2, \ldots, v_n$, the nonzero vector $w := v_1 \wedge v_2 \wedge \cdots \wedge v_n$ determines a point in the projective space $\mathbb{P}(\bigwedge^n(V))$ (whose points are the lines in $\bigwedge^n(V)$).

Part (3) shows that this point is independent of the basis chosen but depends only on the subspace W, thus we can indicate it by the symbol $[W]$.

Part (2) shows that the subspace W is recovered by the point $[W]$. We get:

Corollary. *The map $W \to [W]$ is a 1-1 correspondence between the set of all n-dimensional subspaces of V and the points in $\mathbb{P}(\bigwedge^n V)$ corresponding to decomposable elements.*

Definition. We denote by $Gr_n(V)$ the set of n-dimensional subspaces of V or its image in $\mathbb{P}(\bigwedge^n(V))$ and call it the *Grassmann variety*.

In order to understand the construction we will be more explicit. Consider the set $S_{n,m}$ of n-tuples $v_1, v_2, \ldots, v_n$ of linearly independent vectors in V.

(3.1.1) $$S_{n,m} := \{(v_1, v_2, \ldots, v_n) \in V^n \mid v_1 \wedge v_2 \wedge \cdots \wedge v_n \neq 0\}.$$

To a given basis $e_1, \ldots, e_m$ of V, we associate the basis $e_{i_1} \wedge e_{i_2} \wedge \cdots \wedge e_{i_n}$ of $\bigwedge^n V$ where $(i_1 < i_2 < \cdots < i_n)$.

Represent in coordinates an n-tuple $v_1, v_2, \ldots, v_n$ of vectors in V as the rows of an $n \times m$ matrix X (of rank n if the vectors are linearly independent).

In the basis $e_{i_1} \wedge e_{i_2} \wedge \cdots \wedge e_{i_n}$ of $\bigwedge^n V$ the coordinates of $v_1 \wedge v_2 \wedge \cdots \wedge v_n$ are then the determinants of the maximal minors of X.

Explicitly, let us denote by $X[i_1, i_2, \dots, i_n]$ or just $[i_1, i_2, \dots, i_n]$ the determinant of the maximal minor of X extracted from the columns $i_1\, i_2 \dots i_n$. Then

$$\text{(3.1.2)} \qquad v_1 \wedge v_2 \wedge \cdots \wedge v_n = \sum_{1 \le i_1 < i_2 \dots < i_n \le m} X[i_1, i_2, \dots, i_n] e_{i_1} \wedge e_{i_2} \wedge \cdots \wedge e_{i_n}.$$

$S_{n,m}$ can be identified with the open set of $n \times m$ matrices of maximal rank, $S_{n,m}$ is called the (algebraic) Stiefel manifold.[127]

Let us indicate by $W(X)$ the subspace of V spanned by the rows of X. The group $Gl(n, F)$ acts by left multiplication on $S_{n,m}$ and if $A \in Gl(n, F)$, $X \in S_{n,m}$, we have:

$$W(X) = W(Y), \text{ if and only if, } Y = AX, \ A \in Gl(n, F)$$

$$Y[i_1, i_2, \dots, i_n] = \det(A) X[i_1, i_2, \dots, i_n].$$

In particular $Gr_n(V)$ can be identified with the set of orbits of $Gl(n, F)$ acting by left multiplication on $S_{n,m}$. We want to understand the nature of $Gr_n(V)$ as variety. We need:

Lemma. *Given a map between two affine spaces $\pi : F^k \to F^{k+h}$, of the form*

$$\pi(x_1, x_2, \dots, x_k) = (x_1, x_2, \dots, x_k, p_1, \dots, p_h)$$

with $p_i = p_i(x_1, x_2, \dots, x_k)$ polynomials, its image is a closed subvariety of F^{k+h} and π is an isomorphism of F^k onto its image.[128]

Proof. The image is the closed subvariety given by the equations

$$x_{k+i} - p_i(x_1, x_2, \dots, x_k) = 0.$$

The inverse of the map π is the projection

$$(x_1, x_2, \dots, x_k, \dots, x_{k+h}) \to (x_1, x_2, \dots, x_k). \qquad \square$$

In order to understand the next theorem let us give a general definition. Suppose we are given an algebraic group G acting on an algebraic variety V and a map $\rho : V \to W$ which is constant on the G-orbits.

We say that ρ *is a principal G-bundle locally trivial in the Zariski topology*[129] if there is a covering of W by Zariski open sets U_i in such a way that for each U_i we have a G-equivariant isomorphism $\phi_i : G \times U_i \to \rho^{-1}(U_i)$ so that the following diagram is commutative:

$$\begin{array}{ccc} G \times U_i & \xrightarrow{\ \phi_i\ } & \rho^{-1}(U_i) \\ {\scriptstyle p_2}\big\downarrow & & {\scriptstyle \rho}\big\downarrow \\ U_i & \xrightarrow{\ 1\ } & U_i \end{array} , \qquad p_2(g, u) := u.$$

[127] The usual Stiefel manifold is, over $\mathbb{C}$, the set of n-tuples $v_1, v_2, \dots, v_n$ of orthonormal vectors in $\mathbb{C}^m$. It is homotopic to $S_{n,m}$.

[128] π is the graph of a polynomial map.

[129] Usually the bundles one encounters are locally trivial only in more refined topologies.

We can now state and prove the main result of this section:

Theorem.

(1) The Grassmann variety $Gr_n(V)$ is a smooth projective subvariety of $\mathbb{P}(\bigwedge^n(V))$.
(2) The map $X \to W[X]$ from $S_{n,m}$ to $Gr_n(V)$ is a principal $Gl(n, F)$ bundle (locally trivial in the Zariski topology).

Proof. In order to prove that a subset S of projective space is a subvariety one has to show that intersecting S with each of the open affine subspaces U_i, where the ith coordinate is nonzero, one obtains a Zariski closed set $S_i := S \cap U_i$ in U_i. To prove furthermore that S is smooth one has to check that each S_i is smooth.

The proof will in fact show something more. Consider the affine open set U of $\mathbb{P}(\bigwedge^n(V))$ where one of the projective coordinates is not 0 and intersect it with $Gr_n(V)$. We claim that $U \cap Gr_n(V)$ is closed in U and isomorphic to an $n(m-n)$-dimensional affine space and that on this open set the bundle of (2) is trivial.

To prove this let us assume for simplicity of notation that U is the open set where the coordinate of $e_1 \wedge e_2 \wedge \ldots \wedge e_n$ is not 0. We use in this set the affine coordinates obtained by setting the corresponding projective coordinate equal to 1.

The condition that $W(X) \in U$ is clearly $X[1, 2, \ldots, n] \neq 0$, i.e., that the submatrix A of X formed from the first n columns is invertible. Since we have selected this particular coordinate it is useful to display the elements of $S_{n,m}$ in block form as $X = (A\ T)$, (A, T being respectively $n \times n$, $n \times m - n$ matrices).

Consider the matrix $Y = A^{-1}X = (1_n\ Z)$ with Z an $n \times m - n$ matrix and $T = AZ$. It follows that the map $i : Gl(n, F) \times M_{n,m}(F) \to S_{n,m}$ given by $i(A, Z) = (A\ AZ)$ is an isomorphism of varieties to the open set $S^0_{n,m}$ of $n \times m$ matrices X such that $W(X) \in U$. Its inverse is $j : S^0_{n,m} \to Gl(n, F) \times M_{n,m}(F)$ given by $j(A\ T) = (A, A^{-1}T)$.

Thus we have that the set of matrices of type $(1_n\ Z)$ is a set of representatives for the $Gl(n, F)$-orbits of matrices X with $W(X) \in U$. In other words, in a vector space W such that $[W] \in U$, there is a unique basis which in matrix form is of type $(1_n\ Z)$. i, j also give the required trivialization of the bundle.

Let us now understand in affine coordinates the map from the space of $n \times (m-n)$ matrices to $U \cap Gr_n(V)$. It is given by computing the determinants of the maximal minors of $X = (1_n\ Z)$. A simple computation shows that:

(3.1.3)

$$X[1\ 2 \ldots i-1\ n+k\ i+1 \ldots\ n] = \begin{vmatrix} 1 & 0 & \ldots & 0 & z_{1\,k} & 0 & 0 & \ldots & 0 \\ 0 & 1 & \ldots & 0 & z_{2\,k} & 0 & 0 & \ldots & 0 \\ \ldots & \ldots & \ldots & \ldots & \ldots & \ldots & \ldots & \ldots & \\ \ldots & \ldots & \ldots & \ldots & \ldots & \ldots & \ldots & \ldots & \\ 0 & 0 & \ldots & 1 & z_{i-1\,k} & 0 & 0 & \ldots & 0 \\ 0 & 0 & \ldots & 0 & z_{i\,k} & 0 & 0 & \ldots & 0 \\ 0 & 0 & \ldots & 0 & z_{i+1\,k} & 1 & 0 & \ldots & 0 \\ \ldots & \ldots & \ldots & \ldots & \ldots & \ldots & \ldots & \ldots & \\ \ldots & \ldots & \ldots & \ldots & \ldots & \ldots & \ldots & \ldots & \\ 0 & 0 & \ldots & 0 & z_{n\,k} & 0 & 0 & \ldots & 1 \end{vmatrix}$$

This determinant is z_{ik}. Thus Z maps to a point in U in which $n \times (m-n)$ of the coordinates are, up to sign, the coordinates z_{ik}. The remaining coordinates are instead polynomials in these variables. Now we can invoke the previous lemma and conclude that $Gr_n(V) \cap U$ is closed in U and it is isomorphic to the affine space $F^{n(m-n)}$. □

3.2 Schubert Cells

We now display a matrix $X \in S_{n,m}$ as a sequence $(w_1, w_2, \dots, w_m)$ of *column vectors* so that if A is an invertible matrix, $AX = (Aw_1, Aw_2, \dots, Aw_m)$.

If $i_1 < i_2 < \cdots < i_k$ are indices, the property that the corresponding columns in X are linearly independent is invariant in the $Gl(n, F)$-orbit, and depends only on the space $W(X)$ spanned by the rows. In particular we will consider the sequence $i_1 < i_2 < \dots < i_n$ defined inductively in the following way: w_{i_1} is the first nonzero column and inductively $w_{i_{k+1}}$ is the first column vector which is linearly independent from $w_{i_1}, w_{i_2}, \dots, w_{i_k}$.

For an n-dimensional subspace W we will set $s(W)$ to be the sequence thus constructed from a matrix X for which $W = W(X)$. We set

$$(3.2.1) \quad C_{i_1,i_2,\dots,i_n} = \{W \in Gr_n(V) |\, s(W) = i_1, i_2, \dots, i_n\}, \quad \text{a Schübert cell.}$$

$C_{i_1,i_2,\dots,i_n}$ is contained in the open set $U_{i_1,i_2,\dots,i_n}$ of $Gr_n(V)$ where the Plücker coordinate $[i_1, i_2, \dots, i_n]$ is not zero. In 3.1, we have seen that this open set can be identified to the set of $n \times (m-n)$ matrices X for which the submatrix extracted from the columns $i_1 < i_2 < \dots < i_n$ is the identity matrix. We wish thus to represent our set $C_{i_1,i_2,\dots,i_n}$ by these matrices.

By definition, we have now that the columns $i_1, i_2, \dots, i_n$ are the columns of the identity matrix, the columns before i_1 are 0 and the columns between i_k, i_{k+1} are vectors in which all coordinates greater that k are 0. We will refer to such a matrix as a *canonical representative*. For example, $n = 4$, $m = 11$, $i_1 = 2$, $i_2 = 6$, $i_3 = 9$, $i_4 = 11$. Then a canonical representative is

$$(3.2.2) \quad \begin{vmatrix} 0 & 1 & a_1 & a_2 & a_3 & 0 & b_{11} & b_{12} & 0 & c_{11} & 0 \\ 0 & 0 & 0 & 0 & 0 & 1 & b_{33} & b_{34} & 0 & c_{33} & 0 \\ 0 & 0 & 0 & 0 & 0 & 0 & 0 & 0 & 1 & c_{31} & 0 \\ 0 & 0 & 0 & 0 & 0 & 0 & 0 & 0 & 0 & 0 & 1 \end{vmatrix}.$$

Thus $C_{i_1,i_2,\dots,i_n}$ is an affine subspace of $U_{i_1,i_2,\dots,i_n}$ given by the vanishing of certain coordinates. Precisely the free parameters appearing in the columns between i_k, i_{k+1} are displayed in a $k \times (i_{k+1} - i_k - 1)$ matrix, and the ones in the columns after i_n in an $n \times (m - i_n)$ matrix. Thus:

Proposition. *$C_{i_1,i_2,\dots,i_n}$ is a closed subspace of the open set $U_{i_1,i_2,\dots,i_n}$ of the Grassmann variety called a Schübert cell. Its dimension is*

$$\dim(C_{i_1,i_2,\dots,i_n}) = \sum_{k=1}^{n-1} k(i_{k+1} - i_k - 1) + n(m - i_n)$$

(3.2.3)
$$= nm - \frac{n(n-1)}{2} - \sum_{j=1}^{n} i_j.$$

3.3 Plücker equations

Let us make an important remark. By definition of the indices $i_1, i_2, \dots, i_n$ associated to a matrix X, we have that, given a number $j < i_k$, the submatrix formed by the first j columns has rank at most $k-1$. This implies immediately that if we give indices $j_1, j_2, \dots, j_n$ for which the corresponding Plücker coordinate is nonzero, then $i_1, i_2, \dots, i_n \leq j_1, j_2, \dots, j_n$. In other words:

Proposition. *$C_{i_1,i_2,\dots,i_n}$ is the subset of $Gr_n(V)$ where $i_1, i_2, \dots, i_n$ is nonzero and all Plücker coordinates $[j_1, j_2, \dots, j_n]$ which are not greater than or equal to $[i_1, i_2, \dots, i_n]$ vanish.*

Proof. We have just shown one implication. We must see that if at a point of the Grassmann variety all Plücker coordinates $[j_1, j_2, \dots, j_n]$ which are not greater than or equal to $[i_1, i_2, \dots, i_n]$ vanish and $[i_1, i_2, \dots, i_n]$ is nonzero, then this point is in the cell $C_{i_1,i_2,\dots,i_n}$. Take as representative the matrix X which has the identity in the columns $i_1, i_2, \dots, i_n$. We must show that if $i_k < i < i_{k+1}$ the entries $x_{i,j}$, $j > k$, of this matrix are 0. We can compute this entry up to sign as the Plücker coordinate $[i_1, i_2, \dots i_{j-1}, i, i_{j+1}, \dots, i_n]$ (like in 3.1.3). Finally, reordering, we see that this coordinate is $[i_1, i_2, \dots, i_k, i, i_{k+1}, \dots, i_{j-1}, i_{j+1}, \dots, i_n]$ which is strictly less than $[i_1, i_2, \dots, i_n]$, hence 0 by hypothesis. □

We have thus decomposed the Grassmann variety into cells, indexed by the elements $[i_1, i_2, \dots, i_n]$. We have already seen that this set of indices has a natural total ordering and we wish to understand this order in a geometric fashion. Let us indicate by $P_{n,m}$ this partially ordered set. Let us visualize $P_{2,5}$ (see the diagram on p. 514):

First, let us make a simple remark based on the following:

Definition. In a partially ordered set P we will say that 2 elements a, b are *adjacent* if

$$a < b, \text{ and if } a \leq c \leq b, \text{ then } a = c, \text{ or } c = b.$$

Remark. The elements adjacent to $i_1, i_2, \dots, i_n$ are obtained by selecting any index i_k such that $i_k + 1 < i_{k+1}$ and replacing it by $i_k + 1$ (if $k = n$ the condition is $i_k < m$).

Proof. The proof is a simple exercise left to the reader. □

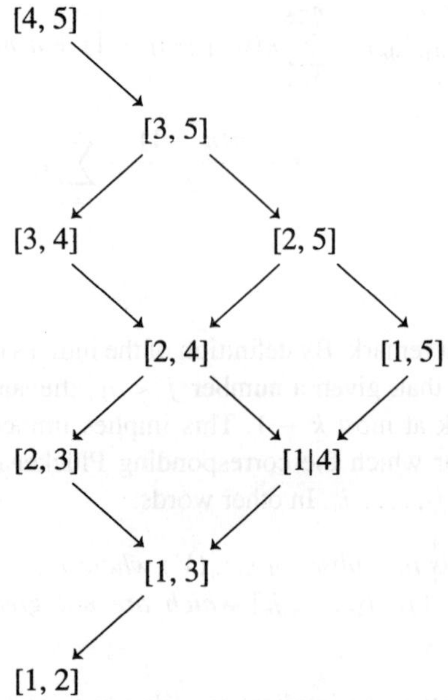

3.4 Flags

There is a geometric meaning of the Schubert cells related to the relative position with respect to a *standard flag*.

Definition. A flag in a vector space V is an increasing sequence of subspaces:

$$F_1 \subset F_2 \subset \cdots \subset F_k.$$

A complete flag in an m-dimensional space V is a flag

$$0 \subset F_1 \subset F_2 \subset \cdots \subset F_{m-1} \subset F_m = V \tag{3.4.1}$$

with $\dim(F_i) = i,\ i = 1, \ldots, m$.

Sometimes it is better to use a projective language, so that F_i gives rise to an $i-1$-dimensional linear subspace in the projective space $\mathbb{P}(V)$.

A complete flag in an m-dimensional projective space is a sequence: $\pi_0 \subset \pi_1 \subset \pi_2 \ldots \subset \pi_m$ with π_i a linear subspace of dimension i.[130]

We fix as standard flag $F_1 \subset F_2 \subset \cdots \subset F_k$ with F_i the set of vectors with the first $m-i$ coordinates equal to 0, spanned by the last i vectors of the basis $e_1, \ldots, e_m$.

Given a space $W \in C_{i_1, i_2, \ldots, i_n}$ let $v_1, \ldots, v_n$ be the corresponding normalized basis as rows of an $n \times m$ matrix X for which the submatrix extracted from the columns

[130] The term flag comes from a simple drawing in 3-dimensional projective space. The base of a flagpole is a point, the pole is a line, and the flag is a plane.

$i_1, i_2, \ldots, i_n$ is the identity matrix. Therefore a linear combination $\sum_{k=1}^{n} c_k v_k$ has the number c_k as the i_k coordinate $1 \leq k \leq n$. Thus for any i we see that

(3.4.2) $$W \cap F_i = \left\{ \sum_{k=1}^{n} c_k v_k | c_k = 0, \text{ if } i_k < m - i \right\}.$$

We deduce that for every $1 \leq i \leq m$,

$$d_i := \dim(F_i \cap W) = n - k \text{ if and only if } i_k < m - i \leq i_{k+1}.$$

In other words, the sequence of d_i's is completely determined, and determines the numbers $\underline{i} := i_1 < i_2 < \cdots < i_n$. Let us denote by $\underline{d}[\underline{i}]$ the sequence thus defined; it has the properties:

$$d_m = n, \; d_1 \leq 1, d_i \leq d_{i+1} \leq d_i + 1.$$

The numbers $m - i_k + 1$ are the ones in which the sequence jumps by 1. For the example given in (3.2.2) we have the sequence

$$1, 1, 2, 2, 2, 3, 3, 3, 3, 4, 4.$$

We observe that given two sequences

$$\underline{i} := i_1 < i_2 < \ldots < i_n, \quad \underline{j} := j_1 < j_2 < \ldots < j_n,$$

we have

$$\underline{i} \leq \underline{j} \text{ iff } \underline{d}[\underline{i}] \leq \underline{d}[\underline{j}].$$

3.5 B-orbits

We pass now to a second fact:

Definition. Let:

$$S_{i_1, i_2, \ldots, i_n} := \{W \mid \dim(F_i \cap W) \leq d_i[\underline{i}], \; \forall i\}.$$

From the previous remarks:

$$C_{i_1, i_2, \ldots, i_n} := \{W \mid \dim(F_i \cap W) = d_i[\underline{i}], \; \forall i\}, \qquad S_{\underline{i}} = \cup_{\underline{j} \geq \underline{i}} C_{\underline{j}}.$$

We need now to interpret these notions in a group-theoretic way.

We define T to be the subgroup of $GL(m, F)$ of diagonal matrices. Let $I_{i_1, i_2, \ldots, i_n}$ be the $n \times m$ matrix with the identity matrix in the columns $i_1, i_2, \ldots, i_n$ and 0 in the other columns. We call this the *center* of the Schubert cell.

Lemma. *The $\binom{m}{n}$ decomposable vectors associated to the matrices $I_{i_1, i_2, \ldots, i_n}$ are the vectors $e_{i_1} \wedge e_{i_2} \wedge \cdots \wedge e_{i_n}$. These are a basis of weight vectors for the group T acting on $\bigwedge^n F^m$. The corresponding points in projective space $P(\bigwedge^n F^m)$ are the fixed points of the action of T, and the corresponding subspaces are the only T-stable subspaces of F^m.*

Proof. Given an action of a group G on a vector space, the fixed points in the corresponding projective space are the stable 1-dimensional subspaces. If the space has a basis of weight vectors of distinct weights, any G-stable subspace is spanned by a subset of these vectors. The lemma follows. □

Remark. When $F = \mathbb{C}$, the space $\bigwedge^n \mathbb{C}^m$ is an irreducible representation of $SL(m, \mathbb{C})$ and a fundamental representation. It has a basis of weight vectors of distinct weights and they are one in orbit under the symmetric group. A representation with this property is called *minuscule*. For general Lie groups few fundamental representations are minuscule.[131]

We define B to be the subgroup of $GL(m, F)$ which stabilizes the standard flag. A matrix $X \in B$ if and only if, for each i, Xe_i is a linear combination of the elements e_j with $j \geq i$. This means that B is the group of lower triangular matrices, usually denoted by B^-. From the definitions we have clearly that the sets $C_{i_1,i_2,\dots,i_n}$, $S_{i_1,i_2,\dots,i_n}$ are stable under the action of B. In fact we have:

Theorem. *$C_{i_1,i_2,\dots,i_n}$ is a B-orbit.*

Proof. Represent the elements of $C_{i_1,i_2,\dots,i_n}$ by their matrices whose rows are the canonical basis. Consider, for any such matrix X, an associated matrix $\tilde{X}$ which has the i_k row equal to the kth row of X and otherwise the rows of the identity matrix. For instance, if X is the matrix of 3.2.2 we have

$$(3.5.1)\qquad \tilde{X} = \begin{vmatrix} 1 & 0 & 0 & 0 & 0 & 0 & 0 & 0 & 0 & 0 & 0 \\ 0 & 1 & a_1 & a_2 & a_3 & 0 & b_{11} & b_{12} & 0 & c_{11} & 0 \\ 0 & 0 & 1 & 0 & 0 & 0 & 0 & 0 & 0 & 0 & 0 \\ 0 & 0 & 0 & 1 & 0 & 0 & 0 & 0 & 0 & 0 & 0 \\ 0 & 0 & 0 & 0 & 1 & 0 & 0 & 0 & 0 & 0 & 0 \\ 0 & 0 & 0 & 0 & 0 & 1 & b_{33} & b_{34} & 0 & c_{33} & 0 \\ 0 & 0 & 0 & 0 & 0 & 0 & 1 & 0 & 0 & 0 & 0 \\ 0 & 0 & 0 & 0 & 0 & 0 & 0 & 1 & 0 & 0 & 0 \\ 0 & 0 & 0 & 0 & 0 & 0 & 0 & 0 & 1 & c_{31} & 0 \\ 0 & 0 & 0 & 0 & 0 & 0 & 0 & 0 & 0 & 1 & 0 \\ 0 & 0 & 0 & 0 & 0 & 0 & 0 & 0 & 0 & 0 & 1 \end{vmatrix}.$$

We have

$$X = I_{i_1,i_2,\dots,i_n} \tilde{X},$$

and $\tilde{X}^t \in B$. This implies the theorem. □

Finally we have:

[131] In this chapter the minuscule property is heavily used to build the standard monomial theory. Nevertheless there is a rather general standard monomial theory due to Lakshmibai–Seshadri (cf [L-S]) and Littelmann for all irreducible representations of semisimple algebraic groups (cf. [Lit], [Lit2]).

Proposition. *$S_{i_1,i_2,\dots,i_n}$ is the Zariski closure of $C_{i_1,i_2,\dots,i_n}$.*

Proof. $S_{i_1,i_2,\dots,i_n}$ is defined by the vanishing of all Plücker coordinates not greater or equal to $i_1, i_2, \dots, i_n$, hence it is closed and contains $C_{i_1,i_2,\dots,i_n}$.

Since $C_{i_1,i_2,\dots,i_n}$ is a B-orbit, its closure is a union of B orbits and hence a union of Schubert cells. To prove the theorem it is enough, by 3.3, to show that, if for some k we have $i_k + 1 < i_{k+1}$, then $I_{i_1,i_2,\dots,i_{k-1},i_k+1,i_{k+1},i_n}$ is in the closure of $C_{i_1,i_2,\dots,i_n}$.

For this consider the matrix $I_{i_1,i_2,\dots,i_n}(b)$ which differs from $I_{i_1,i_2,\dots,i_n}$ only in the $i_k + 1$ column. This column has 0 in all entries except b in the k row.

The space defined by this matrix lies in $C_{i_1,i_2,\dots,i_n}$ and equals the one defined by the matrix obtained from $I_{i_1,i_2,\dots,i_n}(b)$ dividing the k row by b.

This last matrix equals $I_{i_1,i_2,\dots,i_{k-1},i_k+1,i_{k+1},i_n}$ except in the i_k column, which has 0 in all entries except b^{-1} in the k row. The limit as $b \to \infty$ of this matrix tends to $I_{i_1,i_2,\dots,i_{k-1},i_k+1,i_{k+1},i_n}$. For example,

$$W\left(\begin{vmatrix} 0&1&0&0&0&0&0&0&0&0\\ 0&0&0&0&1&b&0&0&0&0\\ 0&0&0&0&0&0&0&1&0&0\\ 0&0&0&0&0&0&0&0&0&1 \end{vmatrix}\right) = W\left(\begin{vmatrix} 0&1&0&0&0&0&0&0&0&0\\ 0&0&0&0&b^{-1}&1&0&0&0&0\\ 0&0&0&0&0&0&0&1&0&0\\ 0&0&0&0&0&0&0&0&0&1 \end{vmatrix}\right);$$

$$\lim_{b\to\infty} W\left(\begin{vmatrix} 0&1&0&0&0&0&0&0&0&0\\ 0&0&0&0&b^{-1}&1&0&0&0&0\\ 0&0&0&0&0&0&0&1&0&0\\ 0&0&0&0&0&0&0&0&0&1 \end{vmatrix}\right) = W\left(\begin{vmatrix} 0&1&0&0&0&0&0&0&0&0\\ 0&0&0&0&0&1&0&0&0&0\\ 0&0&0&0&0&0&0&1&0&0\\ 0&0&0&0&0&0&0&0&0&1 \end{vmatrix}\right).$$

□

3.6 Standard Monomials

We want to apply to standard monomials the theory developed in the previous sections. We have seen that the Schubert variety $S_{i_1,i_2,\dots,i_n} = S_{\underline{i}}$ is the intersection of the Grassmann variety with the subspace where the coordinates $\underline{j}$ which are not greater than or equal to $\underline{i}$ vanish.

Definition. We say that a standard monomial is standard on $S_{\underline{i}}$ if it is a product of Plücker coordinates greater or equal than $\underline{i}$.[132]

Theorem. *The monomials that are standard on $S_{\underline{i}}$ are a basis of the projective coordinate ring of $S_{\underline{i}}$.*

Proof. The monomials that are not standard on $S_{\underline{i}}$ vanish on this variety, hence it is enough to show that the monomials standard on $S_{\underline{i}}$, restricted to this variety, are linearly independent. Assume by contradiction that some linear combination $\sum_{k=1}^{a} c_k T_k$ vanishes on $S_{\underline{i}}$, and assume that the degree of this relation is minimal.

[132] This definition and the corresponding approach to standard monomials is due to Seshadri ([Seh]).

Let us consider, for each monomial T_k, its minimal coordinate p_k and write $T_k = p_k T_k'$; then select, among the Plücker coordinates p_k, a maximal coordinate $p_{\underline{j}}$ and decompose the sum as

$$\sum_{k=1}^{b} c_k p_k T_k' + p_{\underline{j}}\left(\sum_{k=b+1}^{a} c_k T_k'\right),$$

where the sum $\sum_{k=1}^{b} c_k p_k T_k'$ collects all terms which start from a coordinate p_k different from p_j. By hypothesis $\underline{i} \leq \underline{j}$. Restricting the relation to $S_{\underline{j}}$, all the standard monomials which contain coordinates not greater than $\underline{j}$ vanish, so, by choice of $\underline{j}$, we have that $p_{\underline{j}}\,(\sum_{k=m+1}^{n} c_k T_k')$ vanishes on $S_{\underline{j}}$. Since $S_{\underline{j}}$ is irreducible and $p_{\underline{j}}$ is nonzero on $S_{\underline{j}}$, we must have that $(\sum_{k=m+1}^{n} c_k T_k')$ vanishes on $S_{\underline{j}}$. This relation has a lower degree and we reach a contradiction by induction. □

Of course this theorem is more precise than the standard monomial theorem for the Grassmann variety.

4 Double Tableaux

4.1 Double Tableaux

We return now to the polynomial ring $\mathbb{Z}[x_{i,j}] : 1 \leq i \leq n;\ 1 \leq j \leq m$ of §1.1, which we think of as polynomial functions on the space of $n \times m$ matrices.

In this ring we will study the relations among the special polynomials obtained as determinants of minors of the matrix X. We use the notations (1.1.2) of Section 1.1.

Consider the Grassmann variety $Gr_n(m+n)$ and in it the open set A where the Plücker coordinate extracted from the last n columns is nonzero. In §2 we have seen that this open set can be identified with the space $M_{n,m}$ of $n \times m$ matrices. The identification is defined by associating to a matrix X the space spanned by the rows of $(X\ 1_n)$.

Remark. In more intrinsic terms, given two vector spaces V and W, we identify $\hom(V, W)$ with an open set of the Grassmannian in $V \oplus W$ by associating to a map $f : V \to W$ its graph $\Gamma(f) \subset V \oplus W$. The fact that the first projection of $\Gamma(f)$ to V is an isomorphism is expressed by the nonvanishing of the corresponding Plücker coordinate.

The point 0 thus corresponds to the unique 0-dimensional Schubert cell, which is also the only closed Schubert cell. Thus every Schubert cell has a nonempty intersection with this open set.[133]

[133] One can consider that in the Grassman variety we can construct two different cellular decompositions using the two opposite Borel subgroups B^+, B^-. Thus here we are considering the intersection of the open cell relative to B^- with the cells relative to B^+.

We use as coordinates in X the variables x_{ij} but we display them as

$$X' := \begin{vmatrix} x_{n1} & x_{n2} & \dots & x_{n,m-1} & x_{nm} \\ x_{n-1,1} & x_{n-1,2} & \dots & x_{n-1,m-1} & x_{n-1,m} \\ \dots & \dots & \dots & \dots & \dots \\ \dots & \dots & \dots & \dots & \dots \\ x_{11} & x_{12} & \dots & x_{1,m-1} & x_{1m} \end{vmatrix}.$$

Let us compute a Plücker coordinate $[i_1, i_2, \dots, i_n]$ for $X'\ 1_n$. We must distinguish, among the indices i_k appearing, the ones $\leq m$, say $i_1, i_2, \dots, i_h$ and the ones bigger than m, that is $i_{h+t} = m + j_t$ where $t = 1, \dots, n-h$; $1 \leq j_t \leq n$.

The last $n-h$ columns of the submatrix of $(X'\ 1_n)$ extracted from the columns $i_1, i_2, \dots, i_n$ are thus the columns of indices $j_1, j_2, \dots, j_{n-h}$ of the identity matrix.

Let first Y be an $n \times (n-1)$ matrix and e_i the ith column of the identity matrix.

The determinant $\det(Ye_i)$ of the $n \times n$ matrix, obtained from Y by adding e_i as the last column, equals $(-1)^{n+i}\det(Y_i)$, where Y_i is the $(n-1) \times (n-1)$ matrix extracted from Y by deleting the ith row. When we repeat this construction we erase successive rows.

In our case, therefore, we obtain that $[i_1, i_2, \dots, i_h, m+j_1, \dots, m+j_{n-h}]$ equals, up to sign, the determinant $(u_1, u_2, \dots, u_h | i_1, i_2, \dots, i_h)$ of X, where the indices $u_1, u_2, \dots, u_h$ are complementary, in $1, 2, \dots, n$, to the indices $n+1-j_1$, $n+1-j_2, \dots, n+1-j_{n-h}$.

We have defined a bijective map between the set of Plücker coordinates $[i_1, i_2, \dots, i_n]$ in $1, 2, \dots, n+m$ distinct from the last coordinate and the minors of the $n \times m$ matrix.

4.2 Straightening Law

Since the Plücker coordinates are naturally partially ordered, we want to understand the same ordering transported to the minors. It is enough to do it for adjacent elements. We must distinguish various cases.

Suppose thus that we are given a coordinate $[i_1, i_2, \dots, i_h, m+j_1, \dots, m+j_{n-h}]$ corresponding to the minor $(v_h, \dots, v_2, v_1 | i_1, i_2, \dots, i_s, \dots, i_h)$, and consider

$$\begin{aligned} &[i_1, i_2, \dots, i_s, \dots, i_h, m+j_1, \dots, m+j_{n-h}] \\ &\quad \leq [i_1, i_2, \dots, i_s+1, \dots, i_h, m+j_1, \dots, m+j_{n-h}]. \end{aligned}$$

This gives

$$(v_h, \dots, v_2, v_1 | i_1, i_2, \dots, i_s, \dots, i_h) \leq (v_h, \dots, v_2, v_1 | i_1, i_2, \dots, i_s+1, \dots, i_h).$$

Similarly

$$\begin{aligned} &[i_1, \dots, i_h, m+j_1, \dots, m+j_s, \dots, m+j_{n-h}] \\ &\quad \leq [i_1, \dots, i_h, m+j_1, \dots, m+j_s+1, \dots, m+j_{n-h}] \end{aligned}$$

gives, for $v_t := n - j_s - 1$,

$$(v_h, \dots, v_t, \dots, v_2, v_1 | i_1, i_2, \dots, i_h)$$
$$\leq (v_h, \dots, v_t + 1, \dots, v_2, v_1 | i_1, i_2, \dots, i_h).$$

Finally we have the case in which the number of indices $\leq m$ decrease, i.e.,

$$[i_1, \dots, i_{h-1}, i_h = m, m + j_1, \dots, m + j_{n-h}]$$
$$\leq [i_1, \dots, i_{h-1}, m + 1, m + j_1, \dots, m + j_{n-h}].$$

This gives $n = v_h,\ j_1 > 1$ and

$$(n, v_{h-1}, \dots, v_2, v_1 | i_1, i_2, \dots, i_{h-1}, m) \leq (v_{h-1}, \dots, v_2, v_1 | i_1, i_2, \dots, i_{h-1}).$$

In particular we see that a $k \times k$ determinant can be less than an $h \times h$ determinant only if $k \geq h$.

The formal implication is that a standard product of Plücker coordinates, interpreted (up to sign) as a product of determinants of minors, appears as a double tableau, in which the shape of the left side is the reflection of the shape on the right. The columns are non-decreasing. The rows are strictly increasing in the right tableau and strictly decreasing in the left. As example, let $n = 3, m = 5$, and consider a tableau:

$$\begin{vmatrix} 1 & 2 & 3 \\ 1 & 2 & 4 \\ 1 & 4 & 7 \\ 2 & 4 & 8 \\ 2 & 6 & 8 \\ 3 & 7 & 8 \end{vmatrix}.$$

To this corresponds the double tableau:

$$\begin{array}{rrr|lll} 3 & 2 & 1 & 1 & 2 & 3 \\ 3 & 2 & 1 & 1 & 2 & 4 \\ & 3 & 1 & 1 & 4 & \\ & 3 & 2 & 2 & 4 & \\ & & 2 & 2 & & \\ & & 3 & 3 & & \end{array}$$

We will call such a tableau a *double standard tableau.*[134]

Of course, together with the notion of double standard tableau we also have that of double tableau or *bi-tableau*, which can be either thought of as a product of determinants of minors of decreasing sizes or as a pair of tableaux, called left (or row) and right (or column) tableau of the same size.

---

[134] The theory of double standard tableaux was introduced by Doubilet, Rota and Stein [DRS]. The treatment here is due to Seshadri [L-S].

If one takes the second point of view, which is useful when analyzing formally the straightening laws, one may think that the space of one-line tableaux of size k is a vector space M_k with basis the symbols $(v_h, \dots, v_2, v_1 | i_1, i_2, \dots, i_h)$. The right indices run between 1 amd m, while the left indices run between 1 and n. The symbols are assumed to be separately antisymmetric in the left and right indices. In particular, when two indices on the right or on the left are equal the symbol is 0.

For a partition $\lambda := m_1 \geq m_2 \geq \cdots \geq m_t$ the tableaux of shape λ can be thought of as the tensor product $M_{m_1} \otimes M_{m_2} \otimes \ldots \otimes M_{m_t}$. When we evaluate a formal tableau as a product of determinants we have a map with nontrivial kernel (the space spanned by the straightening laws).

We now want to interpret the theory of tableaux in terms of representation theory. For this we want to think of the space of $n \times m$ matrices as $\hom(V, W) = W \otimes V^*$ where V is m-dimensional and W is n-dimensional (as free $\mathbb{Z}$-modules if we work over $\mathbb{Z}$). The algebra R of polynomial functions on $\hom(V, W)$ is the symmetric algebra on $W^* \otimes V$.

$$R = S(W^* \otimes V). \tag{4.2.1}$$

The two linear groups $GL(V),\ GL(W)$ act on the space of matrices and on R.

Over $\mathbb{Z}$ we no longer have the decomposition 6.3.2 of Chapter 9 so our theory is a replacement, and in a way, also a refinement of that decomposition.

In matrix notations the action of an element $(A, B) \in GL(m) \times GL(n)$ on an $n \times m$ matrix Y is BYA^{-1}. If $e_i,\ i = 1, \dots, n$, is a basis of W and $f_j,\ j = 1, \dots, m$, one of V under the identification $R = S(W^* \otimes V) = \mathbb{Z}[x_{ij}]$, the element $e^i \otimes f_j$ corresponds to x_{ij}:

$$\langle e^i \otimes f_j | X \rangle := \langle e^i | X f_j \rangle = \Big\langle e^i \,|\, \sum_h x_{hj} e_h \Big\rangle = x_{ij}.$$

Geometrically we can think as follows. On the Grassmannian $G_{m,m+n}$ acts the linear group $GL(m+n)$. The action is induced by the action on $n \times (m+n)$ matrices Y by $YC^{-1}, C \in GL(m+n)$.

The space of $n \times m$ matrices is identified with the cell $(X \ 1_n)$ and is stable under the diagonal subgroup $GL(m) \times GL(n)$. Thus if $C = \begin{vmatrix} A & 0 \\ 0 & B \end{vmatrix}$ we have

$$(X \ 1_n) C^{-1} = (XA^{-1} \ B^{-1}) \equiv (BXA^{-1} \ 1_n). \tag{4.2.2}$$

If now we want to understand the dual action on polynomials, we can use the standard dual form $(gf)(u) = f(g^{-1}u)$ for the action on a vector space as follows:

Remark. The transforms of the coordinate functions x_{ij} under A, B are the entries of $B^{-1}XA$, where $X = (x_{ij})$ is the matrix having as entries the variables x_{ij}.

Let us study the subspace $M_k \subset R$ of the ring of polynomials spanned by the determinants of $(v_k, \dots, v_2, v_1 | i_1, i_2, \dots, i_s, \dots, i_k)$, i.e., the $k \times k$ minors.

Given an element $A \in \hom(V, W)$, it induces a map $\wedge^k A : \bigwedge^k V \to \bigwedge^k W$. Thus the formula , $i_k(\phi \otimes u)(A) := \langle \phi | \wedge^k Au \rangle$ defines a map

$$i_k : \hom\left(\bigwedge^k V, \bigwedge^k W\right)^* = \bigwedge^k W^* \otimes \bigwedge^k V \to R = S(V^* \otimes W).$$

It is clear that $i_k(f_{v_k} \wedge \dots f_{v_2} \wedge f_{v_1} | e_{i_1} \wedge e_{i_2} \dots \wedge e_{i_k}) = (v_k, \dots, v_2, v_1 | i_1, i_2, \dots, i_s, \dots, i_k)$ and thus M_k is the image of i_k.

Lemma. *M_h is isomorphic to* $\hom(\bigwedge^h V, \bigwedge^h W)^* = \bigwedge^h V \otimes (\bigwedge^h W)^*$ *in a* $GL(V) \times GL(W)$ *equivariant way.*

Proof. Left to the reader as in Chapter 9, 7.1. □

The action of the two linear groups on rows and columns induces, in particular, an action of the two groups of diagonal matrices, and a double tableau is clearly a weight vector under both groups. Its weight (or double weight) is read off from the row and column indices appearing.

We may encode the number of appearances of each index on the row and column tableaux as two sequences

$$1^{h_1} 2^{h_2} \dots n^{h_n};\ 1^{k_1} 2^{k_2} \dots m^{k_m}.$$

When one wants to stress the combinatorial point of view one calls these two sequences the *content* of the double tableau.

According to the definition of the action of a group on functions, we see that the weight of a diagonal matrix in $GL(n)$ with entries b_i acting on rows is $\prod_{i=1}^n b^{-h_i}$ while the weight of a diagonal matrix in $GL(m)$ with entries a_i acting on columns is $\prod_{i=1}^m a^{k_i}$.

We come now to the main theorem:

Theorem. *The double standard tableaux are a $\mathbb{Z}$-basis of $\mathbb{Z}[x_{i,j}]$.*

Proof. The standard monomials in the Plücker coordinates are a basis of $\mathbb{Z}[[i_1, i_2, \dots, i_n]]$, so we have that the double standard tableaux span the polynomial algebra $\mathbb{Z}[x_{i,j}]$ over $\mathbb{Z}$.

We need to show that they are linearly independent. One could give a proof in the same spirit as for the ordinary Plücker coordinates, or one can argue as follows.

We have identified the space of $n \times m$ matrices with the open set of the Grassmann variety where the Plücker coordinate $p = [m+1, m+2, \dots, m+n]$ is nonzero.

There are several remarks to be made:

(1) The coordinate p is the maximal element of the ordered set of coordinates, so that if T is a standard monomial, so is Tp.
(2) Since a $\mathbb{Z}$-basis of $\mathbb{Z}[[i_1, i_2, \dots, i_n]]$ is given by the tableaux Tp^k, where T is a standard tableau not containing p, we have that these tableaux not containing p are a basis over the polynomial ring $\mathbb{Z}[p]$.
(3) The algebra $\mathbb{Z}[x_{i,j}]$ equals the quotient algebra $\mathbb{Z}[[i_1, i_2, \dots, i_n]]/(p-1)$.

From (2) and (3) it follows that the image in $\mathbb{Z}[x_{i,j}]$ of the standard monomials which do not end with p are a $\mathbb{Z}$-basis. But the images of these monomials are the double standard tableaux and the theorem follows. To finish the proof, it remains to check (1), (2), (3).

Points (1) and (2) are clear.

Point (3) is a general fact on projective varieties. If $W \subset P^n$ is a projective variety and A is its homogeneous coordinate ring, the coordinate ring of the affine part of W where a coordinate x is not zero is $A/(x-1)$. □

4.3 Quadratic Relations

We need to analyze now the straightening algorithm for double tableaux. To begin, we consider a basic quadratic relation for a two-line tableau. We have thus to understand the quadratic relation 2.1.4 for a product of two Plücker coordinates $|i_1, \dots, i_n||j_1, \dots, j_n|$ in terms of double tableaux. We may assume without loss of generality that the two coordinates give a double tableau with two rows of length $a \geq b$. There are two possibilities for the point $i_k > j_k$ where the violation occurs: either the two indices i_k, j_k are both column indices or both row indices. Let us treat the first case, the other is similar. In this case all indices $j_1, \dots, j_k$ are column indices while among the $i_k, \dots, i_n$ there can be also row indices.

In each summand of 2.1.4 some top indices are exchanged with bottom indices, so we can separate the sum into two contributions, the first in which no row indices are exchanged and the second with the remaining terms. Thus in the first we have a sum of tableaux of type a, b, while in the second the possible types are $a+t, b-t$, $t > 0$.

Summarizing,

Proposition. *A straightening law on the column indices for a product*

$$T = (u_a \dots u_1 | i_1 \dots i_a)(v_b \dots v_1 | j_1 \dots j_b)$$

of 2 determinants of sizes $a \geq b$ is the sum of two terms $T_1 + T_2$, where T_2 is a sum of tableaux of types $a+t, b-t,\ t > 0$, and T_1 is the sum of the tableaux obtained from T by selecting an index i_k such that $i_k > j_k$ and performing all possible exchanges among $i_k \dots i_a$ and $j_1 \dots j_k$, while leaving fixed the row indices and summing with the sign of the exchange:

$$(4.3.1) \qquad \sum \epsilon \begin{matrix} u_a, u_{a-1}, \dots, u_2, u_1 | i_1, i_2, \dots, \underline{i_k, \dots\dots, i_a} \\ v_b, \dots, v_2, v_1 | \underline{j_1, j_2, \dots, j_k}, \dots, j_b \end{matrix} + T_2.$$

All the terms of the quadratic relation have the same double weight. There is a similar statement for row straightening.

For our future analysis, it is not necessary to make more explicit the terms T_2, which are in any case encoded formally in the identity 2.1.4.

Remark. (1) There is an important special case to be noticed: when the row indices $u_m, \ldots, u_2, u_1$ are all contained in the row indices $i_k \ldots i_1$. In this case the terms T_2 do not appear, since raising a row index creates a determinant with two equal rows.

(2) The shapes of tableaux appearing in the quadratic equations are closely connected with a special case of Pieri's formula (in characteristic 0).

$$a \geq b, \qquad \bigwedge^a V \otimes \bigwedge^b V = \bigoplus_{t=0}^{b} S_{a+t,b-t}(V).$$

Regarding (1) we define:

Definition 1. A tableau A is said to be extremal if, for every $i > 1$, the indices of the ith row are contained in the indices of the $(i-1)^{st}$ row.

Let us take a double tableau $A|B$, where A, B represent the two tableaux of row and column indices. Let us apply sequentially straightening relations on the column indices. We see that again we have two contributions $A|B = T_1 + T_2$, where in T_1 we have tableaux of the same shape while in T_2 the shape has changed (we will see how in a moment).

Lemma. *The contribution from the first part of the sum is of type*

$$T_1 = \sum_C c_{B,C}\, A|C, \tag{4.3.2}$$

where the coefficients $c_{B|C}$ are independent of A. If A is an extremal tableau, $T_2 = 0$. There is an analogous statement for row relations.

We can now use the previous straightening relations to transform a double tableau into a sum of double standard tableaux. For this we have to remark that, starting from a product of determinants of sizes $a_1 \geq a_2 \geq \cdots \geq a_i$ and applying a quadratic relation, we may replace two successive sizes $a \geq b$ with some $a+t, b-t$. In this way the product does not appear as a product of determinants of decreasing sizes. We have thus to reorder the terms of the product to make the sizes decreasing. To understand how the shapes of tableaux behave with respect to this operation, we give the following:

Definition 2. The dominance order for sequences of real numbers is

$$(a_1, \ldots, a_n) \geq (b_1, \ldots, b_n) \quad \text{iff} \quad \sum_{i=1}^{h} a_i \geq \sum_{i=1}^{h} b_i \qquad \forall h = 1, \ldots, n.$$

In particular we obtain a (partial) ordering on partitions.

Remark. If we take a vector $(b_1, \ldots, b_n)$ and construct $(a_1, \ldots, a_n)$ by reordering the entries in decreasing order, then $(a_1, \ldots, a_n) \geq (b_1, \ldots, b_n)$.

Corollary. *Given a double tableau of shape λ, by the straightening algorithm it is expressed as a linear combination of standard tableaux of shapes $\geq \lambda$ in the dominance order and of the same double weight.*

5 Representation Theory

5.1 U Invariants

Consider the root subgroups, which we denote by $a + \lambda b$, acting on matrices by adding to the ath column the bth column multiplied by λ. This is the result of the multiplication $X(1+\lambda e_{ba})$. A single determinant of a minor $D := (i_1, \ldots, i_k | j_1, \ldots, j_k)$ is transformed according to the following rule (cf. §2.3):

If a does not appear among the elements j_s, or if both a, b appear among these elements, D is left invariant.

If $a = j_s$ and b does not appear, D is transformed into $D + \lambda D'$ where D' is obtained from D by substituting a in the column indices with b.

Of course a similar analysis is valid for the row action.

This implies a combinatorial description of the group action of $G = GL(m) \times GL(n)$ on the space of tableaux. In particular we can apply it when the base ring is $\mathbb{Z}$ or a field F, so that the special linear group over F or $\mathbb{Z}$ is generated by the elements $a + \lambda b$. We have described the action of such an element on a single determinant, which then extends by multiplication and straightening algorithm.

An argument similar to the one performed in §2.3 shows that: Given a linear combination $C := \sum_i c_i T_i$ of double standard tableaux, apply to it the transformation $2 + \lambda 1$ and obtain a polynomial in λ. The degree k of this polynomial is the maximum of the number of occurrences of 2 in a tableau T_i as a column index not preceded by 1, i.e., 2 occurs on the first column.

Its leading term is of the form $\sum c_i T_i'$ where the sum extends to all the indices of tableaux T_i where 2 appears in the first column k times and T_i' is obtained from T_i by replacing 2 with 1 in these positions. It is clear that to distinct tableaux T_i correspond distinct tableaux T_i' and thus this leading coefficient is nonzero. It follows that:

The element C is invariant under $2 + \lambda 1$ if and only if in the column tableau, 2 appears only on the second column.

Let us indicate by $A^{1,2}$ this ring of invariant elements under $2 + \lambda 1$.

We can now repeat the argument using $3 + \lambda 1$ on the elements of $A^{1,2}$ and see that

An element $C \in A^{1,2}$ is invariant under $3+\lambda 1$ if and only if in the column tableau each occurrence of 3 is preceded by 1.

By induction we can define $A^{1,k}$ to be the ring of invariants under all the root subgroups $i + \lambda 1, \ i \leq k$.

$A^{1,k}$ is spanned by the elements such that in the column tableau no element $1 < i \leq k$ appears in the first column.

Now when $k = m$, all tableaux which span this space have only 1's on the first column of the right tableau.

Next we can repeat the argument on $A^{1,m}$ using the root subgroups $i + \lambda 2, \ i \leq k$. We thus define $A^{2,k}$ to be the ring of invariants under all the root subgroups $i + \lambda 1$ and all the root subgroups $i + \lambda 2, \ i \leq k$.

$A^{2,k}$ is spanned by the elements with 1 on the first column of the right tableau and no element $2 < i \leq k$ appearing on the second column.

In general, given $i < j \le m$ consider the subgroup $U_{i,j}$ of upper triangular matrices generated by the root subgroups

$$b + \lambda a,\ a \le i-1,\ b \le m;\ b + \lambda i,\ \ b \le j$$

and denote by $A^{i,j}$ the corresponding ring of invariants. Then:

Proposition 1. *$A^{i,j}$ is spanned by the elements in which the first $i-1$ columns of the right tableau are filled, respectively, with the numbers $1, 2, \dots, i-1$, while no number $i < k \le j$ is in the i column.*

Corollary 1. *The ring of polynomial invariants under the full group U^+ of upper triangular matrices, acting on the columns, is spanned by the double standard tableaux whose column tableau has the ith column filled with i for all i. We call such a tableau right canonical.*

The main remark is that, given a shape λ, there is a *unique* canonical tableau of that given shape characterized by having 1 in the first column, 2 in the second, etc. We denote this canonical tableau by C_λ. For example, for $m = 5$,

$$C_{33211} := \begin{matrix} 1 & 2 & 3 \\ 1 & 2 & 3 \\ 1 & 2 & \\ 1 & & \\ 1 & & \end{matrix}\ , \qquad C_{54211} := \begin{matrix} 1 & 2 & 3 & 4 & 5 \\ 1 & 2 & 3 & 4 & \\ 1 & 2 & & & \\ 1 & & & & \\ 1 & & & & \end{matrix}\ .$$

One could have proceeded similarly starting from the subgroups $m + \lambda i$ and getting:

Corollary 2. *The ring of polynomial invariants under the full group U^- of lower triangular matrices, acting on the columns, is spanned by the double standard tableaux whose column side has the property that each index $i < m$ appearing is followed by $i+1$. We call such a tableau anticanonical.*

Again, given a shape λ, there is a *unique* anticanonical tableau of that given shape, e.g., for $m = 5$,

$$\begin{matrix} 3 & 4 & 5 \\ 3 & 4 & 5 \\ 4 & 5 & \\ 5 & & \\ 5 & & \end{matrix}\ , \qquad \begin{matrix} 1 & 2 & 3 & 4 & 5 \\ 2 & 3 & 4 & 5 & \\ 4 & 5 & & & \\ 5 & & & & \\ 5 & & & & \end{matrix}\ .$$

Observe that a tableau can be at the same time canonical and anticanonical if and only if all its rows have length m (e.g., for $m = 5$):

$$(5.1.1) \qquad \begin{matrix} 1 & 2 & 3 & 4 & 5 \\ 1 & 2 & 3 & 4 & 5 \\ 1 & 2 & 3 & 4 & 5 \\ 1 & 2 & 3 & 4 & 5 \end{matrix}\ .$$

Of course we have a similar statement for the action on rows (the left action) except that the invariants under left action by U^- are left canonical, and under the left action by U^+ action are left anticanonical.

Now we will obtain several interesting corollaries.

Definition. For a partition λ define W^λ (resp. V_λ) to be the span of all double tableaux $A|C_\lambda$ of shape λ with left canonical tableau (resp. $C_\lambda|B$).

From Lemma 4.3, since a canonical tableau is extremal, we have:

Proposition 2. *W^λ has as basis the double standard tableaux $A|C_\lambda$.*
V_λ has as basis the double standard tableaux $C_\lambda|A$.

Theorem 1. *The invariants under the right U^+ action (resp. the left U^- action) decompose as*

$$\bigoplus_\lambda W^\lambda, \quad \textit{resp.} \ \bigoplus_\lambda V_\lambda.$$

If we act by right multiplication with a diagonal matrix t with entry a_i in the ii position, this multiplies the ith column by a_i and thus transforms a double tableau T which is right canonical and of shape λ into $T\prod a_i^{k_i}$, where k_i is the length of the ith column.[135]

If t with entries a_i is the diagonal part of an upper triangular matrix, we can think of $\prod a_i^{k_i}$ as a function on B^+ which is still a *character* denoted by λ. Thus the decomposition $\bigoplus_\lambda W^\lambda$ is a decomposition into weight spaces under the Borel subgroup of upper triangular matrices. We have proved:

Theorem 2. *W^λ is the space of functions which, under the right action of B^+, are weight vectors of character λ. W^λ is a $GL(n)$-submodule, (similar statement for V_λ).*

Proof. The left action by $GL(n)$ commutes with the right action and thus each W^λ is a $GL(n)$ submodule. □

Assume for instance $n \leq m$. The $U^- \times U^+$ invariants are spanned by those tableaux which are canonical on the left and the right and will be called *bicanonical.* These tableaux are the polynomials in the determinants

$$d_k := (k, k-1, \dots, 1|1, 2, \dots, k).$$

A monomial $d_1^{h_1} d_2^{h_2} \dots d_n^{h_n}$ is a bicanonical tableau whose shape λ is determined by the sequence h_i and will be denoted by K_λ.

An argument similar to the previous analysis of U invariants shows that:

[135] We are skimming over a point. When we work over $\mathbb{Z}$ there are not enough diagonal matrices so we should really think that we compute our functions, which are defined over $\mathbb{Z}$, into any larger ring. The identities we are using are valid when we compute in any such extension.

Proposition 3.

(1) Any U^- fixed vector in W^λ is multiple of the bicanonical tableau K_λ of shape λ.
(2) If the base ring is an infinite field every U^- stable subspace of W^λ contains K_λ.
(3) W^λ is an indecomposable U^-- or $GL(n)$-module.
(4) $W^\lambda W^\mu = W^{\lambda+\mu}$ (Cartan multiplication).
(5) When we work over an infinite field F, the $GL(n)$-submodule L_λ generated by K_λ is irreducible and it is the unique irreducible submodule of V_λ.

Proof. (1) and (2) follow from the previous analysis. In fact given any U^- submodule M and an element $\sum_i c_i T_i \in M$, a linear combination of double standard tableaux, apply to it the transformation $2+\lambda 1$. By hypothesis, for all λ this element is in M and so its leading term is also in M. Repeat the argument with the other transformations $i + \lambda j$ as in the previous proof until we get the bicanonical tableau in M.

(3) follows from (2). For (4) we have to specify the meaning of $\lambda + \mu$. Its correct meaning is by interpreting the partitions as weights for the torus. Then it is clear that a product of two weight vectors has as weight the sum of the weights. Thus $W^\lambda W^\mu \subset W^{\lambda+\mu}$. To show equality we observe that a standard tableau of shape $\lambda + \mu$ can be written as the product of two standard tableaux of shapes λ and μ.

(5) If A is a minimal submodule of W^λ it is necessarily irreducible. By (1) it must contain K_λ hence L_λ and this suffices to prove the statement. □

Remark. The above proposition is basically the theory of the highest weight vector in this case. The reader is invited to complete the representation theory of the general linear group in characteristic 0 by this combinatorial approach (as an alternative to the one developed in Chapter 9).

In general the previous theorem is interpreted by saying that W^λ is an *induced representation* of a 1-dimensional representation of B^+. The geometric way of expressing this is by taking the 1-dimensional representation F_λ of B, given by the character λ, forming the line bundle $L_\lambda := G \times_{B^+} F_\lambda$ on the flag variety G/B^+ and interpreting:

$$W^\lambda = H^0(G/B^+, L_\lambda).$$

If the reader knows the meaning of these terms it should not be difficult to prove this statement in our case. One has just to identify the sections of the line bundle with the functions on G which are eigenvectors of B^+ of the appropriate character. But it would take us too far afield to introduce this language in detail to explain it here.

If A denotes the coordinate ring of the linear group $GL(n, F)$ we know that $A = F[x_{ij}][1/d]$, where d is the determinant, and we can extend the previous theorems to study A as a representation. It is enough to remark that d is also a U^+- invariant of weight d itself and every double standard tableau is uniquely a power of d times a double standard tableau of shape λ with $ht(\lambda) \leq n-1$. Thus we obtain (cf. Chapter 9, 8.1.3 and 8.2.1):

Theorem 3. *The space A^{U^+} of functions on $GL(n, F)$ right invariant under U^+ decomposes as*

$$A^{U^+} = \bigoplus_{\lambda} \bigoplus_{k \in \mathbb{Z}} W^{\lambda}[d^k], \quad ht(\lambda) \leq n-1.$$

We also have:

Theorem 4. *Every rational irreducible $GL(n, F)$-module is of the type $L_{\lambda}[d^k]$. These modules are not isomorphic.*

Proof. Given a rational irreducible $GL(n, F)$-module M we can embed M into A. Since we have a filtration of A with factors isomorphic to $W^{\lambda}[d^k]$ we must have a nonzero morphism of M into one of these modules. Now $W^{\lambda}[d^k]$ contains a unique irreducible submodule $L_{\lambda}[d^k]$. Hence M is isomorphic to $L_{\lambda}[d^k]$. The fact that these modules are not isomorphic depends on the fact that each of them contains a unique (up to constant) U^+-invariant vector of weight $d^k\lambda$, and these weights are distinct. □

5.2 Good Filtrations

A one-row double tableau which is right canonical is the determinant of a $i \times i$ minor $u_i, \dots, u_1 | 1, 2, \dots, i$ extracted from the first i columns of X. Let W^i denote the space of these tableaux. As a representation of $GL(n) = GL(W)$, W^i is isomorphic to $\bigwedge^i (W)^*$.

Similarly, a one-row double tableau which is left canonical is the determinant of a $i \times i$ minor $i, \dots, 2, 1 | v_1, \dots, v_i$ extracted from the first i rows of X. Let V_i denote the space of these tableaux. As a representation of $GL(m) = GL(V)$, V_i is isomorphic to $\bigwedge^i (V)$.

If $\lambda = k_1 \geq k_2 \geq \dots \geq k_r$ the tableaux of shape λ can be viewed as the natural tensor product basis of $W^{k_1} \otimes W^{k_2} \dots \otimes W^{k_r}$.

The straightening laws for W^{λ} can be viewed as elements of this tensor product, and we will call the subspace spanned by these elements R_{λ}. Then

$$W^{\lambda} := W^{k_1} \otimes W^{k_2} \dots \otimes W^{k_r} / R^{\lambda}.$$

Similarly, on the rows

$$V_{\lambda} := V_{k_1} \otimes V_{k_2} \dots \otimes V_{k_r} / R_{\lambda}.$$

Quite often, when dealing with right or left canonical tableaux it is better to drop the C_{λ} completely and write the corresponding double tableau as a single tableau (since the row or column indices are completely determined).

We can now reinterpret the straightening algorithm as the existence of a *good filtration*[136] on the algebra R of functions on matrices.

[136] The theory of good filtrations has been developed by Donkin [Do] for semisimple algebraic groups and it is an essential tool for the characteristic free theory.

Theorem 1. *(1) Given a double tableau of shape λ by the straightening algorithm it is expressed as a linear combination of standard tableaux of shapes $\geq \lambda$ and of the same double weight.*

(2) Let S_λ, resp. A_λ, denote the linear span of all tableaux of shape $\geq \lambda$ (resp. of standard tableaux of shape λ). We have

$$S_\mu := \bigoplus_{\lambda \geq \mu,\ |\lambda|=|\mu|} A_\lambda.$$

Denote by $S'_\mu := \bigoplus_{\lambda > \mu,\ |\lambda|=|\mu|} A_\lambda$ (which has as basis the double standard tableaux of shape $> \lambda$ in the dominant ordering).

(3) The space S_μ / S'_μ is a representation of $GL(V) \times GL(W)$ equipped with a natural basis indexed by double standard tableaux $A|B$ of shape μ. When we take an operator $X \in GL(V)$ we have $X(A|B) = \sum_C c_{B,C} A|C$ where C runs over the standard tableaux and the coefficients are independent of A, and similarly for $GL(W)$.

(4) As a $GL(V) \times GL(W)$-representation we have that

$$S_\lambda / S'_\lambda \cong W^\lambda \otimes V_\lambda.$$

Proof. The first fact is Corollary 4.3.

By definition, if $\lambda := k_1, k_2, \ldots, k_i$ is a partition, we have that $T_\lambda := M_{k_1} M_{k_2} \ldots M_{k_i}$ is the span of all double tableaux of shape λ. Thus $S_\mu = \sum_{\lambda \geq \mu,\ |\lambda|=|\mu|} T_\lambda$ by (1).

Parts (3) and (4) follow from Lemma 4.3.2. We establish a combinatorial linear isomorphism j_λ between $W^\lambda \otimes V_\lambda$ and S_λ / S'_λ by setting $j_\lambda(A \otimes B) := A|B$ where A is a standard row tableau (identified to $A|C_\lambda$), B a standard column tableau (identified to $C_\lambda|B$) and $A|B$ the corresponding double tableau. From (1), j_λ is an isomorphism of $GL(m) \times GL(n)$-modules. □

Before computing explicitly we relate our work to Cauchy's formula.

In 4.2 we have identified the subspace M_k of the ring of polynomials spanned by the determinants of the $k \times k$ minors with $\bigwedge^k W^* \otimes \bigwedge^k V$. The previous theorem implies in particular that the span of all tableaux of shapes $\geq \mu$ and some fixed degree p is a quotient of a direct sum of tensor products $\tilde{T}_\lambda := M_{k_1} \otimes M_{k_2} \otimes \ldots \otimes M_{k_i}$, where $\lambda \geq \mu,\ |\lambda| = p$, modulo a subspace which is generated by the straightening relations. In other words we can view the straightening laws as a combinatorial description of a set of generators for the kernel of the map $\bigoplus_{\lambda \geq \mu,\ |\lambda|=p} \tilde{T}_\lambda \to S_\mu$. Thus we have a combinatorial description by generators and relations of the group action on S_μ.

Revert for a moment to characteristic 0. Take a Schur functor associated to a partition λ and define

$$i_\lambda : \hom(V_\lambda, W_\lambda)^* = W_\lambda^* \otimes V_\lambda \to R = S(V^* \otimes W),\ i_\lambda(\phi \otimes u)(A) := \langle \phi | Au \rangle.$$

Set $M_\lambda = i_\lambda(W_\lambda^* \otimes V_\lambda)$. The map i_λ is $GL(V) \times GL(W)$-equivariant, $W_\lambda^* \otimes V_\lambda$ is irreducible and we identify $M_\lambda = W_\lambda^* \otimes V_\lambda$.

To connect with our present theory we shall compute the invariants

$$(W_\lambda^* \otimes V_\lambda)^{U^- \times U^+} = (W_\lambda^*)^{U^-} \otimes (V_\lambda)^{U^+}.$$

From the highest weight theory of Chapter 10 we know that V_λ has a unique U^+-fixed vector of weight λ (cf. Chapter 10, §5.2) while W_λ^* has a unique U^--fixed vector of weight $-\lambda$. It follows that the space $(W_\lambda^*)^{U^-} \otimes (V_\lambda)^{U^+}$ is formed by the multiples of the bicanonical tableau K_λ.

Theorem 2. *In characteristic* 0, *if* $\mu \vdash p$*:*

$$S_\mu = \bigoplus_{\substack{|\lambda| \leq \min(m,n), \\ \mu \leq \lambda,\ \lambda \vdash p}} W_\lambda^* \otimes V_\lambda$$

S_μ / S'_μ *is isomorphic to* $W_\mu^* \otimes V_\mu$.

Proof. We can apply the highest weight theory and remark that the highest weight of $W_\lambda^* \otimes V_\lambda$ under $U^- \times U^+$ is the bicanonical tableau of shape λ (since it is the only $U^- \times U^+$ invariant of the correct weight). Thus to identify the weights λ for which $W_\lambda^* \otimes V_\lambda \subset S_\mu$ it suffices to identify the bicanonical tableaux in S_μ. From the basis by standard tableaux we know that the $U^- \times U^+$ fixed vectors in S_μ are the linear combinations of the bicanonical tableaux K_λ for $|\lambda| \leq \min(m, n)$, $\mu \leq \lambda$, $\lambda \vdash p$. □

Over the integers or in positive characteristic we no longer have the direct sum decomposition. The group $GL(n)$ or $SL(n)$ is not linearly reductive and rational representations do not decompose into irreducibles. Nevertheless, often it is enough to use particularly well behaved filtrations. It turns out that the following is useful:

Definition. Given a polynomial representation P of $GL(m)$ a *good filtration* of P is a filtration by $GL(m)$ submodules such that the quotients are isomorphic to the modules V_λ.

5.3 $SL(n)$

Now we want to apply this theory to the special linear group.

We take double tableaux for an $n \times n$ matrix $X = (x_{ij})$. Call $A := F[x_{ij}]$ and observe that $d = \det(X) = (n, \ldots, 1|1, \ldots, n)$ is the first coordinate, so the double standard tableaux with at most $n - 1$ columns are a basis of A over the polynomial ring $F[d]$. Hence, setting $d = 1$ in the quotient ring $A/(d - 1)$ the double standard tableaux with at most $n-1$ columns are a basis over F. Moreover, d is invariant under the action of $SL(n) \times SL(n)$ and thus $A/(d - 1)$ is an $SL(n) \times SL(n)$-module.

We leave to the reader to verify that $A/(d - 1)$ is the coordinate ring of $SL(n)$ and its $SL(n) \times SL(n)$-module action corresponds to the let and right group actions, and that the image of the V_λ for λ with at most $n - 1$ columns give a decomposition of $A/(d - 1)^{U^+}$ (similarly for W^μ).

We want now to analyze the map $\overline{f}(g) := f(g^{-1})$ which exchanges left and right actions on standard tableaux.

For this remark that the inverse of a matrix X of determinant 1 is the adjugate $\bigwedge^{n-1} X$. More generally consider the pairing $\bigwedge^k F^n \times \bigwedge^{n-k} F^n \to \bigwedge^n F^n = F$ under which

$$\left\langle \bigwedge^k X(u_1 \wedge \cdots \wedge u_k) \middle| \bigwedge^{n-k} X(v_1 \wedge \cdots \wedge v_{n-k}) \right\rangle$$
$$= \bigwedge^n X(u_1 \wedge \cdots \wedge u_k \wedge v_1 \wedge \cdots \wedge v_{n-k})$$
$$= u_1 \wedge \cdots \wedge u_k \wedge v_1 \wedge \cdots \wedge v_{n-k}.$$

If we write everything in matrix notation the pairing between basis elements of the two exterior powers is a diagonal $\binom{n}{k}$ matrix of signs ± 1 that we denote by J_k. We thus have:

Lemma. *There is an identification between $(\bigwedge^k X^{-1})^t$ and $J_k \bigwedge^{n-k} X$.*

Proof. From the previous pairing and compatibility of the product with the operators $\bigwedge X$ we have

$$\left(\bigwedge^k X\right)^t J_k \bigwedge^{n-k} X = 1_{\binom{n}{k}}.$$

Thus

$$\left(\bigwedge^k X^{-1}\right)^t = J_k \bigwedge^{n-k} X.$$

This implies that under the map $f \to \overline{f}$ a determinant $(i_1 \dots i_k | j_1 \dots j_k)$ of a k minor is transformed, up to sign, into the $n-k$ minor with complementary row and column indices. □

Corollary. *$f \to \overline{f}$ maps V_λ isomorphically into W^μ where, if λ has rows $k_1, k_2, \dots, k_r$, then μ has rows $n - k_r, n - k_{r-1}, \dots, n - k_1$.*

5.4 Branching Rules

Let us recover in a characteristic free way the branching rule from $GL(m)$ to $GL(m-1)$ of Chapter 9, §10. Here the branching will not give a decomposition but a good filtration.

Consider therefore the module V_λ for $GL(m)$, with its basis of semistandard tableau of shape λ, filled with the indices $1, \dots, m$. First, we can decompose $V_\lambda = \bigoplus_k V_\lambda^k$ where V_λ^k has as basis the semistandard tableau of shape λ where m appears k-times. Clearly each V_λ^k is $GL(m-1)$-stable. Now take a semistandard tableau of shape λ, in which m appears k times. Erase all the boxes where m appears. We obtain a semistandard tableau filled with the indices $1, \dots, m-1$ of some shape μ, obtained from λ by removing k boxes and at most one box in each row.[137] Let us denote by A_μ the space spanned by these tableaux. Thus we can further decompose

[137] In this chapter the indexing by diagrams is dual.

as $V_\lambda^k = \bigoplus_\mu A_\mu$. When we apply an element of $GL(m-1)$ to such a tableau, we see that we obtain in general a tableau of the same shape, but not semistandard. The straightening algorithm of such a tableau will consist of two terms, $T_1 + T_2$; in T_1 the index m is not moved, while in T_2 the index m is moved to some upper row in the tableau. It is easily seen that this implies that:

Theorem. *V_λ and each V_λ^k have a good filtration for $GL(m-1)$ in which the factors are the modules V_μ for the shapes μ obtained from λ by removing k boxes and at most one box in each row.*

This is the characteristic free analogue of the results of Chapter 9, §10.3. Of course, in characteristic 0, we can split the terms of the good filtration and obtain an actual decomposition.

5.5 $SL(n)$ Invariants

Theorem. *The ring generated by the Plücker coordinates $[i_1, \dots, i_n]$ extracted from an $n \times m$ matrix is the ring of invariants under the action of the special linear group on the columns.*

Proof. If an element is $SL(n)$ invariant, it is in particular both U^-- and U^+-invariant under left action. By the analysis in 5.1 its left tableau must be at the same time canonical and anticanonical. Hence by 5.1.1 it is the tableau defining a product of maximal minors involving all rows, i.e., Plücker coordinates. □

Classically this theorem is used to prove the projective normality of the Grassmann variety and the factoriality of the ring of Plücker coordinates, which is necessary for the definition of the Chow variety.

Let us digress on this application. Given an irreducible variety $V \subset \mathbb{P}^n$ of codimension $k+1$, a generic linear subspace of $\mathbb{P}^n$ of dimension k has empty intersection with V. The set of linear subspaces which have a nonempty intersection is (by a simple dimension count) a hypersurface, of some degree u, in the corresponding Grassmann variety. Thus it can be defined by a single equation, which is a polynomial of degree u in the Plücker coordinates. This in turn can finally be seen as a point in the (large) projective space of lines in the space of standard monomials of degree u in the Plücker coordinates. This is the Chow point associated to V, which is a way to parameterize projective varieties.

6 Characteristic Free Invariant Theory

6.1 Formal Invariants

We have been working in this chapter with varieties defined over $\mathbb{Z}$ without really formalizing this concept. If we have an affine variety V over an algebraically closed field k and a subring $A \subset k$ (in our case either $\mathbb{Z}$ or a finite field), we say that V is defined over A if there is an algebra $A[V]$ such that $k[V] = A[V] \otimes_A k$. Similarly, a map of two varieties $V \to W$ both defined over A is itself defined over A if its comorphism maps $A[W]$ to $A[V]$.

For an algebraic group G to be defined over A thus means that also its group structures Δ, S are defined over A.

When a variety is defined over A one can consider the set $V[A]$ of its A-rational points. Thinking of points as homomorphisms, these are the homomorphisms of $A[V]$ to A. Although the variety can be of large dimension, the set of its A-rational points can be quite small. In any case if V is a group, $V[A]$ is also a group.

As a very simple example, we take the multiplicative group defined over $\mathbb{Z}$, its coordinate ring being $\mathbb{Z}[x, x^{-1}]$. Its $\mathbb{Z}$-rational points are invertible integers, that is, only ± 1.

More generally, if B is any A algebra, the A-homomorphisms of $A[V]$ to B are considered as the *B-rational points* of V or *points with coefficients in B*. Of course one can define a new variety defined over B by the *base change* $B[V] := A[V] \otimes_A B$.[138]

This causes a problem in the definition of invariant. If a group G acts on a variety V and the group, the variety and the action are defined over A, one could consider the invariants just under the action of the A-rational points of G. These usually are not really the invariants one wants to analyze. In order to make the discussion complete, let us go back to the case of an algebraically closed field k, a variety V and a function $f(x)$ on V. Under the G-action we have the function $f(g^{-1}x)$ on $G \times V$; f is invariant if and only if this function is independent of g, equivalently, if $f(gv)$ is independent of g. In the language of comorphism we have the comorphisms

$$\mu : k[V] \to k[G] \otimes k[V], \quad \mu(f)(g, v) := f(gv).$$

So, to say that f is invariant, is equivalent to saying that $\mu(f) = 1 \otimes f$. Furthermore in the language of morphisms, a specific point $g_0 \in G$ corresponds to a morphism $\phi : k[V] \to k$ and the function (of x only) $f(g_0 x)$ is $\phi \otimes 1 \circ \mu(f)$.

Now we leave to the reader to verify the simple:

Proposition 1. *Let G, V and the action be defined over $A \subset k$. For an element $f \in A[V]$ the following are equivalent:*

(1) $\mu(f) = 1 \otimes f$.
(2) For every commutative algebra B the function $f \otimes 1 \in B[V]$ is invariant under the group $G[B]$ of B-rational points.
(3) The function $f \otimes 1 \in k[V]$ is invariant.

If f satisfies the previous properties, then it is called an absolute invariant or just an invariant.

---

[138] Actually, to be precise we should extend our notions to the idea of *affine scheme*; otherwise there are some technical problems with this definition.

One suggestive way of thinking of condition (1) is the following. Since we have defined a rational point of G in an algebra B as a homomorphism $A[G] \to B$, we can in particular consider the identity map $A[G] \xrightarrow{1} A[G]$ as a point of G with coefficients in $A[G]$. This is by definition *the generic point* of G. Thus condition (1) means that f is invariant under the action of a generic group element. The action under any group element $\phi : A[G] \to B$ is obtained by *specializing* the generic action.

It may be useful to see when invariance only under the rational points over A implies invariance. We have:

Proposition 2. *If the points $G[A]$ are Zariski dense in G, then a function invariant under $G[A]$ is an invariant.*

Proof. We have $f(x) = f(gx)$ when $g \in G[A]$. Since for any given x the function $f(gx) - f(x)$ is a regular function on G, if it vanishes on a Zariski dense subset it is identically 0. □

Exercise. Prove that if F is an infinite field and k is its algebraic closure, the rational points $GL(n, F)$ are dense in the group $GL(n, k)$.

Prove the same statement for the groups which can be parameterized by a linear space through the Cayley transform (Chapter 4, §5.1).

A similar discussion applies when we say that a vector is a weight vector under a torus defined over $\mathbb{Z}$ or a finite field. We mean that it is an absolute weight vector under any base change. We leave it to the reader to repeat the formal definition.

6.2 Determinantal Varieties

Consider now the more general theory of standard tableaux on a Schubert variety. We have remarked at the beginning of §4.1 that every Schubert cell intersects the affine set A which we have identified with the space $M_{n,m}$ of $n \times m$ matrices. The intersection of a Schubert variety with A will be called an affine Schubert variety. It is indexed by a minor a of the matrix X and indicated by S_a. The proof given in 4.2 and the remarks on the connection between projective and affine coordinate rings give:

Proposition. *Given a minor a of X the ideal of the variety S_a is generated by the determinants of the minors b which are not greater than or equal to the minor a. Its affine coordinate ring has a basis formed by the standard monomials in the determinants of the remaining minors.*

There is a very remarkable special case of this proposition. Choose the $k \times k$ minor whose row and column indices are the first indices $1, 2, \dots, k$. One easily verifies: A minor b is not greater than or equal to a if and only if it is a minor of rank $> k$. Thus S_a is the determinantal variety of matrices of rank at most k. We deduce:

Theorem. *The ideal I_k generated by the determinants of the $(k+1)\times(k+1)$ minors is prime (in the polynomial ring $A[x_{i,j}]$ over any integral domain A).*

The standard tableaux which contain at least one minor of rank $\geq k+1$ are a basis of the ideal I_k.

The standard tableaux formed by minors of rank at most k are a basis of the coordinate ring $A[x_{i,j}]/I_k$.

Proof. The only thing to be noticed is that a determinant of a minor of rank $s > k+1$ can be expanded, by the Laplace rule, as a linear combination of determinants of $(k+1)\times(k+1)$ minors. So these elements generate the ideal defined by the Plücker coordinates which are not greater than a. □

Over a field the variety defined is the determinantal variety of matrices of rank at most k.

6.3 Characteristic Free Invariant Theory

Now we give the characteristic free proof of the first fundamental theorem for the general linear group.

Let F be an infinite field.[139] We want to show the FFT of the linear group for vectors and forms with coefficients in F.

FFT Theorem. *The ring of polynomial functions on $M_{p,m}(F)\times M_{m,q}(F)$ which are $Gl(m,\mathbb{F})$-invariant is given by the polynomial functions on $M_{p,q}(F)$ composed with the product map, which has as image the determinantal variety of matrices of rank at most m.*

Let us first establish some notation. We display a matrix $A \in M_{p,m}(F)$ as p rows ϕ_i:

$$A := \begin{vmatrix} \phi_1 \\ \phi_2 \\ \dots \\ \phi_p \end{vmatrix}$$

and a matrix $B \in M_{m,q}(F)$ as q columns x_i:

$$B := \begin{vmatrix} x_1 & x_2 & \dots & x_p \end{vmatrix}.$$

The entries of the product are the *scalar products* $\overline{x}_{ij} := \langle \phi_i | x_j \rangle$.

The theory developed for the determinantal variety implies that the double standard tableaux in these elements $\overline{x}_{ij}$ with at most m columns are a basis of the ring A_m generated by these elements.

139 One could relax this by working with formal invariants.

Lemma. *Assume that an element $P := \sum c_i T_i \in A_m$, with T_i distinct double standard tableaux, vanishes when we compute it on the variety C_m formed by those pairs A, B of matrices for which the first m columns x_i of B are linearly dependent; then the column tableau of each T_i starts with the row $1, 2, \dots, m$.*

Similarly if P vanishes when we compute it on the variety R_m formed by those pairs A, B of matrices for which the first m rows ϕ_i of A are linearly dependent, then the row tableau of each T_i starts with the row $m, m-1, \dots, 1$.

Proof. First, it is clear that every double standard tableau with column tableau starting with the row $1, 2, \dots, m$ vanishes on C_m. If we split $P = P_0 + P_1$ with P_0 of the previous type, then also P_1 vanishes on C_m. We can thus assume $P = P_1$, and we must show that $P_1 = 0$.

Decompose $P = Q + R$ where R is the sum of all tableaux which do not contain 1 among the column indices. When we evaluate P in the subvariety of $M_{p,m}(F) \times M_{m,q}(F)$ where $x_1 = 0$, we get that R vanishes identically, hence Q vanishes on this variety. But this variety is just of type $M_{p,m}(F) \times M_{m,q-1}(F)$. We deduce that R is a relation on the double standard tableaux in the indices $1, \dots, p; 2, \dots, q$. Hence $R = 0$ and $P = Q$.

Next, by substituting $x_1 \to x_1 + \lambda x_2$ in P we have a polynomial vanishing identically on C_m. Hence its leading term vanishes on C_m. This leading term is a linear combination of double standard tableaux obtained from some of the T_i by substituting all 1's not followed by 2 with 2's.

Next we perform the substitutions $x_1 + \lambda x_3, \dots, x_1 + \lambda x_m$, and in a similar fashion we deduce a new leading term in which the 1's which are not followed by $2, 3, \dots, m$ have been replaced with larger indices.

Formally this step does not immediately produce a standard tableau, for instance if we have a row $1\,2\,3\,7\,\dots$ and replace 1 by 4 we get $4\,2\,3\,7\,\dots$, but this can be immediately rearranged up to sign to $2\,3\,4\,7\,\dots$.

Since by hypothesis P does not contain any tableau with first row in the right side equal to $1, 2, 3, \dots, m$, at the end of this procedure we must get a nontrivial linear combination of double standard tableaux in which 1 does not appear in the column indices and vanishing on C_m. This, we have seen, is a contradiction. The proof for the rows is identical. □

At this point we are ready to prove the FFT.

Proof. We may assume $p \geq m,\ q \geq m$ and consider

$$d := (m, m-1, \dots, 1 | 1, 2, \dots, m).$$

Let $\mathcal{A}$ be the open set in the variety of matrices of rank $\leq m$ in $M_{p,q}(F)$ where $d \neq 0$. Similarly, let $\mathcal{B}$ be the open set of elements in $M_{p,m}(F) \times M_{m,q}(F)$ which, under multiplication, map to $\mathcal{A}$.

The space $\mathcal{B}$ can be described as pairs of matrices in block form, with multiplication:

$$\begin{vmatrix} A \\ B \end{vmatrix}, \begin{vmatrix} C & D \end{vmatrix} \stackrel{\pi}{\rightarrow} \begin{vmatrix} AC & AD \\ BC & BD \end{vmatrix}$$

and AC invertible.

The complement of $\mathcal{B}$ is formed by those pairs of matrices (A, B) in which either the first m columns x_i of B or the first m rows ϕ_j of A are linearly dependent, i.e., in the notations of the Lemma it is $C_m \cup R_m$.

Finally, setting $\mathcal{B}' := \left\{ \left(\begin{vmatrix} 1_m \\ B \end{vmatrix}, \begin{vmatrix} C & D \end{vmatrix} \right) \right\}$ with C invertible, we get that $\mathcal{B}$ is isomorphic to the product $GL(m, F) \times \mathcal{B}'$ by the map

$$\left(A, \left(\begin{vmatrix} 1_m \\ B \end{vmatrix}, \begin{vmatrix} C & D \end{vmatrix} \right) \right) \mapsto \left(\begin{vmatrix} A^{-1} \\ BA^{-1} \end{vmatrix}, \begin{vmatrix} AC & AD \end{vmatrix} \right).$$

By multiplication we get

$$\begin{vmatrix} 1_m \\ B \end{vmatrix} \begin{vmatrix} C & D \end{vmatrix} = \begin{vmatrix} C & D \\ BC & BD \end{vmatrix}.$$

This clearly implies that the matrices $\mathcal{B}'$ are isomorphic to $\mathcal{A}$ under multiplication and that they form a section of the quotient map π. It follows that the invariant functions on $\mathcal{B}$ are just the coordinates of $\mathcal{A}$. In other words, after inverting d, the ring of invariants is the ring of polynomial functions on $M_{p,q}(F)$ composed with the product map.

We want to use the theory of standard tableaux to show that this denominator can be eliminated. Let then f be a polynomial invariant. By hypothesis f can be multiplied by some power of d to get a polynomial coming from $M_{p,q}(F)$.

Now we take a minimal such power of d and will show that it is 1.

For this we remark that fd^h, when $h \geq 1$, vanishes on the complement of $\mathcal{B}$ and so on the complement of $\mathcal{A}$. Now we only have to show that a polynomial on the determinantal variety that vanishes on the complement of $\mathcal{A}$ is a multiple of d.

By the previous lemma applied to columns and rows we see that each first row of each double standard tableau T_i in the development of fd^h is $(m, m-1, \ldots, 1 | 1, 2, \ldots, m)$, i.e., d divides this polynomial, as desired. □

7 Representations of S_n

7.1 Symmetric Group

We want to recover now, and generalize in a characteristic free way, several points of the theory developed in Chapter 9.

Theorem 1. *If V is a finite-dimensional vector space over a field F with at least $m+1$ elements, then the centralizer of $G := GL(V)$ acting on $V^{\otimes m}$ is spanned by the symmetric group.*

Proof. We start from the identification of $\operatorname{End}_G V^{\otimes m}$ with the invariants $(V^{*\otimes m} \otimes V^{\otimes m})^G$. We can clearly restrict to $G = SL(V)$ getting the same invariants.

Now we claim that the elements of $(V^{*\otimes m} \otimes V^{\otimes m})^G$ are invariants for any extension of the field F, and so are multilinear invariants. Then we have that the multilinear invariants as described by Theorem 6.3 are spanned by the products $\prod_{i=1}^m \langle \alpha_{\sigma(i)} | x_i \rangle$, which corresponds to σ, and the theorem is proved.

To see that an invariant $u \in (V^{*\otimes m} \otimes V^{\otimes m})^{SL(V)}$ remains invariant over any extension field, it is enough to show that u is invariant under all the elementary transformations $1 + \lambda e_{ij}$, $\lambda \in SL(V)$, since these elements generate the group $SL(V)$.

If we write the condition of invariance $u(1+\lambda e_{ij}) = (1+\lambda e_{ij})u$, we see that it is a polynomial in λ of degree $\leq m$ and by hypothesis vanishes on F. By the assumption that F has at least $m+1$ elements it follows that this polynomial is identically 0. □

Next, we have seen in Corollary 4.3 that the space of double tableaux of given double weight has as basis the standard bi-tableaux of the same weight. We want to apply this idea to *multilinear tableaux.*

Let us start with a remark on tensor calculus.

Let V be an n-dimensional vector space. Consider $V^{*\otimes m}$, the space of multilinear functions on V. Let e_i, $i = 1, \ldots, n$, be a basis of V and e^i the dual basis. The elements $e^{i_1} \otimes e^{i_2} \otimes \cdots \otimes e^{i_m}$ form an associated basis of $V^{*\otimes m}$. In functional notation $V^{*\otimes m}$ is the space of multilinear functions $f(x_1, \ldots, x_m)$ in the arguments $x_i \in V$.

Writing $x_i := \sum x_{ji} e_j$ we have

$$(7.1.1) \qquad \langle e^{i_1} \otimes e^{i_2} \otimes \cdots \otimes e^{i_m} | x_1 \otimes \ldots \otimes x_m \rangle = \prod_{h=1}^m x_{i_h h}.$$

Thus the space $V^{*\otimes m}$ is identified to the subspace of the polynomials in the variables x_{ij}, $i = 1, \ldots n$; $j = 1, \ldots, m$, which are *multilinear* in the right indices $1, 2, \ldots, m$.

From the theory of double standard tableaux it follows immediately that:

Theorem 2. *$V^{*\otimes m}$ has as basis the double standard tableaux T of size m which are filled with all the indices $1, 2, \ldots, m$ without repetitions in the column tableau and with the indices from $1, 2, \ldots, n$ (with possible repetitions) in the row tableau.*

To these tableau we can apply the theory of §5.3. One should remark that on $V^{*\otimes m}$ we obviously do not have the full action of $GL(n) \times GL(m)$ but only of $GL(n) \times S_m$, where $S_m \subset GL(m)$ as permutation matrices.

Corollary. *(1) Given a multilinear double tableau of shape λ by the straightening algorithm it is expressed as a linear combination of multilinear standard tableaux of shapes $\geq \lambda$.*

(2) Let S^0_λ, resp. A^0_λ, denote the linear span of all multilinear double standard tableaux tableaux of shape $\geq \lambda$, resp. of multilinear double standard tableaux of shape λ. We have

$$S_\mu^0 := \bigoplus_{\lambda \geq \mu, \atop |\lambda|=|\mu|} A_\lambda^0.$$

Denote by $S_\mu^1 := \bigoplus_{\lambda>\mu,\ |\lambda|=|\mu|} A_\lambda^0$ *(which has as basis the multilinear double standard tableaux of shape* $> \lambda$ *in the dominance ordering).*

(3) The space S_μ^0/S_μ^1 *is a representation of* $GL(n) \times S_m$ *equipped with a natural basis indexed by double standard tableaux* $A|B$ *of shape* μ *and with* B *doubly standard (or multilinear).*

It is isomorphic to the tensor product $V_\lambda \otimes M_\lambda$ *with* V_λ *a representation of* $GL(n)$ *with basis the standard tableaux of shape* λ *and* M_λ *a representation of* S_m *with basis the multilinear standard tableaux of shape* λ.

The proof is similar to 5.2, and so we omit it. In both cases the straightening laws give combinatorial rules to determine the actions of the corresponding groups on the basis of standard diagrams.

7.2 The Group Algebra

Let us consider in $\mathbb{Z}[x_{ij}]$, $i, j = 1, \dots, n$, the space Σ_n spanned by the monomials of degree n multilinear in both the right and left indices.

These monomials have as basis the $n!$ monomials $\prod_{i=1}^n x_{\sigma(i)i} = \prod_{j=1}^n x_{j\sigma^{-1}(j)}$, $\sigma \in S_n$ and also the double standard tableaux which are multilinear or doubly standard both on the left and the right.

Proposition. *The map* $\phi : \mathbb{Z}[S_n] \to \Sigma_n$*, defined by* $\phi : \sigma \to \prod_{i=1}^n x_{\sigma(i)i}$*, is an* $S_n \times S_n$ *linear isomorphism, where on the group algebra* $\mathbb{Z}[S_n] \to \Sigma_n$ *we have the usual left and right actions, while on* Σ_n *we have the two actions on left and right indices.*

Proof. By construction it is an isomorphism of abelian groups and

$$\phi(abc^{-1}) = \prod_{i=1}^n x_{(abc^{-1})(i)i} = \prod_{i=1}^n x_{a(b(i))\,c(i)}. \qquad \square$$

As in the previous theory, we have a filtration by the shape of double standard tableaux (this time multilinear on both sides or *bimultilinear*) which is stable under the $S_n \times S_n$ action. The factors are tensor products $M^\lambda \otimes M_\lambda$. It corresponds, in a characteristic free way, to the decomposition of the group algebra in its simple ideals.

Corollary. *(1) Given a bimultilinear double tableau of shape* λ*, by the straightening algorithm it is expressed as a linear combination of bimultilinear standard tableaux of shapes* $\geq \lambda$.

(2) Let S_λ^{00}*, resp.* A_λ^{00}*, denote the linear span of all bimultilinear tableaux of shape* $\geq \lambda$*, resp. of bimultilinear standard tableaux of shape* λ*. We have*

$$S_\mu^{00} := \bigoplus_{\lambda \geq \mu, \atop |\lambda|=|\mu|} A_\lambda^{00}.$$

Denote by $S^{11}_\mu := \bigoplus_{\lambda>\mu,\ |\lambda|=|\mu|} A^{00}_\lambda$ *(which has as basis the multilinear double standard tableaux of shape* $> \lambda$ *in the dominant ordering).*

(3) The space S^{00}_μ / S^{11}_μ *is a representation of* $S_n \times S_n$ *equipped with a natural basis indexed by the double doubly standard (or bimultilinear) tableaux* $A|B$ *of shape* μ.

It is isomorphic to the tensor product $M^\lambda \otimes M_\lambda$ *with* M^λ *a representation of* S_n *with basis the left multilinear standard tableaux of shape* λ *and* M_λ*, a representation of* S_n *with basis the right multilinear standard tableaux of shape* λ.

The proof is similar to 5.2, and so we omit it.

Again one could completely reconstruct the characteristic 0 theory from this approach.

7.3 Kostka Numbers

Let us consider in the tensor power $V^{*\otimes m}$ the tensors of some given weight $h_1, h_2, \dots, h_m, \sum h_i = m$, i.e., the span of the tensors $e^{i_1} \otimes e^{i_2} \otimes \cdots \otimes e^{i_m}$ in which the indices $i_1, i_2, \dots, i_m$ contain 1 h_1 times, 2 h_2 times, and so on. These tensors are just the S_m orbit of $(e^1)^{h_1} \otimes (e^2)^{h_2} \otimes \cdots (e^m)^{h_m}$ and, as a representation of S_m, they give the permutation representation on $S_m / S_{h_1} \times \ldots \times S_{h_m}$. By the theory of standard tableaux this space has also a basis of double tableaux $A|B$ where A is standard and B semistandard of weight $\mu := h_1, h_2, \dots, h_m$. In characteristic 0 we thus obtain:

Theorem. *The multiplicity of the irreducible representation* M_λ *of* S_m *in the permutation representation on* $S_m / S_{h_1} \times \ldots \times S_{h_m}$ *(Kostka number) is the number of semistandard tableaux* B *of shape* λ *and of weight* μ.

In positive characteristic, we replace the decomposition with a good filtration.

8 Second Fundamental Theorem for GL and S_m

8.1 Second Fundamental Theorem for the Linear Group

Given an m-dimensional vector space V over an infinite field F, the first fundamental theorem for the general linear group states that the ring of polynomial functions on $(V^*)^p \times V^q$ which are $GL(V)$-invariant is generated by the functions $\langle \alpha_i | v_j \rangle$.

Equivalently, the ring of polynomial functions on $M_{p,m} \times M_{m,q}$ which are $Gl(m, F)$-invariant is given by the polynomial functions on $M_{p,q}$ composed with the product map, which has as image the determinantal variety of matrices of rank at most m. Thus Theorem 6.2 can be interpreted as:

Theorem (Second fundamental theorem for the linear group). *Every relation among the invariants* $\langle \alpha_i | v_j \rangle$ *is in the ideal* I_m *generated by the determinants of the* $(m + 1) \times (m + 1)$ *minors of the matrix formed by the* $\langle \alpha_i | v_j \rangle$.

8.2 Second Fundamental Theorem for the Symmetric Group

We have seen that the space of $GL(V)$-endomorphisms of $V^{\otimes n}$ is spanned by the symmetric group S_n. We have a linear isomorphism between the space of operators on $V^{\otimes n}$ spanned by the permutations and the space of multilinear invariant functions.

To a permutation σ corresponds f_σ:

$$f_\sigma(\alpha_1, \alpha_2, \ldots, \alpha_n, v_1, v_2, \ldots, v_n) = \prod_{i=1}^{n} \langle \alpha_{\sigma i} | v_i \rangle.$$

More formally, f_σ is obtained by evaluating the variables x_{hk} in the invariants $\langle \alpha_h | v_k \rangle$ in the monomial $\prod_{i=1}^n x_{\sigma i,i}$. We want to analyze the relations among these invariants. We know that such relations are the intersection of the linear span of the given monomials with the determinantal ideal I_k (cf. §6.2).

Now the span of the multilinear monomials $\prod_{i=1}^n x_{\sigma i,i}$ is the span of the double tableaux with n boxes in which both the right and left tableau are filled with the n distinct integers $1, \ldots, n$.

Theorem. *The intersection of the ideal I_k with the span of the multilinear monomials corresponds to the two-sided ideal of the algebra of the symmetric group S_n generated by the antisymmetrizer $\sum_{\sigma \in S_{k+1}} \epsilon_\sigma \sigma$ in $k+1$ elements.*

Proof. By the previous paragraph it is enough to remark that this antisymmetrizer corresponds to the polynomial

$$(k+1, k, \ldots, 2, 1 | 1, 2, \ldots, k, k+1) \prod_{j=k+2}^{m} (j|j),$$

and then apply the symmetric group on both sets of indices, and the straightening laws. □

8.3 More Standard Monomial Theory

We have seen in Chapter 11, §4 the two plethysm formulas 4.5.1, 4.5.2 for $S(S^2(V))$ and $S(\bigwedge^2(V))$. We want to now give a combinatorial interpretation of these
formulas.

We think of the first algebra over $\mathbb{Z}$ as the polynomial ring $\mathbb{Z}[x_{ij}]$ in a set of variables x_{ij} subject to the symmetry condition $x_{ij} = x_{ji}$, while the second algebra is the polynomial ring $\mathbb{Z}[y_{ij}]$ is a set of variables y_{ij}, $i \neq j$, subject to the skew symmetry condition $y_{ij} = -y_{ji}$.

In the first case we will display the determinant of a $k \times k$ minor extracted from the rows $i_1, i_2, \ldots, i_k$ and columns $j_1, j_2, \ldots, j_k$ as a two-row tableau:

$$\begin{vmatrix} i_1, i_2, \ldots, i_k \\ j_1, j_2, \ldots, j_k \end{vmatrix}. \tag{8.3.1}$$

The main combinatorial identity is this:

Lemma. *If we fix any index a and consider the $k+1$ indices $i_a, i_{a+1}, \dots, i_k, j_1, j_2, \dots, j_a$, then alternating the two-row tableau in these indices produces* 0.

Proof. The proof is by decreasing induction on a. Since this is a formal identity we can work in $\mathbb{Q}[x_{ij}]$. It is convenient to rename these indices $u_{a+1}, u_{a+2}, \dots, u_{k+1}, u_1, u_2, \dots, u_a$.

We start by proving the result for the case $a = k$, which is the identity

$$\begin{vmatrix} i_1, i_2, \dots, i_{k-1}, s \\ j_1, j_2, \dots, j_{k-1}, j_k \end{vmatrix} = \sum_{p=1}^{k} \begin{vmatrix} i_1, \; i_2, \; i_3, \dots\dots, \; i_{k-1}, \; j_p \\ j_1, \dots, j_{p-1}, s, j_{p+1}, \dots, j_k \end{vmatrix}.$$

To prove this identity expand the determinants appearing on the right-hand side with respect to the last row:

$$\sum_{p=1}^{k} \begin{vmatrix} i_1, \; i_2, \; i_3, \dots\dots, \; i_{k-1}, \; j_p \\ j_1, \dots, j_{p-1}, s, j_{p+1}, \dots, j_k \end{vmatrix}$$

$$= \sum_{p=1}^{k} \left(\sum_{u=1}^{p-1} (-1)^{n+u} \begin{vmatrix} j_p \\ j_u \end{vmatrix} \begin{vmatrix} i_1, i_2, \dots\dots\dots\dots\dots\dots\dots, i_{k-2}, i_{k-1} \\ j_1, j_2, \dots, \check{j}_u, \dots, j_{p-1}, s, j_{p+1}, \dots, j_k \end{vmatrix} \right.$$

$$+ (-1)^{n+p} \begin{vmatrix} j_p \\ s \end{vmatrix} \begin{vmatrix} i_1, i_2, \dots\dots\dots\dots, i_{k-1} \\ j_1, j_2, \dots, j_{p-1}, j_{p+1}, \dots, j_k \end{vmatrix}$$

$$\left. + \sum_{u=p+1}^{k} (-1)^{n+u} \begin{vmatrix} j_p \\ j_u \end{vmatrix} \begin{vmatrix} i_1, i_2, \dots\dots\dots\dots\dots\dots\dots, i_{k-1} \\ j_1, j_2, \dots, j_{p-1}, s, j_{p+1}, \dots, \check{j}_u, \dots, j_k \end{vmatrix} \right).$$

The right-hand side of the above equals

$$\sum_{p=1}^{k} \left(\sum_{u=1}^{p-1} (-1)^{n+u} \begin{vmatrix} j_p \\ j_u \end{vmatrix} \begin{vmatrix} i_1, i_2, \dots\dots\dots\dots\dots\dots\dots, i_{k-1} \\ j_1, j_2, \dots, \check{j}_u, \dots, j_{p-1}, s, j_{p+1}, \dots, j_k \end{vmatrix} \right)$$

$$+ \sum_{u=1}^{k} \left(\sum_{p=u+1}^{k} (-1)^{n+p} \begin{vmatrix} j_u \\ j_p \end{vmatrix} \begin{vmatrix} i_1, i_2, \dots\dots\dots\dots\dots\dots\dots, i_{k-1} \\ j_1, j_2, \dots, j_{u-1}, s, j_{u+1}, \dots, \check{j}_p, \dots, j_k \end{vmatrix} \right)$$

$$+ \sum_{p=1}^{k} (-1)^{n+p} \begin{vmatrix} j_p \\ s \end{vmatrix} \begin{vmatrix} i_1, i_2, \dots\dots\dots\dots, i_{k-1} \\ j_1, j_2, \dots, j_{p-1}, j_{p+1}, \dots, j_k \end{vmatrix}.$$

The first two terms of the above expression cancel and the last is the expansion of $\begin{vmatrix} i_1, i_2, \dots, i_{k-1}, s \\ j_1, j_2, \dots\dots, j_k \end{vmatrix}$.

Suppose the lemma is proved for some $a+1$. We want to prove it for a. Compute

$$\sum_{\sigma\in S_{k+1}} \epsilon_\sigma \begin{vmatrix} i_1, i_2, \dots, i_a, u_{\sigma(a+1)}, u_{\sigma(a+2)}, \dots, u_{\sigma(k+1)} \\ u_{\sigma(1)}, u_{\sigma(2)}, \dots, u_{\sigma(a)}, j_{a+1}, \dots, j_{k-1}, j_k \end{vmatrix} = \sum_{\sigma\in S_{k+1}} \epsilon_\sigma$$

$$\times \left(\sum_{b=1}^{a} \begin{vmatrix} i_1, i_2, \dots, i_a, u_{\sigma(a+1)}, u_{\sigma(a+2)}, \dots, u_{\sigma(k)}, u_{\sigma(b)} \\ u_{\sigma(1)}, \dots, u_{\sigma(b-1)}, u_{\sigma(k+1)}, u_{\sigma(b+1)}, \dots, u_{\sigma(a)}, j_{a+1}, \dots, j_{k-1}, j_k \end{vmatrix} \right.$$

$$\left. + \sum_{b=a+1}^{k} \begin{vmatrix} i_1, i_2, \dots, i_a, u_{\sigma(a+1)}, u_{\sigma(a+2)}, \dots, u_{\sigma(k)}, j_b \\ u_{\sigma(1)}, \dots, u_{\sigma(a)}, j_{a+1}, \dots, j_{b-1}, u_{\sigma(k+1)}, j_{b+1}, j_{k-1}, j_k \end{vmatrix} \right)$$

$$= -a \sum_{\sigma\in S_{k+1}} \epsilon_\sigma \begin{vmatrix} i_1, i_2, \dots, i_a, u_{\sigma(a+1)}, u_{\sigma(a+2)}, \dots, u_{\sigma(k+1)} \\ u_{\sigma(1)}, u_{\sigma(2)}, \dots, u_{\sigma(a)}, j_{a+1}, \dots, j_{k-1}, j_k \end{vmatrix} + \sum_{b=a+1}^{k} (-1)^{k-b-1}$$

$$\times \sum_{\sigma\in S_{k+1}} \epsilon_\sigma \begin{vmatrix} i_1, i_2, \dots, i_a, j_b, u_{\sigma(a+1)}, u_{\sigma(a+2)}, \dots, u_{\sigma(k)} \\ u_{\sigma(1)}, \dots, u_{\sigma(a)}, u_{\sigma(k+1)}, j_{a+1}, \dots, j_{b-1}, j_{b+1}, j_{k-1}, j_k \end{vmatrix}.$$

By induction this last sum is 0 and we have

$$(1+a) \sum_{\sigma\in S_{k+1}} \epsilon_\sigma \begin{vmatrix} i_1, i_2, \dots, i_a, u_{\sigma(a+1)}, u_{\sigma(a+2)}, \dots, u_{\sigma(k+1)} \\ u_{\sigma(1)}, u_{\sigma(2)}, \dots, u_{\sigma(a)}, j_{a+1}, \dots, j_{k-1}, j_k \end{vmatrix} = 0. \qquad \square$$

Recall that in order to alternate a function which is already alternating on two sets of variables, it is sufficient to alternate it over the coset representatives of $S_{k+1}/S_{k+1-a} \times S_a$. We will use the previous relation in this form.

Let us take any product of minors of a symmetric matrix displayed as in 8.3.1. We obtain a tableau in which each type of row appearing appears an even number of times. In other words, the columns of the tableau are all even. We deduce:

Theorem. *In the case of symmetric variables $x_{i,j} = x_{j,i}$, the standard tableaux with even columns form a $\mathbb{Z}$-basis of $\mathbb{Z}[x_{ij}]$.*

Proof. A product of variables x_{ij} is a tableau (with just one column). We show first that every tableau is a linear combination of standard ones.

So we look at a violation of standardness in the tableau. This can occur in two different ways since a tableau is a product $d_1 d_2 \dots d_s$ of determinants of minors.

The first case is when the violation appears in two indices $i_a > j_a$ of a minor, displayed as $d_k = \begin{vmatrix} i_1, i_2, \dots, i_k \\ j_1, j_2, \dots, j_k \end{vmatrix}$. The identity in the previous lemma implies immediately that this violation can be removed by replacing the tableau with lexicographically smaller ones. The second case is when the violation occurs between a column index of some d_k and the corresponding row index of d_{k+1}. Here we can use the fact that by symmetry we can exchange the rows with the column indices in a minor and then we can apply the identity on double tableaux discussed in §4.3. The final result is to express the given tableau as a linear combination of tableaux which are either of strictly higher shape or lexicographically inferior to the given one. Thus this straightening algorithm terminates.

In order to prove that the standard tableaux so obtained are linearly independent, one could proceed as in the previous sections. Alternatively, we can observe that since standard tableaux of a given shape are, in characteristic 0, in correspondence with a basis of the corresponding linear representation of the linear group, the proposed basis is in each degree k (by the plethysm formula) of cardinality equal to the dimension of $S^k[S^2(V)]$, and so, being a set of linear generators, it must be a basis. □

8.4 Pfaffians

For the symplectic case $\mathbb{Z}[y_{ij}]$, $i, j = 1, \dots, n$, subject to the skew symmetry, we define, for every sequence $1 \leq i_1 < i_2 < \cdots < i_{2k} \leq n$ formed by an even number of indices, the symbol $|i_1, i_2, \dots, i_{2k}|$ to denote the Pfaffian of the principal minor of the skew matrix $Y = (y_{ij})$. A variable y_{ij}, $i < j$ equals the Pfaffian $|ij|$.

A product of such Pfaffians can be displayed as a tableau with even rows, thus a product of variables y_{ij} is a tableau with two columns. The theorem in this case is:

Theorem 1. *The standard tableaux with even rows form a $\mathbb{Z}$-basis of $\mathbb{Z}[y_{ij}]$.*

The proof of this theorem is similar to the previous one. In order to show that every tableau is a linear combination of standard ones we need an identity between Pfaffians, which produces the straightening algorithm.

Lemma 1. *The a_i, b_j are indices among $1, \dots, n$:*

$$[a_1, \dots, a_p][b_1, \dots, b_m] - \sum_{h=1}^{p}[a_1, \dots, a_{h-1}, b_1, a_{h+1}, \dots a_p][a_h, b_2, \dots, b_m]$$
$$= \sum_{k=2}^{m}(-1)^{k-1}[b_2, \dots, \check{b}_k, \dots, b_m][b_k, b_1, a_1, \dots, a_p].$$

Proof. We use the development of a Pfaffian:

$$[a_1,\dots,a_p][b_1,\dots,b_m]-\sum_{h=1}^{p}[a_1,\dots,a_{h-1},b_1,a_{h+1},\dots a_p][a_h,b_2,\dots,b_m]$$

$$=[a_1,\dots,a_p]\left(\sum_{k=2}^{m}(-1)^k[b_1,b_k][b_2,\dots,\check{b}_k,\dots,b_m]\right)$$

$$-\sum_{h=1}^{p}[a_1,\dots,a_{h-1},b_1,a_{h+1},\dots a_p]\sum_{k=2}^{m}(-1)^k[a_h,b_k][b_2,\dots,\check{b}_k,\dots,b_m]$$

$$=\sum_{k=2}^{m}(-1)^k[b_2,\dots,\check{b}_k,\dots,b_m](-[b_k,b_1][a_1,\dots,a_p]$$

$$+(-1)^{h-1}[b_k,a_h][b_1,a_1,\dots,a_{h-1},a_{h+1},\dots,a_p])$$

$$=\sum_{k=2}^{m}(-1)^{k-1}[b_2,\dots,\check{b}_k,\dots,b_m][b_k,b_1,a_1,\dots,a_p]$$ □

We are ready now to state and prove the basic form of the straightening algorithm. First, we do it in a weak form over $\mathbb{Q}$.

Lemma 2. *a_i, b_j, c_l are indices from $1,\dots,n$:*

$$\sum_{\sigma\in S_{k+i+1}}\epsilon_\sigma[a_1,a_2,\dots,a_i,b_{\sigma(1)},\dots,b_{\sigma(k)}][b_{\sigma(k+1)},\dots,b_{\sigma(k+i+1)},c_1,\dots,c_t]$$

is a linear combination, with rational coefficients, of higher terms

$$[i_1,\dots,i_p][j_1,j_2,\dots,j_r]$$

with $p>i+k$.

Proof. We prove the statement by induction on i. If $i=0$, we can apply the previous lemma. Otherwise, assume the statement true for $i-1$ and use Lemma 1 to deduce:

$$\sum_{\sigma\in S_{k+i+1}}\epsilon_\sigma[a_1,\dots,a_i,b_{\sigma(1)},\dots,b_{\sigma(k)}][b_{\sigma(k+1)},\dots,b_{\sigma(k+i+1)},c_1,\dots,c_t]$$

$$=\sum_{\sigma\in S_{k+i+1}}\epsilon_\sigma\Bigg(\sum_{j=1}^{i}[a_1,\dots,a_{j-1},b_{\sigma(k+1)},\dots,a_{j+1},\dots,a_i,b_{\sigma(1)},\dots,b_{\sigma(k)}]$$

$$\times[a_j,b_{\sigma(k+2)},\dots b_{\sigma(k+i+1)},c_1,\dots,c_t]$$

$$+\sum_{u=1}^{k}[a_1,\dots,a_i,b_{\sigma(1)},\dots,b_{\sigma(u-1)},b_{\sigma(k+1)},b_{\sigma(u+1)},\dots,b_{\sigma(k)}]$$

$$\times[b_{\sigma(k)},b_{\sigma(k+2)},\dots,b_{\sigma(k+i+1)},c_1,\dots,c_t]\Bigg)+R$$

$$=R'-k\sum_{\sigma\in S_{k+i+1}}\epsilon_\sigma[a_1,\dots,a_i,b_{\sigma(1)},\dots,b_{\sigma(k)}]$$

$$\times[b_{\sigma(k+1)},\dots,b_{\sigma(k+i+1)},c_1,\dots,c_t]+R$$

where R are terms of *higher shape* given by Lemma 1, while R' are terms of *higher shape* given by induction. Thus

$$(1+k)\sum_{\sigma\in S_{k+i+1}} \epsilon_\sigma [a_1,\dots,a_i,b_{\sigma(1)},\dots,b_{\sigma(k)}][b_{\sigma(k+1)},\dots,b_{\sigma(k+i+1)},c_1,\dots,c_t]$$

is a sum of higher terms. □

Lemma 3. *The standard tableaux (products of Pfaffians) are a linear basis of* $\mathbb{Q}[y_{i,j}]$.

Proof. The fact that they span $\mathbb{Q}[y_{i,j}]$ comes from the fact that the previous lemma gives a straightening algorithm over $\mathbb{Q}$. The linear independence follows from the Plethysm formula and the fact that, from the representation theory of the linear group, we know that the number of standard tableaux of a given degree equals the dimension of the polynomial ring in that degree. □

We now restate, and prove Theorem 1 in a more precise form.

Theorem 2. *The standard tableaux with even rows form a* $\mathbb{Z}$*-basis of* $\mathbb{Z}[y_{ij}]$. *Moreover*

$$\sum_{\sigma\in S_{k+i+1}/S_k\times S_{i+1}} \epsilon_\sigma [a_1,a_2,\dots,a_i,b_{\sigma(1)},\dots,b_{\sigma(k)}] \times [b_{\sigma(k+1)},\dots,b_{\sigma(k+i+1)},c_1,\dots,c_t]$$

is a linear combination, with integral coefficients, of higher terms $[i_1,\dots,i_n]$ $[j_1,j_2,\dots j_r]$ *with* $n>i+k$ *and gives a straightening algorithm over* $\mathbb{Z}$.

Proof. The proof goes in two steps. In the first step we prove that, taking as coefficients an infinite field F, the given standard tableaux are linearly independent. For this we see that the proof of 2.3 applies with a little change. Namely, here the transformations $i+\lambda j$ are applied to the matrix of variables $Y=(y_{i,j})$ on rows and columns. Y transforms to a matrix whose Pfaffians are multilinear in the indices, so if i appears and j does not appear it creates a term in λ, and we can argue as in that section. Starting from a possible relation we get a relation of type $[1,2,\dots,k]^h=0$ which is not valid.

In the second step we see that, if the standard tableaux with even rows are not a $\mathbb{Z}$-basis of $\mathbb{Z}[y_{ij}]$, since they are a basis over $\mathbb{Q}$, we can specialize at some prime so that they become linearly dependent, contradicting the previous step.

As a final step, the straightening algorithm over $\mathbb{Q}$ in the end expresses a two-line tableau T as a sum of standard tableaux of the same and of higher shape. Since we have seen that the standard tableaux are a basis over $\mathbb{Z}$, this implies that the final step of the straightening algorithm must express T as an integral linear combination of tableaux. □

Exercise. Do the standard monomial theory for Pfaffians as a theory of Schubert cells for the variety of pure spinors. Interpret in this way the quadratic equations satisfied by pure spinors as the basic straightening laws (cf. Chapter 11, §7.2).

8.5 Invariant Theory

We are now going to deduce the first fundamental theorem for invariants of the orthogonal and symplectic group in all characteristics, using a method similar to the one of §6.3 for the linear group. For the second fundamental theorem the argument is like the one of §8.1.

We do first the symplectic group which is simpler.[140] For this we have to prove the usual:

Lemma 1. *If a polynomial in the skew product vanishes on the set where the first* $2n$ *elements are linearly dependent, it is a multiple of* $[1, 2, \dots, 2n]$.

The proof is similar to 5.1 and we omit it.

Theorem 1. *Over any field* F *the ring of invariants of* p *copies of the fundamental representation of* $Sp(2n, F)$ *is generated by the skew products.*

Proof. Take p copies of the fundamental representation of $Sp(2n, F)$. We may assume $p \geq 2n$ is even. We work geometrically and think of the invariants $[v_i, v_j]$ as the coordinates of a map π from $p \times 2n$ matrices to skew-symmetric $p \times p$ matrices, $\pi(T) := TJT^t$. The image is formed by the set D^p_{2n} of skew-symmetric $p \times p$ matrices of rank $\leq 2n$. The first step is thus to consider, in the variety D^p_{2n} of skew-symmetric $p \times p$ matrices of rank $\leq 2n$, the open set U where the Pfaffian $[1, 2, \dots, 2n]$ is different from 0.

The open set $\pi^{-1}(U)$ is the set of p-tuples of vectors $v_1, \dots, v_p$ with the property that the first $2n$ vectors are linearly independent. We claim that the map $\pi : \pi^{-1}(U) \to U$ is a locally trivial fibration. For each point in U there is a neighborhood W with $\pi^{-1}(W)$ equal to the product $Sp(2n, F) \times W$. In other words we want to find a section $s : W \to \pi^{-1}(W)$ so that $\pi s = 1$ and the map $Sp(2n, F) \times W \to \pi^{-1}(W)$, $(g, w) \mapsto g(s(w))$ is the required isomorphism.

In fact let A be the principal $2n \times 2n$ minor of a matrix X in U, an invertible skew-symmetric matrix. If $\pi(v_1, \dots, v_p) = X$, the entries of A are the elements $[v_i, v_j]$, $i, j \leq 2n$.

We want to find the desired section and trivialization by interpreting the algorithm of finding a symplectic basis for the form $u^t A v$. We consider A as the matrix of a symplectic form in some basis $b_1, b_2, \dots, b_{2n}$.

First, let us analyze this algorithm which proceeds stepwise. There are two types of steps. In one type of step, we have determined $e_1, f_1, \dots, e_{i-1}, f_{i-1}$ as linear combinations of the b_i and we have to choose e_i. This is done by choosing any vector orthogonal to the previously determined ones, which in turn involves a solution of the linear system of equations $\sum_{j=1}^n x_j[b_j, e_i] = \sum_{j=1}^n x_j[b_j, f_i] = 0$. The linear system is of maximal rank but in order to solve it explicitly we have to choose an invertible maximal minor by whose determinant we have to divide. This choice depends on the initial value A_0 of A and thus the resulting formula is valid only in some open set. The choice of e_i has coordinates which are rational functions of the entries of A.

140 We correct a mistake in [DC] in the statement of the theorem.

The other type of step consists of completing e_i to f_i which is again the solution of a linear equation. The algorithm furnishes a rational function on some open set W containing any given matrix A_0, which associates to a skew matrix A a symplectic basis S written in terms of the given basis b_i, in other words a matrix $f(A)$ such that $f(A)Af(A)^t = J_{2n}$, the standard matrix of the symplectic form. The rows of $f(A)^{-1}$ define an explicit choice of vectors $v_i(A)$, depending on A through a rational function defined in a neighborhood of a given A_0, with matrix of skew products $[v_i(A), v_j(A)] = a_{i,j}$. Using the full matrix $X \in D^p_{2n}$ of skew products, of which A is a principal minor, we can complete this basis to a full p-tuple with skew products X. Since $v_k(A) = \sum_{j=1}^{2n} z_{k,j} v_j(A)$ can be solved from the identities, $x_{i,k} = [v_i(A), v_k(A)] = \sum_{j=1}^{2n} z_{k,j}[v_i(A), v_j(A)] = \sum_{j=1}^{2n} z_{k,i} a_{i,j}$.

Thus we have constructed a section $s(X) \in M_{p,2n}$ with $s(X)Js(X)^t = X$. From this the trivialization is $(X, Y) \to s(X)Y^{-1},\ X \in U,\ Y \in Sp(2n, F)$.

Once we have proved the local triviality of the map, let us take a function on $\pi^{-1}(U)$ which is invariant under the symplectic group. On each open set $\pi^{-1}(W) = Sp(2n, F) \times W$ the function must necessarily come from a regular function on W. Since the regular functions on an algebraic variety have the *sheaf* property, i.e., a function which is locally regular is regular, we deduce that the invariant comes from a function on U.

At this point we know that if f is an invariant, after eventually multiplying it by a power of the Pfaffian $[1, 2, \ldots, 2n]$ it lies in the subring generated by the skew products with basis the standard tableaux. We now have to do the cancellation. This is a consequence of Lemma 1 as in the case of the general linear group. □

We have already mentioned the fact that the orthogonal group is harder. First, we will work in characteristic $\neq 2$. In characteristic 2 there are various options in defining the orthogonal group, one being to define the orthogonal group by the equations $XX^t = 1$ as a *group scheme*, since in characteristic 2 these equations do not generate a radical ideal.

Apart from the problem of characteristic 2, the difference between the symplectic and the orthogonal group is the following. The map $X \to XJX^t$ from invertible $2n \times 2n$ matrices to invertible skew matrices is a fibration locally trivial in the Zariski topology as we have seen by the algorithm of constructing a symplectic basis. For the orthogonal group $O(V)$, $\dim(V) = n$ the map is $X \mapsto XX^t$, but the theory is not the same. In this case we start as before taking the open set U of matrices in which the determinant of the first principal minor $A := \begin{vmatrix} 1, 2, \ldots, n \\ 1, 2, \ldots, n \end{vmatrix}$ is invertible. We need some algorithm to construct some kind of standard basis for the space with a symmetric form given by a matrix A. In general we may try to find an orthogonal basis, otherwise a hyperbolic basis. In the first case, when we do the standard Gram–Schmidt orthogonalization, if we want to pass to an orthonormal basis (as we do) we have to extract some square roots. In the second case we still have to solve quadratic equations since we have to find isotropic vectors. In any case the formulas we will find when we want to find a section of the fibration π as in the previous case will also involve extracting square roots. The technical way of expressing this is that the fibration is locally trivial in the étale topology. In fact, apart from introducing a new technical notion the proof still works. We need to remark though that regular functions have the sheaf property also with respect to this topology. In fact in our case it is really some simple Galois theory. Alternatively we can work more geometrically.

Lemma 2. *The variety S_n^p of symmetric $p \times p$ matrices of rank $\leq n$ is smooth at the points S^0 of rank exactly n. The map $\pi : M_{p,n} \to S_n^p$, $X \mapsto XX^t$ is smooth on the points $\pi^{-1}(S^0)$.*

Proof. The group $GL(n, F)$ acts on both spaces by $AX, AYA^t, X \in M_{p,n}, Y \in S_n^p$ and the map π is equivariant. Since clearly any symmetric matrix of rank n can be transformed, using the action of $GL(n, F)$, to the open set U where the determinant of the first principal minor $A := \begin{vmatrix} 1, 2, \dots, n \\ 1, 2, \dots, n \end{vmatrix}$ is invertible, it is enough to show that U is smooth and that the map $\pi^{-1}(U) \to U, X \mapsto XX^t$ is smooth.

Let

$$X = \begin{vmatrix} A & B \\ B^t & C \end{vmatrix} \in U, \quad \det(A) \neq 0, \quad \operatorname{rank}(X) = n.$$

Next we claim that U projects isomorphically to the pairs A, B with $\det(A) \neq 0$ and B an $n \times (n-p)$ matrix. In fact the entries of the matrix C are determined and are of the form $f_{i,j}(A, B)/\det(A)$ with $f_{i,j}(A, B)$ polynomials.

To prove this, take the $(n+1) \times (n+1)$ minor where to A we add the row i and the column j. By hypothesis its determinant is 0, but this determinant is $\det(A)c_{i,j} + f_{i,j}(A, B)$ (where $f_{i,j}(A, B)$ are the remaining terms of the expansion of the determinant on the last column).

Using this isomorphism we can see that π is a smooth map. In fact compute the differential at a point X by the method explained in Chapter 8, §7.3, substituting $(X+Y)(X+Y)^t$ collecting linear terms $XY^t + YX^t$. Write both $X = \begin{vmatrix} T \\ V \end{vmatrix}, Y = \begin{vmatrix} U \\ W \end{vmatrix}$ in block form with U, W square $n \times n$ matrices. Now the linear terms of the differential read:

$$\begin{vmatrix} TU^t + UT^t & TW^t + UV^t \\ VU^t + WT^t & VW^t + UV^t \end{vmatrix} \to TU^t + UT^t, TW^t + UV^t$$

at a point in which T is invertible. Let the target matrix be a pair (C, D) with C symmetric.

If the characteristic is different from 2, we can solve $TU^t + UT^t = C, \quad TW^t + UV^t = D$ setting $U := C(T^t)^{-1}/2, W^t = T^{-1}(UV^t - D)$. Thus $d\pi$ is surjective and π is smooth. □

Lemma 3. *Let f be a regular function on $\pi^{-1}(U)$ which is invariant under the orthogonal group. Then f comes from a function on U.*

Proof. Let us consider an invariant function f on the open set $\pi^{-1}(U)$. Let R be the ring $F[U][f]$ in which we add f to the ring $k[U]$. We need to prove in fact that $f \in k[U]$. The ring $F[U][f]$ is a coordinate ring of some variety Y so that we have a factorization of the map $\pi : \pi^{-1}(U) \to Y \xrightarrow{\rho} U$. Since f is an invariant, f is constant on the fibers of π which are all orbits. Thus it follows that the map ρ is bijective. At this point we can conclude as follows. The map ρ is separable and bijective, and U is smooth, hence normal, so, by ZMT, ρ is an isomorphism. In other words f is a function on U. □

Lemma 4. *If a polynomial in the scalar product vanishes on the set where the first n elements are linearly dependent, it is a multiple of* $\begin{vmatrix} 1, 2, \dots, n \\ 1, 2, \dots, n \end{vmatrix}$.

The proof is similar to 5.1 and we omit it.

Theorem 2. *Over any field F of characteristic not 2, the ring of invariants of p copies of the fundamental representation of $O(n, F)$ is generated by the scalar products.*

Proof. Let f be an invariant. From the previous lemmas we know that, after eventually multiplying f by a power of the determinant $\begin{vmatrix} 1, 2, \dots, n \\ 1, 2, \dots, n \end{vmatrix}$, it lies in the subring generated by the scalar products with basis the standard tableaux. We have to do the cancellation as in §5.1, and this is achieved by the previous lemma.

Recently M. Domokos and P. E. Frenkel, in the paper "Mod 2 indecomposable orthogonal invariants" (to appear in *Advances in Mathematics*), have shown that in characteristic 2 there are other indecomposable invariants of degree higher than two. So in this case the invariant theory has to be deeply modified.

14

Hilbert Theory

When the algebraists of the 19th century were studying invariants, one of their main concerns was to find a finite set of invariants from which all the others could be expressed as polynomials. In particular they needed to prove that algebras of invariants are finitely generated. In modern geometric terms this result is needed to construct an algebraic variety whose coordinate ring is the ring of invariants which should parameterize in some sense the orbits of the group. Finiteness of the invariants was first proved by Gordan for binary forms by a complicated induction, and then for invariants of forms in any number of variables by Hilbert, who formulated the proofs in a rather abstract way, founding modern commutative algebra. The proof of Hilbert extends immediately to linearly reductive groups. Hilbert asked, in the 14th problem of his famous list, whether this (or rather a more general statement) is always true. It was only in the 1970s that Nagata produced a counterexample. At the same time interest in invariant theory had resurged and the finiteness theorem was also proved for reductive groups, which in characteristic 0 coincide with linearly reductive groups, but not in positive characteristic.

In this chapter we want to give a rather brief introduction to these ideas. We treat in detail some elementary topics and give some ideas of other more advanced topics.

The subject now goes under the name *Geometric Invariant Theory*. It is now a rather rich and deep topic and there are systematic treatments available at least for parts of it.

1 The Finiteness Theorem

1.1 Finite Generation

As already mentioned, one of the themes of 19th century invariant theory was to show that algebras of invariants are finitely generated. Hilbert proved that this is a consequence of the linear reductivity of the group. The main formal ingredient for such a group is *Reynold's operator*, which we discussed in Chapter 6, §2.4. Since

G is linearly reductive, if M is a rational representation we have a canonical G-equivariant map $r_M : M \to M^G$. If $f : M \to N$ is a G-equivariant map of rational representations, we have the commutative diagram:

$$\begin{array}{ccc} M & \xrightarrow{r_M} & M^G \\ {\scriptstyle f}\downarrow & & {\scriptstyle f}\downarrow \\ N & \xrightarrow{r_N} & N^G \end{array}.$$

In particular if G acts rationally as automorphisms of an algebra R and $a \in R^G$ we have:

(Reynold's identity) $$r_R(ab) = ar_R(b), \quad \forall a \in R^G, \quad \forall b \in R.$$

Theorem. *Let G be a linearly reductive group acting on a finite-dimensional vector space V. Then the algebra of invariants is finitely generated.*

Proof. Denote by R the algebra of polynomials and by $S := R^G$ the invariants. Consider the space S^+ of invariant polynomials without constant term and form the ideal RS^+ of R. By the Hilbert basis theorem, this is finitely generated and $RS^+ = \sum_{i=1}^k Ru_i$, $u_i \in S^+$. We may assume that the u_i are homogeneous. Given $u \in S^+$ we thus have $u = \sum_{i=1}^k x_i u_i$, $x_i \in R$. If we apply now Reynold's identity, we obtain $u = \sum_{i=1}^k r_R(x_i)u_i$. In other words the ideal generated by the elements u_i in S is S^+. Now let $T := F[u_1, \dots, u_k]$. We want to show that $T = S$. We proceed by induction. Assume we know that $T_i = S_i$, $i < N$. We want to prove that $T_N = S_N$. Pick $u \in S_N$ and write it as $u = \sum_{i=1}^k v_i u_i$, $v_i \in S$, comparing the degrees we may assume that $\deg v_i + \deg u_i = N$. In particular $\deg v_i < N$, so by induction $v_i \in T$ and the claim follows. □

2 Hilbert's 14th Problem

2.1 Hilbert's 14th Problem

The 14th problem of Hilbert's famous list of 23 problems, presented in the International Congress of Mathematicians in Paris 1900, asks whether the ring of invariants for any group action is finitely generated. In fact Hilbert formulates a more general question: given a finitely generated domain $F[a_1, a_2, \dots, a_k]$ with quotient field G and an intermediate subfield $F \subset H \subset G$, is the algebra $F[a_1, a_2, \dots, a_k] \cap H$ finitely generated over F?

In fact even the first question has a negative answer, as shown by Nagata. The groups for which one can find counterexamples are in fact rather special, being isomorphic to the additive group of a finite-dimensional vector space $\mathbb{C}^m$. At the moment the best result is due to Mukai who gives an example of infinite generation for $m = 3$. For $m = 1$ an old result of Weizenbock shows that instead we have finite generation, while for $m = 2$ the answer is unknown.

Let us quickly explain Weizenbock's result.

The additive group $\mathbb{C}$ is identified with the subgroup U^+ of $SL(2, \mathbb{C})$ of matrices $\begin{vmatrix} 1 & t \\ 0 & 1 \end{vmatrix}$. From Chapter 7, §1.5 and §3.4 it follows that, in every rational representation $\rho : \mathbb{C} \to GL(n, \mathbb{C})$, the matrices induced by $\mathbb{C}$ are unipotent and there is a nilpotent matrix N such that $\rho(t) = e^{tN}$.

Now N is a direct sum of Jordan blocks of sizes $n_1, \dots, n_k$, but then the direct sum of the irreducible representations of $SL(2, \mathbb{C})$ of dimensions $n_1, n_2, \dots, n_k$ restricted to U^+ give the representation V from which we started. At this point one has to verify that the invariants under U^+ of V coincide with the invariants of $SL(2, \mathbb{C})$ on $V \oplus \mathbb{C}^2$. This argument is essentially what we will develop in the next chapter when we discuss covariants, and so we leave it to the reader.

3 Quotient Varieties

We want to discuss quotients in the simplest sense for affine and projective varieties. For a more systematic treatment see [MF], [DL].

From the results of Chapter 7 we can use the fact that, given an algebraic group G acting on a variety X, each orbit Gx is open in its closure $\overline{Gx}$, which is then a union of orbits. All the orbits in $\overline{Gx}$ different from Gx are then necessarily of dimension strictly less than the dimension of Gx. In particular we will use systematically the fact that an orbit of minimal dimension is necessarily closed.

First, let us extend Theorem 1.1. We assume that we are still working over the complex numbers.

Theorem 1. *(1) If a linearly reductive group G acts on an affine variety V, the ring of invariants $\mathbb{C}[V]^G$ is finitely generated.*

(2) If $W \subset V$ is a G-stable subvariety and $\mathbb{C}[W] = \mathbb{C}[V]/I$, the induced map of coordinate rings $\mathbb{C}[V]^G \to \mathbb{C}[W]^G$ is surjective.

Proof. From the general theory of semisimple representations (Chapter 6), if a linearly reductive group G acts on a space M and N is stable, then $M = N \oplus P$ decomposes as direct sum of subrepresentations. $M^G = N^G \oplus P^G$, and so the projection $M^G \to (M/N)^G = P^G$ is surjective. In particular this proves (2).

Now from Chapter 7, Theorem 1.3, given an action of an affine group G on an affine variety V, there exists a linear representation U of G and a G equivariant embedding of V in U. Therefore we have a surjective mapping $\mathbb{C}[U]^G \to \mathbb{C}[V]^G$. From 2.1 $\mathbb{C}[U]^G$ is finitely generated, hence so is $\mathbb{C}[V]^G$. □

We can now give a basic definition.

Definition. Given an action of an affine group G on an affine variety V the ring $\mathbb{C}[V]^G$ is the coordinate ring of an affine variety denoted $V//G$ and called the *categorical quotient* of V by G.

Of course the fact that $\mathbb{C}[V]^G$ is the coordinate ring of an affine variety depends (from the general theory of affine varieties) on the fact that it is finitely generated.

The previous theorem implies furthermore that if W is a G-stable subvariety of V, we have an inclusion $W//G \subset V//G$ as a subvariety.

The natural inclusion $\mathbb{C}[V]^G \subset \mathbb{C}[V]$ determines a canonical *quotient map* $\pi : V \to V//G$.

Theorem 2. *(1) The mapping π is constant on G-orbits.*

(2) The mapping π is surjective.

(3) Given any point $q \in V//G$, the G-stable subvariety $\pi^{-1}(q)$ of V contains a unique closed G-orbit.

Proof. (1) is clear since the map is given by invariant functions.

(2) Let $q \in V//G$ be a point. It is given by a maximal ideal $m \subset \mathbb{C}[V]^G$. We have to prove that there is a maximal ideal $n \subset \mathbb{C}[V]$ with $m = n \cap \mathbb{C}[V]^G$. Since every ideal is contained in a maximal ideal, it suffices to show that the ideal $m\mathbb{C}[V]$ is a proper ideal, or equivalently that $m = m\mathbb{C}[V] \cap \mathbb{C}[V]^G$.

Then let $a = \sum_i s_i m_i \in \mathbb{C}[V]^G$ with $s_i \in \mathbb{C}[V]$, $m_i \in m$. Apply Reynold's identity $a = R(a) = \sum_i R(s_i)m_i$, and hence $a \in m$ as desired.

(3) Since $\pi^{-1}(q)$ is a G-stable subvariety it contains an orbit of minimal dimension which is then necessarily closed. To complete (3) it is thus sufficient to show that two distinct closed orbits A_1, A_2 map to two distinct points in $V//G$. For this observe that $A_1 \cup A_2$ is an algebraic variety (since the two orbits are closed) and $\mathbb{C}[A_1 \cup A_2] = \mathbb{C}[A_1] \oplus \mathbb{C}[A_2]$ since they are disjoint. Then $\mathbb{C}[A_1 \cup A_2]^G = \mathbb{C}[A_1]^G \oplus \mathbb{C}[A_2]^G = \mathbb{C} \oplus \mathbb{C}$ is the coordinate ring of 2 points which, from the preceding discussion, means exactly that these two orbits map to two distinct points. □

One expresses the meaning of the previous theorem by saying that *$V//G$ parameterizes the closed orbits* of the G-action on V.

Another way of expressing part of the previous theorem is to say that invariants separate closed orbits, that is, given two distinct closed orbits A_1, A_2, there is an invariant with value 1 on A_1 and 0 on A_2.

4 Hilbert–Mumford Criterion

4.1 Projective Quotients

One often wants to apply invariant theory to projective varieties. The setting will be this. We have a linear algebraic group G acting linearly on a vector space V and thus projectively on $P(V)$. Suppose that $W \subset P(V)$ is a projective variety which is stable under the action of G. We would like to define $W//G$ as projective variety. To do it let us first consider the homogeneous coordinate ring $\mathbb{C}[C(W)]$ of the cone of W. The invariant ring $\mathbb{C}[C(W)]^G$ is thus a graded ring. As we have already seen when computing explicit invariants, the generators of this ring can be taken to be homogeneous but not necessarily of the same degree. Thus in order to obtain a projective variety the idea is to take a sufficiently large degree m and consider the space of functions in $\mathbb{C}[C(W)]^G$ which are homogeneous of degree m. Considering these functions as homogeneous coordinates, this space gives a map of W into a projective space. However this map is not defined for the points of W where all the invariants vanish.

It is therefore important to understand from the beginning which are the points in $C(W)$ or even in V where all invariants vanish; these points are called the *unstable points*. Of course in the affine picture these points are just the preimage under the quotient $V \to V//G$ of the image of 0. Therefore we have that:

Proposition 1. *A vector $v \in V$ is unstable, or all invariants without constant term vanish on it, if and only if* 0 *is in the closure of the orbit Gv.*

Proof. An invariant is constant on an orbit and its closure. Thus if 0 is in the closure of the orbit Gv, any homogeneous invariant of degree > 0 vanishes at v. Conversely, assume all such invariants vanish at v. Take in the closure of Gv an orbit C of minimal dimension which is then necessarily closed. If $C \neq 0$, we could find an invariant f with $f(0) = 0$ and $f(C) \neq 0$, which is a contradiction. □

One needs a simpler criterion to see that a vector is unstable, and this is furnished by the Hilbert–Mumford criterion.

Theorem. *A vector v is unstable if and only if there is a* 1*-parameter subgroup $\rho : \mathbb{C}^* \to G$ such that* $\lim_{t\to 0} \rho(t)v = 0$.

In other words 0 is also in the closure of the orbit that one has under a single 1-parameter subgroup.

Let us give an idea of the proof of Hilbert in the case of $GL(n, \mathbb{C})$.

The proof goes essentially in 3 steps.

Step 1. In the first step, using the fact that $0 \in \overline{Gv}$, one constructs an analytic curve $\lambda : D \to \overline{Gv}$ where $D = \{t \in \mathbb{C} \mid |t| < 1\}$ with $\lambda(0) = 0,\ \lambda(t) \in Gv,\ \forall t \neq 0$.

Step 2. Next, by eventually passing to a parameter s with $t = s^k$ one can lift this curve to G for the nonzero values of s, i.e., one finds an analytic map $\mu : D - \{0\} \to GL(n, \mathbb{C})$ with a pole at 0 so that $\mu(t)v = \lambda(t)$.

Step 3. Now consider the matrix $\mu(t)$ with entries $\mu_{i,j}(t)$ some Laurent series. We want to apply a method like the one leading to the elementary divisors to write the function $\mu(t)$ in the form $a(t)\rho(t)b(t)$, where $a(t), b(t)$ are convergent power series (without polar part) with values in $GL(n, \mathbb{C})$, while $\rho(t)$ is a diagonal matrix with entries t^{m_i} for some integers m_i. If we can achieve this, then we see that

$$0 = \lim_{t\to 0} \lambda(t) = \lim_{t\to 0} \rho(t)v = \lim_{t\to 0} \mu(t)v.$$

In more detail, to prove the first two steps we need the following:

Lemma. *(a) Let V be an irreducible variety and U a nonempty open set of V. If $p \in V$, there is an irreducible curve $C \subset V$ so that $p \in C$, $C \cap U \neq \emptyset$.*

(b) Let $\rho : W \to V$ be a dominant map of irreducible varieties and $p \in V$. There is a curve $B \subset W$ with $p \in \overline{\rho(B)}$.

Proof. (a) Let $n = \dim V$. If $n = 1$ there is nothing to prove. Since this is a *local statement,* we can assume $V \subset k^n$ affine. Let $X_1, \dots X_k$ be the irreducible components of $V - U$ which are not points. Choose a point $p_i \in X_i$ for each i and $p_i \neq p$. Then choose a hyperplane H passing through p, which does not contain V and does not pass through any of the p_i. Then by a basic result (cf. [Ha], [Sh]) $H \cap V$ is a union of irreducible varieties V_i of dimension $n - 1$. By our choice, none of them is contained in $V - U$; otherwise it would contain one of the X_j. Take one V_i with $p \in V_i$. We have that $U' := V_i \cap U$ is a nonempty open set. We can thus continue and finish by induction.

(b) We know that the image of ρ contains a nonempty open set U, and by the previous part we can find a curve C in V meeting U and with $p \in C$. Consider $Z := \rho^{-1}(C)$. There is at least one irreducible component Z^0 of Z which maps to C in a nonconstant way. Take any point of Z^0 mapping to some $q \in C \cap U$ and consider $\rho^{-1}(q)$ and the set $T := Z^0 - \rho^{-1}(q)$, open in Z^0. By (a) there is an irreducible curve $B \subset Z^0$ with $q \in B$ and $B \cap T \neq \emptyset$. The map ρ, restricted to B, maps B to C, it is not constant, so it is dominant and satisfies the requirements. □

The next properties we need are specific to curves. We give them without proofs (cf. [Ha], [Sh], [Fu]):

Proposition 2. *Given a map $\rho : C \to B$ of irreducible curves, one can complete $C \subset \overline{C}$, $B \subset \overline{B}$ to two projective curves $\overline{C}, \overline{B}$ so that the map extends to these curves.*

The extended map $\overline{\rho}$ is now surjective, and thus given a point $p \in B$ there is a point $q \in C$ with $\overline{\rho}(q) = p$.

The final property of curves that we need is the *local analytic description.*

Proposition 3. *Given a curve C and a point $p \in C$ there is an analytic map f of a disk D to C such that $f(0) = p$ and f restricted to $D - \{0\}$ is an analytic isomorphism to an open set of C.*[141]

All these statements justify the first two steps of the proof of the H-M criterion.

The analytic coordinate can be replaced in positive characteristic by a formal power series argument which still can be used to prove the theorem.

The third step is an easy Gaussian elimination. We construct the two matrices $a(t), b(t)$ as products of elementary operations on rows and columns as follows. First, permuting rows and columns (i.e., multiplying on the right and left by permutation matrices) we may assume that the order of pole of $\mu_{1,1}(t)$ is the highest of

[141] Now open means in the usual complex topology and **not** in the Zariski topology.

all entries of the matrix. Next, for each $i = 2, \dots, n$, $\mu_{i,1}(t)\mu_{1,1}(t)^{-1}$ is holomorphic, and subtracting from the ith row the first multiplied by $\mu_{i,1}(t)\mu_{1,1}(t)^{-1}$, one can make 0 all the elements of the first column except the first, and similarly for the first row. Next write $\mu_{1,1}(t) = t^{n_1} f(t)$ with $f(t)$ holomorphic and $f(0) \neq 0$. To divide the first row by $f(t)$ is equivalent to multiplying by a diagonal matrix with holomorphic entries $f(t)^{-1}, 1, \dots, 1$.

After this step $\mu(T)$ becomes a block matrix $\begin{vmatrix} t^{n_1} & 0 \\ 0 & \mu_1(t) \end{vmatrix}$, and now we continue by induction.

In order to extend the proof to any linearly reductive group one has to replace the last step of Gaussian elimination with a similar argument. This can be done using the Bruhat decomposition.

The Hilbert–Mumford criterion is quite effective in determining unstable points. For instance Hilbert showed:

Proposition (Hilbert). *Consider the action of $SL(2)$ on homogeneous polynomials of degree n (the binary forms). Such a homogeneous form defines n points (its roots), perhaps with some coincidences and multiplicities, on the projective line. A form is unstable if and only if one of its zeroes has multiplicity $> n/2$.*

Proof. Every 1-parameter group is conjugate to one contained in the standard diagonal torus, and up to conjugacy and reparametrization, we can assume it is the 1-parameter group $t \to \begin{vmatrix} t^{-1} & 0 \\ 0 & t \end{vmatrix}$. This group transforms a form $f(x, y) = \sum_{i=0}^n a_i x^{n-i} y^i$ into the form $\sum_{i=0}^n a_i (tx)^{n-i} (t^{-1}y)^i = \sum_{i=0}^n a_i t^{n-2i} x^{n-i} y^i$. Computing the limit, we have

$$\lim_{t \to 0} \sum_{i=0}^{n} a_i t^{n-2i} x^{n-i} y^i = 0$$

if and only if $a_i = 0$, $\forall n - 2i \leq 0$. This implies that x^{n-k} divides $f(x, y)$, where k is the minimum integer for which $n - 2k > 0$. Hence the point $(0, 1)$ is a root of multiplicity $n - k > n/2$. □

The reader can try to determine the unstable points in the various examples in which we constructed the invariants explicitly. For instance, for the conjugation action of m-tuples of matrices one has:

An m-tuple of matrices is unstable if and only if it can be simultaneously conjugated to an m-tuple of strictly upper triangular matrices. This happens if and only if the given matrices generate a nilpotent subalgebra of the algebra of matrices.

5 The Cohen–Macaulay Property

5.1 Hilbert Series

Let $A = F[a_1, \dots, a_k]$ be a finitely generated algebra over a field F. A theorem which is now standard, but is part of the ideas developed by Hilbert, and known as the *Hilbert basis theorem*, is that A is the quotient $A = F[x_1, \dots, x_k]/J$ of a polynomial ring modulo a finitely generated ideal. In other words once an algebra is finitely generated, then it is also *finitely presented* by generators and relations.

Suppose now that A (as usual with algebras of invariants) is also a graded algebra, $A = \bigoplus_{i=0}^{\infty} A_i$. We assume that $A_0 = F$ and the elements a_i are homogeneous of some degrees $h_i > 0$. In this case we also have that A_i is a finite-dimensional vector space, so if we denote by $d_i := \dim_F A_i$ we have $d_i < \infty$ and we can form the *Hilbert series*:

$$H_A(t) := \sum_{i=0}^{\infty} d_i t^i. \tag{5.1.1}$$

The same construction can be performed when we consider a finitely generated graded module $M = \sum_{i=1}^{k} Au_i$ over a finitely generated graded algebra $A = F[a_1, \dots, a_k]$.

Let us then make the basic remark. Let $f : M \to N$ be a graded morphism of graded modules. Let i be the degree of f, this means that $f(M_k) \subset N_{k+i}$ for all k. We have that $\mathrm{Ker} f$ and $\mathrm{Im} f$ are graded submodules of M, N respectively. We have for all k an exact sequence:

$$0 \to (\mathrm{Ker}\, f)_k \to M_k \xrightarrow{f} N_{k+i} \to (\mathrm{Coker}\, f)_{k+i} \to 0$$

from which

$$\dim((\mathrm{Ker}\, f)_k) - \dim(M_k) + \dim(N_{k+i}) - \dim((\mathrm{Coker}\, f)_{k+i}) = 0.$$

We multiply by t^{k+i} and sum to get

$$H_{\mathrm{Coker} f}(t) - t^i H_{\ker f}(t) = H_N(t) - t^i H_M(t). \tag{5.1.2}$$

Theorem. *Let M be a finitely generated graded module over the graded polynomial ring $F[x_1, \dots, x_m]$ where $\deg x_i = h_i$. Then, for some integeres u, v*

$$H_M(t) = \frac{p(t)}{\prod_{i=1}^{m}(1 - t^{h_i})}, \quad p(t) = \sum_{i=-u}^{v} a_i t^i, \quad a_i \in \mathbb{Z}. \tag{5.1.3}$$

If $M_i = 0$ when $i < 0$, then $p(t)$ is a polynomial in t ($u = 0$).

Proof. By induction on m, if $m = 0$, then M is a finite-dimensional graded vector space and the statements are clear. Assume $m > 0$ and consider the map $f : M \xrightarrow{x_m} M$ given by multiplication by x_m. It is a graded morphism of degree h_m, and by construction, $\mathrm{Ker}\, f$ and $\mathrm{Coker}\, f$ are both finitely generated graded modules annihilated by x_m. In other words they can be viewed as $F[x_1, \dots, x_{m-1}]$ modules, and by induction

$$H_{\ker f}(t) = \frac{p_1(t)}{\prod_{i=1}^{m-1}(1 - t^{h_i})}, \quad H_{\mathrm{Coker} f}(t) = \frac{p_2(t)}{\prod_{i=1}^{m-1}(1 - t^{h_i})}.$$

From 5.1.2,

$$t^{h_m} H_{\ker f}(t) - H_{\mathrm{Coker} f}(t) = (1 - t^{h_m}) H_M(t) \implies H_M(t) = \frac{t^{h_m} p_1(t) - p_2(t)}{\prod_{i=1}^{m}(1 - t^{h_i})}. \quad \square$$

Notice in particular that:

Proposition. *For the graded polynomial ring $A := F[x_1, \dots, x_m]$ where* $\deg x_i = h_i$, *we have*

$$H_A(t) = \frac{1}{\prod_{i=1}^m (1 - t^{h_i})}. \tag{5.1.4}$$

Proof. This follows from the previous proof, or from the fact that if we have two graded vector spaces $M = \oplus M_i,\ N = \oplus N_j$ and set $(M \otimes N)_k := \bigoplus_{i+j=k} M_i \otimes N_j$, then we have

$$H_{M\otimes N}(t) = H_M(t)H_N(t),\ F[x_1, \dots, x_m] = F[x_1] \otimes F[x_2] \otimes \cdots \otimes F[x_m],$$
$$H_{F[x]}(t) = \frac{1}{1 - t^{\deg x}}.$$

5.2 Cohen–Macaulay Property

There are two special conditions on graded algebras which are useful and which appear for algebras of invariants: the *Cohen–Macaulay* and the *Gorenstein* property.

These conditions are not exclusive to graded algebras but for simplicity we formulate them in this important case (cf. [E]).

Definition 1. A finitely generated graded algebra $A = F[a_1, \dots, a_m]$ is said to be Cohen–Macaulay if there exist homogeneous elements $u_1, \dots, u_k \in A$ (a regular system of parameters) with the two properties:

(i) The elements u_i are algebraically independent, i.e., the algebra they generate is a polynomial ring in the u_i.
(ii) The ring A is a finite free module over $B := F[u_1, \dots, u_k]$.

Condition (ii) implies the existence of homogeneous elements $p_1, \dots, p_r$ such that

$$A = \bigoplus_{i=1}^r F[u_1, \dots, u_k]p_i. \tag{5.2.1}$$

If $h_i,\ i = 1, \dots, k$, is the degree of u_i and $\ell_j,\ j = 1, \dots, r$, the degree of p_j, we deduce

$$H_A(t) = \frac{\sum_{j=1}^r t^{\ell_j}}{\prod_{i=1}^k (1 - t^{h_i})}. \tag{5.2.2}$$

When A is C-M, a regular system of parameters is clearly not unique. Nevertheless one can prove (cf. [E]) that if $v_1, \dots, v_s \in A$ and $A/(v_1, \dots, v_s)$ is finite dimensional, then $s \geq k$. If furthermore $s = k$, then $v_1, \dots, v_s$ is a regular system of parameters.

The Gorenstein property is subtler and it is best to explain it first for a finite-dimensional graded algebra. Assume $A = F[a_1, \dots, a_k] = \bigoplus_{i=0}^N A_i$ with $A_N \neq 0$, such that the highest degree component is finite dimensional. For the Gorenstein property we need two conditions:

(i) $\dim A_N = 1$.

Let $A_N = Fu_N$. Define $t : A \to F$ by $t(a) = \begin{cases} 0 & \text{if } a \in A_i, i < N; \\ t(fu_N) = f, & f \in F. \end{cases}$

(ii) The symmetric bilinear form $t(ab)$ on A is nondegenerate.

Definition 2. A finitely generated graded algebra $A = F[a_1, \dots, a_m]$ is said to be Gorenstein if there exists a regular system of parameters $u_1, \dots, u_k \in A$ such that the finite-dimensional algebra $D := F[a_1, \dots, a_m]/(u_1, \dots, u_k)$ is Gorenstein.

One can prove again that if the property is verified for one single system of parameters it is verified by all systems.

For a finite-dimensional algebra A with maximum degree N, the Gorenstein property implies that $t(ab)$ establishes an isomorphism between A_i and A^*_{N-i} for all i. In particular we have a consequence for the Hilbert series of a Gorenstein algebra, it is of the form 5.2.2, with the further restriction that the numerator $p(t)$ is a polynomial with nonnegative integer coefficients, constant term 1 and with the symmetry $p(t) = t^N p(t^{-1})$ where $N = \deg p(t)$.

An important theorem of Hochster and Roberts [HR] (see also [B-H] for a short proof due to Knop) is:

Theorem. *If G is a linearly reductive group and M a rational representation, then the ring of invariants of G acting on M is Cohen–Macaulay. If furthermore G is contained in the special linear group, the ring of invariants is even Gorenstein.*

It is in fact quite interesting to see explicitly these properties for the rings of invariants that we studied in Chapter 11. A lot of interesting combinatorics is associated to this property (cf. [Gar], [Stan]).

15

Binary Forms

We want to finish this book going backwards in history. After having discussed many topics mostly belonging to 20th century mathematics, we go back to where they all began, the old invariant theory of the 19th century in its most complete achievement: the theory of binary forms. We show a few of the many computational ideas which were developed at that time.

For a classical exposition see the book by Grace and Young [GY].

1 Covariants

1.1 Covariants

The theory of binary forms studies the invariants of forms in two variables.

Recall that the space of binary forms of degree m is the space:[142]

$$S_m := \left\{ \sum_{i=0}^{m} a_i \binom{m}{i} x^{m-i} y^i \mid a_i \in \mathbb{C} \right\}. \tag{1.1.1}$$

Of course one can also study binary forms over more general coefficient rings.

We have seen that the spaces S_m are irreducible representations of $SL(2, \mathbb{C})$ and in fact exhaust the list of these representations. The observation that S_m has a non-degenerate $SL(2, \mathbb{C})$ invariant form (Chapter 5, §3.7) implies that S_m and S_m^* are isomorphic representations.

The algebra of invariants of binary forms of degree m is the ring of polynomials in the coordinates $(a_0, a_1, \ldots, a_m)$ which are invariant under the action of $SL(2, \mathbb{C})$. Many papers on this subject appeared in the 19th century, but then the theory disappeared from the mathematical literature, to be discovered again in the last 30 years.

Let us establish some notation. Denote by $S(S_m)$, $P(S_m) = \mathbb{C}[a_0, a_1, \ldots, a_m]$, respectively, the symmetric algebra and the polynomials on the space S_m. These two

142 The normalization of the coefficients is for convenience.

algebras are both representations of $SL(2,\mathbb{C})$ and in fact they are isomorphic as representations due to the observation that S_m has a nondegenerate $SL(2,\mathbb{C})$ invariant form. Since $SL(2,\mathbb{C})$ is the only group in this chapter, we will write for short $SL(2,\mathbb{C}) = S(2)$.

There are several ways to approach the theory, but almost always they pass through the study of a more general problem: the study of *covariants*.

Definition. On the space of forms of degree m, a covariant of degree k and order p is a polynomial map $F : S_m \to S_p$, homogeneous of degree k and $S(2)$-equivariant.

Of course an invariant is just a covariant of order 0.

Remark. The first remark to be made is the obvious one. The identity map $f \to f$ of S_m is a covariant of degree 1 and order m. The form f is a covariant (of itself).

A polynomial map $F : S_m \to S_p$, homogeneous of degree k, is the composition of a linear map $f : S^k[S_m] \to S_p$ with $a \mapsto a^k$, where $S^k[S_m]$ is the k-th symmetric power of S_m. Let $M_p(k,m)$ denote the isotypic component of $S^k[S_m]$ of type S_p. We have that $M_p(k,m) = S_p \otimes V_p(k,m)$, where $\dim V_p(k,m)$ is the multiplicity with which S_p appears in $S^k[S_m]$.

Proposition 1. *The space of covariants of degree k and order p is the dual of* $V_p(k,m)$.

Proof. Since $\hom_{S(2)}(S_p, S_p) = \mathbb{C}$, the explicit identification is the following:

$$\hom_{S(2)}(S^k[S_m], S_p) = \hom_{S(2)}(S_p \otimes V_p(k,m), S_p)$$
$$= \hom_{S(2)}(S_p, S_p) \otimes V_p(k,m)^*. \qquad \square$$

This proposition of course implicitly says that knowing covariants is equivalent to knowing the decomposition of $S(S_m)$ into isotypic components.

There is a different way of understanding covariants which is important. Consider the space $V = \mathbb{C}^2$ on which $S_m = S^m[V^*]$ is identified with the space of homogeneous polynomials of degree m. Although by duality V is isomorphic to S_1, it is important to distinguish the two spaces. On V we have the two coordinates x, y. Consider the $S(2)$ invariant polynomials on the space $S_m \oplus V$. Call the variables $(a_0, a_1, \dots, a_m, x, y)$. Any polynomial invariant $f(a_0, a_1, \dots, a_m, x, y)$ decomposes into the sum of the bihomogeneous parts with respect separately to S_m and V. These components are invariant.

Proposition 2. *The space of covariants of degree k and order p on* S_m *can be identified with the space of invariants on* $S_m \oplus V$ *of bidegree k, p.*

Proof. If $F : S_m \to S_p$ is a covariant and $a \in S_m$, $v \in V$, we can evaluate $F(a)(v)$, and this is clearly an invariant function of a, v. If $\lambda, \mu \in \mathbb{C}$, we have

$$F(\lambda a)(\mu v) = \lambda^k F(\mu v) = \lambda^k \mu^p F(v).$$

Conversely, if $F(a_0, a_1, \dots, a_m, x, y)$ is a bihomogeneous invariant of bidegree k, p, we have that for fixed $a = (a_0, a_1, \dots, a_m)$, the function $F(a_0, a_1, \dots, a_m, x, y)$ is a homogeneous form $f_a \in S_p$. The map $a \to f_a$ is equivariant and of degree k. $\square$

In more explicit terms we write an invariant of bidegree k, p as

$$f(a_0, a_1, \ldots, a_m, x, y) = \sum_{i=0}^{p} f_i(a_0, a_1, \ldots, a_m)x^{p-i}y^i$$

which exhibits its nature as a covariant.

1.2 Transvectants

Of course covariants may be composed and defined also for several binary forms. The main construction between forms and covariants is *transvection.*[143]

Transvection is an interpretation of the Clebsch–Gordan formula, which in representation theoretic language is

$$S_p \otimes S_q = \bigoplus_{i=0}^{\min(p,q)} S_{p+q-2i}. \tag{1.2.1}$$

Definition. Given $f \in S_p,\ g \in S_q$, the ith transvection $(f, g)_i$ of f, g is the projection of $f \otimes g$ to S_{p+q-2i}.

It is interesting and useful to see the transvection explicitly in the case of *decomposable* forms. Since $S_m = S_m(S_1)$ we follow the usual procedure of tensor algebra to make explicit a multilinear function on decomposable tensors.

Let us use the classical notation in which several forms are thought of as depending on different sets of variables, denoted $x, y, z, \ldots$. So we change the notation slightly.

Given a linear form $a = a_x := a_0x_1 + a_1x_2$ the form

$$a_x^m = \sum_{i=0}^{m} a_0^{m-i}a_1^i \binom{m}{i} x_1^{m-i}x_2^i$$

is a typically decomposable form. If $b_y = b_0y_1 + b_1y_2$ is another linear form let us define (a, b) by

$$(a, b) := \det \begin{vmatrix} a_0 & a_1 \\ b_0 & b_1 \end{vmatrix} = a_0b_1 - a_1b_0.$$

Proposition. *We have for the ith transvection:*

$$(a_x^p, b_y^q)_i = (a, b)^i a_x^{p-i} b_x^{q-i}. \tag{1.2.2}$$

Proof. One can make explicit the computations of Chapter 3, §3.1.7, or argue as follows.

Consider the polynomial map:

$$T_i : (a_x, b_y) \to (a, b)^i a_x^{p-i} b_x^{q-i} \in S_{p+q-2i}, \tag{1.2.3}$$

T_i is clearly homogeneous of degree p in a_x and of degree q in b_y, therefore it factors as

$$(a_x, b_y) \mapsto a_x^p \otimes b_y^q \in S_p \otimes S_q \xrightarrow{t_i} S_{p+q-2i}.$$

143 *Übershiebung* in German.

It is clear that T_i is $S(2)$-equivariant, hence also $S_p \otimes S_q \xrightarrow{t_i} S_{p+q-2i}$ is a linear equivariant map. At this point there is only one subtlety to discuss. While the projection to the isotypic component of $S_p \otimes S_q$ of type S_{p+q-2i} is uniquely determined, the explicit isomorphism of this component with S_{p+q-2i} is determined up to a scalar, so we can in fact *define* the transvection so that it is normalized by the formula 1.2.2. □

Given two covariants $F : S_m \to S_p$, $G : S_m \to S_q$ we can then form their transvection $(F, G)_i : S_m \to S_{p+q-2i}$. As in the general theory we may want to polarize a covariant $F : S_m \to S_p$ of degree k by introducing k variables in S_m. We obtain then a multilinear covariant

$$\overline{F} : S_m \otimes S_m \otimes \cdots \otimes S_m \to S_p.$$

Iterating the decomposition 1.2.1 as normalized by 1.2.2 we obtain a decomposition:

$$S_m^{\otimes k} = \bigoplus\nolimits_{i_1, i_2, \ldots, i_{k-1}} S_{km-2(i_1+i_2+\cdots+i_{k-1})},$$

$$i_j \leq \min(m, jm - 2(i_1 + i_2 + \cdots + i_{j-1})).$$

This is an explicit way in which $S_m^{\otimes k}$ can be decomposed into irreducible summands.

On an element $f_1 \otimes f_2 \otimes \cdots \otimes f_k$, $f_i \in S_m$, the projection onto $S_{km-2(i_1+i_2+\cdots+i_{k-1})}$ is the iteration of transvections:

$$(((\ldots((f_1, f_2)_{i_1}, f_3)_{i_2}, \ldots, f_k)_{i_k}.$$

For instance, quadratic covariants of a given form f are only the transvections $(f, f)_i$.

Theorem. *Any covariant of f is a linear combination of covariants obtained from f by performing a sequence of transvections.*

1.3 Source

Given a covariant, thought of as an invariant

$$f(a_0, \ldots, a_m, x, y) = \sum_{i=0}^{p} f_i(a_0, \ldots, a_m) x^{p-i} y^i,$$

let us compute it for the vector of coordinates $x = 1$, $y = 0$, getting

$$f(a_0, \ldots, a_m, 1, 0) = f_0(a_0, a_1, \ldots, a_m).$$

Definition. The value $f_0(a_0, a_1, \ldots, a_m)$ is called the source[144] of the covariant.

Example. For the identity covariant f the source is a_0.

[144] *Quelle* in German.

The main point is:

Theorem 1. *The source of a covariant is invariant under the subgroup* $U^+ := \left\{\begin{vmatrix} 1 & u \\ 0 & 1 \end{vmatrix}\right\}$. *Every* U^+ *invariant function* $f_0(a_0, a_1, \dots, a_m)$ *is the source of a covariant. The map*

$$Q : f(a_0, \dots, a_m, x, y) \to f(a_0, \dots, a_m, 1, 0) = f_0(a_0, a_1, \dots, a_m)$$

is an isomorphism between covariants and U^+ *invariants of* S_m.

Proof. Given a covariant $f(a, v)$, since $(1, 0)$ is a vector invariant under U^+ and f is invariant under $SL(2)$, clearly $f(a, (1, 0))$ is U^+-invariant. For the converse we need an important geometric remark. Let $\mathcal{A} := \{(a, v) \in S_m \times V \mid v \neq 0\}$. $\mathcal{A}$ is an open set and its complement is S_m, so it has codimension 2 in the ambient space. Let $S'_m := \{(a, (1, 0)), a \in S_m\} \subset S_m \oplus V$. Given any element $(a, v) \in \mathcal{A}$ there is an element $g \in SL(2)$ with $gv = (1, 0)$. In other words $SL(2)S'_m = \mathcal{A}$. It follows that any $SL(2)$ invariant function is determined by the values that it takes once restricted to S'_m, and this is exactly the map Q. It remains to prove that, given a U^+ invariant function on S'_m, it extends to an $SL(2)$-invariant function on $S_m \times V$. In fact it is enough to show that it extends to an algebraic function on $\mathcal{A}$, since then it extends everywhere due to the codimension condition. The argument is this: given $\overline{f}$ a U^+-invariant we want to define an extension of f on $\mathcal{A}$. Take a vector (a, v) and a $g \in SL(2)$ with $gv = (1, 0)$ and define $f(a, v) := \overline{f}(ga, (1, 0))$. We need to show that this is well defined. This follows from the U^+-invariance and the fact that it is a regular algebraic function, which is a standard fact of algebraic geometry due to the smoothness of $\mathcal{A}$.

In more concrete terms, given (x, y) if $x \neq 0$ (resp. $y \neq 0$) we construct

$$\begin{vmatrix} x \\ y \end{vmatrix} = \begin{vmatrix} x & 1 \\ y & \frac{y+1}{x} \end{vmatrix} \begin{vmatrix} 1 \\ 0 \end{vmatrix}, \qquad \begin{vmatrix} x \\ y \end{vmatrix} = \begin{vmatrix} x & \frac{x-1}{y} \\ y & 1 \end{vmatrix} \begin{vmatrix} 1 \\ 0 \end{vmatrix}.$$

When we use these two matrices as the g to build the function $f(a, v) := \overline{f}(ga, (1, 0))$ we see that it has an expression as a polynomial in $a_0, \dots, a_m, x, y$ with denominator either a power of x or of y, and hence this denominator can be eliminated. □

Remark. In the classical literature an invariant under the subgroup U^+ is also called a *semi-invariant*.

We need to understand directly from the nature of the semi-invariant what type of covariant it is. For this we introduce the notion of weight. At the beginning the notion is formal, but in fact we will see that it is a weight in the sense of characters for the multiplicative group. By definition we give weight i to the variable a_i and extend (by summing the weights) the notion to the weight of a monomial. A polynomial is usually called *isobaric* if it is a sum of monomials of the same weight. In this case we can talk of the weight of the polynomial.

Theorem 2. *The source of a covariant of degree k and order p, of forms in S_m, is a homogeneous polynomial of degree k which is isobaric of weight $\frac{mk-p}{2}$. In particular an $SL(2)$-invariant on S_m, is a semi-invariant of degree k and weight $\frac{mk}{2}$.*

Proof. Consider the torus $D_t := \begin{vmatrix} t^{-1} & 0 \\ 0 & t \end{vmatrix} \subset SL(2)$. It acts on the space V transforming $x \mapsto t^{-1}x,\ y \mapsto ty$. The action on the forms is

$$(D_t f)(x, y) = f(tx, t^{-1}y), \sum_{i=0}^{m} a_i \binom{m}{i} (tx)^{m-i}(t^{-1}y)^i = \sum_{i=0}^{m} a_i \binom{m}{i} t^{m-2i} x^{m-i} y^i.$$

In other words D_t transforms $a_i \to t^{m-2i} a_i$. A covariant, of degree k and order p, must be also an invariant function of this transformation on $S_m \oplus V$ or

$$f(t^m a_0, \dots, t^{m-2i} a_i, \dots, t^{-m} a_m, t^{-1}x, ty) = f(a_0, \dots, a_m, x, y).$$

For the source we get

$$f_0(t^m a_0, \dots, t^{m-2i} a_i, \dots, t^{-m} a_m)(t^{-1}x)^p = f_0(a_0, \dots, a_m)x^p.$$

A monomial in f_0 of weight w is multiplied by t^{mk-2w}; hence we deduce that for every monomial $mk - 2w - p = 0$, as required. □

In particular notice:[145]

Corollary. *The $SL(2)$-invariants of binary forms of degree k are the semi-invariants of degree k and of weight $\frac{mk}{2}$.*

If $B := \left\{ \begin{vmatrix} t^{-1} & u \\ 0 & t \end{vmatrix} \right\}$, the B-invariants coincide with the $SL(2)$-invariants.

Proof. The first assertion is a special case of the previous theorem. The second follows from the proof: a semi-invariant of degree k is invariant under B if and only if its weight w satisfies $mk - 2w = 0$, and then $p = 0$. □

2 Computational Algorithms

2.1 Recursive Computation

The theory of semi-invariants allows one to develop a direct computational method which works very efficiently to compute covariants of forms of low degree, and then its efficiency breaks down! Denote by A_m the ring of polynomial semi-invariants of binary forms of degree m.

First, a little remark. On binary forms the action of U^+ fixes y and transforms $x \mapsto x + ty$, therefore it commutes with $\frac{\partial}{\partial x}$. Moreover the 1-parameter group U^+ is

[145] The reader experienced in algebraic geometry may see that the geometric reason is in the fact that $SL(2)/B = \mathbb{P}^1$ is compact.

the exponential of the element $y\frac{\partial}{\partial x}$. This element acts as a unique nilpotent Jordan block of size $m+1$ on the space S_m. Thus the theory of semi-invariants is the theory of invariants under the exponential of this unique block.

The previous remark implies that the maps

$$i_m : S_m \to S_{m+1},\ i_m(f(x,y)) := f(x,y)y,$$

$$p_{m+1} : S_{m+1} \to S_m,\ p_{m+1}(f(x,y)) := \frac{\partial}{\partial x} f(x,y)$$

are U^+-equivariant. In particular they induce two maps $i_m^* : A_{m+1} \to A_m,\ p_{m+1}^* : A_m \to A_{m+1}$. We have $p_{m+1}i_m = i_{m-1}p_m$.

We will use in particular $p^* := p_{m+1}^*$ to build A_{m+1} starting from A_m.

If we take the basis $u_{i,m} := \frac{1}{i!}x^i y^{m-i}$ we have

$$p(u_{i,m})) = u_{i-1,m-1},\ 1 \le i \le m,\ p(u_{0,m})) = 0.$$

Therefore it is convenient to write the forms as:[146]

$$\sum_{i=0}^{m} a_i \frac{1}{(m-i)!} x^{m-i} y^i. \tag{2.1.1}$$

so that $y\frac{\partial}{\partial x}\sum_{i=0}^{m} a_i \frac{1}{(m-i)!}x^{m-i}y^i = \sum_{i=0}^{m-1} a_i \frac{1}{(m-i-1)!}x^{m-i-1}y^{i+1}$. Passing to coordinates, the transpose transformation maps $a_i \mapsto a_{i-1}$. Thus the differential operator which generates the induced group of linear transformations on the polynomial ring $F[a_0, \dots, a_m]$ is

$$-\sum_{i=1}^{m} a_{i-1}\frac{\partial}{\partial a_i}.$$

Proposition. *A polynomial $f(a_0, \dots, a_m)$ is a semi-invariant if and only if it satisfies the differential equation:*

$$\sum_{i=1}^{m} a_{i-1}\frac{\partial}{\partial a_i} f(a_0, \dots, a_m) = 0. \tag{2.1.2}$$

The operator $\sum_{i=1}^{m} a_{i-1}\frac{\partial}{\partial a_i}$ maps polynomials of degree k and weight p into polynomials of degree k and weight $p-1$.

Let us take a form $\sum_{i=0}^{m} a_i \frac{1}{(m-i)!}x^{m-i}y^i$ with $a_0 \neq 0$. Under the transformation $x \to x - \frac{a_1}{a_0}y,\ y \to y$, it is transformed into a form with $a_1 = 0$. More formally, let S_m^0 be the set of forms of degree m with $a_0 \neq 0$ and let S_m' be the set of forms of degree m with $a_0 \neq 0,\ a_1 = 0$. The previous remark shows that, acting with U^+, we identify $S_m^0 = U^+ \times S_m'$. Thus we have an identification of the U^+-invariant functions on S_m^0 with the functions on S_m'. The functions on S_m' are the polynomials in their

[146] In fact the algorithm could be easily done in all characteristics, in which case it is important to choose the correct basis. The reader may do some exercises and verify that covariants change when the characteristic is small.

coefficients with a_0 inverted. Let us denote by b_i the U^+-invariant function on S_m^0 which corresponds to the ith coefficient, $i = 2, \dots, m$, of forms in S'_m. Calculating explicitly the functions b_i we have

$$a_0^m \sum_{i=0}^{m} a_i \frac{1}{(m-i)!}\left(x - \frac{a_1}{a_0}y\right)^{m-i} y^i = \sum_{i=0}^{m} a_0^i a_i \frac{1}{(m-i)!}(a_0x - a_1y)^{m-i}y^i$$

$$= \sum_{i=0}^{m} a_i \sum_{j=0}^{m-i} a_0^{m-j}(-a_1)^j \frac{1}{j!}\frac{1}{(m-i-j)!}x^{m-i-j}y^{i+j}$$

$$= \sum_{k=0}^{m}\left[\sum_{j=0}^{k} a_{k-j}a_0^{m-j}(-a_1)^j\frac{1}{j!}\right]\frac{1}{(m-k)!}x^{m-k}y^k \implies$$

$$b_k = a_0^{-m}\left[\sum_{j=0}^{k} a_{k-j}a_0^{m-j}(-a_1)^j\frac{1}{j!}\right].$$

Let

$$(1-k)c_k := (-1)^k k! a_0^{k-1} b_k = \sum_{j=0}^{k}(-1)^{k+j}\frac{k!}{j!}a_0^{k-j-1}a_{k-j}a_1^j$$

$$= (1-k)a_1^k + \sum_{s=2}^{k}\frac{(-1)^s k!}{(k-s)!}a_0^{s-1}a_s a_1^{k-s}.$$

We have thus proved:

Theorem. *The ring* $A_m[a_0^{-1}] = F[c_2, \dots, c_m][a_0, a_0^{-1}]$.

Let us make explicit some of these elements:

$$c_2 = a_1^2 - 2a_0a_2$$

$$c_3 = a_1^3 - 3a_0a_2a_1 + 3a_0^2a_3$$

$$c_4 = a_1^4 - 4a_0a_2a_1^2 + 8a_0^2a_3a_1 - 8a_0^3a_4$$

$$c_5 = a_1^5 - 5a_0a_2a_1^3 + 15a_0^2a_3a_1^2 - 30a_0^3a_4a_1 + 30a_0^4a_5$$

$$c_6 = a_1^6 - 6a_0a_2a_1^4 + 24a_0^2a_3a_1^3 - 72a_0^3a_4a_1^2 + 144a_0^4a_5a_1 - 144a_0^5a_6.$$

We have that c_i is a covariant of degree i and weight i.

If we want to understand covariants from these formulas it is necessary to compute

(2.1.3) $$A_m = F[c_2, \dots, c_m][a_0, a_0^{-1}] \cap F[a_0, \dots, a_m].$$

For each polynomial $F(c_2, \dots, c_m)$ we need to consider its order of vanishing for $a_0 = 0$. Write $F(c_2, \dots, c_m) = a_0^k G(a_0, a_1, a_2, \dots, a_m)$ with $G(0, a_1, a_2, \dots, a_m) \neq 0$. Then we may add $G(a_0, a_1, a_2, \dots, a_m)$ to A_m.

Remark. We have that $A_m \subset A_{m+1}$. Moreover the elements c_i are *independent* of m (the ones which exist).

We may now make a further reduction:

Exercise. If a covariant $G(a_0, a_1, a_2, \dots, a_m)$ satisfies $G(0, a_1, a_2, \dots, a_m) \neq 0$, then we also have $G(0, a_1, 0, \dots, a_m) \neq 0$. It follows that if we define $d_2 := c_2(a_0, a_1, 0, a_3, \dots, a_m)$ and $\overline{A}_m = F[d_2, \dots, d_m][a_0, a_0^{-1}] \cap F[a_0, a_1, a_3, \dots, a_m]$, then the natural map $A_m \to \overline{A}_m$ is an isomorphism.

Let us explain how we proceed. Suppose we have already constructed $A_m = F[t_1, t_2, \dots, t_k]$ and want to compute A_{m+1}. We add c_{m+1} to A_m and then want to find the elements in $F[t_1, t_2, \dots, t_k, c_{m+1}]$ which are divisible by a_0 or a power of it.

For this we evaluate the elements $t_1, t_2, \dots, t_k, c_{m+1}$ for $a_0 = 0$. For these evaluated elements consider their ideal of relations. Each element in this ideal lifts to a polynomial $h[t_1, t_2, \dots, t_k, c_{m+1}]$ which is divisible by a_0. We then start determining a set of generators for the ideal lifting to polynomials $h_i[t_1, t_2, \dots, t_k, c_{m+1}] = a_0^k g_i[a_0, a_1, \dots, a_{m+1}]$ where $g_i[0, a_1, \dots, a_{m+1}] \neq 0$. The elements $g_i[a_0, a_1, \dots, a_{m+1}] \in A_{m+1}$ and we add them. In this way we may not have added all the elements coming from the ideal, but we have that:

Lemma. *Let $a_0 = s_0, s_1, \dots, s_k \in F[a_0, a_1, \dots, a_m]$, and denote by $\overline{s}_i := s_i(0, a_1, \dots, a_m)$. Assume we have a basis $h_i[x_1, x_2, \dots, x_k]$ of the ideal J of relations between the elements $\overline{s}_1, \dots, \overline{s}_k$. Then write*

$$h_i[s_1, s_2, \dots, s_k] = a_0^{k_i} g_i[a_0, a_1, \dots, a_{m+1}],$$

with $g_i[0, a_1, \dots, a_m] \neq 0$. Assume furthermore that for every i we have $g_i[a_0, a_1, \dots, a_{m+1}] \in F[s_0, s_1, \dots, s_k]$.

Then for every $h[x_1, x_2, \dots, x_k] \in J$, such that

$$h[s_0, s_1, s_2, \dots, s_k] = a_0^k g[a_0, a_1, \dots, a_m],$$

we have $g[a_0, a_1, \dots, a_m] \in F[s_0, s_1, \dots, s_k]$.

Proof. Let us work by induction on the order of vanishing k. If $k = 0$, there is nothing to prove, otherwise $h[x_1, x_2, \dots, x_k] \in J$, and let

$$h[x_1, x_2, \dots, x_k] = \sum_i t_i[x_1, x_2, \dots, x_k] h_i[x_1, x_2, \dots, x_k].$$

We substitute for the x_i the s_i, and pull out the powers of a_0 and get

$$a_0^k g[a_0, a_1, \dots, a_m] = \sum_i t_i[s_1, \dots, s_k] a_0^{k_i} g_i[a_0, a_1, \dots, a_m].$$

All the k_i are positive and of course $k \geq \min k_i := h > 0$. We cancel h and

$$a_0^{k-h} g[a_0, a_1, \dots, a_m] = \sum_i t_i[s_1, \dots, s_k] a_0^{k_i - h} g_i[a_0, a_1, \dots, a_m].$$

By hypothesis $a_0, g_i[a_0, a_1, \dots, a_m] \in F[s_0, s_1, \dots, s_k]$, hence we finish by induction. □

The previous lemma gives an algorithm to compute A_m. At each step we have determined some ring $F[s_0, s_1, \dots, s_k]$ and we add to this ring all the elements g_i obtained from a basis of the ideal J. When no new elements are found the algorithm stops.

Of course it is not a priori clear that the algorithm will stop; in fact this is equivalent to the classical theorem that covariants are a finitely generated algebra.

On the other hand, if one keeps track of the steps in the algorithm, one has also a presentation for the covariant algebra by generators, relations and the relations among the relations, i.e., the syzygies.

We can perform this explicitly for $m \leq 6$. For $m \leq 4$ we easily define, following the previous algorithm, covariants A, B, C by

(2.1.4) $$3Aa_0^2 = c_2^3 - c_3^2, \quad c_4 - c_2^2 = 8Ba_0^2, \quad A - 2Bc_2 = a_0 C.$$

Since by standard theory A_m is a ring of dimension m we get that

(2.1.5) $$A_2 = F[a_0, c_2], \; A_3 = F[a_0, c_2, c_3, A], \; A_4 = F[a_0, c_2, c_3, B, C],$$

where in A_2 the elements a_0, c_2 are algebraically independent, while for A_3, A_4, we have one more generator than the dimension and the corresponding equations 2.1.4. One may continue by hand for 5,6 but the computations become more complicated and we leave them out.

By weight considerations and Corollary 1.3, one can identify, inside the covariants, the corresponding rings of invariants.

2.2 Symbolic Method

The symbolic method is the one mostly used in classical computations. It is again based on polarization. Let us denote by V the space of linear binary forms and by $S^m(V)$ their symmetric powers, the binary forms of degree m.

One chooses the coefficients so that a form is written as $A := \sum_{i=0}^m a_i \binom{m}{i} x^{m-i} y^i$. This is convenient since, when the form is a power of a linear form $b_1 x + b_2 y$, we obtain $a_i = b_1^{m-i} b_2^i$.

Let $h(A) = h(a_0, \dots, a_m)$ be a polynomial invariant, homogeneous of degree k, for binary forms of degree m. Polarizing it we deduce a multilinear invariant of k forms $h(A_1, \dots, A_k)$. Evaluating this invariant for $A_i := b_i^m$ with b_i a linear form, we obtain an invariant of the linear forms $b_1, \dots, b_k$ which is symmetric in the b_i and homogeneous of degree m in each b_i. Now one can use the FFT for $SL(2)$ proved in Chapter 11, §1.2 (which in this special case could be proved directly), and argue that any such invariant is a polynomial in the determinants $[b_i, b_j]$. Conversely any invariant of the linear forms $b_1, \dots, b_k$ which is homogeneous of degree m in each $b_i \in V$, factors uniquely through a multilinear invariant on the symmetric powers $S^m(V)$. Under restitution one obtains a homogeneous polynomial invariant of binary forms of degree k. Thus:

Theorem 1. *The space of polynomial invariants, homogeneous of degree k, for binary forms of degree m is linearly isomorphic to the space of invariants of k linear forms, homogeneous of degree m in each variable and symmetric in the k forms.*

In practice this method gives a computational recipe to produce a complete list of linearly generating invariants in each degree. One writes a basis of invariants of k linear forms, homogeneous of degree m, choosing products of elements $[b_i, b_j]$ with the only constraint that each index $i = 1, \dots, k$ appears exactly m times. Any such product is a *symbolic invariant*.

Next, one develops the polynomial using the formula

$$[b_i, b_j] = b_{i1}b_{j2} - b_{j1}b_{i2}$$

and looks at any given monomial which will necessarily, by the homogeneity condition, be of the form

$$\lambda_{h_1,h_2,\dots,h_k} b_{11}^{m-h_1} b_{12}^{h_1} b_{21}^{m-h_2} b_{22}^{h_2} \cdots b_{k1}^{m-h_k} b_{k2}^{h_k}.$$

Under the identifications made the invariant is obtained by the substitutions

$$\lambda_{h_1,h_2,\dots,h_k} b_{11}^{m-h_1} b_{12}^{h_1} b_{21}^{m-h_2} b_{22}^{h_2} \cdots b_{k1}^{m-h_k} b_{k2}^{h_k} \mapsto \lambda_{h_1,h_2,\dots,h_k} a_{h_1} a_{h_2} \cdots a_{h_k}.$$

In the previous discussion we have not assumed that the invariant of linear forms is necessarily symmetric. Let us then see an example. Suppose we want to find invariants of degree 2 for forms of degree m. In this case we prefer to write a, b instead of b_1, b_2 and have only one possible symbolic invariant, i.e., $[a, b]^m$. This is symmetric if and only if m is even; thus we have the existence of a quadratic invariant only for forms of even degree. For these forms we already know the existence of an $SL(2)$-invariant quadratic form which, up to normalization is the invariant of Chapter 5, §3.7.

In a similar way one may describe symbolically the invariants of several binary forms, of possibly different degrees.

Covariants can also be treated by the symbolic method. We can use the fact that according to Chapter 5, §3.7 there is a nondegenerate $SL(2)$-invariant bilinear form $\langle x, y\rangle$ on each space $S^p(V)$. Thus if $F : S^m(V) \to S^p(V)$ is a covariant of degree k and we polarize it to $\overline{F}(A_1, \dots, A_k)$, we obtain an invariant of $A_1, \dots, A_k \in S^m(V)$, $B \in S^p(V)$ by the formula $\langle F(A_1, \dots, A_k), B\rangle$. Next, we substitute as before A_i with b_i^m and B with c^p and get a linear combination of products of either (b_i, b_j) or (b_i, c). Conversely, given such a product, we can associate to it a covariant by taking the product of all the terms (b_i, b_j) in the monomial times the product of the b_j's which are paired with c. The previous discussion implies that all covariants will be obtained this way.

There are several difficulties involved with the symbolic method. One is the difficulty of taking care in a combinatorial way of the symmetry. In fact in Chapter 13, §2.3 we have shown that for invariants of several linear forms $b_1, \dots, b_k$ we have a theory of standard rectangular tableaux with rows of length 2 and filled with the numbers $1, 2, \dots, k$. The constraint that they be homogeneous of degree m in each variable means that each index i appears m times, but the symmetry condition is unclear in the combinatorial description and does not lead to another combinatorial constraint.

The second difficulty arises when one wants to build the invariant ring by generators and relations. It is not at all clear how to determine when a symbolic invariant can be expressed as a polynomial of lower degree symbolic invariants. This point was settled by Gordan in a very involved argument, which was later essentially forgotten once the more general theorems of Hilbert on the finite generation of invariants were proved.

Remark. A different proof of the symbolic method could be gotten directly using transvectants as in §2.

There is a second form in which the symbolic method appears and it is based on the following fact (we work over $\mathbb{C}$).

Let us consider the space of m-tuples of linear forms $(a_1, a_2, \dots, a_m)$ and the map to $S^m(V)$ given by

$$\phi : (a_1, a_2, \dots, a_m) \to \prod_{i=1}^{m} a_i.$$

The map ϕ is invariant under the action of the group N generated by the permutations S_m of the linear forms, and the rescaling of the forms

$$(a_1, a_2, \dots, a_m) \mapsto (\lambda_1 a_1, \lambda_2 a_2, \dots, \lambda_m a_m), \quad \prod_{i=1}^{m} \lambda_i = 1.$$

$N = S_m \ltimes T$ is the semidirect product of S_m with the torus

$$T := \{(\lambda_1, \lambda_2, \dots, \lambda_m)\}, \prod_{i=1}^{m} \lambda_i = 1.$$

The property of unique factorization implies that two m-tuples of linear forms map under ϕ to the same form of degree m if and only if they are in the same orbit under N. We claim then that:

Proposition. *The ring of invariants of the action of N on the space V^m of m-tuples $a_1, \dots, a_m$ of linear forms is the ring of polynomials in the coordinates of $S^m(V)$.*

Proof. Consider the quotient variety $V^m//N$. The map ϕ factors through the quotient as $\phi : V^m \xrightarrow{\pi} V^m//N \xrightarrow{\overline{\phi}} S^m(V)$. By the fact that $V^m//N$ parameterizes closed orbits of N, it follows that $\overline{\phi}$ is bijective. Hence, by ZMT, we have that $\overline{\phi}$ is an isomorphism. □

Now when we look at the invariants of $SL(2)$ on forms in $S^m(V)$ we have:

Theorem 2. *The invariants under $SL(2)$ on $S^m(V)$ coincide with the invariants under $N \times SL(2)$ acting on V^m.*

In particular the invariants under $SL(2)$ on $S^m(V)$ coincide with the invariants under N acting on the $SL(2)$-invariants of V^m.

We know that the $SL(2)$-invariants of V^m are described by standard rectangular tableaux with 2 columns and with entries $1, 2, \dots, m$. It is clear by the definitions that such a tableau U transforms under a torus element $(\lambda_1, \lambda_2, \dots, \lambda_m)$ as

$$\lambda_1^{h_1} \lambda_2^{h_2} \dots \lambda_m^{h_m} U$$

where h_i is the number of occurrences of i in the tableau. Thus U is invariant under the torus T if and only if all the indices $1, \dots, m$ appear exactly the same number of times. Finally on this ring of $SL(2) \times T$-invariants acts the symmetric group S_m and the $SL(2)$-invariants of the binary forms are identified to the S_m-invariants of this action.

3 Hilbert Series

3.1 Hilbert Series

In this section we shall develop a formula for the dimensions of the spaces of invariants and covariants in various degrees.

Given a representation ρ of a reductive group G on a vector space W, let R be the ring of polynomial functions on W.

We want to study the numbers (which we shall organize in a generating series):

$$d_i(\rho) := \dim R_i^G, \quad p_\rho(t) := \sum_{i=0}^{\infty} d_i(\rho) t^i. \tag{3.1.1}$$

We have already remarked in Chapter 8, §6 that a maximal compact subgroup K is Zariski dense; hence the G-invariants coincide with the K-invariants. We can thus apply the theory of characters of compact groups Chapter 8, §1.3.1 and Molien's theorem Chapter 9, §4.3.3 so that

$$p_\rho(t) = \int_K \det(1 - \rho(\gamma)t)^{-1} \, d\gamma$$

where $d\gamma$ is the normalized Haar measure on K.

One can transform this formula by the Weyl integration formula to an integration over a maximal torus T with coordinates z and normalized Haar measure dz:

$$p_\rho(t) = \int_T \det(1 - \rho(z)t)^{-1} A(z) \, dz$$

where $A(z)$ is the Weyl factor (Chapter 11, §9.1). In the case of $G = SL(2)$, $K = SU(2)$ the torus T is 1-dimensional and the Weyl factor is

$$A(z) = \frac{2 - z^2 - z^{-2}}{2}. \tag{3.1.2}$$

Let us start with invariants. We fix m and $S^m(V)$ the space of binary forms and let $R := \mathbb{C}[a_0, \dots, a_m]$, the ring of polynomial functions on $S^m(V)$. Denote by $p_m(t)$ the generating series on invariants and $p_m^c(t)$ the generating series for covariants.

The matrix $\rho(z)$ is diagonal with entries z^{-m+2k}, $k = 0, \dots, m$. Thus setting $z := e^{2\pi i \phi}$, the integral to be computed is

$$\int_0^1 \frac{1}{2}(2 - z^2 - z^{-2}) \prod_{k=0}^{m} (1 - z^{-m+2k}t)^{-1} d\phi,$$

or in the language of complex analysis:

$$p_m(t) = \frac{1}{2\pi i} \oint_T \frac{1}{2}(2 - z^2 - z^{-2}) \prod_{k=0}^{m} (1 - z^{-m+2k}t)^{-1} z^{-1}\, dz.$$

For covariants instead we have the extra factors:

$$p_m^c(t) = \frac{1}{2\pi i} \oint_T \frac{1}{2}(2 - z^2 - z^{-2})$$
$$\times \prod_{k=0}^{m} (1 - z^{-m+2k}t)^{-1} (1 - zt)^{-1} (1 - z^{-1}t)^{-1} z^{-1}\, dz.$$

Let us make the computations for invariants and for $m = 2p$ even. In this case one sees immediately that $\int_0^1 f(z^2)\, d\phi = \int_0^1 f(z)\, d\phi$, so we have

$$p_{2p}(t) = \frac{1}{2\pi i} \oint_T \frac{1}{2}(2 - z - z^{-1}) z^{-1} \prod_{k=-p}^{p} (1 - z^k t)^{-1}\, dz$$
$$= \frac{(-1)^p t^{-p}}{2\pi i (1 - t)} \oint_T \frac{1}{2}(2 - z - z^{-1}) z^{-1+p(p+1)/2}$$
$$\times \prod_{k=1}^{p} (1 - z^k t)^{-1} (1 - z^k t^{-1})^{-1}\, dz.$$

We apply partial fractions and the residue theorem. Set $\zeta_k := e^{2\pi i/k}$; we have

$$\prod_{k=1}^{p} (1 - z^k t)(1 - z^k t^{-1}) = \prod_{k=1}^{p} \prod_{j=0}^{k-1} (1 - \zeta_k^j t^{1/k} z)(1 - \zeta_k^{-j} t^{-1/k} z).$$

Let A denote the inverse of the product on the right-hand side. We expand

$$A = \sum_{k=1}^{p} \sum_{j=0}^{k-1} \frac{b_{kj}}{1 - \zeta_k^j t^{1/k} z} + \frac{c_{kj}}{1 - \zeta_k^{-j} t^{-1/k} z}$$

where the numbers b_{kj}, c_{kj} will be calculated later. For the integral p_{2p}, assuming $|t| < 1$ and $p \geq 2$ ($p = 1$ being trivial):

$$p_{2p} = \frac{(-1)^p}{2(1-t)t^p} \sum_{k=1}^{p} \sum_{j=0}^{k-1} c_{kj} (\zeta_k^j t^{1/k})^{p(p+1)/2} (2 - \zeta_k^j t^{1/k} - \zeta_k^{-j} t^{-1/k}). \tag{3.1.3}$$

We work now on the term

$$I_k := \sum_{k=1}^{p} \sum_{j=0}^{k-1} c_{kj} (\zeta_k^j t^{1/k})^{p(p+1)/2} (2 - \zeta_k^j t^{1/k} - \zeta_k^{-j} t^{-1/k}).$$

Recall that the partial fraction decomposition gives

$$c_{kj}^{-1} = \prod_{h=1}^{p} \prod_{s=0}^{h-1} (1 - \zeta_h^s t^{\frac{1}{h}} \zeta_k^j t^{\frac{1}{k}})(1 - \zeta_h^{-s} t^{\frac{-1}{h}} \zeta_k^j t^{\frac{1}{k}}) \prod_{s=0}^{k-1} (1 - \zeta_k^s t^{\frac{1}{k}} \zeta_k^j t^{\frac{1}{k}})$$

$$\times \prod_{s=0,\ s\neq j}^{k-1} (1 - \zeta_k^{-s} t^{\frac{-1}{k}} \zeta_k^j t^{\frac{1}{k}})$$

$$= \prod_{h=1,\ h\neq k}^{p} (1 - \zeta_k^{jh} t^{1+h/k})(1 - \zeta_k^{jh} t^{-1+h/k})(1 - t^2)k. \tag{3.1.4}$$

In order to simplify the formula let us work in the formal field of Laurent series in $t^{1/N}$, where N is the least common multiple of the numbers $k = 1, \dots, p$. The automorphism $\sigma_N : t^{1/N} \to \zeta_N t^{1/N}$ induces a trace operator $\operatorname{Tr} := \frac{1}{N} \sum_{j=0}^{N-1} \sigma_N^j$ which has the property that, applied to a Laurent series in $t^{1/N}$, picks only the terms in t. In other words, $\operatorname{Tr}(t^a) = 0$ if a is not an integer, and $\operatorname{Tr}(t^a) = t^a$ if a is an integer. On the subfield of series in $t^{1/k}$ the operator $\operatorname{Tr} = \frac{1}{k} \sum_{j=0}^{k-1} \sigma_k^j$. Formulas 3.1.3 and 3.1.4 give $p_{2p} = \operatorname{Tr}(J)$ with

$$J := \frac{(-1)^p}{2(1-t)(1-t^2)t^p} \sum_{k=1}^{p} t^{p(p+1)/2k} (2 - t^{1/k} - t^{-1/k})$$

$$\times \prod_{h=1,\ h\neq k}^{p} (1 - t^{1+h/k})^{-1} (1 - t^{-1+h/k})^{-1}.$$

One can further manipulate this formula. Remark that

$$1 - t^{-1+h/k} = 1 - t^{(h-k)/k} = -t^{(h-k)/k}(1 - t^{(k-h)/k}).$$

If h runs from 1 to p and $h \neq k$, we have that $h + k$ runs from $k + 1$ to $k + p$ except $2k$, and if h runs from $k + 1$ to p, then $h - k$ runs from 1 to $p - k$. Finally when h runs from 1 to $k - 1$, we have that $k - h$ runs from 1 to $k - 1$. Then

$$(1-t)(1-t^2) \prod_{h=1,\ h\neq k}^{p} (1 - t^{1+h/k})(1 - t^{-1+h/k})$$

$$= (-1)^{k-1} t^{(1-k)/2} \prod_{h=1}^{p+k} (1 - t^{h/k}) \prod_{h=1}^{p-k} (1 - t^{h/k}).$$

In conclusion:

$$J = \frac{1}{2}(-1)^p \sum_{k=1}^{p} (-1)^{k-1} t^{E(p,k)} (t^{1/k} - 1)^2 \prod_{h=1}^{p+k} (1 - t^{h/k})^{-1} \prod_{h=1}^{p-k} (1 - t^{h/k})^{-1}$$

where $E(p, k) = (p - k + 2)(p - k - 1)/2k$.

Let us show now how the previous formulas can be used for an effective computation.

Let us write:

$$J = \frac{1}{2}(-1)^p \sum_{k=1}^{p} (-1)^{k-1} J_k, \quad J_k := \frac{H(p,k)}{\prod_{h=2}^{p+k}(1 - t^{h/k}) \prod_{h=2}^{p-k}(1 - t^{h/k})}$$

with $H(p, k) = t^{(p-k+2)(p-k+1)/2k}$ for $1 \leq k \leq p - 1$ and $H(p, p) = (1 - t^{1/p})$. Since

$$(1 - t^{h/k})^{-1} = \left(\sum_{j=0}^{k-1} t^{jh/k} \right) (1 - t^h)^{-1},$$

we can write

$$J_k = \frac{H(p,k) \prod_{h=2}^{p+k} (\sum_{j=0}^{k-1} t^{jh/k}) \prod_{h=2}^{p-k} (\sum_{j=0}^{k-1} t^{jh/k})}{\prod_{h=2}^{p+k} (1 - t^h) \prod_{h=2}^{p-k} (1 - t^h)} = \frac{N_k}{D_k},$$

and get

$$\text{(3.1.5)} \qquad \mathrm{Tr}(J) = \frac{1}{2}(-1)^p \sum_{k=1}^{p} (-1)^{k-1} \frac{\mathrm{Tr}(N_k)}{D_k}.$$

Since N_k is an effectively given polynomial in $t^{1/k}$, we have that $\mathrm{Tr}(N_k)$ is an effectively computable polynomial in t.

We would like to express the Hilbert series in the form given by formula 1.1.3 of Chapter 12. For this we prove:

Proposition. *In formula 3.1.5, if p is even, the polynomial $g_e := \prod_{h=2}^{2p-1}(1 - t^h)$ is a common multiple of the D_k's.*

If p is odd, the polynomial $g_o := (1 + t) \prod_{h=2}^{2p-1}(1 - t^h)$ is a common multiple of the D_k's.

Proof. For p odd note that

$$(1 - t^{-1/p}) \left(\sum_{j=0}^{p-1} t^{2j/p} \right) = -t^{-1/p}(1 - t) \left(\sum_{j=0}^{p-1} (-1)^j t^{j/p} \right).$$

For p even

$$(1-t^{-1/p})\left(\sum_{j=0}^{p-1} t^{2j/p}\right)\left(\sum_{j=0}^{p-1} t^{3j/p}\right)$$

$$= -t^{-1/p}(1-t^2)(1-t^{1/p}+t^{2/p})\left(\sum_{j=0}^{(p-2)/2}(-1)^j t^{6j/p}\right).$$

Hence when p is even:

$$J_p = -t^{-1/p}(1-t^{1/p}+t^{2/p})\left(\sum_{j=0}^{(p-2)/2}(-1)^j t^{6j/p}\right)$$

$$\times \prod_{h=4}^{2p-1}\left(\sum_{j=0}^{p-1} t^{jh/p}\right) \Big/ \prod_{h=2}^{2p-1}(1-t^h),$$

while when p is odd:

$$J_p = -t^{-1/p}\left(\sum_{j=0}^{p-1}(-1)^j t^{j/p}\right)\prod_{h=3}^{2p-1}\left(\sum_{j=0}^{p-1} t^{jh/p}\right) \Big/ (1+t)\prod_{h=2}^{2p-1}(1-t^h).$$

To finish we have to prove that the denominators D_k for $k < p$ divide the denominator found for J_p. So, let Φ_d denote the cyclotomic polynomial for primitive d^{th} roots of 1, and $t^h - 1 = \prod_{d\,|\,h}\Phi_d$. Denote finally by $[a/b]$ the integral part of the fraction $a/b,\ a, b \in \mathbb{N}$. Then

$$D_k = \prod_{h=2}^{p+k}(1-t^h)\prod_{h=2}^{p-k}(1-t^h) = \Phi_1^{2p-2}\prod_{d\geq 2}\Phi_d^{[(p+k)/d]+[(p-k)/d]}$$

$$g_e = \Phi_1^{2p-2}\prod_{d\geq 2}\Phi_d^{[(2p-1)/d]}, \qquad g_o = \Phi_1^{2p-2}\Phi_2^{[(2p-1)/2]}\prod_{d\geq 3}\Phi_d^{[(2p-1)/d]} \qquad \square$$

The numerator of the Hilbert series can at this point be made explicit. Computer work, done for forms of degree ≤ 36, gives large numbers and some interesting facts about which we will comment in the next section.

4 Forms and Matrices

4.1 Forms and Matrices

There is even another method to find explicit invariants of binary forms, discovered by Hilbert.

It is based on the Clebsch–Gordan formula

$$S^m(V)\otimes S^n(V)=\bigoplus_{i=0}^{\min m,n} S^{m+n-2i}(V).$$

If we apply it when $m=n$, we can take advantage of the fact that $S^m(V)\equiv S^m(V)^*$ and thus obtain

$$\mathrm{End}(S^m(V))=S^m(V)\otimes S^m(V)^*\equiv S^m(V)\otimes S^m(V)=\bigoplus_{i=0}^{m} S^{2i}(V).$$

Therefore the forms $S^{2i}(V)$ can be embedded in an $SL(2)$-equivariant way into $\mathrm{End}(S^m(V))$ for all $m\geq i$. In particular, the coefficients of the characteristic polynomial of elements of $\mathrm{End}(S^m(V))$, restricted to $S^{2i}(V)$, are $SL(2)$-invariants. The identification $S^m(V)\equiv S^m(V)^*$ is given by an invariant nondegenerate form which, for m odd, is symplectic while for m even, it is symmetric (Chapter 5, §3.7). In both cases, on $\mathrm{End}(S^m(V))$ we have an $SL(2)$-invariant involution $*$ and $(a\otimes b)^*=\epsilon(b\otimes a)$ where $\epsilon=-1$ in the symplectic and $\epsilon=1$ in the orthogonal case. It is interesting to see how the decomposition $S^m(V)\otimes S^m(V)=\bigoplus_{i=0}^{m} S^{2i}(V)$ behaves under the symmetry $\tau: a\otimes b\to b\otimes a$. We have:

Proposition. $\tau=(-1)^i$ *on* $S^{2(m-i)}(V)$.

Proof. Since τ is $SL(2)$ equivariant and has only eigenvalues ±1 on each space $S^{2i}(V)$ which is an irreducible representation of $SL(2)$, it must be either 1 or -1. To determine the sign, recall the formula 2.1.2 of transvection, which in our case is $(a_x^m,b_y^m)_i=(a,b)^i a_x^{m-i}b_x^{m-i}$ and which can be interpreted as the image of $a_x^m\otimes b_x^m$ in $S^{2(m-i)}(V)$ under the canonical projection. We have by equivariance

$$(-1)^i(a,b)^i a_x^{m-i}b_x^{m-i}=(b,a)^i b_x^{m-i}a_x^{m-i}=(b_x^m,a_y^m)_i=\tau((a,b)^i a_x^{m-i}b_x^{m-i}).\quad\square$$

We obtain:

Corollary. $S^{2(m-i)}(V)\subset S^m(V)$ *is formed by symmetric matrices when m and i are odd or m and i are even, otherwise it is formed by antisymmetric matrices.*

Thus one method to find invariants is to take coefficients of the characteristic polynomials of these matrices. When one starts to do this one discovers that it is very hard to relate invariants given by different matrix representations. In any case these computations and the ones on Hilbert series suggest a general conjecture with which we want to finish this book, showing that even in the most ancient and basic themes there are difficult open problems. The problem is the following. From Proposition 6.1 we have seen that the Hilbert series of binary forms, in case the forms are of degree $4k$, a multiple of 4, is of the form

$$f(t)/\prod_{i=2}^{4k-1}(1-t^i).$$

When one computes them (and one can do quite a lot of computations by computer) one discovers that for $4k\leq 36$, the coefficients of $f(t)$ are nonnegative. Recalling the theory of Cohen–Macaulay algebras this suggests that there may be a regular sequence of invariants of degree i, $2 \le i \le 4k - 1$. One way in which such a sequence could arise is as coefficients of the characteristic polynomial of a corresponding matrix.

In particular when we embed $S^{4k}(V) \subset \mathrm{End}(S^{4k-2}(V))$ we have (since the matrices we obtain have order $4k - 1$ and trace 0), exactly $4k - 2$ coefficients. One may attempt to guess that these coefficients form a regular sequence. There is even a possible geometric way of checking this. Saying that in a Cohen–Macaulay graded ring of dimension h, h elements form a regular sequence is equivalent to saying that their common zeros reduce to the point 0. When we are dealing with invariants this is equivalent to saying that as functions of $S^{4k}(V)$ they define the set of unstable points, that is the set on which all invariants vanish. By Chapter 14, §4.1 a binary form is unstable if and only if one of its zeroes has multiplicity $> n/2$. On the other hand, for a matrix x the coefficients of its characteristic polynomial are 0 if and only if the matrix is nilpotent. So, in the end, the question is: if the endomorphism on $S^{4k-2}(V)$ induced by a form on $S^{4k}(V)$ is nilpotent, is the form unstable?

This seems to be very hard to settle. Even if it is false it may still be true that there is a regular sequence of invariants of degree i, $2 \le i \le 4k - 1$.

Bibliography

[Ab] Abe E. *Hopf Algebras*, Cambridge Tracts in Mathematics,Vol. 74, Cambridge University Press, Cambridge, New York, 1980.

[AD] Abeasis, S., Del Fra, A. Characteristic classes through classical invariant theory, *Math. Z.* **164**(2) (1978), 105–114.

[AH] Ackerman M., Hermann R. *Hilbert's Invariant Theory Papers*, Brookline, 1978.

[A] Adams J.F. *Lectures on Lie Groups*, Benjamin, New York, 1969.

[B] Bourbaki N. *Algèbre Multilinéaire*, Ch. 3, Herman, 1958.

[Bor] Borel A. *Linear Algebraic Groups*, Lecture Notes in Mathematics, Benjamin Press, 1968.

[Bor2] Borel A. *Essays in the History of Lie Groups and Algebraic Groups*, History of Mathematics, Vol. 21. A.M.S., Providence, RI; London Mathematical Society, Cambridge, 2001, xiv+184 pp.

[BB] Bialynicki-Birula A. Some theorems on actions of algebraic groups. *Ann. of Math.*, 98 (1973), 480–497.

[B1] Bourbaki N. *Groupes et Algèbres de Lie*, Chapter 1, Hermann, Paris, 1960.

[B2] Bourbaki N. *Groupes et Algèbres de Lie*, Chapters 4–6, Hermann, Paris, 1968.

[B3] Bourbaki N. *Groupes et Algèbres de Lie*, Chapters 7–8, Hermann, Paris, 1975.

[Br] Brezis H. *Analyse fonctionnelle. Théorie et applications*, Collection Mathématiques Appliquées pour la Maîtrise. Masson, Paris, 1983.

[B-H] Bruns W, Herzog J. *Cohen-Macaulay Rings*, Cambridge Studies in Advanced Mathematics, Vol. 39. Cambridge University Press, Cambridge, 1993. xii+403 pp.

[Ca] Capelli A. *Lezioni sulla teoria delle forme algebriche*, Napoli, 1902.

[Car] Carter R.W. *Simple Groups of Lie Type*, Wiley, London, New York, 1972.

[Ch1] Chevalley C. *Classification des groupes de Lie algébriques*, Séminaire Ecole Normale Supérieure, Paris, 1956–58.

[Ch2] Chevalley C. *Theory of Lie Groups*, Princeton University Press, 1946.

[Ch3] Chevalley C. *Théorie de groupes de Lie*, Hermann, Paris, T. I, 1951, T. II, 1952.

[Ch4] Chevalley C. *The algebraic theory of spinors*, Columbia University Press, New York, 1954.

[Cou] Coutinho S.C. *A Primer in Algebraic D-modules*, London Math. Soc. Student Texts, Vol. 33, 1995.

[CR] Curtis C.W., Reiner I. *Representation Theory of Finite Groups and Associative Algebras*, Pure and Applied Mathematics, Vol. XI, Interscience Publishers, John Wiley & Sons, New York, London, 1962.

[DK] Dadok J., Kac V. Polar representations, *J. Algebra* **92**(2) (1985), 504–524.

[DP] De Concini C, Procesi C. A characteristic free approach to invariant theory, *Adv. in Math.* **21** (1976), 330–354.

[DP2] De Concini C, Procesi C. Symmetric functions, conjugacy classes and the flag variety. *Inv. Mathematicae* **64** (1981), 203–219.

[DG] Demazure M., Grothendieck A. *Schémas en groupes*, Lect. Notes in Math., Vols. 151, 152, 153, Springer-Verlag, 1970.

[Dic] Dickson L.E. *Algebraic Invariants*, Cambridge, 1903.

[Die] Dieudonné J.A. *Sur les Groupes Classiques*, Hermann, Paris, 1948.

[DC] Dieudonné J.A., Carrell J.B. Invariant theory, old and new, *Advances in Math.* **4** (1971) New York.

[Di] Dixmier J. *Les C^*-algèbres et leurs Représentations*, Cahiers Scientifiques XXIX, Gauthier-Villars, Paris, 1964.

[DL] Dolgachev I. *Lectures on Invariant Theory*, 2004.

[Do] Donkin S. A filtration for rational modules, *Math. Z.* **177**(1) (1981), 1–8.

[DRS] Doubilet P, Rota G C, Stein J. *On the foundations of combinatorial theory IX*, Studies in Appl. Math., 108, 18 (1974).

[E] Eisenbud D. *Commutative Algebra*, GTM, Vol. 150, Springer, 1994.

[Fo] Fogarty J. *Invariant Theory*, Benjamin, New York, 1969.

[FH] Fulton W., Harris J. *Representation Theory*, GTM, Vol. 129, Springer 1991.

[Fu] Fulton W. *Algebraic Curves. An Introduction to Algebraic Geometry*, notes written with the collaboration of Richard Weiss. Reprint of 1969 original. Advanced Book Classics. Addison-Wesley Publishing Company, Advanced Book Program, Redwood City, CA, 1989.

[Gar] Garsia A.M. Combinatorial methods in the theory of Cohen-Macaulay rings, *Adv. in Math.* **38**(3) (1980), 229–266.

[GW] Goodman R., Wallach N.R. *Representations and Invariants of the Classical Groups*, Cambridge University Press, 1998, corrected paperback 2003.

[GY] Grace J.H., Young A. *Algebra of Invariants*, Cambridge, 1903.

[Gb] Grove L., Benson C. *Finite Reflection Groups*, GTM, Vol. 99, Springer 1985.

[Gu] Gurevich G.B. *Foundations of the Theory of Algebraic Invariants*, Groningen, 1964.

[Hai] Haiman M. Dual equivalence with applications, including a conjecture of Proctor, *Discrete Mathematics* **99** (1992), 79–113.

[Ha] Hartshorne R. *Algebraic Geometry*, GTM, Vol. 52, Springer-Verlag, 1977.

[Hat] Hatcher A. *Algebraic Topology*, Cambridge University Press, 2002.

[He] Helgason S. *Differential Geometry, Lie Groups and Symmetric Spaces*, Academic Press, New York, 1978.

[Her] Herstein I.N. *Topics in Algebra*, Blaisdell Pub. Co., 1964.

[Hil] Hilbert D. Über die vollen Invariantensysteme, *Math. Annalen* **43** (1893), 313–373.

[Ho] Hochschild G. *The Structure of Lie Groups*, Holden-Day, San Francisco, 1965.

[HR] Hochster M., Roberts J. Rings of invariants of reductive groups acting on regular rings are Cohen Macaulay, *Adv. in Math.* **18** (1974), 115–175.

[Hu1] Humphreys J. *Introduction to Lie Algebras and Representation Theory*, GTM, Vol. 9, Springer-Verlag, 1980.

[Hu2] Humphreys J. *Linear Algebraic Groups*, GTM, Vol. 21, Springer-Verlag, 1975.

[Hu3] Humphreys J. *Reflection Groups and Coxeter Groups*, Cambridge Studies in Advanced Mathematics, Vol. 29, Cambridge University Press, 1990.

[J1] Jacobson N. *Lie Algebras*, Wiley-Interscience, New York, London, 1962.

[J2] Jacobson N. *Exceptional Lie Algebras*, Marcel Dekker, New York, 1971.

[JBA] Jacobson N. *Basic Algebra I, II*, Freeman and Co., San Francisco, 1974.

[Ka] Kac V. *Infinite Dimensional Lie Algebras*, Cambridge University Press, 1990.

[Kap] Kaplansky. *Lie Algebras and Locally Compact Groups*, Chicago University Press, Chicago, 1971.

[Ki] Kirillov A. *Élements de la Théorie des Représentations*, (translation), Éditions MIR, Moscow, 1974.

[Kn] Knapp A. *Lie Groups Beyond an Introduction*, Progress in Mathematics, Vol. 140, Birkhäuser, 1996.

[Knu] Knutson D. *λ-rings and the Representation Theory of the Symmetric Group*, Lecture Notes in Mathematics, Vol. 308, Springer-Verlag, Berlin-New York, 1973.

[Kr] Kraft H. *Geometrische Methoden in der Invariantentheorie*, Braunschweig, 1984.

[KrP] Kraft H., Procesi C. *Classical Invariant Theory: A Primer*, http://www.math.unibas.ch.

[KSS] Kraft H., Slodowy P., Springer T.A. *Algebraic Transformation Groups and Invariant Theory*, Birkhäuser, Basel, 1989.

[L-S] Lakshmibai V., Seshadri C.S. Geometry of G/P. II. The work of de Concini and Procesi and the basic conjectures, *Proc. Indian Acad. Sci. Sect. A* **87**(2) (1978), 1–54.

[Lit] Littelmann P. A Littlewood–Richardson rule for symmetrizable Kac-Moody algebras, *Invent. Math.*, **116**(1–3) (1994), 329–346.

[Lit2] Littelmann P. *Bases for Representations, LS-paths and Verma Flags. A tribute to C.S. Seshadri* (Chennai, 2002). Trends Math., Birkhäuser, Basel, 2003, 323–345.

[Li] Littlewood D.E. *The Theory of Group Characters and Matrix Representations of Groups*, New York, 1940.

[Mac] Macdonald I.G. *Symmetric Functions and Hall Polynomials*, Second Edition. The Clarendon Press, Oxford University Press, New York, 1995.

[M] Meyer W.Fr. Invariantentheorie, *Enzyklopädie der mathematischen Wissenschaften*, IB2, 1892.

[MM] Milnor J.W, Moore J.C. On the structure of Hopf algebras, *Ann. of Math.* **81**(2) (1965), 211–264.

[MZ] Montgomery and Zippin. *Topological Transformation Groups*, Wiley, New York, 1955.

[MF] Mumford D., Fogarty J. *Geometric Invariant Theory*, Ergebnisse der Mathematik 34, Springer-Verlag, New York, 1982.

[Mu] Murnaghan F.D. *The Theory of Group Representations*, Dover, 1938.

[Na1] Nagata M. Invariants of a group in an affine ring, *J. Math. Kyoto Univ.* **8** (1964), 369.

[Na2] Nagata M, *Lectures on the Fourteenth Problem of Hilbert*, Tata Institute of Fundamental Research, Bombay, 1965.

[Ne] Neretin Yu.A. *A Construction of Finite-Dimensional Faithful Representation of Lie Algebra*, preprint series: ESI preprints.

[Ogg] Ogg A. *Modular forms and Dirichlet series*, Benjamin, Amsterdam, 1969.

[OV] Onishchik A.L, Vinberg E., *Lie Groups and Algebraic Groups*, Springer Series in Soviet Math., Springer-Verlag, 1990.

[Pr] Procesi C., *A Primer in Invariant Theory*, Brandeis, 1982.

[PrR] Procesi C., Rogora E. Aspetti geometrici e combinatori della teoria delle rappresentazioni del gruppo unitario, *Quaderni U.M.I.* 36 (1991).

[Ra] Raynaud M. *Anneaux locaux henséliens*, Lecture Notes in Mathematics, Vol. 169, Springer-Verlag, Berlin, New York 1970.

[RA] Regev A. On the codimension of matrix algebras. *Algebra – Some Current Trends* (Varna, 1986), Lecture Notes in Mathematics, Vol. 1352, Springer, Berlin, 1988.

[Ru] Rudin W. *Real and Complex Analysis*, McGraw-Hill, New York, 1966.

[Sa] Sagan B.E. *The Symmetric Group*, Wadsworth & Brooks/Cole Math. Series, 1991.

[Sc] Schur I. *Vorlesüngen über Invariantentheorie*, Springer Verlag 1968, (posthumous).

[Sch] Schützenberger M.P. In Foata D. (ed.), *Combinatoire et Representation du Groupe-Symétrique, Strasbourg 1976*, Springer Lecture Notes in Mathematics 579, 1977; pp. 59–113.

[Sch2] Schützenberger M.P. Quelques remarques sur une construction de Schensted, *Math. Scan.* **12** (1963), 117–128.

[Se1] Serre J.P. *Lie Algebras and Lie Groups*, Benjamin, New York, 1965.

[Se2] Serre J.P. *Algèbres de Lie semi-simple complexes*, Benjamin, New York, 1966.

[Seh] Seshadri C.S. Geometry of G/P. I. Theory of standard monomials for minuscule representations. *C.P. Ramanujam – A Tribute*, Tata Institute of Fundamental Research Studies in Math., 8, Springer, Berlin, New York, 1978, pp. 207–239.

[Sh] Shafarevich I.R. *Basic Algebraic Geometry*, Grundlehren der Math. Wiss., Bd. 213, Springer-Verlag, 1974.

[Sp] Spanier E. *Algebraic Topology*, McGraw-Hill Series in Higher Mathematics, 1966.

[Sp1] Springer T.A. *Invariant Theory*, Springer-Verlag, 1977.

[Sp2] Springer T.A. *Linear Algebraic Groups*, PM, Vol. 9, Birkhäuser-Verlag, 1981.

[Stan] Stanley R.P. *Combinatorics and Commutative Algebra*, Second Edition. Progress in Mathematics, Vol. 41, Birkhäuser-Boston, 1996.

[Si] Sylvester. *Mathematical Papers*, Vol. I, Chelsea, New York, 1973, pp. 511 ff.

[Sw] Sweedler M. *Hopf Algebras*, Mathematics Lecture Note Series, W.A. Benjamin, Inc., New York, 1969.

[Ti] Tits J. *Sur les constantes de structure et le théorème d'existence des algèbres de Lie semi-simples*, Inst. Hautes Études Sci. Publ. Math. No. 31, 1966, 21–58.

[Wa] Warner F. *Foundations of Differentiable Manifolds and Lie Groups*, Scott, Foresman and Co., 1971.

[Wie] Wielandt H. Zum Satz von Sylow. *Math. Z.* **60** (1954), 407–408.

[Wt] Weitzenböck R. *Invariantentheorie*, Noordhoff, Groningen, 1923.

[W] Weyl H. *The Classical Groups. Their Invariants and Representations*, Princeton, 1939.

[Ze] Žhelobenko D.P. *Compact Lie Groups and Their Representations*, Translations of Math. Monog., Vol. 40, AMS, Providence, RI, 1973.

Index of Symbols

Subject Index